Chapter 4 Exponents, Polynomials and Additional Applications

Rules of Exponents

1. $x^m \cdot x^n = x^{m+n}$ **product rule**
2. $\dfrac{x^m}{x^n} = x^{m-n}, x \neq 0$ **quotient rule**
3. $(x^m)^n = x^{m \cdot n}$ **power rule**
4. $x^o = 1, x \neq 0$ **zero exponent rule**
5. $x^{-m} = \dfrac{1}{x^m}, x \neq 0$ **negative exponent rule**
6. $\left(\dfrac{ax}{by}\right)^m = \dfrac{a^m x^m}{b^m y^m}, b \neq 0, y \neq 0$ **expanded power rule**

FOIL method (*F*irst, *O*uter, *I*nner, *L*ast) of multiplying binomials: $(a + b)(c + d) = ac + ad + bc + bd$

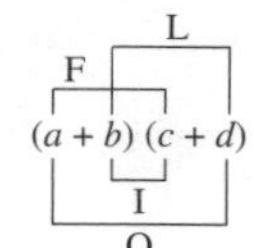

Product of sum and difference of same two terms: $(a + b)(a - b) = a^2 - b^2$

Squares of binomials: $(a + b)^2 = a^2 + 2ab + b^2$

$(a - b)^2 = a^2 - 2ab + b^2$

distance formula: $d = rt$

Chapter 5 Factoring

If $a \cdot b = c$, then a and b are **factors** of c.

Difference of two squares: $a^2 - b^2 = (a + b)(a - b)$

Sum of two cubes: $a^3 + b^3 = (a + b)(a^2 - ab + b^2)$

Difference of two cubes: $a^3 - b^3 = (a - b)(a^2 + ab + b^2)$

To Factor a Polynomial

1. If all the terms of the polynomial have a greatest common factor other than 1, factor it out.
2. If the polynomial has two terms, determine if it is a difference of two squares or a sum or a difference of two cubes. If so, factor using the appropriate formula.
3. If the polynomial has three terms, factor the trinomial using one of the procedures discussed.
4. If the polynomial has more than three terms, try factoring by grouping.
5. As a final step, examine your factored polynomial to see if the terms in any factors listed have a common factor. If you find a common factor, factor it out at this point.

Quadratic equation: $ax^2 + bx + c = 0, a \neq 0.$

Zero-factor Property: If $ab = 0$, then $a = 0$ or $b = 0$.

To Solve a Quadrataic Equation by Factoring

1. Write the equation in standard form with the squared term positive. This will result in one side of the equation being 0.
2. Factor the side of the equation that is not 0.
3. Set each factor containing a variable equal to zero and solve each equation.
4. Check the solution found in step 3 in the original equation.

Chapter 6 Rational Expressions and Equations

To Reduce Rational Expressions

1. Factor both the numerator and denominator as completely as possible.
2. Divide out any common factors, to both the numerator and denominator.

To Multiply Rational Expressions

1. Factor all numerators and denominators completely.
2. Divide out common factors.
3. Multiply the numerators together and multiply the denominators together.

To Add or Subtract Two Rational Expressions

1. Determine the least common denominator (LCD).
2. Rewrite each fraction as an equivalent fraction with the LCD.
3. Add or subtract the numerators while maintaining the LCD.
4. When possible, factor the remaining numerator and reduce the fraction.

To Solve Rational Equations

1. Determine the LCD of all fractions in the equation.
2. Multiply both sides of the equation by the LCD. This will result in every term in the equation being multiplied by the LCD.
3. Remove any parentheses and combine like terms on each side of the equation.
4. Solve the equation.
5. Check your solution in the original equation.

Elementary Algebra for College Students

Elementary Algebra for College Students

Fourth Edition

Allen R. Angel
Monroe Community College

Prentice Hall

Upper Saddle River
New Jersey 07458

f Congress Cataloging-in-Publication Data
en R., 1942–
entary algebra for college students/Allen R. Angel—4th

cm.
Includes index.
ISBN 0-13-324781-3 (hc:alk. paper)
1. Algebra. I. Title.
QA152.2.A54 1996
512.9—dc20 95-17392
CIP

Sponsoring Editor: Melissa S. Acuña
Editor-in-Chief: Jerome Grant
Development Editor: Ellen Credille
Director of Production and Manufacturing: David W. Riccardi
Project Manager: Robert C. Walters
Marketing Manager: Jolene Howard
Copy Editor: Barbara Zeiders
Interior Designer: Geri Davis
Cover Designer: Geri Davis
Creative Director: Paula Maylahn
Art Director: Amy Rosen
Manufacturing Manager: Alan Fischer
Photo Researcher: Rona Tuccillo
Photo Editor: Lorinda Morris-Nantz
Supplements Editor: Audra J. Walsh
Editorial Assistant: April Thrower

Simon & Schuster/A Viacom Company

Upper Saddle River, New Jersey 07458

Printed in the United States of America

10 9 8 7 6 5 4 3 2

ISBN 0-13-324781-3

Prentice-Hall International (UK) Limited, *London*
Prentice-Hall of Australia Pty, Limited, *Sydney*
Prentice-Hall Canada, Inc., *Toronto*
Prentice-Hall Hispanoamericana, S.A., *Mexico*
Prentice-Hall of India Private Limited, *New Delhi*
Prentice-Hall of Japan, Inc., *Tokyo*
Simon & Schuster Asia Pte. Ltd., *Singapore*
Editora Prentice-Hall do Brasil, Ltda., *Rio de Janeiro*

To my wife, Kathy,
and my sons, Robert and Steven

Contents

Preface

This book was written for college students and other adults who have never been exposed to algebra or those who have been exposed but need a refresher course. My primary goal was to write a book that students can read, understand, and enjoy. To achieve this goal I have used short sentences, clear explanations, and many detailed worked-out examples. I have tried to make the book relevant to college students by using practical applications of algebra throughout the text.

Features of the Text

FOUR-COLOR FORMAT: Color is used pedagogically in the following ways:

Important definitions and procedures are color screened.

Color screening or color type is used to make other important items stand out.

Artwork is enhanced and clarified with use of multiple colors.

The four-color format allows for easy identification of important features by students.

The four-color format makes the text more appealing and interesting to students.

READABILITY: One of the most important features of the text is its readability. The book is very readable, even for those with weak reading skills. Short, clear sentences are used and more easily recognized, and easy-to-understand language is used whenever possible.

ACCURACY: Accuracy in a mathematics text is essential. To insure accuracy in this book, mathematicians from around the country have read the galleys carefully for typographical errors and have checked all the answers.

CONNECTIONS: Many of our students do not thoroughly grasp new concepts the first time they are presented. In this text we encourage students to make connections. That is, we introduce a concept, then later in the text briefly reintroduce it and build upon it. Often an important concept is used in many sections of the text. Students are reminded where the material was seen before, or where it will be used again. This also serves to emphasize the importance of the concept. Important concepts are also reinforced throughout the text in the Cumulative Review Exercises and Cumulative Review Test.

PREVIEW AND PERSPECTIVE: This feature at the beginning of each chapter explains to the students why they are studying the material and where this material will be used again in other chapters of the book. This material helps students see the connections between various topics in the book, and the connection to real world situations.

VIDEOTAPE AND SOFTWARE ICONS: At the beginning of each section a videotape and software icon is displayed. These icons tell the student where material in this section can be found on the videotapes, saving your students time when they want to review this material on the videotapes or the tutorial software.

KEYED SECTION OBJECTIVES: Each section opens with a list of skills that the student should learn in that section. The objectives are then keyed to the appropriate portions of the sections with symbols such as 1.

PRACTICAL APPLICATIONS: Practical applications of algebra are stressed throughout the text. Students need to learn how to translate application problems into algebraic symbols. The problem-solving approach used throughout this text gives students ample practice in setting up and solving application problems. The use of practical applications motivates students.

DETAILED WORKED-OUT EXAMPLES: A wealth of examples have been worked out in a step-by-step, detailed manner. Important steps are highlighted in color, and no steps are omitted until after the student has seen a sufficient number of similar examples.

STUDY SKILLS SECTION: Many students taking this course have poor study skills in mathematics. Section 1.1, the first section of this text, discusses the study skills needed to be successful in mathematics. This section should be very beneficial for your students, and should help them to achieve success in mathematics.

HELPFUL HINTS: The helpful hint boxes offer useful suggestions for problem solving and other varied topics. They are set off in a special manner so that students will be sure to read them.

COMMON STUDENT ERRORS: Errors that students often make are illustrated. The reasons why certain procedures are wrong are explained, and the correct procedure for working the problem is illustrated. These common student error boxes will help prevent your students from making those errors we see so often.

CALCULATOR CORNERS: The Calculator Corners, placed at appropriate intervals in the text, are written to reinforce the algebraic topics presented in the section and to give the student pertinent information on using the calculator to solve algebraic problems.

EXERCISE SETS: Each exercise set is graded in difficulty. The early problems help develop the student's confidence, and then students are eased gradually into the more difficult problems. A sufficient number and variety of examples are given in each section for the student to successfully complete even the more difficult exercises. The number of exercises in each section is more than ample for student assignments and practice.

WRITING EXERCISES: Many exercise sets include exercises that require students to write out the answers in words. These exercises improve students' understanding and comprehension of the material. Many of these exercises involve problem solving and help develop better reasoning and critical thinking skills. Writing exercises are indicated by the symbol ✎.

CUMULATIVE REVIEW EXERCISES: All exercise sets (after the first two) contain questions from previous sections in the chapter and from previous chapters. These cumulative review exercises will reinforce topics that were previously covered and help students retain the earlier material, while they are learning the new material. For the students' benefit the Cumulative Review Exercises are keyed to the section where the material is covered.

GROUP ACTIVITY/CHALLENGE PROBLEMS: These exercises, which are part of every exercise set after Section 1.1, provide a variety of problems. Many were written to stimulate student thinking and to lead to interesting group discussions. Others provide additional applications of algebra or present material from future sections of the book so that students can see and learn the material on their own before it is covered in class. Others are more challenging than those in the regular exercise set. These problems may be assigned to your students individually or may be assigned as group exercises.

CHAPTER SUMMARY: At the end of each chapter is a chapter summary which includes a glossary and important chapter facts. The terms in the glossary are keyed to the page where they are first introduced.

REVIEW EXERCISES: At the end of each chapter are review exercises that cover all types of exercises presented in the chapter. The review exercises are keyed to the sections where the material was first introduced.

PRACTICE TESTS: The comprehensive end-of-chapter practice test will enable the students to see how well they are prepared for the actual class test. The Test Item File includes several forms of each chapter test that are similar to the student's practice test. Multiple choice tests are also included in the Test Item File.

CUMULATIVE REVIEW TEST: These tests, which appear at the end of each even-numbered chapter, test the students' knowledge of material from the beginning of the book to the end of that chapter. Students can use these tests for review, as well as for preparation for the final exam. These exams, like the cumulative review exercises, will serve to reinforce topics taught earlier.

ANSWERS: The *odd answers* are provided for the exercise sets. *All answers* are provided for the Cumulative Review Exercises, the Review Exercises, Practice Tests, and the Cumulative Practice Test. *Selected answers* are provided for the Group Activity/Challenge Problem exercises.

National Standards

Recommendations of the *Curriculum and Evaluation Standards for School Mathematics,* prepared by the National Council of Teachers of Mathematics, (NCTM) and *Crossroads in Mathematics: Standards for Introductory College*

Mathematics Before Calculus, prepared by the American Mathematical Association of Two Year Colleges (AMATYC) were incorporated into this section.

Prerequisite

This text assumes no prior knowledge of algebra. However, a working knowledge of arithmetic skills is important. Fractions are reviewed early in the text, and decimals are reviewed in Appendix A.

Modes of Instruction

The format and readability of this book lends itself to many different modes of instruction. The constant reinforcement of concepts will result in greater understanding and retention of the material by your students.

The features of the text and the large variety of supplements available make this text suitable for many types of instructional modes including:

- lecture
- modified lecture
- learning laboratory
- self-paced instruction
- cooperative or group study

Changes in the Fourth Edition

When I wrote the fourth edition I considered the many letters and reviews I got from students and faculty alike. I would like to thank all of you who made suggestions for improving the fourth edition. I would also like to thank the many instructors and students who wrote to inform me of how much they enjoyed and appreciated the text.

Some of the changes made in the fourth edition of the text include:

- *Preview and Perspective* has been added to the beginning of each chapter to give students connections between the material they are studying and the material they have studied or will study, and to associate the material with real life applications.
- Real life applications on current topics have been added throughout the text.
- The section on Study Skills for Success in Mathematics has been updated and enhanced.
- More varied exercises have been added in many exercise sets.
- More challenging problems have been added to the end of the graded exercise sets.
- Additional detailed worked-out examples have been added where needed.
- Current environmental issues are discussed mathematically.
- New and additional Helpful Hints and Common Student Errors have been added.
- More Calculator Corners for scientific calculators have been added.
- More exercises that require written student answers are included.
- *Group Activity/Challenge Problems* have been added to each exercise set. Many of these exercises are appropriate for group work and many will foster creative discussions among the members of the group.

- More illustrations have been added throughout the book to help the visual learner.
- The balance is used as a visual aid when discussing the procedures for solving equations. This helps students understand the material better.
- The section titled *Finding the Least Common Denominator* and the section titled *Addition and Subtraction of Rational Expressions* have been combined into one section.
- Pie graphs, line graphs and bar graphs have been added to the graphing chapter. Many examples and exercises come from current newspaper and magazine articles.
- The section titled *Slope-intercept Form of a Line* and the section titled the *Point-slope Form of a Line* have been combined into one section.
- A section on functions has been added to the graphing chapter.
- The section titled *Introduction* (to solving systems of equations) and the section titled *Solving Systems of Equations Graphically* have been combined into one section.
- The word "optional" has been removed from all sections.
- The text has been fine tuned for greater clarity.

Supplements to the Fourth Edition

For this edition of the book the author has personally coordinated the development of the *Student's Solution Manual* and the *Instructor's Solution, Manual.* Experienced mathematics professors who have prior experience in writing supplements, and whose works have been of superior quality, have been carefully selected for authoring the supplements.

For Instructors

Annotated Instructor's Edition: Includes answers to every exercise on the same page.

Instructor's Solutions Manual: Contains solutions to all exercises not included in the Student's Solution manual.

Test Pro II: Allows users to generate tests by chapter or section number, choosing from thousands of test questions and hundreds of algorithms which generate different numbers for the same item. Editing and graphing capability are included.

Test Item File: Contains five tests per chapter.

For Students

MathPro Software: Carefully keyed section-by-section to the text, this software provides students with interactive feedback and help in solving exercises. Each section contains Warm-up Exercises which allow the student unlimited practice and Practice Problems which are graded and recorded.

Videotapes: Closely tied to the book, these instructional tapes feature a lecture format with worked-out examples and exercises from each section of the book.

Student's Solutions Manual: Includes detailed step-by-step solutions to all exercises whose answers appear in the answer appendix. This includes answers to odd-numbered exercises and answers to **all** Cumulative Review Exercises, Review Exercises, Practice Test, and Cumulative Review Test. *Selected answers* for Group Activity/Challenge Problem exercises are also included.

Student's Study Guide. Includes additional worked-out examples, additional drill problems, and Practice Tests, and their answers. Important concepts are emphasized.

Acknowledgments

Writing a textbook is a long and time-consuming project. Many people deserve thanks for encouraging and assisting me with this project. Most importantly I would like to thank my wife Kathy, and sons, Robert and Steven. Without their constant encouragement and understanding, this project would not have become a reality. In addition, Robert did proofreading and answer checking.

I would like to thank Richard Semmler of Northern Virginia Community College, Larry Clar of Monroe Community College, and Phyllis Barnidge and Lynda Steele of Laurel Technical Services for their conscientiousness and the attention to details they provided in checking galleys, pages, artwork and answers.

I would also like to thank my students and colleagues of Monroe Community College for their suggestions for improving the book. I would like to thank students and faculty from around the country for using the third edition and offering valuable suggestions for the fourth edition. I was especially pleased in receiving so many letters from students informing me how much they enjoyed using the book. Thank you for your kind words.

I would like to thank my editor at Prentice Hall, Melissa Acuña, my production editor Bob Walters, and Ellen Credille for their many valuable suggestions and conscientiousness with this project.

I would like to thank the following reviewers and proofreaders for their thoughtful comments and suggestions.

Celeste Carter, Richland College
Larry Clar, Monroe Community College
Linda Crabtree, Longview Community College
Stephen Drake, Northwestern Michigan College
Pauline Graveline, Canton College of Technology
Frances Leach, Delaware Technical and Community College
Dianne Phelps, Sullivan County Community College
Leela Rakesh, Central Michigan University
Deborah J. Ritchie, Moorpark College
Jack Rotman, Lansing Community College
Richard Semmler, Northern Virginia Community College
Carole Sutphen, Muskegon Community College
Donna M. Szott, Community College of Allegheny County
Mary Vaughn, Texas State Technical College

PREVIEW TO THE NEW EDITION

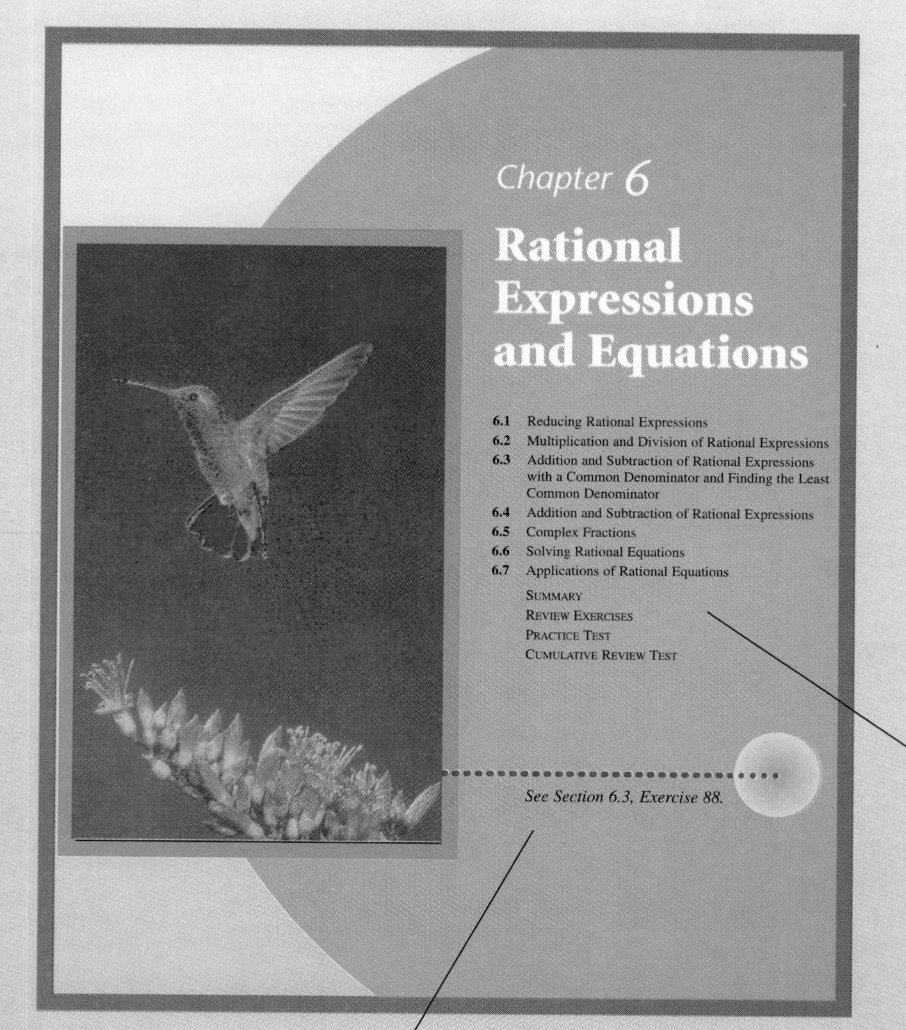

Elementary Algebra for College Students 4/e is a significant revision of the Third Edition. The new design uses color pedagogically and helps increase the readability for the students and the instructors. The exercise sets have been enhanced and new applications have been added. The next few pages will highlight some of the new features of this edition of the text, as well as showing some important features that have made this text successful.

Each chapter opening page contains an outline of the contents of the chapter.

The photo application is taken from an exercise in the chapter, showing how the skills and concepts in the chapter relate to students' lives.

Preview and Perspective

You worked with rational numbers when you discussed fractions in arithmetic. Now you will expand your knowledge to include fractions that contain variables. Fractions that contain variables are often referred to as rational expressions. The same basic procedures you used to reduce, add, subtract, multiply, and divide arithmetic fractions will be used with rational expressions in Sections 6.1 through 6.5. In Section 6.1 we define rational expressions and explain that rational expressions cannot have a denominator of zero. In Section 6.2, multiplication and division of rational expressions are discussed. In Sections 6.3 and 6.4 we explain how to add and subtract rational expressions. Complex fractions are discussed in Section 6.5.

Many real-life problems involve equations that contain rational expressions. Such equations are called rational equations. We discuss solving rational equations in Section 6.6 and give various real-life applications of rational equations in Sections 6.6 and 6.7. Since many formulas contain rational expressions, you will use the material you learn in this chapter in many other courses. In later mathematics courses you may graph rational equations that contain two variables. The material presented here will prepare you for that work.

To be successful in this chapter you need to have a complete understanding of factoring, which was presented in Chapter 5, especially Sections 5.3 and 5.4. You might wish to review reducing, adding, subtracting, multiplying, and dividing numerical fractions discussed in Section 1.2. The material presented in this chapter builds upon the procedures discussed in Section 1.2.

Every chapter begins with a Preview and Perspective, *giving students an overview of the chapter. This feature helps students see the connections between various concepts in the text and how these concepts relate to the real world.*

6.1 Reducing Rational Expressions

Tape 9

1. Determine the values for which a rational expression is defined.
2. Recognize the three signs of a fraction.
3. Reduce rational expressions.
4. Factor a negative 1 from a polynomial.

Each section begins with a list of the objectives for the section. The objectives are keyed to the appropriate sections in the text with icons.

Rational Expressions

1 A **rational expression** (also called an **algebraic fraction**) is an algebraic expression of the form p/q, wh
Examples of rational expressio

$$\frac{4}{5}, \quad \frac{x-6}{x},$$

The denominator of a rational
is not defined. In the expression (x
denominator would then equal 0.
for all real numbers except 0. It is
the value of x cannot be 3 because
of x cannot be used in the expressio
answered correctly. **Whenever w**
variable in the denominator, we
the variable that make the denom

One method that can be used t
that are excluded is to set the deno
equation for the variable.

practice you have at setting up and solving the word problems, the better you will become at solving them.

$l = w + 3$

w

FIGURE 3.8

EXAMPLE 1 Mrs. O'Connor is planning to build a sandbox for her daughter. She has 26 feet of lumber to build the perimeter. What should the dimensions of the rectangular sandbox be if the length is to be 3 feet longer than the width?

Solution: We are asked to find the dimensions of the sandbox.

Let x = width of sandbox
then $x + 3$ = length of sandbox (Fig. 3.8)

From Section 3.1 we know that $P = 2l + 2w$. We have called the width of the sandbox x, and the length $x + 3$. We substitute these expressions into the equation.

$$\begin{aligned} P &= 2l + 2w \\ 26 &= 2(x + 3) + 2x \\ 26 &= 2x + 6 + 2x \\ 26 &= 4x + 6 \\ 20 &= 4x \\ 5 &= x \end{aligned}$$

Thus, the width is 5 feet, and the length is $x + 3 = 5 + 3 = 8$ feet.

Check:
$$\begin{aligned} P &= 2l + 2w \\ 26 &= 2(8) + 2(5) \\ 26 &= 16 + 10 \\ 26 &= 26 \quad \text{true} \end{aligned}$$

Great care has been taken to include practical applications of algebra. The problem-solving approach used in the text gives students needed practice in solving applied problems, and the real-life applications help motivate students.

EXAMPLE 2 The sum of the angles of a triangle measure 180 degrees (180°). If two angles are the same and the third is 30° greater than the other two, find all three angles.

Solution: We are asked to find the three angles.

Let x = each smaller angle
then $x + 30$ = larger angle (Fig. 3.9)

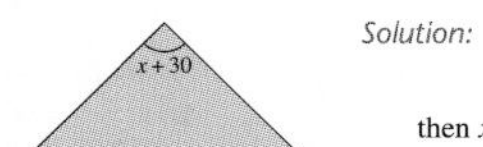

FIGURE 3.9

$$\begin{aligned} \text{Sum of the 3 angles} &= 180 \\ x + x + (x + 30) &= 180 \\ 3x + 30 &= 180 \\ 3x &= 150 \\ x &= \frac{150}{3} = 50° \end{aligned}$$

Therefore, the three angles are 50°, 50°, and 50° + 30° or 80°.

Check:
$$\begin{aligned} 50° + 50° + 80° &= 180° \\ 180° &= 180° \quad \text{true} \end{aligned}$$

Calculator Corners *reinforce the topics taught in the section, and show students necessary keystrokes.*

Calculator Corner

Using Parentheses

When evaluating an expression on a calculator where the order of operations is to be changed, you will need to use parentheses. If you are not sure whether or not they are needed, it will not hurt to add them. Consider $\frac{8}{4-2}$. Since we wish to divide 8 by the difference $4 - 2$, we need to use parentheses.

Evaluate	*Keystrokes*
$\frac{8}{4-2}$	8 ÷ (4 − 2) = 4

What would you obtain if you evaluated 8 ÷ 4 − 2 = on a calculator? Why would you get that result? Here is another example.

Evaluate	*Keystrokes*
$-5(20 - 46) - 12$	5 +/− × (20 − 46) − 12 = 118

...aluate some expressions for given values of the variables.

...uate $7x - 2$ when $x = 2$.

...e 2 for x in the expression.

$7x - 2 = 7(2) - 2 = 14 - 2 = 12$

Insert either >, <, or = in the shaded area to make the statement true.

[1.3] **22.** $-|-6|$ ▒ $|-4|$

23. $|-3|$ ▒ $-|3|$

[1.6] **24.** Evaluate $-6 - (-2) + (-4)$.

[2.1] **25.** Simplify $-6y + x - 3(x - 2) + 2y$.

[3.1] **26.** Solve $2x + 3y = 9$ for y; then find the value of y when $x = 3$.

Group Activity/ Challenge Problems

1. Consider the accompanying figure. **(a)** Write a formula for determining the area of the shaded region. **(b)** Find the area of the shaded region when $S = 9$ inches and $s = 6$ inches.

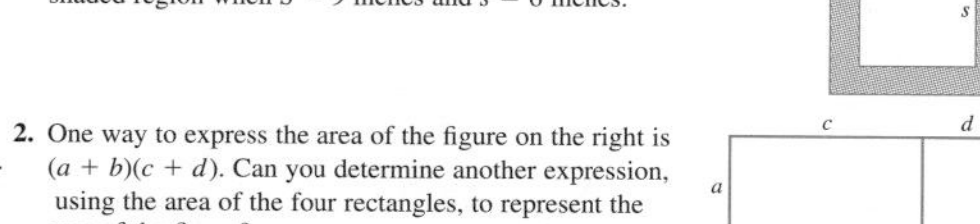

2. One way to express the area of the figure on the right is $(a + b)(c + d)$. Can you determine another expression, using the area of the four rectangles, to represent the area of the figure?

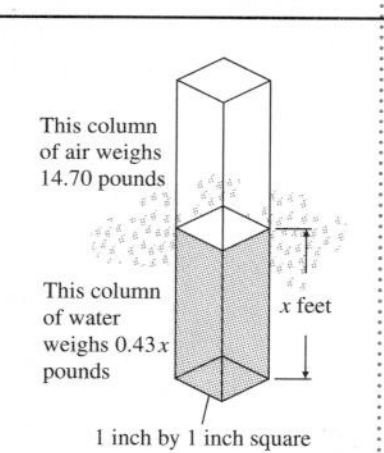

3. The total pressure, P, in pounds per square inch, exerted on an object x feet below sea level is given by the formula $P = 14.70 + 0.43x$. As shown in the accompanying diagram, the 14.70 represents the weight in pounds of the column of air (from sea level to the top of the atmosphere) standing over a 1-inch by 1-inch square of seawater. The $0.43x$ represents the weight, in pounds, of a column of water 1 inch by 1 inch by x feet.

(a) A submarine can w... deep can the submar...

(b) If the pressure gau... square inch, how de...

Group Activity/Challenge Problems, *found in every exercise set, encourage critical thinking. Many exercises can be used as group learning or discussion problems. These exercises show extended applications of the mathematics in the section, and frequently will present material from later sections in the text.*

The solution to Example 2 could be condensed as follows:

$2x - 3 - 5x = 13 + 4x - 2$	
$-3x - 3 = 4x + 11$	Like terms were combined.
$-3 = 7x + 11$	$3x$ was added to both sides of equation.
$-14 = 7x$	11 was subtracted from both sides of equation.
$-2 = x$	Both sides of equation were divided by 7.

We solved Example 2 by moving the terms containing the variable to the right side of the equation. Now rework the problem moving the terms containing the variable to the left side of the equation. You should obtain the same answer.

Example 3 Solve the equation $5.74x + 5.42 = 2.24x - 9.28$.

Solution: We first notice that there are no like terms on the same side of the equal sign that can be combined. We will elect to collect the terms containing the variable on the left side of the equation.

$$5.74x + 5.42 = 2.24x - 9.28$$

Step 3 $5.74x - 2.24x + 5.42 = 2.24x - 2.24x - 9.28$ Subtract $2.24x$ from both sides of equation.

$3.5x + 5.42 = -9.28$

Step 3 $3.5x + 5.42 - 5.42 = -9.28 - 5.42$ Subtract 5.42 from both sides of equation.

$3.5x = -14.7$

Step 4 $\frac{3.5x}{3.5} = \frac{-14.7}{3.5}$ Divide both sides of equation by 3.5.

$x = -4.2$

Example 4 Solve the equation $2(p + 3) = -3p + 10$.

Solution: $2(p + 3) = -3p + 10$

Step 1 $2p + 6 = -3p + 10$ Distributive property was used.

Step 3 $2p + 3p + 6 = -3p + 3p + 10$ Add $3p$ to both sides of equation.

$5p + 6 = 10$

Step 3 $5p + 6 - 6 = 10 - 6$ Subtract 6 from both sides of equation.

$5p = 4$

Step 4 $\frac{5p}{5} = \frac{4}{5}$ Divide both sides of equation by 5.

$p = \frac{4}{5}$

The many worked out examples in the text provide a step-by-step detailed explanaton. Important steps are highlighted in color.

End of the chapter Summary *includes a glossary of key terms from the chapter and a list of the important rules, facts and procedures. This feature helps students review the key concepts and skills from the chapter.*

Summary

Glossary

Binomial *(210):* A two-termed polynomial.

Degree of a polynomial *(211):* The same as the highest-degree term in the polynomial.

Degree of a term *(211):* The exponent on the variable when the polynomial is in one variable.

Descending order, or descending power, of the variable *(210):* Polynomial written so that the exponents on the variable decrease from left to right.

Monomial *(210):* A one-term polynomial.

Polynomial in x *(210):* An expression containing the sum of a finite number of terms of the form ax^n, for any real number a and any whole number n.

Scientific notation *(205):* A number greater than or equal to 1 and less than 10 multiplied by some power of 10.

Trinomial *(210):* A three-termed polynomial.

Important Facts

Rules of Exponents

1. $x^m x^n = x^{m+n}$ product rule
2. $\dfrac{x^m}{x^n} = x^{m-n}$ quotient rule

wrong.

Helpful Hint

Some students start solving equation tion. Sometimes they are not sure tha give up for lack of confidence. You m as you follow the procedure on page tion, even if it takes quite a few ste Our goal is to isolate the variable, an equation you must also do to the oth of the equation equally.

Exercise Set 2.5

(a) *Represent the figure as an equation in the variable* x *and* **(b)** *solve the equation.*

1. $2x = x + 6, x = 6$
2. $x + 6 = 2x + 2, x = 4$
3. $5 + 2x = x + 19, x = 14$
4. $3x = x + 14, x = 7$
5. $5 + x = 2x + 5, x = 0$
6. $3x + 2 = x + 7, x = \frac{5}{2}$
7. $2x + 8 = x + 4, x = -4$
8. $4x + 19 = 2x + 5, x = -7$

Solve each equation. You may wish to use a calculator to solve the equations containing decimal numbers.

9. $4x = 3x + 5$ 5
11. $-4x + 10 = 6x$ 1
13. $5x + 3 = 6$ $\frac{3}{5}$
15. $15 - 3x = 4x - 2x$ 3
17. $2x - 4 = 3x - 6$ 2
19. $3 - 2y = 9 - 8y$ 1
21. $4 - 0.6x = 2.4x - 8.48$ 4.16
23. $5x = 2(x + 6)$ 4
25. $x - 25 = 12x + 9 + 3x$ $-\frac{17}{7}$
27. $2(x + 2) = 4x + 1 - 2x$ no solution
29. $-(w + 2) = -6w + 32$ $\frac{34}{5}$
31. $4 - (2x + 5) = 6x + 31$ -4
33. $0.1(x + 10) = 0.3x - 4$ 25

The exercise sets are carefully developed. They are graded in difficulty to help students gain confidence and succeed with the more difficult exercises. The extensive number of examples in the text help prepare students to understand and work all the exercises.

In mathematics we generally do not write a fraction with a negative sign in the denominator. When a negative sign appears in a denominator, we can move it to the numerator or place it in front of the fraction. For example, the fraction $\frac{5}{-7}$ should be written as either $-\frac{5}{7}$ or $\frac{-5}{7}$.

Example 14 Evaluate $\dfrac{2}{5} \div \dfrac{-8}{15}$.

Solution: $\dfrac{2}{5} \div \dfrac{-8}{15} = \dfrac{\overset{1}{\cancel{2}}}{\underset{1}{\cancel{5}}} \cdot \dfrac{\overset{3}{\cancel{15}}}{\underset{4}{-\cancel{8}}} = \dfrac{1(3)}{1(-4)} = \dfrac{3}{-4} = -\dfrac{3}{4}$

The operations on real numbers are summarized in Table 1.1.

TABLE 1.1 Summary of Operations on Real Numbers

Signs of Numbers	Addition	Subtraction	Multiplication	Division
Both Numbers Are Positive	Sum Is Always Positive	Difference May Be Either Positive or Negative	Product Is Always Positive	Quotient Is Always Positive
Examples 6 and 2 2 and 6	$6 + 2 = 8$ $2 + 6 = 8$	$6 - 2 = 4$ $2 - 6 = -4$	$6 \cdot 2 = 12$ $2 \cdot 6 = 12$	$6 \div 2 = 3$ $2 \div 6 = \frac{1}{3}$
One Number Is Positive and the Other Number Is Negative	Sum May Be Either Positive or Negative	Difference May Be Either Positive or Negative	Product Is Always Negative	Quotient Is Always Negative
Examples 6 and −2 −6 and 2	$6 + (-2) = 4$ $-6 + 2 = -4$	$6 - (-2) = 8$ $-6 - (2) = -8$	$6(-2) = -12$ $-6(2) = -12$	$6 \div (-2) = -3$ $-6 \div 2 = -3$
Both Numbers Are Negative	Sum Is Always Negative	Difference May Be Either Positive or Negative	Product Is Always Positive	Quotient Is Always Positive
Examples −6 and −2 −2 and −6	$-6 + (-2) = -8$ $-2 + (-6) = -8$	$-6 - (-2) = -4$ $-2 - (-6) = 4$	$-6(-2) = 12$ $-2(-6) = 12$	$-6 \div (-2) = 3$ $-2 \div (-6) = \frac{1}{3}$

Helpful Hints *boxes offer suggestions from problem solving and provide students with extra help on many topics.*

Helpful Hint

At this point some students begin confusing problems like $-2 - 3$ with $(-2)(-3)$ and problems like $2 - 3$ with problems like $2(-3)$. If you do not understand the difference between problems like $-2 - 3$ and $(-2)(-3)$, make an appointment to see your instructor as soon as possible.

Subtraction Problems	*Multiplication Problems*
$-2 - 3 = -5$	$(-2)(-3) = 6$
$2 - 3 = -1$	$(2)(-3) = -6$

To the Student

Algebra is a course that cannot be learned by observation. To learn algebra you must become an active participant. You must read the text, pay attention in class, and, most importantly, you must work the exercises. The more exercises you work, the better.

The text was written with you in mind. Short, clear sentences are used, and many examples are given to illustrate specific points. The text stresses useful applications of algebra. Hopefully, as you progress through the course, you will come to realize that algebra is not just another math course that you are required to take, but a course that offers a wealth of useful information and applications.

This text makes full use of color. The different colors are used to highlight important information. Important procedures, definitions, and formulas are placed within colored boxes.

The boxes marked **Helpful Hints** should be studied carefully, for they stress important information. The boxes marked **Common Student Errors** should also be studied carefully. These boxes point out errors that students commonly make, and provide the correct procedures for doing these problems.

Ask your professor early in the course to explain the policy on when the calculator may be used. Pay particular attention to the **Calculator Corners.**

Other questions you should ask your professor early in the course include: What supplements are available for use? Where can help be obtained when the professor is not available? Supplements that may be available include: Student's Study Guide, Student's Solutions Manual, tutorial software, and video tapes, including a tape on the study skills needed for success in mathematics.

You may wish to form a study group with other students in your class. Many students find that working in small groups provides an excellent way to learn the material. By discussing and explaining the concepts and exercises to one another you reinforce your own understanding. Once guidelines and procedures are determined by your group, make sure to follow them.

One of the first things you should do is to read Section 1.1, Study Skills Needed for Success in Mathematics. Read this section slowly and carefully, and pay particular attention to the advice and information given. Occasionally, refer back to this section. This could be the most important section of the book. Carefully read the material on doing your homework and on attending class.

At the end of all exercise sets (after the first two) are **cumulative review exercises.** You should work these problems on a regular basis, even if they are not assigned. These problems are from earlier sections and chapters of the text, and they will refresh your memory and reinforce those topics. If you have a problem when working these exercises, read the appropriate section of the text or study your notes that correspond to that material. The section of the text where the Cumulative Review Exercises were introduced is indicated in brackets, [ ], to the left of the exercise. After reviewing the material, if you still have a problem, make an appointment to see your professor. Working the Cumulative Review Exercises throughout the semester will also help prepare you to take your final exam.

At the end of each exercise set are **Group Activity/Challenge Problem** exercises. These exercises often provide questions that create interesting discussions, present additional applications of algebra, help prepare you for future material or present more challenging questions. You may wish to try some of these questions even if they are not assigned.

At the end of each chapter are a **Summary,** a set of **Review Exercises,** and a **Practice Test.** Before each examination you should review these sections carefully and take the practice test. If you do well on the practice test, you should do well on the class test. The questions in the review exercises are marked to indicate the section in which that material was first introduced. If you have a problem with a review exercise question, reread the section indicated. You may also wish to take the **Cumulative Review Test** that appears at the end of every even-numbered chapter.

In the back of the text there is an **answer section** which contains the answers to the *odd*-numbered exercises, *all* cumulative review exercises, review exercises, practice tests, and cumulative review tests. *Selected answers* are given for the Group Activity/Challenge Problem exercises. The answers should be used only to check your work.

I have tried to make this text as clear and error free as possible. No text is perfect, however. If you find an error in the text, or an example or section that you believe can be improved, I would greatly appreciate hearing from you. If you enjoy the text, I would also appreciate hearing from you.

Allen R. Angel

Elementary Algebra for College Students

Chapter 1
Real Numbers

See Section 1.2, Exercise 73

Preview and Perspective

In this chapter we provide the building blocks for this course and all other mathematics courses you may take.

A review of addition, subtraction, multiplication, and division of numbers containing decimal points is provided in Appendix A. Percents are also discussed in Appendix A. You may wish to review this material now.

For many students, Section 1.1, Study Skills for Success in Mathematics, may be the most important one in this book. Read it carefully and follow the advice given. Following the study skills presented will greatly increase your chance of success in this course and in all other mathematics courses.

In Section 1.2 we discuss fractions, including reducing fractions; addition, subtraction, multiplication, and division of fractions; and mixed numbers. *It is essential that you understand fractions* because we will work with fractions throughout the course. Furthermore, we will soon be discussing algebraic fractions, which we call rational expressions, and the same rules and procedures that apply to arithmetic fractions apply to rational expressions.

In Section 1.3 we introduce the structure of the real number system. In Section 1.4 we introduce inequalities and discuss absolute value informally. Both of these topics are important, especially if you plan on taking another mathematics course. In Sections 1.5 through 1.7 we discuss the operations on the real numbers. *Addition, subtraction, multiplication, and division of real numbers must be clearly understood before you finish this chapter* because you will use real numbers in Chapter 2 and throughout the book.

Exponents are introduced in Section 1.8. We discuss exponents in more depth in Section 4.1. You must master the order of operations to follow when evaluating expressions and formulas. We explain how to do this in Section 1.9.

In Section 1.10, we discuss the properties of the real number system, which we shall use throughout the book.

1.1 Study Skills for Success in Mathematics

Tape 1

1. Recognize the goals of this text.
2. Learn proper study skills.
3. Prepare for and take exams.
4. Learn to manage time and seek help when needed.
5. Purchase a scientific calculator.

This section is extremely important. Take the time to read it carefully and follow the advice given. For many of you this section may be the most important section of the text.

Most of you taking this course fall into one of three categories: (1) those who did not take algebra in high school, (2) those who took algebra in high school but did not understand the material, or (3) those who took algebra in high school and were successful but have been out of school for some time and need a refresher course. Whichever the case, you will need to acquire study skills for mathematics courses.

Before we discuss study skills I will present the goals of this text. These goals may help you realize why certain topics are covered in the text and why they are covered as they are.

Goals of This Text

1 The goals of this text include:

1. Presenting traditional algebra topics
2. Preparing you for more advanced mathematics courses
3. Building your confidence in, and your enjoyment of, mathematics
4. Improving your reasoning and critical thinking skills
5. Increasing your understanding of how important mathematics is in solving real-life problems
6. Encouraging you to think mathematically, so that you will feel comfortable translating real-life problems into mathematical equations, and then solving the problems.

It is important to realize that this course is the foundation for more advanced mathematics courses. A thorough understanding of algebra will make it easier for you to succeed in later mathematics courses.

2 Now we will consider study skills and other items of importance.

Have a Positive Attitude

You may be thinking to yourself, "I hate math," or "I wish I did not have to take this class." You may have heard of "math anxiety" and feel you fit this category. The first thing to do to be successful in this course is to change your attitude to a more positive one. You must be willing to give this course, and yourself, a fair chance.

Based on past experiences in mathematics, you may feel this is difficult. However, mathematics is something you need to work at. Many of you are more mature now than when you took previous mathematics courses. Your maturity and desire to learn are extremely important, and can make a tremendous difference in your ability to succeed in mathematics. I believe you can be successful in this course, but you also need to believe it.

Preparing for and Attending Class

To be prepared for class, you need to do your homework. If you have difficulty with the homework, or some of the concepts, write down questions to ask your instructor. If you were given a reading assignment, read the appropriate material carefully before class. If you were not given a reading assignment, spend a few minutes previewing any new material in the textbook before class. At this point you don't have to understand everything you read. Just get a feeling for the definitions and concepts that will be discussed. This quick preview will help you understand what your instructor is explaining during class.

After the material is explained in class, read the corresponding sections of the text slowly and carefully, word by word.

You should plan to attend every class. Most instructors agree that there is an inverse relationship between absences and grades. That is, the more absences you have, the lower your grade will be. Every time you miss a class, you miss important information. If you must miss a class, contact your instructor ahead of time, and get the reading assignment and homework. If possible, before the next class try to borrow and copy a friend's notes to help you understand the material you missed.

To be successful in this course, you must thoroughly understand the material in this chapter, especially fractions and adding and subtracting real numbers. If you are having difficulty after covering these topics, see your instructor for help.

In algebra and other mathematics courses, the material you learn is cumulative. That is, the new material is built on material that was presented previously. You must understand each section before moving on to the next section, and each chapter before moving on to the next chapter. Therefore, do not let yourself fall behind. Seek help as soon as you need it—do not wait! Make sure that you do all your homework assignments completely and study the text carefully. You will greatly increase your chance of success in this course by following the study skills presented in the next section.

While in class, pay attention to what your instructor is saying. If you don't understand something, ask your instructor to repeat the material. If you have read the assigned material before class and have questions that have not been answered, ask your instructor. If you don't ask questions, your instructor will not know that you have a problem understanding the material.

In class, take careful notes. Write numbers and letters clearly, so that you can read them later. It is not necessary to write down every word your instructor says. Copy the major points and the examples that do not appear in the text. You should not be taking notes so frantically that you lose track of what your instructor is saying. It is a mistake to believe that you can copy material in class without understanding it, and then figure it out later.

Reading the Text

A mathematics text is not a novel. Mathematics textbooks should be read slowly and carefully, word by word. If you don't understand what you are reading, reread the material. When you come across a new concept or definition, you may wish to underline it, so that it stands out. Then it will be easier to find later. When you come across an example, read and follow it line by line. Don't just skim it. Then work out the example on another sheet of paper. Make notes of anything you don't understand to ask your instructor.

This textbook has special features to help you. I suggest that you pay particular attention to these highlighted features, including the Common Student Error boxes, the Helpful Hint boxes, and important procedures and definitions identified by color. The **Common Student Error** boxes point out the most common errors made by students. Read and study this material very carefully and make sure that you understand what is explained. If you avoid making these common errors, your chances of success in this and other mathematics classes will be increased greatly. The **Helpful Hints** offer many valuable techniques for working certain problems. They may also present some very useful information or show an alternative way to work a problem.

Doing Homework

Two very important commitments that you must make to be successful in this course are attending class and doing your homework regularly. Your assignments must be worked conscientiously and completely. Do your homework as soon as possible, so the material presented in class will be fresh in your mind. Research has shown that for mathematics courses, studying and doing homework shortly after learning the material improves retention and performance. Mathematics cannot be learned by observation. You need to practice what you have heard in class. It is through doing homework that you truly learn the material. While working homework you will become aware of the types of problems that you need further help with. If you do not work the assigned exercises, you will not know what questions to ask in class.

When you do your homework, make sure that you write it neatly and carefully. List the exercise number next to each problem and work each problem in a step-by-step manner. Then you can refer to it later and understand what is written. Pay particular attention to copying signs and exponents correctly.

Don't forget to check the answers to your homework assignments. This book contains the answers to the odd-numbered exercises in the back of the book. In addition, the answers to all the cumulative review and end-of-chapter review exercises, practice tests, and cumulative review tests are in the back of the book. Answers to selected Group Activity/Challenge Problems are also provided.

Ask questions in class about homework problems you don't understand. You should not feel comfortable until you understand all the concepts needed to work every assigned problem successfully.

Studying for Class

Study in the proper atmosphere, in an area where you will not be constantly disturbed, so that your attention can be devoted to what you are reading. The area where you study should be well ventilated and well lit. You should have sufficient desk space to spread out all your materials. Your chair should be comfortable. There should be no loud music to distract you from studying.

Before you begin studying, make sure that you have all the materials you need (pencils, markers, calculator, etc.). You may wish to highlight the important points covered in class or in the book.

It is recommended that students study and do homework for at least two hours for each hour of class time. Some students require more time than others. It is important to spread your studying time out over the entire week rather than studying during one large block of time.

When studying, you should not only understand how to work a problem, but also know *why* you follow the specific steps you do to work the problem. If you do not have an understanding of why you follow the specific process, you will not be able to transfer the process to solve similar problems.

This book has **Cumulative Review Exercises** at the end of every section after this section. Even if these exercises are not assigned for homework, I urge you to work them as part of your studying process. These exercises reinforce material presented earlier in the course, and you will be less likely to forget the material if you review it repeatedly throughout the course. They will also help prepare you for the final exam. If you forget how to work one of the Cumulative Review Exercises, turn to the section indicated in blue next to the problem and review that section. Then try the problem again.

Preparing for an Exam

3 If you study a little bit each day you should not need to cram the night before an exam. Begin your studying early. If you wait until the last minute, you may not have time to seek the help you may need if you find you cannot work a problem.

To review for an exam:

1. Read your class notes.
2. Review your homework assignments.
3. Study formulas, definitions, and procedures given in the text.
4. Read the Common Student Error boxes and Helpful Hint boxes carefully.

5. Read the summary at the end of each chapter.
6. Work the review exercises at the end of each chapter. If you have difficulties, restudy those sections. If you still have trouble, seek help.
7. Work the chapter practice test.

Midterm and Final Exams

When studying for a comprehensive midterm or final exam follow the procedures discussed for preparing for an exam. However, also:

1. Study all your previous tests carefully. Make sure that you have learned to work the problems you may have previously missed.
2. Work the cumulative review tests at the end of each even-numbered chapter. These tests cover the material from the beginning of the book to the end of that chapter.
3. If your instructor has given you a worksheet or practice exam, make sure that you complete it. Ask questions on any problems you do not understand.
4. Begin your studying process early so that you can seek all the help you need in a timely manner.

Taking an Exam

Make sure you get sufficient sleep the night before the test. If you studied properly, you should not have to stay up late preparing for a test. Arrive at the exam site early so that you have a few minutes to relax before the exam. If you rush into the exam, you will start out nervous and anxious. After you are given the exam, you should do the following:

1. Carefully write down any formulas or ideas that you need to remember.
2. Look over the entire exam quickly to get an idea of its length. Also make sure that no pages are missing.
3. Read the test directions carefully.
4. Read each question carefully. Answer each question completely, and make sure that you have answered the specific question asked.
5. Work the questions you understand best first; then go back and work those you are not sure of. Do not spend too much time on any one problem or you may not be able to complete the exam. Be prepared to spend more time on problems worth more points.
6. Attempt each problem. You may get at least partial credit even if you do not obtain the correct answer. If you make no attempt at answering the question, you will lose full credit.
7. Work carefully in a step-by-step manner. Copy all signs and exponents correctly when working from step to step, and make sure to copy the original question from the test correctly.
8. Write clearly so that your instructor can read your work. If your instructor cannot read your work, you may lose credit. Also, if your writing is not clear, it is easy to make a mistake when working from one step to the next. When appropriate, make sure that your final answer stands out by placing a box around it.
9. If you have time, check your work and your answers.

10. Do not be concerned if others finish the test before you or if you are the last to finish. Use all your extra time to check your work.

Stay calm when taking your test. Do not get upset if you come across a problem you can't figure out right away. Go on to something else and come back to that problem later.

Time Management

4 As mentioned earlier, it is recommended that students study and do homework for at least two hours for each hour of class time. Finding the necessary time to study is not always easy. Below are some suggestions that you may find helpful.

1. Plan ahead. Determine when you will study and do your homework. Do not schedule other activities for these periods. Try to space these periods evenly over the week.
2. Be organized, so that you will not have to waste time looking for your books, your pen, your calculator, or your notes.
3. If you are allowed to use a calculator, use it for tedious calculations.
4. When you stop studying, clearly mark where you stopped in the text.
5. Try not to take on added responsibilities. You must set your priorities. If your education is a top priority, as it should be, you may have to reduce time spent on other activities.
6. If time is a problem, do not overburden yourself with too many courses. Consider taking fewer credits. If you do not have sufficient time to study, your understanding and all your grades may suffer.

Using Supplements

This text comes with a large variety of supplements. Find out from your instructor early in the semester which supplements are available and might be beneficial for you to use. Supplements should not replace reading the text, but should be used to enhance your understanding of the material.

Seeking Help

Be sure to get help as soon as you need it! Do not wait! In mathematics, one day's material is often based on the previous day's material. So, if you don't understand the material today, you will not be able to understand the material tomorrow.

Where should you seek help? There are often a number of resources on campus. Try to make a friend in the class with whom you can study. Often, you can help one another. You may wish to form a study group with other students in your class. Discussing the concepts and homework with your peers will reinforce your own understanding of the material.

You should know your instructor's office hours, and you should not hesitate to seek help from your instructor when you need it. Make sure that you have read the assigned material and attempted the homework before meeting with your instructor. Come prepared with specific questions to ask.

There are often other sources of help available. Many colleges have a mathematics lab or a mathematics learning center, where tutors are available. Ask your instructor early in the semester where and when tutoring is available. Arrange for a tutor as soon as you need one.

The Calculator

5 I strongly urge you to purchase a scientific calculator as soon as possible. One can be purchased for under $15 and can be used in many courses. Ask your instructor if you may use a calculator in class, on homework, and on tests. If so, you should use your calculator whenever possible to save time. Also ask your instructor if he or she recommends a particular calculator for this or a future mathematics class.

If a calculator contains a [LOG] key or [SIN] key, it is a scientific calculator. You *cannot* use the square root key [$\sqrt{x}$] to identify scientific calculators since both scientific calculators and nonscientific calculators may have this key. You should pay particular attention to the **Calculator Corners** in this book. The Calculator Corners explain how to use your calculator to solve problems.

A Final Word

You can be successful at mathematics if you attend class regularly, pay attention in class, study your text carefully, do your homework daily, review regularly, and seek help as soon as you need it. Good luck in your course.

Exercise Set 1.1

Do you know:

1. Your professor's name and office hours?

2. Your professor's office location and telephone number?

3. Where and when you can obtain help if your professor is not available?

4. The name and phone number of a friend in your class?

5. What supplements are available to assist you in learning?

6. If your instructor is recommending the use of a particular calculator?

7. When you can use your calculator in this course?

If you do not know the answer to any of the questions just asked, you should find out as soon as possible.

8. What are your reasons for taking this course?

9. What are your goals for this course?

10. Are you beginning this course with a positive attitude? It is important that you do!

11. List the things you need to do to prepare properly for class.

12. Explain how a mathematics text should be read.

13. For each hour of class time, how many hours outside of class are recommended for studying and doing homework?

14. When studying, you should not only understand how to work a problem, but also why you follow the specific steps you do. Why is this important?

15. Two very important commitments that you must make to be successful in this course are **(a)** doing homework regularly and completely and **(b)** attending class regularly. Explain why these commitments are necessary.

16. Write a summary of the steps you should follow when taking an exam.

17. Have you given any thought to studying with a friend or a group of friends? Can you see any advantages in doing so? Can you see any disadvantages in doing so?

1.2 Fractions

Tape 1

1. Learn multiplication symbols.
2. Recognize factors.
3. Reduce fractions to lowest terms.
4. Multiply fractions.
5. Divide fractions.
6. Add and subtract fractions.
7. Convert mixed numbers to fractions.

Students taking algebra for the first time often ask, "What is the difference between arithmetic and algebra?" When doing arithmetic, all the quantities used in the calculations are known. In algebra, however, one or more of the quantities are unknown and must be found.

EXAMPLE 1 A recipe calls for 3 cups of flour. Mrs. Clark has 2 cups of flour. How many additional cups does she need?

Solution: The answer is 1 cup.

Although very elementary, this is an example of an algebraic problem. The unknown quantity is the number of additional cups of flour needed.

An understanding of decimal numbers (see Appendix A) and fractions is essential to success in algebra. You will need to know how to reduce a fraction to its lowest terms and how to add, subtract, multiply, and divide fractions. We will review these topics in this section. We will also explain the meaning of factors.

Multiplication Symbols

1 In algebra we often use letters called **variables** to represent numbers. Letters commonly used as variables are **x, y,** and **z.** So that we do not confuse the variable x with the times sign, we use different notation to indicate multiplication.

> **Multiplication Symbols**
>
> If a and b stand for (or represent) any two mathematical quantities, then each of the following may be used to indicate the product of a and b ("a times b").
>
> $$ab \qquad a \cdot b \qquad a(b) \qquad (a)b \qquad (a)(b)$$

Examples

3 times 4 may be written:	3 times x may be written:	x times y may be written:
	$3x$	xy
$3(4)$	$3(x)$	$x(y)$
$(3)4$	$(3)x$	$(x)y$
$(3)(4)$	$(3)(x)$	$(x)(y)$
$3 \cdot 4$	$3 \cdot x$	$x \cdot y$

A word commonly used in algebra is "expression." An **expression** is a general term for any collection of numbers, variables, grouping symbols, such as parentheses ( ) or brackets [ ], and *operations,* such as addition, subtraction, multiplication, and division. Some examples of expressions are $5 - 2$, $x + 7$, $2x - 3y$, and $2(x + 3)$. We will discuss expressions further in Chapter 2.

Factors

2 The numbers or variables multiplied in a multiplication problem are called factors.

> If $a \cdot b = c$, then a and b are **factors** of c.

For example, in $3 \cdot 5 = 15$, the numbers 3 and 5 are factors of the product 15. In $2 \cdot 15 = 30$, the numbers 2 and 15 are factors of the product 30. Note that 30 has many other factors. Since $5 \cdot 6 = 30$, the numbers 5 and 6 are also factors of 30. Since $3x$ means 3 times x, both the 3 and the x are factors of $3x$.

Reduce Fractions

3 Now we have the necessary information to discuss fractions. The top number of a fraction is called the **numerator,** and the bottom number is called the **denominator.** In the fraction $\frac{3}{5}$, the 3 is the numerator and the 5 is the denominator.

A fraction is **reduced to its lowest terms** when the numerator and denominator have no common factors other than 1. To reduce a fraction to its lowest terms, follow these steps.

> **To Reduce a Fraction to Its Lowest Terms**
>
> 1. Find the largest number that will divide into (without remainder) both the numerator and the denominator. This number is called the **greatest common factor** (GCF).
> 2. Then divide both the numerator and the denominator by the greatest common factor.

If you do not remember how to find the greatest common factor of two or more numbers, read Appendix B.

EXAMPLE 2 Reduce $\dfrac{10}{25}$ to its lowest terms.

Solution: The largest number that divides both 10 and 25 is 5. Therefore, 5 is the greatest common factor. Divide both the numerator and the denominator by 5 to reduce the fraction to its lowest terms.

$$\frac{10}{25} = \frac{10 \div 5}{25 \div 5} = \frac{2}{5}$$

EXAMPLE 3 Reduce $\frac{6}{18}$ to its lowest terms.

Solution: Both 6 and 18 can be divided by 1, 2, 3, and 6. The largest of these numbers, 6, is the greatest common factor. Divide both the numerator and the denominator by 6.

$$\frac{6}{18} = \frac{6 \div 6}{18 \div 6} = \frac{1}{3}$$

Note in Example 3 that both the numerator and denominator could have been written with a factor of 6. Then the common factor 6 could be divided out.

$$\frac{6}{18} = \frac{1 \cdot \cancel{6}}{3 \cdot \cancel{6}} = \frac{1}{3}$$

When you work with fractions you should give your answers in lowest terms.

Multiplication of Fractions

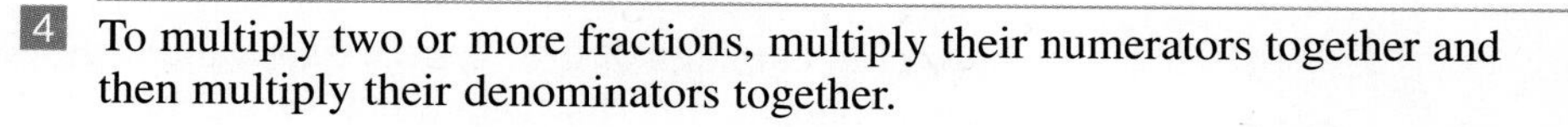

4 To multiply two or more fractions, multiply their numerators together and then multiply their denominators together.

Multiplication of Fractions

$$\frac{a}{b} \cdot \frac{c}{d} = \frac{ac}{bd}$$

EXAMPLE 4 Multiply $\frac{6}{13}$ by $\frac{5}{12}$.

Solution: $\frac{6}{13} \cdot \frac{5}{12} = \frac{6 \cdot 5}{13 \cdot 12} = \frac{30}{156} = \frac{5}{26}$

In Example 4, reducing $\frac{30}{156}$ to its lowest terms, $\frac{5}{26}$, is for many students more difficult than the multiplication itself. Before multiplying fractions, to help avoid having to reduce an answer to its lowest terms, we often divide both a numerator and a denominator by a common factor. **This process can be used only when multiplying fractions; it cannot be used when adding or subtracting fractions.**

EXAMPLE 5 Divide a numerator and a denominator by a common factor and then multiply.

$$\frac{6}{13} \cdot \frac{5}{12}$$

Solution: Since the numerator 6 and the denominator 12 can both be divided by the common factor 6, we divide out the 6 first.

$$\frac{6}{13} \cdot \frac{5}{12} = \frac{\overset{1}{\cancel{6}}}{13} \cdot \frac{5}{\underset{2}{\cancel{12}}} = \frac{1 \cdot 5}{13 \cdot 2} = \frac{5}{26}$$

Note that the answer obtained in Example 5 is identical to the answer obtained in Example 4.

EXAMPLE 6 Multiply $\frac{27}{40} \cdot \frac{16}{9}$.

Solution:
$$\frac{27}{40} \cdot \frac{16}{9} = \frac{\overset{3}{\cancel{27}}}{40} \cdot \frac{16}{\underset{1}{\cancel{9}}} \qquad \text{Divide both 27 and 9 by 9.}$$

$$= \frac{\overset{3}{\cancel{27}}}{\underset{5}{\cancel{40}}} \cdot \frac{\overset{2}{\cancel{16}}}{\underset{1}{\cancel{9}}} \qquad \text{Divide both 40 and 16 by 8.}$$

$$= \frac{3 \cdot 2}{5 \cdot 1} = \frac{6}{5}$$

The numbers 0, 1, 2, 3, 4, . . . are called **whole numbers.** The three dots after the 4 indicate that the whole numbers continue indefinitely in the same manner. Thus the numbers 468 and 5043 are also whole numbers. Whole numbers will be discussed further in Section 1.3. To multiply a whole number by a fraction, write the whole number with a denominator of 1 and then multiply.

EXAMPLE 7 Multiply $5 \cdot \frac{2}{15}$.

Solution: $\frac{5}{1} \cdot \frac{2}{15} = \frac{\overset{1}{\cancel{5}}}{1} \cdot \frac{2}{\underset{3}{\cancel{15}}} = \frac{2}{3}$

Division of Fractions

5 To divide one fraction by another, invert the divisor (the second fraction if written with $\div$) and proceed as in multiplication.

Division of Fractions

$$\frac{a}{b} \div \frac{c}{d} = \frac{a}{b} \cdot \frac{d}{c} = \frac{ad}{bc}$$

EXAMPLE 8 Divide $\frac{3}{5} \div \frac{5}{6}$.

Solution: $\frac{3}{5} \div \frac{5}{6} = \frac{3}{5} \cdot \frac{6}{5} = \frac{3 \cdot 6}{5 \cdot 5} = \frac{18}{25}$

Sometimes, rather than being asked to obtain the answer to a problem by adding, subtracting, multiplying, or dividing, you may be asked to evaluate an expression. To **evaluate** an expression means to obtain the answer to the problem using the operations given.

EXAMPLE 9 Evaluate $\frac{4}{7} \div \frac{5}{12}$.

Solution: $\frac{4}{7} \div \frac{5}{12} = \frac{4}{7} \cdot \frac{12}{5} = \frac{48}{35}$

EXAMPLE 10 Evaluate $\frac{3}{8} \div 9$.

Solution: Write 9 as $\frac{9}{1}$.

$$\frac{3}{8} \div 9 = \frac{3}{8} \div \frac{9}{1} = \frac{\overset{1}{\cancel{3}}}{8} \cdot \frac{1}{\underset{3}{\cancel{9}}} = \frac{1}{24}$$

Addition and Subtraction of Fractions

6 *Only fractions that have the same* (or a common) *denominator can be added or subtracted.* To add (or subtract) fractions with the same denominator, add (or subtract) the numerators and keep the common denominator.

Addition and Subtraction of Fractions

$$\frac{a}{c} + \frac{b}{c} = \frac{a + b}{c} \quad \text{or} \quad \frac{a}{c} - \frac{b}{c} = \frac{a - b}{c}$$

EXAMPLE 11 Evaluate $\frac{9}{15} + \frac{2}{15}$.

Solution: $\frac{9}{15} + \frac{2}{15} = \frac{9 + 2}{15} = \frac{11}{15}$

EXAMPLE 12 Evaluate $\frac{8}{13} - \frac{5}{13}$.

Solution: $\frac{8}{13} - \frac{5}{13} = \frac{8 - 5}{13} = \frac{3}{13}$

To add (or subtract) fractions with unlike denominators, we must first rewrite each fraction with the same, or a common, denominator. The smallest number that is divisible by two or more denominators is called the **least common denominator.** *If you have forgotten how to find the least common denominator, or LCD, review Appendix B now.*

Example 13 Add $\frac{1}{2} + \frac{1}{5}$.

Solution: We cannot add these fractions until we rewrite them with a common denominator. Since the lowest number that both 2 and 5 divide into (without remainder) is 10, we will first rewrite both fractions with the least common denominator of 10.

$$\frac{1}{2} = \frac{1}{2} \cdot \frac{5}{5} = \frac{5}{10} \quad \text{and} \quad \frac{1}{5} = \frac{1}{5} \cdot \frac{2}{2} = \frac{2}{10}$$

Now add.

$$\frac{1}{2} + \frac{1}{5} = \frac{5}{10} + \frac{2}{10} = \frac{7}{10}$$

Note that multiplying both the numerator and denominator by the same number is the same as multiplying by 1. Thus the value of the fraction does not change.

Example 14 Subtract $\frac{3}{4} - \frac{2}{3}$.

Solution: The least common denominator is 12. Therefore, we rewrite both fractions with a denominator of 12.

$$\frac{3}{4} = \frac{3}{4} \cdot \frac{3}{3} = \frac{9}{12} \quad \text{and} \quad \frac{2}{3} = \frac{2}{3} \cdot \frac{4}{4} = \frac{8}{12}$$

Now subtract.

$$\frac{3}{4} - \frac{2}{3} = \frac{9}{12} - \frac{8}{12} = \frac{1}{12}$$

Common Student Error It is important that you realize that dividing out a common factor in the numerator of one fraction and the denominator of a different fraction can be performed only when multiplying fractions. **This process cannot be performed when adding or subtracting fractions.**

Correct	*Incorrect*
Multiplication problems	Addition problems
$\frac{\overset{1}{\cancel{3}}}{5} \cdot \frac{1}{\underset{1}{\cancel{3}}}$	$\frac{\overset{1}{\cancel{3}}}{5} + \frac{\cancel{1}}{\underset{1}{\cancel{3}}}$ (crossed out)
$\frac{\overset{2}{\cancel{8}} \cdot 3}{\underset{1}{\cancel{4}}}$	$\frac{\overset{2}{\cancel{8}} + 3}{\underset{1}{\cancel{4}}}$ (crossed out)

Mixed Numbers

7 Consider the number $5\frac{2}{3}$. This is an example of a **mixed number.** A mixed number consists of a whole number followed by a fraction. The mixed number $5\frac{2}{3}$ means $5 + \frac{2}{3}$. The mixed number $5\frac{2}{3}$ may be changed to a fraction as follows:

$$5\frac{2}{3} = 5 + \frac{2}{3} = \frac{15}{3} + \frac{2}{3} = \frac{17}{3}$$

Any fraction whose numerator is greater than its denominator* may be changed to a mixed number. For example, $\frac{17}{3}$ may be changed to

$$\frac{17}{3} = \frac{15}{3} + \frac{2}{3} = 5 + \frac{2}{3} = 5\frac{2}{3}$$

The procedure used to change from a mixed number to a fraction can be simplified as follows.

To Change a Mixed Number to a Fraction

1. Multiply the denominator of the fraction in the mixed number by the whole number preceding it.
2. Add the numerator of the fraction in the mixed number to the product obtained in step 1. This sum represents the numerator of the fraction we are seeking. The denominator of the fraction we are seeking is the same as the denominator of the fraction in the mixed number.

EXAMPLE 15 Change the mixed number $5\frac{2}{3}$ to a fraction.

Solution: Multiply the denominator, 3, by the whole number, 5, to get a product of 15. To this product add the numerator, 2. This sum, 17, represents the numerator of the fraction. The denominator of the fraction we are seeking is the same as the denominator of the fraction in the mixed number, 3. Thus, $5\frac{2}{3} = \frac{17}{3}$.

$$5\frac{2}{3} = \frac{15 + 2}{3} = \frac{17}{3}$$

EXAMPLE 16 Change $6\frac{5}{9}$ to a fraction.

Solution: Multiply 9 by 6 to get 54; then add 5 to get 59. This is the numerator of the fraction we are seeking.

$$6\frac{5}{9} = \frac{54 + 5}{9} = \frac{59}{9}$$

*A fraction such as $\frac{17}{3}$, whose numerator is greater than its denominator, is sometimes referred to as an *improper fraction.* However, this is misleading because there is nothing "improper" about such fractions. In fact, such fractions are used in all mathematics, and they are generally preferred to mixed numbers.

To Change a Fraction Greater Than 1 to a Mixed Number

1. Divide the numerator by the denominator. Note the quotient and remainder.
2. Write the mixed number. The quotient found in step 1 is the whole number part of the mixed number. The remainder is the numerator of the fraction in the mixed number. The denominator in the fraction of the mixed number will be the same as the denominator in the original fraction.

EXAMPLE 17 Change $\frac{17}{3}$ to a mixed number.

Solution:

$$3\overline{)17}\text{ quotient }5 \leftarrow \text{whole number}$$

Denominator (or divisor) ⟶ 3

$$\begin{array}{r} 5 \\ 3\overline{)17} \\ \underline{15} \\ 2 \end{array}$$

5 ⟵ whole number; 2 ⟵ remainder

$$\frac{17}{3} = 5\frac{2}{3}$$

2 ⟵ remainder
3 ⟵ denominator (or divisor)
5 ⟵ whole number

Thus, $\frac{17}{3}$ changed to a mixed number is $5\frac{2}{3}$.

EXAMPLE 18 Change $\frac{21}{5}$ to a mixed number.

Solution:

$$\begin{array}{r} 4 \\ 5\overline{)21} \\ \underline{20} \\ 1 \end{array} \quad \text{therefore} \quad \frac{21}{5} = 4\frac{1}{5}$$

To add, subtract, multiply, or divide mixed numbers, we often change the mixed numbers to fractions.

EXAMPLE 19 Add $2\frac{1}{4} + \frac{1}{2}$.

Solution: Change $2\frac{1}{4}$ to $\frac{9}{4}$; then add.

$$\begin{aligned} 2\frac{1}{4} + \frac{1}{2} &= \frac{9}{4} + \frac{1}{2} \\ &= \frac{9}{4} + \frac{2}{4} \\ &= \frac{11}{4} \text{ or } 2\frac{3}{4} \end{aligned}$$

EXAMPLE 20 Multiply $\left(3\frac{3}{4}\right)\left(4\frac{3}{5}\right)$.

Solution: Change both mixed numbers to fractions; then multiply.

$$\left(3\frac{3}{4}\right)\left(4\frac{3}{5}\right) = \frac{\overset{3}{\cancel{15}}}{4} \cdot \frac{23}{\underset{1}{\cancel{5}}} = \frac{69}{4} \quad \text{or} \quad 17\frac{1}{4}$$

EXAMPLE 21 Divide $\frac{4}{5} \div 2\frac{5}{8}$.

Solution: Change $2\frac{5}{8}$ to a fraction; then follow the procedure for dividing fractions.

$$\frac{4}{5} \div 2\frac{5}{8} = \frac{4}{5} \div \frac{21}{8}$$

$$= \frac{4}{5} \cdot \frac{8}{21} = \frac{32}{105}$$

Exercise Set 1.2

Reduce each fraction to its lowest terms. If a fraction is already in its lowest terms, so state.

1. $\frac{4}{16}$
2. $\frac{5}{20}$
3. $\frac{10}{15}$
4. $\frac{3}{8}$
5. $\frac{15}{30}$
6. $\frac{9}{30}$
7. $\frac{15}{35}$
8. $\frac{36}{72}$
9. $\frac{40}{64}$
10. $\frac{15}{120}$
11. $\frac{9}{14}$
12. $\frac{6}{42}$
13. $\frac{96}{72}$
14. $\frac{14}{28}$
15. $\frac{50}{35}$
16. $\frac{84}{28}$

Indicate any parts where a common factor can be divided out as a first step in solving the problem. Explain your answer.

17. (a) $\frac{3}{5} \cdot \frac{10}{11}$ (b) $\frac{3}{5} + \frac{10}{11}$ (c) $\frac{3}{5} - \frac{10}{11}$ (d) $\frac{3}{5} \div \frac{10}{11}$

18. (a) $\frac{4}{5} + \frac{1}{4}$ (b) $\frac{4}{5} - \frac{1}{4}$ (c) $\frac{4}{5} \cdot \frac{1}{4}$ (d) $\frac{4}{5} \div \frac{1}{4}$

19. (a) $6 + \frac{5}{12}$ (b) $6 \cdot \frac{5}{12}$ (c) $6 - \frac{5}{12}$ (d) $6 \div \frac{5}{12}$

20. (a) $4 + \frac{3}{4}$ (b) $4 - \frac{3}{4}$ (c) $4 \div \frac{3}{4}$ (d) $4 \cdot \frac{3}{4}$

Find the product or quotient. Write the answers in lowest terms.

21. $\frac{1}{2} \cdot \frac{3}{4}$ **22.** $\frac{3}{5} \cdot \frac{4}{7}$ **23.** $\frac{5}{4} \cdot \frac{2}{7}$ **24.** $\frac{5}{12} \cdot \frac{6}{5}$

25. $\frac{3}{8} \cdot \frac{2}{9}$ **26.** $\frac{15}{16} \cdot \frac{4}{3}$ **27.** $\frac{1}{4} \div \frac{1}{5}$ **28.** $\frac{2}{3} \cdot \frac{3}{5}$

29. $\frac{5}{12} \div \frac{4}{3}$ **30.** $\frac{2}{9} \div \frac{12}{5}$ **31.** $\frac{10}{3} \div \frac{5}{9}$ **32.** $\frac{12}{5} \div \frac{3}{7}$

33. $\frac{5}{12} \cdot \frac{16}{15}$ **34.** $\frac{3}{10} \cdot \frac{5}{12}$ **35.** $\frac{4}{15} \div \frac{13}{12}$ **36.** $\frac{15}{16} \div \frac{1}{2}$

37. $\frac{12}{7} \cdot \frac{19}{24}$ **38.** $\frac{28}{13} \cdot \frac{2}{7}$ **39.** $1\frac{4}{5} \cdot \frac{20}{3}$ **40.** $4\frac{4}{5} \div \frac{8}{15}$

41. $\left(\frac{3}{5}\right)\left(1\frac{2}{3}\right)$ **42.** $\left(\frac{5}{8}\right)\left(3\frac{1}{3}\right)$ **43.** $3\frac{2}{3} \div 1\frac{5}{6}$ **44.** $3\frac{1}{4} \div \frac{5}{6}$

Add or subtract. Write the answers in lowest terms.

45. $\frac{2}{7} + \frac{3}{7}$ **46.** $\frac{3}{10} + \frac{5}{10}$ **47.** $\frac{7}{12} - \frac{5}{12}$ **48.** $\frac{18}{36} - \frac{1}{36}$

49. $\frac{5}{14} + \frac{9}{14}$ **50.** $\frac{9}{10} - \frac{3}{10}$ **51.** $\frac{19}{26} - \frac{5}{26}$ **52.** $\frac{1}{3} + \frac{1}{5}$

53. $\frac{2}{5} + \frac{5}{6}$ **54.** $\frac{1}{9} - \frac{1}{18}$ **55.** $\frac{4}{12} - \frac{2}{15}$ **56.** $\frac{5}{6} - \frac{3}{7}$

57. $\frac{2}{10} + \frac{1}{15}$ **58.** $\frac{5}{8} - \frac{1}{6}$ **59.** $\frac{5}{8} - \frac{4}{7}$ **60.** $\frac{3}{8} + \frac{5}{12}$

61. $\frac{5}{6} + \frac{9}{24}$ **62.** $\frac{7}{15} - \frac{12}{30}$ **63.** $\frac{4}{7} - \frac{1}{4}$ **64.** $3\frac{1}{2} + \frac{1}{4}$

65. $4\frac{1}{4} + \frac{2}{5}$ **66.** $\frac{3}{10} + 2\frac{1}{3}$ **67.** $2\frac{1}{2} + 1\frac{1}{3}$ **68.** $\frac{4}{5} - \frac{2}{7}$

69. $4\frac{2}{3} - 1\frac{1}{5}$ **70.** $3\frac{1}{8} - \frac{3}{4} - \frac{1}{2}$ **71.** $1\frac{4}{5} - \frac{3}{4} + 3$ **72.** $2\frac{2}{3} + 1\frac{3}{5} - \frac{3}{12}$

In many problems you will need to subtract a fraction from 1, where 1 represents "the whole" or the "total amount." Exercises 73–76 are answered by subtracting the fraction given from 1.

73. On Earth, $\frac{1}{6}$ of all fresh water is in the Antarctic. How much of Earth's supply of fresh water is elsewhere?

74. The probability that an event does not occur may be found by subtracting the probability that the event does occur from 1. If the probability that global warming is occurring is $\frac{7}{9}$, find the probability that global warming is not occurring.

75. An article in your local newspaper states that the chance of a person in your tax bracket being audited by the Internal Revenue Service is $\frac{1}{12}$. What is the chance of a person in this tax bracket not being audited?

76. Of all oats grown in the United States, $\frac{19}{20}$ is fed to animals. How much of the total crop is not fed to animals?

Solve.

77. Paul, a drapemaker, wishes to make 3 identical pairs of drapes. If each pair needs $6\frac{3}{4}$ yards of material, how much material will Paul need?

78. A board is $22\frac{1}{2}$ feet long. What is the length of each piece when cut in five equal lengths? (Ignore the thickness of the cuts.)

79. A length of $3\frac{1}{16}$ inches is cut from a piece of wood $16\frac{3}{4}$ inches long. What is the length of the remaining piece of wood?

80. At the beginning of the day a stock was selling for $11\frac{7}{8}$ dollars. At the close of the session it was selling for $13\frac{3}{4}$ dollars. How much did the stock gain that day?

81. Ellen soldered two pieces of pipe measuring $3\frac{3}{8}$ feet and $5\frac{1}{16}$ feet. What is the total length of these two pieces of pipe?

82. A recipe calls for $2\frac{1}{2}$ cups of flour and another $1\frac{1}{3}$ cups of flour to be added later. How much flour does the recipe require?

83. At high tide the water level at a measuring stick is $20\frac{3}{4}$ feet. At low tide the water level dropped to $8\frac{7}{8}$ feet. How much did the water level fall?

84. The inseam on a new pair of pants is 30 inches. If Leland's inseam is $28\frac{3}{8}$ inches, how much will the cuffs need to be shortened?

85. The instructions on a turkey indicate that a 12- to 16-pound turkey should bake at 325°F for about 22 minutes per pound. Donna is planning to bake a $13\frac{1}{2}$ pound turkey. Approximately how long should the turkey be baked?

86. A recipe calls for $\frac{3}{4}$ teaspoon of teriyaki seasoning for each pound of beef. To cook $4\frac{1}{2}$ pounds of beef, how many teaspoons of teriyaki are needed?

87. Dawn cuts a piece of wood measuring $3\frac{1}{8}$ inches into two equal pieces. How long is each piece?

88. Tom wishes to subdivide a $4\frac{5}{8}$-acre lot into three equal-size lots. What will be the acreage of each lot?

89. A nurse must give $\frac{1}{16}$ milligram of a drug for each kilogram of patient weight. If Mr. Duncan weighs (or has a mass of) 80 kilograms, find the amount of the drug Mr. Duncan should be given.

90. Find the total height of the computer desk shown in the figure.

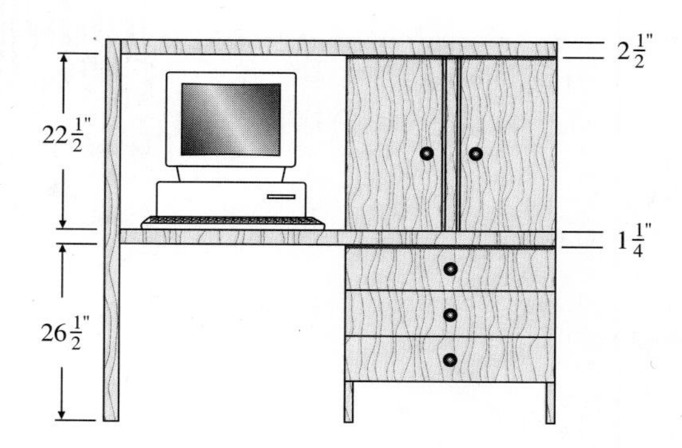

91. Marcinda is considering purchasing a mail order computer. The catalog describes the computer as $7\frac{1}{2}$ inches high and the monitor as $14\frac{3}{8}$ inches high. Marcinda is hoping to place the monitor on top of the computer and to place the computer and monitor together in the opening where the computer is shown in the desk in Exercise 90.

(a) Will there be sufficient room to do this?

(b) If so, how much extra height will she have?

92. A mechanic wishes to use a bolt to fasten a piece of wood $4\frac{1}{2}$ inches thick with a metal tube $2\frac{1}{3}$ inches thick. If the thickness of the nut is $\frac{1}{8}$ inch, find the length of the shaft of the bolt so that the nut fits flush with the end of the bolt (see the figure).

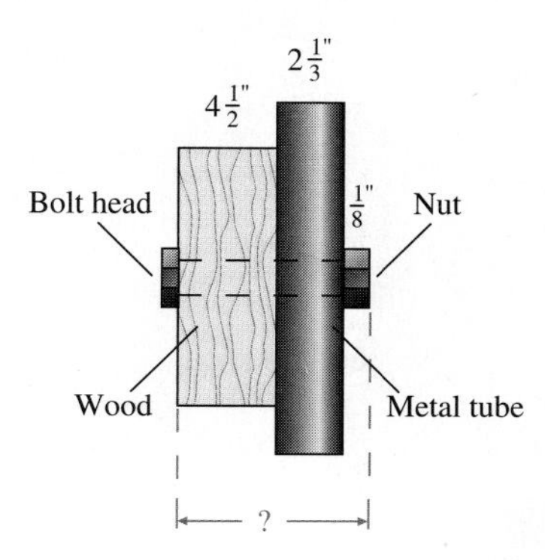

93. Find:

(a) $\frac{5}{6} \cdot \frac{3}{8}$ **(b)** $\frac{5}{6} \div \frac{3}{8}$

(c) $\frac{5}{6} + \frac{3}{8}$ **(d)** $\frac{5}{6} - \frac{3}{8}$

94. What are numbers or variables being multiplied called?

95. What is an expression?

96. In a fraction, what is the name of the **(a)** top number and **(b)** bottom number?

97. Consider parts (a) and (b) below.

(a) $\dfrac{\overset{1}{\cancel{3}}}{5} \cdot \dfrac{1}{\underset{2}{\cancel{6}}}$ **(b)** $\dfrac{\overset{1}{\cancel{3}}}{\underset{2}{\cancel{6}}}$

Which part shows reducing a fraction to its lowest terms? Explain.

98. Explain how to reduce a fraction to its lowest terms.

99. Explain how to multiply fractions.

100. Explain how to divide fractions.

101. Explain how to add or subtract fractions.

102. Explain how to convert a mixed number to a fraction.

103. Explain how to convert a fraction whose numerator is greater than its denominator into a mixed number.

Group Activity/Challenge Problems

1. The directions below show how to make either two or four servings of Minute Rice.

(a) Write down three different methods that can be used to find the amount of each ingredient needed to make three servings of Minute Rice.

(b) Find the amount of each ingredient using each of the three methods.

(c) Do the answers from each of the three methods agree?

AMOUNTS OF RICE AND WATER: USE EQUAL AMOUNTS RICE AND WATER. MINUTE RICE DOUBLES IN VOLUME.

To Make	Rice & Water (equal measures)	Salt	Butter or Margarine (if desired)
2 servings	$\frac{2}{3}$ cup	$\frac{1}{4}$ tsp.	1 tsp.
4 servings	$1\frac{1}{3}$ cups	$\frac{1}{2}$ tsp.	2 tsp.

MINUTE® is a registered trademark of General Foods Corporation, White Plains, N.Y.

2. An allopurinal pill comes in 300-milligram doses. Dr. Duncan wants a patient to get 450 milligrams each day by cutting the pills in half and taking $\frac{1}{2}$ pill three times a day. If she wants to prescribe enough pills for a 6-month period (assume 30 days per month), how many pills should she prescribe?

Evaluate

3. $\dfrac{\frac{1}{2}+\frac{3}{4}}{\frac{3}{4}-\frac{1}{3}}$

4. $\dfrac{\frac{12}{5}-\frac{5}{4}}{\frac{5}{9}\div\frac{2}{3}}$

5. $\left(\frac{5}{12}+\frac{3}{5}\right)\div\left(\frac{5}{7}\cdot\frac{3}{10}\right)$

1.3 The Real Number System

Tape 1

1 Identify some important sets of numbers.
2 Know the structure of the real numbers.

We will be talking about and using various types of numbers throughout the text. This section introduces you to some of those numbers and to the structure of the real number system. This section is a quick overview. Some of the sets of numbers we mention in this section, such as rational and irrational numbers, are discussed in greater depth later in the text.

Sets of Numbers

1 A **set** is a collection of **elements** listed within braces. The set $\{a, b, c, d, e\}$ consists of five elements, namely a, b, c, d, and e. A set that contains no elements is called an **empty set** (or **null set**). The symbol { } or the symbol ϕ is used to represent the empty set.

There are many different sets of numbers. Two important sets are the natural numbers and the whole numbers. The whole numbers were introduced earlier.

Natural numbers: $\{1, 2, 3, 4, 5, \ldots\}$
Whole numbers: $\{0, 1, 2, 3, 4, 5, \ldots\}$

An aid in understanding sets of numbers is the real number line (Fig. 1.1).

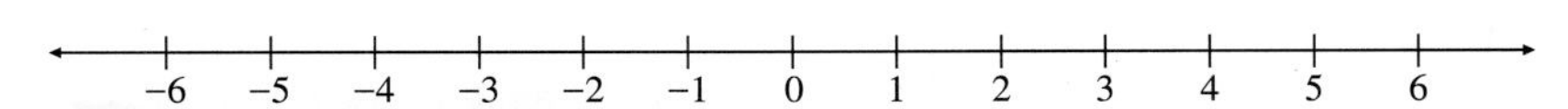

FIGURE 1.1

The real number line continues indefinitely in both directions. The numbers to the right of 0 are positive and those to the left of 0 are negative. Zero is neither positive nor negative (Fig. 1.2).

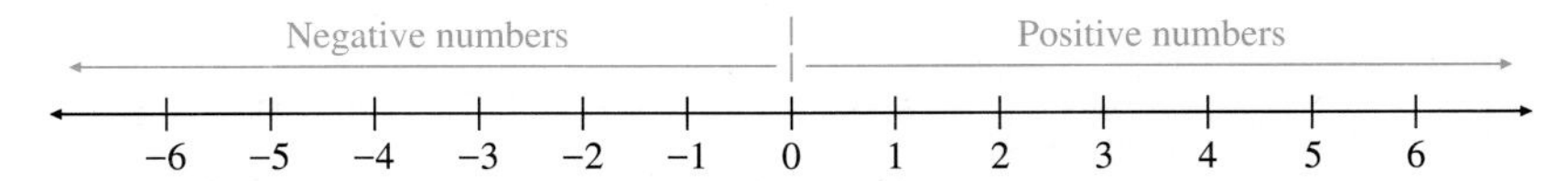

FIGURE 1.2

Figure 1.3 illustrates the natural numbers marked on the number line. The natural numbers are also called the **positive integers** or the **counting numbers.**

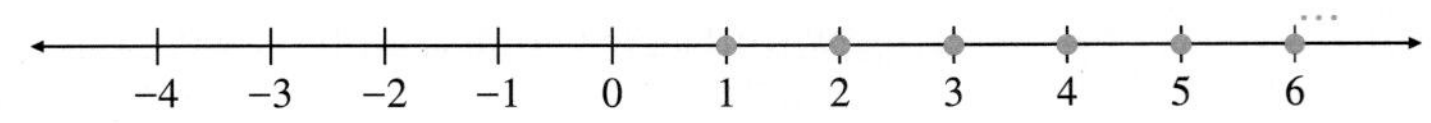

FIGURE 1.3

Another important set of numbers is the integers.

Integers: $\{\underbrace{\ldots, -5, -4, -3, -2, -1}_{\text{negative integers}}, 0, \underbrace{1, 2, 3, 4, 5, \ldots}_{\text{positive integers}}\}$

The integers consist of the negative integers, 0, and the positive integers. The integers are marked on the number line in Figure 1.4.

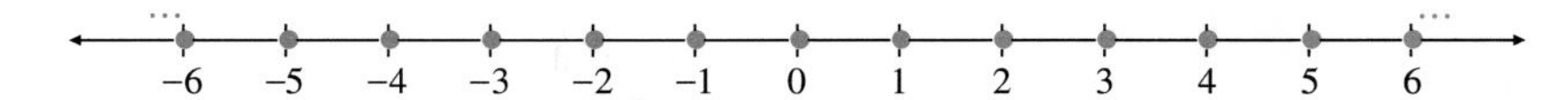

Figure 1.4

Can you think of any numbers that are not integers? You probably thought of "fractions" or "decimal numbers." Fractions and decimal numbers belong to the set of rational numbers. The set of **rational numbers** consists of all the numbers that can be expressed as a quotient of two integers, with the denominator not 0.

Rational numbers: {quotient of two integers, denominator not 0}

The fraction $\frac{1}{2}$ is a quotient of two integers with the denominator not 0. Thus, $\frac{1}{2}$ is a rational number. The decimal number 0.4 can be written $\frac{4}{10}$ and is therefore a rational number. All integers are also rational numbers since they can be written with a denominator of 1: for example, $3 = \frac{3}{1}$, $-12 = \frac{-12}{1}$, and $0 = \frac{0}{1}$. Some rational numbers are illustrated on the number line in Figure 1.5.

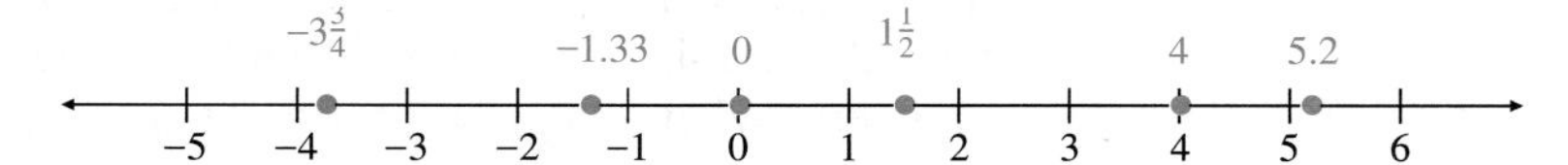

Figure 1.5

Most of the numbers that we use are rational numbers; however, some numbers are not rational. Numbers such as the square root of 2, written $\sqrt{2}$, are not rational numbers. Any number that can be represented on the number line that is not a rational number is called an **irrational number.** The $\sqrt{2}$ is *approximately* 1.41. Some irrational numbers are illustrated on the number line in Figure 1.6. Rational and irrational numbers will be discussed further in later chapters.

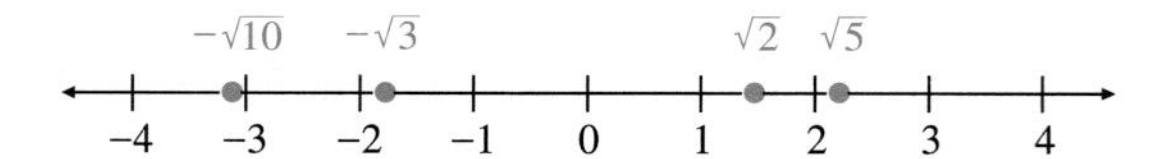

Figure 1.6

Structure of the Real Numbers

2 Notice that many different types of numbers can be illustrated on the number line. Any number that can be represented on the number line is a **real number.**

Real numbers: {all numbers that can be represented on the real number line}

The symbol $\mathbb{R}$ is used to represent the set of real numbers. All the numbers mentioned thus far are real numbers. The natural numbers, the whole numbers, the integers, the rational numbers, and the irrational numbers are all real numbers. There are some types of numbers that are not real numbers, but these numbers are beyond the scope of this book. Figure 1.7 illustrates the relationship between the various sets of numbers within the set of real numbers.

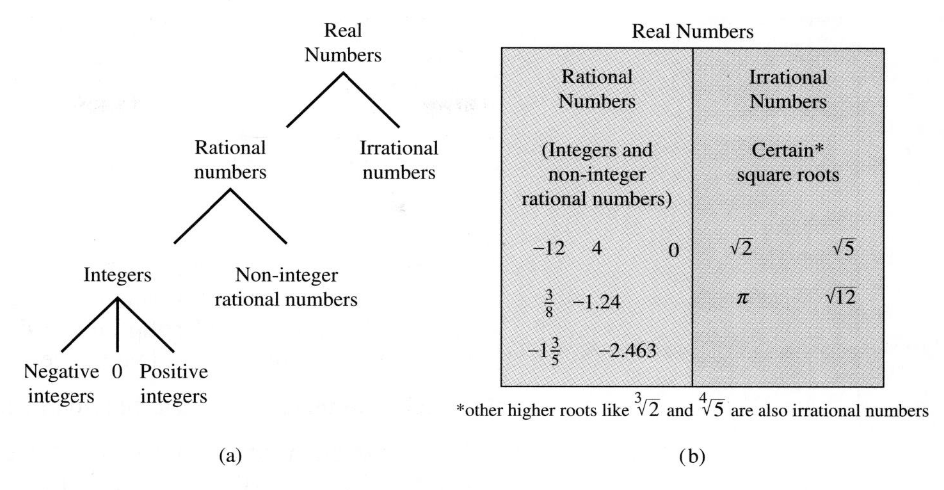

FIGURE 1.7

In Figure 1.7(a), we can see that when we combine the rational numbers and the irrational numbers we get the real numbers. When we combine the integers with the non-integer rational numbers (such as $\frac{1}{2}$ and 0.42), we get the rational numbers. When we combine the positive integers (or natural numbers), 0, and the negative integers, we get the integers.

Consider the positive integer 5. If we follow the positive integer branch in Figure 1.7(a) upward, we see that the number 5 is also an integer, a rational number, and a real number. Now consider the number $\frac{1}{2}$. It belongs to the non-integer rational numbers. If we follow this branch upward, we can see that $\frac{1}{2}$ is also a rational number and a real number.

EXAMPLE 1 Consider the following set of numbers:

$$\left\{-6, -0.5, 4\frac{1}{2}, -96, \sqrt{3}, 0, 9, -\frac{4}{7}, -2.9, \sqrt{7}, -\sqrt{5}\right\}$$

List the elements of the set that are:

(a) Natural numbers. **(b)** Whole numbers.
(c) Integers. **(d)** Rational numbers.
(e) Irrational numbers. **(f)** Real numbers.

Solution: **(a)** 9 **(b)** 0, 9 **(c)** $-6, -96, 0, 9$

(d) $-6, -0.5, 4\frac{1}{2}, -96, 0, 9, -\frac{4}{7}, -2.9$

(e) $\sqrt{3}, \sqrt{7}, -\sqrt{5}$

(f) $-6, -0.5, 4\frac{1}{2}, -96, \sqrt{3}, 0, 9, -\frac{4}{7}, -2.9, \sqrt{7}, -\sqrt{5}$

Exercise Set 1.3

List each set of numbers.

1. Integers.

2. Whole numbers.

3. Natural numbers.

4. Negative integers.

5. Positive integers.

6. Counting numbers.

In Exercises 7–38, state whether each statement is true or false.

7. -1 is a negative integer.

8. 0 is a whole number.

9. 0 is an integer.

10. -1.36 is a real number.

11. $\frac{1}{2}$ is an integer.

12. 0.5 is an integer.

13. $\sqrt{2}$ is a rational number.

14. $\sqrt{7}$ is a real number.

15. $-\frac{3}{5}$ is a rational number.

16. 0 is a rational number.

17. $-4\frac{1}{3}$ is a rational number.

18. -5 is both a rational number and a real number.

19. $-\frac{5}{3}$ is an irrational number.

20. $2\frac{5}{8}$ is an irrational number.

21. 0 is a positive number.

22. When zero is added to the set of counting numbers, the set of whole numbers is formed.

23. The natural numbers, counting numbers, and positive integers are different names for the same set of numbers.

24. When the negative integers, the positive integers, and 0 are combined, the integers are formed.

25. Any number to the left of zero on the number line is a negative number.

26. Every negative integer is a real number.

27. Every integer is a rational number.

28. Every rational number is a real number.

29. Every real number is a rational number.

30. Every negative number is a negative integer.

31. Some real numbers are not rational numbers.

32. Some rational numbers are not real numbers.

33. The symbol $\mathbb{R}$ is used to represent the set of real numbers.

34. Every integer is positive.

35. The symbol ϕ is used to represent the empty set.

36. All real numbers can be represented on the number line.

37. Every number greater than zero is a positive integer.

38. Irrational numbers cannot be represented on the number line.

39. Consider the set of numbers.

$$\left\{-6, 7, 12.4, -\frac{9}{5}, -2\frac{1}{4}, \sqrt{3}, 0, 9, \sqrt{7}, 0.35\right\}$$

List those numbers that are:

(a) Positive integers.

(b) Whole numbers.

(c) Integers.

(d) Rational numbers.

(e) Irrational numbers.

(f) Real numbers.

40. Consider the set of numbers.

$$\left\{-\frac{5}{3}, 0, -2, 5, 5\frac{1}{2}, \sqrt{2}, -\sqrt{3}, 1.63, 207\right\}$$

List those numbers that are:

(a) Positive integers.

(b) Whole numbers.

(c) Integers.

(d) Rational numbers.

(e) Irrational numbers.

(f) Real numbers.

41. Consider the set of numbers

$$\left\{\frac{1}{2}, \sqrt{2}, -\sqrt{2}, 4\frac{1}{2}, \frac{5}{12}, -1.67, 5, -300, -9\frac{1}{2}\right\}$$

List those numbers that are:

(a) Positive integers.

(b) Whole numbers.

(c) Negative integers.

(d) Integers.

(e) Rational numbers.

(f) Irrational numbers.

(g) Real numbers.

In Exercises 42–53, give three examples of numbers that satisfy the conditions.

42. A real number but not an integer.

43. A rational number but not an integer.

44. An integer but not a negative integer.

45. A real number but not a rational number.

46. An irrational number and a positive number.

47. An integer and a rational number.

48. A negative integer and a real number.

49. A negative integer and a rational number.

50. A real number but not a positive rational number.

51. A rational number but not a negative number.

52. An integer but not a positive integer.

53. A real number but not an irrational number.

54. **(a)** What is a rational number?

(b) Explain why every integer is a rational number.

55. Write a paragraph or two describing the structure of the real number system. Explain how whole numbers, counting numbers, integers, rational numbers, irrational numbers, and real numbers are related.

CUMULATIVE REVIEW EXERCISES

2] **56.** Convert $4\frac{2}{3}$ to a fraction.

57. Write $\frac{16}{3}$ as a mixed number.

58. Add $\frac{3}{5} + \frac{5}{8}$.

59. Multiply $\left(\frac{5}{9}\right)\left(4\frac{2}{3}\right)$.

Group Activity/ Challenge Problems

1. **(a)** Does the set of natural numbers have a last number? If so, what is it?

(b) Do you know what a set that has no last number is called?

2. How many decimal numbers are there **(a)** between 1.0 and 2.0; **(b)** between 1.4 and 1.5? Explain your answer.

3. How many fractions are there **(a)** between 1 and 2; **(b)** between $\frac{1}{3}$ and $\frac{1}{5}$? Explain your answer.

Set A **union** *set B, symbolized A∪B, consists of the set of elements that belong to set A or set B (or both sets). Set A* **intersection** *set B, symbolized A∩B, consists of the set of elements common to both Set A and set B. For each of the given pair of sets find A∪B and A∩B.*

4. A = {2, 3, 4, 6, 8, 9} B = {1, 2, 3, 5, 7, 8}

5. A = {a, b, c, d, g, i, j} B = {b, c, d, h, m, p}

6. A = {red, blue, green, yellow} B = {pink, orange, purple}

1.4 Inequalities

Tape 1

1. Determine which is the greater of two numbers.
2. Find the absolute value of a number

Comparing Numbers

1 The number line (Fig. 1.8) can be used to explain inequalities. When comparing two numbers, **the number to the right on the number line is the greater number, and the number to the left is the lesser number.** The symbol > is used to represent the words "is greater than." The symbol < is used to represent the words "is less than."

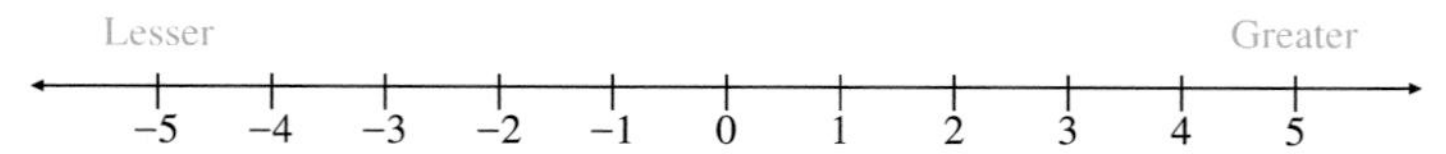

FIGURE 1.8

The statement that the number 3 is greater than the number 2 is written $3 > 2$. Notice that 3 is to the right of 2 on the number line. The statement that the number 0 is greater than the number -1 is written $0 > -1$. Notice that 0 is to the right of -1 on the number line.

Instead of stating that 3 is greater than 2, we could state that 2 is less than 3, written $2 < 3$. Notice that 2 is to the left of 3 on the number line. The statement that the number -1 is less than the number 0 is written $-1 < 0$. Notice that -1 is to the left of 0 on the number line.

EXAMPLE 1 Insert either > or < in the shaded area between the paired numbers to make a true statement.

(a) $-4 \quad -2$ **(b)** $-\frac{3}{2} \quad 2.5$ **(c)** $\frac{1}{2} \quad \frac{1}{4}$ **(d)** $-2 \quad 4$

Solution: The points given are shown on the number line (Fig. 1.9).

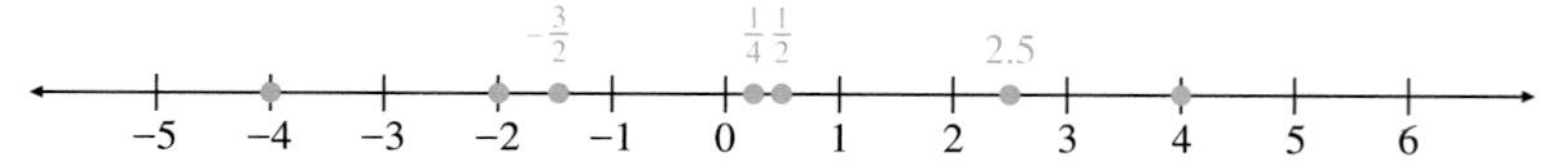

FIGURE 1.9

(a) $-4 < -2$; notice that -4 is to the left of -2.
(b) $-\frac{3}{2} < 2.5$; notice that $-\frac{3}{2}$ is to the left of 2.5.
(c) $\frac{1}{2} > \frac{1}{4}$; notice that $\frac{1}{2}$ is to the right of $\frac{1}{4}$.
(d) $-2 < 4$; notice that -2 is to the left of 4.

EXAMPLE 2 Insert either > or < in the shaded area between the paired numbers to make a true statement.

(a) $-1 \quad -2$ **(b)** $-1 \quad 0$ **(c)** $-2 \quad 2$ **(d)** $-4.09 \quad -4.9$

Solution: The numbers given are shown on the number line (Fig. 1.10).

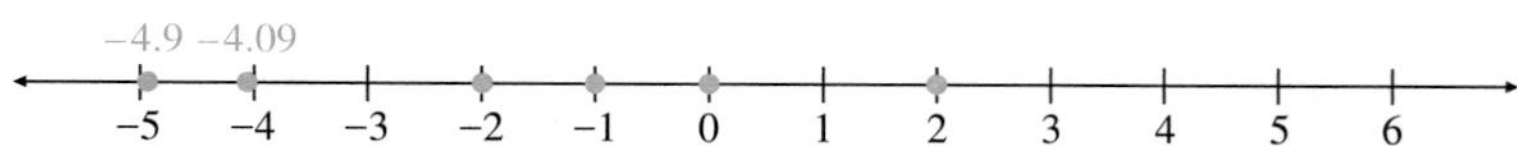

FIGURE 1.10

(a) $-1 > -2$; notice that -1 is to the right of -2.
(b) $-1 < 0$; notice that -1 is to the left of 0.
(c) $-2 < 2$; notice that -2 is to the left of 2.
(d) $-4.09 > -4.9$; notice that -4.09 is to the right of -4.9.

Absolute Value

2 The concept of absolute value can be explained with the help of the number line shown in Figure 1.11. The **absolute value** of a number can be considered the distance between the number and 0 on the number line. Thus, the absolute value of 3, written $|3|$, is 3 since it is 3 units from 0 on the number line. Similarly, the absolute value of -3, written $|-3|$, is also 3 since -3 is 3 units from 0.

$$|3| = 3 \quad \text{and} \quad |-3| = 3$$

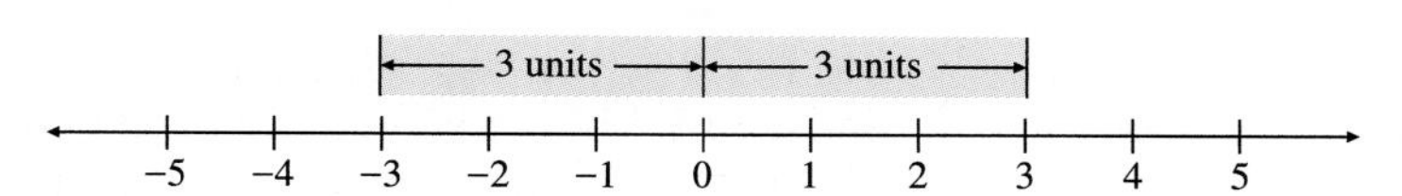

FIGURE 1.11

Since the absolute value of a number measures the distance (without regard to direction) of a number from 0 on the number line, **the absolute value of every number will be either positive or zero.**

Number	*Absolute Value of Number*
6	$\|6\| = 6$
-6	$\|-6\| = 6$
0	$\|0\| = 0$
$-\frac{1}{2}$	$\left\|-\frac{1}{2}\right\| = \frac{1}{2}$

The negative of the absolute value of a nonzero number will always be a negative number.

For example:

$$-|2| = -(2) = -2 \quad \text{and} \quad -|-3| = -(3) = -3$$

EXAMPLE 3 Insert either $>$, $<$, or $=$ in the shaded area to make a true statement.

(a) $|3| \;\square\; 3$ **(b)** $|-2| \;\square\; |2|$ **(c)** $-2 \;\square\; |-4|$
(d) $|-5| \;\square\; 0$ **(e)** $|12| \;\square\; |-18|$

Solution: **(a)** $|3| = 3$.
(b) $|-2| = |2|$, since both $|-2|$ and $|2|$ equal 2.
(c) $-2 < |-4|$, since $|-4| = 4$.
(d) $|-5| > 0$, since $|-5| = 5$.
(e) $|12| < |-18|$, since $|12| = 12$ and $|-18| = 18$.

The concept of absolute value is very important in higher-level mathematics courses. If you take a course in intermediate algebra, you will learn a more formal definition of absolute value. We will use absolute value in Section 1.5 to add and subtract real numbers.

Exercise Set 1.4

Evaluate.

1. $|4|$ **2.** $|-3|$ **3.** $|-15|$ **4.** $|-12|$ **5.** $|0|$

6. $|54|$ **7.** $-|-8|$ **8.** $-|92|$ **9.** $-|65|$ **10.** $-|-34|$

Insert either $<$ or $>$ in the shaded area to make a true statement.

11. $2 \; \square \; 3$ **12.** $4 \; \square \; -2$ **13.** $-3 \; \square \; 0$

14. $-6 \; \square \; -4$ **15.** $\frac{1}{2} \; \square \; -\frac{2}{3}$ **16.** $\frac{3}{5} \; \square \; \frac{4}{5}$

17. $0.2 \; \square \; 0.4$ **18.** $-0.2 \; \square \; -0.4$ **19.** $-\frac{1}{2} \; \square \; -1$

20. $0 \; \square \; -0.9$ **21.** $4 \; \square \; -4$ **22.** $-\frac{3}{4} \; \square \; -1$

23. $-2.1 \; \square \; -2$ **24.** $-1.83 \; \square \; -1.82$ **25.** $\frac{5}{9} \; \square \; -\frac{5}{9}$

26. $-9 \; \square \; -12$ **27.** $-\frac{3}{2} \; \square \; \frac{3}{2}$ **28.** $-4.09 \; \square \; -5.3$

29. $0.49 \; \square \; 0.43$ **30.** $-1.0 \; \square \; -0.7$ **31.** $5 \; \square \; -7$

32. $0.001 \; \square \; 0.002$ **33.** $-0.006 \; \square \; -0.007$ **34.** $\frac{1}{2} \; \square \; -\frac{1}{2}$

35. $-5 \; \square \; -2$ **36.** $\frac{5}{3} \; \square \; \frac{3}{5}$ **37.** $-\frac{2}{3} \; \square \; -3$

38. $\frac{5}{2} \; \square \; \frac{7}{2}$ **39.** $-\frac{1}{2} \; \square \; -\frac{3}{2}$ **40.** $-0.4 \; \square \; -0.5$

Insert either $<$, $>$, or $=$ in the shaded area to make a true statement.

41. $8 \; \square \; |-7|$ **42.** $|-8| \; \square \; |-7|$

43. $|0| \; \square \; \frac{2}{3}$ **44.** $|-4| \; \square \; -3$

45. $|-3| \; \square \; |-4|$ **46.** $|-1.9| \; \square \; -1.8$

47. $4 \; \square \; \left|-\frac{9}{2}\right|$ **48.** $-5 \; \square \; |5|$

49. $\left|-\frac{6}{2}\right| \; \square \; \left|-\frac{2}{6}\right|$ **50.** $\left|\frac{2}{5}\right| \; \square \; |-0.40|$

Insert either >, <, or = in the shaded area to make a true statement.

51. $\frac{2}{3} + \frac{2}{3} + \frac{2}{3} + \frac{2}{3} \quad \square \quad 4 \cdot \frac{2}{3}$

52. $\frac{2}{3} \cdot \frac{2}{3} \quad \square \quad \frac{2}{3} + \frac{2}{3}$

53. $\frac{1}{2} \cdot \frac{1}{2} \quad \square \quad \frac{1}{2} \div \frac{1}{2}$

54. $5 \div \frac{2}{3} \quad \square \quad \frac{2}{3} \div 5$

55. $\frac{5}{8} - \frac{1}{2} \quad \square \quad \frac{5}{8} \div \frac{1}{2}$

56. $2\frac{1}{3} \cdot \frac{1}{2} \quad \square \quad 2\frac{1}{3} + \frac{1}{2}$

57. What numbers are 4 units from 0 on the number line?

58. What numbers are 5 units from 0 on the number line?

59. What numbers are 2 units from 0 on the number line?

60. Are there any real numbers whose absolute value is not a positive number? Explain your answer.

61. What is the absolute value of a number?

CUMULATIVE REVIEW EXERCISES

.2] **62.** Subtract $1\frac{2}{3} - \frac{3}{8}$.

63. List the set of whole numbers.

64. List the set of counting numbers.

.3] **65.** Consider the set of numbers

$$\{5, -2, 0, \tfrac{1}{3}, \sqrt{3}, -\tfrac{5}{9}, 2.3\}.$$

List the numbers in the set that are:

(a) Natural numbers.

(b) Whole numbers.

(c) Integers.

(d) Rational numbers.

(e) Irrational numbers.

(f) Real numbers.

Group Activity/ Challenge Problems

1. A number greater than 0 and less than 1 (or between 0 and 1) is multiplied by itself. Will the product be less than, equal to, or greater than the original number selected? Explain why this is always true.

2. A number between 0 and 1 is divided by itself. Will the quotient be less than, equal to, or greater than the original number selected? Explain why this is always true.

3. What two numbers can be substituted for x to make $|x| = 3$ a true statement?

4. Are there any values for x that would make $|x| = -|x|$ a true statement?

5. **(a)** To what is $|x|$ equal if x represents a real number greater than or equal to 0? **(b)** To what is $|x|$ equal if x represents a real number less than 0? **(c)** Fill in the shaded areas to make a true statement.

$$\text{(c)}\ |x| = \begin{cases} \square, & x \geq 0 \\ \square, & x < 0 \end{cases}$$

1.5 Addition of Real Numbers

Tape 1

1. Add real numbers using the number line.
2. Identify opposites or additive inverses.
3. Add using absolute values.

There are many practical uses for negative numbers. A submarine diving below sea level, a bank account that has been overdrawn, a business spending more than it earns, and a temperature below zero are some examples. In some

European hotels, the floors below the registration lobby are given negative numbers.

The four basic **operations** of arithmetic are addition, subtraction, multiplication, and division. In the next few sections we will explain how to add, subtract, multiply, and divide numbers. We will consider both positive and negative numbers. In this section we discuss the operation of addition.

Add Real Numbers Using the Number Line

1 To add numbers, we make use of the number line. Represent the first number to be added (first *addend*) by an arrow starting at 0. The arrow is drawn to the right if the number is positive. If the number is negative, the arrow is drawn to the left. From the tip of the first arrow, draw a second arrow to represent the second addend. The second arrow is drawn to the right or left, as just explained. The sum of the two numbers is found at the tip of the second arrow. Note that *any number except 0 without a sign in front of it is positive.* For example, 3 means $+3$ and 5 means $+5$.

EXAMPLE 1 Evaluate $3 + (-4)$ using the number line.

Solution: *Always begin at 0.* Since the first addend, the 3, is positive, the first arrow starts at 0 and is drawn 3 units to the right (Fig. 1.12).

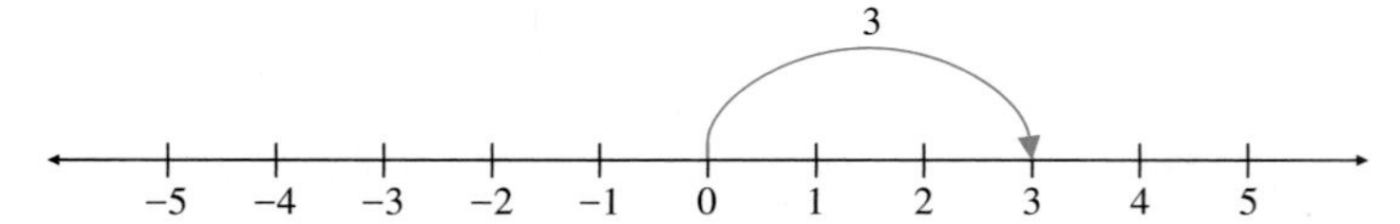

FIGURE 1.12

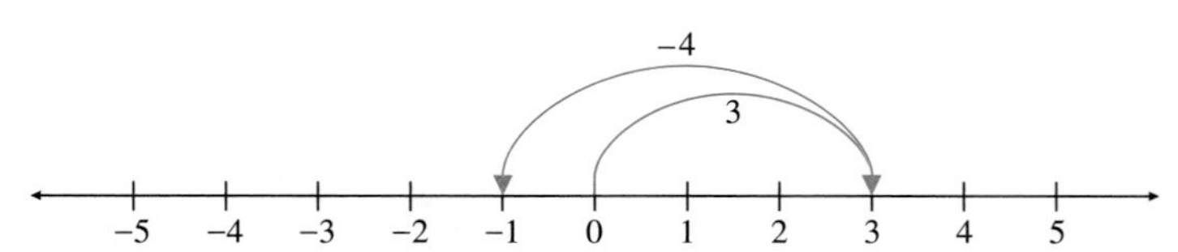

FIGURE 1.13

The second arrow starts at 3 and is drawn 4 units to the left, since the second addend is negative (Fig. 1.13). The tip of the second arrow is at -1. Thus

$$3 + (-4) = -1$$

EXAMPLE 2 Evaluate $-4 + 2$ using the number line.

Solution: Begin at 0. Since the first addend is negative, -4, the first arrow is drawn 4 units to the left. From there, since 2 is positive, the second arrow is drawn 2 units to the right. The second arrow ends at -2 (Fig. 1.14).

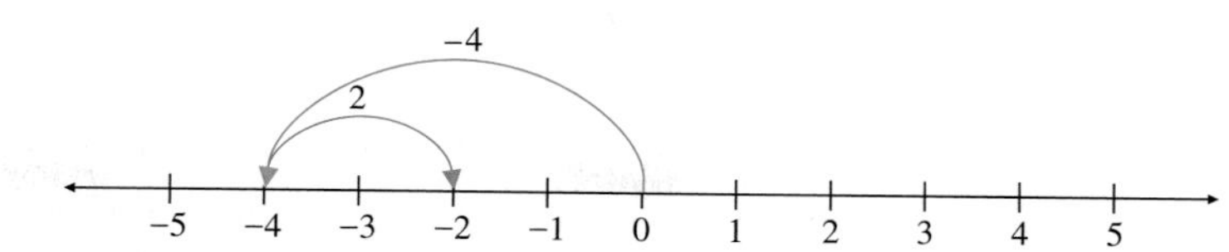

FIGURE 1.14

$$-4 + 2 = -2$$

EXAMPLE 3 Evaluate $-3 + (-2)$ using the number line.

Solution: Start at 0. Since both numbers being added are negative, both arrows will be drawn to the left (Fig. 1.15).

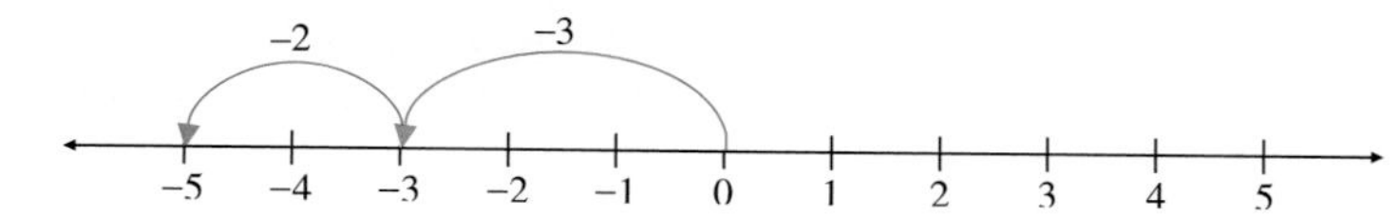

FIGURE 1.15

$$-3 + (-2) = -5$$

In Example 3, we can think of the expression $-3 + (-2)$ as combining a *loss* of 3 and a *loss* of 2 for a total *loss* of 5, or -5.

EXAMPLE 4 Add $5 + (-5)$.

Solution: The first arrow starts at 0 and is drawn 5 units to the right. The second arrow starts at 5 and is drawn 5 units to the left. The tip of the second arrow is at 0. Thus, $5 + (-5) = 0$ (Fig. 1.16).

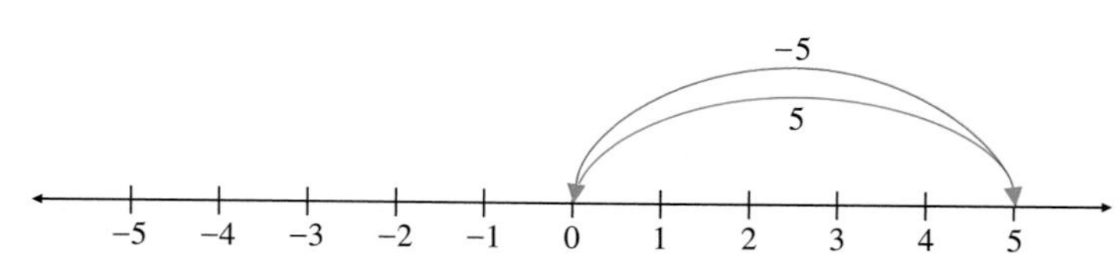

FIGURE 1.16

$$5 + (-5) = 0$$

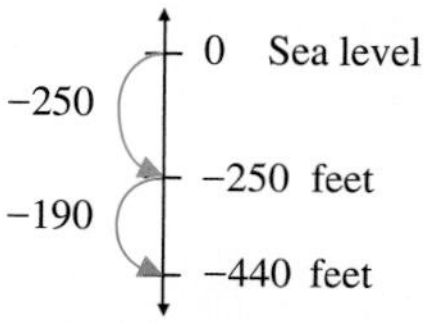

0 Sea level
−250
−250 feet
−190
−440 feet

FIGURE 1.17

EXAMPLE 5 A submarine dives 250 feet. Later it dives an additional 190 feet. Find the depth of the submarine (assume that depths below sea level are indicated by negative numbers).

Solution: A vertical number line (Fig. 1.17) may help you visualize this problem.

$$-250 + (-190) = -440 \text{ feet}$$

Identify Opposites 2

Any two numbers whose sum is zero are said to be **opposites** (or **additive inverses**) of each other. In general, if we let a represent any real number, then its opposite is $-a$ and $a + (-a) = 0$.

In Example 4 the sum of 5 and -5 is zero. Thus -5 is the opposite of 5 and 5 is the opposite of -5.

EXAMPLE 6 Find the opposite of each number.

(a) 3 **(b)** -4

Solution: **(a)** The opposite of 3 is -3, since $3 + (-3) = 0$.
(b) The opposite of -4 is 4, since $-4 + 4 = 0$.

Add Using Absolute Value 3

Now that we have had some practice adding signed numbers on the number line, we give a rule (in two parts) for using absolute value to add signed numbers. Remember that the absolute value of a nonzero number will always be positive. The first part follows.

To add real numbers with the same sign (either both positive or both negative), add their absolute values. The sum has the same sign as the numbers being added.

EXAMPLE 7 Add $4 + 8$.

Solution: Since both numbers have the same sign, both positive, we add their absolute values: $|4| + |8| = 4 + 8 = 12$. Since both numbers being added are positive, the sum is positive. Thus $4 + 8 = 12$.

EXAMPLE 8 Add $-6 + (-9)$.

Solution: Since both numbers have the same sign, both negative, we add their absolute values: $|-6| + |-9| = 6 + 9 = 15$. Since both numbers being added are negative, their sum is negative. Thus $-6 + (-9) = -15$.

The sum of two positive numbers will always be positive and the sum of two negative numbers will always be negative.

To add two signed numbers with different signs (one positive and the other negative), find the difference between the larger absolute value and the smaller absolute value. The answer has the sign of the number with the larger absolute value.

EXAMPLE 9 Add $10 + (-6)$.

Solution: The two numbers being added have different signs; thus we find the difference between the larger absolute value and the smaller: $|10| - |-6| = 10 - 6 = 4$. Since $|10|$ is greater than $|-6|$ and the sign of 10 is positive, the sum is positive. Thus, $10 + (-6) = 4$.

EXAMPLE 10 Add $12 + (-18)$.

Solution: The numbers being added have different signs; thus we find the difference between the larger absolute value and the smaller: $|-18|-|12| = 18 - 12 = 6$. Since $|-18|$ is greater than $|12|$ and the sign of -18 is negative, the sum is negative. Thus, $12 + (-18) = -6$.

EXAMPLE 11 Add $-24 + 19$.

Solution: The two numbers being added have different signs; thus we find the difference between the larger absolute value and the smaller: $|-24|-|19| = 24 - 19 = 5$. Since $|-24|$ is greater than $|19|$, the sum is negative. Therefore, $-24 + 19 = -5$.

The sum of two signed numbers with different signs may be either positive or negative. The sign of the sum will be the same as the sign of the number with the larger absolute value.

Helpful Hint

Architects often make a scale model of a building before starting construction of the building. This "model" helps them visualize the project and often helps them avoid problems.

Mathematicians also construct models. A mathematical *model* may be a physical representation of a mathematical concept. It may be as simple as using tiles or chips to represent specific numbers. For example, below we use a model to help explain addition of real numbers. This may help some of you understand the concepts better.

We let a red chip represent $+1$ and a green chip represent -1.

● $= +1$ ● $= -1$

If we add $+1$ and -1, or a red and a green chip, we get 0.

Now consider the addition problem $3 + (-5)$. We can represent this as

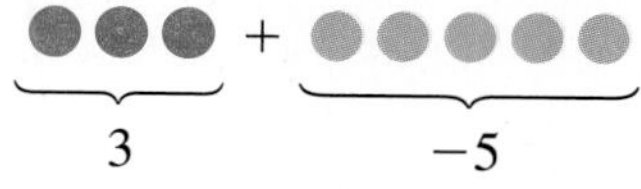

If we remove 3 red chips and 3 green chips, or three zeros, we are left with 2 green chips, which represents a sum of -2. Thus, $3 + (-5) = -2$,

Now consider the problem $-4 + (-2)$. We can represent this as

-4 -2

Since we end up with 6 green chips, and each green chip represents -1, the sum is -6. Therefore, $-4 + (-2) = -6$.

EXAMPLE 12 The ABC Company had a loss of \$4000 for the first 6 months of the year and a profit of \$15,500 for the second 6 months of the year. Find the net profit or loss for the year.

Solution: This problem can be represented as $-4000 + 15{,}500$. Since the two numbers being added have different signs, find the difference between the larger absolute value and the smaller.

$$|15{,}500| - |-4000| = 15{,}500 - 4000 = 11{,}500$$

Thus, the net profit for the year was \$11,500.

Calculator Corner

USE OF A SCIENTIFIC CALCULATOR AND ADDITION OF REAL NUMBERS

It is important that you understand the procedures for adding, subtracting, multiplying, and dividing real numbers *without* using a calculator. To do so, you must understand the basic concepts presented in this and the following two sections. You should not need to rely on a calculator to work problems. You can, however, use the calculator to help save time in difficult calculations. If you have an understanding of the basic concepts, you should be able to tell if you have made an error entering information on the calculator by seeing if the answer is reasonable. In each respective section we show how to add, subtract, multiply, and divide real numbers using a calculator. We strongly urge you to purchase or borrow a *scientific calculator.* Ask your instructor if he or she is recommending a specific calculator. You will find it very useful in this and many other mathematics and science courses. The cost of a scientific calculator is not much more than the cost of the nonscientific calculator. A picture of a typical scientific calculator is shown on the left.

ENTERING NEGATIVE NUMBERS

Scientific calculators contain a +/− key, which is used to enter a negative number. To enter the number −5 we press 5 +/− and a −5 will be displayed. Now we show how to evaluate some addition problems on a calculator.

ADDITION OF REAL NUMBERS

Evaluate	Keystrokes	Answer
$5 + (-2)$	5 + 2 +/− = 3	3
$-3 + (-8)$	3 +/− + 8 +/− = −11	−11
$52 + (-97)$	52 + 97 +/− = −45	−45
$-127 + (-82)$	127 +/− + 82 +/− = −209	−209

Exercise Set 1.5

Write the opposite of each number.

1. 12 **2.** -7 **3.** -40 **4.** 3

5. 0 **6.** 6 **7.** $\frac{5}{3}$ **8.** $-\frac{1}{2}$

9. $\frac{3}{5}$ **10.** -1 **11.** 0.63 **12.** -0.721

13. $3\frac{1}{5}$ **14.** $-5\frac{1}{4}$ **15.** -3.1 **16.** 5.26

Add.

17. $4 + 3$ **18.** $-4 + 3$ **19.** $4 + (-3)$ **20.** $4 + (-2)$

21. $-4 + (-2)$ **22.** $-3 + (-5)$ **23.** $6 + (-6)$ **24.** $-6 + 6$

25. $-4 + 4$ **26.** $-3 + 5$ **27.** $-8 + (-2)$ **28.** $6 + (-5)$

29. $-3 + 3$ **30.** $-8 + 2$ **31.** $-3 + (-7)$ **32.** $0 + (-3)$

33. $0 + 0$ **34.** $0 + (-0)$ **35.** $-6 + 0$ **36.** $-9 + 13$

37. $22 + (-19)$ **38.** $-13 + (-18)$ **39.** $-45 + 36$ **40.** $40 + (-25)$

41. $18 + (-9)$ **42.** $-7 + 7$ **43.** $-14 + (-13)$ **44.** $-27 + (-9)$

45. $-35 + (-9)$ **46.** $34 + (-12)$ **47.** $4 + (-30)$ **48.** $-16 + 9$

49. $-35 + 40$ **50.** $-12 + 17$ **51.** $180 + (-200)$ **52.** $-33 + (-92)$

53. $-105 + 74$ **54.** $183 + (-183)$ **55.** $184 + (-93)$ **56.** $-42 + 129$

57. $-452 + 312$ **58.** $-94 + (-98)$ **59.** $-60 + (-38)$ **60.** $49 + (-63)$

*For each of the following, **(a)** determine by observation whether the sum will be a positive number, zero, or a negative number; **(b)** find the sum using your calculator; and **(c)** examine your answer to part (b) to see if it is reasonable and makes sense.*

61. $463 + (-197)$ **62.** $-140 + (-629)$ **63.** $-84 + (-289)$ **64.** $-593 + 624$

65. $-947 + 495$ **66.** $762 + (-762)$ **67.** $-496 + (-804)$ **68.** $-354 + 1090$

69. $-285 + 263$ **70.** $1035 + (-972)$ **71.** $-1833 + (-2047)$ **72.** $-138 + 648$

73. $4793 + (-6060)$ **74.** $-9095 + (-647)$ **75.** $-1025 + (-1025)$ **76.** $7625 + (-1938)$

77. $-8276 + (-6283)$ **78.** $-4693 + 6773$ **79.** $-9042 + 7827$ **80.** $-4067 + (-3078)$

81. $1046 + (-8540)$ **82.** $-625 + (-9248)$ **83.** $8364 + (-906)$ **84.** $-436 + 8954$

Answer true or false.

85. The sum of two negative numbers is always a negative number.

86. The sum of two positive numbers is never a negative number.

87. The sum of a positive number and a negative number is always a positive number.

88. The sum of a negative number and a positive number is sometimes a negative number.

89. The sum of a positive number and a negative number is always a negative number.

90. The sum of a number and its opposite is always equal to zero.

Write an expression that can be used to solve each problem and then solve.

91. Mr. Yelserp owed \$67 on his bank credit card. He charged another item costing \$107. Find the amount that Mr. Yelserp owed the bank.

92. Mr. Weber charged \$193 worth of goods on his charge card. Find his balance after he made a payment of \$112.

93. Mrs. Petrie paid \$1424 in federal income tax. When she was audited, Mrs. Petrie had to pay an additional \$503. What was her total tax?

94. Mr. Vela hiked 847 meters down the Grand Canyon. He climbed back up 385 meters and then rested. Find his distance from the rim of the canyon.

SEE EXERCISE 94

95. A company is drilling a well. During the first week they drilled 22 feet, and during the second week they drilled another 32 feet before they struck water. How deep is the well?

96. A football team lost 18 yards on one play and then lost 3 yards on the following play. What was the total loss in yardage?

97. In 1992, there were approximately 141 million births and 49 million deaths worldwide. What was the change in the world's population in 1992?

98. An airplane at an altitude of 2400 feet above sea level drops a package into the ocean. The package settles at a point 200 feet below sea level. How far did the object fall?

99. Explain in your own words how to add two numbers with like signs.

100. Explain in your own words how to add two numbers with unlike signs.

CUMULATIVE REVIEW EXERCISES

[1.1] 101. Multiply $\left(\frac{3}{5}\right)\left(1\frac{2}{3}\right)$

102. Subtract $3 - \frac{5}{16}$

[1.3] *Insert either $<$, $>$, or $=$ in the shaded area to make the statement true.*

103. $|-3|$ ▒ 2

104. 8 ▒ $|-7|$

Group Activity/Challenge Problems

Evaluate each exercise by adding the numbers from left to right. We will discuss problems like this shortly.

1. $(-4) + (-6) + (-12)$
2. $5 + (-7) + (-8)$
3. $29 + (-46) + 37$

Find the following sums. Explain how you determined your answer. Hint: pair small numbers with large numbers from the ends inward.

4. $1 + 2 + 3 + \cdots + 10$
5. $1 + 2 + 3 + \cdots + 20$
6. $1 + 2 + 3 + \cdots + 100$
7. $1 + 2 + 3 + \cdots + 5000$

1.6 Subtraction of Real Numbers

1 Subtract real numbers.

2 Subtract real numbers mentally.

Tape 2

Subtraction

1 Any subtraction problem can be rewritten as an addition problem using the additive inverse.

Subtraction of Real Numbers

In general, if a and b represent any two real numbers, then

$$a - b = a + (-b)$$

This rule says that to subtract b from a, add the opposite or additive inverse of b to a.

EXAMPLE 1 Evaluate $9 - (+4)$.

Solution: We are subtracting a positive 4 from 9. To accomplish this we add the opposite of $+4$, which is -4, to 9.

$$9 - (+4) = 9 + (-4) = 5$$

subtract positive 4 add negative 4

We evaluated $9 + (-4)$ using the procedures for *adding* real numbers presented in Section 1.5.

Often in a subtraction problem, when the number being subtracted is a positive number, the + sign preceding the number being subtracted is not illustrated. For example, in the subtraction $9 - 4$,

$$9 - 4 \text{ means } 9 - (+4)$$

Thus, to evaluate $9 - 4$, we must add the opposite of 4 (or $+4$), which is -4, to 9.

$$9 - 4 = 9 + (-4) = 5$$

subtract positive 4 add negative 4

This procedure is illustrated in Example 2.

EXAMPLE 2 Evaluate $5 - 3$.

Solution: We must subtract a positive 3 from 5. To change this problem to an addition problem, add the opposite of 3, which is -3, to 5.

$$\overbrace{5 - 3}^{\text{subtraction problem}} = \overbrace{5 + (-3)}^{\text{addition problem}} = 2$$

subtract positive 3 add negative 3

EXAMPLE 3 Evaluate $4 - 9$.

Solution: Add the opposite of 9, which is -9, to 4.

$$4 - 9 = 4 + (-9) = -5$$

EXAMPLE 4 Evaluate $-4 - 2$.

Solution: Add the opposite of 2, which is -2, to -4.

$$-4 - 2 = -4 + (-2) = -6$$

2

EXAMPLE 5 Evaluate $4 - (-2)$.

Solution: We are asked to subtract a negative 2 from 4. To do this, add the opposite of -2, which is 2, to 4.

$$4 - (-2) = 4 + 2 = 6$$

subtract negative 2 add positive 2

Helpful Hint

By examining Example 5 we see that

$$4 - (-2) = 4 + 2$$

two negative signs together + sign

Whenever we subtract a negative number, we can replace the two negative signs with a plus sign.

EXAMPLE 6 Evaluate $6 - (-3)$.

Solution: Since we are subtracting a negative number, adding the opposite of -3, which is 3, to 6 will result in the two negative signs being replaced by a plus sign.

$$6 - (-3) = 6 + 3 = 9$$

EXAMPLE 7 Evaluate $-15 - (-12)$.

Solution: $-15 - (-12) = -15 + 12 = -3$

Helpful Hint

We will now indicate how we may illustrate subtraction using colored chips. Remember from the preceding section that a red chip represents $+1$ and a green chip -1.

● = +1 ● = −1

Consider the subtraction problem $2 - 5$. If we change this to an addition problem we get $2 + (-5)$. We can then add, as was done in the preceding section. The figure below shows that $2 + (-5) = -3$.

Now consider $-2 - 5$. This means $-2 + (-5)$, which can be represented as follows:

Thus, $-2 - 5 = -7$.

Now consider the problem $-3 - (-5)$. This can be rewritten as $-3 + 5$, which can be represented as follows:

Thus, $-3 - (-5) = 2$.

Some students still have difficulty understanding why when you subtract a negative number you obtain a positive number. Let us look at the problem $3 - (-2)$. This time we will look at it from a little different point of view. Let us start with 3:

From this we wish to subtract a negative two. To the $+3$ shown above we will add two zeros by adding two $+1$ -1 combinations. Remember, $+1$ and -1 sum to 0.

$+3 \qquad 0 \qquad 0$

Now we can subtract or "take away" the two -1's as shown:

From this we see that we are left with $3 + 2$ or 5. Thus, $3 - (-2) = 5$.

EXAMPLE 8 Subtract 12 from 3.

Solution: $3 - 12 = 3 + (-12) = -9$

Helpful Hint

Example 8 asked us to "subtract 12 from 3." Some of you may have expected this to be written as $12 - 3$ since you may be accustomed to getting a positive answer. However, the correct method of writing this is $3 - 12$. Notice that the number following the word "from" is our starting point. That is where the calculation begins. For example:

Subtract 5 from -1, means $-1 - 5$.

Subtract -3 from 2, means $2 - (-3)$.

Subtract a from b, means $b - a$.

From 6 subtract 2, means $6 - 2$.

From -6 subtract -3, means $-6 - (-3)$.

From a subtract b, means $a - b$.

EXAMPLE 9 Subtract 5 from 5.

Solution: $5 - 5 = 5 + (-5) = 0$

EXAMPLE 10 Subtract -6 from 4.

Solution: $4 - (-6) = 4 + 6 = 10$

EXAMPLE 11 Evaluate.

(a) $8 - (-5)$ **(b)** $-3 - (-9)$

Solution: **(a)** $8 - (-5) = 8 + 5 = 13$
(b) $-3 - (-9) = -3 + 9 = 6$

EXAMPLE 12 Mary Jo Morin's checkbook indicated a balance of \$125 before she wrote a check for \$183. Find the balance in her checkbook.

Solution: $125 - 183 = 125 + (-183) = -58$. The negative indicates a deficit. Therefore, Mary Jo is overdrawn by \$58.

EXAMPLE 13 Janet made \$4200 in the stock market, while Mateo lost \$3000. How much further ahead is Janet than Mateo financially?

Solution: Janet's gain is represented as a positive number. Mateo's loss is represented as a negative number.

$$4200 - (-3000) = 4200 + 3000 = 7200$$

Janet is therefore \$7200 ahead of Mateo financially.

EXAMPLE 14 Evaluate.

(a) $12 + (-4)$ **(b)** $-16 - 3$ **(c)** $5 + (-4)$
(d) $6 - (-5)$ **(e)** $-12 - (-3)$ **(f)** $8 - 13$

Solution: Parts (a) and (c) are addition problems, while the other parts are subtraction problems. We can rewrite each subtraction problem as an addition problem to evaluate.

(a) $12 + (-4) = 8$ **(b)** $-16 - 3 = -16 + (-3) = -19$
(c) $5 + (-4) = 1$ **(d)** $6 - (-5) = 6 + 5 = 11$
(e) $-12 - (-3) = -12 + 3 = -9$ **(f)** $8 - 13 = 8 + (-13) = -5$

Subtract Mentally

2 In the previous examples, we changed subtraction problems to addition problems. We did this because we know how to add real numbers. After this chapter, when we work out a subtraction problem, we will not show this step. *You need to practice and thoroughly understand how to add and subtract real numbers. You should understand this material so well that, when asked to evaluate an expression like* $-4 - 6$, *you will be able to compute the answer mentally. You should understand that* $-4 - 6$ *means the same as* $-4 + (-6)$, *but you should not need to write the addition to find the value of the expression,* -10.

Let us evaluate a few subtraction problems without showing the process of changing the subtraction to addition.

EXAMPLE 15 Evaluate.

(a) $-7 - 5$ **(b)** $4 - 12$ **(c)** $18 - 25$ **(d)** $-20 - 12$

Solution: **(a)** $-7 - 5 = -12$ **(b)** $4 - 12 = -8$
(c) $18 - 25 = -7$ **(d)** $-20 - 12 = -32$

In Example 15 we may have reasoned that $-7 - 5$ meant $-7 + (-5)$, which is -12, but we did not need to show it.

In evaluating expressions involving more than one addition and subtraction, work from left to right unless parentheses or other grouping symbols appear.

EXAMPLE 16 Evaluate.

(a) $-6 - 12 - 4$ **(b)** $-3 + 1 - 7$ **(c)** $8 - 10 + 2$

Solution: We work from left to right.

(a) $\underbrace{-6 - 12}{} - 4$ $= -18 - 4$ $= -22$

(b) $\underbrace{-3 + 1}{} - 7$ $= -2 - 7$ $= -9$

(c) $\underbrace{8 - 10}{} + 2$ $= -2 + 2$ $= 0$

After this section you will generally not see an expression like $3 + (-4)$. Instead, the expression will be written as $3 - 4$. Recall that $3 - 4$ means $3 + (-4)$ by our definition of subtraction. **Whenever we see an expression of the form $a + (-b)$, we can write the expression as $a - b$.** For example, $12 + (-15)$ can be written $12 - 15$ and $-6 + (-9)$ can be written $-6 - 9$.

EXAMPLE 17 Evaluate.

(a) $-3 - (-4) + (-10) + (-5)$ **(b)** $-3 - (-4) - 10 - 5$

Solution: **(a)** Again we work from left to right.

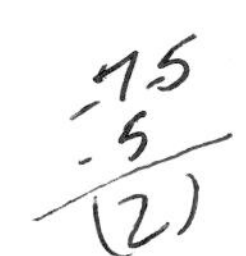

$$\begin{aligned}-3 - (-4) + (-10) + (-5) &= -3 + 4 + (-10) + (-5)\\ &= 1 + (-10) + (-5)\\ &= -9 + (-5)\\ &= -14\end{aligned}$$

(b) This part is really the same problem as part (a), since $+(-10)$ can be written -10 and $+(-5)$ can be written -5.

$$\begin{aligned}-3 - (-4) - 10 - 5 &= -3 + 4 - 10 - 5\\ &= 1 - 10 - 5\\ &= -9 - 5\\ &= -14\end{aligned}$$

Calculator Corner

SUBTRACTION OF REAL NUMBERS

In the calculator corner on page 34 we indicated that the [+/–] key is used to enter a negative number. Below we show some examples of subtracting real numbers on a scientific calculator. The number following the [=] is the answer.

Evaluate	*Keystrokes*
$6 - 10$	6 [–] 10 [=] –4
$-5 - 8$	5 [+/–] [–] 8 [=] –13 *Note:* [+/–] is pressed *after* the number to be made negative.

You can subtract a negative number on the calculator as shown below.

Evaluate	*Keystrokes*
$2 - (-7)$	2 [–] 7 [+/–] [=] 9
$-9 - (-3)$	9 [+/–] [–] 3 [+/–] [=] –6

Exercise Set 1.6

Evaluate.

1. $6 - 3$
2. $-6 - 4$
3. $8 - 9$
4. $5 - 3$
5. $3 - 3$
6. $-4 - 2$
7. $(-7) - (-4)$
8. $-4 - (-3)$
9. $-3 - 3$
10. $-4 - 4$
11. $3 - (-3)$
12. $4 - 4$
13. $0 - 6$
14. $6 - 6$
15. $0 - (-6)$
16. $9 - (-3)$
17. $-3 - 5$
18. $-5 - (-3)$
19. $-5 + 7$
20. $-7 - 9$
21. $5 - 3$
22. $5 - 12$
23. $6 - (-3)$
24. $6 - 10$
25. $8 - 8$
26. $-8 - 8$
27. $-8 - 10$
28. $4 - 12$
29. $-4 - (-2)$
30. $7 - 9$
31. $(-4) - (-4)$
32. $15 - 8$
33. $6 - 6$
34. $(-8) - (-12)$
35. $9 - 9$
36. $-6 - (-2)$
37. $4 - 5$
38. $-9 - 2$
39. $-2 - 3$
40. $9 - (-12)$
41. $-25 - 16$
42. $-20 - (-15)$
43. $37 - 40$
44. $40 - 37$
45. $-100 - 80$
46. $80 - 100$
47. $-20 - 90$
48. $-45 - 37$
49. $-50 - (-40)$
50. $70 - (-70)$
51. $130 - (-90)$
52. $40 - 62$
53. $87 - 87$
54. $93 - (-93)$
55. $-53 - (-7)$
56. $-75 - (-16)$
57. Subtract 3 from -15.
58. Subtract -4 from -5.
59. Subtract 8 from -8.
60. Subtract 10 from -20.
61. Subtract 8 from 18.
62. Subtract 5 from -5.
63. Subtract -3 from -5.
64. Subtract 10 from -3.
65. Subtract -4 from 9.
66. Subtract 18 from -18.
67. Subtract 18 from 18.
68. Subtract 5 from 5.
69. Subtract 12 from 8.
70. Subtract -9 from 12.
71. Subtract -15 from -4.
72. Subtract -12 from 3.
73. Subtract -36 from 45.
74. Subtract 17 from -12.

*For each of the following, **(a)** determine by observation whether the difference will be a positive number, zero, or a negative number; **(b)** find the difference using your calculator; and **(c)** examine your answer to part (b) to see if it is reasonable and makes sense.*

75. $296 - 197$
76. $483 - 569$
77. $102 - 697$
78. $-372 - 195$
79. $349 - (-498)$
80. $843 - (-745)$
81. $950 - (-762)$
82. $575 - (-462)$
83. $-408 - (-604)$
84. $-776 - 358$
85. $-1024 - (-576)$
86. $-1047 - 376$
87. $165.7 - 49.6$
88. $-40.2 - (-12.6)$
89. $-37.2 - (-37.2)$
90. $597.3 - (-64.72)$
91. Subtract 364 from 295.
92. Subtract -387 from -932.
93. Subtract 647 from -1023.
94. Subtract 2432 from -6771.
95. Subtract -7.62 from 89.7.
96. Subtract 16.2 from -87.7.

Evaluate.

97. $6 + 5 - (+4)$
98. $9 - (+6) - (+5)$
99. $-3 + (-4) + 5$
100. $9 - 7 + (-2)$
101. $-13 - (+5) + 3$
102. $7 - (+4) - (-3)$
103. $-9 - (-3) + 4$
104. $15 + (-7) - (-3)$
105. $5 - (+3) + (-2)$
106. $12 + (-5) - (-4)$
107. $25 + (+12) - (-6)$
108. $-7 + 6 - 3$
109. $-4 - 7 + 5$
110. $20 - 4 - 25$
111. $-4 + 7 - 12$
112. $-36 - 5 + 9$
113. $45 - 3 - 7$
114. $-2 + 7 - 9$
115. $-9 - 4 - 8$
116. $25 - 19 + 27$
117. $-4 - 13 + 5$
118. $(-4) + (-3) + 5 - 7$
119. $-9 - 3 - (-4) + 5$
120. $17 + (-3) - 9 - (-7)$
121. $32 + 5 - 7 - 12$
122. $-19 + (-3) - (-5) - (-2)$
123. $6 - 9 - (-3) + 12$
124. $-7 - 4 - 3 + 5$
125. $19 + 4 - 20 - 25$
126. $37 - (-19) + 7 - 12$

127. A Girl Scout troop received 920 boxes of thin mint Girl Scout cookies. They sold 1246 boxes of thin mints. How many more boxes will the troop need to order?

128. An airplane is 2000 feet above sea level. A submarine is 1500 feet below sea level. How far above the submarine is the airplane?

129. Mike made \$750 in the stock market while Kirk lost \$496. What is the difference in their performance?

130. The highest point on Earth, Mt. Everest, is 29,028 feet above sea level. The lowest point on Earth, the Marianas Trench, is 36,198 feet below sea level. How far above the Marianas Trench is the top of Mt. Everest?

See Exercise 130

131. The greatest change in temperature ever recorded within a 24-hour period occurred at Browning, Montana, on January 23, 1916. The temperature fell from 44°F to −56°F. How much did the temperature drop?

132. **(a)** Is the statement $a + (-b) = a - b$ true for all real numbers a and b?

(b) If $a = -3$ and $b = 5$, determine if $a + (-b) = a - b$.

133. **(a)** Explain how to subtract −2 from 6.

(b) Subtract −2 from 6 following the procedure given in part (a).

134. **(a)** Explain how to subtract 6 from −9.

(b) Subtract 6 from −9 using the procedure given in part (a).

135. Two trains start at the same station at the same time. The Amtrak travels 68 miles in 1 hour. The Pacific Express travels 80 miles in 1 hour. **(a)** If the two trains travel in opposite directions, how far apart will they be in 1 hour? Explain.

(b) If the two trains travel in the same direction, how far apart will they be in 1 hour? Explain.

136. Consider the expression $6 - 4 + 3 - 5 - 2$, which contains only additions and subtractions.

(a) State the order we follow to evaluate the expression. Explain why we follow this order.

(b) Evaluate the expression using the order you gave in part (a).

CUMULATIVE REVIEW EXERCISES

[1.3] **137.** List the set of integers.

138. Explain the relationship between the set of rational numbers, the set of irrational numbers, and the set of real numbers.

[1.4] *Insert either >, <, or = in the shaded area to make each statement true.*

139. $|-3| \;\square\; -5$

140. $|-6| \;\square\; |-7|$

Group Activity/ Challenge Problems

Find the sum.

1. $1 - 2 + 3 - 4 + 5 - 6 + 7 - 8 + 9 - 10$

2. $1 - 2 + 3 - 4 + 5 - 6 + \cdots + 99 - 100$

3. $-1 + 2 - 3 + 4 - 5 + 6 - \cdots - 99 + 100$

4. Consider a number line.

(a) What is the distance, in units, between −2 and 5?

(b) Write a subtraction problem to represent this distance (the distance is to be positive).

5. A model rocket is on a hill near the ocean. The hill's height is 62 feet above sea level. When ignited, the rocket climbs upward to 128 feet above sea level, then it falls and lands in the ocean and settles 59 feet below sea level. Find the total distance traveled by the rocket.

Evaluate each of the following.

6. $(-5 - 4 - 3 - 2 - 1) - (-5 - 4 - 3 - 2 - 1)$

7. $-5 + 4 - 3 + 2 - 1 + 5 - 4 + 3 - 2 + 1$

1.7 Multiplication and Division of Real Numbers

Tape 2

1. Multiply real numbers.
2. Divide real numbers.
3. Remove negative signs from denominators.
4. Understand the difference between 0 in the numerator and 0 in the denominator of a fraction.

Multiplication of Real Numbers

1 The following rules are used in determining the sign of the product when two numbers are multiplied.

> **Multiplication of Real Numbers**
>
> 1. The product of two numbers with **like** signs is a **positive** number.
> 2. The product of two numbers with **unlike** signs is a **negative** number.

By this rule, the product of two positive numbers or two negative numbers will be a positive number. The product of a positive number and a negative number will be a negative number.

EXAMPLE 1 Evaluate $3(-5)$.

Solution: Since the numbers have unlike signs, the product is negative.

$$3(-5) = -15$$

EXAMPLE 2 Evaluate $(-6)(7)$.

Solution: Since the numbers have unlike signs, the product is negative.

$$(-6)(7) = -42$$

EXAMPLE 3 Evaluate $(-7)(-5)$.

Solution: Since the numbers have like signs, both negative, the product is positive.

$$(-7)(-5) = 35$$

EXAMPLE 4 Evaluate each expression.

(a) $-6 \quad 3$ **(b)** $(-4)(-8)$ **(c)** $4(-9)$
(d) $0 \quad 4$ **(e)** $0(-2)$ **(f)** $-3(-6)$

Solution: **(a)** $-6 \quad 3 = -18$ **(b)** $(-4)(-8) = 32$ **(c)** $4(-9) = -36$
(d) $0 \quad 4 = 0$ **(e)** $0(-2) = 0$ **(f)** $-3(-6) = 18$

Note that zero multiplied by any real number equals zero.

EXAMPLE 5 Multiply $\left(\frac{-1}{8}\right)\left(\frac{-3}{5}\right)$.

Solution: $\left(\frac{-1}{8}\right)\left(\frac{-3}{5}\right) = \frac{(-1)\cdot(-3)}{8\cdot 5} = \frac{3}{40}$

EXAMPLE 6 Evaluate $\left(\frac{3}{20}\right)\left(\frac{-3}{10}\right)$.

Solution: $\left(\frac{3}{20}\right)\left(\frac{-3}{10}\right) = \frac{3(-3)}{20\,(10)} = \frac{-9}{200}$

Sometimes you may be asked to perform more than one multiplication in a given problem. When this happens, the sign of the final product can be determined by counting the number of *negative* numbers being multiplied. **The product of an even number of negative numbers will always be positive. The product of an odd number of negative numbers will always be negative.** Can you explain why?

EXAMPLE 7 Evaluate $(-2)(3)(-2)(-1)$.

Solution: Since there are three negative numbers (an odd number of negatives), the product will be negative, as illustrated.

$$\begin{aligned}(-2)(3)(-2)(-1) &= (-6)(-2)(-1)\\ &= (12)\,(-1)\\ &= -12\end{aligned}$$

EXAMPLE 8 Evaluate $(-3)(2)(-1)(-2)(-4)$.

Solution: Since there are four negative numbers (an even number), the product will be positive.

$$\begin{aligned}(-3)(2)(-1)(-2)(-4) &= (-6)(-1)(-2)(-4)\\ &= (6)\,(-2)(-4)\\ &= (-12)(-4)\\ &= 48\end{aligned}$$

Division of Real Numbers

2 The rules for dividing numbers are very similar to those used in multiplying numbers.

Division of Real Numbers

1. The quotient of two numbers with **like** signs is a **positive** number.
2. The quotient of two numbers with **unlike** signs is a **negative** number.

Therefore, the quotient of two positive numbers or two negative numbers will be a positive number. The quotient of a positive and a negative number will be a negative number.

EXAMPLE 9 Evaluate $\frac{20}{-5}$.

Solution: Since the numbers have unlike signs, the quotient is negative.

$$\frac{20}{-5} = -4$$

EXAMPLE 10 Evaluate $\frac{-36}{4}$.

Solution: Since the numbers have unlike signs, the quotient is negative.

$$\frac{-36}{4} = -9$$

EXAMPLE 11 Evaluate $\frac{-30}{-5}$.

Solution: Since the numbers have like signs, both negative, the quotient is positive.

$$\frac{-30}{-5} = 6$$

EXAMPLE 12 Evaluate $-16 \div (-2)$.

Solution: $\frac{-16}{-2} = 8$

EXAMPLE 13 Evaluate $\frac{-2}{3} \div \frac{-5}{7}$.

Solution: Invert the *divisor*, $\frac{-5}{7}$, and then multiply.

$$\begin{aligned}\frac{-2}{3} \div \frac{-5}{7} &= \left(\frac{-2}{3}\right)\left(\frac{7}{-5}\right)\\ &= \frac{-14}{-15}\\ &= \frac{14}{15}\end{aligned}$$

Remove Negative Sign in the Denominator

3 We now know that the quotient of a positive and a negative number is a negative number. The fractions $-\frac{3}{4}$, $\frac{-3}{4}$, and $\frac{3}{-4}$ all represent the same negative number, negative three-fourths.

> If a and b represent any real numbers, $b \neq 0$, then
>
> $$\frac{a}{-b} = \frac{-a}{b} = -\frac{a}{b}$$

In mathematics we generally do not write a fraction with a negative sign in the denominator. When a negative sign appears in a denominator, we can move it to the numerator or place it in front of the fraction. For example, the fraction $\frac{5}{-7}$ should be written as either $-\frac{5}{7}$ or $\frac{-5}{7}$.

EXAMPLE 14 Evaluate $\frac{2}{5} \div \frac{-8}{15}$.

Solution: $$\frac{2}{5} \div \frac{-8}{15} = \frac{\overset{1}{\cancel{2}}}{\underset{1}{\cancel{5}}} \cdot \frac{\overset{3}{\cancel{15}}}{\underset{4}{-\cancel{8}}} = \frac{1(3)}{1(-4)} = \frac{3}{-4} = -\frac{3}{4}$$

The operations on real numbers are summarized in Table 1.1.

TABLE 1.1 SUMMARY OF OPERATIONS ON REAL NUMBERS

Signs of Numbers	Addition	Subtraction	Multiplication	Division
Both Numbers Are Positive	Sum Is Always Positive	Difference May Be Either Positive or Negative	Product Is Always Positive	Quotient Is Always Positive
Examples 6 and 2 2 and 6	$6 + 2 = 8$ $2 + 6 = 8$	$6 - 2 = 4$ $2 - 6 = -4$	$6 \cdot 2 = 12$ $2 \cdot 6 = 12$	$6 \div 2 = 3$ $2 \div 6 = \frac{1}{3}$
One Number Is Positive and the Other Number Is Negative	Sum May Be Either Positive or Negative	Difference May Be Either Positive or Negative	Product Is Always Negative	Quotient Is Always Negative
Examples 6 and −2 −6 and 2	$6 + (-2) = 4$ $-6 + 2 = -4$	$6 - (-2) = 8$ $-6 - (2) = -8$	$6(-2) = -12$ $-6(2) = -12$	$6 \div (-2) = -3$ $-6 \div 2 = -3$
Both Numbers Are Negative	Sum Is Always Negative	Difference May Be Either Positive or Negative	Product Is Always Positive	Quotient Is Always Positive
Examples −6 and −2 −2 and −6	$-6 + (-2) = -8$ $-2 + (-6) = -8$	$-6 - (-2) = -4$ $-2 - (-6) = 4$	$-6(-2) = 12$ $-2(-6) = 12$	$-6 \div (-2) = 3$ $-2 \div (-6) = \frac{1}{3}$

Helpful Hint

At this point some students begin confusing problems like $-2 - 3$ with $(-2)(-3)$ and problems like $2 - 3$ with problems like $2(-3)$. If you do not understand the difference between problems like $-2 - 3$ and $(-2)(-3)$, make an appointment to see your instructor as soon as possible.

Subtraction Problems	*Multiplication Problems*
$-2 - 3 = -5$	$(-2)(-3) = 6$
$2 - 3 = -1$	$(2)(-3) = -6$

Helpful Hint

For multiplication and division of real numbers:

$$\left.\begin{array}{ll} (+)(+) = + & \dfrac{(+)}{(+)} = + \\ (-)(-) = + & \dfrac{(-)}{(-)} = + \end{array}\right\} \text{Like signs give positive products and quotients.}$$

$$\left.\begin{array}{ll} (+)(-) = - & \dfrac{(+)}{(-)} = - \\ (-)(+) = - & \dfrac{(-)}{(+)} = - \end{array}\right\} \text{Unlike signs give negative products and quotients}$$

Division Involving 0

4 Now let us look at division involving the number 0. What is $\frac{0}{1}$ equal to? Note that $\frac{6}{3} = 2$ because $3 \cdot 2 = 6$. We can follow the same procedure to determine the value of $\frac{0}{1}$. Suppose that $\frac{0}{1}$ is equal to some number, which we will designate by **?** .

$$\text{If } \frac{0}{1} = \ ? \text{ then } 1 \cdot \ ? = 0$$

Since only $1 \cdot 0 = 0$, the ? must be 0. Thus, $\frac{0}{1} = 0$. Using the same technique, we can show that zero divided by any nonzero number is zero.

If a represents any real number except 0, then

$$\frac{0}{a} = 0$$

Now what is $\frac{1}{0}$ equal to?

$$\text{If } \frac{1}{0} = \ ? \text{ then } 0 \cdot \ ? = 1$$

But since 0 multiplied by any number will be 0, there is no value that can replace ? . We say that $\frac{1}{0}$ is *undefined.* Using the same technique, we can show that any real number, except 0, divided by 0 is undefined.

If a represents any real number except 0, then

$$\frac{a}{0} \text{ is } \textbf{undefined}.$$

What is $\frac{0}{0}$ equal to?

$$\text{If } \frac{0}{0} = \ ? \quad \text{then} \quad 0 \cdot \ ? = 0$$

But since the product of any number and 0 is 0, the ? can be replaced by any real number. Therefore the quotient $\frac{0}{0}$ cannot be determined, and so we will not use it in this course.*

Summary of Division Involving Zero

If a represents any real number except 0, then

$$\frac{0}{a} = 0. \qquad \frac{a}{0} \text{ is undefined.}$$

*At this level, some professors prefer to call $\frac{0}{0}$ *indeterminate* while others prefer to call $\frac{0}{0}$ *undefined*. In higher level mathematics courses $\frac{0}{0}$ is sometimes referred to as the *indeterminate form*.

Calculator Corner

MULTIPLICATION OF REAL NUMBERS

Below we illustrate how real numbers may be multiplied on a calculator.

Evaluate	Keystrokes
$6(-23)$	6 × 23 +/_ = −138
$(-14)(-37)$	14 +/_ × 37 +/_ = 518

Since you know that a positive number multiplied by a negative number will be negative, to obtain the product of $6(-23)$, you can multiply 6×23 and write a negative sign before the answer. Similarly, since you know that a negative number multiplied by a negative number is positive, you can obtain the answer to $(-14)(-37)$ simply by multiplying $(14)(37)$.

DIVISION OF REAL NUMBERS

Below we illustrate how real numbers may be divided on a calculator.

Evaluate	Keystrokes
$\dfrac{-40}{8}$	40 +/_ ÷ 8 = −5
$\dfrac{85}{-5}$	85 ÷ 5 +/_ = −17
$\dfrac{-240}{-16}$	240 +/_ ÷ 16 +/_ = 15

A positive number divided by a negative number, or a negative number divided by a positive number, is a negative number. Therefore, the first two examples could have been done using only positive numbers and then placing a negative sign before the value obtained to get the correct answer. Since a negative number divided by a negative number is positive, the third quotient could have been found by dividing 240/16.

EXAMPLE 15 Indicate whether the quotient is 0 or undefined.

(a) $\frac{0}{2}$ **(b)** $\frac{5}{0}$ **(c)** $\frac{0}{-4}$ **(d)** $\frac{-2}{0}$

Solution: The answer to parts **(a)** and **(c)** is 0. The answer to parts **(b)** and **(d)** is undefined.

Exercise Set 1.7

Find the product.

1. $(-4)(-3)$ **2.** $-4 \cdot 2$ **3.** $3(-3)$ **4.** $6(-2)$

5. $(-4)(8)$ **6.** $(-3)(2)$ **7.** $9(-1)$ **8.** $-1(8)$

9. $-4(-3)$ **10.** $0(4)$ **11.** $-9(-4)$ **12.** $(-12)(-3)$

13. $8(12)$ **14.** $(-5)(-6)$ **15.** $-9(-9)$ **16.** $(15)(-4)$

17. $-2(5)$ **18.** $6(-12)$ **19.** $(-6)(2)(-3)$ **20.** $5(-2)(-8)$

21. $0(3)(8)$ **22.** $2(-3)(7)$ **23.** $(-1)(-1)(-1)$ **24.** $2(4)(-2)(-5)$

25. $-5(-3)(8)(-1)$ **26.** $(-3)(-4)(-5)(-1)$ **27.** $(-4)(3)(-7)(1)$ **28.** $4(3)(1)(-1)$

29. $(-3)(2)(5)(3)$ **30.** $(-1)(3)(0)(-7)$ **31.** $(-5)(-6)(-3)(-4)$ **32.** $(-1)(-1)(9)(8)$

Find the product.

33. $\left(\frac{-1}{2}\right)\left(\frac{3}{5}\right)$ **34.** $\left(\frac{2}{3}\right)\left(\frac{-3}{5}\right)$ **35.** $\left(\frac{-8}{9}\right)\left(\frac{-7}{12}\right)$ **36.** $\left(\frac{-5}{12}\right)\left(\frac{-6}{11}\right)$

37. $\left(\frac{6}{-3}\right)\left(\frac{4}{-2}\right)$ **38.** $\left(\frac{8}{-11}\right)\left(\frac{6}{-5}\right)$ **39.** $\left(\frac{5}{-7}\right)\left(\frac{6}{8}\right)$ **40.** $\left(\frac{9}{10}\right)\left(\frac{7}{-8}\right)$

Find the quotient.

41. $\frac{6}{2}$ **42.** $9 \div (-3)$ **43.** $-16 \div (-4)$ **44.** $\frac{-24}{8}$

45. $\frac{-36}{-9}$ **46.** $-45 \div 5$ **47.** $\frac{-16}{4}$ **48.** $\frac{36}{-2}$

49. $\frac{18}{-1}$ **50.** $\frac{-12}{-1}$ **51.** $-15 \div (-3)$ **52.** $12 \div (-6)$

53. $\frac{-6}{-1}$ **54.** $\frac{60}{-12}$ **55.** $\frac{-25}{-5}$ **56.** $\frac{36}{-4}$

57. $\frac{1}{-1}$ **58.** $\frac{-1}{1}$ **59.** $\frac{-48}{12}$ **60.** $\frac{50}{-5}$

61. $\frac{0}{1}$ **62.** $-40 \div (-8)$ **63.** $-64 \div (-4)$ **64.** $(-120) \div (-120)$

65. Divide 0 by 4. **66.** Divide 20 by -5. **67.** Divide 30 by -10.

68. Divide -30 by -10. **69.** Divide -180 by 30. **70.** Divide -60 by 5.

71. Divide -25 by -5. **72.** Divide 80 by -20.

Find the quotient.

73. $\frac{5}{12} \div \left(\frac{-5}{9}\right)$ **74.** $(-3) \div \frac{5}{19}$ **75.** $\frac{3}{-10} \div (-8)$ **76.** $\frac{-4}{9} \div \left(\frac{-6}{7}\right)$

77. $\frac{-15}{21} \div \left(\frac{-15}{21}\right)$ **78.** $\frac{8}{-15} \div \left(\frac{-9}{10}\right)$ **79.** $(-12) \div \frac{5}{12}$ **80.** $\frac{-16}{3} \div \left(\frac{5}{-9}\right)$

Evaluate.

81. $-6 \cdot 5$ **82.** $-9(-3)$ **83.** $\frac{-18}{-2}$ **84.** $\frac{100}{-5}$

85. $-50 \div (-10)$ **86.** $-3(0)$ **87.** $-5(-12)$ **88.** $56 \div (-8)$

89. $\frac{0}{5}$ **90.** Divide 60 by -2. **91.** $(-1)(-5)(-9)$ **92.** Divide -120 by -10.

93. $-100 \div 5$ **94.** $4(-2)(-1)(-5)$ **95.** Divide 60 by -60. **96.** $(6)(-1)(-3)(4)$

Indicate whether each of the following is 0 or undefined.

97. $0 \div 6$ **98.** $-4 \div 0$ **99.** $\frac{5}{0}$ **100.** $\frac{-2}{0}$

101. $\frac{0}{1}$ **102.** $0 \div (-2)$ **103.** $0 \div 6$ **104.** $6 \div 0$

105. $\frac{0}{-6}$ **106.** $\frac{0}{-1}$ **107.** 3 divided by 0 **108.** 0 divided by 12

For each of the following, (a) determine by observation whether the product or quotient will be a positive number, 0, a negative number, or undefined; (b) find the product or quotient on your calculator (an error message indicates that the quotient is undefined); (c) examine your answer in part (b) to see if it is reasonable and makes sense.

109. $96(-15)$ **110.** $(-212)(-87)$ **111.** $-168 \div 42$ **112.** $204 \div (-17)$

113. $-240/15$ **114.** $190/(-5)$ **115.** $243 \div (-27)$ **116.** $(-323) \div (-17)$

117. $(-15)(-170)$ **118.** $-440 \div 22$ **119.** $(-406)(-42)$ **120.** $(18)(-27)$

121. $(1530)(0)$ **122.** $0 \div 1935$ **123.** $(-19)(10.5)$ **124.** $-86.4 \div (-36)$

125. $7.2 \div 0$ **126.** $-37.74 \div 37$ **127.** $0 \div (-5260)$ **128.** $(4.3)(-2.1)(6.3)$

129. $(-90)(-1.2)(-1.6)$ **130.** $-288.86/1.43$ **131.** $(9.6)(-12.2)(-60)$ **132.** $0.48020/(-19.6)$

In Exercises 133–145, answer true or false.

133. The product of a positive number and a negative number is a negative number.

134. The product of two negative numbers is a negative number.

135. The quotient of two negative numbers is a positive number.

136. The quotient of two numbers with unlike signs is a positive number.

137. The product of an even number of negative numbers is a positive number.

138. The product of an odd number of negative numbers is a negative number.

139. Zero divided by 1 is 1.

140. Six divided by 0 is 0.

141. One divided by 0 is 0.

142. Zero divided by 1 is undefined.

143. Five divided by 0 is undefined.

144. The product of 0 and any real number is 0.

145. Division by 0 does not result in a real number.

146. Write out the rules for determining the sign of the product or quotient of two numbers.

147. Explain why the product of an even number of negative numbers is a positive number.

148. Will the product of $(1)(-2)(3)(-4)(5)(-6)\cdots(33)(-34)$ be a positive number or a negative number? Explain how you determined your answer.

CUMULATIVE REVIEW EXERCISES

[1.1] **149.** Find the quotient $\frac{5}{7} \div \frac{1}{5}$.

[1.5] **150.** Subtract -18 from -20.

Evaluate.

151. $6 - 3 - 4 - 2$

152. $5 - (-2) + 3 - 7$

Group Activity/ Challenge Problems

We will learn in the next section that $2^3 = 2 \cdot 2 \cdot 2$ and $x^m = \underbrace{x \cdot x \cdot x \cdot \cdots \cdot x}_{m \text{ factors of } x}$.

Using this information evaluate each of the following.

1. 3^4 **2.** $(-2)^3$ **3.** $\left(\frac{2}{3}\right)^3$ **4.** 1^{100} **5.** $(-1)^{81}$

Find the quotient.

6. $\dfrac{1 - 2 + 3 - 4 + 5 - \cdots + 99 - 100}{1 - 2 + 3 - 4 + 5 - \cdots + 99 - 100}$

7. $\dfrac{-1 + 2 - 3 + 4 - 5 + \cdots - 99 + 100}{1 - 2 + 3 - 4 + 5 - \cdots + 99 - 100}$

8. $\dfrac{5 \cdot 4 \cdot 3 \cdot 2 \cdot 1}{(-5)(-4)(-3)(-2)(-1)}$

9. $\dfrac{6 \cdot 5 \cdot 4 \cdot 3 \cdot 2 \cdot 1}{(-6)(-5)(-4)(-3)(-2)(-1)}$

10. $\dfrac{(-5)(-4)(-3)(-2)(-1)}{(-1)(2)(-3)(4)(-5)} \cdot (-1)(-2)(-3)$

1.8 An Introduction to Exponents

Tape 2

1. Learning the meaning of exponents.
2. Evaluate expressions containing exponents.
3. Learn the difference between $-x^2$ and $(-x)^2$.

Exponents

1 To understand certain topics in algebra, you must understand exponents. Exponents are introduced in this section and are discussed in more detail in Chapter 4.

In the expression 4^2, the 4 is called the **base,** and the 2 is called the **exponent.** The number 4^2 is read "4 squared" or "4 to the second power" and means

$$\underbrace{4 \cdot 4}_{2 \text{ factors of } 4} = 4^2$$

The number 4^3 is read "4 cubed" or "4 to the third power" and means

$$\underbrace{4 \cdot 4 \cdot 4}_{3 \text{ factors of } 4} = 4^3$$

In general, the number b to the nth power, written b^n, means

$$\underbrace{b \cdot b \cdot b \cdot \cdots \cdot b}_{n \text{ factors of } b} = b^n$$

Thus, $b^4 = b \cdot b \cdot b \cdot b$ or $bbbb$ and $x^3 = x \cdot x \cdot x$ or xxx.

Evaluate Expressions Containing Exponents

2

EXAMPLE 1 Evaluate.

(a) 3^2 **(b)** 2^5 **(c)** 1^5 **(d)** 4^3 **(e)** $(-3)^2$ **(f)** $(-2)^3$ **(g)** $\left(\frac{2}{3}\right)^2$

Solution: **(a)** $3^2 = 3 \cdot 3 = 9$
(b) $2^5 = 2 \cdot 2 \cdot 2 \cdot 2 \cdot 2 = 32$
(c) $1^5 = 1 \cdot 1 \cdot 1 \cdot 1 \cdot 1 = 1$ (1 raised to any power equals 1; why?)
(d) $4^3 = 4 \cdot 4 \cdot 4 = 64$
(e) $(-3)^2 = (-3)(-3) = 9$
(f) $(-2)^3 = (-2)(-2)(-2) = -8$
(g) $\left(\frac{2}{3}\right)^2 = \left(\frac{2}{3}\right)\left(\frac{2}{3}\right) = \frac{4}{9}$

Other examples of exponential notation are:

(a) $x \cdot x \cdot x \cdot x = x^4$ **(b)** $aabbb = a^2b^3$
(c) $x \cdot x \cdot y = x^2y$ **(d)** $aaabb = a^3b^2$
(e) $xyxx = x^3y$ **(f)** $xyzzy = xy^2z^2$
(g) $3 \cdot x \cdot x \cdot y = 3x^2y$ **(h)** $5xyyyy = 5xy^4$
(i) $3 \cdot 3 \cdot x \cdot x = 3^2x^2$ **(j)** $5 \cdot 5 \cdot 5 \cdot xxy = 5^3x^2y$

Notice in parts **(e)** and **(f)** that the order of the factors does not matter.

Helpful Hint

Do you know the difference between **(a)** $x + x + x + x + x + x$ and **(b)** $x \cdot x \cdot x \cdot x \cdot x \cdot x$? Can you write a simplified expression for both parts **(a)** and **(b)**? The simplified expression for part **(a)** is $6x$ and the simplified expression for part **(b)** is x^6. Note that $x + x + x + x + x + x = 6x$ and $x \cdot x \cdot x \cdot x \cdot x \cdot x = x^6$.

It is not necessary to write exponents of 1. Thus, when writing xxy, we write x^2y and not x^2y^1. **Whenever we see a letter or number without an exponent, we always assume that letter or number has an exponent of 1.**

EXAMPLE 2 Write each expression as a product of factors.

(a) x^2y **(b)** xy^3z **(c)** $3x^2yz^3$ **(d)** 2^3xy **(e)** $3^2x^3y^2$

Solution: **(a)** $x^2y = xxy$ **(b)** $xy^3z = xyyyz$ **(c)** $3x^2yz^3 = 3xxyzzz$
(d) $2^3xy = 2 \cdot 2 \cdot 2xy$ **(e)** $3^2x^3y^2 = 3 \cdot 3xxxy$

$-x^2$ and $(-x)^2$

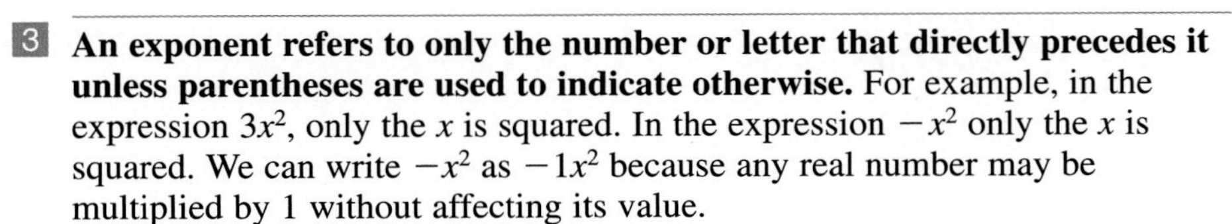

3 **An exponent refers to only the number or letter that directly precedes it unless parentheses are used to indicate otherwise.** For example, in the expression $3x^2$, only the x is squared. In the expression $-x^2$ only the x is squared. We can write $-x^2$ as $-1x^2$ because any real number may be multiplied by 1 without affecting its value.

$$-x^2 = -1x^2$$

By looking at $-1x^2$ we can see that only the x is squared, not the -1. If the entire expression $-x$ was to be squared, we would need to use parentheses and write $(-x)^2$. Note the difference in the following two examples:

$$-x^2 = -(x)(x)$$
$$(-x)^2 = (-x)(-x)$$

Consider the expressions -3^2 and $(-3)^2$. How do they differ?

$$-3^2 = -(3)(3) = -9$$
$$(-3)^2 = (-3)(-3) = 9$$

EXAMPLE 3 Evaluate.

(a) -5^2 **(b)** $(-5)^2$ **(c)** -2^3 **(d)** $(-2)^3$

Solution: **(a)** $-5^2 = -(5)(5) = -25$ **(b)** $(-5)^2 = (-5)(-5) = 25$

(c) $-2^3 = -(2)(2)(2) = -8$ **(d)** $(-2)^3 = (-2)(-2)(-2) = -8$

EXAMPLE 4 Evaluate **(a)** -2^4 and **(b)** $(-2)^4$.

Solution: **(a)** $-2^4 = -(2)(2)(2)(2) = -16$
(b) $(-2)^4 = (-2)(-2)(-2)(-2) = 16$

EXAMPLE 5 Evaluate **(a)** x^2 and **(b)** $-x^2$ for $x = 3$.

Solution: Substitute 3 for x.

(a) $x^2 = 3^2 = 3 \quad 3 = 9$ **(b)** $-x^2 = -3^2 = -(3)(3) = -9$

EXAMPLE 6 Evaluate **(a)** y^2 and **(b)** $-y^2$ for $y = -4$.

Solution: Substitute -4 for y.

(a) $y^2 = (-4)^2 = (-4)(-4) = 16$
(b) $-y^2 = -(-4)^2 = -(-4)(-4) = -16$

Note that $-x^2$ will always be a negative number for any nonzero value of x, and $(-x)^2$ will always be a positive number for any nonzero value of x. Can you explain why? See Exercises 104 and 105.

COMMON STUDENT ERROR The expression $-x^2$ means $-(x^2)$. When asked to evaluate $-x^2$ for any real number x, many students will incorrectly treat $-x^2$ as $(-x)^2$.
Evaluate $-x^2$ when $x = 5$.

Correct	*Incorrect*
$-5^2 = -(5^2) = -(5)(5)$	~~$-5^2 = (-5)(-5)$~~
$= -25$	~~$= 25$~~

Calculator Corner

Use of x^2 and y^x Keys

The x^2 key is used to square a value. For example, to evaluate 5^2 we would press.

$$5 \; x^2 \; 25$$

After the x^2 key is pressed the answer 25 is displayed.
To evaluate $(-5)^2$ on a calculator, we press

$$5 \; {}^{+}\!/\!_{-} \; x^2 \; 25$$

Note that $(-5)^2$ has a value of 25.

To raise a value to a power greater than 2 we use the y^x or x^y key. Some calculators have a y^x key while others have an x^y key. To use these keys you enter the number, then press either the y^x or x^y key, then enter the exponent. After the $=$ key is pressed the answer will be displayed.

Evaluate	*Keystrokes*
2^5	2 y^x 5 $=$ 32
$(13)^4$	13 y^x 4 $=$ 28561

Even when using the parentheses keys *some* scientific calculators cannot directly raise a negative number to a power greater than 2. For example, if you evaluate $(-2)^5$ using the keystrokes (2 +/−) y^x 5 $=$ *some* calculators will give an *error* message. The correct answer is -32. Possibly the easiest way to raise negative numbers to a power may be to raise the positive number to the power and then write a negative sign before the final answer if needed. *A negative number raised to an odd power will be negative, and a negative number raised to an even power will be positive.* Can you explain why this is true?

Evaluate	*Keystrokes*	*Correct Answer*
$(-2)^5$	2 y^x 5 $=$ 32	-32 (since the exponent is odd)
$(-13)^4$	13 y^x 4 $=$ 28561	28561 (since the exponent is even)

Raising Fractions to Powers

To raise fractions to powers we may use the $=$ key before using the x^2 or y^x key, or we may use parentheses.

Evaluate	*Keystrokes*
$\left(\frac{2}{5}\right)^2$	2 $\div$ 5 $=$ x^2 .16 or (2 $\div$ 5) x^2 .16
$\left(\frac{2}{5}\right)^6$	2 $\div$ 5 $=$ y^x 6 $=$ .004096 or (2 $\div$ 5) y^x 6 $=$ .004096

When evaluating $\left(\frac{2}{5}\right)^2$ what would happen if we did not use either the $=$ or the parentheses keys? That is, what would the calculator display if you keyed in 2 $\div$ 5 x^2 and 2 $\div$ 5 x^2 $=$? Try this now on your calculator and explain the results. See Group Activity Exercise 5.

Exercise Set 1.8

Evaluate.

1. 5^2
2. 3^2
3. 2^3
4. 1^5
5. 3^3
6. -5^2
7. 6^3
8. $(-2)^2$
9. $(-2)^3$
10. -3^4
11. $(-1)^3$
12. 6^2
13. 3^3
14. 2^5
15. -6^2
16. 5^3
17. $(-6)^2$
18. $(-3)^3$
19. 2^4
20. $(-3)^4$
21. 4^1
22. -3^2
23. $(-2)^4$
24. -1^4
25. -2^4
26. $(-1)^4$
27. $(-4)^3$
28. $3^2(4)^2$
29. $5^2 \cdot 3^2$
30. $(-1)^4(3)^3$
31. $5(4^2)$
32. $2^3 \cdot 5^1$
33. $2^1 \cdot 4^2$
34. $(-2)^4(-1)^3$
35. $3(-5^2)$
36. $9(-2)^2$

Express in exponential form.

37. $x \cdot x \cdot y \cdot y$
38. $x \cdot y \cdot z \cdot z$
39. $xyyyz$
40. $xxxxz$
41. $yyzzz$
42. $aabbab$
43. $xyxyz$
44. $x \cdot x \cdot y \cdot z \cdot z$
45. $a \cdot x \cdot a \cdot x \cdot y$
46. $x \cdot x \cdot x \cdot y \cdot y$
47. $x \cdot y \cdot y \cdot z \cdot z \cdot z$
48. $xyyyy$
49. $3xyy$
50. $2 \cdot 2 \cdot 2xyyyy$

Express as a product of factors.

51. x^2y
52. y^2z
53. xy^3
54. x^2yz
55. xy^2z^3
56. $2x^2y^2$
57. 3^2yz
58. 2^3y^3
59. 2^3x^3y
60. 3^3xy^3
61. $(-2)^2y^3z$
62. $(-1)^2x^3y^2$

Evaluate **(a)** x^2 *and* **(b)** $-x^2$ *for each of the following values of* x.

63. 3
64. 2
65. 4
66. 1
67. -2
68. 5
69. 7
70. 8
71. -1
72. -5
73. $-\frac{1}{2}$
74. $\frac{3}{4}$

In Exercises 75–94 (a) determine by observation whether the answer should be positive or negative, (b) evaluate the expression on your calculator, and (c) determine if your answer in part (b) is reasonable and makes sense.

75. 3^5
76. 4^6
77. $(-2)^3$
78. 5^4
79. -2^5
80. $(-6)^3$
81. -5^6
82. 10^4
83. $(-6)^4$
84. $(1.3)^3$
85. $(8.4)^3$
86. $(5.3)^4$
87. $(-2.3)^3$
88. $(-4.5)^4$
89. $(-1/2)^4$
90. $-(1/2)^4$
91. $(2/5)^4$
92. $(3/4)^3$
93. $(-2/3)^4$
94. $-\left(\frac{3}{5}\right)^3$

In Exercises 95–103, answer true or false.

95. $(-4)^{20}$ is a negative number.
96. $(-4)^{19}$ is a negative number.
97. $-(-3)^{15}$ is a negative number.
98. $-(-2)^{14}$ is a negative number.
99. x^2y means x^2y^1.
100. $3xy^4$ means $3^1x^1y^4$.
101. $2x^5y$ means $2^1x^5y^1$.
102. When a number is written without an exponent, the exponent on the number is 0.
103. When a variable is written without an exponent, the exponent on the variable is 1.
104. Explain why $-x^2$ will always be a negative number for any nonzero value of x.
105. Explain why $(-x)^2$ will always be a positive number for any nonzero value of x.

106. Will the expression $(-6)^{15}$ be a positive or a negative number? Explain.

107. Will the expression $(-1)^{100}$ be a positive or a negative number? Explain.

108. Will the expression -8^{14} be a positive or a negative number? Explain.

109. We will discuss using zero as an exponent in Section 4.1. On your scientific calculator find the value of 4^0 by using your y^x or x^y key and record its value. Evaluate a few other numbers raised to the zero power. Can you make any conclusions about a real number (other than 0) raised to the zero power?

CUMULATIVE REVIEW EXERCISES

[1.6] **110.** Subtract -6 from 12.

111. Evaluate $-4 - 3 + 9 - 7$.

112. Evaluate $-4672 - 5692$ on your calculator.

[1.7] *Evaluate.*

113. $\left(\frac{-5}{7}\right) \div \left(\frac{-3}{14}\right)$

114. $\frac{0}{4}$

Group Activity/ Challenge Problems

In Exercises 1–4, simplify parts **(a)** *to* **(c)**, *and leave the answer in exponential form. Using parts* **(a)** *to* **(c)**, *and other examples, determine the answer to part* **(d)**.

1. **(a)** $2^2 \cdot 2^3$ **(b)** $3^2 \cdot 3^3$ **(c)** $2^3 \cdot 2^4$ **(d)** $x^m \cdot x^n$

2. **(a)** $\frac{2^3}{2^2}$ **(b)** $\frac{3^4}{3^2}$ **(c)** $\frac{4^5}{4^3}$ **(d)** $\frac{x^m}{x^n}$

3. **(a)** $(2^3)^2$ **(b)** $(3^3)^2$ **(c)** $(4^2)^2$ **(d)** $(x^m)^n$

4. **(a)** $(2x)^2$ **(b)** $(3x)^2$ **(c)** $(4x)^3$ **(d)** $(ax)^m$

General rules that may be used to solve problems like 1 through 4 will be discussed in Chapter 4.

5. In the Calculator Corner in this section we showed that to evaluate $\left(\frac{2}{5}\right)^2$ we press the following keys:

2 ÷ 5 = x^2 .16

Therefore, $(2/5)^2 = 0.16$. Using a scientific calculator, perform the set of keystrokes below. Explain each result.

(a) 2 ÷ 5 x^2 **(b)** 2 ÷ 5 x^2 =

1.9 Use of Parentheses and Order of Operations

1 Learn the order of operations.
2 Learn the use of parentheses.
3 Evaluate expressions for given values of the variable.

Tape 2

Order of Operations

1 Evaluate $2 + 3 \cdot 4$. Is it 20? Is it 14? To answer this question you must know the order of operations to follow when evaluating a mathematical expression. You will often have to evaluate expressions containing multiple operations.

To Evaluate Mathematical Expressions, Use the Following Order

1. First, evaluate the information within **parentheses** ( ), or brackets [ ]. If the expression contains nested parentheses (one pair of parentheses within another pair), evaluate the information in the innermost parentheses first.
2. Next, evaluate all **exponents.**
3. Next, evaluate all **multiplications** or **divisions** in the order in which they occur, working from left to right.
4. Finally, evaluate all **additions** or **subtractions** in the order in which they occur, working from left to right.

We can now evaluate $2 + 3 \cdot 4$. Since multiplications are performed before additions,

$$2 + 3 \cdot 4 \text{ means } 2 + (3 \cdot 4) = 2 + 12 = 14$$

Calculator Corner

We now know that $2 + 3 \cdot 4$ means $2 + (3 \cdot 4)$ and has a value of 14. What will a calculator display if you key in the following?

2 [+] 3 [×] 4 [=]

The answer depends on your calculator. *Scientific calculators* will evaluate an expression following the rules just stated.

Scientific calculator: 2 [+] 3 [×] 4 [=] 14

Nonscientific calculators will perform operations in the order they are entered.

Nonscientific calculator: 2 [+] 3 [×] 4 [=] 20

Remember that in algebra, unless otherwise instructed by parentheses, we always perform multiplications and divisions before additions and subtractions. Is your calculator a scientific calculator?

To calculate $2 + (3 \times 4)$ on a nonscientific calculator, we first enter the multiplication and then the addition, as follows:

3 [×] 4 [+] 2 [=] 14

Scientific calculators are not much more expensive than nonscientific calculators. We recommend that you purchase a scientific calculator, especially if you plan to take more mathematics or science courses.

Use of Parentheses

2 Parentheses or brackets may be used (1) to change the order of operations to be followed in evaluating an algebraic expression or (2) to help clarify the understanding of an expression.

To evaluate the expression $2 + 3 \cdot 4$, we would normally perform the multiplication, $3 \cdot 4$, first. If we wished to have the addition performed before the multiplication, we could indicate this by placing parentheses about the $2 + 3$:

$$(2 + 3) \cdot 4 = 5 \cdot 4 = 20$$

Consider the expression $1 \cdot 3 + 2 \cdot 4$. According to the order, multiplications are to be performed before additions. We can rewrite this expression as $(1 \cdot 3) + (2 \cdot 4)$. Note that the order of operations was not changed. The parentheses were used only to help clarify the order to be followed.

Helpful Hint

If parentheses are not used to change the order of operations, multiplications and divisions are always performed before additions and subtractions. When a problem has only multiplications and divisions, work from left to right. Similarly, when a problem has only additions and subtractions, work from left to right.

EXAMPLE 1 Evaluate $2 + 3 \cdot 5^2 - 7$.

Solution: Color shading is used to indicate the order in which the expression is to be evaluated.

$$\begin{aligned} &2 + 3 \cdot 5^2 - 7 \\ &= 2 + 3 \cdot 25 - 7 \\ &= 2 + 75 - 7 \\ &= 77 - 7 \\ &= 70 \end{aligned}$$

EXAMPLE 2 Evaluate $6 + 3[(12 \div 4) + 5]$.

Solution:

$$\begin{aligned} &6 + 3[(12 \div 4) + 5] \\ &= 6 + 3[3 + 5] \\ &= 6 + 3(8) \\ &= 6 + 24 \\ &= 30 \end{aligned}$$

EXAMPLE 3 Evaluate $(4 \div 2) + 4(5 - 2)^2$.

Solution:

$$\begin{aligned} &(4 \div 2) + 4(5 - 2)^2 \\ &= 2 + 4(3)^2 \\ &= 2 + 4 \cdot 9 \\ &= 2 + 36 \\ &= 38 \end{aligned}$$

EXAMPLE 4 Evaluate $5 + 2^2 \cdot 3 - 3^2$.

Solution:

$$\begin{aligned} &5 + 2^2 \cdot 3 - 3^2 \\ &= 5 + 4 \cdot 3 - 9 \\ &= 5 + 12 - 9 \\ &= 17 - 9 \\ &= 8 \end{aligned}$$

EXAMPLE 5 Evaluate $-8 - 81 \div 9 \cdot 2^2 + 7$.

Solution:

$$\begin{aligned} & 8 - 81 \div 9 \cdot 2^2 + 7 \\ &= -8 - 81 \div 9 \cdot 4 + 7 \\ &= -8 - 9 \cdot 4 + 7 \\ &= -8 - 36 + 7 \\ &= -44 + 7 \\ &= -37 \end{aligned}$$

EXAMPLE 6 Evaluate.

(a) $-4^2 + 6 \div 3$ **(b)** $(-4)^2 + 6 \div 3$

Solution: **(a)**

$$\begin{aligned} & -4^2 + 6 \div 3 \\ &= -16 + 6 \div 3 \\ &= -16 + 2 \\ &= -14 \end{aligned}$$

(b)

$$\begin{aligned} & (-4)^2 + 6 \div 3 \\ &= 16 + 6 \div 3 \\ &= 16 + 2 \\ &= 18 \end{aligned}$$

EXAMPLE 7 Evaluate $\frac{3}{8} - \frac{2}{5} \cdot \frac{1}{12}$.

Solution: First perform the multiplication.

$$\begin{aligned} & \frac{3}{8} - \left(\frac{\overset{1}{\cancel{2}}}{5} \cdot \frac{1}{\underset{6}{\cancel{12}}}\right) \\ &= \frac{3}{8} - \frac{1}{30} \\ &= \frac{45}{120} - \frac{4}{120} \\ &= \frac{41}{120} \end{aligned}$$

EXAMPLE 8 Write the following statements as mathematical expressions using parentheses and brackets and then evaluate: Multiply 5 by 3. To this product add 6. Multiply this sum by 7.

Solution:

$5 \cdot 3$	**Multiply 5 by 3.**
$(5 \cdot 3) + 6$	**Add 6.**
$7[(5 \cdot 3) + 6]$	**Multiply the sum by 7.**

Now evaluate the expression.

$$\begin{aligned} & 7[(5 \cdot 3) + 6] \\ &= 7[15 + 6] \\ &= 7(21) \\ &= 147 \end{aligned}$$

Sometimes brackets are used in place of parentheses to help avoid confusion. If only parentheses had been used, the preceding expression would appear as $7((5 \cdot 3) + 6)$.

EXAMPLE 9 Write the following statements as mathematical expressions using parentheses and brackets and then evaluate: Subtract 3 from 15. Divide this difference by 2. Multiply this quotient by 4.

Solution:

$15 - 3$	Subtract 3 from 15.
$(15 - 3) \div 2$	Divide by 2.
$4[(15 - 3) \div 2]$	Multiply the quotient by 4.

Now evaluate.

$$\begin{aligned} &4[(15 - 3) \div 2] \\ &= 4[12 \div 2] \\ &= 4(6) \\ &= 24 \end{aligned}$$

Calculator Corner

USING PARENTHESES

When evaluating an expression on a calculator where the order of operations is to be changed, you will need to use parentheses. If you are not sure whether or not they are needed, it will not hurt to add them. Consider $\frac{8}{4-2}$. Since we wish to divide 8 by the difference $4 - 2$, we need to use parentheses.

Evaluate	Keystrokes
$\frac{8}{4-2}$	8 ÷ (4 − 2) = 4

What would you obtain if you evaluated 8 ÷ 4 − 2 = on a calculator? Why would you get that result? Here is another example.

Evaluate	Keystrokes
$-5(20 - 46) - 12$	5 +/− × (20 − 46) − 12 = 118

Evaluating Expressions Containing Variables

3 Now we will evaluate some expressions for given values of the variables.

EXAMPLE 10 Evaluate $7x - 2$ when $x = 2$.

Solution: Substitute 2 for x in the expression.

$$7x - 2 = 7(2) - 2 = 14 - 2 = 12$$

EXAMPLE 11 Evaluate $(3x + 1) + 2x^2$ when $x = 4$.

Solution: Substitute 4 for each x in the expression.

$$\begin{aligned}(3x + 1) + 2x^2 &= [3(4) + 1] + 2(4)^2 \\ &= [12 + 1] + 2(4)^2 \\ &= 13 + 2(16) \\ &= 13 + 32 \\ &= 45\end{aligned}$$

EXAMPLE 12 Evaluate $-y^2 + 3(x + 2) - 5$ when $x = -3$ and $y = -2$.

Solution:

$$\begin{aligned}-y^2 + 3(x + 2) - 5 &= -(-2)^2 + 3(-3 + 2) - 5 \\ &= -(-2)^2 + 3(-1) - 5 \\ &= -(4) + 3(-1) - 5 \\ &= -4 - 3 - 5 \\ &= -7 - 5 \\ &= -12\end{aligned}$$

Calculator Corner

EVALUATING EXPRESSIONS

Later in this course you will need to evaluate an expression like $3x^2 - 2x + 5$ for various values of x. Below we show how to evaluate such expressions.

Evaluate	Keystrokes
(a) $3x^2 - 2x + 5$, for $x = 4$ $3(4)^2 - 2(4) + 5$	3 × 4 x^2 − 2 × 4 + 5 = 45
(b) $3x^2 - 2x + 5$, for $x = -6$ $3(-6)^2 - 2(-6) + 5$	3 × 6 +/− x^2 − 2 × 6 +/− + 5 = 125
(c) $-x^2 - 3x - 5$, for $x = -2$ $-(-2)^2 - 3(-2) - 5$	1 +/− × 2 +/− x^2 − 3 × 2 +/− − 5 = −3

Remember in part (c) that $-x^2 = -1x^2$.

Exercise Set 1.9

Evaluate.

1. $3 + 4 \cdot 5$
2. $3 - 5^2 - 2$
3. $6 - 6 + 8$
4. $(6^2 \div 3) - (6 - 4)$
5. $1 + 3 \cdot 2^2$
6. $4 \cdot 3^2 - 2 \cdot 5$
7. $-4^2 + 6$
8. $(-2)^3 + 8 \div 4$
9. $(4 - 3) \cdot (5 - 1)^2$
10. $20 - 6 - 3 - 2$
11. $3 \cdot 7 + 4 \cdot 2$
12. $8 + 5(6 - 1)$
13. $[1 - (4 \cdot 5)] + 6$
14. $[12 - (4 \div 2)] - 5$
15. $4^2 - 3 \cdot 4 - 6$
16. $5 - 3 + 4^2 - 6$
17. $-2[-5 + (3 - 4)]$
18. $(-3)^2 + (3 - 4)^3 - 5$
19. $(6 \div 3)^3 + 4^2 \div 8$
20. $5^2 - 2^2(4 - 2)^2$
21. $-4^2 + 8 \div 2 \cdot 5 + 3$

22. $-4 - (-12 + 4) \div 2 + 1$
23. $3 + (4^2 - 10)^2 - 3$
24. $[-2(2 - 4)^2]^2 - 6$
25. $[6 - (-2 - 3)]^2$
26. $(-2)^2 + 4^2 \div 2^2 + 3$
27. $(3^2 - 1) \div (3 + 1)^2$
28. $-4(5 - 2)^2 + 5$
29. $-[(56 \div 7) - 6 \div 2]$
30. $4[6 + (6 \div 2)^2] - 1$
31. $2[3(8 - 2^2) - 6]$
32. $(13 + 5) - (4 - 2)^2$
33. $10 - [8 - (3 + 4)]^2$
34. $6 - 8 \cdot 2 \div 4 \div 2 + 5$
35. $[4 + ((5 - 2)^2 \div 3)^2]^2$
36. $2[((6 \div 3)^2 + 4)^2 - 3]$
37. $[-3(4 - 2)^2]^2 - [-3(3 - 5)^2]$
38. $[7 - [3(8 \div 4)]^2 + 9 \cdot 4]^2$
39. $(14 \div 7 \cdot 7 \div 7 - 7)^2$
40. $2.5 + 7.56 \div 2.1 + (9.2)^2$
41. $(8.4 + 3.1)^2 - (3.64 - 1.2)$
42. $2[1.63 + 5(4.7)] - 3.15$
43. $(4.3)^2 + 2(5.3) - 3.05$
44. $\frac{2}{3} + \frac{3}{8} \cdot \frac{4}{5}$
45. $\left(\frac{2}{7} + \frac{3}{8}\right) - \frac{3}{112}$
46. $\left(\frac{5}{6} \cdot \frac{4}{5}\right) + \left(\frac{2}{3} \cdot \frac{5}{8}\right)$
47. $\frac{3}{4} - 4 \cdot \frac{5}{40}$
48. $\frac{2}{3} + 4 \div 3^2$
49. $2\left(3 + \frac{2}{5}\right) \div \left(\frac{3}{5}\right)^2$
50. $64 \cdot \frac{1}{2} \div 8 + \frac{3}{4}$

Write the following statements as mathematical expressions using parentheses and brackets and then evaluate.

51. Multiply 6 by 3. From this product, subtract 4. From this difference, subtract 2.
52. Add 4 to 9. Divide this sum by 2. Add 10 to this quotient.
53. Divide 20 by 5. Add 12 to this quotient. Subtract 8 from this sum. Multiply this difference by 9.
54. Multiply 6 by 3. To this product, add 27. Divide this sum by 8. Multiply this quotient by 10.
55. Add $\frac{4}{5}$ to $\frac{3}{7}$. Multiply this sum by $\frac{2}{3}$.
56. Multiply $\frac{3}{8}$ by $\frac{4}{5}$. To this product, add $\frac{7}{120}$. From this sum, subtract $\frac{1}{60}$.

Evaluate for the values given.

57. $x + 4$, when $x = -2$.
58. $2x - 4x + 5$, when $x = 1$.
59. $3x - 2$, when $x = 4$.
60. $3(x - 2)$, when $x = 5$.
61. $x^2 - 6$, when $x = -3$.
62. $x^2 + 4$, when $x = 5$.
63. $-3x^2 - 4$, when $x = 1$.
64. $2x^2 + x$, when $x = 3$.
65. $-4x^2 - 2x + 5$, when $x = -3$.
66. $-3x^2 + 6x + 5$, when $x = 5$.
67. $3(x - 2)^2$, when $x = 7$.
68. $4(x + 1)^2 - 6x$, when $x = 5$.
69. $2(x - 3)(x + 4)$, when $x = 1$.
70. $3x^2(x - 1) + 5$, when $x = -4$.
71. $-6x + 3y$, when $x = 2$ and $y = 4$.
72. $6x + 3y^2 - 5$, when $x = 1$ and $y = -3$.
73. $x^2 - y^2$, when $x = -2$, and $y = -3$.
74. $x^2 - y^2$, when $x = 2$ and $y = -4$.
75. $4(x + y)^2 + 4x - 3y$, when $x = 2$ and $y = -3$.
76. $(4x - 3y)^2 - 5$, when $x = 4$ and $y = -2$.
77. $3(a + b)^2 + 4(a + b) - 6$, when $a = 4$ and $b = -1$.
78. $4xy - 6x + 3$, when $x = 5$ and $y = 2$.
79. $x^2y - 6xy + 3x$, when $x = 2$ and $y = 3$.
80. $\dfrac{6x^2}{3} + \dfrac{2x^2}{2}$, when $x = 2$.
81. $6x^2 + 3xy - y^2$, when $x = 2$ and $y = -3$.
82. $3(x - 4)^2 - (3x - 4)^2$, when $x = -1$.
83. $5(2x - 3)^2 - 4(6 - y)^2$, when $x = -2$ and $y = -1$.
84. $[2(x - 3) + (y + 2)]^2 - 6x^2$, when $x = 3$ and $y = -2$.

*Later in the text we will need to evaluate expressions like $ax^2 + bx + c$ where a, b, and c are real numbers for various values of the variable x. In Exercises 85–100, determine the value of the expression for the value of the variable given **(a)** without using a calculator and **(b)** using a scientific calculator. If parts **(a)** and **(b)** do not agree, determine why. (See the Calculator Corner on page 62.)*

85. $x^2 + 3x - 5, x = 2$
86. $2x^2 - 5x + 3, x = 1$
87. $x^2 - 4x + 7, x = -3$
88. $3x^2 - 6x - 4, x = 2$
89. $-x^2 + 6x - 5, x = 3$
90. $4x^2 - 5x, x = -6$
91. $-x^2 - 2x - 5, x = -3$
92. $-3x^2 - 12, x = -3$
93. $2x^2 - 4x - 10, x = 5$
94. $6x^2 - 3x + 2, x = 4$
95. $-x^2 - 6x + 8, x = 5$
96. $-5x^2 - 3x + 12, x = -3$
97. $x^2 - 16x + 5, x = 5$
98. $4x^2 + 5x - 3, x = -2$
99. $x^2 + 8x - 10, x = 4$
100. $x^2 - 4x + 12, x = 6$

101. In your own words, write the order of operations to follow to evaluate a mathematical expression.

102. (a) Write in your own words the procedure you would use to evaluate $[9 - (8 \div 2)]^2 - 6^3$.

(b) Evaluate the expression in part (a).

103. (a) Write in your own words the procedure you would use to evaluate the expression $-4x^2 + 3x - 6$ when x is 5.

(b) Evaluate the expression in part (a) when $x = 5$.

CUMULATIVE REVIEW EXERCISES

[1.7] 104. Evaluate $(-2)(-4)(6)(-1)(-3)$.

[1.8] 105. When $x = -5$ evaluate (a) x^2 and (b) $-x^2$.

Evaluate.

106. $(-2)^4$

107. -2^4

Group Activity/Challenge Problems

Evaluate for the values given.

1. $4([3(x-2)]^2 + 4)$, when $x = 4$.
2. $[(3-6)^2 + 4]^2 + 3 \cdot 4 - 12 \div 3$.
3. $-2[(3x^2+4)^2 - (3x^2-2)^2]$, when $x = -2$.

Insert one pair of parentheses to make the statement true.

4. $14 + 6 \div 2 \times 4 = 40$
5. $12 - 4 - 6 + 10 = 24$
6. $24 \div 6 \div 2 + 2 = 1$
7. $30 + 15 \div 5 + 10 \div 2 = 38$
8. $18 \div 3^2 - 3 + 5 = 8$

1.10 Properties of the Real Number System

Tape 3

1 Identify the commutative property.
2 Identify the associative property.
3 Identify the distributive property.

Here, we introduce various properties of the real number system. We will use these properties throughout the text.

The Commutative Property

1 The *commutative property of addition* states that the order in which any two real numbers are added does not matter.

Commutative Property of Addition

If a and b represent any two real numbers, then

$$a + b = b + a$$

Notice the commutative property involves a change in *order.* For example,

$$4 + 3 = 3 + 4$$
$$7 = 7$$

The *commutative property of multiplication* states that the order in which any two real numbers are multiplied does not matter.

Commutative Property of Multiplication

If a and b represent any two real numbers, then

$$a \cdot b = b \cdot a$$

For example,

$$\begin{aligned} 6 \cdot 3 &= 3 \cdot 6 \\ 18 &= 18 \end{aligned}$$

The commutative property **does not hold** *for subtraction or division.* For example, $4 - 6 \neq 6 - 4$ and $6 \div 3 \neq 3 \div 6$.

The Associative Property

2 The *associative property of addition* states that, in the addition of three or more numbers, parentheses may be placed around any two adjacent numbers without changing the results.

Associative Property of Addition

If a, b, and c represent any three real numbers, then

$$(a + b) + c = a + (b + c)$$

Notice that the associative property involves a change of *grouping*. For example,

$$\begin{aligned} (3 + 4) + 5 &= 3 + (4 + 5) \\ 7 + 5 &= 3 + 9 \\ 12 &= 12 \end{aligned}$$

In this example the 3 and 4 are grouped together on the left, and the 4 and 5 are grouped together on the right.

The *associative property of multiplication* states that, in the multiplication of three or more numbers, parentheses may be placed around any two adjacent numbers without changing the results.

Associative Property of Multiplication

If a, b, and c represent any three real numbers, then

$$(a \cdot b) \cdot c = a \cdot (b \cdot c)$$

For example,

$$\begin{aligned} (6 \cdot 2) \cdot 4 &= 6 \cdot (2 \cdot 4) \\ 12 \cdot 4 &= 6 \cdot 8 \\ 48 &= 48 \end{aligned}$$

Notice that the associative property involves a change of grouping. When the associative property is used, the content within the parentheses changes.

The associative property **does not hold** *for subtraction or division.* For example, $(4 - 1) - 3 \neq 4 - (1 - 3)$ and $(8 \div 4) \div 2 \neq 8 \div (4 \div 2)$.

The Distributive Property

3 A very important property of the real numbers is the *distributive property of multiplication over addition.*

Distributive Property

If a, b, and c represent any three real numbers, then

$$a(b + c) = ab + ac$$

For example, if we let $a = 2$, $b = 3$, and $c = 4$, then

$$2(3 + 4) = (2 \cdot 3) + (2 \cdot 4)$$
$$2 \cdot 7 = 6 + 8$$
$$14 = 14$$

Therefore, we may either add first and then multiply, or multiply first and then add. The distributive property will be discussed in more detail in Chapter 2.

Helpful Hint

The *commutative property* changes *order.*

The *associative property* changes *grouping.*

The *distributive property* involves two operations, multiplication and addition.

The following are additional illustrations of the commutative, associative, and distributive properties. If we assume that x represents any real number, then:

$x + 4 = 4 + x$ by the commutative property of addition.
$x \cdot 4 = 4 \cdot x$ by the commutative property of multiplication.
$(x + 4) + 7 = x + (4 + 7)$ by the associative property of addition.
$(x \cdot 4) \cdot 6 = x \cdot (4 \cdot 6)$ by the associative property of multiplication.
$3(x + 4) = (3 \cdot x) + (3 \cdot 4)$ or $3x + 12$ by the distributive property.

EXAMPLE 1 Name the properties.

(a) $4 + (-2) = -2 + 4$ **(b)** $x + y = y + x$
(c) $x \cdot y = y \cdot x$ **(d)** $(-12 + 3) + 4 = -12 + (3 + 4)$

Solution: **(a)** Commutative property of addition
(b) Commutative property of addition
(c) Commutative property of multiplication
(d) Associative property of addition

EXAMPLE 2 Name the properties.

(a) $2(x + 2) = (2 \cdot x) + (2 \cdot 2) = 2x + 4$
(b) $4(x + y) = (4 \cdot x) + (4 \cdot y) = 4x + 4y$
(c) $3x + 3y = (3 \cdot x) + (3 \cdot y) = 3(x + y)$
(d) $(3 \cdot 6) \cdot 5 = 3 \cdot (6 \cdot 5)$

Solution: **(a)** Distributive property
(b) Distributive property
(c) Distributive property (in reverse order)
(d) Associative property of multiplication

EXAMPLE 3 Name the properties.

(a) $(3 + 4) + 5 = (4 + 3) + 5$
(b) $(2 + 3) + (4 + 5) = (4 + 5) + (2 + 3)$
(c) $3(x + 4) = 3(4 + x)$ **(d)** $3(x + 4) = (x + 4)3$

Solution: **(a)** Commutative property of addition. The $3 + 4$ was changed to $4 + 3$. The same numbers remain within parentheses; therefore, it is not the associative property.
(b) Commutative property of addition. The order of parentheses was changed; however, the same numbers remain within the parentheses.
(c) Commutative property of addition. $x + 4$ was changed to $4 + x$.
(d) Commutative property of multiplication. The expression within parentheses is not changed.

Helpful Hint

Do not confuse the distributive property with the associative property of multiplication. Make sure you understand the difference.

Distributive Property	*Associative Property of Multiplication*
$3(4 + x) = 3 \cdot 4 + 3 \cdot x$	$3(4 \cdot x) = (3 \cdot 4)x$
$= 12 + 3x$	$= 12x$

For the distributive property to be used, there must be two *terms,* separated by a plus or minus sign, within the parentheses, as in $3(4 + x)$.

EXAMPLE 4 Name the property used to go from one step to the next.

(a) $9 + 4(x + 5)$
(b) $= 9 + 4x + 20$
(c) $= 9 + 20 + 4x$
(d) $= 29 + 4x$ addition facts
(e) $= 4x + 29$

Solution: **(a to b)** Distributive property
(b to c) Commutative property of addition; $4x + 20 = 20 + 4x$
(d to e) Commutative property of addition; $29 + 4x = 4x + 29$

The distributive property can be expanded in the following manner:

$$a(b + c + d + \cdots + n) = ab + ac + ad + \cdots + an$$

For example, $3(x + y + 5) = 3x + 3y + 15$.

Exercise Set 1.10

Name the property illustrated.

1. $4(3 + 5) = 4(3) + 4(5)$

2. $3 + y = y + 3$

3. $5 \cdot y = y \cdot 5$

4. $1(x + 3) = (1)(x) + (1)(3) = x + 3$

5. $2(x + 4) = 2x + 8$

6. $3(4 + x) = 12 + 3x$

7. $x \cdot (y \cdot z) = (x \cdot y) \cdot z$

8. $1(x + 4) = x + 4$

9. $1(x + 3) = x + 3$

10. $3 + (4 + x) = (3 + 4) + x$

Complete using the property given.

11. $3 + 4 =$
commutative property of addition

12. $-3 + 4 =$
commutative property of addition

13. $-6 \cdot (4 \cdot 2) =$
associative property of multiplication

14. $-4 + (5 + 3) =$
associative property of addition

15. $(6)(y) =$
commutative property of multiplication

16. $4(x + 3) =$
distributive property

17. $1(x + y) =$
distributive property

18. $6(x + y) =$
distributive property

19. $4x + 3y =$
commutative property of addition

20. $3(x + y) =$
distributive property

21. $5x + 5y =$
distributive property (in reverse order)

22. $(3 + x) + y =$
associative property of addition

23. $(x + 2)3 =$
commutative property of multiplication

24. $2x + 2z =$
distributive property (in reverse order)

25. $(3x + 4) + 6 =$
associative property of addition

26. $3(x + y) =$
commutative property of addition

27. $3(x + y) =$
commutative property of multiplication

28. $(3x)y =$
associative property of multiplication

29. $4(x + y + 3) =$
distributive property

30. $3(x + y + 2) =$
distributive property

Name the property illustrated to go from one step to the next. See Example 4.

31. $(3 + x) + 4 = (x + 3) + 4$
32. $\quad = x + (3 + 4)$
$\quad = x + 7$

33. $6 + 5(x + 3) = 6 + 5x + 15$
34. $\quad = 6 + 15 + 5x$
$\quad = 21 + 5x$

35. $\quad = 5x + 21$

36. $(x + 4)5 = 5(x + 4)$
37. $\quad = 5x + 20$
38. $\quad = 20 + 5x$

In Exercises 39–42, indicate if the given processes are commutative. That is, does changing the order in which the items are done result in the same final outcome? Explain.

39. Putting sugar and then cream in coffee; putting cream and then sugar in coffee.

40. Applying suntan lotion and then sunning yourself; sunning yourself and then applying suntan lotion.

41. Putting on your socks and then your shoes; putting on your shoes and then your socks.

42. Brushing your teeth and then washing your face; washing your face and then brushing your teeth.

43. Explain how you can tell the difference between the associative property of multiplication and the distributive property.

CUMULATIVE REVIEW EXERCISES

[1.2] **44.** Add $2\frac{3}{5} + \frac{2}{3}$.

45. Subtract $3\frac{5}{8} - 2\frac{3}{16}$.

[1.9] *Evaluate.*

46. $12 - 24 \div 8 + 4 \cdot 3^2$

47. $-4x^2 + 6xy + 3y^2$, when $x = 2$ and $y = -3$

Group Activity/ Challenge Problems

Indicate if the property displayed is the commutative, associative, or distributive property. Explain.

1. $2 + (3 + 4) = (3 + 4) + 2$

2. $(a + b) + (c + d) = (c + d) + (a + b)$

3. $3 + (x + y) + z = 3 + x + (y + z)$

Summary

GLOSSARY

Absolute value (27): The distance between a number and 0 on the number line. The absolute value of any nonzero number will be positive.

Additive inverses or opposites (32): Two numbers whose sum is zero.

Denominator (10): The bottom number of a fraction.

Empty Set (21): A set that contains no elements.

Evaluate (13): To evaluate an expression means to find its value.

Expression (10): An expression is any collection of numbers, letters, grouping symbols, and operations.

Factor (10): If $a \cdot b = c$, then a and b are factors of c.

Greatest common factor (GCF) (10): The largest number that divides into two or more numbers.

Least common denominator (LCD) (13): The smallest number divisible by two or more denominators.

Numerator (10): The top number of a fraction.

Operation (30): The basic operations of arithmetic are addition, subtraction, multiplication, and division.

Reduced to its lowest terms (10): A fraction is reduced to its lowest terms when its numerator and denominator have no common factor other than 1.

Set (21): A collection of elements listed within braces.

Variable (9): A letter used to represent a number.

IMPORTANT FACTS

Fractions: $\frac{a}{c} + \frac{b}{c} = \frac{a+b}{c}$ $\qquad \frac{a}{c} - \frac{b}{c} = \frac{a-b}{c}$

$\frac{a}{b} \cdot \frac{c}{d} = \frac{ac}{bd}$ $\qquad \frac{a}{b} \div \frac{c}{d} = \frac{a}{b} \cdot \frac{d}{c} = \frac{ad}{bc}$

Sets of Numbers

Natural numbers: {1, 2, 3, 4, . . .}

Whole numbers: {0, 1, 2, 3, 4, . . .}

Integers: {. . . , −3, −2, −1, 0, 1, 2, 3, . . .}

Rational numbers: {quotient of two integers, denominator not 0}

Irrational numbers: {real numbers that are not rational numbers}

Real numbers: {all numbers that can be represented on the number line}

Operations on the Real Numbers

To *add real numbers with the same sign,* add their absolute values. The sum has the same sign as the numbers being added.

To *add real numbers with different signs,* find the difference between the larger absolute value and the smaller absolute value. The answer has the sign of the number with the larger absolute value.

To *subtract b from a,* add the opposite of b to a.

$$a - b = a + (-b)$$

The *products* and *quotients* of numbers with *like signs* will be *positive.* The *products* and *quotients* of numbers with *unlike signs* will be *negative.*

Division Involving 0

If a represents any real number except 0, then

$$\frac{0}{a} = 0$$

$$\frac{a}{0} \text{ is undefined}$$

Exponents

$$b^n = \underbrace{b \cdot b \cdot b \cdot \cdots \cdot b}_{n \text{ factors of } b}$$

Order of Operations

1. Evaluate expressions within parentheses.
2. Evaluate all expressions with exponents.
3. Perform multiplications or divisions working left to right.
4. Perform additions or subtractions working left to right.

PROPERTIES OF THE REAL NUMBER SYSTEM

Property	Addition	Multiplication
Commutative	$a + b = b + a$	$ab = ba$
Associative	$(a + b) + c = a + (b + c)$	$(ab)c = a(bc)$
Distributive	$a(b + c) = ab + ac$	

Review Exercises

[1.2] *Perform the operations indicated. Reduce answers to lowest terms.*

1. $\frac{3}{5} \cdot \frac{5}{6}$
2. $\frac{2}{5} \div \frac{10}{9}$
3. $\frac{5}{12} \div \frac{3}{5}$
4. $\frac{5}{6} + \frac{1}{3}$
5. $\frac{3}{8} - \frac{1}{9}$
6. $2\frac{1}{3} - 1\frac{1}{5}$

[1.3] 7. List the set of natural numbers.

8. List the set of whole numbers.
9. List the set of integers.
10. Describe the set of rational numbers.
11. Describe the set of real numbers.
12. Consider the set of numbers

$$\left\{3, -5, -12, 0, \frac{1}{2}, -0.62, \sqrt{7}, 426, -3\frac{1}{4}\right\}$$

List those that are
(a) Positive integers.
(b) Whole numbers.
(c) Integers.
(d) Rational numbers.
(e) Irrational numbers.
(f) Real numbers.

13. Consider the set of numbers

$$\left\{-2.3, -8, -9, 1\frac{1}{2}, \sqrt{2}, -\sqrt{2}, 1, -\frac{3}{17}\right\}$$

List those that are
(a) Natural numbers.
(b) Whole numbers.
(c) Negative integers.
(d) Integers.
(e) Rational numbers.
(f) Real numbers.

[1.4] *Insert either $<$, $>$, or $=$ in the shaded area to make a true statement.*

14. $-3 \;\square\; -5$
15. $-2 \;\square\; 1$
16. $0.6 \;\square\; -1.3$
17. $-2.6 \;\square\; -3.6$
18. $0.50 \;\square\; 0.509$
19. $4.6 \;\square\; 4.06$
20. $-3.2 \;\square\; -3.02$
21. $5 \;\square\; |-3|$
22. $-3 \;\square\; |-7|$
23. $|-2.5| \;\square\; \left|\frac{5}{2}\right|$

[1.5–1.6] *Evaluate.*

24. $-3 + 6$
25. $-4 + (-5)$
26. $-6 + 6$
27. $4 + (-9)$
28. $0 + (-3)$
29. $-10 + 4$
30. $-8 - (-2)$
31. $-9 - (-4)$
32. $4 - (-4)$
33. $0 - 2$
34. $-8 - 1$
35. $2 - 12$
36. $7 - 2$
37. $2 - 7$
38. $0 - (-4)$
39. $-7 - 5$

Evaluate.

40. $6 - 4 + 3$
41. $-5 + 7 - 6$
42. $-5 - 4 - 3$
43. $-2 + (-3) - 2$
44. $-(-4) + 5 - (+3)$
45. $7 - (+4) - (-3)$
46. $5 - 2 - 7 + 3$
47. $4 - (-2) + 3$

[1.7] *Evaluate.*

48. $-4(7)$
49. $(-9)(-3)$
50. $4(-9)$
51. $-2(3)$
52. $\left(\frac{3}{5}\right)\left(\frac{-2}{7}\right)$
53. $\left(\frac{10}{11}\right)\left(\frac{3}{-5}\right)$
54. $\left(\frac{-5}{8}\right)\left(\frac{-3}{7}\right)$
55. $0 \cdot \frac{4}{9}$
56. $4(-2)(-6)$
57. $(-1)(-3)(4)$
58. $-5(2)(7)$
59. $(-3)(-4)(-5)$
60. $-1(-2)(3)(-4)$
61. $(-4)(-6)(-2)(-3)$

Evaluate.

62. $15 \div (-3)$ **63.** $6 \div (-2)$ **64.** $-20 \div 5$

65. $-36 \div (-2)$ **66.** $0 \div 4$ **67.** $0 \div (-4)$

68. $72 \div (-9)$ **69.** $-40 \div (-8)$

70. $-4 \div \left(\frac{-4}{9}\right)$ **71.** $\frac{15}{32} \div (-5)$ **72.** $\frac{3}{8} \div \left(\frac{-1}{2}\right)$

73. $\frac{28}{-3} \div \frac{9}{-2}$ **74.** $\frac{14}{3} \div \left(\frac{-6}{5}\right)$ **75.** $\left(\frac{-5}{12}\right) \div \left(\frac{-5}{12}\right)$

Indicate whether each of the following is 0 *or undefined.*

76. $0 \div 4$ **77.** $0 \div (-6)$ **78.** $8 \div 0$

79. $-4 \div 0$ **80.** $\frac{8}{0}$ **81.** $\frac{0}{-5}$

[1.5–1.7, 1.9] *Evaluate.*

82. $-4(2 - 8)$ **83.** $2(4 - 8)$ **84.** $(3 - 6) + 4$

85. $(-4 + 3) - (2 - 6)$ **86.** $[4 + 3(-2)] - 6$ **87.** $(-4 - 2)(-3)$

88. $[4 + (-4)] + (6 - 8)$ **89.** $9[3 + (-4)] + 5$ **90.** $-4(-3) + [4 \div (-2)]$

91. $(-3 \cdot 4) \div (-2 \cdot 6)$ **92.** $(-3)(-4) + 6 - 3$ **93.** $[-2(3) + 6] - 4$

[1.8] *Evaluate.*

94. 4^2 **95.** 6^2 **96.** 9^3 **97.** 1^5

98. 3^4 **99.** 2^4 **100.** $(-3)^3$ **101.** $(-1)^9$

102. $(-2)^5$ **103.** $\left(\frac{2}{7}\right)^2$ **104.** $\left(\frac{-3}{5}\right)^2$ **105.** $\left(\frac{2}{5}\right)^3$

Express in exponential form.

106. xxy **107.** xyy **108.** $xxyyx$ **109.** $yyzz$

110. $2 \cdot 2 \cdot 3 \cdot 3 \cdot 3xyy$ **111.** $5 \cdot 7 \cdot 7 \cdot xxy$ **112.** $xyxyz$

Express as a product of factors.

113. x^2y **114.** xz^3 **115.** y^3z **116.** $2x^3y^2$

Evaluate for the values given.

117. $-x^2$, when $x = 3$ **118.** $-x^2$, when $x = -4$ **119.** $-x^3$, when $x = 3$ **120.** $-x^4$, when $x = -2$

[1.9] *Evaluate.*

121. $3 + 5 \cdot 4$ **122.** $7 - 3^2$ **123.** $3 \cdot 5 + 4 \cdot 2$

124. $(3 - 7)^2 + 6$ **125.** $6 + 4 \cdot 5$ **126.** $8 - 36 \div 4 \cdot 3$

127. $6 - 3^2 \cdot 5$ **128.** $2 - (8 - 3)$ **129.** $[6 - (3 \cdot 5)] + 5$

130. $3[9 - (4^2 + 3)] \cdot 2$ **131.** $(-3^2 + 4^2) + (3^2 \div 3)$ **132.** $2^3 \div 4 + 6 \cdot 3$

133. $(4 \div 2)^4 + 4^2 \div 2^2$ **134.** $(15 - 2^2)^2 - 4 \cdot 3 + 10 \div 2$ **135.** $4^3 \div 4^2 - 5(2 - 7) \div 5$

Evaluate for the values given.

136. $4x - 6$, when $x = 5$ **137.** $8 - 3x$, when $x = 2$

138. $6 - 4x$, when $x = -5$ **139.** $x^2 - 5x + 3$, when $x = 6$

140. $5y^2 + 3y - 2$, when $y = -1$ **141.** $-x^2 + 2x - 3$, when $x = 2$

142. $-x^2 + 2x - 3$, when $x = -2$ **143.** $-3x^2 - 5x + 5$, when $x = 1$

144. $3xy - 5x$, when $x = 3$ and $y = 4$ **145.** $-x^2 - 8x - 12$, when $x = -3$

−1.9] (a) *Use a scientific calculator to evaluate the expression and* (b) *check to see if your answer is reasonable.*

146. $158 + (-493)$

147. $324 - (-29.6)$

148. $\frac{-17.28}{6}$

149. $(-62)(-1.9)$

150. 5^7

151. $(-3)^6$

152. $-(4.2)^3$

153. $3x^2 - 4x + 3$, when $x = 5$

154. $-2x^2 - 6x - 3$, when $x = -2$

0] *Name the property illustrated.*

155. $(4 + 3) + 9 = 4 + (3 + 9)$

156. $6 \cdot x = x \cdot 6$

157. $4(x + 3) = 4x + 12$

158. $(x + 4)3 = 3(x + 4)$

159. $6x + 3x = 3x + 6x$

160. $(x + 7) + 4 = x + (7 + 4)$

161. $-6x + 3 = 3 + (-6x)$

Practice Test

1. Consider the set of numbers

$$\left\{-6, 42, -3\frac{1}{2}, 0, 6.52, \sqrt{5}, \frac{5}{9}, -7, -1\right\}$$

List those that are:

(a) Natural numbers.

(b) Whole numbers.

(c) Integers.

(d) Rational numbers.

(e) Irrational numbers.

(f) Real numbers.

Insert either <, >, or = in the shaded area to make a true statement.

2. $-6 \quad \square \quad -3$

3. $|-3| \quad \square \quad |-2|$

Evaluate.

4. $-4 + (-8)$

5. $-6 - 5$

6. $4 - (-12)$

7. $5 - 12 - 7$

8. $(-4 + 6) - 3(-2)$

9. $(-4)(-3)(2)(-1)$

10. $\left(\frac{-2}{9}\right) \div \left(\frac{-7}{8}\right)$

11. $\left(-12 \cdot \frac{1}{2}\right) \div 3$

12. $3 \cdot 5^2 - 4 \cdot 6^2$

13. $(4 - 6^2) \div [4(2 + 3) - 4]$

14. $-6(-2 - 3) \div 5 \cdot 2$

15. $(-3)^4$

16. $\left(\frac{3}{5}\right)^3$

17. Write $2 \cdot 2 \cdot 5 \cdot 5 \cdot yyzzz$ in exponential form.

18. Write $2^2 3^3 x^4 y^2$ as a product of factors.

Evaluate for the values given.

19. $2x^2 - 6$, when $x = -4$

20. $6x - 3y^2 + 4$, when $x = 3$ and $y = -2$

21. $-x^2 - 6x + 3$, when $x = -2$

Name the property illustrated.

22. $x + 3 = 3 + x$

23. $4(x + 9) = 4x + 36$

24. $(2 + x) + 4 = 2 + (x + 4)$

25. $5(x + y) = (x + y)5$

Chapter 2

Solving Linear Equations and Inequalities

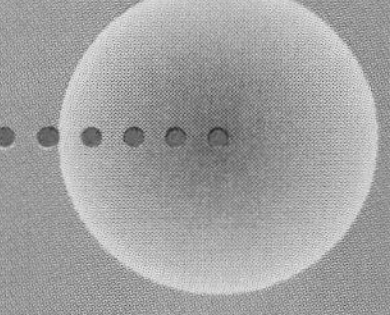

See Section 2.6, Exercise 43

Preview and Perspective

When many students describe algebra they use the words "solving equations." Solving equations is an important part of algebra and of most other mathematics courses you may take. The major emphasis of this chapter is to teach you how to solve linear equations. We will be using the principles learned in this chapter throughout the book. For example, in Chapter 3, when we study applications of algebra, we will write the applications as linear equations and then solve them. Later in the book we will discuss other types of equations, such as quadratic equations. The procedures presented in this chapter will also be used when solving quadratic equations.

To be successful in solving linear equations, you need to have a thorough understanding of adding, subtracting, multiplying, and dividing real numbers. This material was discussed in Chapter 1. The material presented in the first four sections of this chapter are the building blocks for solving linear equations. In Section 2.5 we combine the material presented previously to solve a variety of linear equations.

In Section 2.6 we discuss ratios and proportions and how to set up and solve them. For many students, proportions may be the most common type of equation used to solve real-life application problems. Once you learn to set up and solve proportions you will see that they have a tremendous number of real-life applications.

In Section 2.7 we discuss solving linear inequalities. Solving linear inequalities is an extension to solving linear equations. The procedure used to solve linear inequalities is basically the same as the procedure used to solve linear equations. The material presented in this section will be helpful in Section 8.6 when we graph linear inequalities in two variables. Linear and other types of inequalities will be discussed in more depth in an intermediate algebra course.

2.1 Combining Like Terms

Tape 3

1. Identify terms.
2. Identify like terms.
3. Combine like terms.
4. Use the distributive property to remove parentheses.
5. Remove parentheses when they are preceded by a plus or minus sign.
6. Simplify an expression.

Identify Terms

1 In Section 1.2 and other sections of the text, we indicated that letters called **variables** (or **literal numbers**) are used to represent numbers.

As was indicated in Chapter 1, an **expression** (sometimes referred to as an **algebraic expression**) is a collection of numbers, variables, grouping symbols, and operation symbols. Examples of expressions are:

$$5, \qquad x^2 - 6, \qquad 4x - 3, \qquad 2(x + 5) + 6, \qquad \frac{x + 3}{4}$$

When an algebraic expression consists of several parts, the parts that are added or subtracted are called the **terms** of the expression. The expression $2x - 3y - 5$ has three terms: $2x$, $-3y$, and -5. The expression

$$3x + 2xy + 5(x + y)$$

also has three terms: $3x$, $2xy$, and $5(x + y)$.

The + and − signs that break the expression into terms are a part of the term. However, when listing the terms of an expression, it is not necessary to list the + sign at the beginning of a term.

Expression	*Terms*
$-2x + 3y - 8$	$-2x, \quad 3y, \quad -8$
$3y - 2x + \frac{1}{2}$	$3y, \quad -2x, \quad \frac{1}{2}$
$7 + x + 4 - 5x$	$7, \quad x, \quad 4, \quad -5x$
$3(x - 1) - 4x + 2$	$3(x - 1), \quad -4x, \quad 2$
$\frac{x + 4}{3} - 5x + 3$	$\frac{x + 4}{3}, \quad -5x, \quad 3$

The numerical part of a term is called its **numerical coefficient** or simply its **coefficient.** In the term $6x$, the 6 is the numerical coefficient. Note that $6x$ means the variable x is multiplied by 6.

Term	*Numerical Coefficient*
$3x$	3
$-\frac{1}{2}x$	$-\frac{1}{2}$
$4(x - 3)$	4
$\frac{2x}{3}$	$\frac{2}{3}$, since $\frac{2x}{3}$ means $\frac{2}{3}x$
$\frac{x + 4}{3}$	$\frac{1}{3}$, since $\frac{x + 4}{3}$ means $\frac{1}{3}(x + 4)$

Whenever a term appears without a numerical coefficient, we assume that the numerical coefficient is 1.

Examples

x means $1x$	$-x$ means $-1x$
x^2 means $1x^2$	$-x^2$ means $-1x^2$
xy means $1xy$	$-xy$ means $-1xy$
$(x + 2)$ means $1(x + 2)$	$-(x + 2)$ means $-1(x + 2)$

If an expression has a term that is a number (without a variable), we refer to that number as a **constant term,** or simply a **constant.** In the expression $x^2 + 3x - 4$, the -4 is a constant term, or a constant.

Identify Like Terms

2 **Like terms** are terms that have the same variables with the same exponents. The following are examples of like terms and unlike terms. Note that if two terms are like terms, only their numerical coefficients may differ.

Like Terms	*Unlike Terms*	
$3x, \quad -4x$	$3x, \quad 2$	(One term has a variable, the other is a constant)
$4y, \quad 6y$	$3x, \quad 4y$	(Variables differ)
$5, \quad -6$	$x, \quad 3$	(One term has a variable, the other is a constant)
$3(x + 1), \quad -2(x + 1)$	$2x, \quad 3xy$	(Variables differ)
$3x^2, \quad 4x^2$	$3x, \quad 4x^2$	(Exponents differ)

EXAMPLE 1 Identify any like terms.

(a) $2x + 3x + 4$ **(b)** $2x + 3y + 2$ **(c)** $x + 3 + y - \frac{1}{2}$

Solution: **(a)** $2x$ and $3x$ are like terms.
(b) There are no like terms.
(c) 3 and $-\frac{1}{2}$ are like terms.

EXAMPLE 2 Identify any like terms.

(a) $5x - x + 6$ **(b)** $3 - 2x + 4x - 6$ **(c)** $12 + x + 7$

Solution: **(a)** $5x$ and $-x$ (or $-1x$) are like terms.
(b) 3 and -6 are like terms; $-2x$ and $4x$ are like terms.
(c) 12 and 7 are like terms.

Combine Terms

3 We often need to simplify expressions by combining like terms. **To combine like terms** means to add or subtract the like terms in an expression. To combine like terms, we can use the procedure that follows.

To Combine Like Terms

1. Determine which terms are like terms.
2. Add or subtract the coefficients of the like terms.
3. Multiply the number found in step 2 by the common variables.

Examples 3 through 9 illustrate this procedure.

EXAMPLE 3 Combine like terms: $4x + 3x$.

Solution: $4x$ and $3x$ are like terms with the common variable x. Since $4 + 3 = 7$, then $4x + 3x = 7x$.

EXAMPLE 4 Combine like terms: $\frac{3}{5}x - \frac{2}{3}x$.

Solution: Since $\frac{3}{5} - \frac{2}{3} = \frac{9}{15} - \frac{10}{15} = -\frac{1}{15}$, then $\frac{3}{5}x - \frac{2}{3}x = -\frac{1}{15}x$.

EXAMPLE 5 Combine like terms: $5.23a - 7.45a$.

Solution: Since $5.23 - 7.45 = -2.22$, then $5.23a - 7.45a = -2.22a$.

EXAMPLE 6 Combine like terms: $3x + x + 5$.

Solution: The $3x$ and x are like terms.

$$3x + x + 5 = 3x + 1x + 5 = 4x + 5$$

EXAMPLE 7 Combine like terms: $12 + x + 7$.

Solution: The 12 and 7 are like terms. We can rearrange the terms to get

$$x + 12 + 7 \quad \text{or} \quad x + 19$$

EXAMPLE 8 Combine like terms: $3y + 4x - 3 - 2x$.

Solution: The only like terms are $4x$ and $-2x$.

$$\text{Rearranging terms: } 4x - 2x + 3y - 3$$
$$\text{Combine like terms: } 2x + 3y - 3$$

EXAMPLE 9 Combine like terms: $-2x + 3y - 4x + 3 - y + 5$.

Solution: $-2x$ and $-4x$ are like terms.
$3y$ and $-y$ are like terms.
3 and 5 are like terms.

Grouping the like terms together gives

$$-2x - 4x + 3y - y + 3 + 5$$
$$-6x \quad + \quad 2y \quad + \quad 8$$

The commutative and associative properties were used to rearrange the terms in Examples 7, 8, and 9. The order of the terms in the answer is not critical. Thus $2y - 6x + 8$ is also an acceptable answer to Example 9. When writing answers, we generally list the terms containing variables in alphabetical order from left to right, and list the constant term on the right.

COMMON STUDENT ERROR Students often misinterpret the meaning of a term like $3x$. What does $3x$ mean?

Correct	*Incorrect*
$3x = x + x + x$	~~$3x = x \cdot x \cdot x$~~

Just as $2 + 2 + 2$ can be expressed as $3 \cdot 2$, $x + x + x$ can be expressed as $3 \cdot x$ or $3x$. Note that when we combine like terms in $x + x + x$ we get $3x$. Also note that $x \cdot x \cdot x = x^3$, not $3x$.

Distributive Property

4 We introduced the distributive property in Section 1.10. Because this property is so important, we will study it again. But before we do, let us go back briefly to the subtraction of real numbers. Recall from Section 1.6 that

$$6 - 3 = 6 + (-3)$$

For any real numbers a and b,

$$a - b = a + (-b)$$

We will use the fact that $a + (-b)$ means $a - b$ in discussing the distributive property.

Distributive Property

For any real numbers a, b, and c,

$$a(b + c) = ab + ac$$

EXAMPLE 10 Use the distributive property to remove parentheses.

(a) $2(x + 4)$ **(b)** $-2(x + 4)$

Solution: **(a)** $2(x + 4) = 2x + 2(4) = 2x + 8$

(b) $-2(x + 4) = -2x + (-2)(4) = -2x + (-8) = -2x - 8$

Note in part (b) that, instead of leaving the answer $-2x + (-8)$, we wrote it as $-2x - 8$, which is the proper form of the answer.

EXAMPLE 11 Use the distributive property to remove parentheses.

(a) $3(x - 2)$ **(b)** $-2(4x - 3)$

Solution: **(a)** By the definition of subtraction, we may write $x - 2$ as $x + (-2)$.

$$\begin{aligned} 3(x - 2) = 3[x + (-2)] &= 3x + 3(-2) \\ &= 3x + (-6) \\ &= 3x - 6 \end{aligned}$$

(b) $-2(4x - 3) = -2[4x + (-3)] = -2(4x) + (-2)(-3) = -8x + 6$

The distributive property is used often in algebra, so you need to understand it well. You should understand it so well that you will be able to simplify an expression using the distributive property without having to write down all the steps that we listed in working Examples 10 and 11. Study closely the Helpful Hint on the top of page 81.

The distributive property can be expanded as follows:

$$\mathbf{a(b + c + d + \cdots + n) = ab + ac + ad + \cdots + an}$$

Examples of the expanded distributive property are

$$\begin{aligned} 3(x + y + z) &= 3x + 3y + 3z \\ 2(x + y - 3) &= 2x + 2y - 6 \end{aligned}$$

Helpful Hint

With a little practice, you will be able to eliminate some of the intermediate steps when you use the distributive property. When using the distributive property, there are eight possibilities with regard to signs. Study and learn the eight possibilities that follow.

Positive Coefficient	*Negative Coefficient*
$2(x) = 2x$	$(-2)(x) = -2x$
(a) $2(x + 3) = 2x + 6$	**(e)** $-2(x + 3) = -2x - 6$
$2(+3) = +6$	$(-2)(+3) = -6$
$2(x) = 2x$	$(-2)(x) = -2x$
(b) $2(x - 3) = 2x - 6$	**(f)** $-2(x - 3) = -2x + 6$
$2(-3) = -6$	$(-2)(-3) = +6$
$2(-x) = -2x$	$(-2)(-x) = 2x$
(c) $2(-x + 3) = -2x + 6$	**(g)** $-2(-x + 3) = 2x - 6$
$2(+3) = +6$	$(-2)(+3) = -6$
$2(-x) = -2x$	$(-2)(-x) = 2x$
(d) $2(-x - 3) = -2x - 6$	**(h)** $-2(-x - 3) = 2x + 6$
$2(-3) = -6$	$(-2)(-3) = +6$

EXAMPLE 12 Use the distributive property to remove parentheses.

(a) $4(x - 3)$ **(b)** $-2(2x - 4)$ **(c)** $-\frac{1}{2}(4x + 5)$ **(d)** $-2(3x - 2y + 4z)$

Solution:

(a) $4(x - 3) = 4x - 12$

(b) $-2(2x - 4) = -4x + 8$

(c) $-\frac{1}{2}(4x + 5) = -2x - \frac{5}{2}$

(d) $-2(3x - 2y + 4z) = -6x + 4y - 8z$

The distributive property can also be used from the right, as in Example 13.

EXAMPLE 13 Use the distributive property to remove parentheses from the expression $(2x - 8y)4$.

Solution: We distribute the 4 on the right side of the parentheses over the terms within the parentheses.

$$\begin{aligned}(2x - 8y)4 &= 2x(4) - 8y(4)\\ &= 8x - 32y\end{aligned}$$

Example 13 could have been rewritten as $4(2x - 8y)$ by the commutative property of multiplication, and then the 4 could have been distributed from the left to obtain the same answer, $8x - 32y$.

Plus or Minus Sign Before Parentheses

5 In the expression $(4x + 3)$, how do we remove parentheses? Recall that the coefficient of a term is assumed to be 1 if none is shown. Therefore, we may write

$$\begin{aligned}(4x + 3) &= 1(4x + 3)\\ &= 1(4x) + (1)(3)\\ &= 4x + 3\end{aligned}$$

Note that $(4x + 3) = 4x + 3$. **When no sign or a plus sign precedes parentheses, the parentheses may be removed without having to change the expression inside the parentheses.**

Examples

$$(x + 3) = x + 3$$
$$(2x - 3) = 2x - 3$$
$$+(2x - 5) = 2x - 5$$
$$+(x + 2y - 6) = x + 2y - 6$$

Now consider the expression $-(4x + 3)$. How do we remove parentheses? Here, the number in front of the parentheses is -1, and we write

$$\begin{aligned}-(4x + 3) &= -1(4x + 3)\\ &= -1(4x) + (-1)(3)\\ &= -4x + (-3)\\ &= -4x - 3\end{aligned}$$

Thus, $-(4x + 3) = -4x - 3$. **When a negative sign precedes parentheses, the signs of all the terms within the parentheses are changed when the parentheses are removed.**

Examples

$$-(x + 4) = -x - 4$$
$$-(-2x + 3) = 2x - 3$$
$$-(5x - y + 3) = -5x + y - 3$$
$$-(-2x - 3y - 5) = 2x + 3y + 5$$

Simplify an Expression

6

To Simplify an Expression

1. Use the distributive property to remove any parentheses.
2. Combine like terms.

EXAMPLE 14 Simplify $6 - (2x + 3)$.

Solution: $6 - (2x + 3) = 6 - 2x - 3$ Use the distributive property.
$= -2x + 3$ Combine like terms.

Note: $3 - 2x$ is the same as $-2x + 3$; however, we generally write the term containing the variable first.

EXAMPLE 15 Simplify $6x + 4(2x + 3)$.

Solution: $6x + 4(2x + 3) = 6x + 8x + 12$ Use the distributive property.
$= 14x + 12$ Combine like terms.

EXAMPLE 16 Simplify $2(x - 1) + 9$.

Solution: $2(x - 1) + 9 = 2x - 2 + 9$ Use the distributive property.
$= 2x + 7$ Combine like terms.

EXAMPLE 17 Simplify $2(x + 3) - 3(x - 2) - 4$.

Solution: $2(x + 3) - 3(x - 2) - 4 = 2x + 6 - 3x + 6 - 4$ Use the distributive property.
$= 2x - 3x + 6 + 6 - 4$ Rearrange terms.
$= -x + 8$ Combine like terms.

Helpful Hint

It is important for you to have a clear understanding of the concepts of *term* and *factor*. When two or more expressions are **multiplied,** each expression is a **factor** of the product. For example, since $4 \cdot 3 = 12$, the 4 and the 3 are factors of 12. Since $3 \cdot x = 3x$, the 3 and the x are factors of $3x$. Similarly, in the expression $5xyz$, the 5, x, y, and z are all factors.

In an expression, the parts that are **added or subtracted** are the **terms** of the expression. For example, the expression $2x^2 + 3x - 4$, has three terms, $2x^2$, $3x$, and -4. Note that the terms of an expression may have factors. For example, in the term $2x^2$, the 2 and the x^2 are factors because they are multiplied.

Exercise Set 2.1

Combine like terms when possible.

1. $2x + 3x$
2. $3x + 6$
3. $4x - 5x$
4. $4x + 3y$
5. $12 + x - 3$
6. $-2x - 3x$
7. $-2x + 5x$
8. $4x - 7x + 4$
9. $x + 3x - 7$
10. $3 + 2x - 5$
11. $6 - 3 + 2x$
12. $2 + 2x + 3x$
13. $-4 + 5x + 12$
14. $-2x - 3x - 2 - 3$
15. $5x + 2y + 3 + y$

16. $-x + 2 - x - 2$
17. $4x - 2x + 3 - 7$
18. $x - 4x + 3$
19. $5 + x + 3$
20. $x + 2x + y + 2$
21. $-3x + 2 - 5x$
22. $x + 4 - 6$
23. $5 + 2x - 4x + 6$
24. $3x + 4x - 2 + 5$
25. $x - 2 - 4 + 2x$
26. $2x + 4 - 3 + x$
27. $2 - 3x - 2x + 1$
28. $3x - x + 4 - 6$
29. $2y + 4y + 6$
30. $6 - x - x$
31. $x - 6 + 3x - 4$
32. $-2x + 4x - 3$
33. $4 - x + 4x - 8$
34. $x + 4 + \frac{3}{5}$
35. $x + \frac{3}{4} - \frac{1}{3}$
36. $5.23x + 1.42 - 4.61x$
37. $68.2x - 19.7x + 8.3$
38. $\frac{1}{2}x + 3y + 1$
39. $x + \frac{1}{2}y - \frac{3}{8}y$
40. $2x + 3 + 4x + 5$
41. $-4x - 3.1 - 5.2$
42. $-x + 2x + y$
43. $1 + x + 6 - 3x$
44. $2x - 7 - 5x + 2$
45. $3x - 7 - 9 + 4x$
46. $x - y - 2y + 3$
47. $4x + 6 + 3x - 7$
48. $-y - 6 - 3y - y$
49. $-4 + x - 6 + 2$
50. $x - 3y + 2x + 4$
51. $-19.36 + 40.02x + 12.25 - 18.3x$
52. $52x - 52x - 63.5 - 63.5$
53. $\frac{3}{5}x - 3 - \frac{7}{4}x - 2$
54. $\frac{1}{5}y + 3x - 2x - \frac{2}{3}y$

Use the distributive property to remove parentheses.

55. $2(x + 6)$
56. $3(x - 2)$
57. $5(x + 4)$
58. $-2(x + 3)$
59. $-2(x - 4)$
60. $3(-x + 5)$
61. $-\frac{1}{2}(2x - 4)$
62. $-4(x + 6)$
63. $1(-4 + x)$
64. $4(y + 3)$
65. $\frac{1}{4}(x - 12)$
66. $5(x + y + 4)$
67. $-0.6(3x - 5)$
68. $-(x - 3)$
69. $\frac{1}{2}(-2x + 6)$
70. $-2(x + y - z)$
71. $0.4(2x - 0.5)$
72. $-(x + 4y)$
73. $-(-x + y)$
74. $(3x + 4y - 6)$
75. $-(2x - 6y + 8)$
76. $-(-2x + 6 - y)$
77. $4.6(3.1x - 2.3y + 1.8)$
78. $-2(-x + 3y + 5)$
79. $2\left(\frac{1}{2}x - 4y + \frac{1}{4}\right)$
80. $2\left(3 - \frac{1}{2}x + 4y\right)$
81. $(x + 3y - 9)$
82. $(-x + 5 - 2y)$
83. $-(-x + 4 + 2y)$
84. $2.3(1.6x - 5.9y + 4.8)$

Simplify when possible.

85. $4(x - 2) - x$
86. $2 - (x + 3)$
87. $-2(3 - x) + 1$
88. $-(2x + 3) + 5$
89. $6x + 2(4x + 9)$
90. $3(x + y) + 2y$
91. $2(x - y) + 2x + 3$
92. $6 + (x - 8) + 2x$
93. $(2x + y) - 2x + 3$
94. $4 - (2x + 3) + 5$
95. $8x - (x - 3)$
96. $-(x - 5) - 3x + 4$
97. $2(x - 3) - (x + 3)$
98. $3y - (2x + 2y) - 6x$
99. $4(x - 3) + 2(x - 2) + 4$
100. $4(x + 3) - 2x$
101. $2(x - 4) - 3x + 6$
102. $6 - 2(x + 3) + 5x$
103. $-3(x - 4) + 2x - 6$
104. $-(x + 2) + 3x - 6$
105. $4(x - 3) + 4x - 7$
106. $-3(x + 2y) + 3y + 4$
107. $0.4 + (x + 5) - 0.6 + 2$
108. $4 - (2 - x) + 3x$
109. $9 - (-3x + 4) - 5$
110. $2y - 6(y - 2) + 3$
111. $4(x + 2) - 3(x - 4) - 5$
112. $4 - (y - 5) + 2x + 3$
113. $-0.2(2 - x) + 4(y + 0.2)$
114. $-5(-y + 2) + 3(2 - x) - 4$
115. $-6x + 3y - (6 + x) + (x + 3)$
116. $(x + 3) + (x - 4) - 6x$
117. $-(x + 3) + (2x + 4) - 6$
118. $\frac{1}{2}(x + 3) + \frac{1}{3}(3x + 6)$
119. $\frac{2}{3}(x - 2) - \frac{1}{2}(x + 4)$

120. When no sign or a plus sign precedes an expression within parentheses, explain how to remove the parentheses.

121. When a minus sign precedes an expression within parentheses, explain how to remove the parentheses.

122. Explain the difference between a factor and a term.

123. Consider the expression $2x^2 + 3x - 5$.

(a) List the terms of this expression. Explain why each is a term.

(b) List the positive factors of the term $2x^2$. Explain why each is a factor of the term.

CUMULATIVE REVIEW EXERCISES

[1.4] *Evaluate.*

124. $|-7|$

125. $-|-16|$

[1.9] **126.** Write a paragraph explaining the order of operations.

127. Evaluate $-x^2 + 5x - 6$ when $x = -1$.

Group Activity/ Challenge Problems

Simplify.

1. $4x + 5y + 6(3x - 5y) - 4x + 3$

2. $2x^2 - 4x + 8x^2 - 3(x + 2) - x^2 - 2$

3. $x^2 + 2y - y^2 + 3x + 5x^2 + 6y^2 + 5y$

4. $2[3 + 4(x - 5)] - [2 - (x - 3)]$

5. Consider the expression $3x^2 - 10x + 8$. **(a)** List each term of the expression. **(b)** List the positive factors of $3x^2$, and **(c)** List all factors of the constant 8, including the negative factors. Note that a term may have many factors.

2.2 The Addition Property of Equality

Tape 3

1. Identify linear equations.
2. Check solutions to equations.
3. Identify and define equivalent equations.
4. Use the addition property to solve equations.
5. Solve equations doing some steps mentally.

Linear Equations

1 A statement that shows two algebraic expressions are equal is called an **equation**. For example, $4x + 3 = 2x - 4$ is an equation. In this chapter we learn to solve **linear equations in one variable.**

> A **linear equation** in one variable is an equation that can be written in the form
>
> $$ax + b = c$$
>
> for real numbers a, b, and c, $a \neq 0$.

Examples of linear equations in one variable are

$$x + 4 = 7$$
$$2x - 4 = 6$$

Check Solutions to Equations

2 The **solution of an equation** is the number or numbers that make the equation a true statement. For example, the solution to $x + 4 = 7$ is 3. We will shortly learn how to find the solution to an equation, or to **solve an equation.** But before we do this we will learn how to *check* the solution of an equation.

The solution to an equation may be **checked** by substituting the value that is believed to be the solution into the original equation. If the substitution results in a true statement, your solution is correct. If the substitution results in a false statement, then either your solution or your check is incorrect, and you need to go back and find your error. Try to check all your solutions.

To check whether 3 is the solution to $x + 4 = 7$, we substitute 3 for each x in the equation.

Check: $x = 3$

$$\begin{aligned} x + 4 &= 7 \\ 3 + 4 &= 7 \\ 7 &= 7 \quad \text{true} \end{aligned}$$

Since the check results in a true statement, 3 is a solution.

EXAMPLE 1 Consider the equation $2x - 4 = 6$. Determine whether

(a) 3 is a solution.

(b) 5 is a solution.

Solution: **(a)** To determine whether 3 is a solution to the equation, substitute 3 for x.

Check: $x = 3$

$$\begin{aligned} 2x - 4 &= 6 \\ 2(3) - 4 &= 6 \\ 6 - 4 &= 6 \\ 2 &= 6 \quad \text{false} \end{aligned}$$

Since we obtained a false statement, 3 is not a solution.

(b) Substitute 5 for x in the equation.

Check: $x = 5$

$$\begin{aligned} 2x - 4 &= 6 \\ 2(5) - 4 &= 6 \\ 10 - 4 &= 6 \\ 6 &= 6 \quad \text{true} \end{aligned}$$

Since the value 5 checks, 5 is a solution to the equation.

We can use the same procedures to check more complex equations, as shown in Examples 2 and 3.

EXAMPLE 2 Determine whether 18 is a solution to the equation $3x - 2(x + 3) = 12$.

Solution: To determine whether 18 is a solution, substitute 18 for each x in the equation. If the substitution results in a true statement, then 18 is the solution.

$$\begin{aligned} 3x - 2(x + 3) &= 12 \\ 3(18) - 2(18 + 3) &= 12 \\ 54 - 2(21) &= 12 \\ 54 - 42 &= 12 \\ 12 &= 12 \quad \text{true} \end{aligned}$$

Since we obtain a true statement, 18 is the solution.

EXAMPLE 3 Determine whether $-\frac{3}{2}$ is a solution to the equation

$$3(x + 3) = 6 + x.$$

Solution: Substitute $-\frac{3}{2}$ for each x in the equation.

$$3(x + 3) = 6 + x$$
$$3\left(-\frac{3}{2} + 3\right) = 6 + \left(-\frac{3}{2}\right)$$
$$3\left(-\frac{3}{2} + \frac{6}{2}\right) = \frac{12}{2} - \frac{3}{2}$$
$$3\left(\frac{3}{2}\right) = \frac{9}{2}$$
$$\frac{9}{2} = \frac{9}{2} \quad \text{true}$$

Thus, $-\frac{3}{2}$ is the solution.

Calculator Corner

CHECKING SOLUTIONS

Calculators can be used to check solutions to equations. For example, to check to see if $\frac{-10}{3}$ is a solution to the equation $2x + 3 = 5(x + 3) - 2$, we perform the following steps:

1. Substitute $\frac{-10}{3}$ for each x.

$$2x + 3 = 5(x + 3) - 2$$
$$2\left(\frac{-10}{3}\right) + 3 = 5\left(\frac{-10}{3} + 3\right) - 2$$

2. Evaluate each side of the equation separately using your calculator. If you obtain the same value on both sides, your solution checks. The procedures for evaluating the left and right sides of the equation depend on whether or not your calculator is a scientific calculator (see page 34). If you do not have a scientific calculator, remember that you need to work within parentheses first, and then do your multiplications and divisions from left to right before your additions and subtractions. In the following steps we assume that you have a scientific calculator.

To evaluate the left side of the equation, $2\left(\frac{-10}{3}\right) + 3$, press the following keys:

2 [×] [(] 10 [+/−] [÷] 3 [)] [+] 3 [=] −3.6666667

To evaluate the right side of the equation, $5\left(\frac{-10}{3} + 3\right) - 2$, press the following keys:

5 [×] [(] 10 [+/−] [÷] 3 [+] 3 [)] [−] 2 [=] −3.6666667

Since both sides give the same value the solution checks. Note that because calculators differ in their electronics, sometimes the last digit of a calculation will differ.

Equivalent Equations

3 Now that we know how to check a solution to an equation we will discuss solving equations. Complete procedures for solving equations will be given shortly. For now, you need to understand that **to solve an equation, it is necessary to get the variable alone on one side of the equal sign. We say that we isolate the variable.** To isolate the variable, we make use of two properties: the addition and multiplication properties of equality. Look first at Figure 2.1.

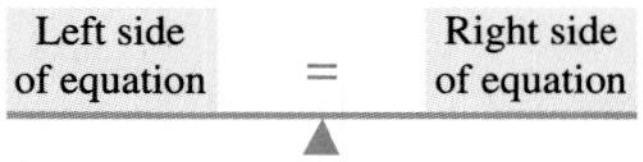

Left side of equation = Right side of equation

FIGURE 2.1

Think of an equation as a balanced statement whose left side is balanced by its right side. When solving an equation, we must make sure that the equation remains balanced at all times. That is, both sides must always remain equal. **We ensure that an equation always remains equal by doing the same thing to both sides of the equation.** For example, if we add a number to the left side of the equation, we must add exactly the same number to the right side. If we multiply the right side of the equation by some number, we must multiply the left side by the same number.

When we add the same number to both sides of an equation or multiply both sides of an equation by the same nonzero number, we do not change the solution to the equation, just the form. Two or more equations with the same solution are called **equivalent equations.** The equations $2x - 4 = 2$, $2x = 6$, and $x = 3$ are equivalent, since the solution to each is 3.

Check: $x = 3$

$2x - 4 = 2$	$2x = 6$	$x = 3$
$2(3) - 4 = 2$	$2(3) = 6$	$3 = 3$ true
$6 - 4 = 2$	$6 = 6$ true	
$2 = 2$ true		

When solving an equation, we use the addition and multiplication properties to express a given equation as simpler equivalent equations until we obtain the solution.

Before stating the addition property of equality, I would like to give you an intuitive and visual introduction to solving equations using the addition property. The multiplication property of equality will be discussed in Section 2.3.

We stated that both sides of an equation must always stay balanced. That is, what you do to one side of the equation you must also do to the other side. Consider Figure 2.2.

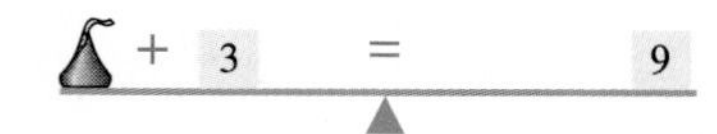

FIGURE 2.2

In this figure and in other figures involving a balance we use the symbol to represent a chocolate "kiss," which is a piece of candy. We could have used a box, a tree, a whale, or any other symbol in place of the kiss. The symbol used is not important in understanding the addition property of equality. The kiss is a symbol that is being used to represent some number. Earlier we stated that in algebra we use letters called variables, such as x and y, to represent numbers. Therefore, if you wished you could replace the kiss with an x or y or some other letter. Whenever we use more than one kiss on a balance the kisses all represent the same value. The value of the kiss may change with each example. We will use the balance to show how the addition property of equalities may be used to solve equations.

In Figure 2.2, can you determine the number the kiss represents if both sides of the equation are to be balanced? If you answered 6, you answered correctly.

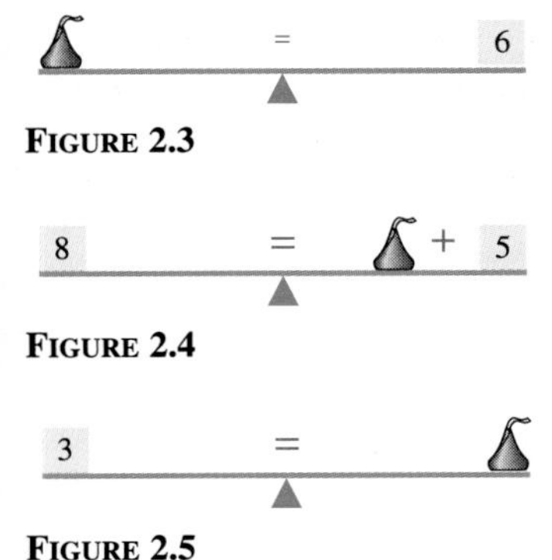

FIGURE 2.3

FIGURE 2.4

FIGURE 2.5

How did you determine your answer? Note that one kiss plus 3 equals 9. If you subtract 3 from both sides of the equation (or scale) you will be left with one kiss equals 6 (Fig. 2.3).

Now consider Figure 2.4.

What is the value of the kiss if both sides are to be balanced? If you answered 3, you are correct. How did you determine your answer? If you subtract 5 from both sides of the balance (Fig. 2.5) you see that one kiss equals 3.

In both of these problems we have actually used the addition property of equality to solve equations. Below we show the balances from Figures 2.2 and 2.4 and the equations that can be determined from the figures. We also indicate the solution to the equations. In the equations we have used the letter x to represent the value of a kiss. However, we could have used k, for kiss, or any other letter. In algebra, we often use the letter x to represent the value of the unknown quantity. To find the value of a kiss, we subtract the amount on the same side of the balance as the kiss from both sides of the balance. To solve each equation for the variable x, we subtract the amount on the same side of the equal sign as the variable from both sides of the equation.

Figure	*Equation*	*Solution*
+ 3 = 9	$x + 3 = 9$	$x = 6$
8 = + 5	$8 = x + 5$	$x = 3$

Use the Addition Property to Solve Equations

4 Now that we have provided an informal introduction, let us define the addition property of equality.

> **Addition Property of Equality**
>
> If $a = b$, then $a + c = b + c$ for any real numbers a, b, and c.

This property implies that the same number can be added to both sides of an equation without changing the solution. **The addition property is used to solve equations of the form $x + a = b$.** To isolate the variable x in equations of this form, add the opposite or additive inverse of a, $-a$, to both sides of the equation.

To isolate the variable when solving equations of the form $x + a = b$, we use the addition property to eliminate the number **on the same side of the equal sign as the variable.** (This is like isolating the kiss on the balance.) Study the following examples carefully.

Equation	*To Solve, Use the Addition Property to Eliminate the Number*
$x + 8 = 10$	8
$x - 7 = 12$	-7
$5 = x - 12$	-12
$-4 = x + 9$	9

Now let us work some problems.

EXAMPLE 4 Solve the equation $x - 4 = 3$.

Solution: To isolate the variable, x, we must eliminate the -4 from the left side of the equation. To do this we add 4, the opposite of -4, to *both sides* of the equation.

$$x - 4 = 3$$
$$x - 4 + 4 = 3 + 4 \quad \text{Add 4 to both sides of the equation.}$$
$$x + 0 = 7$$
$$x = 7$$

Note how the process helps to isolate x.

Check:
$$x - 4 = 3$$
$$7 - 4 = 3$$
$$3 = 3 \quad \text{true}$$

EXAMPLE 5 Solve the equation $y - 3 = -5$.

Solution: To solve this equation, we must isolate the variable, y. To eliminate the -3 from the left side of the equation, we add its opposite, 3, to *both sides* of the equation.

$$y - 3 = -5$$
$$y - 3 + 3 = -5 + 3 \quad \text{Add 3 to both sides of the equation.}$$
$$y + 0 = -2$$
$$y = -2$$

Note that we did not check the solution to Example 5. Space limitations prevent us from showing all checks. However, *you should check all of your answers.*

EXAMPLE 6 Solve the equation $x + 5 = 9$.

Solution: To solve this equation, we must isolate the variable, x. Therefore, we must eliminate the 5 from the left side of the equation. To do this, we add the opposite of 5, -5, to both sides of the equation.

$$x + 5 = 9$$
$$x + 5 + (-5) = 9 + (-5) \quad \text{Add } -5 \text{ to both sides of the equation.}$$
$$x + 0 = 4$$
$$x = 4$$

In Example 6 we added -5 to both sides of the equation. From Section 1.6 we know that $5 + (-5) = 5 - 5$. Thus, we can see that adding a negative 5 to both sides of the equation is equivalent to subtracting a 5 from both sides of the equation. According to the addition property, the same number may be *added* to both sides of an equation. **Since subtraction is defined in terms of addition, the addi-**

tion property also allows us to *subtract* the same number from both sides of the equation. Thus, Example 6 could have also been worked as follows:

$$x + 5 = 9$$
$$x + 5 - 5 = 9 - 5 \quad \text{Subtract 5 from both sides of the equation.}$$
$$x + 0 = 4$$
$$x = 4$$

In this text, unless there is a specific reason to do otherwise, rather than adding a negative number to both sides of the equation, we will subtract a number from both sides of the equation.

EXAMPLE 7 Solve the equation $x + 7 = -3$.

Solution:

$$x + 7 = -3$$
$$x + 7 - 7 = -3 - 7 \quad \text{Subtract 7 from both sides of the equation.}$$
$$x + 0 = -10$$
$$x = -10$$

Check:

$$x + 7 = -3$$
$$-10 + 7 = -3$$
$$-3 = -3 \quad \text{true}$$

Helpful Hint

Some students may not fully understand which number to add or subtract when solving equations. Remember that our goal in solving equations is to get the variable alone on one side of the equation. To do this, we add or subtract **the number on the same side of the equation as the variable** to both sides of the equation.

Equation	*Must Eliminate*	*Number to Add (or Subtract) to (or from) Both Sides of the Equation*	*Correct Results*	*Solution*
$x - 5 = 8$	-5	add 5	$x - 5 + 5 = 8 + 5$	$x = 13$
$x - 3 = -12$	-3	add 3	$x - 3 + 3 = -12 + 3$	$= -9$
$2 = x - 7$	-7	add 7	$2 + 7 = x - 7 + 7$	$9 = x$
$x + 12 = -5$	$+12$	subtract 12	$x + 12 - 12 = -5 - 12$	$x = -17$
$6 = x + 4$	$+4$	subtract 4	$6 - 4 = x + 4 - 4$	$2 = x$
$13 = x + 9$	$+9$	subtract 9	$13 - 9 = x + 9 - 9$	$4 = x$

Notice that under the *Correct Results* column, when the equation is simplified by combining terms, the x will become isolated because the sum of a number and its opposite is 0, and $x + 0$ equals x.

EXAMPLE 8 Solve the equation $4 = x - 5$.

Solution: The variable x is on the right side of the equation. To isolate the x, we must eliminate the -5 from the right side of the equation. This can be accomplished by adding 5 to both sides of the equation.

$$4 = x - 5$$

$$4 + 5 = x - 5 + 5 \quad \text{Add 5 to both sides of the equation.}$$

$$9 = x + 0$$

$$9 = x$$

Thus, the solution is 9.

EXAMPLE 9 Solve the equation $-6.25 = x + 12.78$.

Solution: The variable is on the right side of the equation. Subtract 12.78 from both sides of the equation to isolate the variable.

$$-6.25 = x + 12.78$$

$$-6.25 - 12.78 = x + 12.78 - 12.78 \quad \text{Subtract 12.78 from both sides of the equation.}$$

$$-19.03 = x + 0$$

$$-19.03 = x$$

The solution is -19.03.

COMMON STUDENT ERROR When solving equations, our goal is to get the variable alone on one side of the equal sign. Consider the equation $x + 3 = -4$. How do we solve it?

Correct

Remove the 3 from the left side of the equation.

$$x + 3 = -4$$

$$x + 3 - 3 = -4 - 3$$

$$x = -7$$

Variable is now isolated.

Wrong

Remove the -4 from the right side of the equation.

$$x + 3 = -4$$

$$x + 3 + 4 = -4 + 4$$

$$x + 7 = 0$$

Variable is not isolated.

Remember, use the addition property to **remove the number that is on the same side of the equation as the variable.**

Performing Some Steps Mentally

5 Consider the following two problems:

(a)

$$x - 5 = 12$$

$$x - 5 + 5 = 12 + 5$$

$$x + 0 = 12 + 5$$

$$x = 17$$

(b)

$$15 = x + 3$$

$$15 - 3 = x + 3 - 3$$

$$15 - 3 = x + 0$$

$$12 = x$$

Note how the number on the same side of the equal sign as the variable is transferred to the opposite side of the equal sign when the addition property is used.

Also note that the sign of the number changes when transferred from one side of the equal sign to the other.

When you feel comfortable using the addition property, you may wish to do some of the steps mentally to reduce some of the written work. For example, the preceding two problems may be shortened as follows.

(a)

$$x - 5 = 12$$
$$x - 5 + 5 = 12 + 5 \longleftarrow \text{Do this step mentally.}$$
$$x = 12 + 5$$
$$x = 17$$

Shortened Form

$$x - 5 = 12$$
$$x = 12 + 5$$
$$x = 17$$

(b)

$$15 = x + 3$$
$$15 - 3 = x + 3 - 3 \longleftarrow \text{Do this step mentally.}$$
$$15 - 3 = x$$
$$12 = x$$

Shortened Form

$$15 = x + 3$$
$$15 - 3 = x$$
$$12 = x$$

Exercise Set 2.2

By checking, determine if the number following the equation is a solution to the equation.

1. $2x - 3 = 5, 4$

2. $2x + 1 = x - 5, -6$

3. $2x - 5 = 5(x + 2), -3$

4. $2(x - 3) = 3(x + 1), 1$

5. $3x - 5 = 2(x + 3) - 11, 0$

6. $-2(x - 3) = -5x + 3 - x, -2$

7. $5(x + 2) - 3(x - 1) = 4, 2.3$

8. $x + 3 = 3x + 2, \frac{1}{2}$

9. $4x - 4 = 2x - 3, \frac{1}{2}$

10. $3x + 4 = -2x + 9, \frac{1}{2}$

11. $3(x + 2) = 5(x - 1), \frac{11}{2}$

12. $-(x + 3) - (x - 6) = 3x - 4, 5$

(a) *In Exercises 13–20, represent each figure as an equation using x, and* **(b)** *solve the equation. Refer to page 89 for examples.*

13. + 5 = 8

14. + 2 = 9

15. 12 = + 3

16. 15 = + 4

17. 10 = + 7

18. + 9 = 10

19. + 6 = 4 + 11

20. 12 + 4 = + 3

Solve each equation and check your solution.

21. $x + 2 = 6$

22. $x - 4 = 9$

23. $x + 7 = -3$

24. $x - 4 = -8$

25. $x + 4 = -5$

26. $x - 16 = 36$

27. $x + 43 = -18$

28. $6 + x = 9$

29. $-8 + x = 14$

30. $7 = 9 + x$
31. $27 = x - 16$
32. $-9 = x - 25$
33. $-13 = x - 1$
34. $4 = 11 + x$
35. $29 = -43 + x$
36. $-18 = -14 + x$
37. $7 + x = -19$
38. $9 + x = 9$
39. $x + 29 = -29$
40. $4 + x = -9$
41. $9 = x - 3$
42. $5 + x = 12$
43. $x + 7 = -5$
44. $6 = 4 + x$
45. $9 + x = 12$
46. $-4 = x - 3$
47. $-5 = 4 + x$
48. $12 = 16 + x$
49. $40 = x - 13$
50. $15 + x = -5$
51. $x - 12 = -9$
52. $x + 6 = -12$
53. $4 + x = 9$
54. $-6 = 9 + x$
55. $-8 = -9 + x$
56. $-12 = 8 + x$
57. $5 = x - 12$
58. $2 = x + 9$
59. $-50 = x - 24$
60. $-29 + x = -15$
61. $16 + x = -20$
62. $-25 = 18 + x$
63. $40.2 + x = -7.3$
64. $-27.23 + x = 9.77$
65. $-37 + x = 9.5$
66. $7.2 + x = 7.2$
67. $x - 8.42 = -30$
68. $6.2 + x = 5.7$
69. $9.75 = x + 9.75$
70. $139 = x - 117$
71. $600 = x - 120$

72. What is an equation?

73. **(a)** What is meant by the "solution of an equation"? **(b)** What does it mean to "solve an equation"?

74. Explain how the solution to an equation may be checked.

75. In your own words explain the Addition Property of Equality.

76. What are equivalent equations?

77. To solve an equation we "isolate the variable." **(a)** Explain what this means, and **(b)** explain how to isolate the variable in the equations discussed in this section.

78. When solving the equation $x - 4 = 6$, would you add 4 to both sides of the equation or subtract 6 from both sides of the equation? Explain.

79. When solving the equation $5 = x + 3$, would you subtract 5 from both sides of the equation or subtract 3 from both sides of the equation? Explain.

CUMULATIVE REVIEW EXERCISES

[1.9] *Evaluate.*

80. $3x + 4(x - 3) + 2$ when $x = 4$.

81. $6x - 2(2x + 1)$ when $x = -3$.

[2.1] *Simplify.*

82. $4x + 3(x - 2) - 5x - 7$.

83. $-(x - 3) + 7(2x - 5) - 3x$.

Group Activity/Challenge Problems

1. By checking, determine which of the following are solutions to the equation $2(x + 3) = 2x + 6$.

(a) -1 **(b)** 5 **(c)** $\frac{1}{2}$

(d) Select any number not given in parts **(a)**, **(b)**, or **(c)** and determine if that number is a solution to the equation.

(e) Will every real number be a solution to this equation? Explain.

2. By checking, determine which of the following are solutions to $2x^2 - 7x + 3 = 0$.

(a) 3 **(b)** 2 **(c)** $\frac{1}{2}$

3. In the next section we introduce the multiplication property. When discussing the multiplication property we will use a figure like the one that follows.

(a) Write an equation using the variable x, that can be used to represent this figure.

(b) Solve the equation.

Follow the instructions in Exercise 3 for the following figures.

4. **5.**

2.3 The Multiplication Property of Equality

Tape 3

1. Identify reciprocals.
2. Use the multiplication property of equality to solve equations.
3. Solve equations of the form $-x = a$.
4. Do some steps mentally when solving equations.

Identify Reciprocals

1 Before we discuss the multiplication property, let us discuss what is meant by the **reciprocal** of a number. Two numbers are reciprocals of each other when their product is 1. Some examples of numbers and their reciprocals follow.

Number	*Reciprocal*	*Product*
3	$\frac{1}{3}$	$(3)\left(\frac{1}{3}\right) = 1$
$-\frac{3}{5}$	$-\frac{5}{3}$	$\left(-\frac{3}{5}\right)\left(-\frac{5}{3}\right) = 1$
-1	-1	$(-1)(-1) = 1$

The reciprocal of a positive number is a positive number and the reciprocal of a negative number is a negative number. Note that 0 has no reciprocal.

In general, if a represents any number, its reciprocal is $\frac{1}{a}$. For example, the reciprocal of 3 is $\frac{1}{3}$ and the reciprocal of -2 is $\frac{1}{-2}$ or $-\frac{1}{2}$. The reciprocal of $-\frac{3}{5}$ is $\frac{1}{-\frac{3}{5}}$, which can be written as $1 \div \left(-\frac{3}{5}\right)$. Simplifying, we get $\left(\frac{1}{1}\right)\left(-\frac{5}{3}\right) = -\frac{5}{3}$. Thus, the reciprocal of $-\frac{3}{5}$ is $-\frac{5}{3}$.

Use the Multiplication Property to Solve Equations

2 In Section 2.2 we used the addition property of equality to solve equations of the form $x + a = b$ where a and b represent real numbers. In this section we will solve equations of the form $ax = b$, where a and b represent real numbers. Equations of the form $ax = b$ are solved using the multiplication property of equality. It is important that you recognize the difference between equations like $x + 2 = 8$ and $2x = 8$. In $x + 2 = 8$ the 2 is being added to x, so we use the addition property to solve the equation. In $2x = 8$ the 2 is multiplying the x, so we use the multiplication property to solve the equation. The multiplication property is used to solve linear equations where the coefficient of the x term is a number other than 1. Below we give a visual interpretation of the difference between the equations. To write the equation we have used x to represent the value of a kiss.

Figure	*Equation*	*Property to Use to Solve Equation*	*Solution*
	$x + 2 = 8$	Addition (the equation contains only one x)	6
	$x + x = 8$ or $2x = 8$	Multiplication (the left side of the equation contains more than one x)	4
	$15 = x + x + x$ or $15 = 3x$	Multiplication (the right side of the equation contains more than one x).	5

To help you understand the multiplication property of equality we will give a visual interpretation of the property before stating it.

Consider Figure 2.6. To find the value of one kiss, we need to redraw the balance with only one kiss on one side of the balance. We need to eliminate one of the two kisses on the left side of the balance. We can accomplish this by either multiplying the two kisses by $\frac{1}{2}$ to get $\frac{1}{2}(2) = 1$, or by dividing the two kisses by 2 to get $\frac{2}{2} = 1$. The two processes are equivalent. We must remember that whatever we do to one side of the balance we must do to the other side. Thus, if we multiply the two kisses by $\frac{1}{2}$, we need to multiply the 8 by $\frac{1}{2}$ to get $\frac{1}{2}(8) = 4$. If we divide the two kisses by 2, we need to divide the 8 by 2 to get $\frac{8}{2} = 4$. Either procedure results in the balance shown in Figure 2.7, where we can see that the value of a kiss is 4.

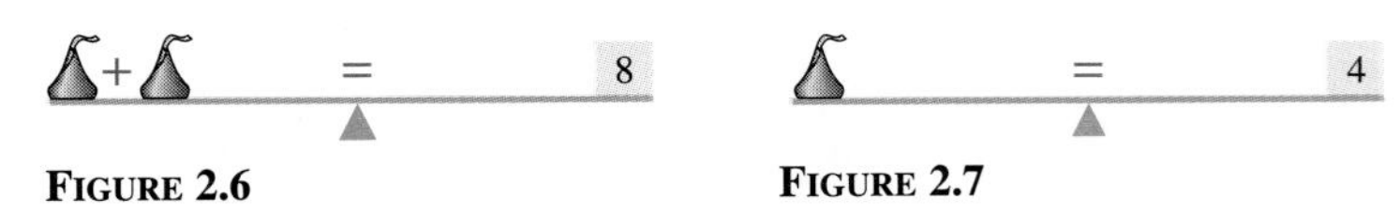

FIGURE 2.6 FIGURE 2.7

FIGURE 2.8

Now consider the balance in Figure 2.8. There are 4 kisses of equal value on the right side of the balance. To find the value of one kiss we need to redraw the balance so that only one kiss appears on the right side. We can do this by multiplying the 4 kisses by $\frac{1}{4}$ to get $4(\frac{1}{4}) = 1$, or by dividing the 4 kisses by 4 to get $\frac{4}{4} = 1$. If we multiply the kisses on the right side of the balance by $\frac{1}{4}$, we need to multiply the 9 on the left side of the balance by $\frac{1}{4}$. This gives $\frac{1}{4}(9) = \frac{9}{4}$. If we divide the kisses on the right side of the balance by 4, we need to divide the 9 on the left side of the balance by 4. This gives $\frac{9}{4}$. Either method results in the kiss on the right side of the balance having a value of $\frac{9}{4}$ (Fig. 2.9).

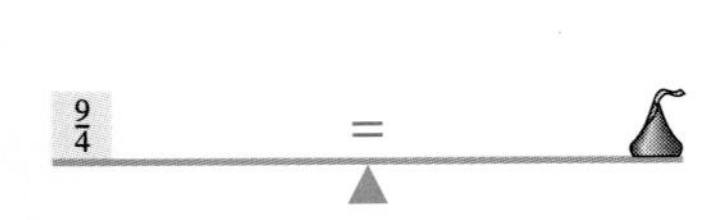

FIGURE 2.9

Below we illustrate how Figure 2.8 may be represented as an equation if we let x represent the value of a kiss.

Figure	*Equation*	*Solution*
9 = (four kisses)	$9 = x + x + x + x$ or $9 = 4x$	$\frac{9}{4}$

To solve the equation $9 = 4x$ we perform a similar process used in finding the value of one kiss on the balance. To find the value of x in the equation $9 = 4x$ we can *multiply* both sides of the equation *by the reciprocal of the number of x's that appear.* Since the right side of the equation contains 4 x's we can multiply both sides of the equation by $\frac{1}{4}$. We can also solve the equation by *dividing* both sides of the equation *by the number of x's that appear,* 4. Using either method we find that $x = \frac{9}{4}$.

The information presented above may help you in solving equations using the multiplication property of equality. Now we present the multiplication property of equality.

Multiplication Property of Equality

If $a = b$, then $a \cdot c = b \cdot c$ for any numbers a, b, and c.

The multiplication property implies that both sides of an equation can be multiplied by the same number without changing the solution. **The multiplication property can be used to solve equations of the form $ax = b$.** We can isolate the variable in equations of this form by multiplying both sides of the equation by the reciprocal of a, which is $\frac{1}{a}$. By doing so the numerical coefficient of the variable, x, becomes 1, which can be omitted when we write the variable. By following this process, we say that we *eliminate* the coefficient from the variable.

Equation	*To Solve, Use the Multiplication Property to Eliminate the Coefficient*
$4x = 9$	4
$-5x = 20$	-5
$15 = \frac{1}{2}x$	$\frac{1}{2}$
$7 = -9x$	-9

Now let us work some problems.

EXAMPLE 1 Solve the equation $3x = 6$.

Solution: To isolate the variable, x, we must eliminate the 3 from the left side of the equation. To do this, we multiply both sides of the equation by the reciprocal of 3, which is $\frac{1}{3}$.

$$3x = 6$$

$$\frac{1}{3} \cdot 3x = \frac{1}{3} \cdot 6 \quad \text{Multiply both sides of the equation by } \frac{1}{3}.$$

$$\frac{1}{\cancel{3}_1} \cdot \overset{1}{\cancel{3}}x = \frac{1}{\cancel{3}_1} \cdot \overset{2}{\cancel{6}} \quad \text{Divide out the common factors.}$$

$$1x = 2$$

$$x = 2$$

Notice in Example 1 that $1x$ is replaced by x in the next step. Usually we do this step mentally. How would you represent and solve the equation $3x = 6$ using a balance? Try this now.

EXAMPLE 2 Solve the equation $\frac{x}{2} = 4$.

Solution: Since dividing by 2 is the same as multiplying by $\frac{1}{2}$, the equation $\frac{x}{2} = 4$ is the same as $\frac{1}{2}x = 4$. We will therefore multiply both sides of the equation by the reciprocal of $\frac{1}{2}$, which is 2.

$$\frac{x}{2} = 4$$

$$\overset{1}{\cancel{2}}\left(\frac{x}{\cancel{2}_1}\right) = 2 \cdot 4 \quad \text{Multiply both sides of the equation by 2.}$$

$$x = 2 \cdot 4$$

$$x = 8$$

Check:

$$\frac{x}{2} = 4$$

$$\frac{8}{2} = 4$$

$$4 = 4 \quad \text{true}$$

EXAMPLE 3 Solve the equation $\frac{2}{3}x = 6$.

Solution: The reciprocal of $\frac{2}{3}$ is $\frac{3}{2}$. Multiply both sides of the equation by $\frac{3}{2}$.

$$\frac{2}{3}x = 6$$

$$\frac{3}{2} \cdot \frac{2}{3}x = \frac{3}{2} \cdot 6$$

$$1x = 9$$

$$x = 9$$

Check: $\frac{2}{3}x = 6$

$$\frac{2}{3}(9) = 6$$

$$6 = 6 \quad \text{true}$$

In Example 1, $3x = 6$, we multiplied both sides of the equation by $\frac{1}{3}$ to isolate the variable. We could have also isolated the variable by dividing both sides of the equation by 3, as follows:

$$3x = 6$$

$$\frac{\overset{1}{\cancel{3}}x}{\underset{1}{\cancel{3}}} = \frac{\overset{2}{\cancel{6}}}{\underset{1}{\cancel{3}}} \quad \text{Divide both sides of the equation by 3.}$$

$$x = 2$$

We can do this because dividing by 3 is equivalent to multiplying by $\frac{1}{3}$. **Since division can be defined in terms of multiplication ($\frac{a}{b}$ means $a \cdot \frac{1}{b}$), the multiplication property also allows us to divide both sides of an equation by the same nonzero number.** This process is illustrated in Examples 4 through 6.

EXAMPLE 4 Solve the equation $8p = 5$.

Solution: $8p = 5$

$$\frac{8p}{8} = \frac{5}{8} \quad \text{Divide both sides of the equation by 8.}$$

$$p = \frac{5}{8}$$

EXAMPLE 5 Solve the equation $-12 = -3x$.

Solution: In this equation the variable, x, is on the right side of the equal sign. To isolate x, we divide both sides of the equation by -3.

$$-12 = -3x$$

$$\frac{-12}{-3} = \frac{-3x}{-3} \quad \text{Divide both sides of the equation by } -3.$$

$$4 = x$$

EXAMPLE 6 Solve the equation $0.32x = 1.28$.

Solution: We begin by dividing both sides of the equation by 0.32 to isolate the variable x.

$$0.32x = 1.28$$

$$\frac{0.32x}{0.32x} = \frac{1.28}{0.32} \quad \text{Divide both sides of the equation by 0.32.}$$

$$x = 4$$

Working problems involving decimal numbers on a calculator will probably save you time.

> ***Helpful Hint***
>
> When solving an equation of the form $ax = b$, we can isolate the variable by
>
> **1.** Multiplying both sides of the equation by the reciprocal of a, $\frac{1}{a}$, as was done in Examples 1, 2, and 3, or
>
> **2.** Dividing both sides of the equation by a, as was done in Examples 4, 5, and 6.
>
> Either method may be used to isolate the variable. However, if the equation contains a fraction, or fractions, you will arrive at a solution more quickly by multiplying by the reciprocal of a. This is illustrated in Examples 7 and 8.

EXAMPLE 7 Solve the equation $-2x = \frac{3}{5}$.

Solution: Since this equation contains a fraction, we will isolate the variable by multiplying both sides of the equation by $-\frac{1}{2}$, which is the reciprocal of -2.

$$-2x = \frac{3}{5}$$

$$\left(-\frac{1}{2}\right)(-2x) = \left(-\frac{1}{2}\right)\left(\frac{3}{5}\right) \qquad \text{Multiply both sides of the equation by } \frac{1}{2}.$$

$$1x = \left(-\frac{1}{2}\right)\left(\frac{3}{5}\right)$$

$$x = -\frac{3}{10}$$

In Example 7, if you wished to solve the equation by dividing both sides of the equation by -2, you would have to divide the fraction $\frac{3}{5}$ by -2.

EXAMPLE 8 Solve the equation $-6 = -\frac{3}{5}x$.

Solution: Since this equation contains a fraction, we will isolate the variable by multiplying both sides of the equation by the reciprocal of $-\frac{3}{5}$, which is $-\frac{5}{3}$.

$$-6 = -\frac{3}{5}x$$

$$(-6)\left(-\frac{5}{3}\right) = \left(-\frac{5}{3}\right)\left(-\frac{3}{5}x\right)$$

$$10 = x$$

In Example 8 the equation was written as $-6x = -\frac{3}{5}x$. This equation is equivalent to the equations $-6 = \frac{-3}{5}x$ and $-6 = \frac{3}{-5}x$. Can you explain why? All three equations have the same solution, 10.

Solve Equations of the Form $-x = a$

3 When solving an equation we may obtain an equation like $-x = 7$. This is not a solution since $-x = 7$ means $-1x = 7$. The solution to an equation is of the form $x =$ some number. When an equation is of the form $-x = 7$, we can solve for x by multiplying both sides of the equation by -1, as illustrated in the following example.

EXAMPLE 9 Solve the equation $-x = 7$.

Solution: $-x = 7$ means that $-1x = 7$. We are solving for x, not $-x$. We can multiply both sides of the equation by -1 to get x on the left side of the equation.

$$-x = 7$$
$$-1x = 7$$
$$(-1)(-1x) = (-1)(7) \quad \text{Multiply both sides of the equation by } -1.$$
$$1x = -7$$
$$x = -7$$

Check:

$$-x = 7$$
$$-(-7) = 7$$
$$7 = 7 \quad \text{true}$$

Thus, the solution is -7.

Example 9 may also be solved by dividing both sides of the equation by -1. Try this now and see that you get the same solution. Whenever we have the opposite (or negative) of a variable equal to a quantity, as in Example 9, we can solve for the variable by multiplying (or dividing) both sides of the equation by -1.

EXAMPLE 10 Solve the equation $-x = -5$.

Solution:

$$-x = -5$$
$$-1x = -5$$
$$(-1)(-1x) = (-1)(-5) \quad \text{Multiply both sides of the equation by } -1.$$
$$1x = 5$$
$$x = 5$$

Helpful Hint

For any real number a, $a \neq 0$,
If $-x = a$, then $x = -a$

Examples:

$$-x = 7 \qquad\qquad -x = -2$$
$$x = -7 \qquad\qquad x = -(-2)$$
$$x = 2$$

Performing Some Steps Mentally

4 When you feel comfortable using the multiplication property, you may wish to do some of the steps mentally to reduce some of the written work. Now we present two examples worked out in detail, along with their shortened form.

EXAMPLE 11 Solve the equation $-3x = -21$.

Solution:

$$-3x = -21$$

$$\frac{-3x}{-3} = \frac{-21}{-3} \longleftarrow \text{Do this step mentally.}$$

$$x = \frac{-21}{-3}$$

$$x = 7$$

Shortened Form

$$-3x = -21$$

$$x = \frac{-21}{-3}$$

$$x = 7$$

EXAMPLE 12 Solve the equation $\frac{1}{3}x = 9$.

Solution:

$$\frac{1}{3}x = 9$$

$$3\left(\frac{1}{3}x\right) = 3(9) \longleftarrow \text{Do this step mentally.}$$

$$x = 3(9)$$

$$x = 27$$

Shortened Form

$$\frac{1}{3}x = 9$$

$$x = 3(9)$$

$$x = 27$$

In Section 2.2 we discussed the addition property and in this section we discussed the multiplication property. It is important that you understand the difference between the two. The following Helpful Hint should be studied carefully.

Helpful Hint

The **addition property** is used to solve equations of the form $\boldsymbol{x + a = b}$. The *addition property* is used when a number is *added to or subtracted from* a variable.

$$x + 3 = -6 \qquad\qquad x - 5 = -2$$

$$x + 3 - 3 = -6 - 3 \qquad\qquad x - 5 + 5 = -2 + 5$$

$$x = -9 \qquad\qquad x = 3$$

The **multiplication property** is used to solve equations of the form $\boldsymbol{ax = b}$. It is used when a variable is *multiplied or divided by a number.*

$$3x = 6 \qquad\qquad \frac{x}{2} = 4 \qquad\qquad \frac{2}{5}x = 12$$

$$\frac{3x}{3} = \frac{6}{3} \qquad\qquad 2\left(\frac{x}{2}\right) = 2(4) \qquad\qquad \left(\frac{5}{2}\right)\left(\frac{2}{5}x\right) = \left(\frac{5}{2}\right)(12)$$

$$x = 2 \qquad\qquad x = 8 \qquad\qquad x = 30$$

Exercise Set 2.3

(a) *Express the figure as an equation in the variable x, and,* **(b)** *solve the equation.*

1. 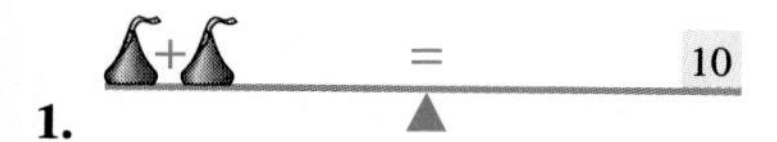

2.

3. 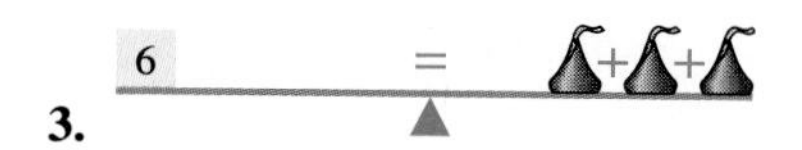

4.

5.

6.

7.

8.

Solve each equation and check your solution.

9. $2x = 6$

10. $5x = 20$

11. $\frac{x}{2} = 4$

12. $\frac{x}{3} = 12$

13. $-4x = 8$

14. $8 = 16y$

15. $\frac{x}{6} = -2$

16. $\frac{x}{3} = -2$

17. $\frac{x}{5} = 1$

18. $-2x = 12$

19. $-32x = -96$

20. $16 = -4y$

21. $-6 = 4z$

22. $\frac{x}{8} = -3$

23. $-x = -6$

24. $-x = 9$

25. $-2 = -y$

26. $-3 = \frac{x}{5}$

27. $-\frac{x}{7} = -7$

28. $4 = \frac{x}{9}$

29. $4 = -12x$

30. $12y = -15$

31. $-\frac{x}{3} = -2$

32. $-\frac{a}{8} = -7$

33. $13x = 10$

34. $-24x = -18$

35. $-4.2x = -8.4$

36. $-3.72 = 1.24y$

37. $7x = -7$

38. $3x = \frac{3}{5}$

39. $5x = -\frac{3}{8}$

40. $-2b = -\frac{4}{5}$

41. $15 = -\frac{x}{5}$

42. $\frac{x}{16} = -4$

43. $-\frac{x}{5} = -25$

44. $-x = -\frac{5}{9}$

45. $\frac{x}{5} = -7$

46. $-3r = -18$

47. $5 = \frac{x}{4}$

48. $-3 = \frac{x}{-5}$

49. $6d = -30$

50. $\frac{2}{7}x = 7$

51. $\frac{y}{-2} = -6$

52. $-2x = \frac{3}{5}$

53. $\frac{-3}{8}w = 6$

54. $-x = \frac{4}{7}$

55. $\frac{1}{3}x = -12$

56. $6 = \frac{3}{5}x$

57. $-4 = -\frac{2}{3}z$

58. $-8 = \frac{-4}{5}x$

59. $-1.4x = 28.28$

60. $-0.42x = -2.142$

61. $2x = -\frac{5}{2}$

62. $6x = \frac{8}{3}$

63. $\frac{2}{3}x = 6$

64. $-\frac{1}{2}x = \frac{2}{3}$

65. **(a)** If $-x = a$, where a represents any real number, what does x equal?

(b) If $-x = 5$, what is x?

(c) If $-x = -5$, what is x?

66. When solving the equation $3x = 5$, would you divide both sides of the equation by 3 or by 5? Explain.

67. When solving the equation $-2x = 5$, would you add 2 to both sides of the equation or divide both sides of the equation by -2? Explain.

68. Consider the equation $\frac{2}{3}x = 4$. This equation could be solved by multiplying both sides of the equation by $\frac{3}{2}$, the reciprocal of $\frac{2}{3}$, or by dividing both sides of the equation by $\frac{2}{3}$. Which method do you feel would be easier? Explain your answer. Find the solution to the equation.

69. Consider the equation $4x = \frac{3}{5}$. Would it be easier to solve this equation by dividing both sides of the equation by 4 or by multiplying both sides of the equation by $\frac{1}{4}$, the reciprocal of 4? Explain your answer. Find the solution to the problem.

70. Consider the equation $\frac{3}{7}x = \frac{4}{5}$. Would it be easier to solve this equation by dividing both sides of the equation by $\frac{3}{7}$ or by multiplying both sides of the equation by $\frac{3}{7}$, the reciprocal of $\frac{3}{7}$? Explain your answer. Find the solution to the equation.

CUMULATIVE REVIEW EXERCISES

[1.6] **71.** Subtract -4 from -8.

72. Evaluate $6 - (-3) - 5 - 4$.

[2.1] **73.** Simplify $-(x + 3) - 5(2x - 7) + 6$.

[2.2] **74.** Solve the equation $-48 = x + 9$.

Group Activity/ Challenge Problems

In the next section we will solve equations using both the addition and multiplication properties. We can use figures like those in Exercises 1–4 to illustrate such problems. For each exercise,

(a) Find the value of a kiss. Hint: Use the addition property first to get the kisses by themselves on one side of the balance. Then use the multiplication property to find the value of a kiss.

(b) Write an equation in variable x that can be used to represent the figure.

(c) Solve the equation and find the value of x. (The value of x should be the same as the value of a kiss. Follow the hint presented in part (a).)

1. + + 6 = 14

2. 9 = + + + 9

3. 6 = + 4 +

4. 7 = + + 2 +

2.4 Solving Linear Equations with a Variable on Only One Side of the Equation

1 Solve linear equations that contain a variable on only one side of the equal sign.

Tape 4

Solve Linear Equations

1 In this section we discuss how to solve linear equations using **both** the addition and multiplication properties of equality when a variable appears on only one side of the equal sign. In Section 2.5 we will discuss how to solve linear equations using both properties when a variable appears on both sides of the equal sign.

Below we show some illustrations that have been represented as equations. In the equations we represented the value of a kiss with the letter x. In each of these equations the variable appears on only one side of the equal sign. We also give the solution to each equation. You probably cannot determine the solutions yet. Do not worry about this. The purpose of this section is to teach you the procedures for finding the solution to such problems.

Figure	*Equation*	*Solution*
+ + 4 = 14	$2x + 4 = 14$	5
+ + + 6 = 27	$3x + 6 = 27$	7
12 = + + 5	$12 = 2x + 5$	$\frac{7}{2}$

The general procedure we use to solve equations is to "isolate the variable." That is, we must get the variable, x, alone on one side of the equal sign. If you consider the balance, we will need to eliminate all the numbers from the same side of the balance as the kisses. No one method is the "best" to solve all linear equations. Following is a general procedure that can be used to solve linear equations when the variable appears on only one side of the equation and the equation does not contain fractions.

To Solve Linear Equations with a Variable on Only One Side of the Equal Sign

1. Use the distributive property to remove parentheses.
2. Combine like terms on the same side of the equal sign.
3. Use the addition property to obtain an equation with the term containing the variable on one side of the equal sign and a constant on the other side. This will result in an equation of the form $ax = b$.
4. Use the multiplication property to isolate the variable. This will give a solution of the form $x = \frac{b}{a}\left(\text{or } 1x = \frac{b}{a}\right)$.
5. Check the solution in the *original* equation.

When solving an equation you should always check your solution, as is indicated in step 5. We will not show all checks because of lack of space. We solved some equations containing fractions in Section 2.3. More complex equations containing fractions will be solved by a different procedure in Section 6.6.

When solving an equation remember that our goal is to obtain the variable alone on one side of the equation.

To help you visualize the boxed procedure consider the figure and corresponding equation below.

Figure	*Equation*
+ + 4 = 10	$2x + 4 = 10$

The equation $2x + 4 = 10$ contains no parentheses and no like terms on the same side of the equal sign. Therefore, we start with step 3, using the addition property. Remember from Section 2.2 that the addition property allows us to add (or subtract) the same quantity to (or from) both sides of an equation without changing its solution. Here we subtract 4 from both sides of the equation to isolate the term containing the variable.

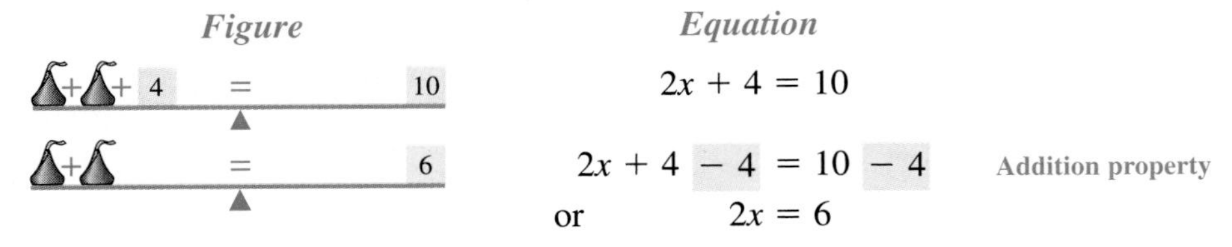

Notice how the term containing the variable, $2x$, is now by itself on one side of the equal sign. Now we use the multiplication property, step 4, to isolate the variable and obtain the solution. Remember from Section 2.3 that the multiplication property allows us to multiply or divide both sides of the equation by the same nonzero number without changing its solution. Here we divide both sides of the equation by 2, the coefficient of the term containing the variable, to obtain the solution, 3.

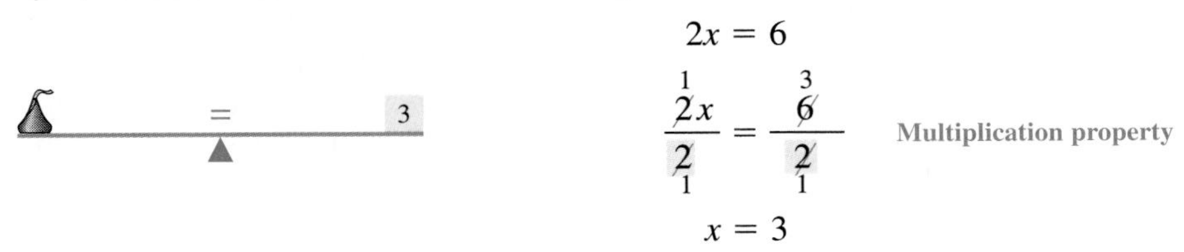

EXAMPLE 1 Solve the equation $2x - 5 = 9$.

Solution: We will follow the procedure outlined for solving equations. Since the equation contains no parentheses and since there are no like terms to be combined, we start with step 3.

Step 3

$$2x - 5 = 9$$

$$2x - 5 + 5 = 9 + 5 \quad \text{Add 5 to both sides of the equation.}$$

$$2x = 14$$

Step 4

$$\frac{2x}{2} = \frac{14}{2} \quad \text{Divide both sides of the equation by 2.}$$

$$x = 7$$

Step 5 *Check:*

$$2x - 5 = 9$$

$$2(7) - 5 = 9$$

$$14 - 5 = 9$$

$$9 = 9 \quad \text{true}$$

Since the check is true, the solution is 7. Note that after completing step 3 we obtain $2x = 14$, which is an equation of the form $ax = b$. After completing step 4 we obtain the answer in the form $x =$ a real number.

Helpful Hint

When solving an equation that does not contain fractions, **the addition property (step 3) is to be used before the multiplication property (step 4).** If you use the multiplication property before the addition property, it is still possible to obtain the correct answer. However, you will usually have to do more work, and you may end up working with fractions. What would happen if you tried to solve Example 1 using the multiplication property before the addition property?

EXAMPLE 2 Solve the equation $-2x - 6 = -3$.

Solution: $-2x - 6 = -3.$

Step 3 $-2x - 6 + 6 = -3 + 6$ — Add 6 to both sides of the equation.

$$-2x = 3$$

Step 4 $\dfrac{-2x}{-2} = \dfrac{3}{-2}$ — Divide both sides of the equation by -2.

$$x = -\frac{3}{2}$$

Step 5 *Check:*

$$-2x - 6 = -3$$

$$-2\left(-\frac{3}{2}\right) - 6 = -3$$

$$3 - 6 = -3$$

$$-3 = -3 \quad \text{true}$$

The solution is $-\dfrac{3}{2}$.

Note that checks are always made with the original equation. In some of the following examples the check will be omitted to save space.

EXAMPLE 3 Solve the equation $16 = 4x + 6 - 2x$.

Solution: Again we must isolate the variable x. Since the right side of the equation has two like terms containing the variable x, we will first combine these like terms.

Step 2 $16 = 4x + 6 - 2x$

$16 = 2x + 6$ — Like terms were combined.

Step 3 $16 - 6 = 2x + 6 - 6$ — Subtract 6 from both sides of equation.

$$10 = 2x$$

Step 4 $\dfrac{10}{2} = \dfrac{2x}{2}$ — Divide both sides of equation by 2.

$$5 = x$$

The preceding solution can be condensed as follows.

$16 = 4x + 6 - 2x$

$16 = 2x + 6$ — Like terms were combined.

$10 = 2x$ — 6 was subtracted from both sides of equation.

$5 = x$ — Both sides of equation were divided by 2.

In Chapter 3 we will be solving many equations that contain decimal numbers. To solve such equations we follow the same procedure as outlined earlier. Example 4, on page 109, illustrates the solution to an equation that contains decimal numbers.

Helpful Hint

In the first two chapters you have been introduced to a variety of mathematics terms. Some of the most commonly used terms are "evaluate," "simplify," "solve," and "check." Make sure you understand what each term means and when each term is used.

Evaluate: To *evaluate an expression* means to find its numerical value.

Evaluate:
$$\begin{aligned} &16 \div 2^2 + 36 \div 4 \\ &= 16 \div 4 + 36 \div 4 \\ &= 4 + 36 \div 4 \\ &= 4 + 9 \\ &= 13 \end{aligned}$$

Evaluate:
$$\begin{aligned} &-x^2 + 3x - 2 \text{ when } x = 4 \\ &= -4^2 + 3(4) - 2 \\ &= -16 + 12 - 2 \\ &= -4 - 2 \\ &= -6 \end{aligned}$$

Simplify: To *simplify an expression* means to perform the operations and combine like terms.

Simplify: $3(x - 2) - 4(2x + 3)$
$$\begin{aligned} 3(x - 2) - 4(2x + 3) &= 3x - 6 - 8x - 12 \\ &= -5x - 18 \end{aligned}$$

Note that when you simplify an expression containing variables you do not generally end up with just a numerical value unless all the variable terms happen to add to zero.

Solve: To *solve an equation* means to find the value or the values of the variables that make the equation a true statement.

Solve:
$$\begin{aligned} 2x + 3(x + 1) &= 18 \\ 2x + 3x + 3 &= 18 \\ 5x + 3 &= 18 \\ 5x &= 15 \\ x &= 3 \end{aligned}$$

Check: To *check an equation,* we substitute the value believed to be the solution into the original equation. If this substitution results in a true statement, then we say the answer checks. For example, to check the solution of the equation just solved, we substitute 3 for x in the equation.

Check:
$$\begin{aligned} 2x + 3(x + 1) &= 18 \\ 2(3) + 3(3 + 1) &= 18 \\ 6 + 3(4) &= 18 \\ 6 + 12 &= 18 \\ 18 &= 18 \quad \text{true} \end{aligned}$$

Since we obtained a true statement, the 3 checks.

It is important to realize that *expressions may be evaluated or simplified* (depending on the type of problem) and *equations are solved and then checked.*

EXAMPLE 4 Solve the equation $x + 1.24 - 0.07x = 4.96$.

Solution: $x + 1.24 - 0.07x = 4.96$

$0.93x + 1.24 = 4.96$ Like terms were combined, $1x - 0.07x = 0.93x$.

$0.93x + 1.24 - 1.24 = 4.96 - 1.24$ Subtract 1.24 from both sides of equation.

$0.93x = 3.72$

$\frac{0.93x}{0.93} = \frac{3.72}{0.93}$ Divide both sides of equation by 0.93.

$x = 4$

EXAMPLE 5 Solve the equation $2(x + 4) - 5x = -3$.

Solution: $2(x + 4) - 5x = -3$.

$2x + 8 - 5x = -3$ The distributive property was used.

$-3x + 8 = -3$ Like terms were combined.

$-3x + 8 - 8 = -3 - 8$ Subtract 8 from both sides of equation.

$-3x = -11$

$\frac{-3x}{-3} = \frac{-11}{-3}$ Divide both sides of equation by -3.

$x = \frac{11}{3}$

The solution to Example 5 can be condensed as follows:

$2(x + 4) - 5x = -3$

$2x + 8 - 5x = -3$ The distributive property was used.

$-3x + 8 = -3$ Like terms were combined.

$-3x = -11$ 8 was subtracted from both sides of equation.

$x = \frac{11}{3}$ Both sides of equation were divided by -3.

EXAMPLE 6 Solve the equation $2x - (x + 2) = 6$.

Solution: $2x - (x + 2) = 6$

$2x - x - 2 = 6$ The distributive property was used.

$x - 2 = 6$ Like terms were combined.

$x = 8$ 2 was added to both sides of equation.

Exercise Set 2.4

(a) *Represent the figure as an equation in the variable x and* **(b)** *solve the equation.*

1. + + 4 = 16

2. + + + 8 = 20

3. 30 = + + 12

4. 27 = + + + 9

5. + + 10 + = 4

6. 3 = + 6 + +

7. 5 + + + = 12

8. 9 = + + 12

Solve each equation. You may wish to use a calculator for equations containing decimal numbers.

9. $2x + 4 = 10$
10. $2x - 4 = 8$
11. $-2x - 5 = 7$
12. $-4x + 5 = -3$
13. $5x - 6 = 19$
14. $6 - 3x = 18$
15. $5x - 2 = 10$
16. $-9x + 3 = 15$
17. $-x - 4 = 8$
18. $6 = 2x - 3$
19. $12 - x = 9$
20. $-3x - 3 = -12$
21. $8 + 3x = 19$
22. $-2x + 7 = -10$
23. $16x + 5 = -14$
24. $19 = 25 + 4x$
25. $-4.2 = 2x + 1.6$
26. $-24 + 16x = -24$
27. $6x - 9 = 21$
28. $-x + 4 = -8$
29. $12 = -6x + 5$
30. $15 = 7x + 1$
31. $-2x - 7 = -13$
32. $-2 - x = -12$
33. $x + 0.05x = 21$
34. $x + 0.07x = 16.05$
35. $2.3x - 9.34 = 6.3$
36. $-2.3 = -1.4 + 0.6x$
37. $28.8 = x - 0.10x$
38. $32.76 = 2.45x - 8.75x$
39. $3(x + 2) = 6$
40. $3(x - 2) = 12$
41. $4(3 - x) = 12$
42. $-2(x + 3) = -9$
43. $-4 = -(x + 5)$
44. $-3(2 - 3x) = 9$
45. $12 = 4(x + 3)$
46. $-2(x + 4) + 5 = 1$
47. $5 = 2(3x + 6)$
48. $-2 = 5(3x + 1) - 12x$
49. $2x + 3(x + 2) = 11$
50. $4 = -2(x + 3)$
51. $x - 3(2x + 3) = 11$
52. $3(4 - x) + 5x = 9$
53. $5x + 3x - 4x - 7 = 9$
54. $-(x + 2) = 4$
55. $0.7(x + 3) = 4.2$
56. $12 + (x + 9) = 7$
57. $1.4(5x - 4) = -1.4$
58. $0.1(2.4x + 5) = 1.7$
59. $3 - 2(x + 3) + 2 = 1$
60. $2(3x - 4) - 4x = 12$
61. $1 - (x + 3) + 2x = 4$
62. $5x - 2x - 7x = -20$
63. $4.22 - 6.4x + 9.60 = 0.38$
64. $-4(x + 2) - 3x = 20$
65. $5.76 - 4.24x - 1.9x = 27.864$

66. **(a)** In your own words, write the general procedure for solving an equation where the variable appears on only one side of the equal sign.
(b) Refer to page 105 to see if you omitted any steps.

67. When solving equations that do not contain fractions, do we normally use the addition or multiplication property first in the process of isolating the variable? Explain your answer.

68. **(a)** Explain, in a step-by-step manner, how to solve the equation $2(3x + 4) = -4$.
(b) Solve the equation by following the steps listed in part (a).

69. **(a)** Explain, in a step-by-step manner, how to solve the equation $4x - 2(x + 3) = 4$.
(b) Solve the equation by following the steps listed in part (a).

CUMULATIVE REVIEW EXERCISES

[1.2] **70.** Add $\frac{5}{8} + \frac{3}{5}$.

[1.9] **71.** Evaluate $[5(2 - 6) + 3(8 \div 4)^2]^2$.

[2.2] **72.** To solve an equation, what do you need to do to the variable?

[2.3] **73.** To solve the equation $7 = -4x$, would you add 4 to both sides of the equation or divide both sides of the equation by -4? Explain your answer.

Group Activity/ Challenge Problems

Solve each equation.

1. $3(x - 2) - (x + 5) - 2(3 - 2x) = 18.$

2. $-6 = -(x - 5) - 3(5 + 2x) - 4(2x - 4).$

3. $4[3 - 2(x + 4)] - (x + 3) = 13.$

In Chapter 3 we will discuss procedures for writing application problems as equations. Let us see if you can figure out how to write an equation for Exercises 4–6. For each exercise, ***(a)*** *draw a balance that represents the problem;* ***(b)*** *using the balance in part (a), write an equation that represents the problem;* ***(c)*** *solve the equation,* ***(d)*** *check your answer.*

4. John and Mary purchased 2 large chocolate kisses and a birthday card. The birthday card cost \$2. The total amount for the 3 items cost \$8. What was the price of a single chocolate kiss?

5. Eduardo purchased 3 boxes of stationary. He also purchased wrapping paper and thank you cards. If the wrapping paper and thank you cards together cost \$6, and the total he paid was \$42, find the cost of a box of stationary.

6. Mahandi purchased three rolls of peppermint candies and the local newspaper. The newspaper cost 50 cents. He paid \$2.75 in all. What did a roll of candies cost?

In Section 2.5 we will solve equations in which the variable appears on both sides of the equation. In Exercises 7 – 10, ***(a)*** *express the figure as an equation,* ***(b)*** *solve the equation, and* ***(c)*** *explain how you determined the solution. We will give step-by-step procedures for solving such equations in the next section, but you should start thinking about this now. Hint: To solve use the addition property to get all terms containing the variable on one side of the equation and all constant terms on the other side of the equation. Then use the multiplication property.*

7. + = + 3

8. + + 6 = + + + 4

9. + + 3 = + + + + 2

10. + + + 1 = + 4

2.5 Solving Linear Equations with the Variable on Both Sides of the Equation

Tape 4

1. Solve equations when the variable appears on both sides of the equal sign.
2. Identify identities and contradictions.

Variable on Both Sides of Equation

1 Below we show some figures that have been represented as equations. We have once again let the value of a kiss be represented by the letter x. In each of these equations the variable appears on both sides of the equation. We also give the solution to each equation. At this time you probably cannot determine the solutions. Do not worry about this. In this section we will teach you the procedure to solve equations of this type.

Figure	*Equation*	*Solution*
+ + 3 = + 6	$2x + 3 = x + 6$	3
+ 4 = + + 2	$x + 4 = 2x + 2$	2
+ + + 5 = + 20	$3x + 5 = x + 20$	$\frac{15}{2}$

Following is a general procedure, similar to the one outlined in Section 2.4, to solve linear equations with the variable on both sides of the equal sign.

To Solve Linear Equations with the Variable on Both Sides of the Equal Sign

1. Use the distributive property to remove parentheses.
2. Combine like terms on the same side of the equal sign.
3. Use the addition property to rewrite the equation with all terms containing the variable on one side of the equal sign and all terms not containing the variable on the other side of the equal sign. It may be necessary to use the addition property twice to accomplish this goal. You will eventually get an equation of the form $ax = b$.
4. Use the multiplication property to isolate the variable. This will give a solution of the form $x =$ some number.
5. Check the solution in the original equation.

The steps listed here are basically the same as the steps listed in the boxed procedure on page 105, except that in step 3 you may need to use the addition property more than once to obtain an equation of the form $ax = b$.

Remember that our goal in solving equations is to isolate the variable, that is, to get the variable alone on one side of the equation. To help you visualize the boxed procedure, consider the figure and corresponding equation that follow.

Figure	*Equation*
+ + + 4 = + 12	$3x + 4 = x + 12$

The equation $3x + 4 = x + 12$ contains no parentheses and no like terms on the same side of the equal sign. Therefore, we start with step 3, the addition property. We will use the addition property twice in order to obtain an equation where the variable appears on only one side of the equal sign. We begin by subtracting x from both sides of the equation to get all the terms containing the variable on the left side of the equation. This will give the following:

Figure	*Equation*	
+ + + 4 = + 12	$3x + 4 = x + 12$	
	$3x - x + 4 = x - x + 12$	Addition property
+ + 4 = 12	or $2x + 4 = 12$	

Notice that the variable, x, now appears on only one side of the equation. However, $+4$ still appears on the same side of the equal sign as the $2x$. We use the addition property a second time to get the term containing the variable by itself on one side of the equation. Subtracting 4 from both sides of the equation gives $2x = 8$, which is an equation of the form $ax = b$.

Figure	*Equation*	
⯁+⯁+ 4 = 12	$2x + 4 = 12$	
⯁+⯁ = 8	$2x + 4 - 4 = 12 - 4$ $2x = 8$	Addition property

Now that the $2x$ is by itself on one side of the equation we can use the multiplication property to isolate the variable and solve the equation for x. Divide both sides of the equation by 2 to isolate the variable and solve the equation.

$$2x = 8$$

$$\frac{\overset{1}{\cancel{2}}x}{\underset{1}{\cancel{2}}} = \frac{\overset{4}{\cancel{8}}}{\underset{1}{\cancel{2}}} \quad \text{Multiplication property}$$

⯁ = 4

$$x = 4$$

The solution to the equation is 4.

EXAMPLE 1 Solve the equation $4x + 6 = 2x + 4$.

Solution: Remember that our goal is always to get all terms with the variable on one side of the equal sign and all terms without the variable on the other side. The terms with the variable may be collected on either side of the equal sign. Many methods can be used to isolate the variable. We will illustrate two. In method 1, we will isolate the variable on the left side of the equation. In method 2, we will isolate the variable on the right side of the equation. In both methods, we will follow the steps given in the box on page 112. Since this equation does not contain parentheses and there are no like terms on the same side of the equal sign, we begin with step 3.

Method 1: Isolating the variable on the left

$$4x + 6 = 2x + 4$$

Step 3 $\quad 4x - 2x + 6 = 2x - 2x + 4 \quad$ **Subtract $2x$ from both sides of the equation.**

$$2x + 6 = 4$$

Step 3 $\quad 2x + 6 - 6 = 4 - 6 \quad$ **Subtract 6 from both sides of the equation.**

$$2x = -2$$

Step 4 $\quad \dfrac{2x}{2} = \dfrac{-2}{2} \quad$ **Divide both sides of the equation by 2.**

$$x = -1$$

Method 2: Isolating the variable on the right

$$4x + 6 = 2x + 4$$

Step 3 $$4x - 4x + 6 = 2x - 4x + 4$$ **Subtract 4x from both sides of the equation.**

$$6 = -2x + 4$$

Step 3 $$6 - 4 = -2x + 4 - 4$$ **Subtract 4 from both sides of the equation.**

$$2 = -2x$$

Step 4 $$\frac{2}{-2} = \frac{-2x}{-2}$$ **Divide both sides of the equation by −2.**

$$-1 = x$$

The same answer is obtained whether we isolate the variable on the left or right.

Step 5 *Check:*

$$4x + 6 = 2x + 4$$
$$4(-1) + 6 = 2(-1) + 4$$
$$-4 + 6 = -2 + 4$$
$$2 = 2 \quad \text{true}$$

EXAMPLE 2 Solve the equation $2x - 3 - 5x = 13 + 4x - 2$.

Solution: We will choose to collect the terms containing the variable on the right side of the equation. Since there are like terms *on the same side of the equal sign,* we will begin by combining these like terms.

Step 2 $$2x - 3 - 5x = 13 + 4x - 2$$

$$-3x - 3 = 4x + 11$$ **Like terms were combined.**

Step 3 $$-3x + 3x - 3 = 4x + 3x + 11$$ **Add 3x to both sides of the equation.**

$$-3 = 7x + 11$$

Step 3 $$-3 - 11 = 7x + 11 - 11$$ **Subtract 11 from both sides of the equation.**

$$-14 = 7x$$

Step 4 $$\frac{-14}{7} = \frac{7x}{7}$$ **Divide both sides of the equation by 7.**

$$-2 = x$$

Step 5 *Check:*

$$2x - 3 - 5x = 13 + 4x - 2$$
$$2(-2) - 3 - 5(-2) = 13 + 4(-2) - 2$$
$$-4 - 3 + 10 = 13 - 8 - 2$$
$$-7 + 10 = 5 - 2$$
$$3 = 3 \quad \text{true}$$

Since the check is true, the solution is -2.

The solution to Example 2 could be condensed as follows:

$2x - 3 - 5x = 13 + 4x - 2$	
$-3x - 3 = 4x + 11$	Like terms were combined.
$-3 = 7x + 11$	$3x$ was added to both sides of equation.
$-14 = 7x$	11 was subtracted from both sides of equation.
$-2 = x$	Both sides of equation were divided by 7.

We solved Example 2 by moving the terms containing the variable to the right side of the equation. Now rework the problem moving the terms containing the variable to the left side of the equation. You should obtain the same answer.

EXAMPLE 3 Solve the equation $5.74x + 5.42 = 2.24x - 9.28$.

Solution: We first notice that there are no like terms on the same side of the equal sign that can be combined. We will elect to collect the terms containing the variable on the left side of the equation.

$$5.74x + 5.42 = 2.24x - 9.28$$

Step 3 $5.74x - 2.24x + 5.42 = 2.24x - 2.24x - 9.28$ Subtract $2.24x$ from both sides of equation.

$$3.5x + 5.42 = -9.28$$

Step 3 $3.5x + 5.42 - 5.42 = -9.28 - 5.42$ Subtract 5.42 from both sides of equation.

$$3.5x = -14.7$$

Step 4 $\dfrac{3.5x}{3.5} = \dfrac{-14.7}{3.5}$ Divide both sides of equation by 3.5.

$$x = -4.2$$

EXAMPLE 4 Solve the equation $2(p + 3) = -3p + 10$.

Solution: $2(p + 3) = -3p + 10$

Step 1 $2p + 6 = -3p + 10$ Distributive property was used.

Step 3 $2p + 3p + 6 = -3p + 3p + 10$ Add $3p$ to both sides of equation.

$$5p + 6 = 10$$

Step 3 $5p + 6 - 6 = 10 - 6$ Subtract 6 from both sides of equation.

$$5p = 4$$

Step 4 $\dfrac{5p}{5} = \dfrac{4}{5}$ Divide both sides of equation by 5.

$$p = \frac{4}{5}$$

The solution to Example 4 could be condensed as follows:

$$2(p + 3) = -3p + 10$$

$2p + 6 = -3p + 10$ Distributive property was used.

$5p + 6 = 10$ $3p$ was added to both sides of equation.

$5p = 4$ 6 was subtracted from both sides of equation.

$p = \frac{4}{5}$ Both sides of equation were divided by 5.

Helpful Hint

After the distributive property was used in step 1, Example 4, we obtained the equation $2p + 6 = -3p + 10$. Then we had to decide whether to collect terms with the variable on the left or the right side of the equal sign. If we wish the sum of the terms containing a variable to be positive, we use the addition property to eliminate the variable, with the *smaller* numerical coefficient from one side of the equation. Since -3 is smaller than 2, we added $3p$ to both sides of the equation. This eliminated $-3p$ from the right side of the equation and resulted in the sum of the variable terms on the left side of the equation, $5p$, being positive.

EXAMPLE 5 Solve the equation $2(x - 5) + 3 = 3x + 9$.

Solution: $2(x - 5) + 3 = 3x + 9$

Step 1 $2x - 10 + 3 = 3x + 9$ Distributive property was used.

Step 2 $2x - 7 = 3x + 9$ Like terms were combined.

Step 3 $-7 = x + 9$ $2x$ was subtracted from both sides of equation.

Step 3 $-16 = x$ 9 was subtracted from both sides of equation.

EXAMPLE 6 Solve the equation $7 - 2x + 5x = -2(-3x + 4)$.

Solution: $7 - 2x + 5x = -2(-3x + 4)$

Step 1 $7 - 2x + 5x = 6x - 8$ Distributive property was used.

Step 2 $7 + 3x = 6x - 8$ Like terms were combined.

Step 3 $7 = 3x - 8$ $3x$ was subtracted from both sides of equation.

Step 3 $15 = 3x$ 8 was added to both sides of equation.

Step 4 $5 = x$ Both sides of equation were divided by 3.

The solution is 5.

Identities and Contradictions

2 Thus far all the equations we have solved have had a single value for a solution. Equations of this type are called **conditional equations,** for they are only true under specific conditions. Some equations, as in Example 7, are true for all values of x. Equations that are true for all values of x are called **identities.** A third type of equation, as in Example 8, has no solution and is called a **contradiction.**

EXAMPLE 7 Solve the equation $2x + 6 = 2(x + 3)$.

Solution:

$$2x + 6 = 2(x + 3)$$
$$2x + 6 = 2x + 6$$

Since the same expression appears on both sides of the equal sign, the statement is true for all values of x. If we continue to solve this equation further, we might obtain

$$2x = 2x \quad \text{6 was subtracted from both sides of equation.}$$
$$0 = 0 \quad 2x \text{ was subtracted from both sides of equation.}$$

Note: The solution process could have been stopped at $2x + 6 = 2x + 6$. Since one side is identical to the other side, the equation is true for all values of x. **Therefore, the solution to this equation is all real numbers.**

COMMON STUDENT ERROR Some students confuse combining like terms with using the addition property. Remember that *when combining terms you work on only one side of the equal sign at a time,* as in

$$3x + 4 - x = 4x - 8$$
$$2x + 4 = 4x - 8 \quad \text{The } 3x \text{ and } -x \text{ were combined.}$$

When using the addition property, you add (or subtract) the same quantity to (from) ***both sides of the equation,*** *as shown below.*

Correct

$$2x + 4 = 4x - 8$$
$$2x - 2x + 4 = 4x - 2x - 8 \quad 2x \text{ was subtracted from } \textit{both sides of} \text{ } \textit{equation.}$$
$$4 = 2x - 8$$
$$4 + 8 = 2x - 8 + 8 \quad \text{8 was added to } \textit{both sides of equation.}$$
$$12 = 2x$$
$$x = 6$$

Incorrect

$$3x + 4 - x = 4x - 8$$
$$3x + x + 4 - x + x = 4x - 8$$

Wrong use of the addition property; note x was not added to *both* sides of the equation.

Ordinarily, when solving an equation, combining like terms is done before using the addition property.

EXAMPLE 8 Solve the equation $-3x + 4 + 5x = 4x - 2x + 5$.

Solution:

$$-3x + 4 + 5x = 4x - 2x + 5$$
$$2x + 4 = 2x + 5 \quad \text{Like terms were combined.}$$
$$2x - 2x + 4 = 2x - 2x + 5 \quad \text{Subtract } 2x \text{ from both sides of equation.}$$
$$4 = 5 \quad \text{false}$$

When solving an equation, if you obtain an obviously false statement, as in this example, the equation has no solution. No value of x will make the equation a true statement. **Therefore, when giving the answer to this problem, you should use the words *no solution.*** An answer left blank may be marked wrong.

> *Helpful Hint*
>
> Some students start solving equations correctly but do not complete the solution. Sometimes they are not sure that what they are doing is correct and they give up for lack of confidence. You must have confidence in yourself. As long as you follow the procedure on page 112 you should obtain the correct solution, even if it takes quite a few steps. Remember two important things: (1) Our goal is to isolate the variable, and (2) whatever you do to one side of the equation you must also do to the other side. That is, you must treat both sides of the equation equally.

Exercise Set 2.5

(a) *Represent the figure as an equation in the variable* x *and* **(b)** *solve the equation.*

1. 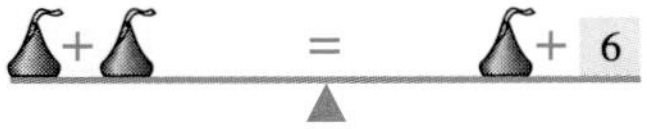 + = + 6

2. + 6 = + + 2

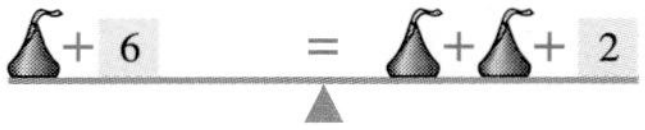

3. 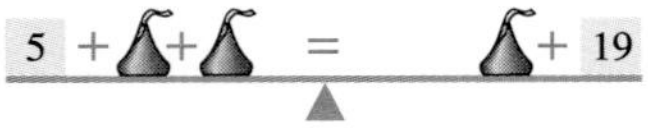 5 + + = + 19

4. 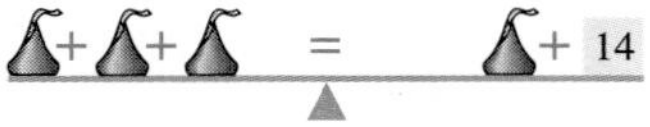 + + = + 14

5. 5 + = + + 5

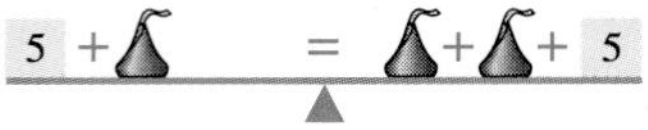

6. 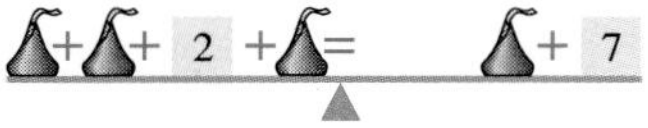 + + 2 + = + 7

7. 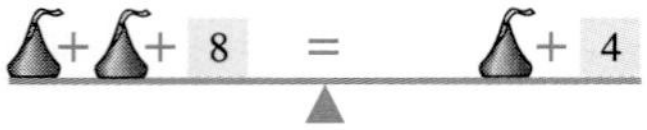 + + 8 = + 4

8. + + + + 19 = + 5 +

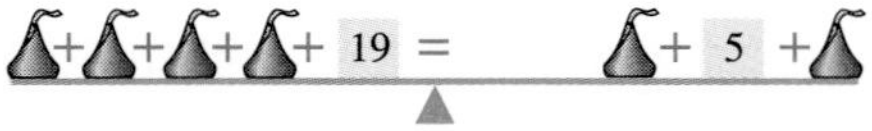

Solve each equation. You may wish to use a calculator to solve the equations containing decimal numbers.

9. $4x = 3x + 5$

10. $x + 6 = 2x - 4$

11. $-4x + 10 = 6x$

12. $6x = 4x + 8$

13. $5x + 3 = 6$

14. $-6x = 2x + 16$

15. $15 - 3x = 4x - 2x$

16. $8 - 6x = 4x + 10$

17. $2x - 4 = 3x - 6$

18. $-5x = -4x + 9$

19. $3 - 2y = 9 - 8y$

20. $124.8 - 9.4x = 4.8x + 32.5$

21. $4 - 0.6x = 2.4x - 8.48$

22. $8 + y = 2y - 6 + y$

23. $5x = 2(x + 6)$

24. $8x - 4 = 3(x - 2)$

25. $x - 25 = 12x + 9 + 3x$

26. $5y + 6 = 2y + 3 - y$

27. $2(x + 2) = 4x + 1 - 2x$

28. $4r = 10 - 2(r - 4)$

29. $-(w + 2) = -6w + 32$

30. $15(4 - x) = 5(10 + 2x)$

31. $4 - (2x + 5) = 6x + 31$

32. $4(2x - 3) = -2(3x + 16)$

33. $0.1(x + 10) = 0.3x - 4$

34. $3y - 6y + 2 = 8y + 6 - 5y$

35. $2(x + 4) = 4x + 3 - 2x + 5$

36. $5(2.9x - 3) = 2(x + 4)$

37. $9(-y + 3) = -6y + 15 - 3y + 12$

38. $-4(-y + 3) = 12y + 8 - 2y$

39. $-(3 - p) = -(2p + 3)$

40. $12 - 2x - 3(x + 2) = 4x + 6 - x$

41. $-(x + 4) + 5 = 4x + 1 - 5x$

42. $19x + 3(4x + 9) = -6x - 38$

43. $35(2x + 12) = 7(x - 4) + 3x$

44. $10(x - 10) + 5 = 5(2x - 20)$

45. $0.4(x + 0.7) = 0.6(x - 4.2)$

46. $3(x - 4) = 2(x - 8) + 5x$

47. $-(x - 5) + 2 = 3(4 - x) + 5x$

48. $1.2(6x - 8) = 2.4(x - 5)$

49. $2(x - 6) + 3(x + 1) = 4x + 3$

50. $-2(-3x + 5) + 6 = 4(x - 2)$

51. $5 + 2x = 6(x + 1) - 5(x - 3)$

52. $4 - (6x + 3) = -(-2x + 3)$

53. $5 - (x - 5) = 2(x + 3) - 6(x + 1)$

54. $12 - 6x + 3(2x + 3) = 2x + 5$

55. **(a)** In your own words, write the general procedure for solving an equation that does not contain fractions where the variable appears on both sides of the equation. **(b)** Refer to page112 to see if you omitted any steps.

56. When solving an equation, how will you know if the equation is an identity?

57. When solving an equation, how will you know if the equation has no real solution?

58. **(a)** Explain, in a step-by-step manner, how to solve the equation $4(x + 3) = 6(x - 5)$.

(b) Solve the equation by following the steps listed in part (a).

59. **(a)** Explain, in a step-by-step manner, how to solve the equation $4x + 3(x + 2) = 5x - 10$.

(b) Solve the equation by following the steps listed in part (a).

CUMULATIVE REVIEW EXERCISES

3] **60.** Evaluate $\left(\frac{2}{3}\right)^5$ on your calculator.

1] **61.** Explain the difference between a factor and a term.

[2.4] **62.** Simplify $2(x - 3) + 4x - (4 - x)$.

63. Solve $2(x - 3) + 4x - (4 - x) = 0$.

64. Solve $(x + 4) - (4x - 3) = 16$.

Group Activity/ Challenge Problems

1. Solve $-2(x + 3) + 5x = -3(5 - 2x) + 3(x + 2) + 6x$.

2. Solve $4(2x - 3) - (x + 7) - 4x + 6 = 5(x - 2) - 3x + 7(2x + 2)$.

3. Solve $4 - [5 - 3(x + 2)] = x - 3$.

In the next chapter we will be discussing procedures for writing application problems as equations. Can you write equations for Exercises 4–6? For each exercise ***(a)*** *make a sketch using a balance like those in Exercises 1–8 that represents the problem;* ***(b)*** *represent the sketch as an equation in the variable x.* ***(c)*** *solve the equation and* ***(d)*** *check your answer to make sure that it makes sense.*

4. Mary Kay purchased 2 large chocolate kisses. The total cost of the two kisses was equal to the cost of 1 kiss plus \$6. Find the cost of one chocolate kiss.

5. Three identical boxes are weighed. Their total weight is the same as (or equals) the weight of one of the boxes plus 20 pounds. Find the weight of a box.

6. Isaac purchased 4 gallons of skim milk. The price of the 4 gallons of milk is the same as the price of 2 gallons of milk plus some other groceries that cost \$5.20. What is the price of a gallon of skim milk?

2.6 Ratios and Proportions

Tape 4

1. Understand ratios.
2. Solve proportions using cross multiplication.
3. Solve practical application problems.
4. Use proportions to change units.
5. Use proportions in geometric problems.

Ratios

1 A **ratio** is a quotient of two quantities. Ratios provide a way to compare two numbers or quantities. The ratio of the number a to the number b may be written

$$a \text{ to } b, \qquad a:b, \qquad \text{or} \qquad \frac{a}{b}$$

where *a* and *b* are called the **terms** of the ratio.

EXAMPLE 1 An algebra class consists of 11 males and 15 females.

(a) Find the ratio of males to females.

(b) Find the ratio of females to the entire class.

Solution: **(a)** 11:15 **(b)** 15:26

In Example 1, part (a) could also have been written $\frac{11}{15}$ or "11 to 15." Part (b) could also have been written $\frac{15}{26}$ or "15 to 26."

EXAMPLE 2 There are two types of cholesterol, low-density lipoprotein, (LDL—considered the harmful type of cholesterol) and high-density lipoprotein (HDL—considered the healthful type of cholesterol). Some doctors recommend that the ratio of low- to high-density cholesterol be less than or equal to 4:1. Mr. Kane's cholesterol test showed that his low-density cholesterol measured 167 milligrams per deciliter, and his high-density cholesterol measured 40 milligrams per deciliter. Is Mr. Kane's ratio of low- to high-density cholesterol less than or equal to the recommended 4:1 ratio?

Solution: The ratio of low- to high-density cholesterol is $\frac{167}{40}$. If we divide 167 by 40, we obtain 4.175. Thus, Mr. Kane's ratio is equivalent to 4.175:1. Therefore, his ratio is not less than or equal to the desired 4:1 ratio.

EXAMPLE 3 Find the ratio of 8 feet to 20 yards.

Solution: To express this as a ratio, both quantities must be in the same units. Since 1 yard equals 3 feet, 20 yards equals 60 feet. Thus, the ratio is $\frac{8}{60}$. The ratio in lowest terms is $\frac{2}{15}$ (or 2:15).

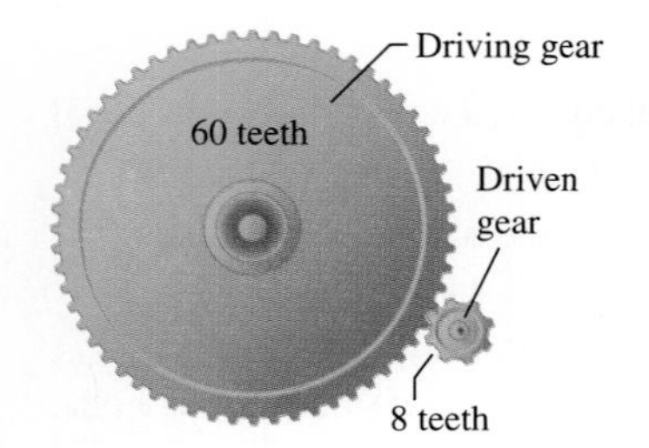

FIGURE 2.10

EXAMPLE 4 The *gear ratio* of two gears is defined as

$$\text{gear ratio} = \frac{\text{number of teeth on the driving gear}}{\text{number of teeth on the driven gear}}$$

Find the gear ratio of the gears shown in Figure 2.10.

Solution: Gear ratio $= \dfrac{60}{8} = \dfrac{15}{2}$. Thus, the gear ratio is 15:2. Gear ratios are generally given as some quantity to 1. If we divide both parts of the ratio by the second number, we will obtain a ratio of some number to 1. Dividing both 15 and 2 by 2 gives a gear ratio of 7.5:1. (A typical first gear ratio on a passenger car may be 3.545:1).

Solve Proportions

2 A **proportion** is a special type of equation. It is a statement of equality between two ratios. One way of denoting a proportion is $a:b = c:d$, which is read "a is to b as c is to d." In this text we write proportions as

$$\frac{a}{b} = \frac{c}{d}$$

The a and d are referred to as the **extremes,** and the b and c are referred to as the **means** of the proportion. One method that can be used in evaluating proportions is **cross-multiplication:**

Cross-Multiplication

$$\text{If } \frac{a}{b} = \frac{c}{d} \text{ then } ad = bc.$$

Note that the product of the means is equal to the product of the extremes.

If any three of the four quantities of a proportion are known, the fourth quantity can easily be found.

EXAMPLE 5 Solve for x by cross-multiplying $\dfrac{x}{3} = \dfrac{25}{15}$.

Solution:

$$\frac{x}{3} = \frac{25}{15}$$
$$x \cdot 15 = 3 \cdot 25$$
$$15x = 75$$
$$x = \frac{75}{15} = 5$$

Check:

$$\frac{x}{3} = \frac{25}{15}$$
$$\frac{5}{3} = \frac{25}{15}$$
$$\frac{5}{3} = \frac{5}{3} \quad \text{true}$$

EXAMPLE 6 Solve for x by cross-multiplying $\dfrac{-8}{3} = \dfrac{64}{x}$.

Solution:

$$\frac{-8}{3} = \frac{64}{x}$$
$$-8 \cdot x = 3 \cdot 64$$
$$-8x = 192$$
$$\frac{-8x}{-8} = \frac{192}{-8}$$
$$x = -24$$

Check:

$$\frac{-8}{3} = \frac{64}{x}$$
$$\frac{-8}{3} = \frac{\overset{8}{\cancel{64}}}{\underset{3}{\cancel{-24}}}$$
$$\frac{-8}{3} = \frac{8}{-3}$$
$$\frac{-8}{3} = \frac{-8}{3} \quad \text{true}$$

Applications

3 Often, practical applications can be solved using proportions. To solve such problems, use the following procedure.

To Solve Problems Using Proportions

1. Represent the unknown quantity by a variable (a letter).
2. Set up the proportion by listing the given ratio on the left side of the equal sign, and the unknown and the other given quantity on the right side of the equal sign. When setting up the right side of the proportion, the same respective quantities should occupy the same respective positions on the left and the right. For example, an acceptable proportion might be

$$\text{Given ratio }\left\{\frac{\text{miles}}{\text{hour}}\right. = \frac{\text{miles}}{\text{hour}}$$

3. Once the proportion is correctly written, drop the units and cross-multiply.
4. Solve the resulting equation.
5. Answer the questions asked.

Note that the two ratios* must have the same units. For example, if one ratio is given in miles/hour and the second ratio is given in feet/hour, one of the ratios must be changed before setting up the proportion.

EXAMPLE 7 A 30-pound bag of fertilizer will cover an area of 2500 square feet.

(a) How many pounds are needed to cover an area of 16,000 square feet?

(b) How many bags of fertilizer are needed?

Solution: **(a)** The given ratio is 30 pounds per 2500 square feet. The unknown quantity is the number of pounds necessary to cover 16,000 square feet.

Step 1 Let x = number of pounds.

Step 2 Given ratio $\left\{\frac{30\text{ pounds}}{2500\text{ square feet}}\right. = \frac{x\text{ pounds}}{16{,}000\text{ square feet}}$ ← Unknown (numerator), ← Given quantity (denominator)

Note how the pounds and the area are given in the same relative positions.

Step 3

$$\frac{30}{2500} = \frac{x}{16{,}000}$$

$$30(16{,}000) = 2500x$$

*Strictly speaking, a quotient of two quantities with different units, such as $\frac{6\textit{ miles}}{1\textit{ hour}}$, is called a *rate*. However, few books make the distinction between ratios and rates when discussing proportions.

Step 4

$$480{,}000 = 2500x$$
$$\frac{480{,}000}{2500} = x$$
$$192 = x$$

Step 5 One hundred ninety-two pounds of fertilizer are needed.

(b) Since each bag weighs 30 pounds, the number of bags is found by division.

$$192 \div 30 = 6.4 \text{ bags}$$

The number of bags needed is therefore 7, since you must purchase whole bags.

EXAMPLE 8 In Washington County the property tax rate is $8.065 per $1000 of assessed value. If a house and its property have been assessed at $124,000, find the tax the owner will have to pay.

Solution: The unknown quantity is the tax the property owner must pay. Let us call this unknown x.

$$\frac{\text{tax}}{\text{assesed value}} = \frac{\text{tax}}{\text{assessed value}}$$

Given tax rate

$$\frac{8.065}{1000} = \frac{x}{124{,}000}$$
$$(8.065)(124{,}000) = 1000x$$
$$1{,}000{,}060 = 1000x$$
$$\$1000.06 = x$$

The owner will have to pay $1000.06 tax.

EXAMPLE 9 A doctor asks a nurse to give a patient 250 milligrams of the drug simethicone. The drug is available only in a solution whose concentration is 40 milligrams of simethicone per 0.6 milliliter of solution. How many milliliters of solution should the nurse give the patient?

Solution: We can set up the proportion using the medication on hand as the given ratio and the number of milliliters needed to be given as the unknown.

Given ratio (medication on hand)

$$\frac{40 \text{ milligrams}}{0.6 \text{ milliliter}} = \frac{250 \text{ milligrams}}{x \text{ milliliters}}$$

← Desired medication (numerator); ← Unknown (denominator)

Now solve for x.

$$\frac{40}{0.6} = \frac{250}{x}$$
$$40x = 0.6(250)$$
$$40x = 150$$
$$x = \frac{150}{40} = 3.75$$

Thus, the nurse should administer 3.75 milliliters of the simethicone solution.

COMMON STUDENT ERROR When you set up a proportion the same units should not be multiplied by themselves during cross-multiplication.

Correct	*Incorrect*
$\frac{\text{miles}}{\text{hour}} = \frac{\text{miles}}{\text{hour}}$	~~$\frac{\text{miles}}{\text{hour}} = \frac{\text{hour}}{\text{miles}}$~~

Conversions

4 Proportions can also be used to convert from one quantity to another. For example, you can use a proportion to convert a measurement in feet to a measurement in meters, or to convert from U.S. dollars to Mexican pesos. The following examples illustrate converting units.

EXAMPLE 10 Convert 18.36 inches to feet.

Solution: We know that 1 foot is 12 inches. We use this known fact in one ratio of our proportion. In the second ratio we set the quantities with the same units in the same respective positions.

$$\text{Known ratio}\left\{\frac{1 \text{ foot}}{12 \text{ inches}}\right. = \frac{x \text{ feet}}{18.36 \text{ inches}}$$

Since we are given 18.36 inches, we place this quantity in the denominator of the second ratio. The unknown quantity is the number of feet, which we will call x. Note that both numerators contain the same units and both denominators contain the same units. Now drop the units and solve for x by cross-multiplying.

$$\frac{1}{12} = \frac{x}{18.36}$$
$$1(18.36) = 12x$$
$$18.36 = 12x$$
$$\frac{18.36}{12} = \frac{12x}{12}$$
$$1.53 = x$$

Thus, 18.36 inches equals 1.53 feet.

EXAMPLE 11 One kilogram is equal to 2.2 pounds.

(a) Find the weight in pounds of a poodle that weighs 7.48 kilograms.

(b) Mary Jo weighs 121 pounds. How many kilograms does she weigh?

Solution: **(a)** We use the fact that 1 kilogram = 2.2 pounds for our known ratio. The unknown quantity is the number of pounds. We will call this quantity x.

$$\text{Known ratio}\left\{\frac{1 \text{ kilogram}}{2.2 \text{ pounds}}\right. = \frac{7.48 \text{ kilograms}}{x \text{ pounds}}$$
$$\frac{1}{2.2} = \frac{7.48}{x}$$
$$1x = (2.2)(7.48)$$
$$x = 16.456$$

Thus, the poodle weighs 16.456 pounds.

(b) The unknown quantity is the number of kilograms. We will call the unknown quantity x.

$$\frac{1 \text{ kilogram}}{2.2 \text{ pounds}} = \frac{x \text{ kilograms}}{121 \text{ pounds}}$$
$$\frac{1}{2.2} = \frac{x}{121}$$
$$1(121) = 2.2x$$
$$121 = 2.2x$$
$$\frac{121}{2.2} = x$$
$$55 = x$$

Thus, Mary Jo weighs 55 kilograms.

EXAMPLE 12 Marisa exchanged 15 U.S. dollars for 46.75 Mexican pesos at a bank in Cancun, Mexico.

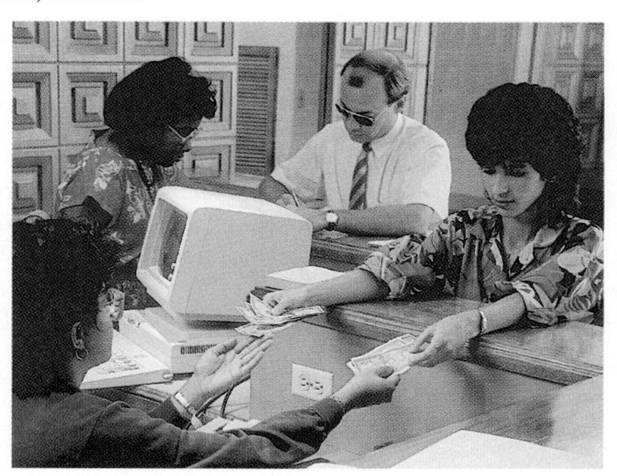

(a) What is the conversion rate per U.S. dollar (that is, what is 1 U.S. dollar worth in pesos)?

(b) At the straw market in downtown Cancun, Marisa purchased a handmade Mayan calendar for 385 pesos. What is the cost of the calendar in U.S. dollars?

Solution: **(a)** We know that 15 U.S. dollars equals 46.75 Mexican pesos. We use this fact in our proportion. The unknown quantity is the number of pesos equal to 1 U.S. dollar.

Known ratio $\left\{\frac{15 \text{ dollars}}{46.75 \text{ pesos}}\right. = \frac{1 \text{ dollar}}{x \text{ pesos}}$

$$15x = 46.75$$
$$x = \frac{46.75}{15} = 3.1167$$

Thus, \$1 can be converted to 3.1167 pesos.

(b) We need to convert 385 pesos to U.S. dollars. We will call the amount of U.S. dollars x.

$$\frac{15 \text{ dollars}}{46.75 \text{ pesos}} = \frac{x \text{ dollars}}{385 \text{ pesos}}$$
$$(15)(385) = 46.75x$$
$$5775 = 46.75x$$
$$\frac{5775}{46.75} = x$$
$$123.53 = x$$

Thus, the cost of the Mayan calendar is 123.53 U.S. dollars or \$123.53.

Helpful Hint

Some of the problems we have just worked using proportions could have been done without using proportions. However, when working problems of this type, students often have difficulty in deciding whether to multiply or divide to obtain the correct answer. By setting up a proportion, you may be better able to understand the problem and have more success at obtaining the correct answer.

Similar Figures

5 Proportions can also be used to solve problems in geometry and trigonometry. The following examples illustrate how proportions may be used to solve problems involving **similar figures.** Two figures are said to be similar when their corresponding angles are equal and their corresponding sides are in proportion.

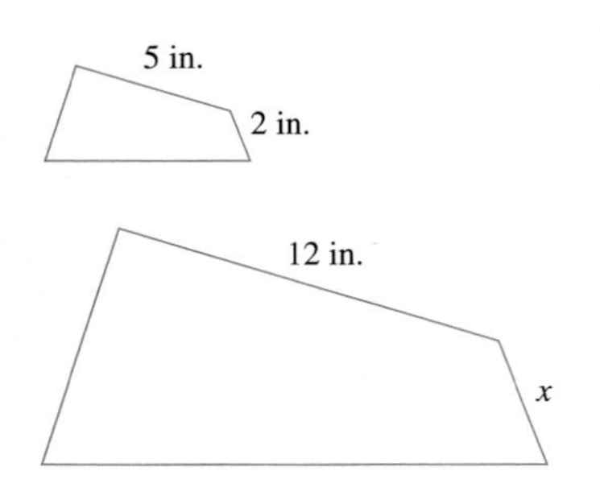

EXAMPLE 13 The figures to the left are similar. Find the length of the side indicated by the x.

Solution: We set up a proportion of corresponding sides to find the length of side x.

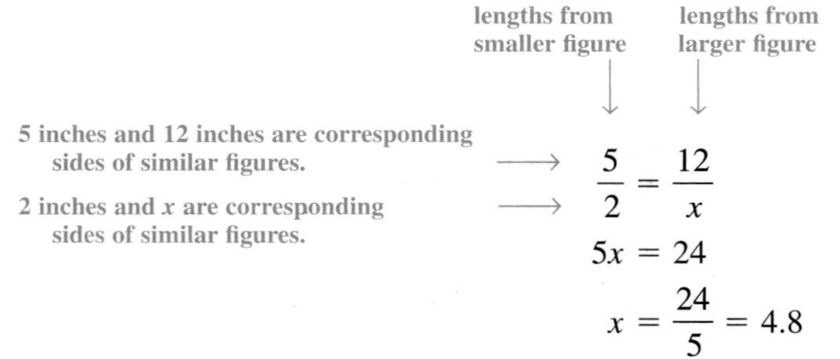

$$\frac{5}{2} = \frac{12}{x}$$
$$5x = 24$$
$$x = \frac{24}{5} = 4.8$$

Thus, the side is 4.8 inches in length.

Note in Example 13 that the proportion could have also been set up as

$$\frac{5}{12} = \frac{2}{x}$$

because one pair of corresponding sides is in the numerator and another pair is in the denominator.

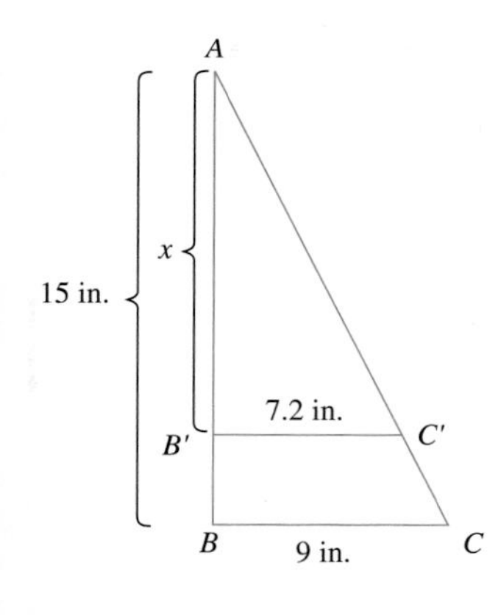

EXAMPLE 14 Triangles ABC and $AB'C'$ are similar triangles. Use a proportion to find the length of side AB'.

Solution: We set up a proportion of corresponding sides to find the length of side AB'. We will let x represent the length of side AB'. One proportion we can use is

$$\frac{\text{length of } AB}{\text{length of } BC} = \frac{\text{length of } AB'}{\text{length of } B'C'}$$

Now insert the proper values and solve for the variable x.

$$\frac{15}{9} = \frac{x}{7.2}$$

$$(15)(7.2) = 9x$$

$$108 = 9x$$

$$12 = x$$

Thus, the length of side AB' is 12 inches.

Exercise Set 2.6

The results of an English examination are 5 A's, 6 B's, 8 C's, 4 D's, and 2 F's. Write the following ratios.

1. A's to C's.
2. A's to total grades.
3. D's to F's.
4. Grades better than C to total grades.
5. Total grades to D's.
6. Grades better than C to grades less than C.

Determine the following ratios. Write each ratio in lowest terms.

7. 5 feet to 3 feet.
8. 60 dollars to 80 dollars.
9. 20 hours to 60 hours.
10. 100 people to 80 people.
11. 4 hours to 40 minutes.
12. 6 feet to 4 yards.
13. 26 ounces to 4 pounds.
14. 7 dimes to 12 nickels.

Find the gear ratio. See Example 4.

15. Driving gear, 40 teeth; driven gear, 5 teeth.
16. Driven gear, 8 teeth; driving gear, 30 teeth.

In Exercises 17–20 **(a)** *determine the indicated ratio, and* **(b)** *write the ratio as some quantity to 1*

17. In 1970 the world population was 3.6 billion people. In 1990 the world population had increased to 5.3 billion. What is the ratio of the 1990 world population to the 1970 world population?
18. In 1970 in the United States, 72,700 metric tons of aluminum was used for soft-drink and beer containers. In 1990 this amount had increased to 1,251,900 metric tons. Find the ratio of the amount of aluminum used for beer and soft-drink containers in 1990 to the amount used in 1970.
19. The Department of Agriculture has estimated that an acre of land in the United States can produce about 20,000 pounds of potatoes. The same acre of land, if used to raise cattle feed, can produce about 165 pounds of beef. Find the ratio of pounds of potatoes to pounds of beef that can be produced on an acre of land.
20. The U.S. population is about 256,560,000. About 635,000 of these persons are medical doctors. The population of China is 1,169,620,000, including about 1,810,000 medical doctors. Find the ratio of persons to doctors for the United States and for China.

Solve for the variable by cross-multiplying.

21. $\frac{4}{x} = \frac{5}{20}$ **22.** $\frac{x}{4} = \frac{12}{48}$ **23.** $\frac{5}{3} = \frac{75}{x}$ **24.** $\frac{x}{32} = \frac{-5}{4}$

25. $\frac{90}{x} = \frac{-9}{10}$ **26.** $\frac{-3}{8} = \frac{x}{40}$ **27.** $\frac{1}{9} = \frac{x}{45}$ **28.** $\frac{y}{6} = \frac{7}{42}$

29. $\frac{3}{z} = \frac{2}{-20}$ **30.** $\frac{3}{12} = \frac{-1.4}{z}$ **31.** $\frac{15}{20} = \frac{x}{8}$ **32.** $\frac{12}{3} = \frac{x}{-100}$

Write a proportion that can be used to solve the problem, then solve the problem. Use a calculator where appropriate.

33. A car can travel 32 miles on 1 gallon of gasoline. How far can it travel on 12 gallons of gasoline?

34. A quality control worker can check 12 parts in 2.5 minutes. How long will it take her to check 60 parts?

35. A gallon of paint covers 825 square feet. How much paint is needed to cover a house with a surface area of 5775 square feet?

36. If 100 feet of wire has an electrical resistance of 7.3 ohms, find the electrical resistance of 40 feet of wire.

37. A blueprint of a shopping mall is in the scale of 1:150. Thus 1 foot on a blueprint represents 150 feet of actual length. One part of the mall is to be 190 feet long. How long will it appear on the blueprint?

38. The property tax in the town of Plainview, Texas, is \$8.235 per \$1000 of assessed value. If the Litton's house is assessed at \$122,000, how much property tax will they owe?

39. A photograph shows a boy standing next to a tall cactus. If the boy, who is actually 48 inches tall, measures 0.6 inch in the photograph, how tall is the cactus that measures 3.25 inches in the photo?

See Exercise 39.

40. If a 40-pound bag of fertilizer covers 5000 square feet, how many pounds of fertilizer are needed to cover an area of 26,000 square feet?

41. The instructions on a bottle of liquid insecticide say "use 3 teaspoons of insecticide per gallon of water." If your sprayer has an 8-gallon capacity, how much insecticide should be used to fill the sprayer?

42. A recipe for McGillicutty stew calls for $4\frac{1}{2}$ pounds of beef. If the recipe is for 20 servings, how much beef is needed to make 12 servings?

43. Every ton of recycled paper saves approximately 17 trees. (It also saves dumping cost, landfill space, about 7000 gallons of water, and 4100 kilowatt-hours of electricity that would be used in new paper products. Furthermore, to collect and recycle paper provides five times as many jobs as to harvest virgin timber.) If your college recycles 20 tons of paper in a year, how many trees has it saved?

44. A recipe for pancake mix calls for two eggs for each 6 cups of pancake mix. The Mesa County Fire Department is planning a Sunday brunch for the community. How many eggs will they use if they plan to use 120 cups of the pancake mix?

45. A nurse must administer 220 micrograms of atropine sulfate. The drug is available in solution form. The concentration of the atropine sulfate solution is 400 micrograms per milliliter. How many milliliters should be given?

46. A doctor asks a nurse to administer 0.7 gram of meprobamate per square meter of body surface. The patient's body surface is 0.6 square meter. How much meprobamate should be given?

47. While on fast forward, the counter of your VCR goes from 0 to 250 in 30 seconds. Your videocassette tape contains two movies. The second movie starts at 800 on the VCR counter. If you are at the beginning of the tape, approximately how long will you keep the VCR on fast forward to reach the beginning of the second movie?

48. Mary read 40 pages of a novel in 30 minutes. **(a)** If she continues reading at the same rate, how long will it take her to read the entire 760-page book? **(b)** How long will it take her to finish the book from page 31?

49. In the United States in 1990, the birth rate was 16 per one thousand people. In the United States in 1990, there were approximately 4,179,200 births. What was the U.S. population in 1990?

50. It is estimated that each year 1 in every 15,000 (1:15,000) people are born with a genetic disorder called Prader–Willi syndrome. If there were approximately 4,179,200 births in the United States in 1990, approximately how many children were born with Prader–Willi syndrome?

Solve using a proportion. Round your answers to two decimal places.

51. Convert 57 inches to feet.

52. Convert 17,952 feet to miles (5280 feet = 1 mile).

53. Convert 26.1 square feet to square yards (9 square feet = 1 square yard).

54. Convert 146.4 ounces to pounds.

55. One inch equals 2.54 centimeters. Find the length of a book in inches if it measures 26.67 centimeters.

56. One liter equals approximately 1.06 quarts. Find the volume in quarts of a 5-liter container.

57. One mile equals approximately 1.6 kilometers. Find the distance in miles of a 25-kilometer kangaroo crossing.

58. One mile equals approximately 1.6 kilometers. Find the distance in kilometers from San Diego, California, to San Francisco, California, a distance of 520 miles.

59. If gold is selling for \$400 per 480 grains (a troy ounce), what is the cost per grain?

60. In chemistry we learn that 100 torr (a unit of measurement) equals 0.13 atmosphere. Find the number of torr in 0.39 atmosphere.

61. In a statistics course, we find that for one particular set of scores 16 points equals 3.2 standard deviations. How many points equals 1 standard deviation?

62. When Fong visited the United States from Canada, he exchanged 10 Canadian dollars for 7.40 U.S. dollars. If he exchanges his remaining 2000 Canadian dollars for U.S. dollars, how much more will he receive?

63. Antonio, who is visiting the United States from Italy, wishes to obtain U.S. currency. If one Italian lira can be converted to 0.00059 U.S. dollar, how many lire will be need to obtain 1200 U.S. dollars?

64. Ms. Johnson spent an evening in a hotel in London, England. When she checked out, she was charged 90 pounds. What was the U.S. dollar equivalence of her hotel bill if 1 English pound could be converted to 1.64 U.S. dollars?

The figures below are similar. For each pair, find the length of the side indicated with an x.

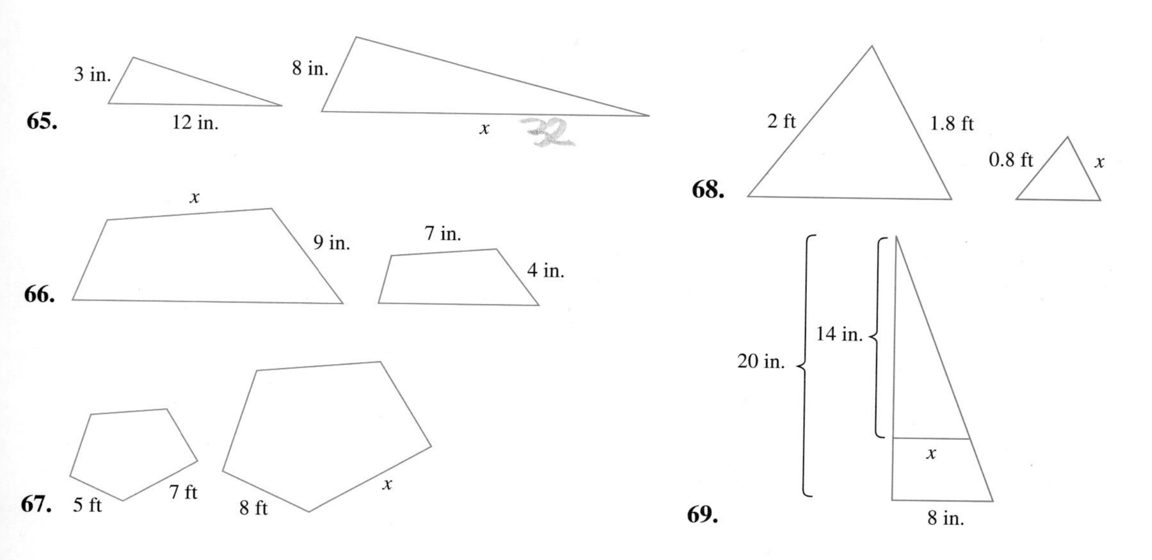

70.

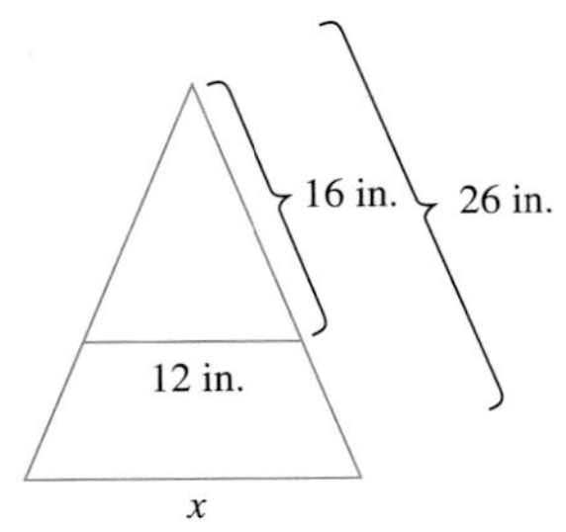

71. Mrs. Sanchez's low-density cholesterol level is 127 milligrams per deciliter (mg/dL). Her high-density cholesterol level is 60 mg/dL. Is Mrs. Sanchez's ratio of low-density to high-density cholesterol level less than or equal to the 4:1 recommended level? See Example 2.

72. **(a)** Another ratio used by some doctors when measuring cholesterol level is the ratio of total cholesterol to high-density cholesterol.* Is this ratio increased or decreased if the total cholesterol remains the same but the high-density level is increased? Explain.

(b) Doctors recommend that the ratio of total cholesterol to high-density cholesterol be less than or equal to 4.5:1. If Mike's total cholesterol is 220 mg/dL and his high-density cholesterol is 50 mg/dL, is his ratio less than or equal to the 4.5:1? Explain.

73. **(a)** Find the ratio of your height to your arm span when your arms are extended horizontally outward. You will need help in getting these measurements.

(b) If a box were to be drawn about your body with your arms extended, would the box be a square or a rectangle? If a rectangle, would the larger length be your arm span or your height measurement? Explain.

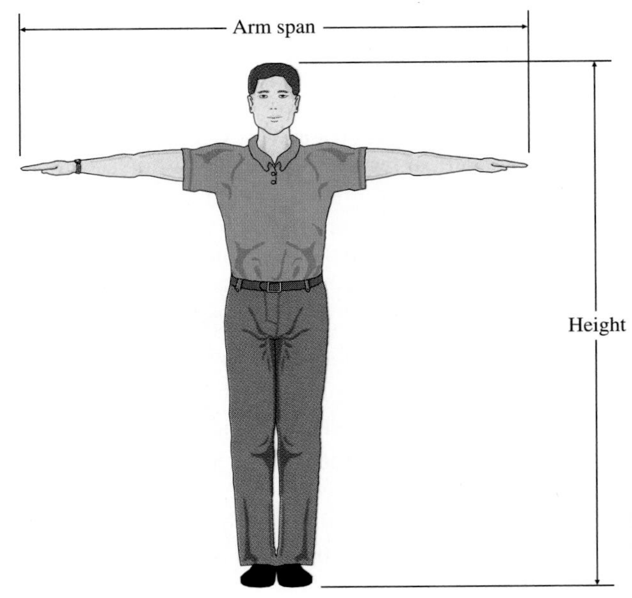

See Exercise 7.3

74. A special ratio in mathematics is called the golden ratio. Do research in a history of mathematics book, or another book recommended by your professor, and explain what the golden ratio is and why it is important.

75. As you have learned, proportions can be used to solve a wide variety of problems. What information is needed for a problem to be set up and solved using a proportion?

CUMULATIVE REVIEW EXERCISES

[1.10] *Name the properties illustrated.*

76. $x + 3 = 3 + x$

77. $3(xy) = (3x)y$

78. $2(x - 3) = 2x - 6$

[2.5] **79.** Solve $-(2x + 6) = 2(3x - 6)$.

* Total cholesterol includes both low- and high-density cholesterol, plus other types of cholesterol.

Group Activity/ Challenge Problems

1. A GE Soft White A-19 incandescent bulb has an average life of about 750 hours. A fluorescent bulb, the Phillips SL 18, produces equivalent light. Its average life is 10,000 hours. **(a)** Express the ratio of the life of the incandescent bulb to the life of the fluorescent bulb. **(b)** Express this ratio as some number to 1.
2. The recipe for the filling for an apple pie calls for:

12 cups sliced apples	$\frac{1}{4}$ teaspoon salt
$\frac{1}{2}$ cup flour	2 tablespoons butter or margarine
1 teaspoon nutmeg	$1\frac{1}{2}$ cups sugar
1 teaspoon cinnamon	

   Determine the amount of each of the other ingredients that should be used if only 8 cups of apples are available.
3. Insulin comes in 10-cubic-centimeter (cc) vials labeled in the number of units of insulin per cubic centimeter. Thus a vial labeled U40 means there are 40 units of insulin per cubic centimeter of fluid. If a patient needs 25 units of insulin, how many cubic centimeters of fluid should be drawn up into a syringe from the U40 vial?
4. In 1938, about 12,000 nesting pairs of wood storks lived in the Everglades. By 1988, 50 years later, their numbers had decreased to about 1200 pairs, partly because of the loss of peripheral wetlands. If the decrease in population continues at the present rate, in what year would the wood storks become extinct in the Everglades?

5. In 1992, according to the world population data sheet, the world population was 5420 million people. The world birth rate that year was 26 per 1000 people and the world death rate was 9 per 1000 people. **(a)** Find the number of births worldwide in 1992. **(b)** Find the number of deaths worldwide in 1992. **(c)** Find the world population increase in 1992.
6. An important concept that we will discuss in Chapter 7 is "slope." The slope of a straight line may be defined as *a ratio* of the vertical change to the horizontal change between any two points on the line. Determine the slope of the line below.

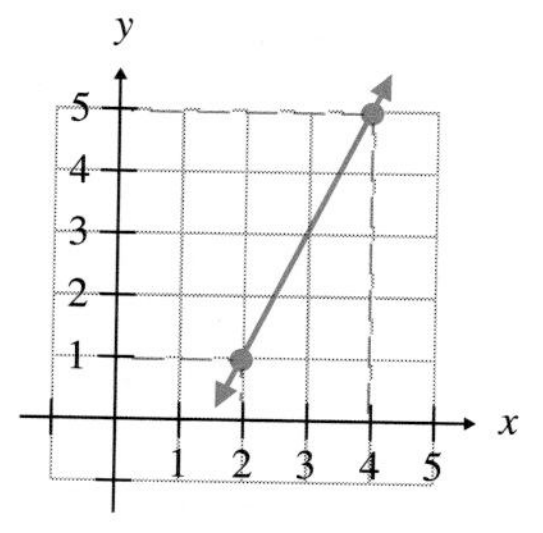

7. Recent legislation requires new nutritional information on food packages. This information can be used to calculate the *percent of calories from fat* in a product. In all foods, each gram of carbohydrates has 4 calories, each gram of protein has 4 calories, and each gram of fat has 9 calories. The percent of calories from fat is found by taking the *ratio* of the calories from fat to the total calories of the food and multiplying this value, in decimal form, by 100. For example, if a food product contains 7 grams of carbohydrates, 2 grams of protein, and 5 grams of fat, we compute the percent of calories from fat as follows:

$$\begin{aligned} \text{calories from carbohydrates} &= 7 \times 4 = 28 \\ \text{calories from protein} &= 2 \times 4 = \ \ 8 \\ \text{calories from fat} &= 5 \times 9 = \underline{45} \\ \text{total calories} &= \qquad\quad\ 81 \end{aligned}$$

$$\begin{aligned} \text{percent of calories from fat} &= \frac{\text{calories from fat}}{\text{total calories}} \times 100 \\ &= \frac{45}{81}(100) = 0.556(100) = 55.6\% \end{aligned}$$

Thus, this product contains about 56% calories from fat. Most medical associations recommend that human diets contain no more than 30% calories from fat.

According to its label, a 4-ounce serving of Healthy Choice Rocky Road Frozen Dairy Dessert contains less than 160 calories, 3 grams of protein, 32 grams of carbohydrates, and 2 grams of fat. It also indicates that the percent of calories from fat is 11%. Use the procedure presented above to determine if the manufacturer's claim regarding the number of calories and percent of calories from fat is correct.

2.7 Inequalities in One Variable

1. Solve linear inequalities.
2. Solve linear inequalities that have all real numbers as their solution or have no solution.

Tape 4

1 Solve Linear Inequalities

The greater-than symbol, $>$, and less-than symbol, $<$, were introduced in Section 1.4. The symbol $\geq$ means greater than or equal to and $\leq$ means less than or equal to. A mathematical statement containing one or more of these symbols is called an **inequality.** The direction of the symbol is sometimes called the **sense** or **order** of the inequality.

Examples of Inequalities in One Variable

$$x + 3 < 5, \qquad x + 4 \geq 2x - 6, \qquad 4 > -x + 3$$

To solve an inequality, we must get the variable by itself on one side of the inequality symbol. To do this, we make use of properties very similar to those used to solve equations. Here are four properties used to solve inequalities. Later in this section we will introduce two additional properties.

Properties Used to Solve Inequalities

For real numbers, a, b, and c:

1. If $a > b$, then $a + c > b + c$.
2. If $a > b$, then $a - c > b - c$.
3. If $a > b$ **and $c > 0$,** then $ac > bc$.
4. If $a > b$ **and $c > 0$,** then $\frac{a}{c} > \frac{b}{c}$.

Property 1 says the same number may be added to both sides of an inequality. Property 2 says the same number may be subtracted from both sides of an inequality. Property 3 says the same *positive* number may be used to multiply both sides of an inequality. Property 4 says the same *positive* number may be used to divide both sides of an inequality. When any of these four properties is used, the direction of the inequality symbol does not change.

EXAMPLE 1 Solve the inequality $x - 4 > 7$, and graph the solution on the real number line.

Solution: To solve this inequality, we need to isolate the variable, x. Therefore, we must eliminate the -4 from the left side of the inequality. To do this, we add 4 to both sides of the inequality.

$$x - 4 > 7$$

$$x - 4 + 4 > 7 + 4 \qquad \text{Add 4 to both sides of the inequality.}$$

$$x > 11$$

The solution is all real numbers greater than 11. We can illustrate the solution on the number line by placing an open circle at 11 on the number line and drawing an arrow to the right (Fig. 2.11).

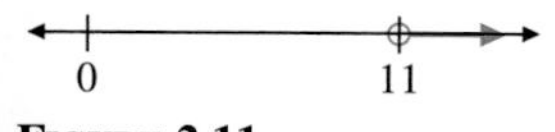

FIGURE 2.11

The open circle at the 11 indicates that the 11 is *not* part of the solution. The arrow going to the right indicates that all the values greater than 11 are solutions to the inequality.

EXAMPLE 2 Solve the inequality $2x + 6 \leq -2$, and graph the solution on the real number line.

Solution: To isolate the variable, we must eliminate the $+6$ from the left side of the inequality. We do this by subtracting 6 from both sides of the inequality.

$$2x + 6 \leq -2$$

$$2x + 6 - 6 \leq -2 - 6 \qquad \text{Subtract 6 from both sides of the inequality.}$$

$$2x \leq -8$$

$$\frac{2x}{2} \leq \frac{-8}{2} \qquad \text{Divide both sides of the inequality by 2.}$$

$$x \leq -4$$

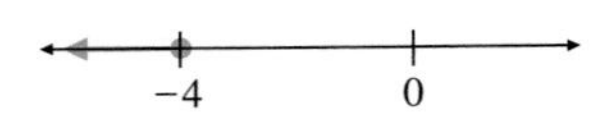

FIGURE 2.12

The solution is all real numbers less than or equal to -4. We can illustrate the solution on the number line by placing a closed, or darkened, circle at -4 and drawing an arrow to the left (Fig. 2.12).

The darkened circle at -4 indicates that -4 *is* a part of the solution. The arrow going to the left indicates that all the values less than -4 are also solutions to the inequality.

Notice in properties 3 and 4 that we specified that $c > 0$. What happens when an inequality is multiplied or divided by a negative number? Examples 3 and 4 will illustrate this.

EXAMPLE 3 Multiply both sides of the inequality $8 > -4$ by -2.

Solution:

$$8 > -4$$
$$-2(8) < -2(-4)$$
$$-16 < 8$$

EXAMPLE 4 Divide both sides of the inequality $8 > -4$ by -2.

Solution:

$$8 > -4$$
$$\frac{8}{-2} < \frac{-4}{-2}$$
$$-4 < 2$$

Examples 3 and 4 illustrate that **when an inequality is multiplied or divided by a negative number, the direction of the inequality symbol changes.**

Additional Properties Used to Solve Inequalities

5. If $a > b$ and $c < 0$, then $ac < bc$.

6. If $a > b$ and $c < 0$, then $\frac{a}{c} < \frac{b}{c}$.

EXAMPLE 5 Solve the inequality $-2x > 6$, and graph the solution on the real number line.

Solution: To isolate the variable, we must eliminate the -2 on the left side of the inequality. To do this, we can divide both sides of the inequality by -2. When we do this, however, we must remember to change the direction of the inequality symbol.

$$-2x > 6$$
$$\frac{-2x}{-2} < \frac{6}{-2} \quad \text{Divide both sides of the inequality by } -2 \text{ and change the direction of the inequality symbol.}$$
$$x < -3$$

FIGURE 2.13

The solution is all real numbers less than -3. The solution is graphed on the number line in Figure 2.13.

EXAMPLE 6 Solve the inequality $4 \geq -5 - x$, and graph the solution on the real number line.

Solution: *Method 1:*

$$4 \geq -5 - x$$

$$4 + 5 \geq -5 + 5 - x \quad \text{Add 5 to both sides of the inequality.}$$

$$9 \geq -x$$

$$-1(9) \leq -1(-x) \quad \text{Multiply both sides of the inequality by } -1 \text{ and change the direction of the inequality symbol.}$$

$$-9 \leq x$$

The inequality $-9 \leq x$ can also be written $x \geq -9$.

Method 2:

$$4 \geq -5 - x$$

$$4 + x \geq -5 - x + x \quad \text{Add } x \text{ to both sides of the inequality.}$$

$$4 + x \geq -5$$

$$4 - 4 + x \geq -5 - 4 \quad \text{Subtract 4 from both sides of the inequality.}$$

$$x \geq -9$$

FIGURE 2.14

The solution is graphed on the number line in Figure 2.14. Other methods could also be used to solve this problem.

Notice in Example 6, method 1, we wrote $-9 \leq x$ as $x \geq -9$. Although the solution $-9 \leq x$ is correct, it is customary to write the solution to an inequality with the variable on the left. One reason we write the variable on the left is that it often makes it easier to graph the solution on the number line. How would you graph $-3 > x$? How would you graph $-5 \leq x$? If you rewrite these inequalities with the variable on the left side, the answer becomes clearer.

$$-3 > x \quad \text{means} \quad x < -3$$

$$\text{and} \quad -5 \leq x \quad \text{means} \quad x \geq -5$$

Notice that you can change an answer from a greater-than statement to a less-than statement or from a less-than statement to a greater-than statement. When you change the answer from one form to the other, remember that the inequality symbol must point to the letter or number to which it was pointing originally.

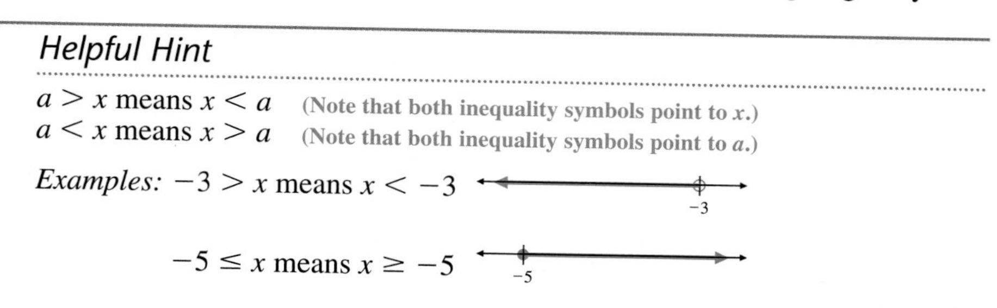

Helpful Hint

$a > x$ means $x < a$ (Note that both inequality symbols point to x.)
$a < x$ means $x > a$ (Note that both inequality symbols point to a.)

Examples: $-3 > x$ means $x < -3$

$-5 \leq x$ means $x \geq -5$

Let us now solve inequalities where the variable appears on both sides of the inequality symbol. To solve these inequalities we use the same basic procedure that we used to solve equations. However, we must remember that whenever we multiply or divide both sides of an inequality by a negative number, we must change the direction of the inequality symbol.

EXAMPLE 7 Solve the inequality $2x + 4 < -x + 12$, and graph the solution on the real number line.

Solution: $2x + 4 < -x + 12$

$2x + x + 4 < -x + x + 12$ Add x to both sides of the inequality.

$3x + 4 < 12$

$3x + 4 - 4 < 12 - 4$ Subtract 4 from both sides of the inequality.

$3x < 8$

$\frac{3x}{3} < \frac{8}{3}$ Divide both sides of the inequality by 3.

$x < \frac{8}{3}$

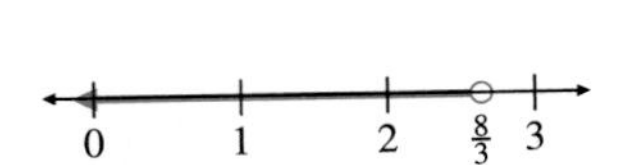

FIGURE 2.15

The solution is graphed on the number line in Figure 2.15.

EXAMPLE 8 Solve the inequality $-5x + 9 < -2x + 6$, and graph the solution on the real number line.

Solution: $-5x + 9 < -2x + 6$

$-5x < -2x - 3$ 9 was subtracted from both sides of the inequality.

$-3x < -3$ $2x$ was added to both sides of the inequality.

$x > 1$ Both sides of inequality were divided by -3 and direction of inequality symbol was changed.

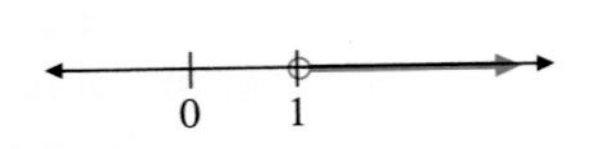

FIGURE 2.16

The solution is graphed in Figure 2.16.

Special Cases

2 In Examples 9 and 10 we illustrate two special types of inequalities. Example 9 is an inequality that is always true for all real numbers, and Example 10 is an inequality that is never true for any real number.

EXAMPLE 9 Solve the inequality $2(x + 3) \leq 5x - 3x + 8$, and graph the solution on the real number line.

Solution: $2(x + 3) \leq 5x - 3x + 8$

$2x + 6 \leq 5x - 3x + 8$ Distributive property was used.

$2x + 6 \leq 2x + 8$ Like terms were combined.

$2x - 2x + 6 \leq 2x - 2x + 8$ Subtract $2x$ from both sides of the inequality.

$6 \leq 8$

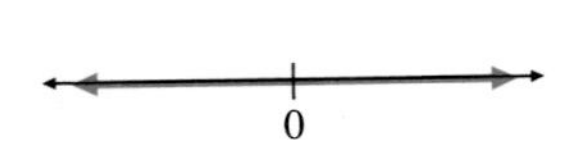

FIGURE 2.17

Since 6 is always less than or equal to 8, the solution is **all real numbers** (Fig. 2.17).

EXAMPLE 10 Solve the inequality $4(x + 1) > x + 5 + 3x$, and graph the solution on the real number line.

Solution: $4(x + 1) > x + 5 + 3x$

$4x + 4 > x + 5 + 3x$ Distributive property was used.

$4x + 4 > 4x + 5$ Like terms were combined.

$4x - 4x + 4 > 4x - 4x + 5$ Subtract $4x$ from both sides of the inequality.

$4 > 5$

0

FIGURE 2.18

Since 4 is never greater than 5, the answer is **no solution** (Fig. 2.18). There is no real number that makes the statement true.

Exercise Set 2.7

Solve each inequality and graph the solution on the real number line.

1. $x + 3 > 7$

2. $x - 4 > -3$

3. $x + 5 \geq 3$

4. $4 - x \geq 3$

5. $-x + 3 < 8$

6. $4 < 3 + x$

7. $6 > x - 4$

8. $-4 \leq -x - 3$

9. $8 \leq 4 - x$

10. $2x < 4$

11. $-2x < 3$

12. $6 \geq -3x$

13. $2x + 3 \leq 5$

14. $-4x - 3 > 5$

15. $12x + 24 < -12$

16. $3x - 4 \leq 9$

17. $4 - 6x > -5$

18. $8 < 4 - 2x$

19. $15 > -9x + 50$

20. $3x - 4 < 5$

21. $4 < 3x + 12$

22. $-4x > 2x + 12$

23. $6x + 2 \leq 3x - 9$

24. $-2x - 4 \leq -5x + 12$

25. $x - 4 \leq 3x + 8$

26. $-3x - 5 \geq 4x - 29$

27. $-x + 4 < -3x + 6$

28. $2(x - 3) < 4x + 10$

29. $-3(2x - 4) > 2(6x - 12)$

30. $-(x + 3) \leq 4x + 5$

31. $x + 3 < x + 4$

32. $x + 5 \geq x - 2$

33. $6(3 - x) < 2x + 12$

34. $2(3 - x) + 4x < -6$

35. $-21(2 - x) + 3x > 4x + 4$

36. $-(x + 3) \geq 2x + 6$

37. $4x - 4 < 4(x - 5)$

38. $-2(-5 - x) > 3(x + 2) + 4 - x$

39. $5(2x + 3) \geq 6 + (x + 2) - 2x$

40. $-3(-2x + 12) < -4(x + 2) - 6$

41. When solving an inequality, if you obtain the result $3 < 5$, what is the solution?

42. When solving an inequality, if you obtain the result $4 \geq 2$, what is the solution?

43. When solving an inequality, if you obtain the result $5 < 2$, what is the solution?

44. When solving an inequality, if you obtain the result $-4 \geq -2$, what is the solution?

45. When solving an inequality, under what conditions will it be necessary to change the direction of the inequality symbol?

46. List the six rules used to solve inequalities.

CUMULATIVE REVIEW EXERCISES

8] **47.** Evaluate $-x^2$ for $x = 3$.

48. Evaluate $-x^2$ for $x = -5$.

5] **49.** Solve $4 - 3(2x - 4) = 5 - (x + 3)$.

[2.6] **50.** The Milford electric company charges $0.174 per kilowatt-hour of electricity. The Cisneros's monthly electric bill was $87 for the month of July. How many kilowatt-hours of electricity did the Cisneros use in July?

Group Activity/ Challenge Problems

1. Solve the inequality
   $3(2 - x) - 4(2x - 3) \leq 6 + 2x - 6(x - 5) + 2x.$
2. Solve the inequality
   $-(x + 4) + 6x - 5 > -4(x + 3) + 2(x + 6) - 5x.$
3. The inequality symbols discussed so far are $<$, $\leq$, $>$, and $\geq$. Can you name an inequality symbol that we have not mentioned in this section?
4. Reproduced below is a portion of the Florida Individual and Joint Intangible Tax Return for 1995.

TAX CALCULATION WORKSHEET

(COMPLETE ONLY ONE (1) COLUMN BELOW)

FILING STATUS (Step 1)	INDIVIDUAL		JOINT	
IF YOUR TAXABLE ASSETS FROM SCHEDULE A LINE 5 ARE:	BOX A $100,000 or LESS	BOX B GREATER than $100,000	BOX C $200,000 or LESS	BOX D GREATER than $200,000
6A. TAXABLE ASSETS (SCHEDULE A, LINE 5) (Step 2)	$______	$______	$______	$______
6B. TIMES TAX RATE	x .001	x .002	x .001	x .002
6C. GROSS TAX (Step 3) (MULTIPLY LINE 6A x LINE 6B)	$______	$______	$______	$______
6D. LESS EXEMPTION	- $20.00	-$120.00	-$40.00	-$240.00
6E. TOTAL TAX DUE (Step 4) (SUBTRACT LINE 6D FROM LINE 6C IF LESS THAN ZERO ENTER 0)	$______	$______	$______	$______
CARRY TOTAL TAX DUE AMOUNT TO SCHEDULE A, LINE 6				

Use the Tax Calculation Worksheet to determine your total tax due (line 6e) if your total taxable assets from Schedule A line 5 are as follows.

(a) \$30,000 and your filing status is individual.
(b) \$175,000 and your filing status is individual.
(c) \$200,000 and your filing status is joint.
(d) \$300,000 and your filing status is joint.

Summary

GLOSSARY

Algebraic expression (76): A collection of numbers, variables, grouping symbols, and operation symbols.

Check (85): A procedure where the value believed to be the solution to an equation is substituted back into the equation.

Coefficient or numerical coefficient (77): The numerical part of a term.

Constant or constant term (77): A term in an expression that does not contain a variable.

Equation (85): A statement that two algebraic expressions are equal.

Equivalent equations (88): Two or more equations that have the same solution.

Identity (116): An equation that is true for all values of the variable.

Inequality (132): A mathematical statement containing one or more inequality symbols ($>$, $\geq$, $<$, $\leq$).

Like terms (77): Terms that have the same variables with the same exponents.

Proportion (121): A statement of equality between two ratios.

Ratio (120): A quotient of two quantities with the same units.

Reciprocal of real number a (95): $\frac{1}{a}$, $a \neq 0$.

Similar figures (126): Two figures are similar when their corresponding angles are equal and their corresponding sides are in proportion.

Simplify (82): To simplify an expression means to combine like terms in the expression.

Solution (85): The value or values of the variable that make an equation a true statement.

Solve (85): To find the solution to an equation.

Term (76): The parts that are added or subtracted in an algebraic expression.

Important Facts

Distributive property:

$$a(b + c) = ab + ac.$$

Addition property:

If $a = b$, then $a + c = b + c$.

Multiplication property:

If $a = b$, then $a \cdot c = b \cdot c$.

Cross-multiplication:

If $\frac{a}{b} = \frac{c}{d}$, then $ad = bc$.

Properties used to solve inequalities

1. If $a > b$, then $a + c > b + c$.
2. If $a > b$, then $a - c > b - c$.
3. If $a > b$ and $c > 0$, then $ac > bc$.
4. If $a > b$ and $c > 0$, then $\frac{a}{c} > \frac{b}{c}$.
5. If $a > b$ and $c < 0$, then $ac < bc$.
6. If $a > b$ and $c < 0$, then $\frac{a}{c} < \frac{b}{c}$.

Review Exercises

1] *Use the distributive property to simplify.*

1. $2(x + 4)$
2. $3(x - 2)$
3. $2(4x - 3)$
4. $-2(x + 4)$
5. $-(x + 2)$
6. $-(x - 2)$
7. $-4(4 - x)$
8. $3(6 - 2x)$
9. $4(5x - 6)$
10. $-3(2x - 5)$
11. $6(6x - 6)$
12. $4(-x + 3)$
13. $-3(x + y)$
14. $-2(3x - 2)$
15. $-(3 + 2y)$
16. $-(x + 2y - z)$
17. $3(x + 3y - 2z)$
18. $-2(2x - 3y + 7)$

Simplify where possible.

19. $2x + 3x$
20. $4y + 3y + 2$
21. $4 - 2y + 3$
22. $1 + 3x + 2x$
23. $6x + 2y + y$
24. $-2x - x + 3y$
25. $2x + 3y + 4x + 5y$
26. $6x + 3y + 2$
27. $2x - 3x - 1$
28. $5x - 2x + 3y + 6$
29. $x + 8x - 9x + 3$
30. $-4x - 8x + 3$
31. $3(x + 2) + 2x$
32. $-2(x + 3) + 6$
33. $2x + 3(x + 4) - 5$
34. $4(3 - 2x) - 2x$
35. $6 - (-x + 3) + 4x$
36. $2(2x + 5) - 10 - 4$
37. $-6(4 - 3x) - 18 + 4x$
38. $6 - 3(x + y) + 6x$
39. $3(x + y) - 2(2x - y)$
40. $3x - 6y + 2(4y + 8)$
41. $3 - (x - y) + (x - y)$
42. $(x + y) - (2x + 3y) + 4$

-2.5] *Solve.*

43. $2x = 4$
44. $x + 3 = -5$
45. $x - 4 = 7$
46. $\frac{x}{3} = -9$
47. $2x + 4 = 8$
48. $14 = 3 + 2x$
49. $8x - 3 = -19$
50. $6 - x = 9$
51. $-x = -12$

52. $2(x + 2) = 6$
53. $-3(2x - 8) = -12$
54. $4(6 + 2x) = 0$
55. $3x + 2x + 6 = -15$
56. $4 = -2(x + 3)$
57. $27 = 46 + 2x - x$
58. $4x + 6 - 7x + 9 = 18$
59. $4 + 3(x + 2) = 12$
60. $-3 + 3x = -2(x + 1)$
61. $3x - 6 = -5x + 30$
62. $-(x + 2) = 2(3x - 6)$
63. $2x + 6 = 3x + 9$
64. $-5x + 3 = 2x + 10$
65. $3x - 12x = 24 - 6x$
66. $2(x + 4) = -3(x + 5)$
67. $4(2x - 3) + 4 = 9x + 2$
68. $6x + 11 = -(6x + 5)$
69. $2(x + 7) = 6x + 9 - 4x$
70. $-5(3 - 4x) = -6 + 20x - 9$
71. $4(x - 3) - (x + 5) = 0$
72. $-2(4 - x) = 6(x + 2) + 3x$

[2.6] *Determine the following ratios. Write each ratio in lowest terms.*

73. 15 feet to 20 feet.
74. 80 ounces to 12 pounds.
75. 32 ounces to 2 pounds.

Solve each proportion.

76. $\frac{x}{9} = \frac{6}{18}$
77. $\frac{15}{10} = \frac{x}{20}$
78. $\frac{3}{x} = \frac{15}{45}$
79. $\frac{20}{45} = \frac{15}{x}$
80. $\frac{6}{5} = \frac{-12}{x}$
81. $\frac{x}{9} = \frac{8}{-3}$
82. $\frac{-4}{9} = \frac{-16}{x}$
83. $\frac{x}{-15} = \frac{30}{-5}$

Each of the following pairs of figures are similar. Find the length of the side indicated by x.

84.

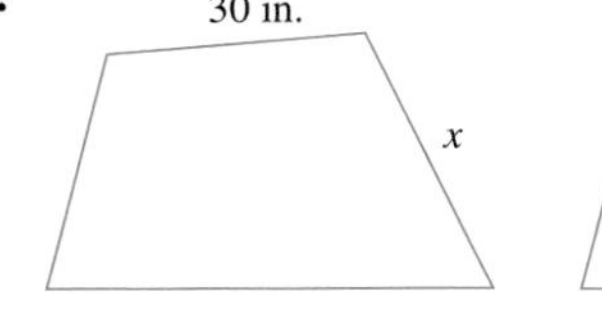

85.

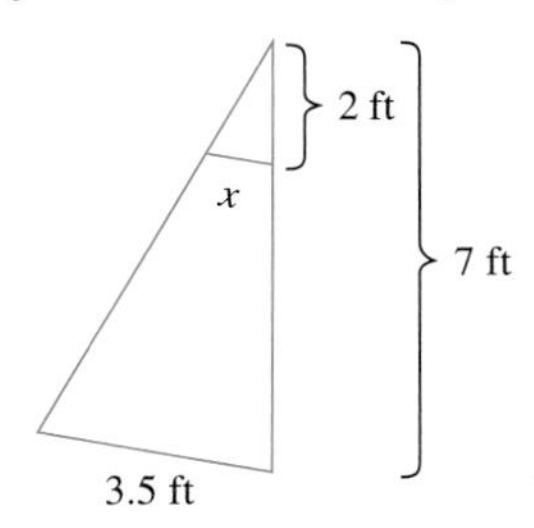

[2.7] *Solve each inequality, and graph the solution on the real number line.*

86. $2x + 4 \geq 8$
87. $6 - 2x > 4x - 12$
88. $6 - 3x \leq 2x + 18$
89. $2(x + 4) \leq 2x - 5$
90. $2(x + 3) > 6x - 4x + 4$
91. $x + 6 > 9x + 30$
92. $x - 2 \leq -4x + 7$
93. $-(x + 2) < -2(-2x + 5)$
94. $2(x + 3) < -(x + 3) + 4$
95. $-6x - 3 \geq 2(x - 4) + 3x$
96. $-2(x - 4) \leq 3x + 6 - 5x$
97. $2(2x + 4) > 4(x + 2) - 6$

[2.6] *Set up a proportion and solve each problem.*

98. If a 4-ounce piece of cake has 160 calories, how many calories does a 6-ounce piece of that cake have?

99. If a copy machine can copy 5 pages per minute, how many pages can be copied in 22 minutes?

100. If the scale of a map is 1 inch to 60 miles, what distance on the map represents 380 miles?

101. Bryce builds a model car to a scale of 1 inch to 0.9 feet. If the completed model is 10.5 inches, what is the size of the actual car?

102. If one U.S. dollar can be exchanged for 3.1165 Mexican pesos, find the value of 1 peso in terms of U.S. dollars.

103. If 3 radians equal 171.9 degrees, find the number of degrees in 1 radian.

104. If a machine can fill and cap 80 bottles of catsup in 50 seconds, how many bottles of catsup can it fill and cap in 2 minutes?

Practice Test

Use the distributive property to simplify.

1. $-2(4 - 2x)$
2. $-(x + 3y - 4)$

Simplify.

3. $3x - x + 4$
4. $4 + 2x - 3x + 6$
5. $y - 2x - 4x - 6$
6. $x - 4y + 6x - y + 3$
7. $2x + 3 + 2(3x - 2)$

Solve.

8. $2x + 4 = 12$
9. $-x - 3x + 4 = 12$
10. $4x - 2 = x + 4$
11. $3(x - 2) = -(5 - 4x)$
12. $2x - 3(-2x + 4) = -13 + x$
13. $3x - 4 - x = 2(x + 5)$
14. $-3(2x + 3) = -2(3x + 1) - 7$
15. $\dfrac{9}{x} = \dfrac{3}{-15}$

Solve and graph the solution on the real number line.

16. $2x - 4 < 4x + 10$
17. $3(x + 4) \geq 5x - 12$
18. $4(x + 3) + 2x < 6x - 3$

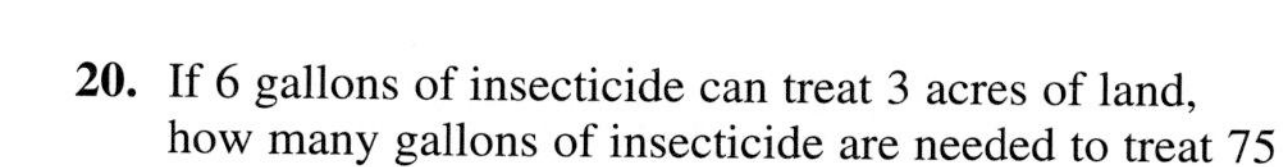

19. The following figures are similar figures. Find the length of side x.

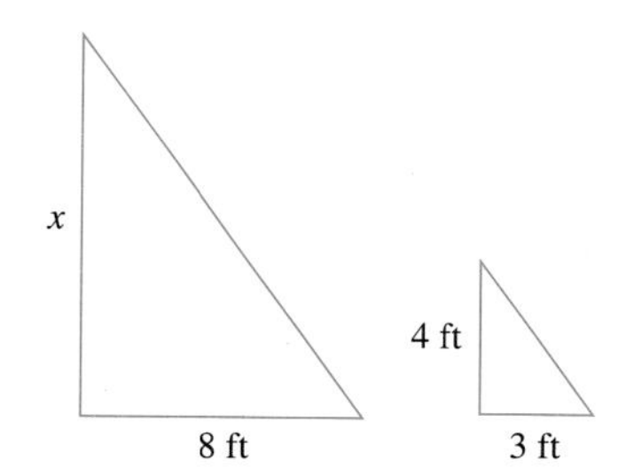

20. If 6 gallons of insecticide can treat 3 acres of land, how many gallons of insecticide are needed to treat 75 acres?

Cumulative Review Test

1. Multiply $\dfrac{16}{20} \cdot \dfrac{4}{5}$.
2. Divide $\dfrac{8}{24} \div \dfrac{2}{3}$.
3. Insert <, >, or = in the shaded area to make a true statement: $|-2|$ ▒ 1.
4. Evaluate $-6 - (-3) + 5 - 8$.
5. Subtract -4 from -12.
6. Evaluate $16 - 6 \div 2 \cdot 3$.
7. Evaluate $3[6 - (4 - 3^2)] - 30$.
8. Evaluate $-3x^2 - 4x + 5$ when $x = -2$.
9. Name the property illustrated: $(x + 4) + 6 = x + (4 + 6)$.

Simplify.

10. $6x + 2y + 4x - y$
11. $3x - 2x + 16 + 2x$

Solve.

12. $4x - 2 = 10$
13. $\dfrac{1}{4}x = -10$

14. $6x + 5x + 6 = 28$

15. $3(x - 2) = 5(x - 1) + 3x + 4$

16. $\frac{15}{30} = \frac{3}{x}$

Solve, and graph the solution on the number line.

17. $x - 4 > 6$

18. $2x - 7 \leq 3x + 5$

19. A 36-pound bag of fertilizer can fertilize an area of 5000 square feet. How many pounds of fertilizer will Marisa need to fertilize her 22,000-square-foot lawn?

20. If Samuel earns $10.50 after working for 2 hours scrubbing boats at the marina, how much does he earn after 8 hours?

Chapter 3

Formulas and Applications of Algebra

See Section 3.2, Group Activity 2

Preview and Perspective

The major goal of this chapter is to teach you the terminology and techniques to write real-life application problems as equations. The equations are then solved using the techniques taught in Chapter 2. For mathematics to be relevant it must be useful. In this chapter we explain and illustrate the tremendous number of real-life applications of algebra. Since this is such an important topic, and since we want you to learn it well and feel comfortable applying mathematics to real-world situations, we cover this chapter especially slowly. You need to have confidence in your work, and you need to do all your homework. The more problems you attempt, the better you will become at setting up and solving application (or word) problems. After completing this chapter, whenever a real-life application of mathematics comes up, try to set it up and solve it algebraically.

We begin this chapter with a discussion of formulas. We explain how to evaluate a formula and how to solve for a variable in a formula. We will be using formulas throughout this course, so you must understand their use. You probably realize that most mathematics and science courses use a wide variety of formulas. Formulas are also used in many other disciplines, including the arts, business and economics, medicine, and technology, to name just a few.

In Section 3.2 we explain how to write real-life applications as equations. In Section 3.3 we continue writing application problems as equations. We then solve the equations to determine the answers to the problems. The general procedures discussed in Sections 3.2 and 3.3 are then applied to specific geometric applications in Section 3.4. Since application problems are so important, we cover them in many different sections throughout the book. For example, we discuss applications of motion problems and mixture problems in Section 4.7.

3.1 Formulas

Tape 5

1. Use the simple interest formula.
2. Use geometric formulas.
3. Solve for a variable in a formula.

A **formula** is an equation commonly used to express a specific relationship mathematically. For example, the formula for the area of a rectangle is

$$\text{area} = \text{length} \cdot \text{width} \quad \text{or} \quad A = lw$$

To **evaluate a formula,** substitute the appropriate numerical values for the variables and perform the indicated operations.

Simple Interest Formula

1 A formula used in banking is the *simple interest formula.*

Simple Interest Formula

$$\text{interest} = \text{principal} \cdot \text{rate} \cdot \text{time} \quad \text{or} \quad i = prt$$

This formula is used to determine the simple interest, i, earned on some savings accounts, or the simple interest an individual must pay on certain loans. In the simple interest formula $i = prt$, p is the principal (the amount invested or borrowed), r is the interest rate in decimal form, and t is the amount of time of the investment or loan.

EXAMPLE 1 Avery borrows \$2000 from a bank for 3 years. The bank charges 12% simple interest per year for the loan. How much interest will Avery owe the bank?

Solution: The principal, p, is \$2000, the rate, r, is 12% or 0.12 in decimal form, and the time, t, is 3 years. Substituting these values in the simple interest formula gives

$$i = prt$$
$$i = 2000(0.12)(3) = 720$$

The simple interest is \$720. After 3 years when Avery repays his loan he will pay the principal, \$2000, plus the interest, \$720, for a total of \$2720.

EXAMPLE 2 Amber invests \$5000 in a savings account which earns simple interest for 2 years. If the interest earned from the account is \$800, find the rate.

Solution: We use the simple interest formula, $i = prt$. We are given the principal, p, the time, t, and the interest, i. We are asked to find the rate, r. We substitute the given values in the simple interest formula and solve the resulting equation for r.

$$i = prt$$
$$800 = 5000(r)(2)$$
$$800 = 10{,}000r$$
$$\frac{800}{10{,}000} = \frac{10{,}000r}{10{,}000}$$
$$0.08 = r$$

Thus, the simple interest rate is 0.08, or 8% per year.

Geometric Formulas

2 The **perimeter,** P, is the sum of the lengths of the sides of a figure. Perimeters are measured in the same common unit as the sides. For example, perimeter may be measured in centimeters, inches, or feet. The **area,** A, is the total surface within the figure's boundaries. Areas are measured in square units. For example, area may be measured in square centimeters, square inches, or square feet. Table 3.1 on page 146 gives the formulas for finding the areas and perimeters of triangles and quadrilaterals. **Quadrilateral** is a general name for a four-sided figure.

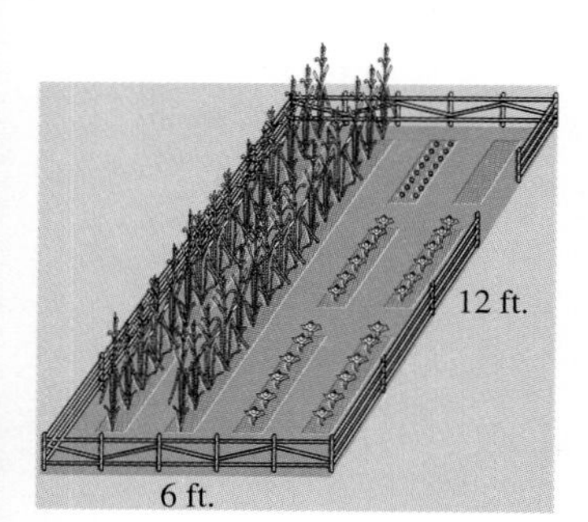

FIGURE 3.1

EXAMPLE 3 Kim's rectangular vegetable garden is 12 feet long and 6 feet wide (Fig. 3.1).

(a) If Kim wants to put fencing around the garden to keep the animals out, how much fencing will she need?

(b) What is the area of Kim's garden?

Solution: **(a)** To find the amount of fencing needed we need to find the perimeter of the garden. Substitute 12 for l and 6 for w in the formula for the perimeter of a rectangle.

$$P = 2l + 2w$$
$$P = 2(12) + 2(6) = 24 + 12 = 36 \text{ feet}$$

Thus, Kim will need 36 feet of fencing.

(b) Substitute 12 for l and 6 for w in the formula for the area of a rectangle.

$$A = lw$$
$$A = (12)(6) = 72 \text{ square feet (or } 72 \text{ ft}^2)$$

Kim's vegetable garden has an area of 72 square feet.

TABLE 3.1 FORMULAS FOR AREAS AND PERIMETERS OF QUADRILATERALS AND TRIANGLES

Figure	Sketch	Area	Perimeter
Square	s	$A = s^2$	$P = 4s$
Rectangle	w, l	$A = lw$	$P = 2l + 2w$
Parallelogram	h, w, l	$A = lh$	$P = 2l + 2w$
Trapezoid	d, a, h, c, b	$A = \frac{1}{2}h(b + d)$	$P = a + b + c + d$
Triangle	a, c, h, b	$A = \frac{1}{2}bh$	$P = a + b + c$

EXAMPLE 4 Find the length of a rectangle whose perimeter is 22 inches and whose width is 3 inches.

Solution: Substitute 22 for P and 3 for w in the formula $P = 2l + 2w$; then solve for the length, l.

$$P = 2l + 2w$$
$$22 = 2l + 2(3)$$
$$22 = 2l + 6$$
$$22 - 6 = 2l + 6 - 6$$
$$16 = 2l$$
$$\frac{16}{2} = \frac{2l}{2}$$
$$8 = l$$

The length is 8 inches.

EXAMPLE 5 The area of a triangle is 30 square feet and its base is 12 feet (Fig. 3.2). Find its height.

Solution:

$$A = \frac{1}{2}bh$$

$$30 = \frac{1}{2}(12)h$$

$$30 = 6h$$

$$\frac{30}{6} = \frac{6h}{6}$$

$$5 = h$$

FIGURE 3.2

The height of the triangle is 5 feet.

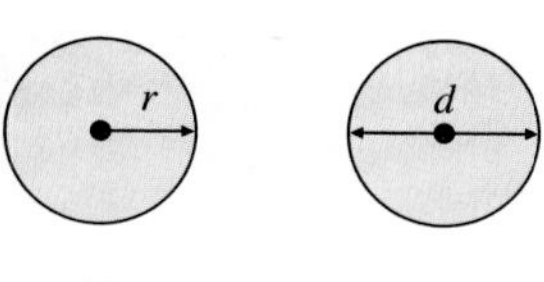

(a) (b)

FIGURE 3.3

Another figure that we see and use daily is the circle. The **circumference,** C, is the length (or perimeter) of the curve that forms a circle. The **radius,** r, is the line segment from the center of the circle to any point on the circle (Fig. 3.3a). The **diameter** of a circle is a line segment through the center whose end points both lie on the circle (Fig. 3.3b). *Note that the length of the diameter is twice the length of the radius.*

The formulas for both the area and the circumference of a circle are given in Table 3.2.

TABLE 3.2 FORMULAS FOR CIRCLES

Circle	Area	Circumference
(circle with radius r)	$A = \pi r^2$	$C = 2\pi r$

The value of pi, symbolized by the Greek lowercase letter π, is *approximately* 3.14.

Most scientific calculators have a key for the value of π. If you press $\boxed{\pi}$, your calculator will display 3.1415927. This is only an approximation of π since π is an irrational number. If you own a scientific calculator, use the π key instead of using 3.14 when working problems that involve π. In this book we will use 3.14 for π since not every student owns a scientific calculator. If you use the π key, your answers will be slightly more accurate than ours, but still approximate.

EXAMPLE 6 Determine the area and circumference of a circle whose diameter is 16 inches.

Solution: The radius is half the diameter, so $r = \frac{16}{2} = 8$ inches.

$$A = \pi r^2 \qquad C = 2\pi r$$
$$A = 3.14(8)^2 \qquad C = 2(3.14)(8)$$
$$A = 3.14(64) \qquad C = 50.24 \text{ inches}$$
$$A = 200.96 \text{ square inches}$$

If you used a scientific calculator and used the π key, your answer for the area would be 201.06193 and for the circumference would be 50.265482.

Table 3.3 gives formulas for finding the volume of certain three-dimensional figures. Volume is measured in cubic units, such as cubic centimeters or cubic feet.

TABLE 3.3 FORMULAS FOR VOLUMES OF THREE-DIMENSIONAL FIGURES

Figure	Sketch	Volume
Rectangular solid		$V = lwh$
Right circular cylinder		$V = \pi r^2 h$
Right circular cone		$V = \frac{1}{3}\pi r^2 h$
Sphere		$V = \frac{4}{3}\pi r^3$

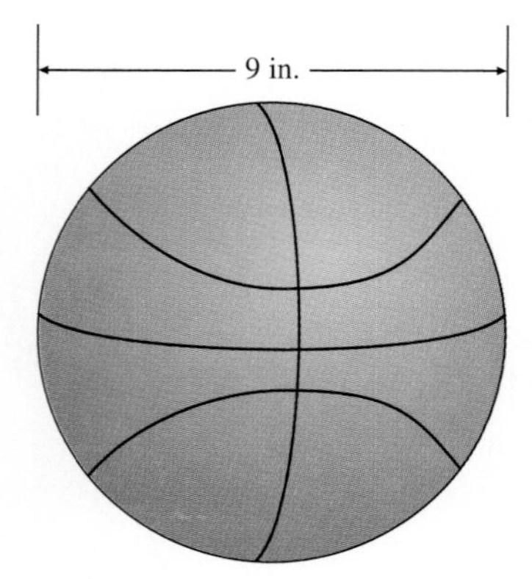

FIGURE 3.4

EXAMPLE 7 Find the volume of a basketball if its diameter is 9 inches (Fig. 3.4).

Solution: Since its diameter is 9 inches, its radius is 4.5 inches.

$$V = \frac{4}{3}\pi r^3$$

$$V = \frac{4}{3}(3.14)(4.5)^3 = \frac{4}{3}(3.14)(91.125) = 381.51$$

Therefore, a basketball has a volume of 381.51 cubic inches. If you used the π key on your calculator, your answer would be 381.7035074.

EXAMPLE 8 Find the height of a right circular cylinder if its volume is 904.32 cubic inches and its radius is 6 inches (Fig. 3.5)

Solution:

$$V = \pi r^2 h$$
$$904.32 = (3.14)(6)^2 h$$
$$904.32 = (3.14)(36)h$$
$$904.32 = 113.04h$$
$$\frac{904.32}{113.04} = \frac{113.04h}{113.04}$$
$$8 = h$$

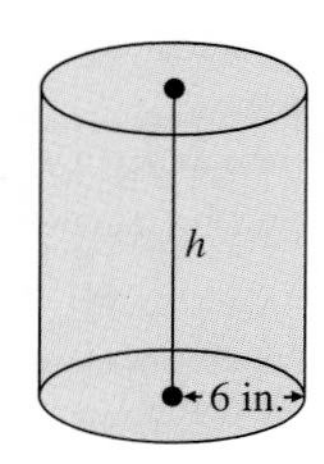

FIGURE 3.5

Thus, the height is 8 inches.

Let us do one more problem that involves evaluating a formula.

EXAMPLE 9 The number of diagonals, d, in a polygon of n sides is given by the formula $d = \frac{1}{2}n^2 - \frac{3}{2}n$.

(a) How many diagonals does a quadrilateral (4 sides) have?

(b) How many diagonals does an octagon (8 sides) have?

Check: for **(a)**

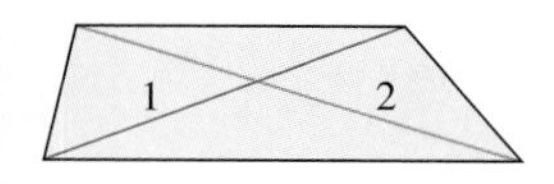

Solution: **(a)** $n = 4$

$$d = \frac{1}{2}(4)^2 - \frac{3}{2}(4)$$
$$= \frac{1}{2}(16) - 6$$
$$= 8 - 6 = 2$$

(b) $n = 8$

$$d = \frac{1}{2}(8)^2 - \frac{3}{2}(8)$$
$$= \frac{1}{2}(64) - 12$$
$$= 32 - 12 = 20$$

Solving for a Variable in a Formula or Equation

3 Often in this course and in other mathematics and science courses, you will be given an equation or formula solved for one variable and have to solve it for a different variable. We will now learn how to do this. This material will reinforce what you learned about solving equations in Chapter 2. We will use the procedures learned here to solve problems in many other sections of the text.

To solve for a variable in a formula, treat each of the quantities, except the one you are solving for, as if they were constants. Then solve for the desired variable by isolating it on one side of the equation, as you did in Chapter 2.

EXAMPLE 10 Solve the formula $A = lw$ for w.

Solution: We must get w by itself on one side of the equation. We begin by removing the l from the right side of the equation to isolate the w.

$$A = lw$$

$$\frac{A}{l} = \frac{\cancel{l}w}{\cancel{l}} \qquad \text{Divide both sides of the equation by } l.$$

$$\frac{A}{l} = w$$

EXAMPLE 11 Solve the formula $P = 2l + 2w$ for l.

Solution: We must get l all by itself on one side of the equation. We begin by removing the $2w$ from the right side of the equation to isolate the term containing the l.

$$P = 2l + 2w$$

$$P - 2w = 2l + 2w - 2w \qquad \text{Subtract } 2w \text{ from both sides of the equation.}$$

$$P - 2w = 2l$$

$$\frac{P - 2w}{2} = \frac{2l}{2} \qquad \text{Divide both sides of the equation by 2.}$$

$$\frac{P - 2w}{2} = l \qquad \left(\text{or } l = \frac{P}{2} - w\right)$$

EXAMPLE 12 An equation we use in Chapter 7 is $y = mx + b$. Solve for m.

Solution: We must get the m all by itself on one side of the equal sign.

$$y = mx + b$$

$$y - b = mx + b - b \qquad \text{Subtract } b \text{ from both sides of the equation.}$$

$$y - b = mx$$

$$\frac{y - b}{x} = \frac{mx}{x} \qquad \text{Divide both sides of the equation by } x.$$

$$\frac{y - b}{x} = m \qquad \left(\text{or } m = \frac{y}{x} - \frac{b}{x}\right)$$

EXAMPLE 13 Solve the simple interest formula $i = prt$ for p.

Solution: We must isolate the p. Since p is multiplied by both r and t, we divide both sides of the equation by rt.

$$i = prt$$

$$\frac{i}{rt} = \frac{prt}{rt}$$

$$\frac{i}{rt} = p$$

In Chapter 7 when discussing graphing we will need to solve many equations for the variable y. This procedure is illustrated in Example 14.

EXAMPLE 14 **(a)** Solve the equation $2x + 3y = 12$ for y.

(b) Find the value of y when $x = 6$.

Solution: **(a)** Begin by isolating the term containing the variable y.

$$2x + 3y = 12$$
$$2x - 2x + 3y = 12 - 2x \quad \text{Subtract } 2x \text{ from both sides of equation.}$$
$$3y = 12 - 2x$$
$$\frac{3y}{3} = \frac{12 - 2x}{3} \quad \text{Divide both sides of equation by 3.}$$
$$y = \frac{12 - 2x}{3} \quad \left(\text{or } y = \frac{12}{3} - \frac{2x}{3} = 4 - \frac{2}{3}x\right)$$

(b) To find the value of y when x is 6, substitute 6 for x in the equation solved for y in part (a).

$$y = \frac{12 - 2x}{3}$$
$$y = \frac{12 - 2(6)}{3} = \frac{12 - 12}{3} = \frac{0}{3} = 0$$

We see that when $x = 6$, $y = 0$.

Some formulas contain fractions. When a formula contains a fraction, we can eliminate the fraction by multiplying both sides of the equation by the denominator, as illustrated in Example 15.

EXAMPLE 15 Solve the formula $A = \dfrac{m + n}{2}$ for m.

Solution: We begin by multiplying both sides of the equation by 2 to eliminate the fraction. Then we isolate the variable m.

$$A = \frac{m + n}{2}$$
$$2A = 2\left(\frac{m + n}{2}\right) \quad \text{Multiply both sides of equation by 2.}$$
$$2A = m + n$$
$$2A - n = m + n - n \quad \text{Subtract } n \text{ from both sides of equation.}$$
$$2A - n = m$$

Thus, $m = 2A - n$.

Exercise Set 3.1

Use the formula to find the value of the variable indicated. Use a calculator to save time. Round answers off to hundredths.

1. $A = s^2$; find A when $s = 5$.
2. $P = a + b + c$; find P when $a = 4$, $b = 3$, and $c = 7$.
3. $P = 2l + 2w$; find P when $l = 6$ and $w = 5$.
4. $A = \frac{1}{2}bh$; find A when $b = 12$ and $h = 8$

5. $A = \frac{1}{2}h(b + d)$; find A when $h = 6$, $b = 18$, and $d = 24$
6. $A = \pi r^2$; find A when $r = 6$. Use $\pi = 3.14$.
7. $C = 2\pi r$; find C when $r = 2$. Use $\pi = 3.14$.
8. $p = i^2 r$; find r when $p = 4000$ and $i = 2$.
9. $A = \frac{1}{2}bh$; find h when $A = 30$ and $b = 6$.
10. $V = \frac{1}{3}Bh$; find h when $V = 40$ and $B = 12$.
11. $V = lwh$; find l when $V = 18$, $w = 1$ and $h = 3$.
12. $T = \frac{RS}{R + S}$; find T when $R = 50$ and $S = 50$.
13. $A = P(1 + rt)$; find A when $P = 1000$, $r = 0.08$, and $t = 1$.
14. $P = 2l + 2w$; find l when $P = 28$ and $w = 6$.
15. $M = \frac{a + b}{2}$; find b when $M = 36$ and $a = 16$.
16. $\mathrm{F} = \frac{9}{5}\mathrm{C} + 32$; find F when $\mathrm{C} = 10$.
17. $\mathrm{C} = \frac{5}{9}(\mathrm{F} - 32)$; find C when $\mathrm{F} = 41$.
18. $z = \frac{x - m}{s}$; find z when $x = 115$, $m = 100$, and $s = 15$.
19. $z = \frac{x - m}{s}$; find x when $z = 2$, $m = 50$, and $s = 5$.
20. $z = \frac{x - m}{s}$; find s when $z = 3$, $x = 80$, and $m = 59$.
21. $K = \frac{1}{2}mv^2$; find m when $K = 288$ and $v = 6$.
22. $A = P(1 + rt)$; find r when $A = 1500$, $t = 1$, and $P = 1000$.
23. $V = \pi r^2 h$; find h when $V = 678.24$, and $r = 6$. Use $\pi = 3.14$.
24. $V = \frac{4}{3}\pi r^3$; find V when $r = 6$. Use $\pi = 3.14$.

Solve each equation for y; then find the value of y for the given value of x. See Example 14.

25. $2x + y = 8$, when $x = 2$.
26. $6x + 2y = -12$, when $x = -3$.
27. $2x = 6y - 4$, when $x = 10$.
28. $-3x - 5y = -10$, when $x = 0$.
29. $2y = 6 - 3x$, when $x = 2$.
30. $15 = 3y - x$, when $x = 3$.
31. $-4x + 5y = -20$, when $x = 4$.
32. $3x - 2y = -18$, when $x = -1$.
33. $-3x = 18 - 6y$, when $x = 0$.
34. $-12 = -2x - 3y$, when $x = -2$.
35. $-8 = -x - 2y$, when $x = -4$.
36. $2x + 5y = 20$, when $x = -5$.

Solve for the variable indicated.

37. $d = rt$, for t
38. $d = rt$, for r
39. $i = prt$, for p
40. $i = prt$, for r
41. $C = \pi d$, for d
42. $V = lwh$, for w
43. $A = \frac{1}{2}bh$, for b
44. $E = IR$, for I
45. $P = 2l + 2w$, for w
46. $PV = KT$, for T
47. $4n + 3 = m$, for n
48. $3t - 4r = 25$, for t
49. $y = mx + b$, for b
50. $y = mx + b$, for x
51. $I = P + Prt$, for r
52. $A = \frac{m + d}{2}$, for m
53. $A = \frac{m + 2d}{3}$, for d
54. $R = \frac{l + 3w}{2}$, for w
55. $d = a + b + c$, for b
56. $A = \frac{a + b + c}{3}$, for b
57. $ax + by = c$, for y
58. $ax + by + c = 0$, for y
59. $V = \pi r^2 h$, for h
60. $V = \frac{1}{3}\pi r^2 h$, for h

Use the formula in Example 9, $d = \frac{1}{2}n^2 - \frac{3}{2}n$, *to find the number of diagonals in a figure with the given number of sides.*

61. 10 sides

62. 6 sides

Use the formula $C = \frac{5}{9}(F - 32)$ *to find the Celsius temperature* (C) *equivalent to the given Fahrenheit temperature* (F).

63. $F = 50°$

64. $F = 86°$

Use the formula $F = \frac{9}{5}C + 32$, *to find the Fahrenheit temperature* (F) *equivalent to the given Celsius temperature* (C).

65. $C = 35°$

66. $C = 10°$

In chemistry the ideal gas law is $P = KT/V$, *where P is pressure, T is temperature, V is volume, and K is a constant. Find the missing quantity.*

67. $T = 10, K = 1, V = 1$

68. $T = 30, P = 3, K = 0.5$

69. $P = 80, T = 100, V = 5$

70. $P = 100, K = 2, V = 6$

The sum of the first n *even numbers can be found by the formula* $S = n^2 + n$. *Find the sum of the numbers indicated.*

71. First 5 even numbers.

72. First 10 even numbers.

In Exercises 73 through 76, use the simple interest formula. See Examples 1 and 2.

73. Mr. Thongsophaporn borrowed \$4000 for 3 years at 12% simple interest per year. How much interest did he pay?

74. Ms. Rodriguez lent her brother \$4000 for a period of 2 years. At the end of the 2 years, her brother repaid the \$4000 plus \$640 interest. What simple interest rate did her brother pay?

75. Ms. Levy invested a certain amount of money in a savings account paying 7% simple interest per year. When she withdrew her money at the end of 3 years, she received \$1050 in interest. How much money did Ms. Levy place in the savings account?

76. Mr. O'Connor borrowed \$6000 at $7\frac{1}{2}\%$ simple interest per year. When he withdrew his money, he received \$1800 in interest. How long had he left his money in the account?

Use the formulas given in Tables 3.1, 3.2, and 3.3 to work Exercises 77–90. See Examples 3–8.

77. Find the perimeter of a triangle whose sides are 5 inches, 12 inches, and 13 inches.

78. Find the area of a rectangle whose length is 9 inches and whose width is 4 inches.

79. Find the area of a triangle whose base is 6 centimeters and whose height is 8 centimeters.

80. Find the perimeter of a rectangle whose length is 5 meters and whose width is 3 meters.

81. Find the area of a circle whose radius is 4 inches. Use 3.14 for π.

82. Find the area of a circle whose diameter is 6 centimeters.

83. Find the circumference of a circle whose diameter is 8 inches.

84. Find the area of a trapezoid whose height is 2 feet and whose bases are 6 feet and 4 feet.

85. The area of the smallest post office in America (in Ochopee, Florida) is 48 square feet. If the length of the post office is 6 feet, find the width of the post office.

86. A sail on a sailboat is in the shape of a triangle. If the area of the sail is 36 square feet and the height of the sail is 12 feet, find the base of the sail.

87. The largest banyon tree in the continental United States is at the Edison House in Fort Myers, Florida. The circumference of the aerial roots of the tree is 390 feet. Find **(a)** the radius of the aerial roots to the nearest tenth of a foot, and **(b)** the diameter of the aerial roots to the nearest tenth of a foot.

See Exercise 87.

88. Donovan's garden is in the shape of a trapezoid. If the height of the trapezoid is 12 meters, one base is 15 meters, and the area is 126 square meters, find the length of the other base..

89. An oil drum has a height of 4 feet and a diameter of 22 inches. Find the volume of the drum in cubic inches.

90. Find the volume of an ice cream cone (cone only) if its diameter is 3 inches and its height is 5 inches.

91. By using any formula for area, explain why area is measured in square units.

92. By using any formula for volume, explain why volume is measured in cubic units.

93. **(a)** Consider the formula for the circumference of a circle, $C = 2\pi r$. If you solve this formula for π, what will you obtain?

(b) If you take the ratio of the circumference of a circle to its diameter, about what numerical value will you obtain? Explain how you determined your answer.

(c) Carefully draw a circle, make it at least 4 inches in diameter. Use a piece of string and a ruler to determine the circumference and diameter of the circle. Find the ratio of the circumference to the diameter. When you divide the circumference by the diameter, what value do you obtain?

CUMULATIVE REVIEW EXERCISES

[1.9] **94.** Evaluate $[4(12 \div 2^2 - 3)^2]^2$.

[2.6] **95.** A stable has 4 Morgan and 6 Arabian horses. Find the ratio of Arabians to Morgans.

96. It takes 3 minutes to siphon 25 gallons of water out of a swimming pool. How long will it take to empty a 13,500-gallon swimming pool by siphoning? Write a proportion that can be used to solve the problem, and then find the desired value.

[2.7] **97.** Solve $2(x - 4) \geq 3x + 9$

Group Activity/ Challenge Problems

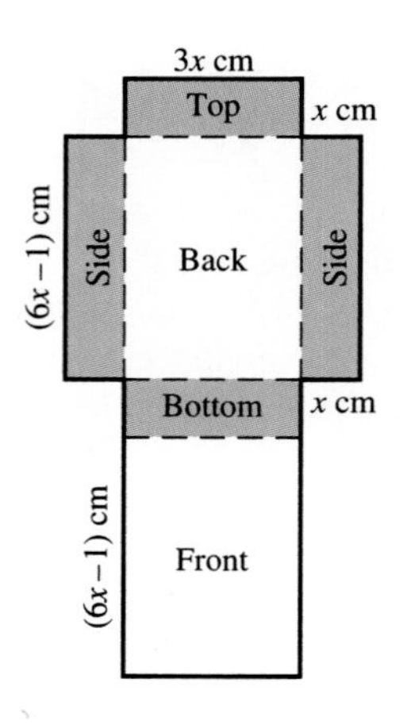

1. **(a)** Using the formulas presented in this section, write an equation in d that can be used to find the shaded area in the figure shown.

(b) Find the shaded area when $d = 4$ feet.

(c) Find the shaded area when $d = 6$ feet.

2. A cereal box is to be made by folding the cardboard along the dashed lines as shown in the figure on the left.

(a) Using the formula

$$\text{volume} = \text{length} \cdot \text{width} \cdot \text{height}$$

write an equation for the volume of the box.

(b) Find the volume of the box when $x = 7$ cm.

(c) Write an equation for the surface area of the box.

(d) Find the surface area when $x = 7$ cm.

3. Earth's diameter is 3963 miles and the moon's diameter is 2160 miles. The moon travels in an elliptical orbit around the Earth. From the center of Earth to the center of the moon the minimum distance is 221,463 miles and the maximum distance is 252,710 miles. Assuming that Earth and the moon are spheres, find **(a)** the nearest approach of their surfaces, **(b)** the farthest approach of their surfaces, and **(c)** the circumference of the moon.

4. The Pantheon is an ancient building in Rome constructed about A.D. 126. It is shaped like a circular cylinder with a dome on top. The outside circumference of the cylinder is about 446 feet.

(a) Find the radius and diameter of the cylindrical part of the Pantheon.

(b) If the walls of the Pantheon are 4 feet thick, find the inside diameter of the floor of the Pantheon.

(c) Find the surface area of the marble floor inside the Pantheon.

(d) If the height of the cylindrical part of the Pantheon (excluding the domed portion) is 120 feet, find its inside volume.

3.2 Changing Application Problems into Equations

Tape 5

1. Translate phrases into mathematical expressions.
2. Write expressions involving percent.
3. Express the relationship between two related quantities.
4. Write expressions involving multiplication.
5. Translate application problems into equations.

Translate Phrases into Mathematical Expressions

1 One practical advantage of knowing algebra is that you can use it to solve everyday problems involving mathematics. For algebra to be useful in solving everyday problems, you must first be able to transform application problems into mathematical language. The purpose of this section is to help you take a verbal or word problem and write it as a mathematical equation.

Often the most difficult part of solving an application problem is translating it into an equation. Before you can translate a problem into an equation, you must understand the meaning of certain statements and how they are expressed mathematically. Here are examples of statements represented as algebraic expressions.

Verbal	*Algebraic*
5 more than a number	$x + 5$
a number increased by 3	$x + 3$
7 less than a number	$x - 7$
a number decreased by 12	$x - 12$
twice a number	$2x$
the product of 6 and a number	$6x$
one-eighth of a number	$\frac{1}{8}x$ or $\frac{x}{8}$
a number divided by 3	$\frac{1}{3}x$ or $\frac{x}{3}$
4 more than twice a number	$2x + 4$
5 less than three times a number	$3x - 5$
3 times the sum of a number and 8	$3(x + 8)$
twice the difference of a number and 4	$2(x - 4)$

To give you more practice with the mathematical terms, we will also convert some algebraic expressions into verbal expressions. Often an algebraic expression can be written in several different ways. Following is a list of some of the possible verbal expressions that can be used to represent the given algebraic expression.

Algebraic	*Verbal*
$2x + 3$	Three more than twice a number
	The sum of twice a number and three
	Twice a number, increased by three
	Three added to twice a number
$3x - 4$	Four less than three times a number
	Three times a number, decreased by four
	The difference of three times a number and four
	Four subtracted from three times a number

EXAMPLE 1 Express each phrase as an algebraic expression.

(a) The distance, d, increased by 10 miles.

(b) 6 less than twice the area.

(c) 3 pounds more than four times the weight.

(d) Twice the sum of the height plus 3 feet.

Solution: **(a)** $d + 10$ **(b)** $2a - 6$

(c) $4w + 3$ **(d)** $2(h + 3)$

In Example 1, the letter x (or any other letter) could have been used in place of those selected.

EXAMPLE 2 Write three different verbal statements to represent the following expressions.

(a) $5x - 2$ **(b)** $2x + 7$

Solution: **(a) 1.** Two less than five times a number.

2. Five times a number, decreased by two.

3. The difference of five times a number and two.

(b) 1. Seven more than twice a number.

2. Two times a number, increased by seven.

3. The sum of two times a number and seven.

EXAMPLE 3 Write a verbal statement to represent each expression.

(a) $3x - 4$ **(b)** $3(x - 4)$

Solution: **(a)** One possible statement is: four less than three times a number.

(b) Three times the difference of a number and four.

Write Expressions Involving Percent

2 Since percents are used so often, you must have a clear understanding of how to write expressions involving percent. Study Example 4 carefully.

EXAMPLE 4 Express each phrase as an algebraic expression.

(a) The cost increased by 6%.

(b) The population decreased by 12%.

Solution: **(a)** When shopping we may see a "25% off" sales sign. We assume that this means 25% off *the original cost,* even though this is not stated. This question asks for the cost increased by 6%. We assume that this means the cost increased by 6% of the original cost, and write

$$c + 0.06c$$

original cost (c) — increased by ($+$) — 6% of the original cost ($0.06c$)

Thus, the answer is $c + 0.06c$.

(b) Using the same reasoning as in part **(a)**, the answer is $p - 0.12p$.

COMMON STUDENT ERROR In Example 4(a) we asked you to represent a cost increased by 6%. Note the answer is $c + 0.06c$. Often, students write the answer to this question as $c + 0.06$. It is important to realize that a percent of a quantity must always be a percent multiplied by some number or letter. Some phrases involving the word percent and the correct and incorrect interpretations follow.

Phrase	*Correct*	*Incorrect*
A $7\frac{1}{2}$% sales tax on c dollars	$0.075c$	0.075 (crossed out)
The cost, c, increased by a $7\frac{1}{2}$% sales tax	$c + 0.075c$	$c + 0.075$ (crossed out)
The cost, c, reduced by 25%	$c - 0.25c$	$c - 0.25$ (crossed out)

Express Relationships between Two Related Quantities

3 Sometimes in a problem, two numbers are related to each other in a certain way. We often represent the simplest, or most basic number that needs to be expressed, as a variable, and the other as an expression containing that variable. Some examples follow.

Verbal	*One Number*	*Second Number*
two numbers differ by 3	x	$x + 3$
John's age now and John's age in 6 years	x	$x + 6$
one number is six times the other number	x	$6x$
one number is 12% less than the other	x	$x - 0.12x$

Note that often more than one pair of expressions can be used to represent the two numbers. For example, "two numbers differ by 3" can also be expressed as x and $x - 3$. Let us now look at two more verbal statements.

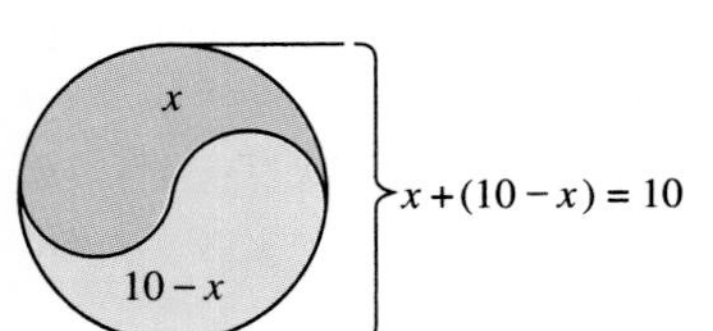

FIGURE 3.6

Verbal	*One Number*	*Second Number*
the sum of two numbers is 10	x	$10 - x$
a 25-foot length of wood cut in two pieces	x	$25 - x$

It may not be obvious why in "the sum of two numbers is 10" the two numbers are represented as x and $10 - x$. Suppose that one number is 2; what is the other number? Since the sum is 10, the second number must be $10 - 2$ or 8. Suppose that one number is 6; the second number must be $10 - 6$, or 4. In general, if the first number is x, the second number must be $10 - x$. Note that the sum of x and $10 - x$ is 10 (Fig. 3.6).

Consider the phrase "a 25-foot length of wood cut in two pieces." If we call one length x, then the other length must be $25 - x$. For example, if one length is 6 feet, the other length must be $25 - 6$ or 19 feet, (Fig. 3.7).

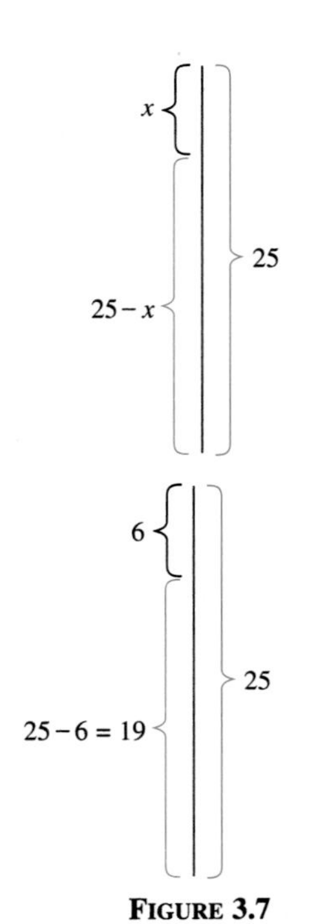

FIGURE 3.7

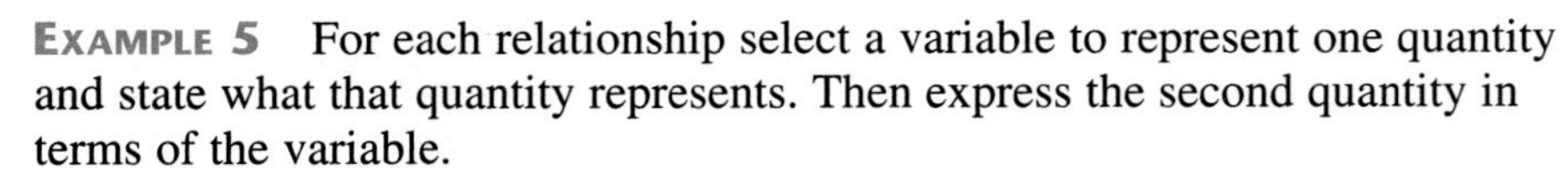

EXAMPLE 5 For each relationship select a variable to represent one quantity and state what that quantity represents. Then express the second quantity in terms of the variable.

(a) A boy is 15 years older than his brother.

(b) The speed of one car is 1.4 times the speed of another.

(c) Two business partners share $75.

(d) John has $5 more than three times the amount that Dee has.

(e) The length of a rectangle is 3 feet less than four times its width.

(f) A number is increased by 6%.

Solution:

(a) Let x be the age of the younger brother; then $x + 15$ is the age of the older brother.

(b) Let s be the speed of the slower car; then $1.4s$ is the speed of the faster car.

(c) Let d be the amount in dollars one partner receives; then $75 - d$ is the amount in dollars the other partner receives.

(d) Let d be Dee's money in dollars; then $3d + 5$ is John's money in dollars.

(e) Let w be the width of the rectangle; then $4w - 3$ is the length of the rectangle.

(f) Let n be the number. Then that number increased by 6% is $n + 0.06n$.

Write Expressions Involving Multiplication

4 Consider the statement "the cost of 3 items at \$5 each." How would you represent this quantity using mathematical symbols? You would probably reason that the cost would be 3 times \$5 and write $3 \cdot 5$ or 3(5).

Now consider the statement "the cost of x items at \$5 each." How would you represent this statement using mathematical symbols? If you use the same reasoning, you might write $x \cdot 5$ or $x(5)$. Another way to write this product is $5x$. Thus, the cost of x items at \$5 each could be represented as $5x$.

Finally, consider the statement "the cost of x items at y dollars each." How would you represent this statement using mathematical symbols? Following the reasoning used in the previous two illustrations, you might write $x \cdot y$ or $x(y)$. Since these products can be written as xy, the cost of x items at y dollars each can be represented as xy.

EXAMPLE 6 Write each of the following as an algebraic expression.

(a) The cost of purchasing x items at \$2 each.

(b) A 5% commission on x dollars in sales.

(c) The number of calories in x potato chips, where each potato chip has 8 calories.

(d) The increase in population in n years for a city growing at a rate of 300 per year.

(e) The distance traveled in t hours when 55 miles are traveled each hour.

Solution:

(a) We can reason like this: one item would cost 1(2) dollars, two items would cost 2(2) dollars, three items 3(2), four items 4(2), and so on. Continuing this reasoning process, we can see that x items would cost $x(2)$ or $2x$ dollars.

(b) A 5% commission on \$1 sales would be 0.05(1), on \$2 sales 0.05(2), on \$3 sales 0.05(3), on \$4 sales 0.05(4), and so on. Therefore, the commission on sales of x dollars would be $0.05(x)$ or $0.05x$.

(c) $8x$

(d) $300n$

(e) $55t$

EXAMPLE 7 A slice of white bread contains 65 calories and a slice of whole-wheat bread contains 55 calories. Write an algebraic expression to represent the total number of calories in x slices of white and y slices of whole-wheat bread.

Solution: x slices of white bread contain $65x$ calories.

y slices of whole-wheat bread contain $55y$ calories.

Together they contain $65x + 55y$ calories.

EXAMPLE 8 Write an algebraic expression for each phrase.

(a) The number of ounces in x pounds.

(b) The number of cents in a dimes and b nickels.

(c) The number of seconds in x hours, y minutes, and z seconds (3600 seconds = 1 hour).

Solution:

(a) Since each pound contains 16 ounces, x pounds is $16 \cdot x$ or $16x$ ounces.

(b) $10a + 5b$

(c) $3600x + 60y + z$

Some terms that we will be using are consecutive integers, consecutive even integers, and consecutive odd integers. **Consecutive integers** are integers that differ by 1 unit. For example, the integers 6 and 7 are consecutive integers. Two consecutive integers may be represented as x and $x + 1$. **Consecutive even integers** are even integers that differ by 2 units. For example, 6 and 8 are consecutive even integers. **Consecutive odd integers** are odd integers that differ by 2 units. For example, 7 and 9 are consecutive odd integers. Two consecutive even integers, or two consecutive odd integers, may be represented as x and $x + 2$.

Translate Application Problems into Equations

5 The word *is* in a verbal problem often means *is equal to* and is represented by an equal sign. Some examples of verbal problems written as equations follow:

Verbal	*Equation*
6 more than twice a number *is* 4	$2x + 6 = 4$
a number decreased by 4 *is* 3 more than twice the number	$x - 4 = 2x + 3$
the product of two consecutive integers *is* 56	$x(x + 1) = 56$
one number is 4 more than three times the other number; their sum *is* 60	$x + (3x + 4) = 60$
a number increased by 15% *is* 120	$x + 0.15x = 120$
the sum of two consecutive odd integers *is* 24	$x + (x + 2) = 24$

Now let us translate some equations into verbal statements. Some examples of equations written as verbal statements follow. We will write only two verbal statements for each equation, but remember there are other ways these equations can be written.

Equation	*Verbal*
$3x - 4 = 4x + 3$	Four less than three times a number *is* three more than four times the number.
	Three times a number, decreased by four *is* four times the number, increased by three.
$3(x - 2) = 6x - 4$	Three times the difference of a number and two *is* four less than six times the number.
	The product of three and the difference of a number and two *is* six times the number, decreased by four.

EXAMPLE 9 Write two verbal statements to represent the equation $x - 2 = 3x - 5$.

Solution:

1. A number decreased by two *is* five less than three times the number.
2. The difference of a number and two *is* the difference of three times the number and five.

EXAMPLE 10 Write a verbal statement to represent the equation $x + 2(x - 4) = 6$.

Solution: The sum of a number and twice the difference of the number and four *is* six.

EXAMPLE 11 Write each problem as an equation.

(a) One number is four less than twice the other. Their sum is 14.

(b) For two consecutive integers, the sum of the smaller and three times the larger is 23.

Solution:

(a) First, express the two numbers in terms of the variable.

$$\text{Let } x = \text{one number}$$
$$\text{then } 2x - 4 = \text{second number}$$

Now we write the equation using the information given.

$$\text{First number} + \text{second number} = 14$$
$$x + (2x - 4) = 14$$

(b) First, express the two consecutive integers in terms of the variable.

$$\text{Let } x = \text{smaller consecutive integer}$$
$$\text{then } x + 1 = \text{larger consecutive integer}$$

Now we write the equation using the information given.

$$\text{Smaller} + \text{three times the larger} = 23$$
$$x + 3(x + 1) = 23$$

EXAMPLE 12 Write the following problem as an equation. One train travels 3 miles more than twice the distance another train travels. The total distance traveled by both trains is 800 miles.

Solution: First express the distance traveled by each train in terms of the variable.

Let x = distance traveled by one train

then $2x + 3$ = distance traveled by second train

Now write the equation using the information given.

distance of train 1 + distance of train 2 = total distance

$$x + (2x + 3) = 800$$

EXAMPLE 13 Express each of the following as an equation.

(a) The cost of renting a snow blower for x days at \$12 per day is \$60.

(b) The population of the town of Newton is increasing at a rate of 500 people per year. The increase in population in t years is 2500.

(c) The distance Dawn and Jack traveled for x days at 600 miles per day is 1500 miles.

(d) The number of cents in d dimes is 120.

Solution: **(a)** $12x = 60$ **(b)** $500t = 2500$

(c) $600x = 1500$ **(d)** $10d = 120$

Exercise Set 3.2

Write as an algebraic expression.

1. Five more than a number.
2. Seven less than a number.
3. Four times a number.
4. The product of a number and eight
5. 70% of a number x.
6. 8% of a number y.
7. A 10% sales tax on a piano costing c dollars.
8. A $7\frac{1}{2}$% sales tax on a car costing p dollars.
9. The 16% of the U.S. population, p, who do not receive adequate nourishment.
10. Only 7% of all U.S. tires, t, are recycled.
11. Three less than six times a number.
12. Six times the diffference of a number and 3.
13. Seven plus three-fourths of a number.
14. Four times a number, decreased by two.
15. Twice the sum of a number and 8.
16. Seventeen decreased by x.
17. The cost of purchasing x rolls of electrical tape at \$4 each.
18. The rental fee for subscribing to Home Box Office for x months at \$12 per month.
19. The cost in dollars of traveling x miles at 23 cents per mile.
20. The distance traveled in t hours when traveling 30 miles per hour.
21. The cost of renting a mailbox for b months at a cost of \$8.20 per month.
22. The cost of waste disposal for y months at \$16 per month.
23. The population growth of a city in n years if the city is growing at a rate of 300 persons per year.
24. The number of calories in x grams of carbohydrates if each gram of carbohydrate contains 4 calories.
25. The number of cents in x quarters.
26. The number of cents in x quarters and y dimes.
27. The number of inches in x feet.

28. The number of inches in x feet and y inches.

29. The number of ounces in c pounds.

30. The number of ounces in c pounds and d ounces.

31. An average chicken egg contains 275 milligrams (mg) of cholesterol and an ounce of chicken contains about 25 mg of cholesterol. Write an expression for the amount of cholesterol in x chicken eggs and y ounces of chicken.

32. According to U.S. guidelines, each gram of carbohydrates contains 4 calories, each gram of protein contains 4 calories, and each gram of fat contains 9 calories. Write an expression for the number of calories in a serving of a product that contains x grams of carbohydrates, y grams of protein, and z grams of fat.

Express as a verbal statement. (There are many acceptable answers.)

33. $x - 6$

34. $x + 3$

35. $4x + 1$

36. $3x - 4$

37. $5x - 7$

38. $2x - 3$

39. $4x - 2$

40. $5 - x$

41. $2 - 3x$

42. $4 + 6x$

43. $2(x - 1)$

44. $3(x + 2)$

Select a variable to represent one quantity and state what that variable represents. Then express the second quantity in terms of the variable.

45. Eileen's salary is $45 more than Martin's salary.

46. A boy is 12 years older than his brother.

47. A number is one-third of another.

48. Two consecutive integers.

49. Two consecutive even integers.

50. One hundred dollars divided between two people.

51. Two numbers differ by 12.

52. A number is 5 less than four times another number.

53. A number is 3 more than one-half another number.

54. A Cadillac costs 1.7 times as much as a Ford.

55. A number is 4 less than three times another number.

56. An 80-foot tree cut into two pieces.

57. Two consecutive odd integers.

58. A number and the number increased by 12%.

59. A number and the number decreased by 15%.

60. The cost of an item and the cost increased by a 7% sales tax.

Use the given variable to represent one quantity. Express the second quantity in terms of the first.

61. The cost of an item, c, and the cost decreased by 10%.

62. The president's salary, s, and the president's salary increased by 3%.

63. The pollution level, p, and the pollution level reduced by 50%.

64. The federal deficit, d, and the federal deficit reduced by 5%.

65. The world population, w, and double the world population.

66. The sales of compact disks, s, and the sales cut in half.

67. The mileage of a car, m, and the mileage increased by 15%.

68. The number of students, n, earning a grade of A in this course, and that number increased by 100%.

Express as an equation.

69. One number is five times another. The sum of the two numbers is 18.

70. Marie is 6 years older than Denise. The sum of their ages is 48.

71. The sum of two consecutive integers is 47.

72. The product of two consecutive even integers is 48.

73. Twice a number, decreased by 8 is 12.

74. For two consecutive integers, the sum of the smaller and twice the larger is 29.

75. One-fifth of the sum of a number and 10 is 150.

76. One train travels six times as far as another. The total distance traveled by both trains is 700 miles.

77. One train travels 8 miles less than twice the other. The total distance traveled by both trains is 1000 miles.

78. One number is 3 greater than six times the other. Their product is 408.
79. A number increased by 8% is 92.
80. The cost of a car plus a 7% tax is $13,600.
81. The cost of a jacket at a 25%-off sale is $65.
82. The cost of a meal plus a 15% tip is $18.
83. The cost of a videocassette recorder reduced by 20% is $215.
84. The product of a number and the number plus 5% is 120.
85. One number is 3 less than twice another number. Their sum is 21.
86. The cost of renting a phone at a cost of $2.37 per month for x months is $27.
87. The distance traveled by a car going 40 miles per hour for t hours is 180 miles.
88. The cost of traveling x miles at 23 cents per mile is $12.80.
89. The number of calories in y french fried potatoes at 15 calories per french fry is 215.
90. Milltown is increasing at a rate of 200 per year. The increase in population in t years is 2400.
91. The number of cents in q quarters is 150.
92. The number of ounces in p pounds is 64.

In Exercises 93–104, express each equation as a verbal statement. (There are many acceptable answers.)

93. $x + 3 = 6$
94. $x - 5 = 2x$
95. $3x - 1 = 2x + 4$
96. $x - 3 = 2x + 3$
97. $4(x - 1) = 6$
98. $3x + 2 = 2(x - 3)$
99. $5x + 6 = 6x - 1$
100. $x - 3 = 2(x + 1)$
101. $x + (x + 4) = 8$
102. $x + (2x + 1) = 5$
103. $2x + (x + 3) = 5$
104. $2x - (x + 3) = 6$
105. Explain why the cost of purchasing x items at 6 dollars each is represented as $6x$.
106. Explain why the cost of purchasing x items at y dollars each is represented as xy.

CUMULATIVE REVIEW EXERCISES

[2.6] *Write a proportion that can be used to solve each problem. Solve each problem and find the desired values.*

107. A recipe for chicken stew calls for $\frac{1}{2}$ teaspoon of thyme for each pound of poultry. If the poultry for the stew weighs 6.7 pounds, how much thyme should be used?
108. Melinda mixes water with dry cat chow for her cat Max. If the directions say to mix 1 cup of water with every 3 cups of dry cat chow, how much water will Melinda add to $\frac{1}{2}$ cup of dry cat chow?

[3.1] 109. $P = 2l + 2w$; find l when $P = 40$ and $w = 5$
110. Solve $3x - 2y = 6$ for y. Then find the value of y when x has a value of 6.

Group Activity/ Challenge Problems

1. (a) Write an algebraic expression for the number of seconds in d days, h hours, m minutes, and s seconds.
   (b) Use the expression found in part (a) to determine the number of seconds in 4 days, 6 hours, 15 minutes, and 25 seconds.
2. At the time of this writing, the toll for southbound traffic on the Golden Gate Bridge is $1.50 per vehicle axle (there is no toll for northbound traffic).
   (a) If the number of 2-, 3-, 4-, 5-, and 6-axle vehicles are represented with the letters r, s, t, u, and v, respectively, write an *expression* that represents the daily revenue of the Golden Gate Bridge Authority.

(b) Write an *equation* that can be used to determine the daily revenue, d.

*In Exercises 3–7, **(a)** write down the quantity you are being asked to find and represent this quantity with a varible, and **(b)** write an equation containing that variable that can be used to solve the problem. Do not solve the equation. We will discuss problems like those that follow in the next section.*

3. An average bath uses 30 gallons of water and an average shower uses 6 gallons of water per minute. How long a shower would result in the same water usage as a bath?

4. A subway token costs $1.20. It takes one token to go to work and one token to return from work daily. A monthly subway pass costs $50 and provides for unlimited use of the subway. How long would it take for the daily subway cost to equal the cost of the monthly pass?

5. The average American produces 40,000 pounds of carbon dioxide each year by driving a car, running air conditioners, lighting, and using appliances and other items that require the burning of fossil fuels. How long would it take for the average American to produce 1,000,000 pounds of carbon dioxide?

6. An employee has a choice of two salary plans. Plan A provides a weekly salary of $200 plus a 5% commission on the employee's sales. Plan B provides a weekly salary of $100 plus an 8% commission on the employee's sales. What must be the weekly sales for the two plans to give the same weekly salary?

7. The cost of renting an 18-foot truck from Mertz is $20 a day plus 60 cents a mile. The cost of renting a similar truck from U-Hail is $30 a day plus 45 cents a mile. How far would you have to drive the rental truck in 1 day for the total cost to be the same with both companies?

3.3 Solving Application Problems

Tape 5

1. Set up and solve verbal problems.
2. Selecting a mortgage.
3. Solving application problems containing large numbers.

There are many types of application problems that can be solved using algebra. In this section we introduce several types. In Section 3.4, we introduce additional types of application problems. Application problems are also presented in many other sections and exercise sets throughout the book. Your instructor may not have time to cover all the applications given in this book. If not, you may still wish to spend a little time on your own reading those problems just to get a feel for the types of applications presented.

To be prepared for this section, you must understand the material presented in Section 3.2. The best way to learn to set up a verbal or word problem is to practice. The more verbal problems you study and attempt, the easier it will become to solve them.

Set Up and Solve Verbal Problems

1 We often transform verbal problems into mathematical terms without realizing it. For example, if you need 3 cups of milk for a recipe and the measuring cup holds only 2 cups, you reason that you need 1 additional cup of milk after the initial 2 cups. You may not realize it, but when you do this simple operation, you are using algebra.

Let x = number of additional cups of milk needed

Thought process: (initial 2 cups) + $\left(\begin{array}{c}\text{number of}\\\text{additional cups}\end{array}\right)$ = total milk needed

Equation to represent problem: $2 + x = 3$

When we solve for x, we get 1 cup of milk.

You probably said to yourself: Why do I have to go through all this when I know that the answer is $3 - 2$ or 1 cup? When you perform this subtraction, you have mentally solved the equation $2 + x = 3$.

$$
\begin{aligned}
2 + x &= 3\\
2 - 2 + x &= 3 - 2\\
x &= 3 - 2\\
x &= 1
\end{aligned}
$$

Let's look at another example.

EXAMPLE 1 Suppose that you are at a supermarket, and your purchases so far total \$13.20. In addition to groceries, you wish to purchase as many packages of gum as possible, but you have a total of only \$18. If a package of gum costs \$1.15, how many can you purchase?

Solution: How can we represent this problem as an equation? We might reason as follows. We need to find the number of packages of gum. Let us call this unknown quantity x.

Let x = number of packages of gum

Thought process: cost of groceries + cost of gum = total cost

Substitute \$13.20 for the cost of groceries and \$18 for the total cost to get

13.20 + cost of gum = 18

At this point you might be tempted to replace the cost of gum with the letter x. But look at what x represents. The variable x represents the *number* of packages of gum, *not the cost of the gum.* In Section 3.2 we learned that the cost of x packages of gum at \$1.15 per package is $1.15x$. Now substitute the cost of the x packages of gum, $1.15x$, into the equation to obtain

Equation to represent problem: $13.20 + 1.15x = 18$

When we solve this equation, we obtain $x = 4.2$ packages (to the nearest tenth). Since you cannot purchase a part of a pack of gum, only 4 packages of gum can be purchased.

Now let us look at the procedure for setting up and solving a word problem.

To Solve a Word Problem

1. Read the question carefully.
2. If possible, draw a sketch to help visualize the problem.
3. Determine which quantity you are being asked to find. Choose a letter to represent this unknown quantity. Write down exactly what this letter represents. If there is more than one unknown quantity, represent all unknown quantities in terms of this variable.

4. Write the word problem as an equation.
5. Solve the equation for the unknown quantity.
6. Answer the question or questions asked.
7. Check the solution in the original stated problem.

Let us now set up and solve some word problems using this procedure.

EXAMPLE 2 Two subtracted from four times a number is 10. Find the number.

Solution: We are asked to find the number. We designate the unknown number by the letter x.
Let x = unknown number.

2 subtracted from 4 times a number is 10

Write the equation:	$4x - 2 = 10$
Solve the equation:	$4x = 12$
Answer the question:	$x = 3$

Check: Substitute 3 for the number in the original problem. Two subtracted from four times a number is 10.

$$4(3) - 2 = 10$$
$$10 = 10 \quad \text{true}$$

Since the solution checks, the unknown number is 3.

EXAMPLE 3 The sum of two numbers is 17. Find the two numbers if the larger is five more than twice the smaller number.

Solution: We are asked to find *two* numbers. We will call the smaller number x. Then we will represent the larger number in terms of x.

Let x = smaller number

then $2x + 5$ = larger number

The sum of the two numbers is 17. Therefore, we write the equation

$$\text{smaller number} + \text{larger number} = 17$$
$$x + (2x + 5) = 17$$

Now solve the equation.

$$3x + 5 = 17$$
$$3x = 12$$
$$x = 4$$

Answer the questions:

$$\begin{aligned} \text{smaller number} &= 4 \\ \text{larger number} &= 2x + 5 \\ &= 2(4) + 5 = 13 \end{aligned}$$

Check: The sum of the two numbers = 17

$$\begin{aligned} 4 + 13 &= 17 \\ 17 &= 17 \quad \text{true} \end{aligned}$$

EXAMPLE 4 Doug and Mila's roller blade company manufactures 10,000 pairs of roller blades each year. They wish to increase production by 1250 pairs of roller blades each year until their yearly production is 25,000 pairs. How long will it take for them to reach their goal?

Solution: We are asked to find the number of years.

$$\begin{aligned} \text{Let } n &= \text{number of years} \\ \text{then } 1250n &= \text{increase in production over } n \text{ years} \end{aligned}$$

$$(\text{present production}) + \begin{pmatrix}\text{increased production} \\ \text{over } n \text{ years}\end{pmatrix} = \text{future production}$$

$$\begin{aligned} 10{,}000 + 1250n &= 25{,}000 \\ 1250n &= 15{,}000 \\ n &= \frac{15{,}000}{1250} \\ n &= 12 \text{ years} \end{aligned}$$

A check will show that 12 years is the correct answer.

EXAMPLE 5 A 32-fluid-ounce container of fruit punch contains 3.84 fluid ounces of pure fruit juice. Find the percent by volume of pure fruit juice in the punch.

Solution: We are asked to find the percent of pure juice.

Let x = percent of pure juice

We use the formula

total volume × percent of pure juice = amount of pure juice

Since 32 is the total volume, x is the percent of pure juice, and 3.84 is the amount of pure juice, the equation is $32x = 3.84$.

$$\begin{aligned} 32x &= 3.84 \\ \frac{32x}{32} &= \frac{3.84}{32} \\ x &= 0.12 \quad \text{or} \quad 12\% \end{aligned}$$

Therefore, the punch is 12% pure fruit juice by volume.

Example 6 John is considering buying a copier for his small business run from his home. He currently pays 8 cents a copy at his local copy center. A new copier is on sale for \$360. How many copies would John need to make at the copy center for the copying cost to equal the cost of the new copier?

Solution: We are asked to find the number of copies.

$$\text{Let } x = \text{number of copies}$$

$$\text{then } 0.08x = \text{cost for making } x \text{ copies}$$

$$\text{cost for making } x \text{ copies} = \text{cost of the new copier}$$

$$0.08x = 360$$

$$x = \frac{360}{0.08} = 4500$$

Thus, after 4500 copies are made the cost would be equal. Of course, other factors, such as the cost of paper, convenience, and distance to the copy center, must be considered when deciding whether to purchase the copier.

Example 7 The cost of renting a Me-Haul 24-foot truck for local moving is \$50 a day plus 40 cents a mile. Find the maximum distance that Mrs. Ahmed can travel if she has only \$80.

Solution: We are asked to find the number of miles Mrs. Ahmed can drive.

$$\text{Let } x = \text{number of miles}$$

$$\text{then } 0.40x = \text{cost of driving } x \text{ miles}$$

$$\text{daily cost} + \text{mileage cost} = \text{total cost}$$

$$50 + 0.40x = 80$$

$$0.40x = 30$$

$$\frac{0.40x}{0.40} = \frac{30}{0.40}$$

$$x = 75$$

Therefore, Mrs. Ahmed can drive 75 miles in one day.

Example 8 At A^+ Auto the new owner is giving her sales staff a choice of salary plans. Plan 1 is a \$200 per week base salary plus a 2% commission on sales. Plan 2 is a straight 8% commission on sales.

(a) Maria must select one of the plans but is not sure of the sales needed for her weekly salary to be the same under the two plans. Can you determine it?

(b) If Maria is certain that she can make \$6000 sales per week, which plan should she select?

Solution: **(a)** We are asked to find the dollar sales.

$$\text{Let } x = \text{dollar sales}$$

$$\text{then } 0.02x = \text{commission from plan 1 sales}$$

$$\text{and } 0.08x = \text{commission from plan 2 sales}$$

$$\begin{aligned}
\text{salary from plan 1} &= \text{salary from plan 2}\\
\text{base salary} + 2\%\text{ commission} &= 8\%\text{ commission}\\
200 + 0.02x &= 0.08x\\
200 &= 0.06x\\
0.06x &= 200\\
\frac{0.06x}{0.06} &= \frac{200}{0.06}\\
x &= 3333.33
\end{aligned}$$

If Maria's sales are about $3333 dollars, both plans would give her about the same salary.

(b) If Maria's sales are $6000, she would earn more weekly by working on straight commission, plan 2. Check this out yourself by computing Maria's salary under both plans and comparing them.

EXAMPLE 9 Allied Airlines wishes to keep its airfare, including a 7% tax, between Dallas, Texas, and Los Angeles, California, at exactly $160. Find the cost of the ticket before tax.

Solution: We are asked to find the cost of the ticket before tax.

$$\begin{aligned}
\text{Let } x &= \text{cost of the ticket before tax}\\
\text{then } 0.07x &= \text{tax on the ticket}
\end{aligned}$$

$$\begin{aligned}
\begin{pmatrix}\text{cost of ticket}\\ \text{before tax}\end{pmatrix} + \begin{pmatrix}\text{tax on}\\ \text{the ticket}\end{pmatrix} &= 160\\
x + 0.07x &= 160\\
1.07x &= 160\\
x &= \frac{160}{1.07}\\
x &= 149.53
\end{aligned}$$

Thus, if Allied prices the ticket at $149.53, the total cost including a 7% tax will be $160.

EXAMPLE 10 According to a will an estate is to be divided among two grandchildren and two charities. The two grandchildren, Alisha and Rayanna, are each to receive twice as much as each of the two charities, the Red Cross and the Salvation Army. If the estate is valued at $240,000, how much will each grandchild and each charity receive?

Solution: We are asked how much each grandchild and each charity will receive. Since the grandchildren will each receive twice as much as the charities, we will let x represent the amount each charity receives.

$$\begin{aligned}
\text{Let } x &= \text{amount each charity receives}\\
\text{then } 2x &= \text{amount each grandchild receives}
\end{aligned}$$

The total received by the two grandchildren and two charities is $240,000. Thus, the equation we use is:

$$\underbrace{x + x}_{\text{each charity receives } x} + \underbrace{2x + 2x}_{\text{each grandchild receives } 2x} = 240{,}000$$
$$6x = 240{,}000$$
$$x = 40{,}000$$

Thus, the American Red Cross and Salvation Army will receive $40,000, and each grandchild will receive 2(40,000), or $80,000.

EXAMPLE 11 Mr. and Mrs. Frank plan to install a security system in their house. They have narrowed down their choices to two security dealers: Moneywell and Doile security systems. Moneywell's system costs $3360 to install and their monitoring fee is $17 per month. Doile's equivalent system costs only $2260 to install, but their monitoring fee is $28 per month.

(a) Assuming that their monthly monitoring fees do not change, in how many months would the total cost of Moneywell's and Doile's system be the same?

(b) If both dealers guarantee not to raise monthly fees for 10 years, and if you plan to use the system for 10 years, which system would be the least expensive?

Solution: **(a)** Doile's system has a smaller initial cost ($2260 vs. $3360); however, their monthly monitoring fees are greater ($28 vs. $17). We are asked to find the number of months after which the total cost of the two systems will be the same.

Let n = number of months
then $17n$ = monthly monitoring cost for Moneywell's system for n months
and $28n$ = monthly monitoring cost for Doile's system for n months

$$\text{total cost of Moneywell} = \text{total cost of Doile}$$
$$\begin{pmatrix}\text{initial}\\\text{cost}\end{pmatrix} + \begin{pmatrix}\text{monthly cost}\\\text{for } n \text{ months}\end{pmatrix} = \begin{pmatrix}\text{initial}\\\text{cost}\end{pmatrix} + \begin{pmatrix}\text{monthly cost}\\\text{for } n \text{ months}\end{pmatrix}$$
$$3360 + 17n = 2260 + 28n$$
$$1100 + 17n = 28n$$
$$1100 = 11n$$
$$100 = n$$

The total cost would be the same in 100 months or about 8.3 years.

(b) Over a 10-year period Moneywell's system would be less expensive. After 8.3 years Moneywell will be less expensive because of their lower monthly cost. Determine the cost for both Moneywell and Doile for 10 years of use now.

Selecting a Mortgage

2 Many of you will purchase a house. Choosing the wrong mortgage can cost you thousands of extra dollars. Table 3.4 is used to determine monthly mortgage payments of principal and interest. The table gives the monthly mortgage payment per $1000 of mortgage at different mortgage rates for various terms of the loan. For example, for a mortgage for 30 years at 7.5%, the monthly payment of principal and interest is $7.00 per $1000 borrowed (circled in table). Thus, for a $50,000 mortgage for 30 years at 7.5% the monthly mortgage payment would be 50 times $7.00 or $350.

$$50(7.00) = \$350$$

TABLE 3.4 ANY BANK, USA: EQUAL MONTHLY PAYMENT TO AMORTIZE A LOAN OF $1,000

Rate	Payment for a Mortgage Period (years) of:				Rate	Payment for a Mortgage Period (years) of:			
(%)	15	20	25	30	(%)	15	20	25	30
4.500	7.65	6.33	5.56	5.07	8.625	9.93	8.76	8.14	7.78
4.625	7.71	6.39	5.63	5.14	8.750	10.00	8.84	8.23	7.87
4.750	7.78	6.46	5.70	5.22	8.875	10.07	8.92	8.31	7.96
4.875	7.84	6.53	5.77	5.29	9.000	10.15	9.00	8.40	8.05
5.000	7.91	6.60	5.85	5.37	9.125	10.22	9.08	8.48	8.14
5.125	7.97	6.67	5.92	5.44	9.250	10.30	9.16	8.57	8.23
5.250	8.04	6.73	6.00	5.52	9.375	10.37	9.24	8.66	8.32
5.375	8.10	6.81	6.07	5.60	9.500	10.45	9.33	8.74	8.41
5.500	8.17	6.88	6.14	5.68	9.625	10.52	9.41	8.83	8.50
5.625	8.24	6.95	6.22	5.76	9.750	10.60	9.49	8.92	8.60
5.750	8.30	7.02	6.29	5.84	9.875	10.67	9.57	9.00	8.69
5.875	8.37	7.09	6.37	5.92	10.000	10.75	9.66	9.09	8.78
6.000	8.44	7.16	6.44	6.00	10.125	10.83	9.74	9.18	8.87
6.125	8.51	7.24	6.52	6.08	10.250	10.90	9.82	9.27	8.97
6.250	8.57	7.31	6.60	6.16	10.375	10.98	9.90	9.36	9.06
6.375	8.64	7.38	6.67	6.24	10.500	11.06	9.99	9.45	9.15
6.500	8.71	7.46	6.75	6.32	10.625	11.14	10.07	9.54	9.25
6.625	8.78	7.53	6.83	6.40	10.750	11.21	10.16	9.63	9.34
6.750	8.85	7.60	6.91	6.49	10.875	11.29	10.24	9.72	9.43
6.875	8.92	7.68	6.99	6.57	11.000	11.37	10.33	9.81	9.53
7.000	8.99	7.76	7.07	6.66	11.125	11.45	10.41	9.90	9.62
7.125	9.06	7.83	7.15	6.74	11.250	11.53	10.50	9.99	9.72
7.250	9.13	7.91	7.23	6.83	11.375	11.61	10.58	10.08	9.81
7.375	9.20	7.98	7.31	6.91	11.500	11.69	10.67	10.17	9.91
7.500	9.28	8.06	7.39	7.00	11.625	11.77	10.76	10.26	10.00
7.625	9.35	8.14	7.48	7.08	11.750	11.85	10.84	10.35	10.10
7.750	9.42	8.21	7.56	7.17	11.875	11.93	10.93	10.44	10.20
7.875	9.49	8.29	7.64	7.26	12.000	12.01	11.02	10.54	10.29
8.000	9.56	8.37	7.72	7.34	12.125	12.09	11.10	10.63	10.39
8.125	9.63	8.45	7.81	7.43	12.250	12.17	11.19	10.72	10.48
8.250	9.71	8.53	7.89	7.52	12.375	12.25	11.28	10.82	10.58
8.375	9.78	8.60	7.97	7.61	12.500	12.33	11.37	10.91	10.68
8.500	9.85	8.68	8.06	7.69					

This payment does not include taxes or insurance, which are sometimes added to the mortgage payment and sometimes paid separately. Also, these figures may be slightly inaccurate because of round-off error.

Sometimes banks charge "points" when they give a loan. One point is 1% of the mortgage. Thus for a \$50,000 mortgage, one point is 0.01(50,000) = \$500 and 3 points is 0.03(50,000) = \$1500.

EXAMPLE 12 Kristen needs a 30-year \$50,000 mortgage. Banc One is charging 7.5% with no points and Collier's Bank is charging 7.125% with 3.00 points.

(a) How long would it take for the total cost of both mortgages to be the same?

(b) If Kristen is planning to sell the house after 5 years, which mortgage should she select?

(c) If Kristen is planning on paying off the loan in 30 years, how much will she save by selecting the 7.125% mortgage?

Solution: **(a)** With the 7.5% mortgage, Kristen's monthly payment is 50(7) = \$350. With the 7.125% mortgage, Kristen's monthly payment is 50(6.74) = \$337. In addition, Kristen must pay 0.03(50,000) = \$1500 because of the 3 points.

Let x = number of months when total payments from both mortgages are equal
then $350x$ = monthly payments for x months with 7.5% loan
and $337x$ = monthly payments for x months with 7.125% loan

Now set up an equation and solve.

$$\begin{pmatrix}\text{monthly payments}\\ \text{for 7.5\% mortgage}\end{pmatrix} = \begin{pmatrix}\text{monthly payments}\\ \text{for 7.125\% mortgage}\end{pmatrix} + \begin{pmatrix}\text{points}\end{pmatrix}$$

$$\begin{aligned} 350x &= 337x + 1500 \\ 13x &= 1500 \\ x &= 115.4 \text{ months} \end{aligned}$$

Thus, the two mortgages would be the same after about 115 months or about 9 years 7 months.

(b) Because of the lower initial cost Banc One's total cost will be lower until about 9 years 7 months. After this, Collier's Bank will have the lower total cost because of their lower monthly payment. For five years Kristen should select the one with the lower initial cost. That is the 7.5% Banc One mortgage.

(c) Over 30 years (360 months) the total cost of each plan is as follows:

7.5%	*7.125%*
350(360) = \$126,000	337(360) + 1500 = 121,320 + 1500
	= 122,820

Thus, over 30 years Kristen would save 126,000 − 122,820 = \$3180 with the 7.125% Collier's Bank mortgage.

Large Numbers

3 **EXAMPLE 13** In 1990 there were about 800 million hectares of rain forest on the earth (about 32 million square miles). That same year 17 million hectares of rain forest were destroyed (an area about twice the size of Ireland). If the deforestation were to continue at the same rate, when would the rain forest be destroyed completely?

Solution: We are asked to find the number of years when the rain forest would be destroyed.

Let x = number of years

then $17x$ = millions of hectares of rain forest lost in x years

area of rain forest in 1990 − area lost in x years = area remaining

$$800 - 17x = 0$$
$$800 = 17x$$
$$\frac{800}{17} = \frac{17x}{17}$$
$$47 = x$$

Thus, at the current rate the rain forest would be destroyed in about 47 years from 1990, which is the year 2037.

Notice in Example 13 that the numbers given were 800 million and 17 million. Since both numbers were given in millions it was not necessary to write the numbers as 800,000,000 and 17,000,000, respectively, to set up and solve the equation. Had we written the equation as $800{,}000{,}000 - 17{,}000{,}000x = 0$ and solved this equation we would have obtained the same answer.

Exercise Set 3.3

For Exercises 1–45, set up an equation that can be used to solve the problem. Solve the equation and answer the question asked. Use a calculator where you feel it is appropriate.

1. The sum of two consecutive integers is 45. Find the numbers.

2. The sum of two consecutive even integers is 106. Find the numbers.

3. The sum of two consecutive odd numbers is 68. Find the numbers.

4. One number is 3 more than twice a second number. Their sum is 27. Find the numbers.

5. One number is 5 less than three times a second number. Their sum is 43. Find the numbers.

6. The sum of three consecutive integers is 39. Find the three integers.

7. The sum of three consecutive odd integers is 87. Find the three integers.

8. The sum of the two facing page numbers in an open book is 149. What are the page numbers?

9. The larger of two integers is 8 less than twice the smaller. When the smaller number is subtracted from the larger, the difference is 17. Find the two numbers.

10. The sum of three integers is 29. Find the three numbers if one number is twice the smallest and the third number is 4 more than twice the smallest.

11. In 1993 the most fuel-efficient car on highways was the Geo Metro xFi and the least efficient was the Vector Acromotive V8. The Geo gets 3 miles per gallon (mpg) more than 5 times that of the Vector. Find the number of miles per gallon of each car if the sum of the miles per gallon for the two cars is 69 mpg.

12. The total number of medals won by Norway in the 1994 Winter Olympic games was twice the amount the United States won. If both countries together won 39 medals, how many medals did each team win?

13. The number of operating nuclear reactors in Canada in 1994 is 6 more than 15 times the number in Mexico. If the sum of the nuclear reactors in 1994 in these two countries is 22, find the number of reactors in each country. (The United States presently has 109 of the world's 424 reactors in use today.)

14. A small town has a population of 4000. If its population increases by 200 per year, how long will it be before the population reaches 5800?

15. Caldwell Banker sold a house for Mrs. Sanchez. The amount that Mrs. Sanchez received after the real estate broker subtracted her 6% commission was $65,800. Find the selling price of the house.

16. A 4-ounce glass of table wine has an alcohol content of 0.48 ounce (14 grams). What is the percent alcohol by volume of the wine?

17. An 18-karat gold bracelet weighing 20 grams contains 15 grams of pure gold. What is the percent of pure gold by weight contained in 18-karat gold?

18. It cost Teshanna $5.75 a week to wash and dry her clothing at the corner laundry. If a washer and dryer cost a total of $747.50, how many weeks would it take for Teshanna's laundry cost to equal the cost of purchasing a washer and dryer?

19. The rock group Purple Finger received $1500 plus $2.00 per head for their performance at the Big Mac Arena. If the total they received for their performance was $3100, how many people were in attendance?

20. The cost of renting a truck is $39 a day plus 40 cents per mile. How far can Milt drive in one day if he has only $75?

21. A tennis star was hired to sign autographs at a convention. She was paid $2000 plus 2% of all admission fees collected at the door. If the total amount she received for the day was $2400, find the total amount collected at the door.

22. State College reimburses its employees $40 a day plus 21 cents a mile when they use their own vehicles on college business. If Professor Kohn takes a 1-day trip and is reimbursed $103 how far did she travel?

23. At Ace Warehouse, for a yearly fee of $60 you save 8% of the price of all items purchased in the store. What would be the total Mary would need to spend during the year so that her savings equal the yearly fee?

24. At a one-day 20%-off sale, Tan purchased a hat for $15.99. What is the regular price of the hat?

25. During the 1995 contract negotiations the city school board approved an 8% pay increase for its teachers effective in 1996. If Paul, a first-grade teacher, projects his 1996 annual salary to be $37,800, what is his present salary?

26. Mr. Murphy receives a weekly salary of $210. He also receives a 6% commission on the total dollar volume of all sales he makes. What must his dollar volume be in a week if he is to make a total of $450?

27. Eunice, a hot dog vendor, wishes to price her hot dogs such that the total cost of the hot dog, including a 7% tax, is $1.50. What will be the price of a hot dog before tax?

28. Essex County has an 8% sales tax. How much does Karita's car cost before tax if the total cost of the car plus its sales tax is $12,800?

29. Mr. Wironowski left his estate of $210,000 to his two children and his favorite charity. If each child is to receive three times the amount left to his favorite charity, how much did each child and the charity receive?

30. Ninety-one hours of overtime must be split among four workers. The two younger workers are to be assigned the same number of hours. The third worker is to be assigned twice as much as each of the younger workers. The fourth worker is to be assigned three times as much as each of the younger workers. How much overtime should be assigned to each worker?

31. Presently only about 4000 landfills are operating in the United States (down from 14,000 in 1978). This number is expected to drop by 300 per year. How long will it take for the number of landfills to drop to 2000?

32. Gary Gutchell worked a 55-hour week last week. He is not sure of his hourly rate, but knows that he is paid $1\frac{1}{2}$ times his regular hourly rate for all hours over a 40-hour week. His pay last week was \$400. What was his hourly rate?

33. Installing a water-saving showerhead saves about 60% of the water when you shower. If a 10-minute shower with a water-saving showerhead uses 24 gallons of water, how much water would a 10-minute shower without the special showerhead use?

34. The Midtown Tennis Club has two payment plans for its members. Plan 1 has a monthly fee of \$20 plus \$8 per hour court rental time. Plan 2 has no monthly fee, but court time is \$16.25 per hour. If court time is rented in 1-hour intervals, how many hours would you have to play per month so that plan 1 becomes a better buy?

35. The fine for speeding in Boomtown is \$5 per mile per hour over the speed limit plus a \$15 administrative charge. Michelle received a speeding ticket and had to pay a fine of \$65. How many miles per hour over the speed limit was she traveling?

36. Miss Dunn is on vacation and stops at a store to buy some film and other items. Her total bill before tax was \$22. After tax the bill came to \$23.76. Find the local sales tax rate.

37. The *World Almanac* identifies Punta Gorda, Florida as the fastest-growing metropolitan area in the United States from 1980 to 1990. From 1980 to 1990 the population grew about 90%. The almanac gives the 1990 population as 110,000 people but does not give the 1980 population. Can you find the 1980 population?

38. The city with the largest population loss from 1980 to 1990 according to the *World Almanac* was Gary, Indiana. Gary lost about 23% of its population during this period. If Gary's 1990 population was about 116,000, find the population of Gary, Indiana in 1980.

39. After Mrs. Egan is seated in a restaurant, she realizes that she has only \$20. If from this \$20 she must pay a 7% tax and she wishes to leave a 15% tip, what is the maximum price for a meal that she can afford to pay?

40. During the first week of a going-out-of-business sale, the Alpine ski shop reduced the price of all items by 20%. During the second week of the sale, they reduced the price of all items over \$100 by an additional \$25. If Helga purchases a pair of Head skis during the second week for \$231, what is the regular price of the skis?

41. The Holiday Health Club has reduced its annual membership fee by 10%. In addition, if you sign up on a Monday, they will take an additional \$20 off the already reduced price. If Jorge purchases a year's membership on a Monday and pays \$250, what is the regular membership fee?

42. Refer to Example 11. Assume that Moneywell's security system cost \$4200 and their monthly monitoring fee is \$16. Also assume that Doile's system cost \$2500 and their monthly monitoring fee is \$25. When will the total cost of these systems be the same?

43. A chain saw uses a mixture of gasoline and oil. For each part oil, you need 15 parts gasoline. If a total of 4 gallons of the oil–gas mixture is to be made, how much oil and how much gas will need to be mixed?

44. The number of divorces granted in 1992 was 2% higher than in 1991. If 1.215 million divorces were granted in 1992, approximately how many divorces were granted in 1991?

45. The number of marriages performed in the United States in 1992 was 1% lower than in 1991. If the number of marriages in 1992 was 2.362 million, approximately how many marriages were performed in 1991? (See Example 13).

Solve the following problems

46. The SavUmor mail order prescription drug suppliers provide two membership plans. Under plan A you pay an annual \$200 membership plus 50% of the manufacturer's list price of each drug. Under plan B you pay a \$50 annual membership fee plus 75% of the manufacturer's list price of any drug.

(a) How much per year must a family's drug bill total for the two plans to result in the same cost?

(b) If Mr. Renaud's allopurinol treatment has a manufacturer's list price of \$7.50 for a month's supply, and Mrs. Renaud's monthly treatment of Premarin and Medroxyprogesterone together have a list price of \$48, and they expect to average an additional \$10 per month for other prescriptions, which plan would be the least expensive?

47. The Yearstons are considering two banks for a 20-year $50,000 mortgage. M&T is charging 9.50% interest with no points, and Citibank is charging 8.00% interest with 4 points.

(a) How long would it take for the total cost of the two mortgages to be the same?

(b) If they plan to live in their house for the 20 years, which mortgage would be the least expensive?

48. Song is considering two banks for a 30-year $60,000 mortgage. Marine Midland is charging 8.25% interest with no points and Chase Bank is charging 8.00% interest with 3 points.

(a) How long would it be for the total cost of the two mortgages to be the same?

(b) If she plans to sell her house in 10 years, which mortgage would be the less expensive?

49. John is considering two banks for a 20-year $100,000 mortgage. Key Mortgage Corp. is charging 9.00% interest with no points, and Countrywide Mortgage Corp. is charging 8.875% interest, also with no points. However, the credit check and application fee at Countrywide is $150 greater than that at Key Mortgage Corp.

(a) How long would it take for the total cost of the two mortgages to be the same?

(b) If John plans to live in the house for 10 years, which mortgage would be less expensive?

50. Lisa is considering two banks for a 30-year $75,000 mortgage. NationsBank is charging 9.5% interest with 1 point and Bank America is charging 9.25% interest with 2 points. The Bank America application fee is also $150 greater than NationsBank.

(a) How long would it take for the total cost of the two mortgages to be the same?

(b) If Lisa plans to sell her house in 8 years, which mortgage would be less expensive?

51. Because interest rates are low, the Appletons are considering refinancing their house. They presently have $50,000 of their mortgage remaining and they are making monthly mortgage payments of prinicpal and interest of $740. The bank they are considering will refinance their $50,000 mortgage for 20 years at 7.875% interest with no points. However, the closing cost for refinancing the house is $3000 (the closing costs are paid by the borrower when refinancing).

(a) If the Appletons refinance, how long will it take for the money they save from the lower payment to equal the closing cost?

(b) How much lower will their monthly payments be?

52. The Smiths are considering refinancing their house. They presently have $40,000 of their mortgage remaining and they are making monthly mortgage payments of $450. The bank they are considering will refinance their house for a 30-year period at 9.25% interest with 2 points. The closing cost is $2500.

(a) If they refinance, how long will it take for the money they save from the lower monthly payments to equal the closing cost and points?

(b) If they plan to live in the house for only 6 more years, does it pay for them to refinance?

53. Make up your own realistic word problem that can be solved using algebra. Express the problem as an equation and solve the equation. Make sure you answer the question that was asked in the problem.

CUMULATIVE REVIEW EXERCISES

54. Evaluate $\frac{1}{4} + \frac{3}{4} \div \frac{1}{2} - \frac{1}{3}$.

Name the property used.

[1.10] **55.** $(x + y) + 5 = x + (y + 5)$

56. $xy = yx$

57. $x(x + y) = x^2 + xy$

[2.6] **58.** At a firemen's chicken barbecue, the chef estimates that he will need $\frac{1}{2}$ pound of coleslaw for each 5 people. If he expects 560 residents to attend, how many pounds of coleslaw will he need?

[3.1] **59.** Solve the formula $M = \frac{a + b}{2}$ for b.

Group Activity/ Challenge Problems

1. To find the **average** of a set of values, you find the sum of the values and divide the sum by the number of values. **(a)** If Paul's first three test grades are 74, 88, and 76, write an equation that can be used to find the grade that Paul must get on his fourth exam to have an 80 average. **(b)** Solve the equation from part (a) and determine the grade Paul must receive.

2. At a basketball game Duke University scored 78 points. Duke made 12 free throws (1 point each). Duke also made 4 times as many 2-point field goals as 3-point field goals (field goals made from more than 18 feet from the basket). How many 2-point field goals and how many 3-point field goals did Duke make?

3. Pick any number, say 9. $\quad 9$

Multiply the number by 4: $\quad 9 \cdot 4 = 36$

Add 6 to the product: $\quad 36 + 6 = 42$

Divide the sum by 2: $\quad 42 \div 2 = 21$

Subtract 3 from the quotient: $\quad 21 - 3 = 18$

The solution is twice the number you started with. Show that when you select n to represent the given number the solution will always be $2n$.

For Exercises 4–6, ***(a)*** *set up an equation that can be used to solve the problem and* ***(b)*** *solve the equation and answer the question asked.*

4. In 1992 14,000 metric tons of uranium waste were produced. By the year 2000 this amount is expected to increase to 40,000. Find the percent increase from 1992 to 2000.

5. A driver education course costs \$45 but saves those under age 25 10% of their annual insurance premiums until they reach age 25. Dan has just turned 18, and his insurance costs \$600 per year.

(a) How long will it take for the amount saved from insurance to equal the price of the course?

(b) When Dan turns 25, how much will he have saved?

6. The world's most densely populated country (measured in people per square mile) is Hong Kong. Hong Kong has one of the lowest population growth rates in Southeast Asia, 0.8%. If the population density of Hong Kong was 253,488 people per square mile in 1994, find its 1993 population density. (New York City has a population density of about 11,480 people per square mile and Tokyo has a population density of about 25,019 people per square mile.)

3.4 Geometric Problems

Tape 5

1 Solve geometric problems.

Geometric Problems

1 This section serves two purposes. One is to reinforce the geometric formulas introduced in Section 3.1. The second is to reinforce procedures for setting up and solving word problems discussed in Sections 3.2 and 3.3. The more

practice you have at setting up and solving the word problems, the better you will become at solving them.

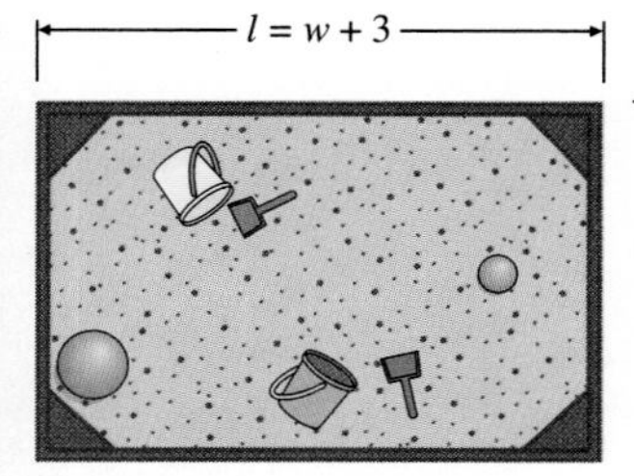

FIGURE 3.8

EXAMPLE 1 Mrs. O'Connor is planning to build a sandbox for her daughter. She has 26 feet of lumber to build the perimeter. What should the dimensions of the rectangular sandbox be if the length is to be 3 feet longer than the width?

Solution: We are asked to find the dimensions of the sandbox.

Let x = width of sandbox

then $x + 3$ = length of sandbox (Fig. 3.8)

From Section 3.1 we know that $P = 2l + 2w$. We have called the width of the sandbox x, and the length $x + 3$. We substitute these expressions into the equation.

$$\begin{aligned} P &= 2l + 2w \\ 26 &= 2(x + 3) + 2x \\ 26 &= 2x + 6 + 2x \\ 26 &= 4x + 6 \\ 20 &= 4x \\ 5 &= x \end{aligned}$$

Thus, the width is 5 feet, and the length is $x + 3 = 5 + 3 = 8$ feet.

Check:
$$\begin{aligned} P &= 2l + 2w \\ 26 &= 2(8) + 2(5) \\ 26 &= 16 + 10 \\ 26 &= 26 \quad \text{true} \end{aligned}$$

EXAMPLE 2 The sum of the angles of a triangle measure 180 degrees (180°). If two angles are the same and the third is 30° greater than the other two, find all three angles.

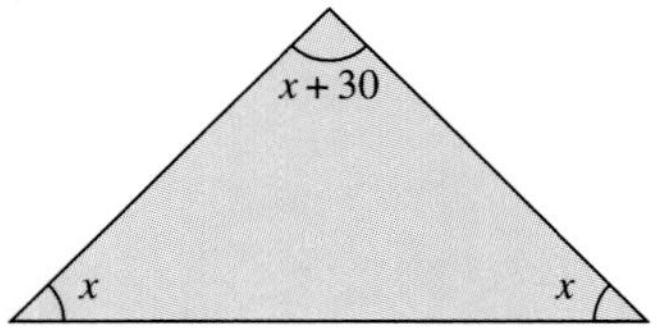

Solution: We are asked to find the three angles.

Let x = each smaller angle

then $x + 30$ = larger angle (Fig. 3.9)

x + 30

x x

FIGURE 3.9

$$\begin{aligned} \text{Sum of the 3 angles} &= 180 \\ x + x + (x + 30) &= 180 \\ 3x + 30 &= 180 \\ 3x &= 150 \\ x &= \frac{150}{3} = 50° \end{aligned}$$

Therefore, the three angles are 50°, 50°, and 50° + 30° or 80°.

Check:
$$\begin{aligned} 50° + 50° + 80° &= 180° \\ 180° &= 180° \quad \text{true} \end{aligned}$$

Recall from Section 3.1 that a quadrilateral is a four-sided figure. Quadrilaterals include squares, rectangles, parallelograms, and trapezoids. The sum of the measures of the angles of any quadrilateral is 360°. We use this information in Example 3.

EXAMPLE 3 In a parallelogram the opposite angles have the same measures. If each of the two larger angles in a parallelogram is 20° less than three times the smaller angles, find the measure of each angle.

Solution: Let x = the measure of each of the two smaller angles

then $3x - 20$ = the measure of each of the two larger angles

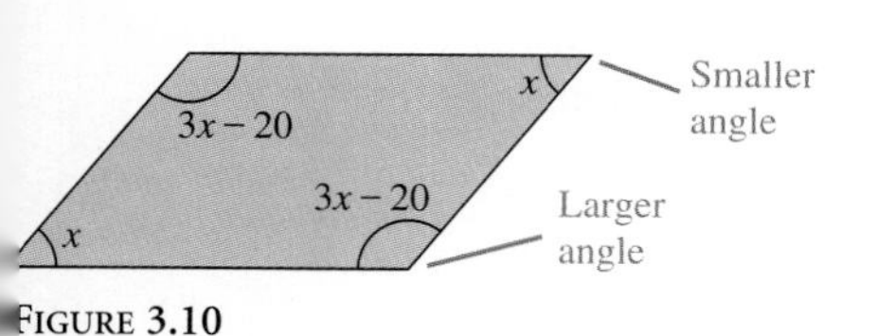

FIGURE 3.10

A diagram of the parallelogram is given in Figure 3.10.

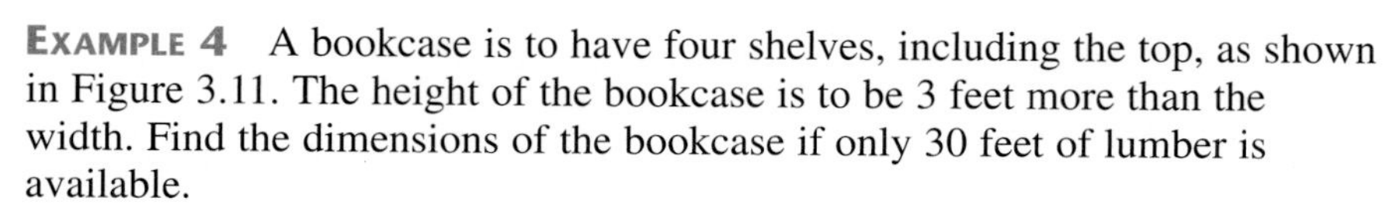

$$\begin{pmatrix}\text{measure of the}\\ \text{two smaller angles}\end{pmatrix} + \begin{pmatrix}\text{measure of the}\\ \text{two larger angles}\end{pmatrix} = 360$$

$$x + x + (3x - 20) + (3x - 20) = 360$$
$$x + x + 3x - 20 + 3x - 20 = 360$$
$$8x - 40 = 360$$
$$8x = 400$$
$$x = 50$$

Thus, each of the two smaller angles is 50° and each of the two larger angles is $3x - 20 = 3(50) - 20 = 130°$. As a check, $50° + 50° + 130° + 130° = 360°$.

EXAMPLE 4 A bookcase is to have four shelves, including the top, as shown in Figure 3.11. The height of the bookcase is to be 3 feet more than the width. Find the dimensions of the bookcase if only 30 feet of lumber is available.

Solution: We are asked to find the dimensions of the bookcase.

Let x = length of a shelf

then $x + 3$ = height of bookcase

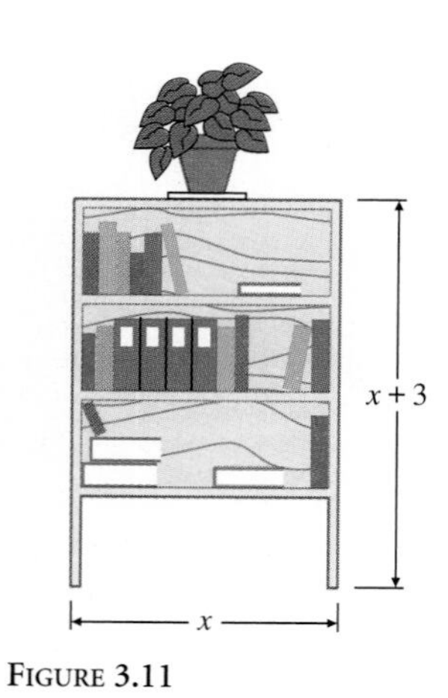

FIGURE 3.11

4 shelves + 2 sides = total lumber available

$$4x + 2(x + 3) = 30$$
$$4x + 2x + 6 = 30$$
$$6x + 6 = 30$$
$$6x = 24$$
$$x = 4$$

The length of each shelf is 4 feet and the height of the bookcase is 4 + 3 or 7 feet.

Check: $4 + 4 + 4 + 4 + 7 + 7 = 30$

$30 = 30$ true

Exercise Set 3.4

Solve the following geometric problems.

1. An **equilateral triangle** is a triangle that has three sides of the same length. If the perimeter of an equilateral triangle is 28.5 inches, find the length of each side. Equilateral triangles are discussed in Appendix C.
2. Two angles are **complementary angles** if the sum of their measures is 90°. If angle A and angle B are complementary angles, and angle A is 21° more than twice angle B, find the measures of angle A and angle B. Complementary angles are discussed in Appendix C.
3. Two angles are **supplementary angles** if the sum of their measures is 180°. If angle A and angle B are supplementary angles, and angle B is 8° less than three times angle A, find the measures of angle A and angle B. Supplementary angles are discussed in Appendix C.
4. If one angle of a triangle is 20° larger than the smallest angle, and the third angle is six times as large as the smallest angle, find the measures of the three angles.
5. If one angle of a triangle is 10° greater than the smallest angle, and the third angle is 30° less than twice the smallest angle, find the measures of the three angles.
6. The length of a rectangle is 8 feet more than its width. What are the dimensions of the rectangle if the perimeter is 48 feet?
7. In an **isosceles triangle,** two sides are equal. The third side is 2 meters less than each of the other sides. Find the length of each side if the perimeter is 10 meters. Isosceles triangles are discussed in Appendix C.
8. The perimeter of a rectangle is 120 feet. Find the length and width of the rectangle if the length is twice the width.
9. The perimeter of a basement floor of a house is 240 feet. Find the length and width of the rectangular floor if the length is 24 feet less than twice the width.
10. If the two smaller angles of a parallelogram have equal measures and the two larger angles are each 30° larger than each smaller angle, find the measure of each angle.
11. If the two smaller angles of a parallelogram have equal measures and the two larger angles each measure 27° less than twice each smaller angle, find the measure of each angle.
12. The measure of one angle of a quadrilateral is 10° greater than the smallest angle, the third angle is 14° greater than twice the smallest angle, and the fourth angle is 21° greater than the smallest angle. Find the measures of the four angles of the quadrilateral.
13. A bookcase is to have four shelves as shown. The height of the bookcase is to be 2 feet more than the width, and only 20 feet of lumber is available. What should be the dimensions of the bookcase?

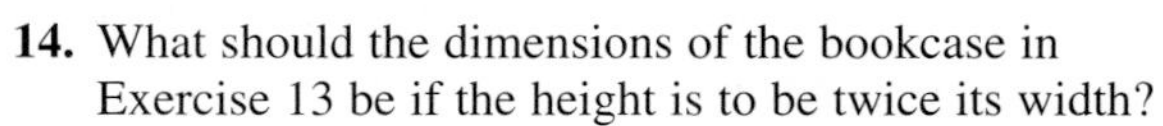

14. What should the dimensions of the bookcase in Exercise 13 be if the height is to be twice its width?
15. Betty McKane plans to build storage shelves as shown. If she has only 45 feet of lumber for the entire unit and wishes the length to be 3 times the height, find the length and height of the unit.

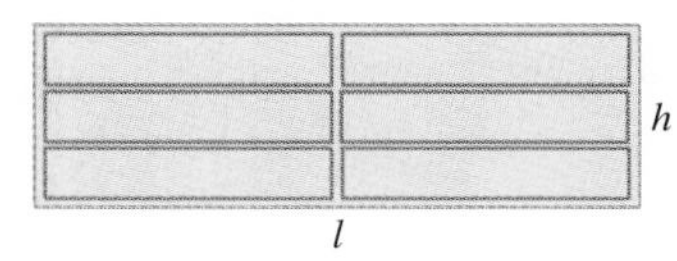

16. An area is to be fenced in along a straight river bank as illustrated. If the length of the fenced-in area is to be 4 feet greater than the width, and the total amount of fencing used is 64 feet, find the width and length of the fenced-in area.

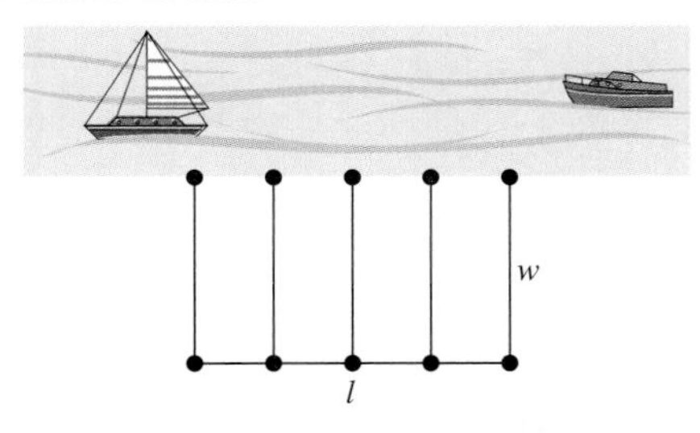

17. In the equation $A = l \cdot w$, what happens to the area if the length is doubled and the width is halved? Explain your answer.
18. In the equation $A = s^2$, what happens to the area if the length of a side, s, is doubled? Explain your answer.
19. In the equation $V = l \cdot w \cdot h$, what happens to the volume if the length, width, and height are all doubled? Explain your answer.
20. In the equation $V = \frac{4}{3}\pi r^3$, what happens to the volume if the radius is tripled? Explain your answer.
21. Create your own realistic geometric word problem that can be solved using algebra. Write the problem as an equation and solve it. Answer the question asked in the original problem.

CUMULATIVE REVIEW EXERCISES

Insert either >, <, or = in the shaded area to make the statement true.

[1.3] **22.** $-|-6| \quad |-4|$

23. $|-3| \quad -|3|$

[1.6] **24.** Evaluate $-6 - (-2) + (-4)$.

[2.1] **25.** Simplify $-6y + x - 3(x - 2) + 2y$.

[3.1] **26.** Solve $2x + 3y = 9$ for y; then find the value of y when $x = 3$.

Group Activity/Challenge Problems

1. Consider the accompanying figure. **(a)** Write a formula for determining the area of the shaded region. **(b)** Find the area of the shaded region when $S = 9$ inches and $s = 6$ inches.

2. One way to express the area of the figure on the right is $(a + b)(c + d)$. Can you determine another expression, using the area of the four rectangles, to represent the area of the figure?

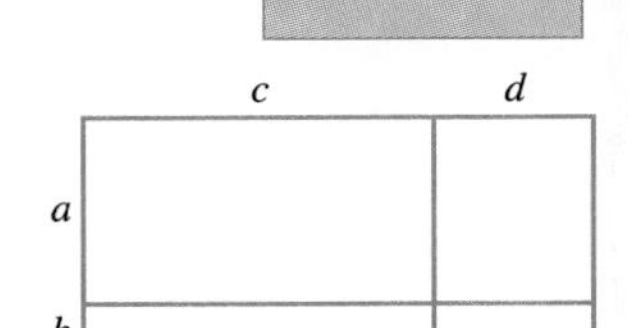

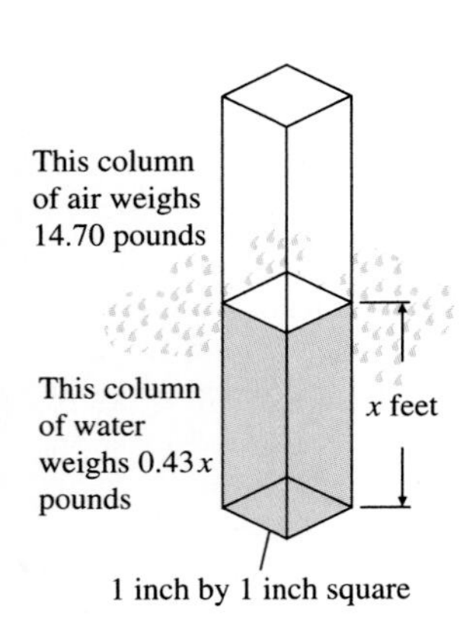

3. The total pressure, P, in pounds per square inch, exerted on an object x feet below sea level is given by the formula $P = 14.70 + 0.43x$. As shown in the accompanying diagram, the 14.70 represents the weight in pounds of the column of air (from sea level to the top of the atmosphere) standing over a 1-inch by 1-inch square of seawater. The $0.43x$ represents the weight, in pounds, of a column of water 1 inch by 1 inch by x feet.

(a) A submarine can withstand a total pressure of 162 pounds per square inch. How deep can the submarine go?

(b) If the pressure gauge in the submarine registers a pressure of 97.26 pounds per square inch, how deep is the submarine?

Summary

GLOSSARY

Area (145): The total surface area within a figure's boundaries.

Circumference (147): The length of the curve that forms a circle.

Complementary angles (181): Two angles whose measures sum to 90°.

Diameter (147): A line segment through the center whose endpoints both lie on the circle.

Equilateral triangle (181): A triangle whose three sides have the same length.

Formula (144): An equation commonly used to express a specific physical concept mathematically.

Perimeter (145): The sum of the lengths of the sides of a figure.

Quadrilateral (145): A four-sided figure.

Radius (147): A line segment from the center of a circle to any point on the circle.

Supplementary angles (181): Two angles whose measures sum to 180°.

IMPORTANT FACTS

Simple interest formula: $i = prt$

The sum of the measures of the angles in any triangle is 180°.

The sum of the measures of the angles of a quadrilateral is 360°.

To Solve a Word Problem

1. Read the question carefully.
2. If possible, draw a sketch to help visualize the problem.
3. Determine which quantity you are being asked to find. Choose a letter to represent this unknown quantity; write down exactly what this letter represents. If there is more than one unknown quantity, express all unknown quantities in terms of the variable selected.
4. Write the word problem as an equation.
5. Solve the equation for the unknown quantity.
6. Answer the question or questions asked.
7. Check the solution in the original problem.

Review Exercises

[3.1] *Find the value of the variable indicated.*

1. $C = \pi d$; find C when $d = 4$. Use $\pi = 3.14$.

2. $A = \frac{1}{2}bh$; find A when $b = 12$ and $h = 8$.

3. $P = 2l + 2w$; find P when $l = 6$ and $w = 4$.

4. $i = prt$; find i when $p = 1000$, $r = 15\%$, and $t = 2$.

5. $E = IR$; find E when $I = 0.12$ and $R = 2000$.

6. $A = \pi r^2$; find A when $r = 3$. Use $\pi = 3.14$.

7. $V = \frac{4}{3}\pi r^3$; find V when $r = 3$. Use $\pi = 3.14$.

8. $Fd^2 = km$; find k when $F = 60$, $m = 12$, and $d = 2$.

9. $y = mx + b$; find b when $y = 15$, $m = 3$, and $x = -2$.

10. $2x + 3y = -9$; find y when $x = 12$.

11. $4x - 3y = 15 + x$; find y when $x = -3$.

12. $2x = y + 3z + 4$; find y when $x = 5$ and $z = -3$.

13. $IR = E + Rr$; find r when $I = 5$, $E = 100$, and $R = 200$.

Solve the equation for y; then find the value of y for the given value of x.

14. $2x - y = 12, x = 10$

15. $3x - 2y = -4, x = 2$

16. $3x = 5 + 2y, x = -3$

17. $-6x - 2y = 20, x = 0$

18. $6 = -3x - 2y, x = -6$

19. $3y - 4x = -3, x = 2$

Solve for the variable indicated.

20. $F = ma$, for m

21. $A = \frac{1}{2}bh$, for h

22. $i = prt$, for t

23. $P = 2l + 2w$, for w

24. $2x - 3y = 6$, for y

25. $A = \frac{B + C}{2}$, for B

26. $V = \pi r^2 h$, for h

Solve.

27. How much interest will Karen pay if she borrows \$600 for 2 years at 15% simple interest? (Use $i = prt$.)

[3.3] **28.** The perimeter of a rectangle is 16 inches. Find the length of the rectangle if the width is 2 inches.

[3.2, 3.3] *Solve each problem.*

29. One number is 4 more than the other. Find the two numbers if their sum is 62.

30. The sum of two consecutive integers is 255. Find the two integers.

31. The larger of two integers is 3 more than five times the smaller integer. Find the two numbers if the smaller subtracted from the larger is 31.

32. What is the cost of a car before tax if the total cost including a 5% tax is \$8400?

33. In Paul's present position as a salesman he receives a base salary of \$500 per week plus a 3% commission on all sales he makes. He is considering moving to another company where he would sell the same goods. His base salary would be only \$400 per week, but his commission would be 8% on all sales he makes. What weekly dollar sales would he have to make for the total salary of each company to be the same?

34. During the first week of a going-out-of-business sale, all prices are reduced by 20%. During the second week of the sale, all prices that still cost more than \$100 are reduced by an additional \$25. During the second week of the sale, Kathy purchased a camcorder for \$495. What was the original price of the camcorder?

35. The Johnsons are considering two banks for a 30-year \$60,000 mortgage. Comerica Bank is offering 8.875% interest with no points and Mellon Bank is offering 8.625% interest with 3 points.

(a) How long would it take for the total payments from each bank to be the same?

(b) If the Johnsons plan on keeping the house for 20 years, to which bank should they apply?

36. Debra Houy is considering refinancing her house. The present balance on her mortgage is \$70,000 and her monthly payment of principal and interest is \$750. First Chicago Corporation is offering her a 20-year \$70,000 mortgage with 8.50% interest and one point. The closing cost, in addition to the one point, is \$3200.

(a) How long would it take for the money she saves in monthly payments to equal the cost of the point and the closing cost?

(b) If she plans to live in the house for only 10 more years, does it pay for her to refinance?

[3.4] **37.** If one angle of a triangle measures 10° greater than the smallest angle, and the third angle measures 10° less than twice the smallest angle, find the measures of the three angles.

38. One angle of a trapezoid measures 10° greater than the smallest angle. A third angle measures five times the smallest angle. The fourth angle measures 20° greater than four times the smallest angle. Find the measure of the four angles.

39. Mrs. Appleby wants a garden whose length is 4 feet longer than its width. The perimeter of the garden is to be 70 feet. What will be the dimensions of the garden?

[3.1–3.4]

40. The sum of two consecutive odd integers is 208. Find the two integers.

41. What is the cost of a television set before tax if the total cost, including a 6% tax, is \$477?

42. Mr. McAdams sells water softeners. He receives a weekly salary of \$300 plus a 5% commission on the sales he makes. If Mr. McAdams earned \$900 last week, what were his sales in dollars?

43. One angle of a triangle is 8° greater than the smallest angle. The third angle is 4° greater than twice the smallest angle. Find the measure of the three angles of the triangle.

44. Dreyel Company plans to increase its number of employees by 25 per year. If the company presently has 427 employees, how long will it take before they reach 627 employees?

45. If the two larger angles of a parallelogram each measure 40° greater than the two smaller angles, find the measure of the four angles.

46. Two copy centers across the street from one another are competing for business and both have made special offers. Under Copy King's plan, for a monthly fee of \$20 each copy made in that month costs only 4 cents. King Kopie charges a monthly fee of \$25 plus 3 cents a copy.

(a) How many copies made in a month would result in both copy centers charging the same amount?

(b) If you belong to both plans, and intend to make 1000 copies of an advertisement for your band, which center would cost less, and by how much?

Practice Test

1. *Use* $P = 2l + 2w$ to find P when $l = 6$ feet and $w = 3$ feet.
2. Use $A = P + Prt$ to find A when $P = 100$, $r = 0.15$, and $t = 3$.
3. Use $V = \frac{1}{3}\pi r^2 h$ to find V when $r = 4$ and $h = 6$. Use $\pi = 3.14$.

Solve for the variable indicated.

4. $P = IR$, for R
5. $3x - 2y = 6$, for y
6. $A = \dfrac{a + b}{3}$, for a
7. $D = R(c + a)$, for c
8. The sum of two integers is 158. Find the two integers if the larger is 10 less than twice the smaller.
9. The sum of three consecutive integers is 42. Find the three integers.
10. Mr. Herron has only $20. If he wishes to leave a 15% tip and must pay 7% tax, find the price of the most expensive meal that he can order.
11. A triangle has a perimeter of 75 inches. Find the three sides if one side is 15 inches larger than the smallest side, and the third side is twice the smallest side.
12. The sum of the angles of a parallelogram is 360°. If the two smaller angles are equal and the two larger angles are each 30° greater than twice the smaller angles, find the measure of each angle.

Chapter 4

Exponents, Polynomials, and Additional Applications

See Section 4.7, Exercise 5

Preview and Perspective

In this chapter we discuss exponents and polynomials. In Sections 4.1 and 4.2 we discuss the rules of exponents. Most of the rules are given in Section 4.1. Make sure that you understand this section before you cover the negative exponent rule in Section 4.2. The rules of exponents are used and expanded upon in Section 9.7 when we discuss the use of fractional exponents. When discussing scientific notation in Section 4.3 we use the rules of exponents to solve real-life application problems that involve very large or very small numbers. A knowledge of scientific notation may also help you in your science courses and when reading certain magazines.

In Section 4.4 we define a polynomial, and in Sections 4.4 through 4.6 we explain how to add, subtract, multiply, and divide polynomials. To be successful with this material you must understand the rules of exponents presented in the first two sections of this chapter. *To understand factoring, which is covered in Chapter 5, you need to understand polynomials, especially multiplication of polynomials.* As you will learn, factoring is the reverse of multiplying polynomials. We will also be working with polynomials throughout Chapter 6.

In Section 4.7 we discuss motion and mixture problems. These are two additional applications of algebra. The material presented here will build on, and reinforce, the material on applications presented in Chapter 3. Motion and mixture problems are used and expanded upon in Sections 6.8 and 8.5. You will find that motion and mixture problems have many real-life applications.

4.1 Exponents

Tape 6

1. Review exponents.
2. Learn the rules of exponents.
3. Learn to simplify an expression prior to using the expanded power rule.

Exponents

1 To use polynomials, we need to expand our knowledge of exponents. Exponents were introduced in Section 1.8. Let us review the fundamental concepts. In the expression x^n, x is referred to as the **base** and n is called the **exponent.** x^n is read "x to the nth power."

$$x^2 = \underbrace{x \cdot x}_{\text{2 factors of } x}$$

$$x^4 = \underbrace{x \cdot x \cdot x \cdot x}_{\text{4 factors of } x}$$

$$x^m = \underbrace{x \cdot x \cdot x \cdot \cdots \cdot x}_{m \text{ factors of } x}$$

EXAMPLE 1 Write $xxxxyy$ using exponents.

Solution: $\underbrace{x\ x\ x\ x}_{\text{4 factors of } x}\ \underbrace{y\ y}_{\text{2 factors of } y} = x^4y^2$

Remember, when a term containing a variable is given without a numerical coefficient, the numerical coefficient of the term is assumed to be 1. For example $x = 1x$ and $x^2y = 1x^2y$.

Also recall that when a variable or numerical value is given without an exponent, the exponent of that variable or numerical value is assumed to be 1. For example, $x = x^1$, $xy = x^1y^1$, $x^2y = x^2y^1$, and $2xy^2 = 2^1x^1y^2$.

Rules of Exponents

2 Now we will learn the rules of exponents.

EXAMPLE 2 Multiply $x^4 \cdot x^3$.

Solution:

$$\underbrace{x \cdot x \cdot x \cdot x}_{x^4} \cdot \underbrace{x \cdot x \cdot x}_{x^3} = x^7$$

Example 2 illustrates that when multiplying expressions with the same base we keep the base and *add* the exponents. This is the product rule for exponents.

Product Rule for Exponents

$$x^m \cdot x^n = x^{m+n}$$

In Example 2 we showed that $x^4 \cdot x^3 = x^7$. This problem could also be done using the product rule: $x^4 \cdot x^3 = x^{4+3} = x^7$.

EXAMPLE 3 Multiply using the product rule.

(a) $3^2 \cdot 3$ **(b)** $2^4 \cdot 2^2$ **(c)** $x \cdot x^4$ **(d)** $x^2 \cdot x^5$ **(e)** $y^4 \cdot y^7$

Solution:

(a) $3^2 \cdot 3 = 3^2 \cdot 3^1 = 3^{2+1} = 3^3$ or 27

(b) $2^4 \cdot 2^2 = 2^{4+2} = 2^6$ or 64

(c) $x \cdot x^4 = x^1 \cdot x^4 = x^{1+4} = x^5$

(d) $x^2 \cdot x^5 = x^{2+5} = x^7$

(e) $y^4 \cdot y^7 = y^{4+7} = y^{11}$

COMMON STUDENT ERROR Note in Example 3(a) that $3^2 \cdot 3^1$ is 3^3 and not 9^3. When multiplying powers of the same base, *do not multiply the bases.*

Correct	*Incorrect*
$3^2 \cdot 3^1 = 3^3$	~~$3^2 \cdot 3^1 = 9^3$~~

Example 4 will help you understand the quotient rule for exponents.

EXAMPLE 4 Divide $x^5 \div x^3$.

Solution: $$\frac{x^5}{x^3} = \frac{\not{x} \cdot \not{x} \cdot \not{x} \cdot x \cdot x}{\not{x} \cdot \not{x} \cdot \not{x}} = \frac{1x^2}{1} = x^2$$

When dividing expressions with the same base, keep the base and *subtract* the exponent in the denominator from the exponent in the numerator.

Quotient Rule for Exponents

$$\frac{x^m}{x^n} = x^{m-n}, \qquad x \neq 0$$

In Example 4 we showed that $x^5/x^3 = x^2$. This problem could also be done using the quotient rule: $x^5/x^3 = x^{5-3} = x^2$.

EXAMPLE 5 Divide each expression.

(a) $\frac{3^5}{3^2}$ **(b)** $\frac{5^4}{5}$ **(c)** $\frac{x^{12}}{x^5}$ **(d)** $\frac{x^9}{x^5}$ **(e)** $\frac{x^7}{x}$

Solution:

(a) $\frac{3^5}{3^2} = 3^{5-2} = 3^3$ or 27

(b) $\frac{5^4}{5} = \frac{5^4}{5^1} = 5^{4-1} = 5^3$ or 125

(c) $\frac{x^{12}}{x^5} = x^{12-5} = x^7$

(d) $\frac{x^9}{x^5} = x^{9-5} = x^4$

(e) $\frac{x^7}{x} = \frac{x^7}{x^1} = x^{7-1} = x^6$

COMMON STUDENT ERROR Note in Example 5(a) that $3^5/3^2$ is 3^3 and not 1^3. When dividing powers of the same base, *do not divide out the bases.*

Correct	*Incorrect*
$\frac{3^3}{3^1} = 3^2$ or 9	~~$\frac{3^3}{3^1} = 1^2$~~

The answer to Example 5(c), x^{12}/x^5, is x^7. We obtained this answer using the quotient rule. This answer could also be obtained by dividing out the common factor in both the numerator and denominator as follows.

$$\frac{x^{12}}{x^5} = \frac{(\cancel{x} \cdot \cancel{x} \cdot \cancel{x} \cdot \cancel{x} \cdot \cancel{x}) \cdot x \cdot x \cdot x \cdot x \cdot x \cdot x \cdot x}{(\cancel{x} \cdot \cancel{x} \cdot \cancel{x} \cdot \cancel{x} \cdot \cancel{x})} = x^7$$

We divided out the product of five x's, which is x^5. We can indicate this process in shortened form as follows:

$$\frac{x^{12}}{x^5} = \frac{\cancel{x^5} \cdot x^7}{\cancel{x^5}} = x^7$$

In this section, to simplify an expression when the numerator and denominator have the same base and the exponent in the denominator is greater than the exponent in the numerator, we divide out common factors. For example, x^5/x^{12} can be simplified by dividing out the common factor, x^5, as follows:

$$\frac{x^5}{x^{12}} = \frac{\cancel{x^5}}{\cancel{x^5} \cdot x^7} = \frac{1}{x^7}$$

We will now simplify some expressions by dividing out common factors.

EXAMPLE 6 Simplify by dividing out a common factor in both the numerator and denominator.

(a) $\frac{x^9}{x^{12}}$ **(b)** $\frac{y^4}{y^9}$

Solution:

(a) Since the numerator is x^9 write the denominator with a factor of x^9. Since $x^9 \cdot x^3 = x^{12}$, we break x^{12} up into $x^9 \cdot x^3$.

$$\frac{x^9}{x^{12}} = \frac{\cancel{x^9}}{\cancel{x^9} \cdot x^3} = \frac{1}{x^3}$$

(b) $\frac{y^4}{y^9} = \frac{\cancel{y^4}}{\cancel{y^4} \cdot y^5} = \frac{1}{y^5}$

In the next section, we will show another way to evaluate expressions like $\frac{x^9}{x^{12}}$ by using the negative exponent rule.

Example 7 leads us to our next rule, the zero exponent rule.

EXAMPLE 7 Divide $\frac{x^3}{x^3}$.

Solution: By the quotient rule,

$$\frac{x^3}{x^3} = x^{3-3} = x^0$$

However,

$$\frac{x^3}{x^3} = \frac{\cancel{x} \cdot \cancel{x} \cdot \cancel{x}}{\cancel{x} \cdot \cancel{x} \cdot \cancel{x}} = 1$$

Since $x^3/x^3 = x^0$ and $x^3/x^3 = 1$, then x^0 must equal 1.

Zero Exponent Rule

$$x^0 = 1, \qquad x \neq 0$$

By the zero exponent rule, any real number, except 0, raised to the zero power equals 1. Note that 0^0 is not a real number.

EXAMPLE 8 Simplify each expression.

(a) 3^0 **(b)** x^0 **(c)** $3x^0$ **(d)** $(3x)^0$

Solution:

(a) $3^0 = 1$

(b) $x^0 = 1$

(c) $3x^0 = 3(x^0)$
$= 3 \cdot 1 = 3$

(d) $(3x)^0 = 1$

Remember, the exponent refers only to the immediately preceding symbol unless parentheses are used.

COMMON STUDENT ERROR An expression raised to the zero power is not equal to 0; it is equal to 1.

Correct	*Incorrect*
$x^0 = 1$	~~$x^0 = 0$~~
$5^0 = 1$	~~$5^0 = 0$~~

The power rule will be explained with the aid of Example 9.

EXAMPLE 9 Simplify $(x^3)^2$.

Solution: $(x^3)^2 = \underbrace{x^3 \cdot x^3}_{\text{2 factors of } x^3} = x^{3+3} = x^6$

Power Rule for Exponents

$$(x^m)^n = x^{m \cdot n}$$

The power rule indicates that when we raise an exponential expression to a power, we keep the base and *multiply* the exponents. Example 9 could also be simplified using the power rule.

$$(x^3)^2 = x^{3 \cdot 2} = x^6$$

Note that the answers are the same.

EXAMPLE 10 Simplify each term.
(a) $(x^3)^4$ **(b)** $(3^4)^2$ **(c)** $(y^3)^8$

Solution:
(a) $(x^3)^4 = x^{3 \cdot 4} = x^{12}$
(b) $(3^4)^2 = 3^{4 \cdot 2} = 3^8$
(c) $(y^3)^8 = y^{3 \cdot 8} = y^{24}$

Helpful Hint

Students often confuse the product and power rules. Note the difference carefully.

Product Rule	*Power Rule*
$x^m \cdot x^n = x^{m+n}$	$(x^m)^n = x^{m \cdot n}$
$2^3 \cdot 2^5 = 2^{3+5} = 2^8$	$(2^3)^5 = 2^{3 \cdot 5} = 2^{15}$

Example 11 will help us in explaining the expanded power rule. As the name suggests, this rule is an expansion of the power rule.

EXAMPLE 11 Simplify $\left(\dfrac{ax}{by}\right)^4$.

Solution: $\left(\frac{ax}{by}\right)^4 = \frac{ax}{by} \cdot \frac{ax}{by} \cdot \frac{ax}{by} \cdot \frac{ax}{by}$

$$= \frac{a \cdot a \cdot a \cdot a \cdot x \cdot x \cdot x \cdot x}{b \cdot b \cdot b \cdot b \cdot y \cdot y \cdot y \cdot y} = \frac{a^4 \cdot x^4}{b^4 \cdot y^4} = \frac{a^4x^4}{b^4y^4}$$

Example 11 illustrates the expanded power rule.

Expanded Power Rule for Exponents

$$\left(\frac{ax}{by}\right)^m = \frac{a^m x^m}{b^m y^m}, \qquad b \neq 0,\ y \neq 0$$

The expanded power rule illustrates that every factor within parentheses is raised to the power outside the parentheses when the expression is simplified.

EXAMPLE 12 Simplify each expression.

(a) $(2x)^2$ **(b)** $(-x)^3$ **(c)** $(2xy)^3$ **(d)** $\left(\frac{-3x}{2y}\right)^2$

Solution:

(a) $(2x)^2 = 2^2x^2 = 4x^2$

(b) $(-x)^3 = (-1x)^3 = (-1)^3x^3 = -1x^3 = -x^3$

(c) $(2xy)^3 = 2^3x^3y^3 = 8x^3y^3$

(d) $\left(\frac{-3x}{2y}\right)^2 = \frac{(-3)^2x^2}{2^2y^2} = \frac{9x^2}{4y^2}$

Simplify before Using the Power Rule

3 Whenever we have an expression raised to a power, it helps to simplify the expression in parentheses before using the expanded power rule. This procedure is illustrated in Examples 13 and 14.

EXAMPLE 13 Simplify $\left(\frac{8x^3y^2}{4xy^2}\right)^3$.

Solution: We first simplify the expression within parentheses by dividing out common factors.

$$\left(\frac{8x^3y^2}{4xy^2}\right)^3 = \left(\frac{8}{4} \cdot \frac{x^3}{x} \cdot \frac{y^2}{y^2}\right)^3 = (2x^2)^3$$

Now use the expanded power rule to simplify further.

$$(2x^2)^3 = 2^3(x^2)^3 = 8x^6$$

Thus, $\left(\frac{8x^3y^2}{4xy^2}\right)^3 = 8x^6$.

EXAMPLE 14 Simplify $\left(\frac{25x^4y^3}{5x^2y^7}\right)^4$.

Solution: Begin by simplifying the expression within parentheses.

$$\left(\frac{25x^4y^3}{5x^2y^7}\right)^4 = \left(\frac{25}{5} \cdot \frac{x^4}{x^2} \cdot \frac{y^3}{y^7}\right)^4 = \left(\frac{5x^2}{y^4}\right)^4$$

Now use the expanded power rule to simplify further.

$$\left(\frac{5x^2}{y^4}\right)^4 = \frac{5^4x^8}{y^{16}} = \frac{625x^8}{y^{16}}.$$

Thus, $\left(\frac{25x^4y^3}{5x^2y^7}\right)^4 = \frac{625x^8}{y^{16}}.$

COMMON STUDENT ERROR Students sometimes make errors in simplifying expressions containing exponents. It is very important that you have a thorough understanding of exponents. One of the most common errors made by students follows. Study this error carefully to make sure you do not make this error.

Correct	*Incorrect*
$\frac{4}{2x} = \frac{\overset{2}{\cancel{4}}}{\underset{1}{\cancel{2}x}} = \frac{2}{x}$	$\frac{4}{x+2} = \frac{\overset{2}{\cancel{4}}}{x+\underset{1}{\cancel{2}}} = \frac{2}{x+1}$
$\frac{x}{xy} = \frac{\overset{1}{\cancel{x}}}{\underset{1}{\cancel{x}}y} = \frac{1}{y}$	$\frac{x}{x+y} = \frac{\overset{1}{\cancel{x}}}{\underset{1}{\cancel{x}}+y} = \frac{1}{1+y}$
$\frac{x^3y^2}{y^2} = \frac{x^3\cancel{y^2}}{\cancel{y^2}} = x^3$	$\frac{x^3+y^2}{y^2} = \frac{x^3+\overset{1}{\cancel{y^2}}}{\underset{1}{\cancel{y^2}}} = x^3+1$

The simplifications on the right side are not correct because only common **factors** can be divided out (remember, factors are multiplied together). In the first denominator on the right, $x + 2$, the x and 2 are terms, not factors, since they are being added. Similarly, in the second denominator, $x + y$, the x and the y are terms, not factors, since they are being added. Also, in the numerator $x^3 + y^2$, the x^3 and y^2 are terms, not factors. No common factors can be divided out in the fractions on the right.

Summary of the Rules of Exponents Presented in This Section

1. $x^m \cdot x^n = x^{m+n}$		**product rule**
2. $\frac{x^m}{x^n} = x^{m-n}$,	$x \neq 0$	**quotient rule**
3. $x^0 = 1$,	$x \neq 0$	**zero exponent rule**
4. $(x^m)^n = x^{m \cdot n}$		**power rule**
5. $\left(\frac{ax}{by}\right)^m = \frac{a^mx^m}{b^my^m}$,	$b \neq 0, y \neq 0$	**expanded power rule**

EXAMPLE 15 Simplify $(2x^2y^3)^4(xy^2)$.

Solution: First simplify $(2x^2y^3)^4$ by using the expanded power rule.

$$(2x^2y^3)^4 = 2^4x^{2\cdot 4}y^{3\cdot 4} = 16x^8y^{12}$$

Now use the product rule to simplify further.

$$\begin{aligned}(2x^2y^3)^4(xy^2) &= (16x^8y^{12})(x^1y^2)\\ &= 16\cdot x^8\cdot x^1\cdot y^{12}\cdot y^2\\ &= 16x^9y^{14}\end{aligned}$$

Thus, $(2x^2y^3)^4(xy^2) = 16x^9y^{14}$.

Exercise Set 4.1

Simplify.

1. $x^2\cdot x^4$
2. $x^5\cdot x^4$
3. $y\cdot y^2$
4. $4^2\cdot 4$
5. $3^2\cdot 3^3$
6. $x^4\cdot x^2$
7. $y^3\cdot y^2$
8. $x^3\cdot x^4$
9. $y^4\cdot y$
10. $\dfrac{x^4}{x^3}$
11. $\dfrac{x^{10}}{x^3}$
12. $\dfrac{y^3}{y}$
13. $\dfrac{5^4}{5^2}$
14. $\dfrac{3^5}{3^2}$
15. $\dfrac{y^2}{y}$
16. $\dfrac{x^3}{x^5}$
17. $\dfrac{x^2}{x^2}$
18. $\dfrac{x^{13}}{x^4}$
19. $\dfrac{y^{12}}{y^{11}}$
20. $\dfrac{3^4}{3^4}$
21. x^0
22. 5^0
23. $3x^0$
24. $-2x^0$
25. $(3x)^0$
26. $-(4x)^0$
27. $(-4x)^0$
28. $(x^2)^3$
29. $(x^5)^2$
30. $(x^2)^2$
31. $(x^5)^5$
32. $(x^4)^2$
33. $(x^3)^1$
34. $(x^3)^2$
35. $(x^4)^3$
36. $(x^5)^4$
37. $(x^5)^3$
38. $(2x)^2$
39. $(1.3x)^2$
40. $(-3x)^2$
41. $(-x)^2$
42. $(-x)^3$
43. $(4x^2)^3$
44. $(2.5x^3)^2$
45. $(-3x^3)^3$
46. $(xy)^4$
47. $(2x^2y)^3$
48. $(4x^3y^2)^3$
49. $(8.6x^2y^5)^2$
50. $(2xy^4)^3$
51. $(-6x^3y^2)^3$
52. $(7x^2y^4)^2$
53. $(-x^4y^5z^6)^3$
54. $(-2x^4y^2z)^3$
55. $\left(\dfrac{x}{y}\right)^2$
56. $\left(\dfrac{x}{3}\right)^2$
57. $\left(\dfrac{x}{4}\right)^3$
58. $\left(\dfrac{2}{x}\right)^3$
59. $\left(\dfrac{y}{x}\right)^5$
60. $\left(\dfrac{3}{y}\right)^4$
61. $\left(\dfrac{6}{x}\right)^3$
62. $\left(\dfrac{2x}{y}\right)^3$
63. $\left(\dfrac{3x}{y}\right)^3$
64. $\left(\dfrac{5x^2}{y}\right)^2$
65. $\left(\dfrac{3x}{5}\right)^2$
66. $\left(\dfrac{3x^4}{2}\right)^3$
67. $\left(\dfrac{4y^3}{x}\right)^3$
68. $\left(\dfrac{-4x^2}{5}\right)^2$
69. $\left(\dfrac{-3x^3}{4}\right)^3$
70. $\left(\dfrac{-x^5}{y^2}\right)^3$
71. $\dfrac{x^3y^2}{xy^6}$
72. $\dfrac{x^2y^6}{x^4y}$
73. $\dfrac{x^5y^7}{x^{12}y^3}$
74. $\dfrac{x^4y^5}{x^7y^{12}}$
75. $\dfrac{10x^3y^8}{2xy^{10}}$
76. $\dfrac{5x^{12}y^2}{10xy^9}$

77. $\dfrac{4xy}{16x^3y^2}$ **78.** $\dfrac{20x^4y^6}{5xy^9}$ **79.** $\dfrac{35x^4y^7}{10x^9y^{12}}$ **80.** $\dfrac{20x^8y^{12}}{5x^8y^7}$

81. $\dfrac{-36xy^9z}{12x^4y^5z^2}$ **82.** $\dfrac{4x^4y^7z^3}{32x^5y^4z^9}$ **83.** $\dfrac{-6x^2y^7z^5}{2x^5y^9z^6}$ **84.** $\dfrac{-25x^4y^{10}}{30x^3y^7z}$

85. $\left(\dfrac{4x^4}{2x^6}\right)^3$ **86.** $\left(\dfrac{4x^4}{8x^8}\right)^3$ **87.** $\left(\dfrac{6y^7}{2y^3}\right)^3$ **88.** $\left(\dfrac{125y^4}{25y^{10}}\right)^3$

89. $\left(\dfrac{27x^9}{30x^5}\right)^0$ **90.** $\left(\dfrac{18y^6}{24y^{10}}\right)^3$ **91.** $\left(\dfrac{x^4y^3}{x^2y^5}\right)^2$ **92.** $\left(\dfrac{2x^7y^2}{4xy^2}\right)^3$

93. $\left(\dfrac{9y^2z^7}{18y^7z}\right)^4$ **94.** $\left(\dfrac{y^7z^5}{y^8z^4}\right)^{10}$ **95.** $\left(\dfrac{4x^2y^5}{y^2}\right)^3$ **96.** $\left(\dfrac{-64xy^6}{32xy^9}\right)^4$

97. $\left(\dfrac{-x^4y^6}{x^2}\right)^2$ **98.** $\left(\dfrac{-x^3y^5}{xy^7}\right)^3$ **99.** $\left(\dfrac{-12x}{16x^7y^2}\right)^2$ **100.** $\left(\dfrac{-x^4z^7}{x^2z^5}\right)^4$

101. $(3xy^4)^2$ **102.** $(2x^4y)(-y^5)$ **103.** $(-6xy^5)(3x^2y^4)$ **104.** $(-2xy)(3xy)$

105. $(2x^4y^2)(4xy^6)$ **106.** $(5x^2y)(3xy^5)$ **107.** $(5xy)(2xy^6)$

108. $(3x^2y)^2(xy)$ **109.** $(2xy)^2(3xy^2)^0$ **110.** $(3x^2)^4(2xy^5)$

111. $(x^4y^6)^3(3x^2y^5)$ **112.** $(4x^2y)(3xy^2)^3$ **113.** $(2x^2y^5)(3x^5y^4)^3$

114. $(5x^4y^7)(2x^3y)^3$ **115.** $(x^7y^5)(xy^2)^4$ **116.** $(xy^4)(xy^4)^3$

117. $(3x^4y^{10})^2(2x^2y^8)$ **118.** $(3x^6y)^2(4xy^8)$

Study the Common Student Error on page 194. Simplify the following expressions by dividing out common factors. If the expression cannot be simplified by dividing out common factors, state so.

119. $\dfrac{x + y}{x}$ **120.** $\dfrac{xy}{x}$ **121.** $\dfrac{x^2 + 2}{x}$

122. $\dfrac{2x^2}{2}$ **123.** $\dfrac{x + 4}{2}$ **124.** $\dfrac{2x}{2}$

125. $\dfrac{x^2y^2}{x^2}$ **126.** $\dfrac{x^2 + y^2}{x^2}$ **127.** $\dfrac{x}{x + 1}$

128. $\dfrac{1}{x + 1}$ **129.** $\dfrac{x^4}{x^2y}$ **130.** $\dfrac{y^2}{x^2 + y}$

131. For what value of x is $x^0 \neq 1$?

132. Explain the difference between the product rule and power rule. Give an example of each.

133. Consider the expression $(-x^5y^7)^9$. When the power rule is used to simplify the expression, will the *sign* of the simplified expression be positive or negative? Explain how you determined your answer.

134. Consider the expression $(-9x^4y^6)^8$. When the power rule is used to simplify the expression, what will be the *sign* of the simplified expression? Explain how you determined your answer.

135. Consider the expression $(-8x^5y^7)^6$. When simplified, what will be the *sign* of the simplified expression? Explain how you determined your answer.

136. In your own words, discuss **(a)** the product rule. **(b)** the quotient rule, **(c)** the zero exponent rule, **(d)** the power rule, and **(e)** the expanded power rule.

Cumulative Review Exercises

Answer each question in your own words.

[2.2] **137.** What is a linear equation?

138. What is a conditional linear equation?

139. What is an identity?

[3.1] **140.** Find the circumference and area of the circle shown.

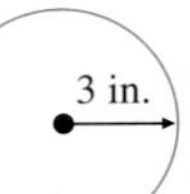

141. Solve the equation $2x - 5y = 6$ for y.

Group Activity/ Challenge Problems

Simplify.

1. $\left(\dfrac{3x^4y^5}{6x^6y^8}\right)^3\left(\dfrac{9x^7y^8}{3x^3y^5}\right)^2$

2. $(2xy^4)^3\left(\dfrac{6x^2y^5}{3x^3y^4}\right)^3(3x^2y^4)^2$

Determine the number (or numbers) that when substituted in the shaded area (or areas) makes the statement true.

3. $x^{\square} \cdot x^5 = x^{20}$

4. $\dfrac{x^8}{x^{\square}} = x^2$

5. $(2x^{\square})^4 = 16x^{12}$

6. $(3x^5)^{\square} = 27x^{15}$

7. $\dfrac{x^{\square}y^8}{x^3y^{\square}} = \dfrac{y^3}{x}$

8. $(x^2y^{\square})^2(x^{\square}y^6) = x^7y^{14}$

9. $\left(\dfrac{x^6y^{\square}}{x^{\square}y^8}\right)^3 = x^6y^9$

10. $\left(\dfrac{\square x^2y^9}{5x^{\square}y^{\square}}\right)^2 = \dfrac{4}{y^2}$

11. $(\square x^{\square}y^{\square})^2(5x^3y^5) = 45x^7y^{13}$

4.2 Negative Exponents

Tape 6

1 Understand the negative exponent rule.

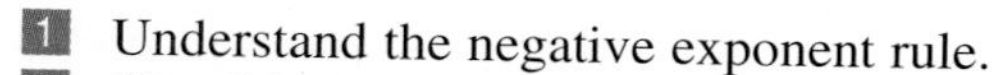

2 Simplify expressions containing negative exponents.

The Negative Exponent Rule

1 One additional rule that involves exponents is the negative exponent rule. You will need to understand negative exponents to be successful with scientific notation in the next section.

The negative exponent rule will be developed using the quotient rule illustrated in Example 1.

EXAMPLE 1 Simplify x^3/x^5 by (a) using the quotient rule, and (b) dividing out common factors.

Solution: **(a)** By the quotient rule,

$$\frac{x^3}{x^5} = x^{3-5} = x^{-2}$$

(b) By dividing out common factors,

$$\frac{x^3}{x^5} = \frac{\cancel{x} \cdot \cancel{x} \cdot \cancel{x}}{\cancel{x} \cdot \cancel{x} \cdot \cancel{x} \cdot x \cdot x} = \frac{1}{x^2}$$

In Example 1 we see that x^3/x^5 is equal to both x^{-2} and $1/x^2$. Therefore, x^{-2} must equal $1/x^2$. That is, $x^{-2} = 1/x^2$. This is an example of the negative exponent rule.

Negative Exponent Rule

$$x^{-m} = \frac{1}{x^m}, \quad x \neq 0$$

When a variable or number is raised to a negative exponent, the expression may be rewritten as 1 divided by the variable or number to that positive exponent.

Examples

$$x^{-3} = \frac{1}{x^3} \qquad 4^{-2} = \frac{1}{4^2} = \frac{1}{16}$$

$$y^{-7} = \frac{1}{y^7} \qquad 5^{-3} = \frac{1}{5^3} = \frac{1}{125}$$

COMMON STUDENT ERROR Students sometimes believe that a negative exponent automatically makes the value of the expression negative. This is not true.

Expression	*Correct*	*Incorrect*	*Also Incorrect*
3^{-2}	$\frac{1}{3^2} = \frac{1}{9}$	~~-3^2~~	~~$-\frac{1}{3^2}$~~
x^{-3}	$\frac{1}{x^3}$	~~$-x^3$~~	~~$-\frac{1}{x^3}$~~

To help you see that the negative exponent rule makes sense, consider the following sequence of exponential expressions and their corresponding values.

$2^4 = 16$	
$2^3 = 8$	(One-half of 16 is 8.)
$2^2 = 4$	(One-half of 8 is 4.)
$2^1 = 2$	(One-half of 4 is 2.)
$2^0 = 1$	(One-half of 2 is 1.)
$2^{-1} = \frac{1}{2^1}$ or $\frac{1}{2}$	(One-half of 1 is $\frac{1}{2}$.)
$2^{-2} = \frac{1}{2^2}$ or $\frac{1}{4}$	(One-half of $\frac{1}{2}$ is $\frac{1}{4}$.)
$2^{-3} = \frac{1}{2^3}$ or $\frac{1}{8}$	(One-half of $\frac{1}{4}$ is $\frac{1}{8}$.)
$2^{-4} = \frac{1}{2^4}$ or $\frac{1}{16}$	(One-half of $\frac{1}{8}$ is $\frac{1}{16}$.)

Note that each time the exponent decreases by 1 the value of the expression is halved. For example, when we go from 2^4 to 2^3, the value of the expression goes from 16 to 8. If we continue decreasing the exponents beyond $2^0 = 1$, the next exponent in the pattern is -1. And if we take half of 1 we get $\frac{1}{2}$. This pattern illustrates that $x^{-m} = \frac{1}{x^m}$

Simplify Expressions Containing Negative Exponents

2 Generally, when you are asked to simplify an exponential expression **your final answer should contain no negative exponents.** You may simplify exponential expressions using the negative exponent rule and the rules of exponents presented in the previous section. The following examples indicate how exponential expressions containing negative exponents may be simplified.

EXAMPLE 2 Use the negative exponent rule to write each expression with positive exponents.

(a) x^{-2} (b) y^{-4} (c) 3^{-2} (d) 5^{-1}

Solution:

(a) $x^{-2} = \frac{1}{x^2}$ (b) $y^{-4} = \frac{1}{y^4}$

(c) $3^{-2} = \frac{1}{3^2} = \frac{1}{9}$ (d) $5^{-1} = \frac{1}{5}$

EXAMPLE 3 Use the negative exponent rule to write each expression with positive exponents.

(a) $\frac{1}{x^{-2}}$ (b) $\frac{1}{4^{-1}}$

Solution: First use the negative exponent rule on the denominator. Then simplify further.

(a) $\frac{1}{x^{-2}} = \frac{1}{1/x^2} = \frac{1}{1} \cdot \frac{x^2}{1} = x^2$ (b) $\frac{1}{4^{-1}} = \frac{1}{1/4} = \frac{1}{1} \cdot \frac{4}{1} = 4$

Helpful Hint

From Examples 2 and 3 we can see that when a factor is moved from the denominator to the numerator or from the numerator to the denominator, the sign of the *exponent* changes.

$$x^{-4} = \frac{1}{x^4} \qquad \frac{1}{x^{-4}} = x^4$$

$$3^{-5} = \frac{1}{3^5} \qquad \frac{1}{3^{-5}} = 3^5$$

Now let's look at additional examples that combine two or more of the rules presented so far.

EXAMPLE 4 Simplify.

(a) $(y^{-3})^8$ **(b)** $(4^2)^{-3}$

Solution:

(a) $(y^{-3})^8 = y^{(-3)(8)}$ by the power rule

$= y^{-24}$

$= \frac{1}{y^{24}}$ by the negative exponent rule

(b) $(4^2)^{-3} = 4^{(2)(-3)}$ by the power rule

$= 4^{-6}$

$= \frac{1}{4^6}$ by the negative exponent rule

EXAMPLE 5 Simplify.

(a) $x^3 \cdot x^{-5}$ (b) $3^{-4} \cdot 3^{-7}$

Solution:

(a) $x^3 \cdot x^{-5} = x^{3+(-5)}$ **by the product rule**

$= x^{-2}$

$= \frac{1}{x^2}$ **by the negative exponent rule**

(b) $3^{-4} \cdot 3^{-7} = 3^{-4+(-7)}$ **by the product rule**

$= 3^{-11}$

$= \frac{1}{3^{11}}$ **by the negative exponent rule**

COMMON STUDENT ERROR What is the sum of $3^2 + 3^{-2}$? Look carefully at the correct solution.

Correct	*Incorrect*
$3^2 + 3^{-2} = 9 + \frac{1}{9}$	~~$3^2 + 3^{-2} = 0$~~
$= 9\frac{1}{9}$	

Note that $3^2 \cdot 3^{-2} = 3^{2+(-2)} = 3^0 = 1$.

EXAMPLE 6 Simplify.

(a) $\frac{x^{-4}}{x^{10}}$ (b) $\frac{5^{-7}}{5^{-4}}$

Solution:

(a) $\frac{x^{-4}}{x^{10}} = x^{-4-10}$ **by the quotient rule**

$= x^{-14}$

$= \frac{1}{x^{14}}$ **by the negative exponent rule**

(b) $\frac{5^{-7}}{5^{-4}} = 5^{-7-(-4)}$ **by the quotient rule**

$= 5^{-7+4}$

$= 5^{-3}$

$= \frac{1}{5^3}$ or $\frac{1}{125}$ **by the negative exponent rule**

Helpful Hint

Consider a division problem where a variable has a negative exponent in either its numerator or its denominator. To simplify the expression, we can move the variable with the negative exponent from the numerator to the denominator, or from the denominator to the numerator, and change the sign of the exponent. For example,

$$\frac{x^{-4}}{x^5} = \frac{1}{x^5 \cdot x^4} = \frac{1}{x^{5+4}} = \frac{1}{x^9}$$

$$\frac{y^3}{y^{-7}} = y^3 \cdot y^7 = y^{3+7} = y^{10}$$

Now consider a division problem where the variable has a negative exponent in both its numerator and denominator. To simplify such an expression, we move the variable with the more negative exponent from the numerator to the denominator, or from the denominator to the numerator, and change the sign of the exponent from negative to positive. For example,

$$\frac{x^{-8}}{x^{-3}} = \frac{1}{x^8 \cdot x^{-3}} = \frac{1}{x^{8-3}} = \frac{1}{x^5} \qquad \text{Note that } -8 < -3.$$

$$\frac{y^{-4}}{y^{-7}} = y^7 \cdot y^{-4} = y^{7-4} = y^3 \qquad \text{Note that } -7 < -4.$$

EXAMPLE 7 Simplify.

(a) $4x^2(5x^{-5})$ (b) $\dfrac{8x^3y^{-2}}{4xy^2}$ (c) $\dfrac{2x^2y^5}{8x^7y^{-3}}$

Solution:

(a) $4x^2(5x^{-5}) = 4 \cdot 5 \cdot x^2 \cdot x^{-5} = 20x^{-3} = \dfrac{20}{x^3}$

(b) $\dfrac{8x^3y^{-2}}{4xy^2} = \dfrac{8}{4} \cdot \dfrac{x^3}{x} \cdot \dfrac{y^{-2}}{y^2}$

$= 2 \cdot x^2 \cdot \dfrac{1}{y^4} = \dfrac{2x^2}{y^4}$

(c) $\dfrac{2x^2y^5}{8x^7y^{-3}} = \dfrac{2}{8} \cdot \dfrac{x^2}{x^7} \cdot \dfrac{y^5}{y^{-3}}$

$= \dfrac{1}{4} \cdot \dfrac{1}{x^5} \cdot y^8 = \dfrac{y^8}{4x^5}$

In Example 7(b), the variable with the negative exponent was moved from the numerator to the denominator. In Example 7(c), the variable with the negative exponent was moved from the denominator to the numerator. In each case, the sign of the exponent was changed from negative to positive when the variable factor was moved.

EXAMPLE 8 Simplify $(4x^{-3})^{-2}$.

Solution: Begin by using the expanded power rule.

$$(4x^{-3})^{-2} = 4^{-2}x^{(-3)(-2)}$$
$$= 4^{-2}x^{6}$$
$$= \frac{1}{4^2}x^6$$
$$= \frac{x^6}{16}$$

COMMON STUDENT ERROR Can you explain why the simplification on the right is incorrect?

Correct

$$\frac{x^3y^{-2}}{w} = \frac{x^3}{wy^2}$$

Incorrect

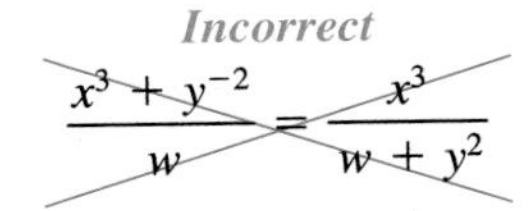

The simplification on the right is incorrect because in the numerator $x^3 + y^{-2}$ the y^{-2} *is not a factor;* it is a term. We will learn how to simplify expressions like this when we study complex fractions in Section 6.5.

Summary of Rules of Exponents

1. $x^m \cdot x^n = x^{m+n}$		**product rule**
2. $\frac{x^m}{x^n} = x^{m-n}$,	$x \neq 0$	**quotient rule**
3. $x^0 = 1, x \neq 0$		**zero exponent rule**
4. $(x^m)^n = x^{m \cdot n}$		**power rule**
5. $\left(\frac{ax}{by}\right)^m = \frac{a^mx^m}{b^my^m}$,	$b \neq 0, y \neq 0$	**expanded power rule**
6. $x^{-m} = \frac{1}{x^m}$,	$x \neq 0$	**negative exponent rule**

Exercise Set 4.2

Simplify.

1. x^{-2}
2. y^{-5}
3. 4^{-1}
4. 5^{-2}
5. $\frac{1}{x^{-3}}$
6. $\frac{1}{x^{-2}}$
7. $\frac{1}{x^{-1}}$
8. $\frac{1}{y^{-4}}$
9. $\frac{1}{4^{-2}}$
10. $\frac{1}{6^{-3}}$
11. $(x^{-2})^3$
12. $(x^{-4})^2$
13. $(y^{-7})^3$
14. $(y^3)^{-8}$
15. $(x^4)^{-2}$
16. $(x^{-9})^{-2}$

17. $(2^{-3})^{-2}$

18. $(2^{-3})^2$

19. $x^4 \cdot x^{-1}$

20. $x^{-3} \cdot x^1$

21. $x^7 \cdot x^{-5}$

22. $x^{-3} \cdot x^{-2}$

23. $3^{-2} \cdot 3^4$

24. $5^3 \cdot 5^{-4}$

25. $\frac{x^8}{x^{10}}$

26. $\frac{x^2}{x^{-1}}$

27. $\frac{y^0}{y^{-3}}$

28. $\frac{x^{-2}}{x^5}$

29. $\frac{x^{-7}}{x^{-3}}$

30. $\frac{x^{-8}}{x^{-3}}$

31. $\frac{3^2}{3^{-1}}$

32. $\frac{2^6}{2^{-1}}$

33. 3^{-3}

34. x^{-7}

35. $\frac{1}{z^{-9}}$

36. $\frac{1}{4^{-3}}$

37. $(x^5)^{-5}$

38. $(x^{-3})^{-4}$

39. $(y^{-2})^{-3}$

40. $x^9 \cdot x^{-12}$

41. $x^3 \cdot x^{-7}$

42. $x^{-3} \cdot x^{-5}$

43. $x^{-8} \cdot x^{-7}$

44. $4^{-3} \cdot 4^3$

45. $\frac{x^{-3}}{x^5}$

46. $\frac{y^6}{y^{-8}}$

47. $\frac{y^9}{y^{-1}}$

48. $\frac{3^{-4}}{3}$

49. $\frac{2^{-3}}{2^{-3}}$

50. $(5^{-2}x^3)^0$

51. $(2^{-1} + 3^{-1})^0$

52. 5^0y^3

53. $\frac{1}{1^{-5}}$

54. $(z^{-7})^{-3}$

55. $(x^{-4})^{-1}$

56. $(x^{-3})^0$

57. $(x^0)^{-3}$

58. $(2^{-2})^{-1}$

59. $2^{-3} \cdot 2$

60. $6^4 \cdot 6^{-2}$

61. $6^{-4} \cdot 6^2$

62. $\frac{x^{-5}}{x^{-9}}$

63. $\frac{x^{-1}}{x^{-4}}$

64. $\frac{x^{-4}}{x^{-1}}$

65. $(3^2)^{-1}$

66. $(5^{-2})^{-2}$

67. $\frac{5}{5^{-2}}$

68. $\frac{x^6}{x^7}$

69. $\frac{2^{-4}}{2^{-2}}$

70. $x^{-12} \cdot x^8$

71. $\frac{7^{-1}}{7^{-1}}$

72. $\frac{3^{-2}}{3^{-1}}$

73. $2x^{-1}y$

74. $(6x^2)^{-2}$

75. $(3x^3)^{-1}$

76. $3x^{-2}y^2$

77. $5x^4y^{-1}$

78. $5x^{-5}y^{-2}$

79. $(3x^2y^3)^{-2}$

80. $(4x^2y^{-3})^{-2}$

81. $(x^5y^{-3})^{-3}$

82. $(3x^{-3}y^4)^{-1}$

83. $3x(5x^{-4})$

84. $(4x^{-2})(5x^{-3})$

85. $2x^5(3x^{-6})$

86. $6x^4(-2x^{-2})$

87. $(9x^5)(-3x^{-7})$

88. $(4x^2y)(3x^3y^{-1})$

89. $(2x^{-3}y^{-2})(x^4y^0)$

90. $(5^0x^{-7})(4x^{-2}y)$

91. $(3y^{-2})(5x^{-1}y^3)$

92. $(4x^{-2})(6x^{-2}y^{-1})$

93. $\frac{8x^4}{4x^{-1}}$

94. $\frac{3x^5}{6x^{-2}}$

95. $\frac{12x^{-2}}{3x^5}$

96. $\frac{2y^{-6}}{6y^4}$

97. $\frac{5x^{-2}}{25x^{-5}}$

98. $\frac{36x^{-4}}{9x^{-2}}$

99. $\frac{12x^{-2}y^0}{2x^3y^2}$

100. $\frac{3x^4y^{-2}}{6y^3}$

101. $\frac{16x^{-7}y^{-2}}{4x^5y^2}$

102. $\frac{32x^4y^{-2}}{4x^{-2}y^{-3}}$

103. $\frac{9x^4y^{-7}}{18x^{-1}y^{-1}}$

104. $\frac{32x^4y^{-3}}{4x^{-3}y^{-3}}$

105. **(a)** Does $a^{-1}b^{-1} = \frac{1}{ab}$? Explain your answer.

(b) Does $a^{-1} + b^{-1} = \frac{1}{a + b}$? Explain your answer:

106. **(a)** Does $\frac{x^{-1}y^2}{z} = \frac{y^2}{xz}$? Explain your answer.

(b) Does $\frac{x^{-1} + y^2}{z} = \frac{y^2}{x + z}$? Explain your answer.

107. In your own words, describe the negative exponent rule.

CUMULATIVE REVIEW EXERCISES

[1.9] **108.** Evaluate $2[6 - (4 - 5)] \div 2 - 5^2$.

109. Evaluate $\dfrac{-3^2 \cdot 4 \div 2}{\sqrt{9} - 2^2}$

[2.6] **110.** According to the instructions on a bottle of concentrated household cleaner, 8 ounces of the cleaner should be mixed with 3 gallons of water. If your bucket holds only 2.5 gallons of water, how much cleaner should you use?

[3.3] **111.** The larger of two integers is one more than three times the smaller. If the sum of the two integers is 37, find the two integers.

Group Activity/ Challenge Problems

1. Often problems involving exponents can be done in more than one way. Simplify $\left(\dfrac{3x^2y^3}{z}\right)^{-2}$

(a) By first using the expanded power rule.

(b) By first using the negative exponent rule.

2. For any nonzero real number a, if $a^{-1} = x$, describe each of the following in terms of x.

(a) $-a^{-1}$ **(b)** $\dfrac{1}{a^{-1}}$

Determine the number (or numbers) that when placed in the shaded area (or areas) make the statement true.

3. $5^{\square} = \dfrac{1}{25}$

4. $\dfrac{1}{3^{\square}} = 9$

5. $(x^{\square}y^3)^{-2} = \dfrac{x^4}{y^6}$

6. $(\square x^{\square}y^{-2})^3 = \dfrac{8}{x^9y^6}$

7. $(x^4y^{-3})^{\square} = \dfrac{y^9}{x^{12}}$

8. $\left(\dfrac{x^{\square}y^{-3}}{x^{-4}y^5}\right)^{\square} = y^{16}$

Evaluate.

9. $3^0 - 3^{-1}$

10. $4^{-1} - 3^{-1}$

11. $2 \cdot 4^{-1} + 4 \cdot 3^{-1}$

12. $-\left(-\dfrac{2}{5}\right)^{-1}$

13. Consider $(3^{-1} + 2^{-1})^0$. We know this is equal to 1 by the zero exponent rule. Can you determine the error in the following calculation? Explain your answer.

$$\begin{aligned}\text{Does } (3^{-1} + 2^{-1})^0 &= (3^{-1})^0 + (2^{-1})^0?\\ &= 3^{-1(0)} + 2^{-1(0)}\\ &= 3^0 + 2^0\\ &= 1 + 1 = 2\end{aligned}$$

4.3 Scientific Notation

Tape 6

1. Convert numbers to and from scientific notation.
2. Recognize numbers in scientific notation with a coefficient of 1.
3. Do calculations with numbers in scientific notation.

Converting Numbers

1 When working with scientific problems, we often deal with very large and very small numbers. For example, the distance from Earth to the sun is about 93,000,000 miles. The wavelength of yellow light is about 0.0000006 meter. Because it is difficult to work with many zeros, scientists often express such numbers with exponents. For example, the number 93,000,000 might be written 9.3×10^7 and the number 0.0000006 might be written 6.0×10^{-7}.

Numbers such as 9.3×10^7 and 6.0×10^{-7} are in a form called **scientific notation.** Each number written in scientific notation is written as a number greater than or equal to 1 and less than 10 ($1 \le a < 10$) multiplied by some power of 10.

Examples of Numbers in Scientific Notation

$$1.2 \times 10^6$$
$$3.762 \times 10^3$$
$$8.07 \times 10^{-2}$$
$$1 \times 10^{-5}$$

Below we change the number 68,400 to scientific notation.

$$68{,}400 = 6.84 \times 10{,}000$$
$$= 6.84 \times 10^4 \qquad \text{Note that } 10{,}000 = 10 \cdot 10 \cdot 10 \cdot 10 = 10^4.$$

Therefore, $68{,}400 = 6.84 \times 10^4$. To go from 68,400 to 6.84 the decimal point was moved 4 places to the left. Note that the exponent on the 10, the 4, is the same as the number of places the decimal point was moved to the left. Here is a simplified procedure for writing a number in scientific notation:

To Write a Number in Scientific Notation

1. Move the decimal point in the original number to the right of the first nonzero digit. This will give a number greater than or equal to 1 and less than 10.
2. Count the number of places you moved the decimal to obtain the number in step 1. If the original number was 10 or greater, the count is to be considered positive. If the original number was less than 1, the count is to be considered negative.
3. Multiply the number obtained in step 1 by 10 raised to the count (power) found in step 2.

EXAMPLE 1 Write the following numbers in scientific notation.

(a) 10,700 **(b)** 0.000386 **(c)** 972,000 **(d)** 0.0083

Solution:

(a) 10,700 means 10,700.

$10{,}700. = 1.07 \times 10^4$

4 places

The original number is greater than 10; therefore, the exponent is positive.

(b) $0.000386 = 3.86 \times 10^{-4}$

4 places

The original number is less than 1; therefore, the exponent is negative.

(c) $972{,}000. = 9.72 \times 10^5$

5 places

(d) $0.0083 = 8.3 \times 10^{-3}$

3 places

Now we explain how to write a number in scientific notation as a number without exponents.

To Convert a Number from Scientific Notation to Decimal Form

1. Observe the exponent of the power of 10.
2. (a) If the exponent is positive, move the decimal point in the number (greater than or equal to 1 and less than 10) to the right the same number of places as the exponent. It may be necessary to add zeros to the number. This will result in a number greater than or equal to 10.

   (b) If the exponent is 0, do not move the decimal point. Drop the factor 10^0 since it equals 1. This will result in a number greater than or equal to 1 but less than 10.

   (c) If the exponent is negative, move the decimal point in the number to the left the same number of places as the exponent (dropping the negative sign). It may be necessary to add zeros. This will result in a number less than 1.

EXAMPLE 2 Write each number without exponents.

(a) 3.2×10^4 **(b)** 6.28×10^{-3} **(c)** 7.95×10^8

Solution: **(a)** Move the decimal point four places to the right.

$$3.2 \times 10^4 = 3.2 \times 10{,}000 = 32{,}000$$

(b) Move the decimal point three places to the left.

$$6.28 \times 10^{-3} = 0.00628$$

(c) Move the decimal point eight places to the right.

$$7.95 \times 10^8 = 795{,}000{,}000$$

Coefficient of 1

2 When reading scientific magazines you will sometimes see numbers written as powers of 10, but without a numerical coefficient. For example in Figure 4.1 we see that the frequency of FM radio is about 10^8 hertz (or cycles per

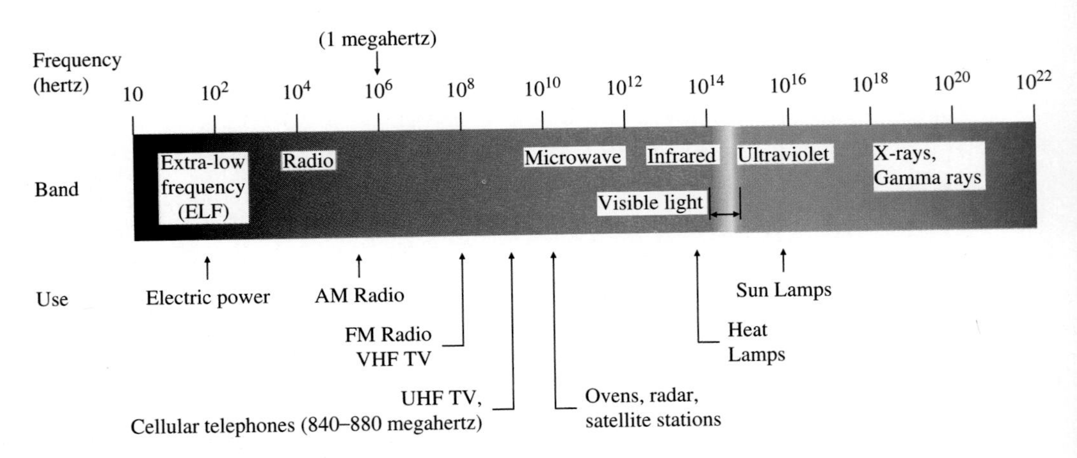

FIGURE 4.1

second). If no numerical coefficient is indicated, the numerical coefficient is always assumed to be 1. Thus, the frequency of FM radio is $10^8 = 1.0 \times 10^8 = 100{,}000{,}000$ hertz (or 100 megahertz since 1 megahertz $= 10^6$ hertz).

Now write the frequency of microwave ovens, 10^{10}, as a frequency in hertz. Your answer should be 10,000,000,000 (which equals 10,000 megahertz).

Calculations Using Scientific Notation

3 We can use the rules of exponents presented in Sections 4.1 and 4.2 when working with numbers written in scientific notation.

EXAMPLE 3 Multiply $(4.2 \times 10^6)(2 \times 10^{-4})$.

Solution:
$$(4.2 \times 10^6)(2 \times 10^{-4}) = (4.2 \times 2)(10^6 \times 10^{-4})$$
$$8.4 \times 10^{6+(-4)}$$
$$= 8.4 \times 10^2 \quad \text{or} \quad 840$$

EXAMPLE 4 Divide $\dfrac{6.2 \times 10^{-5}}{2 \times 10^{-3}}$.

Solution:
$$\frac{6.2 \times 10^{-5}}{2 \times 10^{-3}} = \left(\frac{6.2}{2}\right)\left(\frac{10^{-5}}{10^{-3}}\right)$$
$$= 3.1 \times 10^{-5-(-3)}$$
$$= 3.1 \times 10^{-5+3}$$
$$= 3.1 \times 10^{-2} \quad \text{or} \quad 0.031$$

Calculator Corner

What will your calculator show when you multiply very large or very small numbers? The answer depends on whether your calculator has the ability to display an answer in scientific notation. On calculators without the ability to express numbers in scientific notation, you will probably get an error message because the answer will be too large or too small for the display.

Example: On a calculator without scientific notation:

8000000 [×] 600000 [=] Error

Example: On a calculator that uses scientific notation:

8000000 [×] 600000 [=] 4.8 12

This 4.8 12 means 4.8×10^{12}.

Example: On a calculator that uses scientific notation:

.0000003 [×] .004 [=] 1.2 −9

This 1.2 −9 means 1.2×10^{-9}.

EXAMPLE 5 If a computer can do a calculation in 0.000008 second (eight millionths of a second), how long, in seconds, would it take the computer to do 6 billion (6,000,000,000) calculations?

Solution: The computer could do 1 calculation in 1(0.000008) seconds, 2 calculations in 2(0.000008) seconds, 3 calculations in 3(0.000008) seconds, and 6,000,000,000 calculations in 6,000,000,000 (0.000008) seconds. We will multiply by converting each number to scientific notation.

$$\begin{aligned}(6{,}000{,}000{,}000)(0.000008) &= (6 \times 10^9)(8 \times 10^{-6})\\ &= (6 \times 8)(10^9 \times 10^{-6})\\ &= 48 \times 10^3\\ &= 48{,}000 \text{ seconds}\end{aligned}$$

Thus, the computer can do the calculations in 48,000 seconds, or about $13\frac{1}{3}$ hours.

Exercise Set 4.3

Express each number in scientific notation.

1. 42,000 **2.** 3,610,000 **3.** 900 **4.** 0.00062

5. 0.053 **6.** 0.0000462 **7.** 19,000 **8.** 5,260,000,000

9. 0.00000186 **10.** 0.0003 **11.** 0.00000914 **12.** 37,000

13. 107 **14.** 0.02 **15.** 0.153 **16.** 416,000

Express each number without exponents.

17. 4.2×10^3 **18.** 1.63×10^{-4} **19.** 4×10^7 **20.** 6.15×10^5

21. 2.13×10^{-5} **22.** 9.64×10^{-7} **23.** 3.12×10^{-1} **24.** 4.6×10^1

25. 9×10^6 **26.** 7.3×10^4 **27.** 5.35×10^2 **28.** 1.04×10^{-2}

29. 3.5×10^4 **30.** 2.17×10^{-6} **31.** 1×10^4 **32.** 1×10^{-3}

Perform the indicated operation and express each number without exponents.

33. $(4 \times 10^2)(3 \times 10^5)$ **34.** $(2 \times 10^{-3})(3 \times 10^2)$ **35.** $(5.1 \times 10^1)(3 \times 10^{-4})$

36. $(1.6 \times 10^{-2})(4 \times 10^{-3})$ **37.** $(6.2 \times 10^4)(1.5 \times 10^{-2})$ **38.** $(9 \times 10^8)(1.2 \times 10^{-4})$

39. $\dfrac{6.4 \times 10^5}{2 \times 10^3}$ **40.** $\dfrac{8 \times 10^{-3}}{2 \times 10^1}$ **41.** $\dfrac{8.4 \times 10^{-6}}{4 \times 10^{-3}}$

42. $\dfrac{25 \times 10^3}{5 \times 10^{-2}}$ **43.** $\dfrac{4 \times 10^5}{2 \times 10^4}$ **44.** $\dfrac{16 \times 10^3}{8 \times 10^{-3}}$

Perform the indicated operation by first converting each number to scientific notation. Write the answer in scientific notation.

45. (700,000)(6,000,000) **46.** (0.0006)(5,000,000) **47.** (0.003)(0.00015)

48. (230,000)(3000) **49.** $\dfrac{1{,}400{,}000}{700}$ **50.** $\dfrac{20{,}000}{0.0005}$

51. $\dfrac{0.00004}{200}$ **52.** $\dfrac{0.0012}{0.000006}$ **53.** $\dfrac{150{,}000}{0.0005}$

54. List the following numbers from smallest to largest: 4.8×10^5, 3.2×10^{-1}, 4.6, 8.3×10^{-4}

55. List the following numbers from smallest to largest: 9.2×10^{-5}, 8.4×10^3, 1.3×10^{-1}, 6.2×10^4

In Exercises 56–63, write the answer without exponents unless asked to do otherwise.

56. The distance from Earth to the planet Jupiter is approximately 4.5×10^8 miles. If a spacecraft travels at a speed of 25,000 miles per hour, how long, in hours, would it take the spacecraft to travel from Earth to Jupiter? Use distance = rate × time.

57. If a computer can do a calculation in 0.000004 second, how long, in seconds, would it take the computer to do 8 trillion (8,000,000,000,000) calculations? Write your answer in scientific notation.

58. The half-life of a radioactive isotope is the time required for half the quantity of the isotope to decompose. The half-life of uranium 238 is 4.5×10^9 years, and the half-life of uranium 234 is 2.5×10^5 years. How many times greater is the half-life of uranium 238 than uranium 234?

59. A treaty between the United States and Canada requires that during the tourist season a minimum of 100,000 cubic feet of water per second flows over Niagara Falls (another 130,000 to 160,000 cubic feet/sec is diverted for power generation). Find the minimum volume of water that will flow over the falls in a 24-hour period during the tourist season.

60. The numer of people in the world in 1992 was approximately 5.42×10^9 and the number of people living in China was 1.17×10^9. How many people lived outside China?

61. In the United States only 5% of the 4.0×10^9 pounds of plastic produced annually is recycled. How many pounds of plastic is recycled annually? Write your answer **(a)** in scientific notation and **(b)** as a number without exponents. (Plastic makes up about 25% of all U.S. solid waste and takes up about 8% of all landfill space.)

62. In 1970, 3.0×10^{12} cubic feet of natural gas was produced worldwide. In 1990, 6.6×10^{12} cubic feet of natural gas was produced. How many more cubic feet of natural gas was produced in 1990 than in 1970? Write your answer **(a)** in scientific notation and **(b)** as a number without exponents.

63. Laid end to end the 18 billion disposable diapers thrown away in the United States each year would reach the moon and back 7 times. (7 round trips)
(a) Write 18 billion as a number in scientific notation.
(b) If the distance from Earth to the moon is 2.38×10^5 miles, what is the length of all these diapers placed end to end? Write your answer in scientific notation and as a number without exponents.

64. In Example 2(b) we showed that $6.28 \times 10^{-3} = 0.00628$. Show that you can obtain the same results by using the negative exponent rule.

65. Consider the problem $\dfrac{(4 \cdot 10^3)(6 \cdot 10^{\square})}{24 \cdot 10^{-5}} = 1$. Find the exponent that when placed in the shaded area makes this statement true. Explain how you found your answer.

66. Consider the problem $\dfrac{25 \cdot 10^8}{(5 \cdot 10^{-3})(5 \cdot 10^{\square})} = 1$. Find the exponent that when placed in the shaded area makes this statement true. Explain how you found your answer.

CUMULATIVE REVIEW EXERCISES

[1.9] **67.** Evaluate $4x^2 + 3x + \dfrac{x}{2}$ when $x = 0$.

[2.3] **68.** **(a)** If $-x = -\dfrac{3}{2}$, what is the value of x?
(b) If $5x = 0$, what is the value of x?

[2.5] **69.** Solve the equation $2x - 3(x - 2) = x + 2$.

[4.1] **70.** Simplify $\left(\dfrac{-2x^5y^7}{8x^8y^3}\right)^3$

Group Activity/ Challenge Problems

1. Many students have no idea of the difference in size between a million (1,000,000), a billion (1,000,000,000), and a trillion (1,000,000,000,000).

(a) Write a million, a billion, and a trillion in scientific notation.

(b) Determine how long it would take to spend a million dollars if you spent $1000 a day.

(c) Repeat part (b) for a billion dollars.

(d) Repeat part (b) for a trillion dollars.

(e) How many times greater is a billion dollars than a million dollars?

2. It took all of human history for the world's population to reach 5.4×10^9 in 1992. At current rates the world population will double in about 40 years.

(a) Estimate the world's population in 2032.

(b) Assuming 365 days in a year, estimate the average number of people added to Earth's population each day between 1992 and 2032.

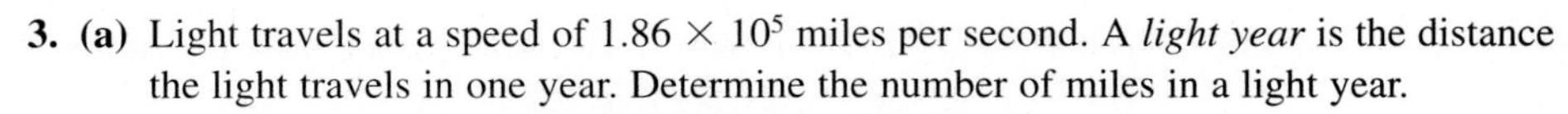

3. (a) Light travels at a speed of 1.86×10^5 miles per second. A *light year* is the distance the light travels in one year. Determine the number of miles in a light year.

(b) The Earth is approximately 93,000,000 miles from the sun. How long does it take light from the sun to reach the Earth?

4.4 Addition and Subtraction of Polynomials

Tape 6

1. Identify polynomials.
2. Add polynomials.
3. Subtract polynomials.

Identify Polynomials

1 A **polynomial in x** is an expression containing the sum of a finite number of terms of the form ax^n, for any real number a and any whole number n.

Examples of Polynomials	*Not Polynomials*	
$2x$	$4x^{1/2}$	**(fractional exponent)**
$\frac{1}{3}x - 4$	$3x^2 + 4x^{-1} + 5$	**(negative exponent)**
$x^2 - 2x + 1$	$4 + \frac{1}{x}$	$\left(\frac{1}{x} = x^{-1}\text{, negative exponent}\right)$

A polynomial is written in **descending order** (or **descending powers) of the variable** when the exponents on the variable decrease from left to right.

Example of Polynomial in Descending Order

$$2x^4 + 4x^2 - 6x + 3$$

Note in the example that the constant term 3 is last because it can be written as $3x^0$. Remember that $x^0 = 1$.

A polynomial of one term is called a **monomial.** A **binomial** is a two-termed polynomial. A **trinomial** is a three-termed polynomial. Polynomials containing more than three terms are not given special names. The prefix "poly" means "many." The chart that follows summarizes this information.

Type of Polynomial	*Number of Terms*	*Examples*
Monomial	One	$5,\quad 4x,\quad -6x^2$
Binomial	Two	$x + 4,\quad x^2 - 6,\quad 2x^2 - 5x$
Trinomial	Three	$x^2 - 2x + 3,\quad 5x^2 - 6x + 7$

The **degree of a term** of a polynomial in one variable is the exponent on the variable in that term.

Term	*Degree of Term*	
$4x^2$	Second	
$2y^5$	Fifth	
$-5x$	First	($-5x$ can be written $-5x^1$.)
3	Zero	(3 can be written $3x^0$.)

The **degree of a polynomial** in one variable is the same as that of its highest-degree term.

Polynomial	*Degree of Polynomial*	
$8x^3 + 2x^2 - 3x + 4$	Third	($8x^3$ is highest-degree term.)
$x^2 - 4$	Second	(x^2 is highest-degree term.)
$2x - 1$	First	($2x$ or $2x^1$ is highest-degree term.)
4	Zero	(4 or $4x^0$ is highest-degree term.)

Addition of Polynomials

2 In Section 2.1 we stated that like terms are terms having the same variables and the same exponents. That is, like terms differ only in their numerical coefficients.

Examples of Like Terms

$3,\quad -5$

$2x,\quad x$

$-2x^2,\quad 4x^2$

$3y^2,\quad 5y^2$

$3xy^2,\quad 5xy^2$

To add polynomials, combine the like terms of the polynomials.

EXAMPLE 1 Simplify $(4x^2 + 6x + 3) + (2x^2 + 5x - 1)$.

Solution:

$$(4x^2 + 6x + 3) + (2x^2 + 5x - 1)$$

$$= 4x^2 + 6x + 3 + 2x^2 + 5x - 1 \qquad \text{Remove parentheses.}$$

$$= \underbrace{4x^2 + 2x^2}_{} + \underbrace{6x + 5x}_{} + \underbrace{3 - 1}_{} \qquad \text{Rearrange terms.}$$

$$= 6x^2 + 11x + 2 \qquad \text{Combine like terms.}$$

EXAMPLE 2 Simplify $(4x^2 + 3x + y) + (x^2 - 6x + 3)$.

Solution:

$$\begin{aligned} &(4x^2 + 3x + y) + (x^2 - 6x + 3) \\ &= 4x^2 + 3x + y + x^2 - 6x + 3 && \text{Remove parentheses.} \\ &= \underbrace{4x^2 + x^2}_{} + \underbrace{3x - 6x}_{} + y + 3 && \text{Rearrange terms.} \\ &= 5x^2 - 3x + y + 3 && \text{Combine like terms.} \end{aligned}$$

EXAMPLE 3 Simplify $(3x^2y - 4xy + y) + (x^2y + 2xy + 3y)$.

Solution:

$$\begin{aligned} &(3x^2y - 4xy + y) + (x^2y + 2xy + 3y) \\ &= 3x^2y - 4xy + y + x^2y + 2xy + 3y && \text{Remove parentheses.} \\ &= \underbrace{3x^2y + x^2y}_{} \underbrace{- 4xy + 2xy}_{} \underbrace{+ y + 3y}_{} && \text{Rearrange terms.} \\ &= 4x^2y - 2xy + 4y && \text{Combine like terms.} \end{aligned}$$

Usually when we add polynomials we will do so as in Examples 1 through 3. That is, we will list the polynomials horizontally. However, in Section 4.6, when we divide polynomials, there will be steps where we add polynomials in columns.

To Add Polynomials in Columns

1. Arrange polynomials in descending order one under the other with like terms in the same columns.
2. Add the terms in each column.

EXAMPLE 4 Add $4x^2 - 2x + 2$ and $-2x^2 - x + 4$ using columns.

Solution:

$$\begin{array}{r} 4x^2 - 2x + 2 \\ -2x^2 - x + 4 \\ \hline 2x^2 - 3x + 6 \end{array}$$

EXAMPLE 5 Add $(3x^3 + 2x - 4)$ and $(2x^2 - 6x - 3)$ using columns.

Solution: Since the polynomial $3x^3 + 2x - 4$ does not have an x^2 term, we will add the term $0x^2$ to the polynomial. This procedure sometimes helps in aligning like terms.

$$\begin{array}{r} 3x^3 + 0x^2 + 2x - 4 \\ 2x^2 - 6x - 3 \\ \hline 3x^3 + 2x^2 - 4x - 7 \end{array}$$

3 Subtraction of Polynomials

To Subtract Polynomials

1. Remove parentheses. (This will have the effect of changing the sign of *every* term within the parentheses of the polynomial being subtracted.)
2. Combine like terms.

EXAMPLE 6 Simplify $(3x^2 - 2x + 5) - (x^2 - 3x + 4)$.

Solution:

$$(3x^2 - 2x + 5) - (x^2 - 3x + 4)$$
$$= 3x^2 - 2x + 5 - x^2 + 3x - 4$$ Remove parentheses (change the sign of each term being subtracted).
$$= \underbrace{3x^2 - x^2}\underbrace{- 2x + 3x}\underbrace{+ 5 - 4}$$ Rearrange terms.
$$= 2x^2 + x + 1$$ Combine like terms.

EXAMPLE 7 Subtract $(-x^2 - 2x + 3)$ from $(x^3 + 4x + 6)$.

Solution:

$$(x^3 + 4x + 6) - (-x^2 - 2x + 3)$$
$$= x^3 + 4x + 6 + x^2 + 2x - 3$$ Remove parentheses.
$$= x^3 + x^2 \underbrace{+ 4x + 2x}\underbrace{+ 6 - 3}$$ Rearrange terms.
$$= x^3 + x^2 + 6x + 3$$ Combine like terms.

COMMON STUDENT ERROR One of the most common mistakes made by students occurs when subtracting polynomials. When subtracting one polynomial from another, **the sign of each term in the polynomial being subtracted must be changed, not just the sign of the first term.**

Correct

$$6x^2 - 4x + 3 - (2x - 3x + 4)$$
$$= 6x^2 - 4x + 3 - 2x^2 + 3x - 4$$
$$= 4x^2 - x - 1$$

Incorrect

$$6x^2 - 4x + 3 - (2x^2 - 3x + 4)$$
$$= 6x^2 - 4x + 3 - 2x^2 - 3x + 4$$
$$= 4x^2 - 7x + 7$$

Do not make this mistake!

Subtraction of Polynomials in Columns

To Subtract Polynomials in Columns

1. Write *the polynomial being subtracted* below the polynomial from which it is being subtracted. List like terms in the same column.
2. **Change the sign of each term** in the polynomial being subtracted. (This step can be done mentally, if you like.)
3. Add the terms in each column.

EXAMPLE 8 Subtract $(x^2 - 6x + 5)$ from $(3x^2 + 5x + 7)$ using columns.

Solution: Align like terms in columns (step 1).

$$\begin{array}{r} 3x^2 + 5x + 7 \\ \underline{-(x^2 - 6x + 5)} \end{array}$$ Align like terms.

Change *all* signs in the second row (step 2); then add (step 3).

$$\begin{array}{rrr} 3x^2 & + \ 5x & + \ 7 \\ -x^2 & + \ 6x & - \ 5 \\ \hline 2x^2 & + \ 11x & + \ 2 \end{array}$$

Change all signs.

Add.

EXAMPLE 9 Using columns, subtract $(2x^2 - 6)$ from $(-3x^3 + 4x - 3)$.

Solution: To help align like terms, we will write each expression in descending order. If any power of x is missing, we will write that term with a numerical coefficient of 0.

$$-3x^3 + 4x - 3 = -3x^3 + 0x^2 + 4x - 3$$
$$2x^2 - 6 = 2x^2 + 0x - 6$$

Align like terms.

$$\begin{array}{r} -3x^3 + 0x^2 + 4x - 3 \\ -(2x^2 + 0x - 6) \\ \hline \end{array}$$

Change all signs in the second row; then add the terms in each column.

$$\begin{array}{r} -3x^3 + 0x^2 + 4x - 3 \\ - 2x^2 - 0x + 6 \\ \hline -3x^3 - 2x^2 + 4x + 3 \end{array}$$

Note: Many of you will find that you can change the signs mentally and can therefore align and change the signs in one step.

Exercise Set 4.4

Indicate the expressions that are polynomials. If the polynomial has a specific name—for example, monomial or binomial—give that name.

1. $6x$
2. $2x^2 - 6x + 7$
3. -8
4. $4x^{-2}$
5. $-4x - 6x^2$
6. $4x^3 - 8$
7. $6x^2 - 2x + 8$
8. $x - 3$
9. $3x^{1/2} + 2x$
10. $-2x^2 + 5x^{-1}$
11. $2x + 5$
12. $4 - 3x$
13. 7
14. $x^{1/3} + x^{2/3}$
15. $3x^3 - 6x^2 + 4x - 5$
16. $x^2 - 3$
17. $5 - x^2 - 6x$
18. $-3x$
19. $2x^{-2}$
20. $6x^2 + 3x - 5$

Express each polynomial in descending order. If the polynomial is already in descending order, so state. Give the degree of each polynomial.

21. $4x$
22. 5
23. $2x^2 - 6 + x$
24. $-4 + x^2 - 2x$
25. $-8 - 4x - x^2$
26. $2x + 4 - x^2$
27. $x^3 - 6$
28. $-x - 1$
29. $2x^2 + 5x - 8$
30. $3x^3 - x + 4$
31. $4 - 6x^3 + x^2 - 3x$
32. $-4 + x - 3x^2 + 4x^3$
33. $-2x + 5x^2 - 4$
34. $1 - x^3 + 3x$
35. $5x + 3x^2 - 6 - 2x^3$
36. $4 - 2x - 3x^2 + 5x^4$

Add.

37. $(2x + 3) + (4x - 2)$
38. $(5x - 6) + (2x - 3)$
39. $(-4x + 8) + (2x + 3)$
40. $(-5x - 3) + (-2x + 3)$
41. $(5x + 8) + (-6x - 10)$
42. $(-8x + 4) + (3x - 12)$
43. $(9x - 12) + (12x - 9)$
44. $(3x - 8) + (-8x + 5)$
45. $(x^2 + 2.6x - 3) + (4x + 3.8)$
46. $(-2x^2 + 3x - 9) + (-2x - 3)$
47. $(5x - 7) + (2x^2 + 3x + 12)$
48. $(-3x + 8) + (-2x^2 - 3x - 5)$
49. $(3x^2 - 4x + 8) + (2x^2 + 5x + 12)$
50. $(x^2 - 6x + 7) + (-x^2 + 3x + 5)$
51. $(-x^2 - 4x + 8) + (5x - 2x^2 + \frac{1}{2})$
52. $(9x^2 + 3x - 12) + (5x^2 - \frac{1}{3}x - 3)$

53. $(8x^2 + 4) + (-2.6x^2 - 5x)$

54. $(8x^3 + 4x^2 + 6) + (0.2x^2 + 5x)$

55. $(-7x^3 - 3x^2 + 4) + (4x + 5x^3 - 7)$

56. $(9x^3 - 2x^2 + 4x - 7) + (2x^3 - 6x^2 - 4x + 3)$

57. $(x^2 + xy - y^2) + (2x^2 - 3xy + y^2)$

58. $(x^2y + 6x^2 - 3xy^2) + (-x^2y - 12x^2 + 4xy^2)$

59. $(2x^2y + 2x - 3) + (3x^2y - 5x + 5)$

60. $(x^2y + x - y) + (2x^2y + 2x - 6y + 3)$

Add in columns.

61. Add $3x - 6$ and $4x + 5$.

62. Add $-2x + 5$ and $-3x - 5$.

63. Add $x^2 - 2x + 4$ and $3x + 12$.

64. Add $4x^2 - 6x + 5$ and $-2x - 8$.

65. Add $-2x^2 + 4x - 12$ and $-x^2 - 2x$.

66. Add $5x^2 + x + 9$ and $2x^2 - 12$.

67. Add $3x^2 + 4x - 5$ and $4x^2 + 3x - 8$.

68. Add $-5x^2 - 3$ and $x^2 + 2x - 9$.

69. Add $2x^3 + 3x^2 + 6x - 9$ and $7 - 4x^2$.

70. Add $-3x^3 + 3x + 9$ and $2x^2 - 4$.

71. Add $6x^3 - 4x^2 + x - 9$ and $-x^3 - 3x^2 - x + 7$.

72. Add $4x^3 + 7$ and $-2x^3 - 4x - 1$.

73. Add $xy + 6x + 4$ and $2xy - 3x - 1$.

74. Add $x^2y - 6x + 3$ and $-2x^2y - 4x - 8$.

Subtract.

75. $(3x - 4) - (2x + 2)$

76. $(6x + 3) - (4x - 2)$

77. $(-2x - 3) - (-5x - 7)$

78. $(12x - 3) - (-2x + 7)$

79. $(-x + 4) - (-x + 9)$

80. $(4x + 8) - (3x + 9)$

81. $(6 - 12x) - (3 - 5x)$

82. $(4x^2 - 6x + 3) - (3x + 7)$

83. $(9x^2 + 7x - 5) - (3x^2 + 3.5)$

84. $(-2x^2 + 4x - 5.2) - (5x^2 + 3x + 7.5)$

85. $(5x^2 - x - 1) - (-3x^2 - 2x - 5)$

86. $(5x^2 - 7) - (4x - 3)$

87. $(5x^2 - x + 12) - (5 + x)$

88. $(-5x^2 - 2x) - (2x^2 - 7x + 9)$

89. $(7x - 0.6) - (-2x^2 + 4x - 8)$

90. $(8x^3 + 5x^2 - 4) - (4x - 3 + 6x^2)$

91. $(2x^3 - 4x^2 + 5x - 7) - (3x + \frac{3}{5}x^2 - 5)$

92. $(-3x^2 + 4x - 7) - (x^3 + 4x^2 - 8x + 5)$

93. $(9x^3 - \frac{1}{5}) - (x^2 + 5x)$

94. $(3x^3 - 6x^2 + 5x) - (4x^3 - 2x^2 + 5)$

95. Subtract $(4x - 6)$ from $(3x + 5)$.

96. Subtract $(-4x + 7)$ from $(-3x - 9)$.

97. Subtract $(5x - 6)$ from $(2x^2 - 4x + 8)$.

98. Subtract $(2x^2 - 6x + 4)$ from $(5x^2 + 6x + 8)$.

99. Subtract $(4x^3 - 6x^2)$ from $(3x^3 + 5x^2 + 9x - 7)$.

100. Subtract $(-4x^2 + 8x - 7)$ from $(-5x^3 - 6x^2 + 7)$.

Perform each subtraction using columns.

101. Subtract $(2x - 7)$ from $(5x + 10)$.

102. Subtract $(6x + 8)$ from $(2x - 5)$.

103. Subtract $(-9x - 4)$ from $(-5x + 3)$.

104. Subtract $(-3x + 8)$ from $(6x^2 - 5x + 3)$.

105. Subtract $(4x^2 - 7)$ from $(9x^2 + 7x - 9)$.

106. Subtract $(4x^2 + 7x - 9)$ from $(x^2 - 6x + 3)$.

107. Subtract $(-4x^2 + 6x)$ from $(x - 6)$.

108. Subtract $(x^2 - 6)$ from $(x^2 + 4x)$.

109. Subtract $(x^2 + 6x - 7)$ from $(4x^3 - 6x^2 + 7x - 9)$.

110. Subtract $(2x^3 + 4x^2 - 9x)$ from $(-5x^3 + 4x - 12)$.

111. In your own words describe a polynomial.

112. **(a)** What is a monomonial? Make up three examples.
(b) What is a binomial? Make up three examples.
(c) What is a trinomial? Make up three examples.

113. **(a)** Explain how to find the degree of a term in one variable.
(b) Explain how to find the degree of a polynomial in one variable.

114. Make up your own fifth degree polynomial with three terms. Explain why it is a fifth degree polynomial with three terms.

115. Explain how to write a polynomial in descending order of the variable.

116. Make up your own addition problem where the sum of two binomials is $-2x + 4$.

117. Make up your own addition problem where the sum of two trinomials is $2x^2 + 5x - 6$.

118. Make up your own subtraction problem where the difference of two trinomials is $x - 2$.

119. Make up your own subtraction problem where the difference of two trinomials is $-x^2 + 4x - 5$.

CUMULATIVE REVIEW EXERCISES

[1.4] **120.** Insert either >, <, or = in the shaded area to make the statement true. $|-4|$ ▒ $|-6|$.

[1.7] *Answer true or false.*

121. The product of two negative numbers is always a negative number.

122. The sum of two negative numbers is always a negative number.

123. The difference of two negative numbers is always a negative number.

124. The quotient of two negative numbers is always a negative number.

[4.1] **125.** Simplify $\left(\frac{3x^4y^5}{6x^7y^4}\right)^3$.

Group Activity/ Challenge Problems

Simplify.

1. $(3x^2 - 6x + 3) - (2x^2 - x - 6) - (x^2 + 7x - 9)$
2. $3x^2y - 6xy - 2xy + 9xy^2 - 5xy + 3x$
3. $4(x^2 + 2x - 3) - 6(2 - 4x - x^2) - 2x(x + 2)$
4. By what degree polynomial must a second-degree polynomial be multiplied if the product is to be a sixth-degree polynomial? Explain your answer.
5. By what degree polynomial must an eighth-degree polynomial be divided if its quotient is to be a second-degree polynomial? Explain your answer.
6. Make up an addition and subtraction problem where when a binomial is added to a trinomial and a different trinomial is then subtracted from their sum the answer is 0.

4.5 Multiplication of Polynomials

Tape 7

1. Multiply a monomial by a monomial.
2. Multiply a polynomial by a monomial.
3. Multiply binomials using the distributive property.
4. Multiply binomials using the FOIL method.
5. Multiply polynomials using formulas for special products.
6. Multiply any two polynomials.

Multiplying a Monomial by a Monomial

1 We begin our discussion of multiplication of polynomials by multiplying a monomial by a monomial. To multiply two monomials, multiply their coefficients and use the product rule of exponents to determine the exponents on the variables. Problems of this type were done in Section 4.1.

EXAMPLE 1 Multiply $(3x^2)(5x^5)$.

Solution: $(3x^2)(5x^5) = 3 \cdot 5 \cdot x^2 \cdot x^5 = 15x^{2+5} = 15x^7$

EXAMPLE 2 Multiply $(-2x^6)(3x^4)$.

Solution: $(-2x^6)(3x^4) = (-2)(3) \cdot x^6 \cdot x^4 = -6x^{6+4} = -6x^{10}$

EXAMPLE 3 Multiply $(5x^2y)(8x^5y^4)$.

Solution: Remember that when a variable is given without an exponent we assume that the exponent on the variable is 1.

$$(5x^2y)(8x^5y^4) = 40x^{2+5}y^{1+4} = 40x^7y^5$$

EXAMPLE 4 Multiply $6xy^2z^5(-3x^4y^7z)$.

Solution: $6xy^2z^5(-3x^4y^7z) = -18x^5y^9z^6$

EXAMPLE 5 Multiply $(-4x^4z^9)(-3xy^7z^3)$.

Solution: $(-4x^4z^9)(-3xy^7z^3) = 12x^5y^7z^{12}$

Multiplying a Polynomial by a Monomial

2 To multiply a polynomial by a monomial, we use the distributive property presented earlier.

$$a(b + c) = ab + ac$$

The distributive property can be expanded to

$$a(b + c + d + \cdots + n) = ab + ac + ad + \cdots + an$$

EXAMPLE 6 Multiply $2x(3x^2 + 4)$.

Solution:
$$\begin{aligned} 2x(3x^2 + 4) &= (2x)(3x^2) + (2x)(4) \\ &= 6x^3 + 8x \end{aligned}$$

Notice that the use of the distributive property results in monomials being multiplied by monomials. If we study Example 6, we see that the $2x$ and $3x^2$ are both monomials, as are the $2x$ and 4.

EXAMPLE 7 Multiply $-3x(4x^2 - 2x - 1)$.

Solution:
$$\begin{aligned} -3x(4x^2 - 2x - 1) &= (-3x)(4x^2) + (-3x)(-2x) + (-3x)(-1) \\ &= -12x^3 + 6x^2 + 3x \end{aligned}$$

EXAMPLE 8 Multiply $3x^2(4x^3 - 2x + 7)$.

Solution:
$$\begin{aligned} 3x^2(4x^3 - 2x + 7) &= (3x^2)(4x^3) + (3x^2)(-2x) + (3x^2)(7) \\ &= 12x^5 - 6x^3 + 21x^2 \end{aligned}$$

EXAMPLE 9 Multiply $2x(3x^2y - 6xy + 5)$.

Solution:
$$\begin{aligned} 2x(3x^2y - 6xy + 5) &= (2x)(3x^2y) + (2x)(-6xy) + (2x)(5) \\ &= 6x^3y - 12x^2y + 10x \end{aligned}$$

EXAMPLE 10 Multiply $(3x^2 - 2xy + 3)4x$.

Solution:
$$\begin{aligned} (3x^2 - 2xy + 3)4x &= (3x^2)(4x) + (-2xy)(4x) + (3)(4x) \\ &= 12x^3 - 8x^2y + 12x \end{aligned}$$

This problem could be written as $4x(3x^2 - 2xy + 3)$ by the commutative property of multiplication, and then simplified as in Examples 6 through 9.

Multiplying a Binomial by a Binomial

3 Now we will discuss multiplying a binomial by a binomial. Before we explain how to do this, consider the multiplication problem $43 \cdot 12$.

$$\begin{array}{rcrcl} & & 43 & \longleftarrow & \text{multiplicand} \\ & & \underline{12} & \longleftarrow & \text{multiplier} \\ 2(4) & \longrightarrow & 86 & \longleftarrow & 2(3) \\ 1(4) & \longrightarrow & \underline{43} & \longleftarrow & 1(3) \\ & & 516 & \longleftarrow & \text{Product} \end{array}$$

Note how the 2 multiplies both the 3 and the 4 and the 1 also multiplies both the 3 and the 4. That is, every digit in the multiplier multiplies every digit in the multiplicand. We can also illustrate the multiplication process as follows.

$$\begin{aligned} (43)(12) &= (40 + 3)(10 + 2) \\ &= (40 + 3)(10) + (40 + 3)(2) \\ &= (40)(10) + (3)(10) + (40)(2) + (3)(2) \\ &= 400 + 30 + 80 + 6 \\ &= 516 \end{aligned}$$

Whenever any two polynomials are multiplied the same process must be followed. That is, **every term in one polynomial must multiply every term in the other polynomial.**

Consider multiplying $(a + b)(c + d)$. Treating $(a + b)$ as a single term and using the distributive property, we get

$$(a + b)(c + d) = (a + b)c + (a + b)d$$

Using the distributive property a second time gives

$$= ac + bc + ad + bd$$

Notice how each term of the first polynomial was multiplied by each term of the second polynomial, and all the products were added to obtain the answer.

EXAMPLE 11 Multiply $(3x + 2)(x - 5)$.

Solution:
$$\begin{aligned} (3x + 2)(x - 5) &= (3x + 2)x + (3x + 2)(-5) \\ &= 3x(x) + 2(x) + 3x(-5) + 2(-5) \\ &= 3x^2 + \underbrace{2x - 15x}_{} - 10 \\ &= 3x^2 - 13x - 10 \end{aligned}$$

Note that after performing the multiplication like terms must be combined.

EXAMPLE 12 Multiply $(x - 4)(y + 3)$.

Solution:
$$\begin{aligned} (x - 4)(y + 3) &= (x - 4)y + (x - 4)3 \\ &= xy - 4y + 3x - 12 \end{aligned}$$

FOIL Method

4 A common method used to multiply two binomials is the *FOIL method.* This procedure also results in each term of one binomial being multiplied by each term in the other binomial. Students often prefer to use this method when multiplying two binomials.

The FOIL Method

Consider $(a + b)(c + d)$

F stands for **first**—multiply the first terms of each binomial together:

$$\overset{\text{F}}{(a + b)(c + d)} \qquad \text{product } ac$$

O stands for **outer**—multiply the two outer terms together:

$$\overset{\text{O}}{(a + b)(c + d)} \qquad \text{product } ad$$

I stands for **inner**—multiply the two inner terms together:

$$\overset{\text{I}}{(a + b)(c + d)} \qquad \text{product } bc$$

L stands for **last**—multiply the last terms together:

$$\overset{\text{L}}{(a + b)(c + d)} \qquad \text{product } bd$$

The product of the two binomials is the sum of these four products.

$$(a + b)(c + d) = ac + ad + bc + bd$$

The FOIL method is not actually a different method used to multiply binomials, but rather an acronym to help students remember to correctly apply the distributive property. We could have used IFOL or any other arrangement of the four letters. However, FOIL is easier to remember than the other arrangements.

EXAMPLE 13 Using the FOIL method, multiply $(2x - 3)(x + 4)$.

Solution:

$$(2x - 3)(x + 4)$$

F, O, I, L

$$\begin{aligned} &\overset{\text{F}}{(2x)(x)} + \overset{\text{O}}{(2x)(4)} + \overset{\text{I}}{(-3)(x)} + \overset{\text{L}}{(-3)(4)} \\ &= 2x^2 + 8x - 3x - 12 \\ &= 2x^2 + 5x - 12 \end{aligned}$$

Thus, $(2x - 3)(x + 4) = 2x^2 + 5x - 12$.

EXAMPLE 14 Multiply $(4 - 2x)(6 - 5x)$.

Solution:

$$(4 - 2x)(6 - 5x)$$

F, O, I, L

$$\begin{aligned} &\quad\ \ \overset{F}{4(6)} + \overset{O}{4(-5x)} + \overset{I}{(-2x)(6)} + \overset{L}{(-2x)(-5x)} \\ &= 24 - 20x - 12x + 10x^2 \\ &= 10x^2 - 32x + 24 \end{aligned}$$

Thus, $(4 - 2x)(6 - 5x) = 10x^2 - 32x + 24$.

EXAMPLE 15 Multiply $(x - 5)(x + 5)$.

Solution:

$$\begin{aligned} &\quad\ \ \overset{F}{(x)(x)} + \overset{O}{(x)(5)} + \overset{I}{(-5)(x)} + \overset{L}{(-5)(5)} \\ &= x^2 + 5x - 5x - 25 \\ &= x^2 - 25 \end{aligned}$$

Thus, $(x - 5)(x + 5) = x^2 - 25$.

EXAMPLE 16 Multiply $(2x + 3)(2x - 3)$.

Solution:

$$\begin{aligned} &\quad\ \ \overset{F}{(2x)(2x)} + \overset{O}{(2x)(-3)} + \overset{I}{(3)(2x)} + \overset{L}{(3)(-3)} \\ &= 4x^2 - 6x + 6x - 9 \\ &= 4x^2 - 9 \end{aligned}$$

Thus, $(2x + 3)(2x - 3) = 4x^2 - 9$.

Special Products

5 Examples 15 and 16 are examples of a special product, the product of the sum and difference of the same two terms.

Product of the Sum and Difference of the Same Two Terms

$$(a + b)(a - b) = a^2 - b^2$$

In the special product above a represents one term and b the other term. Then $(a + b)$ is the sum of the terms and $(a - b)$ is the difference of the terms. This special product is also called the **difference-of-squares formula** because the expression on the right side of the equal sign is the difference of two squares.

EXAMPLE 17 Use the rule for finding the product of the sum and difference of two quantities to multiply each expression.

(a) $(x + 3)(x - 3)$ **(b)** $(2x + 4)(2x - 4)$ **(c)** $(3x + 2y)(3x - 2y)$

Solution: **(a)** If we let $x = a$ and $3 = b$, then

$$\begin{array}{l} (a + b)(a - b) = a^2 - b^2 \\ \downarrow \quad \downarrow \quad \downarrow \quad \downarrow \qquad \downarrow \qquad \downarrow \\ (x + 3)(x - 3) = (x)^2 - (3)^2 \\ \qquad\qquad\qquad = x^2 - 9 \end{array}$$

(b)

$$\begin{array}{l} (a + b)(a - b) = a^2 - b^2 \\ \downarrow \quad \downarrow \quad \downarrow \quad \downarrow \qquad \downarrow \qquad \downarrow \\ (2x + 4)(2x - 4) = (2x)^2 - (4)^2 \\ \qquad\qquad\qquad = 4x^2 - 16 \end{array}$$

(c)

$$\begin{array}{l} (a + b)(a - b) = a^2 - b^2 \\ \downarrow \quad \downarrow \quad \downarrow \quad \downarrow \qquad \downarrow \qquad \downarrow \\ (3x + 2y)(3x - 2y) = (3x)^2 - (2y)^2 \\ \qquad\qquad\qquad = 9x^2 - 4y^2 \end{array}$$

This problem could also be done using the FOIL method.

EXAMPLE 18 Using the FOIL method, find $(x + 3)^2$.

Solution: $(x + 3)^2 = (x + 3)(x + 3)$

$$\begin{array}{l} \quad \text{F} \qquad \text{O} \qquad \text{I} \qquad \text{L} \\ x(x) + x(3) + 3(x) + (3)(3) \\ = x^2 + 3x + 3x + 9 \\ = x^2 + 6x + 9 \end{array}$$

Example 18 is an example of the square of a binomial, another special product.

Square of Binomial Formulas

$$(a + b)^2 = (a + b)(a + b) = a^2 + 2ab + b^2$$
$$(a - b)^2 = (a - b)(a - b) = a^2 - 2ab + b^2$$

EXAMPLE 19 Use the square of a binomial formula to multiply each expression.

(a) $(x + 5)^2$ **(b)** $(2x - 4)^2$
(c) $(3x + 2y)(3x + 2y)$ **(d)** $(x - 3)(x - 3)$

Solution: **(a)** If we let $x = a$ and $5 = b$, then

$$\begin{array}{l} (a + b)(a + b) = a^2 + 2a\,b + b^2 \\ \downarrow \quad \downarrow \quad \downarrow \quad \downarrow \qquad \downarrow \qquad \downarrow \downarrow \qquad \downarrow \\ (x + 5)^2 = (x + 5)(x + 5) = (x)^2 + 2(x)(5) + (5)^2 \\ \qquad\qquad\qquad = x^2 + 10x + 25 \end{array}$$

(b) $(a - b)(a - b) = a^2 - 2ab + b^2$

$$(2x - 4)(2x - 4) = (2x)^2 - 2(2x)(4) + (4)^2$$
$$= 4x^2 - 16x + 16$$

(c) $(a + b)(a + b) = a^2 + 2ab + b^2$

$$(3x + 2y)(3x + 2y) = (3x)^2 + 2(3x)(2y) + (2y)^2$$
$$= 9x^2 + 12xy + 4y^2$$

(d) $(a - b)(a - b) = a^2 - 2ab + b^2$

$$(x - 3)(x - 3) = (x)^2 - 2(x)(3) + (3)^2$$
$$= x^2 - 6x + 9$$

This problem could also be done using the FOIL method.

COMMON STUDENT ERROR

Correct	*Incorrect*
$(a + b)^2 = a^2 + 2ab + b^2$	~~$(a + b)^2 = a^2 + b^2$~~
$(a - b)^2 = a^2 - 2ab + b^2$	~~$(a - b)^2 = a^2 - b^2$~~

Do not forget middle term when you square a binomial.

$$(x + 2)^2 \neq x^2 + 4$$
$$(x + 2)^2 = (x + 2)(x + 2)$$
$$= x^2 + 4x + 4$$

Multiplying a Polynomial by a Polynomial

6 When multiplying a binomial by a binomial, we saw that every term in the first binomial was multiplied by every term in the second binomial. When multiplying two polynomials, each term of one polynomial must be multiplied by each term of the other polynomial. In the multiplication $(3x + 2)(4x^2 - 5x - 3)$, we use the distributive property as follows:

$$(3x + 2)(4x^2 - 5x - 3)$$
$$= 3x(4x^2 - 5x - 3) + 2(4x^2 - 5x - 3)$$
$$= 12x^3 - 15x^2 - 9x + 8x^2 - 10x - 6$$
$$= 12x^3 - 7x^2 - 19x - 6$$

Thus, $(3x + 2)(4x^2 - 5x - 3) = 12x^3 - 7x^2 - 19x - 6$.

Multiplication problems can be performed by using the distributive property, as we just illustrated. However, many students prefer to multiply a polynomial by a polynomial using a vertical procedure. On page 218 we showed that when multiplying the number 43 by the number 12, we multiply each digit in the number 43 by each digit in the number 12. Review that example now. We can follow a similar procedure when multiplying a polynomial by a polynomial, as illustrated in the following examples. We must be careful, however, to align like terms in the same columns when performing the individual multiplications.

EXAMPLE 20 Multiply $(3x + 4)(2x + 5)$.

Solution: First write the polynomials one beneath the other.

$$\begin{array}{r} 3x + 4 \\ \underline{2x + 5} \end{array}$$

Next, multiply each term in $(3x + 4)$ by 5.

$$\begin{array}{rr} & 3x + 4 \\ & \underline{2x + 5} \\ 5(3x + 4) \longrightarrow & 15x + 20 \end{array}$$

Next, multiply each term in $(3x + 4)$ by $2x$ and align like terms.

$$\begin{array}{rr} & 3x + 4 \\ & \underline{2x + 5} \\ & 15x + 20 \\ 2x(3x + 4) \longrightarrow & \underline{6x^2 + 8x } \\ & 6x^2 + 23x + 20 \end{array}$$

Add like terms in columns.

The same answer is obtained using the FOIL method.

EXAMPLE 21 Multiply $(5x - 2)(2x^2 + 3x - 4)$.

Solution: For convenience we place the shorter expression on the bottom, as illustrated.

$$\begin{array}{rrrr} & 2x^2 & + 3x & - 4 \\ & & 5x & - 2 \\ \hline & -4x^2 & - 6x & + 8 \\ 10x^3 & + 15x^2 & - 20x & \\ \hline 10x^3 & + 11x^2 & - 26x & + 8 \end{array}$$

Multiply top polynomial by -2.

Multiply top polynomial by $5x$; align like terms.

Add like terms in columns.

EXAMPLE 22 Multiply $x^2 - 3x + 2$ by $2x^2 - 3$.

Solution:

$$\begin{array}{rrrrr} & & x^2 & - 3x & + 2 \\ & & & 2x^2 & - 3 \\ \hline & & -3x^2 & + 9x & - 6 \\ 2x^4 & - 6x^3 & + 4x^2 & & \\ \hline 2x^4 & - 6x^3 & + x^2 & + 9x & - 6 \end{array}$$

Multiply top polynomial by -3.

Multiply top polynomial by $2x^2$; align like terms.

Add like terms in columns.

EXAMPLE 23 Multiply $(3x^3 - 2x^2 + 4x + 6)(x^2 - 5x)$.

Solution:

$$\begin{array}{rrrrrr} & 3x^3 & - 2x^2 & + 4x & + 6 & \\ & & & x^2 & - 5x & \\ \hline & -15x^4 & + 10x^3 & - 20x^2 & - 30x & \\ 3x^5 & - 2x^4 & + 4x^3 & + 6x^2 & & \\ \hline 3x^5 & - 17x^4 & + 14x^3 & - 14x^2 & - 30x & \end{array}$$

Multiply top polynomial by $-5x$.

Multiply top polynomial by x^2.

Add like terms in columns.

Exercise Set 4.5

Multiply.

1. $x^2 \cdot 3xy$
2. $6xy^2 \cdot 3xy^4$
3. $5x^4y^5(6xy^2)$
4. $-5x^2y^3(6x^2y^5)$
5. $4x^4y^6(-7x^2y^9)$
6. $12x^6y^2(2x^9y^7)$
7. $9xy^6 \cdot 6x^5y^8$
8. $(3x^5y^6)(5y^7)$
9. $(6x^2y)(\frac{1}{2}x^4)$
10. $\frac{2}{3}x(6x^2y^3)$
11. $(1.5x^4y^3)(6xy)$
12. $(2.3x^5)(4.1x^2y^4)$

Multiply.

13. $3(x + 4)$
14. $3(x - 4)$
15. $2x(x - 3)$
16. $-5x(x + 2)$
17. $-4x(-2x + 6)$
18. $-x(3x - 5)$
19. $2x(x^2 + 3x - 1)$
20. $-x(2x^2 - 6x + 5)$
21. $-2x(x^2 - 2x + 5)$
22. $-3x(-2x^2 + 5x - 6)$
23. $5x(-4x^2 + 6x - 4)$
24. $x(x^2 - x + 1)$
25. $(3x^2 + 4x - 5)8x$
26. $1.2x(3x^2 + y)$
27. $0.3x(2xy + 5x - 6y)$
28. $-\frac{1}{2}x(2x^2 + 4x - 6y^2)$
29. $(x - y - 3)y$
30. $\frac{1}{3}y(y^2 - 6y + 3x)$

Multiply.

31. $(x + 3)(x + 4)$
32. $(2x - 3)(x + 5)$
33. $(2x + 5)(3x - 6)$
34. $(x + 5)(x - 5)$
35. $(2x - 4)(2x + 4)$
36. $(4 + 3x)(2 - x)$
37. $(5 - 3x)(6 + 2x)$
38. $(4 + 6x)(x - 3)$
39. $(-x + 3)(2x + 5)$
40. $(6x - 1)(-2x + 5)$
41. $(x + 4)(x + 3)$
42. $(x - 3)(x + 5)$
43. $(x + 4)(x - 2)$
44. $(2x + 3)(x + 5)$
45. $(3x + 4)(2x + 5)$
46. $(3x - 6)(4x - 2)$
47. $(3x + 4)(2x - 3)$
48. $(4x + 4)(x + 1)$
49. $(x - 1)(x + 1)$
50. $(3x - 8)(2x + 3)$
51. $(2x - 3)(2x - 3)$
52. $(6x - 1)(2x + 1)$
53. $(4 - x)(3 + 2x)$
54. $(6 - 2x)(5x - 3)$
55. $(2x + 3)(4 - 2x)$
56. $(5 - 6x)(2x - 7)$
57. $(x + y)(x - y)$
58. $(x + 2y)(2x - 3)$
59. $(2x - 3y)(3x + 2y)$
60. $(x + 3)(2y - 5)$
61. $(4x - 3y)(2y - 3)$
62. $(2x - 0.1)(x + 2.4)$
63. $(x + 0.6)(x + 0.3)$
64. $(3x - 6)(x + \frac{1}{3})$
65. $(2x + 4)(x + \frac{1}{2})$
66. $(x + 4)(x - \frac{1}{2})$

Multiply using a special product formula.

67. $(x + 4)(x - 4)$
68. $(x + 3)^2$
69. $(2x - 1)(2x + 1)$
70. $(x + 2)(x + 2)$
71. $(x + y)^2$
72. $(2x - 3)(2x + 3)$
73. $(x - 0.2)^2$
74. $(x + 3y)(x + 3y)$
75. $(3x + 5)(3x - 5)$
76. $(5x + 4)(5x - 4)$
77. $(0.4x + y)^2$
78. $(x - \frac{1}{2}y)^2$

Multiply.

79. $(x + 3)(2x^2 + 4x - 1)$
80. $(2x + 3)(4x^2 - 5x + 6)$
81. $(5x + 4)(x^2 - x + 4)$
82. $(2x - 5)(3x^2 - 4x + 7)$
83. $(-2x^2 - 4x + 1)(7x - 3)$
84. $(4x^2 + 9x - 2)(x - 2)$
85. $(-3x + 9)(-6x^2 + 5x - 3)$
86. $(4x^2 + 1)(2x + 1)$

87. $(3x^2 - 2x + 4)(2x^2 + 3x + 1)$

88. $(x^2 - 2x + 3)(x^2 - 4)$

89. $(x^2 - x + 3)(x^2 - 2x)$

90. $(-3x^2 - 2x + 4)(2x^2 - 4x + 3)$

91. $(3x^3 + 2x^2 - x)(x - 3)$

92. $(-x^3 + x^2 - 6x + 3)(x^2 + 2x)$

93. $(a + b)(a^2 - ab + b^2)$

94. $(a - b)(a^2 + ab + b^2)$

95. Will the product of a monomial and a monomial always be a monomial? Explain your answer.

96. Will the product of a monomial and a binomial ever be a trinomial? Explain your answer.

97. Will the product of two binomials after like terms are combined always be a trinomial? Explain your answer.

98. Consider the multiplication
$3x^2(2x^{\square} - 5x^{\square} + 3x^{\square}) = 6x^8 - 15x^5 + 9x^3$.
Determine the exponents to be placed in the shaded area to make a true statement. Explain how you determined your answer.

99. Suppose that one side of a rectangle is represented as $x + 2$ and a second side is represented as $2x + 1$.
(a) Express the area of the rectangle in terms of x.
(b) Find the area if $x = 4$ feet.
(c) What value of x, in feet, would result in the rectangle being a square? Explain how you determined your answer to part (c).

CUMULATIVE REVIEW EXERCISES

3] **100.** The cost of a taxi ride is $2.00 for the first mile and $1.50 for each additional mile or part thereof. Find the maximum distance Heidi can ride in the taxi if she has only $20.

1] **101.** Simplify $\left(\frac{4x^8y^5}{8x^8y^6}\right)^4$

[4.1–4.2] **102.** Evaluate each of the following:
(a) -6^3 **(b)** 6^{-3}

[4.4] **103.** Subtract $5x^2 - 4x - 3$ from $-x^2 - 6x + 5$.

Group Activity/Challenge Problems

Perform each polynomial multiplication.

1. $\sqrt{5}x\left(2x^2 + \sqrt{5}x - \frac{1}{2}\right)$ *Hint:*$(\sqrt{5})2 = 2\sqrt{5}$ and $\sqrt{5} \cdot \sqrt{5} = \sqrt{25} = 5$.

2. $\left(\frac{x}{2} + \frac{2}{3}\right)\left(\frac{2x}{3} - \frac{2}{5}\right)$

3. $(2x^3 - 6x^2 + 5x - 3)(3x^3 - 6x + 4)$

Find the expression that when placed in the shaded area makes the statement true.

4. $\square(3x^2 + x - 4) = 6x^2 + 2x - 8$

5. $\square(4x^2 - 2x + 5) = 12x^4 - 6x^3 + 15x^2$

6. $\square(3x^2 - 6x + 8) = -12x^3y^3 + 24x^2y^3 - 32xy^3$

7. $(\square)(x + 3) = x^2 + 5x + 6$

8. $(x + 4)(\square) = x^2 + 2x - 8$

9. $(2x + 1)(\square) = 2x^2 - 5x - 3$

Find two binomials that when placed in the shaded areas make the statement true. The product of the two binomials must equal the trinomial.

10. $x^2 + 3x + 2 = (\square)(\square)$

11. $x^2 + x - 6 = (\square)(\square)$

12. $2x^2 + 7x + 3 = (\square)(\square)$

4.6 Division of Polynomials

Tape 7

1. Divide a polynomial by a monomial.
2. Divide a polynomial by a binomial.
3. Check division of polynomial problems.
4. Write polynomials in descending order when dividing.

1 Dividing a Polynomial by a Monomial

To divide a polynomial by a monomial, divide each term of the polynomial by the monomial.

EXAMPLE 1 Divide: $\frac{2x + 16}{2}$.

Solution: $\frac{2x + 16}{2} = \frac{2x}{2} + \frac{16}{2} = x + 8$

EXAMPLE 2 Divide: $\frac{4x^2 - 8x}{2x}$

Solution: $\frac{4x^2 - 8x}{2x} = \frac{4x^2}{2x} - \frac{8x}{2x} = 2x - 4$

COMMON STUDENT ERROR

Correct	*Incorrect*
$\frac{x + 2}{2} = \frac{x}{2} + \frac{2}{2} = \frac{x}{2} + 1$	$\frac{x + \cancel{2}^{1}}{\cancel{2}_{1}} = \frac{x + 1}{1} = x + 1$
$\frac{x + 2}{x} = \frac{x}{x} + \frac{2}{x} = 1 + \frac{2}{x}$	$\frac{\cancel{x}^{1} + 2}{\cancel{x}_{1}} = \frac{1 + 2}{1} = 3$

Can you explain why the procedures on the right are not correct?

EXAMPLE 3 Divide: $\frac{4x^5 - 6x^4 + 8x - 3}{2x^2}$

Solution: $\frac{4x^5 - 6x^4 + 8x - 3}{2x^2} = \frac{4x^5}{2x^2} - \frac{6x^4}{2x^2} + \frac{8x}{2x^2} - \frac{3}{2x^2}$

$= 2x^3 - 3x^2 + \frac{4}{x} - \frac{3}{2x^2}$

EXAMPLE 4 Divide: $\dfrac{3x^3 - 6x^2 + 4x - 1}{-3x}$

Solution: A negative sign appears in the denominator. Usually it is easier to divide if the divisor is positive. We can multiply both numerator and denominator by -1 to get a positive denominator.

$$\frac{(-1)(3x^3 - 6x^2 + 4x - 1)}{(-1)(-3x)} = \frac{-3x^3 + 6x^2 - 4x + 1}{3x}$$

$$= \frac{-3x^3}{3x} + \frac{6x^2}{3x} - \frac{4x}{3x} + \frac{1}{3x}$$

$$= -x^2 + 2x - \frac{4}{3} + \frac{1}{3x}$$

Dividing a Polynomial by a Binomial

2 We divide a polynomial by a binomial in much the same way as we perform long division. This procedure will be explained in Example 5.

EXAMPLE 5 Divide: $\dfrac{x^2 + 6x + 8}{x + 2}$ $\begin{array}{l} \longleftarrow \text{dividend} \\ \longleftarrow \text{divisor} \end{array}$

Solution: Rewrite the division problem in the following form:

$$x + 2 \overline{\smash{)}\, x^2 + 6x + 8}$$

Divide x^2 (the first term in the dividend) by x (the first term in the divisor).

$$\frac{x^2}{x} = x$$

Place the quotient, x, above the like term containing x in the dividend.

$$\begin{array}{r} x \\ x + 2 \overline{\smash{)}\, x^2 + 6x + 8} \end{array}$$

Next, multiply the x by $x + 2$ as you would do in long division and place the terms of the product under their like terms.

times

$$\begin{array}{r} x \\ x + 2 \overline{\smash{)}\, x^2 + 6x + 8} \\ x^2 + 2x \end{array} \longleftarrow x(x + 2)$$

equals

Now subtract $x^2 + 2x$ from $x^2 + 6x$. When subtracting, remember to change the sign of the terms being subtracted and then add the like terms.

$$\begin{array}{r} x \\ x + 2 \overline{\smash{)}\, x^2 + 6x + 8} \\ \underline{-x^2 \mp 2x} \\ 4x \end{array}$$

Next, bring down the 8, the next term in the dividend.

$$\begin{array}{r} x \\ x+2\overline{)\,x^2+6x+8} \\ \underline{x^2+2x} \\ 4x+8 \end{array}$$

Now divide $4x$, the first term at the bottom, by x, the first term in the divisor.

$$\frac{4x}{x} = +4$$

Write the $+4$ in the quotient above the constant in the dividend.

$$\begin{array}{r} x+4 \\ x+2\overline{)\,x^2+6x+8} \\ \underline{x^2+2x} \\ 4x+8 \end{array}$$

Multiply the $x + 2$ by 4 and place the terms of the product under their like terms.

times

$$\begin{array}{r} x+4 \\ x+2\overline{)\,x^2+6x+8} \\ \underline{x^2+2x} \\ 4x+8 \\ 4x+8 \leftarrow 4(x+2) \end{array}$$

equals

Now subtract.

$$\begin{array}{r} x+4 \leftarrow \text{quotient} \\ x+2\overline{)\,x^2+6x+8} \\ \underline{x^2+2x} \\ 4x+8 \\ \underline{\overset{-}{}4x\overset{-}{\not+}8} \\ 0 \leftarrow \text{remainder} \end{array}$$

Thus,

$$\frac{x^2 + 6x + 8}{x + 2} = x + 4$$

There is no remainder.

EXAMPLE 6 Divide: $\dfrac{6x^2 - 5x + 5}{2x + 3}$

Solution:

$$\frac{6x^2}{2x} \quad \frac{-14x}{2x}$$

$$\begin{array}{r@{}l} & 3x - 7 \\ 2x + 3 \,\big)\!\! & \overline{6x^2 - 5x + 5} \\ & \underline{6x^2 + 9x} \quad \leftarrow 3x(2x + 3) \\ & -14x + 5 \\ & \underline{-14x - 21} \quad \leftarrow -7(2x + 3) \\ & 26 \quad \leftarrow \text{remainder} \end{array}$$

When there is a remainder, as in this example, list the quotient, plus the remainder above the divisor. Thus,

$$\frac{6x^2 - 5x + 5}{2x + 3} = 3x - 7 + \frac{26}{2x + 3}$$

Checking Polynomial Division

3 The answer to a division problem can be checked. Consider the division problem $13 \div 5$.

$$\begin{array}{r} 2 \\ 5\,\overline{)\,13} \\ \underline{10} \\ 3 \end{array}$$

Note that the divisor times the quotient, plus the remainder, equals the dividend:

$$(\text{Divisor} \times \text{quotient}) + \text{remainder} = \text{dividend}$$

$$(5 \cdot 2) + 3 = 13$$

$$10 + 3 = 13$$

$$13 = 13 \quad \text{true}$$

This same procedure can be used to check all division problems.

> **To Check Division of Polynomials**
>
> $$(\text{Divisor} \times \text{quotient}) + \text{remainder} = \text{dividend}$$

Let us check the answer to Example 6. The divisor is $2x + 3$, the quotient is $3x - 7$, the remainder is 26, and the dividend is $6x^2 - 5x + 5$.

Check: (divisor × quotient) + remainder = dividend

$$(2x + 3)(3x - 7) + 26 = 6x^2 - 5x + 5$$

$$(6x^2 - 5x - 21) + 26 = 6x^2 - 5x + 5$$

$$6x^2 - 5x + 5 = 6x^2 - 5x + 5 \quad \text{true}$$

Write Polynomials in Descending Order

4 When dividing a polynomial by a binomial, both the polynomial and binomial should be listed in descending order. If a given power term is missing, it is often helpful to include that term with a numerical coefficient of 0. For example, to divide $(6x^2 + x^3 - 4)/(x - 2)$, we begin by writing $(x^3 + 6x^2 + 0x - 4)/(x - 2)$.

EXAMPLE 7 Divide $(-x + 9x^3 - 28)$ by $(3x - 4)$.

Solution: First rewrite the dividend in descending order to get $(9x^3 - x - 28) \div (3x - 4)$. Since there is no x^2 term in the dividend, we will add $0x^2$ to help align like terms.

$$\frac{9x^3}{3x}, \quad \frac{12x^2}{3x}, \quad \frac{15x}{3x}$$

$$\downarrow \qquad \downarrow \qquad \downarrow$$

$$\begin{array}{r l}
 & 3x^2 + 4x + 5 \\
3x - 4 \,\big)\!\! & 9x^3 + 0x^2 - x - 28 \\
 & \underline{9x^3 - 12x^2} \quad \leftarrow 3x^2(3x - 4) \\
 & 12x^2 - x \\
 & \underline{12x^2 - 16x} \quad \leftarrow 4x(3x - 4) \\
 & 15x - 28 \\
 & \underline{15x - 20} \quad \leftarrow 5(3x - 4) \\
 & -8 \quad \leftarrow \text{remainder}
\end{array}$$

Thus, $\dfrac{-x + 9x^3 - 28}{3x - 4} = 3x^2 + 4x + 5 - \dfrac{8}{3x - 4}$.

Check this division yourself using the procedure just discussed.

Exercise Set 4.6

Rewrite each multiplication problem as a division problem. There is more than one correct answer.

1. $x^2 - 2x - 15 = (x + 3)(x - 5)$
2. $x^2 - 6x + 8 = (x - 4)(x - 2)$
3. $(2x + 3)(x + 1) = 2x^2 + 5x + 3$
4. $2x^2 + 5x - 12 = (2x - 3)(x + 4)$
5. $4x^2 - 9 = (2x + 3)(2x - 3)$
6. $(3x + 2)(3x - 2) = 9x^2 - 4$

Divide.

7. $\dfrac{2x + 4}{2}$
8. $\dfrac{4x - 6}{2}$
9. $\dfrac{2x + 6}{2}$
10. $(4x + 3) \div 2$
11. $\dfrac{3x + 8}{2}$
12. $\dfrac{5x - 12}{6}$
13. $\dfrac{-6x + 4}{2}$
14. $\dfrac{-4x + 5}{-3}$
15. $\dfrac{-9x - 3}{-3}$
16. $\dfrac{5x - 4}{-5}$
17. $(3x + 6) \div x$
18. $\dfrac{4x - 3}{2x}$
19. $\dfrac{9 - 3x}{-3x}$
20. $\dfrac{6 - 5x}{-3x}$
21. $(3x^2 + 6x - 9) \div 3x^2$
22. $\dfrac{12x^2 - 6x + 3}{3}$
23. $\dfrac{-4x^5 + 6x + 8}{2x^2}$
24. $\dfrac{5x^2 + 4x - 8}{2}$
25. $\dfrac{x^6 + 4x^4 - 3}{x^3}$
26. $(4x^2 - 6x + 7) \div x$
27. $\dfrac{6x^5 - 4x^4 + 12x^3 - 5x^2}{2x^3}$

28. $\dfrac{8x^2 - 5x + 10}{-2x}$

29. $\dfrac{4x^3 + 6x^2 - 8}{-4x}$

30. $\dfrac{-12x^4 + 6x^2 - 15x + 4}{-3x}$

31. $\dfrac{9x^6 + 3x^4 - 10x^2 - 9}{3x^2}$

32. $\dfrac{-10x^3 - 6x^2 + 15}{-5x^3}$

Divide.

33. $\dfrac{x^2 + 4x + 3}{x + 1}$

34. $(x^2 + 7x + 10) \div (x + 5)$

35. $\dfrac{2x^2 + 13x + 15}{x + 5}$

36. $\dfrac{2x^2 + x - 10}{x - 2}$

37. $\dfrac{6x^2 + 16x + 8}{3x + 2}$

38. $\dfrac{2x^2 + 13x + 15}{2x + 3}$

39. $\dfrac{2x^2 + x - 10}{2x + 5}$

40. $\dfrac{8x^2 - 26x + 15}{4x - 3}$

41. $(2x^2 + 7x - 18) \div (2x - 3)$

42. $\dfrac{x^2 - 25}{x - 5}$

43. $(4x^2 - 9) \div (2x - 3)$

44. $\dfrac{9x^2 - 4}{3x + 2}$

45. $\dfrac{6x + 8x^2 - 25}{4x + 9}$

46. $\dfrac{10x + 3x^2 + 6}{x + 2}$

47. $\dfrac{6x + 8x^2 - 12}{2x + 3}$

48. $\dfrac{x^3 + 3x^2 + 5x + 3}{x + 1}$

49. $\dfrac{4x^3 + 12x^2 + 7x - 3}{2x + 3}$

50. $\dfrac{2x^3 - 3x^2 - 3x + 6}{x - 1}$

51. $\dfrac{9x^3 - 3x^2 - 9x + 4}{3x + 2}$

52. $\dfrac{2x^3 + 6x - 4}{x + 4}$

53. $(x^3 - 8) \div (x - 3)$

54. $\dfrac{x^3 + 8}{x + 2}$

55. $\dfrac{x^3 - 27}{x - 3}$

56. $\dfrac{x^3 + 27}{x + 3}$

57. $\dfrac{4x^3 - 5x}{2x - 1}$

58. $\dfrac{9x^3 - x + 3}{3x - 2}$

59. $\dfrac{-x^3 - 6x^2 + 2x - 3}{x - 1}$

Determine the expression to be placed in the shaded area to make a true statement. Explain how you determined your answer.

60. $\dfrac{16x^4 + 20x^3 - 4x^2 + 12x}{\square} = 4x^3 + 5x^2 - x + 3$

61. $\dfrac{9x^5 - 6x^4 + 3x^2 + 12}{\square} = 3x^3 - 2x^2 + 1 + \dfrac{4}{x^2}$

Determine the exponents to be placed in the shaded areas to make a true statement. Explain how you determined your answer.

62. $\dfrac{8x^{\square} + 4x^{\square} - 20x^{\square} - 5x^{\square}}{2x^2} = 4x^3 + 2x - 10 - \dfrac{5}{2x}$

63. $\dfrac{15x^{\square} + 25x^{\square} + 5x^{\square} + 10x^{\square}}{5x^2} = 3x^5 + 5x^4 + x^2 + 2.$

CUMULATIVE REVIEW EXERCISES

64. Consider the set of numbers $\{2, -5, 0, \sqrt{7}, \frac{2}{5}, -6.3, \sqrt{3}, -23/34\}$. List those that are **(a)** natural numbers, **(b)** whole numbers, **(c)** rational numbers, **(d)** irrational numbers, and **(e)** real numbers.

65. **(a)** To what is 0/1 equal?
(b) How do we refer to an expression like 1/0?

[1.9] 66. Give the order of operations to be followed when evaluating a mathematical expression.

[2.5] 67. Solve the equation $2(x + 3) + 2x = x + 4$.

Group Activity/ Challenge Problems

Divide. The quotients in Exercises 1 and 2 will contain fractions.

1. $\dfrac{4x^3 - 4x + 6}{2x + 3}$ 2. $\dfrac{3x^3 - 5}{3x - 2}$ 3. $\dfrac{3x^2 + 6x - 10}{-x - 3}$

Determine the polynomial that when substituted in the shaded area results in a true statement. Explain how you determined your answer.

4. $\dfrac{\square}{x + 4} = x + 2 + \dfrac{2}{x + 4}$ 5. $\dfrac{\square}{x + 3} = x + 1 - \dfrac{1}{x + 3}$

4.7 Motion and Mixture Problems

Tape 7

1 Solve motion problems involving only one rate.
2 Learn the distance formula.
3 Solve motion problems involving two rates.
4 Solve mixture problems.

In Chapter 2 we discussed applications that could be solved using proportions. In Chapter 3 we gave a general procedure for setting up and solving application problems. Now we will use that procedure to solve two additional types of application problems, motion and mixture problems.

Motion and mixture problems are grouped in the same section because, as you will learn shortly, you use the same general multiplication procedure to solve them. We begin by discussing motion problems.

Motion Problems with Only One Rate

1 A motion problem is one in which an object is moving at a specified rate for a specified period of time. A car traveling at a constant speed, a swimming pool being filled or drained (the water is moving at a specified rate), and spaghetti being cut on a conveyor belt (conveyor belt moving at a specified speed) are all motion problems.

The formula often used to solve motion problems follows.

Motion Problems

$$\text{Amount} = \text{rate} \cdot \text{time}$$

The amount can be a measure of many different quantities, depending on the rate. For example, if the rate is measuring *distance* per unit time, the amount will be *distance.* If the rate is measuring *volume* per unit time, the amount will be *volume;* and so on.

The distance traveled by a car moving at a constant rate for a specific time can be found by the formula distance = rate · time. When a swimming pool is being filled at a constant rate for a specific period of time, the volume of water in the pool can be found by the formula volume = rate · time. Note that these formulas are more specific versions of the formula in the box.

EXAMPLE 1 A swimming pool is being filled at a rate of 10 gallons per minute. How many gallons have been added after 25 minutes?

Solution: Since we are discussing gallons, which measure volume, the formula we will use is volume = rate · time. We are given the rate, 10 gallons per minute, and the time, 25 minutes. We are asked to find the volume.

$$\begin{aligned} \text{volume} &= \text{rate} \cdot \text{time} \\ &= 10 \cdot 25 = 250 \end{aligned}$$

Thus, the volume after 25 minutes is 250 gallons.

Let us look at the units of measurement in Example 1. The rate is given in gallons per minute and the time is given in minutes. If we analyze the units, called *dimensional analysis,* we see that the volume is measured in gallons.

$$\begin{aligned} \text{volume} &= \text{rate} \cdot \text{time} \\ &= \frac{\text{gallons}}{\cancel{\text{minute}}} \cdot \cancel{\text{minutes}} \\ &= \text{gallons} \end{aligned}$$

EXAMPLE 2 A patient is to receive 1200 cubic centimeters of fluid intravenously over an 8-hour period. What should be the average intravenous flow rate?

Solution: Here we are given the volume and the length of time the fluid is to be administered. We are asked to find the rate.

$$\begin{aligned} \text{volume} &= \text{rate} \cdot \text{time} \\ 1200 &= r \cdot 8 \\ \frac{1200}{8} &= r \\ 150 &= r \end{aligned}$$

The fluid should be administered at a rate of 150 cubic centimeters per hour.

Can you explain why the rate in Example 2 must be in cubic centimeters per hour?

The Distance Formula

2 When the "*amount*" in the rate formula is "*distance*" we often refer to the formula as the **distance formula,**

$$\textbf{distance} = \textbf{rate} \cdot \textbf{time} \quad \text{or} \quad d = r \cdot t$$

Examples 3 and 4 illustrate the use of the distance formula.

EXAMPLE 3 An oil-well-drilling device can drill 3 feet per hour. How long will it take to drill to a depth of 1870 feet?

Solution: Since we are given a distance of 1870 feet we will use the distance formula. We are given the distance and the rate and need to solve for the time, t.

$$\text{distance} = \text{rate} \cdot \text{time}$$
$$1870 = 3t$$
$$\frac{1870}{3} = t$$
$$623.33 = t$$
$$\text{or} \qquad t = 623.33 \text{ hours}$$

Helpful Hint

When working motion problems, the units must be consistent with each other. If you are given a problem where the units are not consistent, you will need to change one of the quantities so that the units will agree before you substitute the values into the formula. Example 4 illustrates how this is done.

EXAMPLE 4 A conveyor belt transporting uncut spaghetti moves at a rate of 1.5 feet per second. A cutting blade is activated at regular intervals to cut the spaghetti into proper lengths. At what time intervals should the blade be activated if the spaghetti is to be cut in 9-inch lengths?

Solution: Since lengths are distances, we will use the distance formula. We are asked to find the time, t, at which the blade should be activated. Since the rate is given in *feet* per second and the length is given in *inches*, one of these units must be changed. One foot equals 12 inches; thus, to change from feet per second to inches per second, we multiply the rate by 12.

$$1.5 \text{ feet per second} = (1.5)(12) = 18 \text{ inches per second}$$
$$\text{distance} = \text{rate} \cdot \text{time}$$
$$9 = 18 \cdot t$$
$$\frac{9}{18} = t$$
$$\frac{1}{2} = t$$

The blade should cut at $\frac{1}{2}$-second intervals.

Motion Problems with Two Rates

3 Now we will look at some motion problems that involve *two rates*, such as two trains traveling at different speeds. In these problems, we generally begin by letting the variable represent one of the unknown quantities, and then we represent the second unknown quantity in terms of the first unknown quantity. For example, suppose that one train travels 20 miles per hour faster than another train. We might let r represent the rate of the slower train and $r + 20$ the rate of the faster train.

To solve problems of this type that use the distance formula, we generally add the two distances, or subtract the smaller distance from the larger, or set the two distances equal to each other, depending on the information given in the problem.

Often, when working problems involving two different rates, we construct a table to organize the information. Examples 5 through 8 illustrate the procedures used.

EXAMPLE 5 Two trains leave the same station along parallel tracks going in opposite directions. The train traveling east has a speed of 40 miles per hour. The train traveling west has a speed of 60 miles per hour. In how many hours will they be 500 miles apart?

Solution: We are asked to find the time, t, when the trains are 500 miles apart.

Let t = time for the trains to be 500 miles apart.

Begin by making a sketch to help visualize the problem (Fig. 4.2).

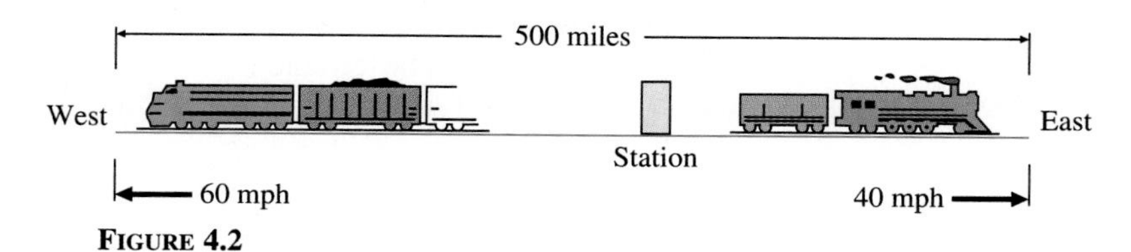

FIGURE 4.2

When the trains are 500 miles apart, each train has traveled for the same amount of time, t. We set up a table to help analyze the problem. The distance is found by multiplying the rate by the time.

Train	Rate	Time	Distance
Eastbound	40	t	$40t$
Westbound	60	t	$60t$

Since the trains are traveling in opposite directions, the sum of their distances must be 500 miles.

$$\begin{pmatrix}\text{distance traveled}\\ \text{by train 1}\end{pmatrix} + \begin{pmatrix}\text{distance traveled}\\ \text{by train 2}\end{pmatrix} = 500 \text{ miles}$$

$$\begin{aligned} 40t + 60t &= 500 \\ 100t &= 500 \\ t &= 5 \end{aligned}$$

The two trains will be 500 miles apart in 5 hours.

EXAMPLE 6 Two cross-country skiers start skiing at the same time on the same trail going in the same direction. The more experienced skier averages 6 miles per hour, while the beginning skier averages 2 miles per hour. After how many hours of skiing will the two skiers be 10 miles apart?

Solution: We are asked to find the time it takes for the skiers to become separated by 10 miles. We will construct a table to aid us in setting up the problem.

Let t = time when skiers are 10 miles apart

Draw a sketch to help visualize the problem (Fig. 4.3).
When the two skiers are 10 miles apart, each has skied for the same number of hours, t.

Rate: 2 mph

Rate: 6 mph

10 miles

Skier	Rate	Time	Distance
Slower	2	t	$2t$
Faster	6	t	$6t$

FIGURE 4.3

Since the skiers are traveling in the same direction, the distance between them is found by subtracting the distance traveled by the slower skier from the distance traveled by the faster skier.

$$\begin{pmatrix}\text{distance traveled}\\ \text{by faster skier}\end{pmatrix} - \begin{pmatrix}\text{distance traveled}\\ \text{by slower skier}\end{pmatrix} = 10 \text{ miles}$$

$$6t - 2t = 10$$

$$4t = 10$$

$$t = 2.5$$

Thus, after 2.5 hours of skiing the two skiers will be 10 miles apart.

EXAMPLE 7 Two construction crews are 20 miles apart working toward each other. Both are laying sewer pipe in a straight line that will eventually be connected together. Both crews will work the same hours. One crew has better equipment and more workers and can lay a greater length of pipe per day. If the faster crew lays 0.4 mile of pipe per day more than the slower crew, and the two pipes are connected after 10 days, find the rate at which each crew lays pipe.

Solution: We are asked to find the two rates. We are told that both crews work for 10 days.

$$\text{Let } r = \text{rate of slower crew}$$

$$\text{then } r + 0.4 = \text{rate of faster crew}$$

Make a sketch (Fig. 4.4) and set up a table of values.

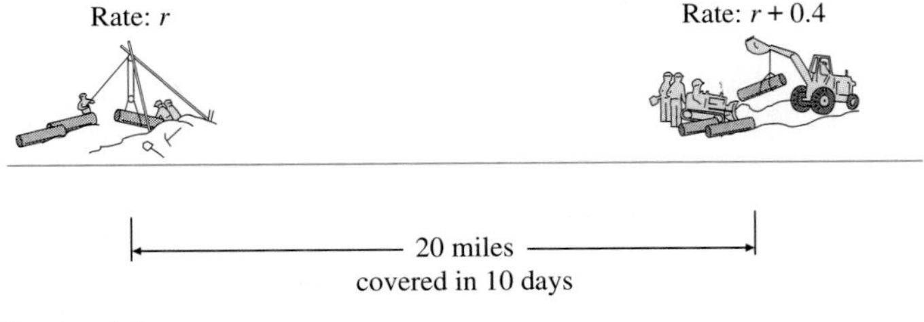

FIGURE 4.4

Crew	Rate	Time	Distance
Slower	r	10	$10r$
Faster	$r + 0.4$	10	$10(r + 0.4)$

The total distance covered by both crews is 20 miles.

$$\begin{pmatrix}\text{distance covered}\\ \text{by slower crew}\end{pmatrix} + \begin{pmatrix}\text{distance covered}\\ \text{by faster crew}\end{pmatrix} = 20 \text{ miles}$$

$$\begin{aligned} 10r + 10(r + 0.4) &= 20 \\ 10r + 10r + 4 &= 20 \\ 20r + 4 &= 20 \\ 20r &= 16 \\ \frac{20r}{20} &= \frac{16}{20} \\ r &= 0.8 \end{aligned}$$

Thus, the slower crew lays 0.8 mile of pipe per day and the faster crew lays $r + 0.4$ or $0.8 + 0.4 = 1.2$ miles of pipe per day.

EXAMPLE 8 A mother and daughter go hiking on the Appalachian Trail. The mother hikes at an average of 4 miles per hour, the daughter at 5 miles per hour. If the daughter begins hiking $\frac{1}{2}$ hour after the mother, and they plan to meet on the trail:

(a) How long will it take for the mother and daughter to meet?

(b) How far from the starting point will they be when they meet?

Solution: **(a)** Since the daughter is the faster hiker, she will cover the same distance in less time. When they meet, each has traveled the same distance. Since the rate is given in miles per hour, the time will be in hours. The daughter begins $\frac{1}{2}$ hour later; therefore, her time will be $\frac{1}{2}$ hour less than her mother's. We are asked to find the time for the mother and daughter to meet.

Let t = time mother is hiking

then $t - \dfrac{1}{2}$ = time daughter is hiking

Make a sketch (Fig. 4.5) and set up a table.

FIGURE 4.5

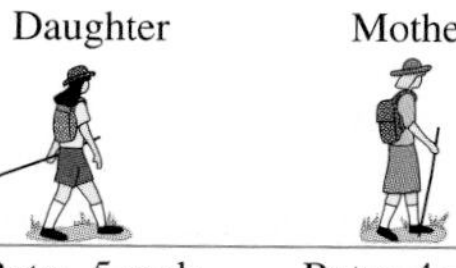

Hiker	Rate	Time	Distance
Mother	4	t	$4t$
Daughter	5	$t - \frac{1}{2}$	$5(t - \frac{1}{2})$

$$\text{mother's distance} = \text{daughter's distance}$$
$$4t = 5\left(t - \frac{1}{2}\right)$$
$$4t = 5t - \frac{5}{2}$$
$$4t - 5t = 5t - 5t - \frac{5}{2}$$
$$-t = -\frac{5}{2}$$
$$t = \frac{5}{2}$$

Thus, they will meet in $\frac{5}{2}$ or $2\frac{1}{2}$ hours.

(b) The distance can be found using either the mother's or daughter's rate and time. We will use the mother's.

$$d = r \cdot t$$
$$= 4 \cdot \frac{5}{2} = \frac{20}{2} = 10 \text{ miles}$$

The mother and daughter will meet 10 miles from the starting point.

Mixture Problems

4 Now we will work some mixture problems. Any problem in which two or more quantities are combined to produce a different quantity or a single quantity is separated into two or more different quantities may be considered a mixture problem. Mixture problems are familiar to everyone, as we can see in the everyday examples that follow.

When solving mixture problems, we often let the variable represent one unknown quantity, and then we represent a second unknown quantity in terms of the first unknown quantity. For example, if we know that when two solutions are mixed they make a total of 80 liters, we may represent the number of liters of one of the solutions as x and the number of liters of the second solution as $80 - x$. Note that when we add x and $80 - x$ we get 80, the total amount.

We generally solve mixture problems by using the fact that the amount (or value) of one part of the mixture plus the amount (or value) of the second part of the mixture is equal to the total amount (or value) of the total mixture.

As we did with motion problems involving two rates, we will use a table to help analyze the problem.

EXAMPLE 9 Deborah wishes to mix coffee worth $7 per pound with 12 pounds of coffee worth $4 per pound.

(a) How many pounds of coffee worth $7 per pound must be mixed to make a mixture worth $6 per pound?

(b) How much of the mixture will be produced?

Solution: **(a)** We are asked to find the number of pounds of \$7 coffee.

Let x = number of pounds of \$7 coffee

Often it is helpful to make a sketch of the situation (Fig. 4.6). Then we construct a table.

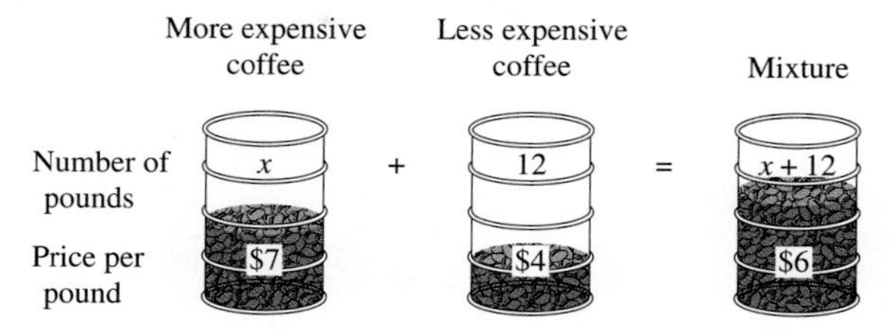

FIGURE 4.6

The value of the coffee is found by multiplying the number of pounds by the price per pound.

Coffee	Price per pound	Number of Pounds	Value of Coffee
More expensive	7	x	$7x$
Less expensive	4	12	4(12)
Mixture	6	$x+12$	$6(x + 12)$

$$\begin{aligned} \text{value of \$7 coffee} + \text{value of \$4 coffee} &= \text{value of mixture} \\ 7x + 4(12) &= 6(x + 12) \\ 7x + 48 &= 6x + 72 \\ x + 48 &= 72 \\ x &= 24 \text{ pounds} \end{aligned}$$

Thus, 24 pounds of the \$7 coffee must be mixed with 12 pounds of the \$4 coffee to make a mixture worth \$6 per pound.

(b) The number of pounds of the mixture is

$$x + 12 = 24 + 12 = 36 \text{ pounds}$$

EXAMPLE 10 Luis invests \$15,000, part at 8% simple interest and the rest at 6% simple interest for 1 year. How much did Luis invest at each rate if his total interest for the year was \$1100?

Solution: We use the simple interest formula that was introduced in Section 3.1 to solve this problem: interest = principal · rate · time.

Let x = amount invested in the 6% account

then $15{,}000 - x$ = amount invested in the 8% account

Account	Principal	Rate	Time	Interest
6%	x	0.06	1	$0.06x$
8%	$15{,}000 - x$	0.08	1	$0.08(15{,}000 - x)$

Since the sum of the interest from the two accounts is $1100, we write the equation

$$\begin{pmatrix}\text{interest from} \\ \text{6\% account}\end{pmatrix} + \begin{pmatrix}\text{interest from} \\ \text{8\% account}\end{pmatrix} = \text{total interest}$$

$$\begin{aligned} 0.06x + 0.08(15{,}000 - x) &= 1100 \\ 0.06x + 0.08(15{,}000) - 0.08(x) &= 1100 \\ 0.06x + 1200 - 0.08x &= 1100 \\ -0.02x + 1200 &= 1100 \\ -0.02x &= -100 \\ x &= \frac{-100}{-0.02} = 5000 \end{aligned}$$

Thus, $5000 was invested at 6% interest. The amount invested at 8% was

$$15{,}000 - x = 15{,}000 - 5000 = 10{,}000$$

Therefore, $5000 was invested at 6% and $10,000 was invested at 8%. The total amount invested is $15,000, which checks with the information given.

In Example 10, we let x represent the amount invested at 6% simple interest. How do you think the answer would change if we let x represent the amount invested at 8% simple interest? Work this problem now letting x represent the amount invested at 8% interest and see what value you obtain for x.

EXAMPLE 11 Melissa Acuña sells both small and large paintings at an art fair. The small paintings sell for $50 each and large paintings sell for $175 each. By the end of the day Melissa lost track of the number of paintings of each size she sold. However by looking at her receipts she knows that she sold a total of 14 paintings for a total of $1200. Determine the number of small and the number of large paintings she sold that day.

Solution: We are asked to find the number of paintings of each size sold.

$$\begin{aligned} \text{Let } x &= \text{number of small paintings sold} \\ \text{then } 14 - x &= \text{number of large paintings sold} \end{aligned}$$

The income received from the sale of the small paintings is found by multiplying the number of small paintings sold by the cost of a small painting. The income received from the sale of the large paintings is found by multiplying the number of large paintings sold by the cost of a large painting. The total income received for the day is the sum of the income from the small paintings and the large paintings.

Painting	Cost	Number of Paintings	Income from Paintings
Small	50	x	$50x$
Large	175	$14 - x$	$175(14 - x)$

$$\begin{aligned}\begin{pmatrix}\text{income from}\\ \text{small paintings}\end{pmatrix} + \begin{pmatrix}\text{income from}\\ \text{large paintings}\end{pmatrix} &= (\text{total income})\\ 50x + 175(14 - x) &= 1200\\ 50x + 2450 - 175x &= 1200\\ -125x + 2450 &= 1200\\ -125x &= -1250\\ x &= \frac{-1250}{-125} = 10\end{aligned}$$

Thus, 10 small paintings and $14 - 10$ or 4 large paintings were sold.

Check: 10 Small Paintings = 500
4 Large Paintings = 700
Total = 1200 true

EXAMPLE 12 How many liters of a 25% acetic acid solution must be added to 80 liters of a 40% acetic acid solution to get a solution that is 30% acetic acid?

Solution: We are asked to find the number of liters of the 25% acetic acid solution.

Let x = number of liters of 25% acetic acid solution

Let us draw a sketch of the solution, (Fig. 4.7).

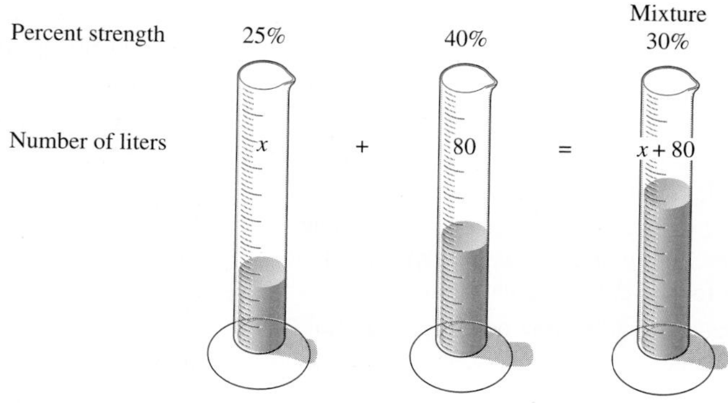

FIGURE 4.7

The amount of pure acid in a given solution is found by multiplying the percent strength by the number of liters.

Solution	Strength	Liters	Amount of Acetic Acid
25%	0.25	x	$0.25x$
40%	0.40	80	$0.40(80)$
mixture	0.30	$x + 80$	$0.30(x + 80)$

$$\begin{pmatrix}\text{amount of pure acid}\\ \text{in 25\% solution}\end{pmatrix} + \begin{pmatrix}\text{amount of pure acid}\\ \text{in 40\% solution}\end{pmatrix} = \begin{pmatrix}\text{amount of pure acid}\\ \text{in 30\% mixture}\end{pmatrix}$$

$$\begin{aligned} 0.25x + 0.40(80) &= 0.30(x + 80) \\ 0.25x + 32 &= 0.30x + 24 \\ 0.25x + 8 &= 0.30x \\ 8 &= 0.05x \\ \frac{8}{0.05} &= x \\ 160 &= x \end{aligned}$$

Therefore, 160 liters of 25% acetic acid solution must be added to the 80 liters of 40% acetic acid solution to get a 30% acetic acid solution. The total number of liters that will be obtained is 160 + 80 or 240.

Exercise Set 4.7

Set up an equation that can be used to solve each problem. Solve the equation, and answer the question. Use a calculator when you feel it is appropriate. See Examples 1–4.

1. A typical shower uses 30 gallons of water and lasts for 6 minutes. How much water is used per minute?
2. How fast must a car travel to cover 150 miles in 3 hours?
3. At what rate must a photocopying machine copy to make 100 copies in 2.5 minutes?
4. A certain laser can cut through steel at 0.2 centimeter per minute. How thick is a steel door if the laser requires 12 minutes to cut through it?
5. *Apollo 11* took approximately 87 hours to reach the moon, a distance of about 238,000 miles. Find the average speed of *Apollo 11*. Give the answer rounded to the nearest mile per hour.
6. On a sunny day a work crew pouring cement on a new highway pours a volume of 1.4 tons per hour. Find the time it will take for the crew to pour 6.3 tons.
7. A patient is to receive 1500 cubic centimeters of an intravenous fluid over a period of 6 hours. What should be the intravenous flow rate?
8. Kilauea Volcano's lava is flowing slowly toward the ocean. If the lava flows a distance of 57 meters in 9.5 hours, find the average rate of flow of the lava.
9. If John walks at 3 miles per hour, how far will he walk in 20 minutes?
10. Morton can check 12 parts *per minute* on the assembly line. How many parts will he be able to check in an 8-*hour* day?

SEE EXERCISE 4.

11. The average American uses 7 gallons of water *per hour.* How much water does the average American use in 2 *days*?
12. A standard T120 VHS videocassette tape contains 246 meters of tape. When played on standard play speed (SP), the tape will play for 2 hours. The tape will play for 4 hours on long play speed (LP) and will play for 6 hours on super long play speed (SLP). Determine the rate of play of the tape at **(a)** SP, **(b)** LP, and **(c)** SLP.

Solve the following motion problems. See Examples 5–8.

13. Two planes leave an airport at the same time. One plane flies north at 500 miles per hour. The other flies south at 650 miles per hour. In how many hours will they be 4025 miles apart?

14. Two joggers, Sonya and Jeri, start from the same point and jog in the same direction. Sonya jogs at 8 miles per hour, and Jeri jogs at 11 miles per hour. In how many hours will Sonya and Jeri be 9 miles apart?

15. Two trains, an Amtrak train and the Santa Fe Special, are 804 miles apart. Both start at the same time and travel toward each other. They meet 6 hours later. If the speed of the Amtrak train is 30 miles per hour faster than the Santa Fe Special, find the speed of each train.

16. Two trains, a Conrail train and the Durango Special, leave South Street Station at the same time traveling in the same direction along parallel tracks. The Conrail train travels 80 miles per hour, and the Durango Special travels 62 miles per hour. In how many hours will the two trains be 144 miles apart?

17. Two Coast Guard cutters are 225 miles apart traveling toward each other, one from the east and the other from the west, searching for a disabled boat. Because of the current, the eastbound cutter travels 5 miles per hour faster than the westbound cutter, **(a)** If the two cutters pass each other after 3 hours, find the average speed of each cutter. **(b)** Speeds at sea are generally measured in knots. One knot is approximately equal to 1.15 miles per hour. Find the speed of each boat in knots.

18. Two sailboats are 9.8 miles apart and sailing toward each other. The larger boat, the *Lorelei,* sails 4 miles per hour faster than the smaller boat, the *Brease Along.* If the two boats pass each other after 0.7 hour, find the speed of each boat.

19. A VHS T120 videocassette tape plays at a rate of 6.72 feet per minute for 2 hours (120 minutes) on standard play speed (SP). What is the rate of the tape on super long play speed (SLP) if the same tape plays for 6 hours on this speed?

20. Two construction crews are laying blacktop on a road. They start at the same time at opposite ends of a 12-mile road and work toward one another. One crew lays blacktop at an average rate of 0.75 mile a day faster than the other crew. If the two crews meet after 3.2 days, find the rate of each crew.

21. Two ranchers are 16.5 miles apart at opposite ends of the Flying W Ranch in Montana. Both start at the same time and travel toward each other on horseback. Chet is walking his horse Chestnut while Annie is trotting on her horse Midnight. If Midnight is traveling at a rate of 3 miles per hour faster than Chestnut, and they meet after 1.5 hours, what is the speed of each horse?

22. Mrs. Weber and her son belong to a health club and exercise together regularly. They start running on two treadmills at the same time. The son's machine is set for 6 miles per hour and the mother's machine is set for 4 miles per hour. When they finish, they compare the distances and find that together they have run a total of 11 miles. How long had they worked out?

23. Serge and Francine go mountain climbing together. Francine begins climbing the mountain 30 minutes before Serge and averages 18 feet per minute. When Serge begins climbing, he averages 20 feet per minute. How far up the mountain will they meet?

24. Two rockets are launched in the same direction 1 hour apart. The first rocket is launched at noon and travels at 12,000 miles per hour. If the second rocket travels at 14,400 miles per hour, at what time will the rockets be the same distance from Earth?

25. The moving walkway in the United Airlines terminal at Chicago's O'Hare International Airport (also called a "travelator" by United Airlines) is like a flat escalator moving along the floor. When Marguerita Vela walked the floor along side the moving walkway at her normal speed of 100 feet per minute, it took her 2.75 minutes to walk the length of the moving walkway. When she returned the same distance, walking at her normal speed on the moving walkway, it took her only 1.25 minutes. **(a)** Find the length of the moving walkway. **(b)** Find the speed of the moving walkway.

26. Tanya starts for school walking at an average rate of 120 feet per minute. A short while later she realizes that she forgot her lunch and turns around and jogs back to the house at 360 feet per minute. If it took her a total of 10 minutes to go and return home, how far had Tanya walked before she turned around?

27. Because of cracks in their foundations, a road leading to a bridge and the bridge must be replaced. The engineer estimates that it will take 20 days for the road crew to tear up and clear the road and an additional 60 days for the same road crew to dismantle and clear the bridge. If the rate of clearing of the road is 1.2 feet per day faster than the rate of clearing of the bridge and the total distance cleared is 124 feet, find the rate for clearing the road and the rate for clearing the bridge.

28. John delivers the Citizen Shopper to households along a 20.4-mile route. It normally takes him 6 hours to do the whole route. One Friday he needs to leave work early, so he gets his friend Brianna to help him. John will start delivering at one end and Brianna will start at the other end 1 hour later, and they will meet somewhere along his route. If Brianna covers 2.6 miles per hour, how long after Brianna begins will the two meet?

Solve the following mixture problems. See Examples 10–13.

29. Mr. Ellis invested \$8900, part at 8% simple interest and the rest at 11% simple interest for a period of 1 year. How much did he invest at each rate if his total annual interest from both investments was \$874? Use interest = principal · rate · time.

30. Linda invested \$6000, part at 9% simple interest and the rest at 10% simple interest for a period of 1 year. If she received a total annual interest of \$562 from both investments, how much did she invest at each rate?

31. Julio invested \$5000, part at 10% simple interest and part at 6% simple interest for a period of 1 year. How much did he invest at each rate if each account earned the same interest?

32. The Clars invested \$12,500, part at 8% simple interest and part at 12% simple interest for a period of 1 year. How much was invested at each rate if each account earned the same interest?

33. Charles knows that his subscription rate for the basic tier of cable television increased from \$13.20 to \$14.50 at some point during the calendar year. If he knew that during the calendar year he paid a total of \$170.10 to the cable company, determine the month of the rate increase.

34. Kathleen has \$3.75 in dimes and nickels. If she has 62 coins, how many of each coin does she have?

35. Phil has 12 bills in his wallet. Some are \$1 bills and some are \$10 bills. The value of the 12 bills is \$39. How many of each type does he have?

36. Mari has a total of 25 quarters and half-dollars. The total value of the coins is \$9.00. How many of each coin does she have?

37. Casey holds two part-time jobs. One job pays \$6.00 per hour, and the other pays \$6.50 per hour. Last week Casey worked a total of 18 hours and earned \$114.00. How many hours did he work on each job?

38. Almonds cost \$6.00 per pound. Walnuts cost \$6.40 per pound. How many pounds of each should Bridget mix to produce a 30-pound mixture that costs \$6.25 per pound?

39. How many pounds of coffee costing \$6.20 per pound must Guido mix with 18 pounds of coffee costing \$5.60 per pound to produce a mixture that costs \$5.80 per pound?

40. There were 600 people at a movie. Adult admission was \$6. Children's admission was \$5. How many adults and how many children attended if the total receipts were \$3250 for the showing?

41. In chemistry class, Ramon has 1 liter of a 20% sulfuric acid solution. How much of a 12% sulfuric acid solution must he mix with the 1 liter of 20% solution to make a 15% sulfuric acid solution?

42. Doug Fisher, a pharmacist, has a 60% solution of the drug sodium iodite. He also has a 25% solution of the same drug. He gets a prescription calling for a 40% solution of the drug. How much of each solution should he mix to make 0.5 liter of the 40% solution?

43. Six quarts of orange juice punch for the class party contains 12% orange juice. If $\frac{1}{2}$ quart of water is added to the punch, find the percent of orange juice in the new mixture.

44. The label on a 12-ounce can of frozen concentrate Hawaiian Punch indicates that when the can of concentrate is mixed with 3 cans of cold water the resulting mixture is 10% juice. Find the percent of pure juice in the concentrate.

45. Scott's Family grass seed sells for \$2.25 per pound and Scott's Spot Filler grass seed sells for \$1.90 per pound. How many pounds of each should be mixed to get a 10-pound mixture that sells for \$2.00 per pound?

46. Citicorp stock is selling at \$37 a share. Mobil Oil stock is selling at \$75 a share. Mr. Abelard has a minimum of \$7800 to invest. He wishes to purchase five times as many shares of Citicorp as of Mobil Oil. How many shares of each can he purchase?

47. United Airlines stock is selling at \$140 per share. Mattel stock is selling at \$22 a share. An investor has \$8000 to invest. She wishes to purchase four times as many shares of United Airlines as Mattel. If only whole shares of stock can be purchased:

(a) How many shares of each will she purchase?

(b) How much money will be left over?

CUMULATIVE REVIEW EXERCISES

2] **48.** **(a)** Divide $2\frac{3}{4} \div 1\frac{5}{8}$.

(b) Add $2\frac{3}{4} + 1\frac{5}{8}$.

5] **49.** Solve the equation $6(x - 3) = 4x - 18 + 2x$.

[2.6] **50.** Solve the proportion $\frac{6}{x} = \frac{72}{9}$.

[2.7] **51.** Solve the inequality $3x - 4 \leq -4x + 3(x - 1)$.

Group Activity/ Challenge Problems

1. Every 20 minutes Americans dump enough cars into junkyards that the cars, if stacked one on top of the other, would reach the top of the Empire State Building, 1472 feet tall.

(a) Write an equation using distance = rate · time, for t in hours, for determining how long it would take for the stack of cars to reach a height of 10 miles. There are 5280 feet in a mile.

(b) Find how long it would take for the stack of cars to reach a height of 10 miles.

2. Sixty-six lines of type will fit on a standard $8\frac{1}{2} \times 11$-inch sheet of paper. A Panasonic printer types 10 characters per inch at a speed of 100 characters per second. How long will it take the printer to type a full page of 66 lines if the page is set for a 1-inch margin on the left and a $\frac{1}{2}$-inch margin on the right?

3. An automatic garage door opener is designed to begin to open when a car is 100 feet from the garage. At what rate will the garage door have to open if it is to raise 6 feet by the time a car traveling at 4 miles per hour reaches it? (1 mile per hour = 1.47 feet per second.)

4. The radiator of Mark's 1968 Dodge Challenger holds 16 quarts. It is presently filled with a 20% antifreeze solution. How many quarts must Mark drain and replace with pure antifreeze for the radiator to contain a 50% antifreeze solution?

Summary

GLOSSARY

Binomial *(210):* A two-termed polynomial.

Degree of a polynomial *(211):* The same as the highest-degree term in the polynomial.

Degree of a term *(211):* The exponent on the variable when the polynomial is in one variable.

Descending order, or descending power, of the variable *(210):* Polynomial written so that the exponents on the variable decrease from left to right.

Monomial *(210):* A one-term polynomial.

Polynomial in x *(210):* An expression containing the sum of a finite number of terms of the form ax^n, for any real number a and any whole number n.

Scientific notation *(205):* A number greater than or equal to 1 and less than 10 multiplied by some power of 10.

Trinomial *(210):* A three-termed polynomial.

IMPORTANT FACTS

Rules of Exponents

1. $x^m x^n = x^{m+n}$ product rule

2. $\frac{x^m}{x^n} = x^{m-n}$ quotient rule

3. $x^0 = 1, x \neq 0$ zero exponent rule

4. $(x^m)^n = x^{mn}$ power rule

5. $\left(\frac{ax}{by}\right)^m = \frac{a^m x^m}{b^m y^m}, b \neq 0, y \neq 0$ expanded power rule

6. $x^{-m} = \frac{1}{x^m}, x \neq 0$ negative exponent rule

Product of sum and difference of the same two quantities (also called the **difference of two squares**):

$$(a + b)(a - b) = a^2 - b^2$$

FOIL Method to Multiply Two Binomials (**F**irst, **O**uter, **I**nner, **L**ast)

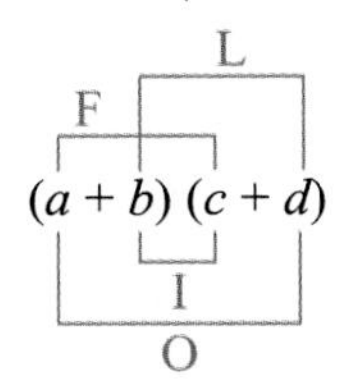

Square of a binomial:

$$(a + b)^2 = a^2 + 2ab + b^2$$
$$(a - b)^2 = a^2 - 2ab + b^2$$

Rate formula: Amount = rate · time
Distance formula: Distance = rate · time

Review Exercises

[4.1] *Simplify.*

1. $x^4 \cdot x^2$ **2.** $x^3 \cdot x^5$ **3.** $3^2 \cdot 3^3$ **4.** $2^4 \cdot 2$

5. $\frac{x^4}{x}$ **6.** $\frac{x^6}{x^6}$ **7.** $\frac{3^5}{3^3}$ **8.** $\frac{4^5}{4^3}$

9. $\frac{x^6}{x^8}$ **10.** $\frac{x^7}{x^2}$ **11.** x^0 **12.** $3x^0$

13. $(3x)^0$ **14.** 4^0 **15.** $(2x)^2$ **16.** $(3x)^3$

17. $(-2x)^2$ **18.** $(-3x)^3$ **19.** $(2x^2)^4$ **20.** $(-x^4)^3$

21. $(-x^3)^4$ **22.** $\left(\frac{2x^3}{y}\right)^2$ **23.** $\left(\frac{3x^4}{2y}\right)^3$ **24.** $6x^2 \cdot 4x^3$

25. $\frac{16x^2y}{4xy^2}$ **26.** $(2x^2y)^2 \cdot 3x$ **27.** $\left(\frac{9x^2y}{3xy}\right)^2$ **28.** $(2x^2y)^3(3xy^4)$

29. $4x^2y^3(2x^3y^4)^2$ **30.** $6x^4(2x^3y^4)^2$ **31.** $\left(\frac{8x^4y^3}{2xy^5}\right)^2$ **32.** $\left(\frac{5x^4y^7}{10xy^{10}}\right)^3$

2] *Simplify.*

33. x^{-3}

34. y^{-7}

35. 5^{-2}

36. $\dfrac{1}{x^{-3}}$

37. $\dfrac{1}{x^{-7}}$

38. $\dfrac{1}{3^{-2}}$

39. $x^3 \cdot x^{-5}$

40. $x^{-2} \cdot x^{-3}$

41. $x^4 \cdot x^{-7}$

42. $x^{13} \cdot x^{-5}$

43. $\dfrac{x^2}{x^{-3}}$

44. $\dfrac{x^5}{x^{-2}}$

45. $\dfrac{x^{-3}}{x^3}$

46. $(3x^4)^{-2}$

47. $(4x^{-3}y)^{-3}$

48. $(-5x^{-2})^3$

49. $-2x^3 \cdot 4x^5$

50. $(3x^{-2}y)^3$

51. $(4x^{-2}y^3)^{-2}$

52. $2x(3x^{-2})$

53. $(5x^{-2}y)(2x^4y)$

54. $4x^5(6x^{-7}y^2)$

55. $2x^{-4}(3x^{-2}y^{-1})$

56. $\dfrac{6xy^4}{2xy^{-1}}$

57. $\dfrac{9x^{-2}y^3}{3xy^2}$

58. $\dfrac{25xy^{-6}}{5y^{-2}}$

59. $\dfrac{36x^4y^7}{9x^5y^{-3}}$

60. $\dfrac{4x^5y^{-2}}{8x^7y^3}$

3] *Express each number in scientific notation.*

61. 364,000

62. 1,640,000

63. 0.00763

64. 0.176

65. 2080

66. 0.000314

Express each number without exponents.

67. 4.2×10^{-3}

68. 1.65×10^4

69. 9.7×10^5

70. 4.38×10^{-6}

71. 9.14×10^{-1}

72. 5.36×10^2

Perform the indicated operation and write each answer without exponents.

73. $(2.3 \times 10^2)(2 \times 10^4)$

74. $(4.2 \times 10^{-3})(3 \times 10^5)$

75. $(6.4 \times 10^{-3})(3.1 \times 10^3)$

76. $\dfrac{6.8 \times 10^3}{2 \times 10^{-2}}$

77. $\dfrac{36 \times 10^4}{4 \times 10^6}$

78. $\dfrac{15 \times 10^{-3}}{5 \times 10^2}$

Convert each number to scientific notation. Then calculate. Express your answer in scientific notation.

79. (60,000)(20,000)

80. (0.00004)(600,000)

81. (0.00023)(40,000)

82. $\dfrac{40{,}000}{0.0002}$

83. $\dfrac{0.000068}{0.02}$

84. $\dfrac{1{,}500{,}000}{0.003}$

85. *Modern Maturity* was the best-selling magazine in the United States in 1992. Its circulation was about 2.29×10^7 copies. The magazine *Sport* was the number 100 best-selling magazine, with 8.17×10^5 copies sold.

(a) How many times greater were the sales of *Modern Maturity* than *Sport*?

(b) How many more copies of *Modern Maturity* were sold? Write your answers without exponents.

86. If one light year is 5.87×10^{12} miles, how far away in miles is the star Alpha Orion (the brightest star in Orion), which is 520 light years from Earth? Write your answer in scientific notation form.

4] *Indicate if each expression is a polynomial. If the polynomial has a specific name, give that name. If the polynomial is not written in descending order, rewrite it in descending order. State the degree of each polynomial.*

87. $x + 3$

88. -2

89. $x^2 - 4 + 3x$

90. $-3 - x + 4x^2$

91. $-5x^2 + 3$

92. $4x^{1/2} - 6$

93. $x - 4x^2$

94. $x^3 + x^{-2} + 3$

95. $x^3 - 2x - 6 + 4x^2$

[4.4–4.6] *Perform the operations indicated.*

96. $(x + 3) + (2x + 4)$

97. $(5x - 5) + (4x + 6)$

98. $(-3x + 4) + (5x - 9)$

99. $(-x^2 + 6x - 7) + (-2x^2 + 4x - 8)$

100. $(12x^2 + 4x - 8) + (-x^2 - 6x + 5)$

101. $(2x - 4.3) - (x + 2.4)$

102. $(-4x + 8) - (-2x + 6)$

103. $(9x^2 - \frac{3}{4}x) - (\frac{1}{2}x - 4)$

104. $(6x^2 - 6x + 1) - (12x + 5)$

105. $(-2x^2 + 8x - 7) - (3x^2 + 12)$

106. $(x^2 + 7x - 3) - (x^2 + 3x - 5)$

107. $x(2x - 4)$

108. $4.5x(x^2 - 3x)$

109. $3x(2x^2 - 4x + 7)$

110. $-x(3x^2 - 6x - 1)$

111. $-4x(-6x^2 + 4x - 2)$

112. $(x + 4)(x + 5)$

113. $(2x + 4)(x - 3)$

114. $(4x + 6)^2$

115. $(6 - 2x)(2 + 3x)$

116. $(x + 4)(x - 4)$

117. $(3x + 1)(x^2 + 2x + 4)$

118. $(x - 1)(3x^2 + 4x - 6)$

119. $(-5x + 2)(-2x^2 + 3x - 6)$

120. $\dfrac{2x + 4}{2}$

121. $\dfrac{4x - 8}{4}$

122. $\dfrac{8x^2 + 4x}{x}$

123. $\dfrac{6x^2 + 9x - 4}{3}$

124. $\dfrac{8x^2 + 6x - 4}{x}$

125. $\dfrac{8x^5 - 4x^4 + 3x^2 - 2}{2x}$

126. $\dfrac{16x - 4}{-2}$

127. $\dfrac{12 + 6x^2 + 3x}{-3x}$

128. $\dfrac{5x^3 + 10x + 2}{2x^2}$

129. $\dfrac{x^2 + x - 12}{x - 3}$

130. $\dfrac{6x^2 - 11x + 3}{3x - 1}$

131. $\dfrac{5x^2 + 28x - 10}{x + 6}$

132. $\dfrac{4x^3 + 12x^2 + x - 12}{2x + 3}$

133. $\dfrac{4x^3 - 5x + 4}{2x - 1}$

[4.7] *Solve the following motion problems.*

134. A train travels at 70 miles per hour. How long will it take the train to travel 280 miles?

135. How fast must a plane fly to travel 3500 miles in 6.5 hours?

136. Two joggers follow the same route. Marty jogs at 8 kilometers per hour and Nick at 6 kilometers per hour. If they leave at the same time, how long will it take for them to be 4 kilometers apart?

Solve the following mixture problems.

137. Kathy Platico wishes to place part of \$12,000 into a savings account earning 8% simple interest and part into a savings account earning $7\frac{1}{4}$% simple interest. How much should she invest in each if she wishes to earn \$900 in interest for the year?

138. A chemist wishes to make 2 liters of an 8% acid solution by mixing a 10% acid solution and a 5% acid solution. How many liters of each should the chemist use?

Solve the following problems.

139. Marty completed the 26-mile Boston Marathon in 4 hours. Find his average speed.

140. Two trains going in opposite directions leave from the same station on parallel tracks. One train travels at 50 miles per hour and the other at 60 miles per hour. How long will it take for the trains to be 440 miles apart?

141. A butcher combined hamburger that cost \$3.50 per pound with hamburger that cost \$4.10 per pound. How many pounds of each were used to make 80 pounds of a mixture that sells for \$3.65 per pound?

142. Joan paid \$12.40 for 40 stamps. Some are 32-cent stamps and some are 22-cent stamps. How many of each type did she purchase?

143. Two brothers who are 230 miles apart start driving toward each other. If the younger brother travels 5 miles per hour faster than the older brother and the brothers meet after 2 hours, find the speed traveled by each brother.

144. How many liters of a 30% acid solution must be mixed with 2 liters of a 12% acid solution to obtain a 15% acid solution?

Practice Test

Simplify each expression.

1. $2x^2 \cdot 3x^4$
2. $(3x^2)^3$
3. $\dfrac{8x^4}{2x}$
4. $\left(\dfrac{3x^2y}{6xy^3}\right)^3$
5. $(2x^3y^{-2})^{-2}$
6. $\dfrac{2x^4y^{-2}}{10x^7y^4}$

Determine whether each expression is a polynomial. If the polynomial has a specific name, give that name.

7. $x^2 - 4 + 6x$
8. -3
9. $x^{-2} + 4$
10. Write the polynomial $-5 + 6x^3 - 2x^2 + 5x$ in descending order, and give its degree.

Perform the operations indicated.

11. $(2x + 4) + (3x^2 - 5x - 3)$
12. $(x^2 - 4x + 7) - (3x^2 - 8x + 7)$
13. $(4x^2 - 5) - (x^2 + x - 8)$
14. $3x(4x^2 - 2x + 5)$
15. $(4x + 7)(2x - 3)$
16. $(6 - 4x)(5 + 3x)$
17. $(2x - 4)(3x^2 + 4x - 6)$
18. $\dfrac{16x^2 + 8x - 4}{4}$
19. $\dfrac{3x^2 - 6x + 5}{-3x}$
20. $\dfrac{8x^2 - 2x - 15}{2x - 3}$
21. Madison can fertilize 0.7 acre per hour. How long will it take him to fertilize a 40-acre farm?
22. Train A travels 60 miles per hour for 4 hours. If train B is to travel the same distance in 3 hours, find its speed.
23. How many liters of 20% salt solution must be added to 60 liters of 40% salt solution to get a solution that is 35% salt?

Convert each number to scientific notation. Then calculate. Express your answer in scientific notation.

24. $(42{,}000)(30{,}000)$
25. $\dfrac{0.0008}{4000}$

Cumulative Review Test

1. Evaluate $16 \div (4 - 6) \cdot 5$.
2. Solve the equation $2x + 5 = 3(x - 5)$.
3. Solve the equation $3(x - 2) - (x + 4) = 2x - 10$.
4. Solve the inequality $2x - 14 > 5x + 1$. Graph the solution on the number line.
5. Solve for w. $v = lwh$.
6. Solve for y. $4x - 3y = 6$.
7. Simplify $(3x^4)(2x^5)$.
8. Simplify $(3x^2y^4)^3(5x^2y)$.
9. Write the polynomial $-2x + 3x^2 - 5$ in descending order, and give its degree.
10. Subtract $6x^2 - 3x + 4$ from $2x^2 - 9x - 7$.

Perform the operations indicated.

11. $(2x^2 + 4x - 3) + (6x^2 - 7x + 12)$.
12. $(4x^2 - 5x - 2) - (3x^2 - 2x + 5)$.
13. $(2x - 3)(3x - 5)$.
14. $(2x^2 + 4x + 8)(x - 5)$.
15. $\dfrac{9x^2 - 6x + 8}{3x}$
16. $\dfrac{2x^2 + x - 6}{x + 2}$
17. At Hsu's Grocery Store, 3 cans of egg drop soup sell for \$1.25. Find the cost of 8 cans.
18. Eleven increased by twice a number is nineteen. Find the number.
19. The length of a rectangle is four more than twice the width. Find the dimensions of the rectangle if its perimeter is 26 feet.
20. Two runners start at the same point and run in opposite directions. One runs at 6 mph and the other runs at 8 mph. In how many hours will they be 28 miles apart?

Chapter 5

Factoring

See Section 5.6, Exercise 61

Preview and Perspective

In this chapter we introduce factoring polynomials, which is the reverse process of multiplying polynomials. When we factor a polynomial, we rewrite it as a product of two or more factors. In Sections 5.1 through 5.4 we present factoring techniques for several types of polynomials. In Section 5.5 we present some special factoring formulas, which can simplify the factoring process for some polynomials. In Section 5.6 we introduce quadratic equations. One of the main uses of factoring is in solving quadratic equations. Material from all previous sections in this chapter will be used when we solve quadratic equations. Quadratic equations can also be solved by other methods, including completing the square and the quadratic formula. Both of these methods for solving quadratic equations are discussed in Chapter 10. However, if the polynomial in a quadratic equation is *factorable,* or can be factored, then factoring may be the easiest way to solve the problem. Factoring will also be used to find the x intercepts when graphing quadratic equations in Chapter 10 and when graphing other equations in intermediate algebra courses.

It is essential that you have a thorough understanding of factoring, especially Sections 5.3 through 5.5, to complete Chapter 6 successfully. The factoring techniques used in this chapter are used throughout Chapter 6.

5.1 Factoring a Monomial from a Polynomial

Tape 7

1. Identify factors.
2. Find the greatest common factor of two or more numbers.
3. Find the greatest common factor of two or more terms.
4. Factor a monomial from a polynomial.

Identify Factors

1 In Chapter 4 you learned how to multiply polynomials. In this chapter we focus on factoring, the reverse process of multiplication. In Section 4.5 we showed that $2x(3x^2 + 4) = 6x^3 + 8x$. In this chapter we start with an expression like $6x^3 + 8x$ and determine that its factors are $2x$ and $3x^2 + 4$, and write $6x^3 + 8x = 2x(3x^2 + 4)$. To **factor an expression** means to write the expression as a product of its factors. Factoring is important because it can be used to solve equations.

If $a \cdot b = c$, then a and b are said to be *factors* of c.

$3 \cdot 5 = 15$; thus 3 and 5 are factors of 15.

$x^3 \cdot x^4 = x^7$; thus x^3 and x^4 are factors of x^7.

$x(x + 2) = x^2 + 2x$; thus x and $(x + 2)$ are factors of $x^2 + 2x$.

$(x - 1)(x + 3) = x^2 + 2x - 3$; thus $(x - 1)$ and $(x + 3)$ are factors of $x^2 + 2x - 3$.

A given number or expression may have many factors. Consider the number 30.

$$1 \cdot 30 = 30, \qquad 2 \cdot 15 = 30, \qquad 3 \cdot 10 = 30, \qquad 5 \cdot 6 = 30$$

Thus, the positive factors of 30 are 1, 2, 3, 5, 6, 10, 15, and 30. Factors can also be negative. Since $(-1)(-30) = 30$, -1 and -30 are also factors of 30. In fact, for each factor a of an expression, $-a$ must also be a factor. Other factors of 30 are therefore $-1, -2, -3, -5, -6, -10, -15$, and -30. When asked to list the factors of an expression with a positive numerical coefficient that contains a variable, we generally list only positive factors.

EXAMPLE 1 List the factors of $6x^3$.

Solution:

$$\overbrace{1 \cdot 6x^3}^{\text{factors}} = 6x^3 \qquad \overbrace{x \cdot 6x^2}^{\text{factors}} = 6x^3$$
$$2 \cdot 3x^3 = 6x^3 \qquad 2x \cdot 3x^2 = 6x^3$$
$$3 \cdot 2x^3 = 6x^3 \qquad 3x \cdot 2x^2 = 6x^3$$
$$6 \cdot x^3 = 6x^3 \qquad 6x \cdot x^2 = 6x^3$$

The factors of $6x^3$ are $1, 2, 3, 6, x, 2x, 3x, 6x, x^2, 2x^2, 3x^2, 6x^2, x^3, 2x^3, 3x^3$, and $6x^3$. The opposite (or negative) of each of these factors is also a factor, but these opposites are generally not listed unless specifically asked for.

Here are examples of multiplying and factoring: Notice that factoring is the reverse process of multiplying.

Multiplying	*Factoring*
$2(3x + 4) = 6x + 8$	$6x + 8 = 2(3x + 4)$
$5x(x + 4) = 5x^2 + 20x$	$5x^2 + 20x = 5x(x + 4)$
$(x + 1)(x + 3) = x^2 + 4x + 3$	$x^2 + 4x + 3 = (x + 1)(x + 3)$

The Greatest Common Factor of Numbers

2 To factor a monomial from a polynomial, we make use of the *greatest common factor (GCF)*. If after studying the following material you wish to see additional material on obtaining the GCF, you may read Appendix B, where one of the topics discussed is finding the GCF.

Recall from Section 1.2 that the **greatest common factor** of two or more numbers is the greatest number that divides into all the numbers. The greatest common factor of the numbers 6 and 8 is 2. Two is the greatest number that divides into both 6 and 8. What is the GCF of 48 and 60? When the GCF of two or more numbers is not easily found, we can find it by writing each number as a product of prime numbers. A **prime number** is an integer greater than 1 that has exactly two factors, itself and one. The first 13 prime numbers are

$$2, 3, 5, 7, 11, 13, 17, 19, 23, 29, 31, 37, 41.$$

A positive integer (other than 1) that is not prime is called **composite.** The number 1 is neither prime nor composite, it is called a **unit.**

To write a number as a product of prime numbers, follow the procedure illustrated in Examples 2 and 3.

EXAMPLE 2 Write 48 as a product of prime numbers.

Solution: Select any two numbers whose product is 48. Two possibilities are $6 \cdot 8$ and $4 \cdot 12$, but there are other choices. Continue breaking down the factors until all the factors are prime, as illustrated in Figure 5.1.

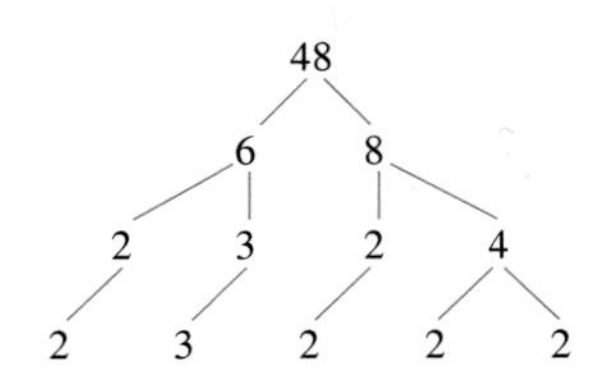

FIGURE 5.1

Note that no matter how you select your initial factors,

$$48 = 2 \cdot 2 \cdot 2 \cdot 2 \cdot 3 = 2^4 \cdot 3$$

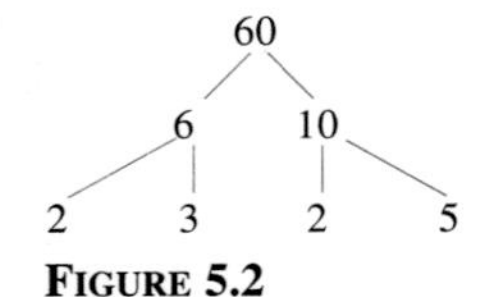

FIGURE 5.2

EXAMPLE 3 Write 60 as a product of prime numbers.

Solution: One way to find the prime factors is shown in Figure 5.2.

Therefore, $60 = 2 \cdot 2 \cdot 3 \cdot 5 = 2^2 \cdot 3 \cdot 5$.

To Find the GCF of Two or More Numbers

1. Write each number as a product of prime factors.
2. Determine the prime factors common to all the numbers.
3. Multiply the common factors found in step 2. The product of these factors is the GCF.

EXAMPLE 4 Find the greatest common factor of 48 and 60.

Solution: From Examples 2 and 3 we know that

$$48 = 2 \cdot 2 \cdot 2 \cdot 2 \cdot 3 = 2^4 \cdot 3$$
$$60 = 2 \cdot 2 \cdot 3 \cdot 5 = 2^2 \cdot 3 \cdot 5$$

Two factors of 2 and one factor of 3 are common to both numbers. The product of these factors is the GCF of 48 and 60:

$$\text{GCF} = 2 \cdot 2 \cdot 3 = 12$$

The GCF of 48 and 60 is 12. Twelve is the greatest number that divides into both 48 and 60.

EXAMPLE 5 Find the GCF of 18 and 24.

Solution: $18 = 2 \cdot 3 \cdot 3 = 2 \cdot 3^2$
$24 = 2 \cdot 2 \cdot 2 \cdot 3 = 2^3 \cdot 3$

One factor of 2 and one factor of 3 are common to both 18 and 24.

$$\text{GCF} = 2 \cdot 3 = 6$$

The Great Common Factor of Terms

3 The GCF of several terms containing variables is easily found. Consider the terms x^3, x^4, x^5, and x^6. The GCF of these terms is x^3, since x^3 is the highest power of x that divides into all four terms. We can illustrate this by writing the terms in factored form.

$$x^3 = x^3 \cdot 1$$
$$x^4 = x^3 \cdot x$$
$$x^5 = x^3 \cdot x^2$$
$$x^6 = x^3 \cdot x^3$$

GCF of all four terms is x^3

EXAMPLE 6 Find the GCF of the terms p^6, p^2, p^7, and p^4.

Solution: The GCF is p^2 because p^2 is the highest power of p that divides into each term. Notice that $\frac{p^6}{p^2} = p^4$, that $\frac{p^2}{p^2} = 1$, that $\frac{p^7}{p^2} = p^5$ and that $\frac{p^4}{p^2} = p^2$.

EXAMPLE 7 Find the GCF of the terms x^2y^3, x^3y^2, and xy^4.

Solution: The highest power of x common to all three terms is x^1 or x. The highest power of y common to all three terms is y^2. So the GCF of the three terms is xy^2.

Greatest Common Factor of Two or More Terms

To find the GCF of two or more terms, take each factor the *fewest* number of times that it appears in any of the terms.

EXAMPLE 8 Find the GCF of the terms xy, x^2y^2, and x^3.

Solution: The GCF is x. The smallest power of x that appears in any of the terms is x. Since the term x^3 does not contain a power of y, the GCF does not contain y.

EXAMPLE 9 Find the GCF of each set of terms.

(a) $18y^2, 15y^3, 21y^5$ **(b)** $-20x^2, 8x, 40x^3$ **(c)** $4x^2, x^2, x^3$

Solution:

(a) The GCF of 18, 15, and 21 is 3. The GCF of the three terms is $3y^2$.
(b) The GCF of -20, 8, and 40 is 4. The GCF of the three terms is $4x$.
(c) The GCF is x^2.

EXAMPLE 10 Find the GCF of each pair of terms.

(a) $x(x + 3)$ and $2(x + 3)$ **(b)** $x(x - 2)$ and $x - 2$
(c) $2(x + y)$ and $3x(x + y)$

Solution:
(a) The GCF is $(x + 3)$.
(b) $x - 2$ can be written as $1(x - 2)$. Therefore, the GCF of $x(x - 2)$ and $1(x - 2)$ is $x - 2$.
(c) The GCF is $(x + y)$.

Factoring a Monomial from a Polynomial

To Factor a Monomial from a Polynomial

1. Determine the greatest common factor of all terms in the polynomial.
2. Write each term as the product of the GCF and its other factor.
3. Use the distributive property to factor out the GCF.

EXAMPLE 11 Factor $6x + 12$.

Solution: The GCF is 6.

$$6x + 12 = 6 \cdot x + 6 \cdot 2$$ Write each term as a product of the GCF and some other factor.

$$= 6(x + 2)$$ Distributive property

To check the factoring process, multiply the factors using the distributive property. If the factoring is correct, the product will be the polynomial you started with. Following is a check of the factoring in Example 11.

Check: $6(x + 2) = 6x + 12$

EXAMPLE 12 Factor $8x - 10$.

Solution: The GCF of $8x$ and -10 is 2.

$$\begin{aligned} 8x - 10 &= 2 \cdot 4x - 2 \cdot 5 \\ &= 2(4x - 5) \end{aligned}$$

Check that the factoring is correct by multiplying.

EXAMPLE 13 Factor $6y^2 + 9y^5$.

Solution: The GCF is $3y^2$.

$$\begin{aligned} 6y^2 + 9y^5 &= 3y^2 \cdot 2 + 3y^2 \cdot 3y^3 \\ &= 3y^2(2 + 3y^3) \end{aligned}$$

Check to see that the factoring is correct.

EXAMPLE 14 Factor $8x^3 + 12x^2 - 16x$.

Solution: The GCF is $4x$.

$$\begin{aligned} 8x^3 + 12x^2 - 16x &= 4x \cdot 2x^2 + 4x \cdot 3x - 4x \cdot 4 \\ &= 4x(2x^2 + 3x - 4) \end{aligned}$$

Check: $4x(2x^2 + 3x - 4) = 8x^3 + 12x^2 - 16x$

EXAMPLE 15 Factor $45x^2 - 30x + 5$.

Solution: The GCF is 5.

$$45x^2 - 30x + 5 = 5 \cdot 9x^2 - 5 \cdot 6x + 5 \cdot 1$$
$$= 5(9x^2 - 6x + 1)$$

EXAMPLE 16 Factor $4x^3 + x^2 + 8x^2y$.

Solution: The GCF is x^2.

$$4x^3 + x^2 + 8x^2y = x^2 \cdot 4x + x^2 \cdot 1 + x^2 \cdot 8y$$
$$= x^2(4x + 1 + 8y)$$

Notice in Examples 15 and 16 that when one of the terms is itself the GCF, we express it in factored form as the product of the term itself and 1.

EXAMPLE 17 Factor $3x(2x - 1) + 4(2x - 1)$.

Solution: The GCF is $(2x - 1)$. Factoring out the GCF gives

$$3x(2x - 1) + 4(2x - 1) = (3x + 4)(2x - 1)$$

The answer may also be expressed as $(2x - 1)(3x + 4)$ by the commutative property of multiplication.

EXAMPLE 18 Factor $x(x + 3) - 5(x + 3)$.

Solution: The GCF is $(x + 3)$. Factoring out the GCF gives

$$x(x + 3) - 5(x + 3) = (x - 5)(x + 3)$$

EXAMPLE 19 Factor $3x^2 + 5xy + 7y^2$.

Solution: The only factor common to all three terms is 1. Whenever the only factor common to all the terms in a polynomial is 1, the polynomial cannot be factored by the method presented in this section.

Whenever you are factoring a polynomial by any of the methods presented in this chapter, the first step will always be to see if there is a common factor (other than 1) to all the terms in the polynomial. If so, factor the greatest common factor from each term using the distributive property.

Helpful Hint

Checking a Factoring Problem

Every factoring problem may be checked by multiplying the factors. The product of the factors should be identical to the expression that was originally factored. You should check all factoring problems.

Exercise Set 5.1

Write each number as a product of prime numbers.

1. 36
2. 60
3. 90
4. 180
5. 200
6. 96

Find the greatest common factor for the two given numbers.

7. 20, 24
8. 45, 27
9. 60, 84
10. 120, 96
11. 72, 90
12. 76, 68

Find the greatest common factor for each set of terms.

13. x^3, x, x^2
14. x^2, x^5, x^7
15. $3x, 6x^2, 9x^3$
16. $6p, 4p^2, 8p^3$
17. x, y, z
18. xy, x, x^2
19. xy, xy^2, xy^3
20. $x^2y, x^3y, 4x$
21. $x^3y^7, x^7y^{12}, x^5y^5$
22. $6x, 12y, 18x^2$
23. $-5, 20x, 30x^2$
24. $18r^4, 6r^2s, 9s^2$
25. $9x^3y^4, 8x^2y^4, 12x^4y^2$
26. $16x^9y^{12}, 8x^5y^3, 20x^4y^2$
27. $40x^3, 27x, 30x^4y^2$
28. $-8x^2, -9x^3, 12xy^3$
29. $2(x + 3), 3(x + 3)$
30. $4(x - 5), 3x(x - 5)$
31. $x^2(2x - 3), 5(2x - 3)$
32. $x(2x + 5), 2x + 5$
33. $3x - 4, y(3x - 4)$
34. $x(x + 7), x + 7$
35. $x + 4, y(x + 4)$
36. $y(x - 2), 3(x - 2)$

Factor the GCF from each term in the expression. If an expression cannot be factored, so state.

37. $3x + 6$
38. $4x + 2$
39. $15x - 5$
40. $12x + 15$
41. $13x + 5$
42. $6x^2 + 3x$
43. $9x^2 - 12x$
44. $24y - 6y^2$
45. $26p^2 - 8p$
46. $8x + 16x^2$
47. $4x^3 - 6x$
48. $7x^5 - 9x^4$
49. $36x^{12} - 24x^8$
50. $45y^{12} + 30y^{10}$
51. $24y^{15} - 9y^3$
52. $38x^4 - 16x^5$
53. $x + 3xy^2$
54. $2x^2y - 6x$
55. $6x + 5y$
56. $3x^2y + 6x^2y^2$
57. $16xy^2z + 4x^3y$
58. $80x^5y^3z^4 - 36x^2yz^3$
59. $34x^2y^2 + 16xy^4$
60. $42xy^6z^{12} - 18y^4z^2$
61. $36xy^2z^3 + 36x^3y^2z$
62. $19x^4y^{12}z^{13} - 8x^5y^3z^9$
63. $14y^3z^5 - 9xy^3z^5$
64. $7x^4y^9 - 21x^3y^7z^5$
65. $3x^2 + 6x + 9$
66. $x^3 + 6x^2 - 4x$
67. $9x^2 + 18x + 3$
68. $4x^2 - 16x + 24$
69. $4x^3 - 8x^2 + 12x$
70. $12x^2 - 9x + 9$
71. $35x^2 - 16x + 10$
72. $5x^3 - 7x^2 + x$
73. $15p^2 - 6p + 9$
74. $35y^3 - 7y^2 + 14y$
75. $24x^6 + 8x^4 - 4x^3$
76. $44x^5y + 11x^3y + 22x^2$
77. $8x^2y + 12xy^2 + 9xy$
78. $52x^2y^2 + 16xy^3 + 26z$
79. $x(x + 4) + 3(x + 4)$
80. $5x(2x - 5) + 3(2x - 5)$
81. $7x(4x - 3) - 4(4x - 3)$
82. $3x(7x + 1) - 2(7x + 1)$
83. $4x(2x + 1) + 1(2x + 1)$
84. $3x(4x - 5) + 1(4x - 5)$
85. $4x(2x + 1) + 2x + 1$
86. $3x(4x - 5) + 4x - 5$

87. What is a factored expression?

88. What is the greatest common factor of two or more numbers?

89. In your own words, explain how to factor a monomial from a polynomial.

90. How may any factoring problem be checked?

CUMULATIVE REVIEW EXERCISES

[2.1] **91.** Simplify $3x - (x - 6) + 4(3 - x)$.

[2.5] **92.** Solve the equation $2(x + 3) - x = 5x + 2$.

[3.1] **93.** If $A = P(1 + rt)$, find r when $A = 1000$, $t = 2$, and $P = 500$.

94. Solve the formula $A = \frac{1}{2}bh$ for h.

Group Activity/ Challenge Problems

1. Factor $4x^2(x - 3)^3 - 6x(x - 3)^2 + 4(x - 3)$.
2. Factor $6x^5(2x + 7) + 4x^3(2x + 7) - 2x^2(2x + 7)$.
3. Factor $x^{7/3} + 5x^{4/3} + 6x^{1/3}$. Begin by factoring $x^{1/3}$ from all three terms.
4. Factor $15x^{1/2} + 5x^{-1/2}$
5. Factor $x^2 + 2x + 3x + 6$. *Hint:* Factor the first two terms, then factor the last two terms, then factor the resulting two terms. We will discuss factoring problems of this type in Section 5.2.
6. Consider the expression

   $1 + 2 - 3 + 4 + 5 - 6 + 7 + 8 - 9 + 10 + 11 - 12 + 13 + 14 - 15$

   (a) Construct groups of three terms (for example, the first group is $1 + 2 - 3$), and write the sum of each group as a product of 3 and another factor [for example, the first group would be 3(0)].

   (b) Factor out the common factor of 3.

   (c) Find the sum of the numbers.

   (d) Use the procedure above to find the sum of the numbers if the process above was continued until . . . $+ 31 + 32 - 33$.

5.2 Factoring by Grouping

Tape 8

1 Factor a polynomial containing four terms by grouping.

Factoring by Grouping

1 It may be possible to factor a polynomial containing four or more terms by removing common factors from groups of terms. This process is called *factoring by grouping.* In Section 5.4 we discuss factoring trinomials. One of the methods we will use requires a knowledge of factoring by grouping. Example 1 illustrates the procedure for factoring by grouping.

EXAMPLE 1 Factor $ax + ay + bx + by$.

Solution: There is no factor (other than 1) common to all four terms. However, a is common to the first two terms and b is common to the last two terms. Factor a from the first two terms and b from the last two terms.

$$ax + ay + bx + by = a(x + y) + b(x + y)$$

This factoring gives two terms, and $(x + y)$ is common to both terms. Proceed to factor $(x + y)$ from each term, as shown below.

$$a(x + y) + b(x + y) = (a + b)(x + y)$$

Notice when $(x + y)$ is factored out we are left with $a + b$, which becomes the other factor, $(a + b)$. Thus, $ax + ay + bx + by = (a + b)(x + y)$.

To Factor a Four-Term Polynomial Using Grouping

1. Determine if there are any factors common to all four terms. If so, factor the greatest common factor from each of the four terms.
2. If necessary, arrange the four terms so that the first two terms have a common factor and the last two have a common factor.
3. Use the distributive property to factor each group of two terms.
4. Factor the greatest common factor from the results of step 3.

EXAMPLE 2 Factor $x^2 + 3x + 4x + 12$ by grouping.

Solution: No factor is common to all four terms. However, you can factor x from the first two terms and 4 from the last two terms.

$$x^2 + 3x + 4x + 12 = x(x + 3) + 4(x + 3)$$

Notice that the factor $(x + 3)$ is common to both terms on the right. Factor out the $(x + 3)$ using the distributive property.

$$x(x + 3) + 4(x + 3) = (x + 4)(x + 3)$$

Thus, $x^2 + 3x + 4x + 12 = (x + 4)(x + 3)$.

In Example 2, the $3x$ and $4x$ are like terms and may be combined. However, since we are explaining how to factor four terms by grouping we will not combine them. Some four-term polynomials, such as in Example 9, have no like terms that can be combined. When we discuss factoring trinomials in Section 5.4 we will sometimes start with a trinomial and rewrite it using four terms. For example, we may start with a trinomial like $2x^2 + 11x + 12$ and rewrite it as $2x^2 + 8x + 3x + 12$. We then factor the resulting four terms by grouping. This is one method that can be used to factor trinomials, as will be explained later.

EXAMPLE 3 Factor $6x^2 + 9x + 8x + 12$ by grouping.

Solution:
$$\begin{aligned} 6x^2 + 9x + 8x + 12 &= 3x(2x + 3) + 4(2x + 3) \\ &= (3x + 4)(2x + 3) \end{aligned}$$

A factoring by grouping problem can be checked by multiplying the factors using the FOIL method. If you have not made a mistake, your result will be the polynomial you began with. Here is a check of Example 3.

Check:
$$\begin{aligned} (3x + 4)(2x + 3) &= \overset{F}{(3x)(2x)} + \overset{O}{(3x)(3)} + \overset{I}{(4)(2x)} + \overset{L}{(4)(3)} \\ &= 6x^2 + 9x + 8x + 12 \end{aligned}$$

Because this is the polynomial we started with, the factoring is correct.

EXAMPLE 4 Factor $6x^2 + 8x + 9x + 12$ by grouping.

Solution: $6x^2 + 8x + 9x + 12 = 2x(3x + 4) + 3(3x + 4)$

$$= (2x + 3)(3x + 4)$$

Notice that Example 4 is the same as Example 3 with the two middle terms switched. The answers to examples 3 and 4 are the same. When factoring by grouping, the two like terms may be switched and the answer will remain the same.

EXAMPLE 5 Factor $x^2 + 3x + x + 3$ by grouping.

Solution: In the first two terms, x is the common factor. Is there a common factor in the last two terms? Yes; remember that 1 is a factor of every term. Factor 1 from the last two terms.

$$\begin{aligned} x^2 + 3x + x + 3 &= x^2 + 3x + 1 \cdot x + 1 \cdot 3 \\ &= x(x + 3) + 1(x + 3) \\ &= (x + 1)(x + 3) \end{aligned}$$

Note that $x + 3$ was expressed as $1(x + 3)$.

EXAMPLE 6 Factor $4x^2 - 2x - 2x + 1$ by grouping.

Solution: When $2x$ is factored from the first two terms, we get

$$4x^2 - 2x - 2x + 1 = 2x(2x - 1) - 2x + 1$$

What should we factor from the last two terms? We wish to factor $-2x + 1$ in such a manner that we end up with an expression that is a multiple of $(2x - 1)$. **Whenever we wish to change the sign *of each term of an expression, we can factor out a negative number from each term.*** In this case we factor out -1.

$$-2x + 1 = -1(2x - 1)$$

Now rewrite $-2x + 1$ as $-1(2x - 1)$.

$$2x(2x - 1) -2x + 1 = 2x(2x - 1) -1(2x - 1)$$

Now factor out the common factor $(2x - 1)$.

$$2x(2x - 1) - 1(2x - 1) = (2x - 1)(2x - 1) \text{ or } (2x - 1)^2$$

EXAMPLE 7 Factor $x^2 + 3x - x - 3$ by grouping.

Solution: $x^2 + 3x - x - 3 = x(x + 3) - x - 3$

$$\begin{aligned} &= x(x + 3) - 1(x + 3) \\ &= (x - 1)(x + 3) \end{aligned}$$

Note that we factored -1 from $-x - 3$ to get $-1(x + 3)$.

EXAMPLE 8 Factor $3x^2 - 6x - 4x + 8$ by grouping.

Solution:
$$\begin{aligned} 3x^2 - 6x - 4x + 8 &= 3x(x - 2) - 4(x - 2) \\ &= (3x - 4)(x - 2) \end{aligned}$$

Note: $-4x + 8 = -4(x - 2)$.

Helpful Hint

When factoring four terms by grouping, if the coefficient of the third term is positive, as in Examples 2 through 5, you will generally factor out a positive coefficient from the last two terms. *If the coefficient of the third term is negative,* as in Examples 6 through 8, *you will generally factor out a negative coefficient from the last two terms.*

In the examples illustrated so far, the two middle terms have been like terms. This need not be the case, as illustrated in Example 9.

EXAMPLE 9 Factor by grouping $xy + 3x - 2y - 6$.

Solution: This problem contains two variables, x and y. The procedure to factor here is basically the same as before. Factor x from the first two terms and -2 from the last two terms.

$$\begin{aligned} xy + 3x - 2y - 6 &= x(y + 3) - 2(y + 3) \\ &= (x - 2)(y + 3) \end{aligned}$$

EXAMPLE 10 Factor $2x^2 + 4xy + 3xy + 6y^2$.

Solution: We will factor out $2x$ from the first two terms and $3y$ from the last two terms.

$$2x^2 + 4xy + 3xy + 6y^2 = 2x(x + 2y) + 3y(x + 2y)$$

Now factor out the common factor $(x + 2y)$ from each term on the right.

$$2x(x + 2y) + 3y(x + 2y) = (2x + 3y)(x + 2y)$$

Check:
$$\begin{aligned} (2x + 3y)(x + 2y) &= \overset{F}{(2x)(x)} + \overset{O}{(2x)(2y)} + \overset{I}{(3y)(x)} + \overset{L}{(3y)(2y)} \\ &= 2x^2 + 4xy + 3xy + 6y^2 \end{aligned}$$

If Example 10 were given as $2x^2 + 3xy + 4xy + 6y^2$, would the results be the same? Try it and see.

EXAMPLE 11 Factor $6r^2 - 9rs + 8rs - 12s^2$.

Solution: Factor $3r$ from the first two terms and $4s$ from the last two terms.

$$\begin{aligned} 6r^2 - 9rs + 8rs - 12s^2 &= 3r(2r - 3s) + 4s(2r - 3s) \\ &= (3r + 4s)(2r - 3s) \end{aligned}$$

EXAMPLE 12 Factor $3x^2 - 15x + 6x - 30$.

Solution: *The first step in any factoring problem is to determine if all the terms have a common factor. If so, we factor out that common factor.* In this polynomial, 3 is common to every term. Therefore, begin by factoring out the 3.

$$3x^2 - 15x + 6x - 30 = 3(x^2 - 5x + 2x - 10)$$

Now we factor the expression in parentheses by grouping. Factor out x from the first two terms and 2 from the last two terms.

$$\begin{aligned} 3(x^2 - 5x + 2x - 10) &= 3[x(x - 5) + 2(x - 5)] \\ &= 3[(x + 2)(x - 5)] \\ &= 3(x + 2)(x - 5) \end{aligned}$$

Thus, $3x^2 - 15x + 6x - 30 = 3(x + 2)(x - 5)$.

Exercise Set 5.2

Factor by grouping.

1. $x^2 + 4x + 3x + 12$
2. $x^2 + 5x + 2x + 10$
3. $x^2 + 2x + 4x + 8$
4. $x^2 - x + 3x - 3$
5. $x^2 + 2x + 5x + 10$
6. $x^2 - 2x + 4x - 8$
7. $x^2 + 3x - 5x - 15$
8. $x^2 + 3x - 2x - 6$
9. $4x^2 + 6x - 6x - 9$
10. $4x^2 - 6x + 6x - 9$
11. $3x^2 + 9x + x + 3$
12. $x^2 + 4x + x + 4$
13. $4x^2 - 2x - 2x + 1$
14. $2x^2 + 6x - x - 3$
15. $8x^2 + 32x + x + 4$
16. $8x^2 - 4x - 2x + 1$
17. $3x^2 - 2x + 3x - 2$
18. $35x^2 + 21x - 40x - 24$
19. $2x^2 - 4x - 3x + 6$
20. $35x^2 - 40x + 21x - 24$
21. $15x^2 - 9x + 25x - 15$
22. $10x^2 - 15x - 8x + 12$
23. $x^2 + 2xy - 3xy - 6y^2$
24. $x^2 - 3xy + 2xy - 6y^2$
25. $6x^2 - 9xy + 2xy - 3y^2$
26. $3x^2 - 18xy + 4xy - 24y^2$
27. $10x^2 - 12xy - 25xy + 30y^2$
28. $12x^2 - 9xy + 4xy - 3y^2$
29. $x^2 + bx + ax + ab$
30. $xy + 3x + 2y + 6$
31. $xy + 4x - 2y - 8$
32. $x^2 - 2x + ax - 2a$
33. $a^2 + 2a + ab + 2b$
34. $2x^2 - 8x + 3xy - 12y$
35. $xy - x + 5y - 5$
36. $y^2 - yb + ya - ab$
37. $12 + 8y - 3x - 2xy$
38. $2y - 6 - xy + 3x$
39. $a^3 + 2a^2 + a + 2$
40. $x^3 - 3x^2 + 2x - 6$
41. $x^3 + 4x^2 - 3x - 12$
42. $y^3 + 2y^2 - 4y - 8$
43. $2x^2 - 12x + 8x - 48$
44. $3x^2 - 3x - 3x + 3$
45. $4x^2 + 8x + 8x + 16$
46. $2x^3 - 5x^2 - 6x^2 + 15x$
47. $6x^3 + 9x^2 - 2x^2 - 3x$
48. $9x^3 + 6x^2 - 45x^2 - 30x$
49. $2x^2 - 4xy + 8xy - 16y^2$
50. $18x^2 + 27xy + 12xy + 18y^2$

Rearrange the terms so that the first two terms have a common factor and the last two terms have a common factor (other than 1). Then factor by grouping. There may be more than one way to rearrange the factors. However, the answer should be the same regardless of the arrangement selected.

51. $3x + 2y + 6 + xy$

52. $5a + 3y + ay + 15$

53. $6x + 5y + xy + 30$

54. $ax - 12 - 3x + 4a$

55. $ax + by + ay + bx$

56. $ax - 8 - 2a + 4x$

57. $cd - 12 - 4d + 3c$

58. $ca - 2b + 2a - cb$

59. $ac - bd - ad + bc$

60. $ax + 2by + 2bx + ay$

61. What is the first step in any factoring by grouping problem?

62. How can you check the solution to a factoring by grouping problem?

63. A polynomial of four terms is factored by grouping and the result is $(x - 2)(x + 4)$. Find the polynomial that was factored, and explain how you determined the answer.

64. A polynomial of four terms is factored by grouping and the result is $(x - 2y)(x - 3)$. Find the polynomial that was factored, and explain how you determined the answer.

CUMULATIVE REVIEW EXERCISES

[3.1] **65.** The diameter of a willow tree grows about 3.5 inches per year. What is the approximate age of a willow tree whose diameter is 25 inches?

[3.3] **66.** Steve takes the Transit Authority bus to and from work each day. The one-way fare is \$1.25. The Transit Authority offers a monthly bus pass for \$52, which provides unlimited rides. How many days would Steve have to travel to and from work in a month to make it worthwhile for him to purchase the monthly bus pass?

[4.6] **67.** Divide $\dfrac{15x^3 - 6x^2 - 9x + 5}{3x}$.

68. Divide $\dfrac{x^2 - 9}{x - 3}$.

SEE EXERCISE 65.

Group Activity/ Challenge Problems

Factor by grouping.

1. $3x^5 - 15x^3 + 2x^3 - 10x$

2. $x^3 + xy - x^2y - y^2$

3. $18a^2 + 3ax^2 - 6ax - x^3$

4. $4a^4b^4 + 12a^3b^2 - 8a^2b^3 - 24ab$

In Section 5.5 we will factor trinomials of the form $ax^2 + bx + c$, $a \neq 0$ using grouping. To do this we rewrite the middle term of the trinomial, bx, as a sum or difference of two terms. Then we factor the resulting polynomial of four terms by grouping. For each of the following (a) rewrite the trinomial as a polynomial of four terms by replacing the bx term with the sum or difference given. (b) Factor the polynomial of four terms. Note that the factors obtained are the factors of the trinomial.

5. $3x^2 + 10x + 8$, $10x = 6x + 4x$

6. $3x^2 + 10x + 8$, $10x = 4x + 6x$

7. $2x^2 - 11x + 15$, $-11x = -6x - 5x$

8. $2x^2 - 11x + 15$, $-11x = -5x - 6x$

9. $4x^2 - 17x - 15$, $-17x = -20x + 3x$

10. $4x^2 - 17x - 15$, $-17x = 3x - 20x$

5.3 Factoring Trinomials with $a = 1$

Tape 8

1. Factor trinomials of the form $ax^2 + bx + c$, where $a = 1$.
2. Understand the trial and error method of factoring trinomials.
3. Remove a common factor from the trinomial before factoring a trinomial.

An Important Note Regarding Factoring Trinomials

Factoring trinomials is important in algebra, higher-level mathematics, physics, and other science courses. Because it is important, you need to study and learn Sections 5.3 and 5.4 well. *To be successful in Chapter 6, you must be able to factor trinomials. You should study and learn Sections 5.3 and 5.4 well, so that when you need to factor trinomials in Chapter 6 you will not have to look back to this chapter to remember how to factor them.*

In this section we learn to factor trinomials of the form $ax^2 + bx + c$, where a, the numerical coefficient of the squared term, is 1. That is, we will be factoring trinomials of the form $x^2 + bx + c$. One example of this type of trinomial is $x^2 + 5x + 6$. Recall that x^2 means $1x^2$.

In Section 5.4 we will learn to factor trinomials of the form $ax^2 + bx + c$, where $a \neq 1$. One example of this type of trinomial is $2x^2 + 7x + 3$.

Factoring Trinomials

1 Now we discuss how to factor trinomials of the form $ax^2 + bx + c$, where a, the numerical coefficient of the squared term, is 1. Examples of such trinomials are

$$x^2 + 7x + 12 \qquad\qquad x^2 - 2x - 24$$
$$a = 1, b = 7, c = 12 \qquad\qquad a = 1, b = -2, c = -24$$

Recall that factoring is the reverse process of multiplication. We can show with the FOIL method that

$$(x + 3)(x + 4) = x^2 + 7x + 12 \quad \text{and} \quad (x - 6)(x + 4) = x^2 - 2x - 24$$

Therefore, $x^2 + 7x + 12$ and $x^2 - 2x - 24$ factor as follows:

$$x^2 + 7x + 12 = (x + 3)(x + 4) \quad \text{and} \quad x^2 - 2x - 24 = (x - 6)(x + 4)$$

Notice that each of these trinomials when factored results in the product of two binomials in which the first term of each binomial is x and the second term is a number (including its sign). In general, when we factor a trinomial of the form $x^2 + bx + c$ we will get a pair of binomial factors as follows:

$$x^2 + bx + c = (x + \square)(x + \square)$$

numbers go here

If, for example, we find that the numbers that go in the shaded areas of the factors are 4 and -6, the factors are written $(x + 4)$ and $(x - 6)$. Notice that instead of listing the second factor as $(x + (-6))$, we list it as $(x - 6)$.

The procedure that follows is one method to determine the numbers to place in the shaded areas when factoring a given trinomial of the form $x^2 + bx + c$.

In Section 4.5 we illustrated how the FOIL method is used to multiply two binomials. Let us multiply $(x + 3)(x + 4)$ using the FOIL method.

$$(x+3)(x+4) \overset{\text{F} \quad \text{O} \quad \text{I} \quad \text{L}}{= x^2 + 4x + 3x + 12}$$
$$= x^2 + \quad 7x \quad + 12$$

We see that $(x + 3)(x + 4) = x^2 + 7x + 12$.

Note that the *sum of the outer and inner terms is 7x* and the *product of the last terms is 12.* To factor $x^2 + 7x + 12$, we look for two numbers whose product is 12 and whose sum is 7. List the factors of 12 first and then list the sum of the factors.

Factors of 12	*Sum of Factors*
$(1)(12) = 12$	$1 + 12 = 13$
$(2)(6) = 12$	$2 + 6 = 8$
$(3)(4) = 12$	$3 + 4 =$
$(-1)(-12) = 12$	$-1 + (-12) = -13$
$(-2)(-6) = 12$	$-2 + (-6) = -8$
$(-3)(-4) = 12$	$-3 + (-4) = -7$

The only factors of 12 whose sum is a positive 7 are 3 and 4. The factors of $x^2 + 7x + 12$ will therefore be $(x + 3)$ and $(x + 4)$.

$$x^2 + 7x + 12 = (x + 3)(x + 4)$$

In the previous illustration all the possible factors of 12 were listed so that you could see them. However, when working a problem, once you find the specific factors you are seeking you need go no further.

To Factor Trinomials of the Form $ax^2 + bx + c$, where $a = 1$.

1. Find two numbers whose product equals the constant, c, and whose sum equals the coefficient, b.

2. Use the two numbers found in step 1, including their signs, to write the trinomial in factored form. The trinomial in factored form will be

$$(x + \text{one number})(x + \text{second number})$$

How do we find the two numbers mentioned in steps 1 and 2? The sign of the constant, c, is a key in finding the two numbers. *The Helpful Hint that follows is very important and useful. Study it carefully.*

Helpful Hint

When asked to factor a trinomial of the form $x^2 + bx + c$, the first thing you should do is to observe the sign of the constant.

(a) If the constant, c, is positive, both numbers in the factors will have the same sign, either both positive or both negative. Furthermore, that common sign will be the same as b, the sign of the coefficient of the x term of the trinomial being factored.

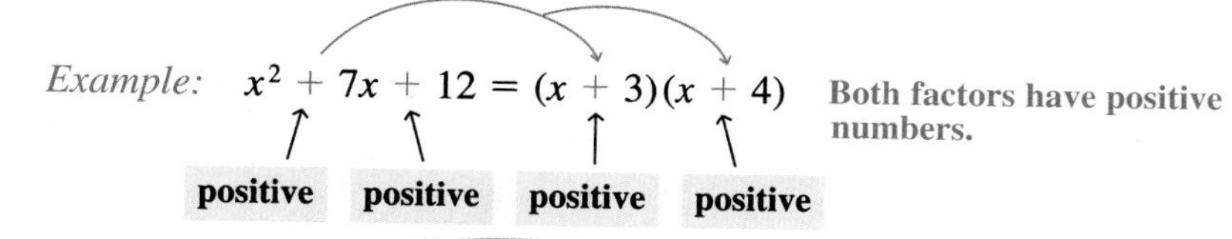

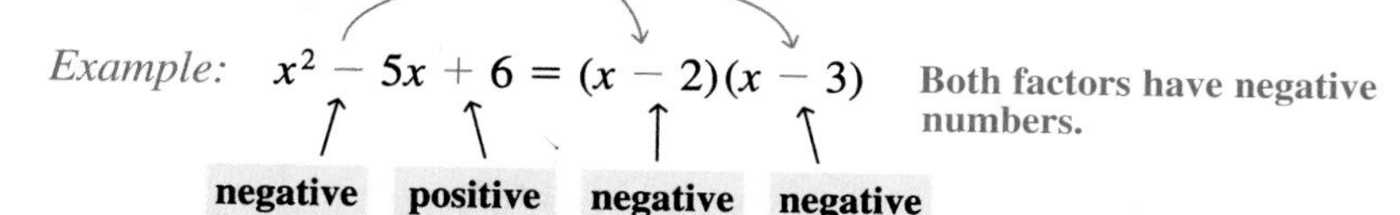

(b) If the constant is negative, the two numbers in the factors will have opposite signs. That is, one number will be positive and the other number will be negative.

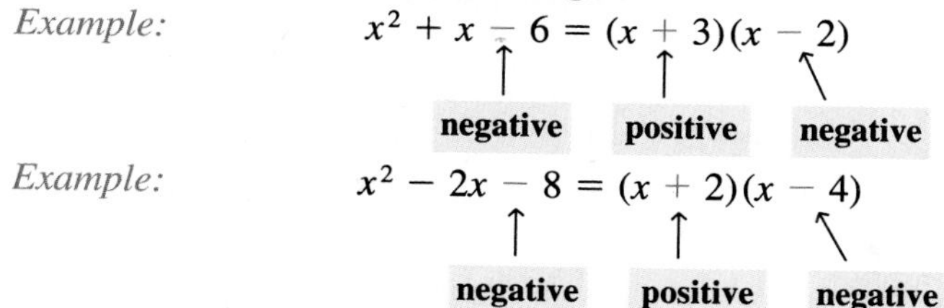

We will use this information as a starting point when factoring trinomials.

EXAMPLE 1 Consider a trinomial of the form $x^2 + bx + c$. Use the signs of b and c given below to determine the signs of the numbers in the factors when the trinomial is factored.

(a) b is negative and c is positive.
(b) b is negative and c is negative.
(c) b is positive and c is negative.
(d) b is positive and c is positive.

Solution:

(a) Since the constant, c, is positive, both numbers must have the same sign. Since the coefficient of the x term, b, is negative, both factors will contain negative numbers.

(b) Since the constant, c, is negative, one factor will contain a positive number and the other will contain a negative number.

(c) Since the constant, c, is negative, one factor will contain a positive number and the other will contain a negative number.

(d) Since the constant, c, is positive, both numbers must have the same sign. Since the coefficient of the x term, b, is positive, both factors will contain positive numbers.

EXAMPLE 2 Factor $x^2 + x - 6$.

Solution: We must find two numbers whose product is the constant, -6, and whose sum is the coefficient of the x term, 1. Remember that x means $1x$. Since the constant is negative, one number must be positive and the other negative. Recall that the product of two numbers with unlike signs is a negative number. We now list the factors of -6 and look for the two factors whose sum is 1.

Factors of −6	*Sum of Factors*
$1(-6) = -6$	$1 + (-6) = -5$
$2(-3) = -6$	$2 + (-3) = -1$
$3(-2) = -6$	$3 + (-2) = 1$
$6(-1) = -6$	$6 + (-1) = 5$

Note that the factors 1 and -6 in the top row are different from the factors -1 and 6 in the bottom row, and their sums are different.

The numbers 3 and -2 have a product of -6 and a sum of 1. Thus, the factors are $(x + 3)$ and $(x - 2)$.

$$x^2 + x - 6 = (x + 3)(x - 2)$$

The order of the factors is not crucial. Therefore, $x^2 + x - 6 = (x - 2)(x + 3)$ is also an acceptable answer.

As mentioned earlier, **trinomial factoring problems can be checked by multiplying the factors using the FOIL method.** If the factoring is correct, the product obtained using the FOIL method will be identical to the original trinomial. Let us check the factors obtained in Example 2.

Check: $(x + 3)(x - 2) = x^2 - 2x + 3x - 6 = x^2 + x - 6$

Since the product of the factors is the original trinomial, the factoring was correct.

EXAMPLE 3 Factor $x^2 - x - 6$.

Solution: The factors of -6 are illustrated in Example 2. The factors whose product is -6 and whose sum is -1 are 2 and -3.

Factors of −6	*Sum of Factors*
$2(-3) = -6$	$2 + (-3) = -1$

Therefore, $x^2 - x - 6 = (x + 2)(x - 3)$.

EXAMPLE 4 Factor $x^2 - 5x + 6$.

Solution: We must find two numbers whose product is 6 and whose sum is -5. Since the constant, 6, is positive, both factors must have the same sign. Since the coefficient of the x term, -5, is negative, both numbers must be negative. Recall that the product of a negative number and a negative number is positive. We now list the negative factors of 6 and look for the pair whose sum is -5.

Factors of 6	*Sum of Factors*
$(-1)(-6)$	$-1 + (-6) = -7$
$(-2)(-3)$	$-2 + (-3) = -5$

The factors of 6 whose sum is -5 are -2 and -3.

$$x^2 - 5x + 6 = (x - 2)(x - 3)$$

EXAMPLE 5 Factor $x^2 + 2x - 8$.

Solution: We must find the factors of -8 whose sum is 2. Since the constant is negative, one factor will be positive and the other factor will be negative.

Factors of −8	*Sum of Factors*
$(1)(-8)$	$1 + (-8) = -7$
$(2)(-4)$	$2 + (-4) = -2$
$(4)(-2)$	$4 + (-2) = 2$
$(8)(-1)$	

Since we have found the two numbers, 4 and -2, whose product is -8 and whose sum is 2, we need go no further.

$$x^2 + 2x - 8 = (x + 4)(x - 2)$$

EXAMPLE 6 Factor $x^2 - 6x + 9$.

Solution: We must find the factors of 9 whose sum is -6. Both factors must be negative. Can you explain why? The two factors whose product is 9 and whose sum is -6 are -3 and -3.

$$\begin{aligned} x^2 - 6x + 9 &= (x - 3)(x - 3) \\ &= (x - 3)^2 \end{aligned}$$

EXAMPLE 7 Factor $x^2 - 4x - 60$.

Solution: We must find two numbers whose product is -60 and whose sum is -4. Since the constant is negative, one factor must be positive and the other negative. The desired factors are -10 and 6 because $(-10)(6) = -60$ and $-10 + 6 = -4$.

$$x^2 - 4x - 60 = (x - 10)(x + 6)$$

EXAMPLE 8 Factor $x^2 + 4x + 12$.

Solution: Let us first find the two numbers whose product is 12 and whose sum is 4. Since both the constant and x term are positive, the two numbers must also be positive.

Factors of 12	*Sum of Factors*
(1)(12)	$1 + 12 = 13$
(2)(6)	$2 + 6 = 8$
(3)(4)	$3 + 4 = 7$

Note that there are no two integers whose product is 12 and whose sum is 4. When two integers cannot be found to satisfy the given conditions, the trinomial cannot be factored by the method presented in this section. Therefore, we write "*cannot be factored*" as our answer.

When factoring a trinomial of the form $x^2 + bx + c$, there is at most one pair of numbers whose product is c and whose sum is b. For example, when factoring $x^2 - 2x - 24$, the two numbers whose product is -24 and whose sum is -2 are -6 and 4. No other pair of numbers will satisfy these specific conditions. Thus, the only factors of $x^2 - 2x - 24$ are $(x - 6)(x + 4)$.

A slightly different type of problem is illustrated in Example 9.

EXAMPLE 9 Factor $x^2 + 2xy + y^2$.

Solution: In this problem the second term contains two variables, x and y, and the last term is not a constant. The procedure used to factor this trinomial is similar to that outlined previously. You should realize, however, that the product of the first terms of the factors we are looking for must be x^2, and the product of the last terms of the factors must be y^2.

We must find two numbers whose product is 1 (from $1y^2$) and whose sum is 2 (from $2xy$). The two numbers are 1 and 1. Thus

$$x^2 + 2xy + y^2 = (x + 1y)(x + 1y) = (x + y)(x + y) = (x + y)^2$$

EXAMPLE 10 Factor $x^2 - xy - 6y^2$.

Solution: Find two numbers whose product is -6 and whose sum is -1. The numbers are -3 and 2. The last terms must be $-3y$ and $2y$ to obtain $-6y^2$.

$$x^2 - xy - 6y^2 = (x - 3y)(x + 2y)$$

Trial and Error Method

2 Another method that can be used to factor trinomials of the form $x^2 + bx + c$ is called the **trial and error method.** With this method we write down factors of the form $(x + \square)(x + \square)$ and then try different sets of factors of the constant, c, in the shaded areas of the parentheses. We multiply each pair of factors using the FOIL method, and continue until we find the pair whose sum of the products of the outer and inner terms is the same as the x term in the trinomial. For example, to factor the trinomial $x^2 - 6x - 16$, we determine the possible factors of -16. Then we try each pair of factors until we obtain a pair whose product from the FOIL method contains $-6x$, the same x term as in the trinomial. We now illustrate how to factor $x^2 - 6x - 16$ using trial and error. Begin by listing the factors of -16. Then list the possible factors of the trinomial and the products of these factors, as indicated in the table below. Finally, determine which, if any, of these products give the correct middle term, $-6x$.

Factor $x^2 - 6x - 16$.

Factors of −16	Possible Factors of Trinomial	Product of Factors
$(16)(-1)$	$(x + 16)(x - 1)$	$x^2 + 15x - 16$
$(8)(-2)$	$(x + 8)(x - 2)$	$x^2 + 6x - 16$
$(4)(-4)$	$(x + 4)(x - 4)$	$x^2 - 16$
$(2)(-8)$	$(x + 2)(x - 8)$	$x^2 - 6x - 16$
$(1)(-16)$	$(x + 1)(x - 16)$	$x^2 - 15x - 16$

In the last column, we find our trinomial in the fourth line. Thus,

$$x^2 - 6x - 16 = (x + 2)(x - 8)$$

Once we found the correct factors we could have stopped and answered the question. All the possible factors were listed here for your benefit.

If you use this method, you should still make use of the information about signs given to you earlier in this section. For example, if the constant is positive, both numbers in the factors will have the same sign, and they will have the sign of the coefficient of the x term. If the constant is negative, one factor will contain a positive number and the other factor will contain a negative number.

When the number of pairs of possible factors of the constant is small, you may wish to use the trial and error method. The trial and error method is discussed further in the next section.

EXAMPLE 11 Factor $x^2 - 14x + 48$ by trial and error.

Solution: Since the constant, 48, is positive and the x term, $-14x$, is negative, both factors of 48 must be negative. Therefore, we will list only the negative factors in the table.

Factor $x^2 - 14x + 48$

Factors of 48	Possible Factors of Trinomial	Product of Factors
$(-1)(-48)$	$(x - 1)(x - 48)$	$x^2 - 49x + 48$
$(-2)(-24)$	$(x - 2)(x - 24)$	$x^2 - 26x + 48$
$(-3)(-16)$	$(x - 3)(x - 16)$	$x^2 - 19x + 48$
$(-4)(-12)$	$(x - 4)(x - 12)$	$x^2 - 16x + 48$
$(-6)(-8)$	$(x - 6)(x - 8)$	$x^2 - 14x + 48$

In the last column, we find our trinomial in the last line. Thus,

$$x^2 - 14x + 48 = (x - 6)(x - 8).$$

EXAMPLE 12 Factor $x^2 + 3xy - 18y^2$ by trial and error.

Solution: Since the last term of the trinomial is negative, one factor will contain a positive number and the other factor will contain a negative number. We try pairs of factors of -18 until we find the pair that gives the correct middle coefficient of 3. We may disregard the second variable, y, at this time and just work with $x^2 + 3x - 18$. Once we obtain the set of factors we are seeking we include the variable y with the factors.

$$x^2 + 3x - 18 = (x + 6)(x - 3)$$

Thus, $x^2 + 3xy - 18y^2 = (x + 6y)(x - 3y)$.

Note that the sum of the products of the outer and inner terms of $(x + 6y)(x - 3y)$ is the $+\,3xy$ we are seeking.

Remove a Common Factor

3 Sometimes each term of a trinomial has a common factor. When this occurs, factor out the common factor first, as explained in Section 5.1. **Whenever the numerical coefficient of the highest-degree term is not 1, you should check for a common factor.** After factoring out any common factor, you should factor the remaining trinomial by one of the methods presented in this section.

EXAMPLE 13 Factor $2x^2 + 2x - 12$.

Solution: Since the numerical coefficient of the squared term is not 1, we check for a common factor. Because 2 is common to each term of the polynomial, we factor it out.

$$2x^2 + 2x - 12 = 2(x^2 + x - 6) \quad \text{Factor out the common factor.}$$

Now factor the remaining trinomial $x^2 + x - 6$ into $(x + 3)(x - 2)$. Thus,

$$2x^2 + 2x - 12 = 2(x + 3)(x - 2).$$

Note that the trinomial $2x^2 + 2x - 12$ is now completely factored into *three* factors: two binomial factors, $x + 3$ and $x - 2$, and a monomial factor, 2. After 2 has been factored out, it plays no part in the factoring of the remaining trinomial.

EXAMPLE 14 Factor $3x^3 + 24x^2 - 60x$.

Solution: We see that $3x$ divides into each term of the polynomial and therefore is a common factor. After factoring out the $3x$, we factor the remaining trinomial.

$$\begin{aligned} 3x^3 + 24x^2 - 60x &= 3x(x^2 + 8x - 20) && \text{Factor out the common factor.} \\ &= 3x(x + 10)(x - 2) && \text{Factor the remaining trinomial.} \end{aligned}$$

Exercise Set 5.3

Factor each expression. If an expression cannot be factored by a method presented in this section, so state.

1. $x^2 + 3x + 2$ **2.** $x^2 - 7x + 12$ **3.** $x^2 + 6x + 8$

4. $x^2 + 7x + 6$ **5.** $x^2 + 7x + 12$ **6.** $x^2 - x - 6$

7. $x^2 - 7x + 9$
8. $y^2 - 6y + 8$
9. $y^2 - 16y + 15$
10. $x^2 + 3x - 28$
11. $x^2 + x - 6$
12. $p^2 - 3p - 10$
13. $r^2 + 2r - 15$
14. $x^2 - 5x + 8$
15. $b^2 - 11b + 18$
16. $x^2 + 11x - 30$
17. $x^2 - 8x - 15$
18. $x^2 - 10x + 21$
19. $a^2 + 12a + 11$
20. $x^2 + 16x + 64$
21. $x^2 + 13x - 30$
22. $x^2 - 30x - 64$
23. $x^2 + 4x + 4$
24. $x^2 - 4x + 4$
25. $k^2 + 6k + 9$
26. $k^2 - 6k + 9$
27. $x^2 + 10x + 25$
28. $x^2 - 10x - 25$
29. $w^2 - 18w + 45$
30. $x^2 - 11x + 10$
31. $x^2 + 22x - 48$
32. $x^2 - 2x + 8$
33. $x^2 - x - 20$
34. $x^2 - 17x - 60$
35. $y^2 - 9y + 14$
36. $x^2 + 15x + 56$
37. $x^2 + 12x - 64$
38. $x^2 - 18x + 80$
39. $x^2 - 14x + 24$
40. $x^2 - 13x + 36$
41. $x^2 - 2x - 80$
42. $x^2 + 18x + 32$
43. $x^2 - 17x + 60$
44. $x^2 - 15x - 16$
45. $x^2 + 30x + 56$
46. $x^2 - 2xy + y^2$
47. $x^2 - 4xy + 4y^2$
48. $x^2 - 6xy + 8y^2$
49. $x^2 + 8xy + 15y^2$
50. $x^2 - 5xy - 14y^2$

Factor completely.

51. $2x^2 - 12x + 10$
52. $3x^2 - 6x - 24$
53. $5x^2 + 20x + 15$
54. $4x^2 + 12x - 16$
55. $2x^2 - 14x + 24$
56. $3y^2 - 33y + 54$
57. $x^3 - 3x^2 - 18x$
58. $x^3 + 11x^2 - 42x$
59. $2x^3 + 6x^2 - 56x$
60. $3x^3 - 36x^2 + 33x$
61. $x^3 + 4x^2 + 4x$
62. $2x^3 - 12x^2 + 10x$

For each trinomial, determine the signs *that will appear in the binomial factors. Explain how you determined your answers.*

63. $x^2 + 180x + 8000$
64. $x^2 - 180x + 8000$
65. $x^2 + 20x - 8000$
66. $x^2 - 20x - 8000$
67. $x^2 - 240x + 8000$
68. $x^2 + 160x - 8000$

Write the trinomial whose factors are listed. Explain how you determined your answer.

69. $(x - 3)(x - 8)$
70. $(x - 3y)(x + 6y)$
71. $2(x - 5y)(x + y)$
72. $3(x + y)(x - y)$

73. How can a trinomial factoring problem be checked?
74. Explain how to determine the factors when factoring a trinomial of the form $x^2 + bx + c$.

CUMULATIVE REVIEW EXERCISES

75. Solve the equation $2(2x - 3) = 2x + 8$.
76. Multiply $(2x^2 + 5x - 6)(x - 2)$.
77. Divide $3x^2 - 10x - 10$ by $x - 4$.
78. Dr. Kaufman, a chemist, mixes 4 liters of an 18% acid solution with 1 liter of a 26% acid solution. Find the strength of the mixture.
79. Factor by grouping: $3x^2 + 5x - 6x - 10$.

SEE EXERCISE 78.

Group Activity/ Challenge Problems

Factor.

1. $x^2 + 0.6x + 0.08$

2. $x^2 - 0.5x - 0.06$

3. $x^2 + \frac{2}{5}x + \frac{1}{25}$

4. $x^2 - \frac{2}{3}x + \frac{1}{9}$

5. $-x^2 - 6x - 8$

6. $-2x^2 - 5x + 3$

7. $x^2 + 5x - 300$

8. $x^2 - 24x - 256$

9–14. Factor Exercises 63–68 on page 273, respectively.

5.4 Factoring Trinomials with $a \neq 1$

Tape 8

1. Factor trinomials of the form $ax^2 + bx + c$, $a \neq 1$, by trial and error.
2. Factor trinomials of the form $ax^2 + bx + c$, $a \neq 1$, by grouping.

An Important Note To Students

In this section we discuss two methods of factoring trinomials of the form $ax^2 + bx + c$, $a \neq 1$. That is, we will be factoring trinomials whose squared term has a numerical coefficient not equal to 1, after removing any common factors. Examples of trinomials with $a \neq 1$ are

$$2x^2 + 11x + 12 \ (a = 2) \qquad 4x^2 - 3x + 1 \ (a = 4)$$

The methods we discuss are (1) **factoring by trial and error** and (2) **factoring by grouping.** We present two different methods for factoring these trinomials because some students, and some instructors, prefer one method, while others prefer the second method. You may use either method unless your instructor asks you to use a specific method. *We will use the same examples to illustrate both methods so that you can make a comparison. Each method is treated independently of the other. So if your teacher asks you to use a specific method, either factoring by grouping or factoring by trial and error, you need only read the material related to that specific method.* Factoring by trial and error is covered on pages 274–281, and factoring by grouping is covered on pages 281–286.

Method 1: Trial and Error

1 Let us now discuss factoring trinomials of the form $ax^2 + bx + c$, $a \neq 1$, by the trial and error method, introduced in Section 5.3. It may be helpful for you to reread that material before going any further.

Recall that factoring is the reverse of multiplying. Consider the product of the following two binomials:

$$\begin{aligned} & \quad\ \ \text{F} \qquad\quad \text{O} \qquad\quad \text{I} \qquad \text{L} \\ (2x + 3)(x + 5) &= 2x(x) + (2x)(5) + 3(x) + 3(5) \\ &= 2x^2 + 10x + 3x + 15 \\ &= 2x^2 + 13x + 15 \end{aligned}$$

Notice that the product of the first terms of the binomials gives the x-squared term of the trinomial, $2x^2$. Also notice that the product of the last terms of the binomials gives the last term, or constant, of the trinomial, $+ 15$. Finally, notice that the sum of the products of the outer terms and inner terms of the binomials gives the middle term of the trinomial, $+ 13x$. When we factor a trinomial using trial and error, we make use of these important facts. Note that $2x^2 + 13x + 15$ in factored form is $(2x + 3)(x + 5)$.

$$2x^2 + 13x + 15 = (2x + 3)(x + 5)$$

When factoring a trinomial of the form $ax^2 + bx + c$ by trial and error, the product of the first terms in the binomial factors must equal the first term of the trinomial, ax^2. Also, the product of the constants in the binomial factors, including their signs, must equal the constant, c, of the trinomial.

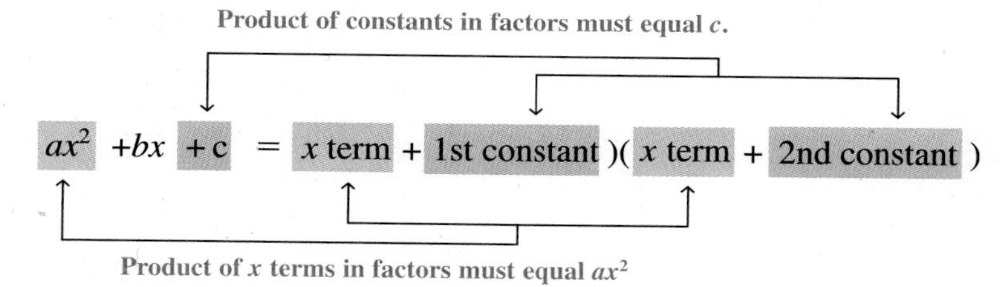

For example, when factoring the trinomial $2x^2 + 7x + 6$, each of the following pairs of factors has a product of the first terms equal to $2x^2$ and a product of the last terms equal to 6.

Trinomial	Possible Factors	Product of First Terms	Product of Last Terms
$2x^2 + 7x + 6$	$(2x + 1)(x + 6)$	$2x(x) = 2x^2$	$1(6) = 6$
	$(2x + 2)(x + 3)$	$2x(x) = 2x^2$	$2(3) = 6$
	$(2x + 3)(x + 2)$	$2x(x) = 2x^2$	$3(2) = 6$
	$(2x + 6)(x + 1)$	$2x(x) = 2x^2$	$6(1) = 6$

Each of these pairs of factors is a possible answer, but only one has the correct factors. How do we determine which is the correct factoring of the trinomial $2x^2 + 7x + 6$? The key lies in the bx term. We know that when we multiply two binomials using the FOIL method the sum of the products of the outer and inner terms gives us the bx term of the trinomial. We use this concept in reverse to determine the correct pair of factors. We need to find the pair of factors whose sum of the products of the outer and inner terms is equal to the bx term of the trinomial.

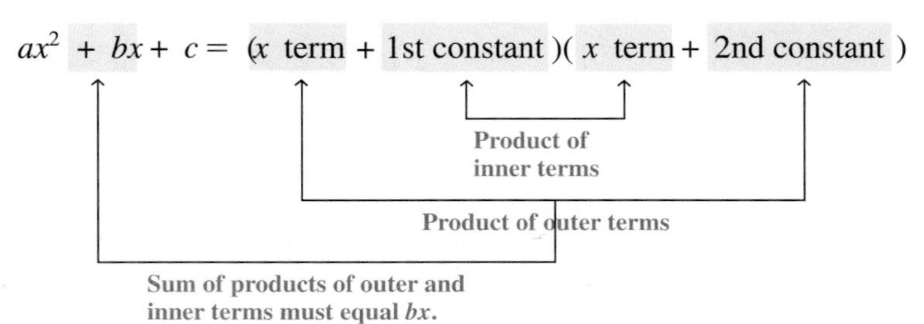

Now look at the possible pairs of factors we obtained for $2x^2 + 7x + 6$ to see if any yield the correct x term, $7x$.

Trinomial	Possible Factors	Product of the First Terms	Product of the Last Terms	Sum of the Products of Outer and Inner Terms
$2x^2 + 7x + 6$	$(2x + 1)(x + 6)$	$2x^2$	6	$2x(6) + 1(x) = 13x$
	$(2x + 2)(x + 3)$	$2x^2$	6	$2x(3) + 2(x) = 8x$
	$(2x + 3)(x + 2)$	$2x^2$	6	$2x(2) + 3(x) = 7x$
	$(2x + 6)(x + 1)$	$2x^2$	6	$2x(1) + 6(x) = 8x$

Since $(2x + 3)(x + 2)$ yields the correct x term, $7x$, the trinomial $2x^2 + 7x + 6$ factors into $(2x + 3)(x + 2)$.

$$2x^2 + 7x + 6 = (2x + 3)(x + 2)$$

We can check this factoring using the FOIL method.

$$\begin{aligned} & \quad\ \ \text{F} \qquad\ \ \text{O} \qquad\ \ \text{I} \qquad \text{L} \\ \textit{Check: } (2x + 3)(x + 2) &= 2x(x) + 2x(2) + 3(x) + 3(2) \\ &= 2x^2 + 4x + 3x + 6 \\ &= 2x^2 + 7x + 6 \end{aligned}$$

Since we obtained the original trinomial, our factoring is correct.

Note in the preceding illustration that $(2x + 1)(x + 6)$ are different factors than $(2x + 6)(x + 1)$, because in one case 1 is paired with $2x$ and in the second case 1 is paired with x. The factors $(2x + 1)(x + 6)$ and $(x + 6)(2x + 1)$ are, however, the same set of factors with their order reversed.

Helpful Hint

When factoring a trinomial of the form $ax^2 + bx + c$, remember that the sign of the constant, c, and the sign of the x term, bx, offer valuable information. When factoring a trinomial by trial and error, first check the sign of the constant. If it is positive, the signs in both factors will be the same as the sign of the bx term. If the constant is negative, one factor will contain a plus sign and the other a negative sign.

Now we outline the procedure to factor trinomials of the form $ax^2 + bx + c$, $a \neq 1$, by trial and error. Keep in mind that the more you practice, the better you will become at factoring.

To Factor Trinomials of the Form $ax^2 + bx + c, a \neq 1$, Using Trial and Error

1. Determine if there is any factor common to all three terms. If so, factor it out.
2. Write all pairs of factors of the coefficient of the squared term, a.
3. Write all pairs of factors of the constant term, c.
4. Try various combinations of these factors until the correct middle term, bx, is found.

When factoring using this procedure, if there is more than one pair of numbers whose product is a, we generally begin with the middle-size pair. We will illustrate the procedure with Examples 1 through 8.

EXAMPLE 1 Factor $2x^2 + 11x + 12$.

Solution: We first determine that all three terms have no common factors other than 1. Since the first term is $2x^2$, one factor must contain a $2x$ and the other an x. Therefore the factors will be of the form $(2x + \quad)(x + \quad)$. Now we must find the numbers to place in the shaded areas. The product of the last terms in the factors must be 12. Since the constant and the coefficient of the x term are both positive, only the positive factors of 12 need be considered. We will list the positive factors of 12, the possible factors of the trinomial, and the sum of the products of the outer and inner terms. Once we find the factors of 12 that yield the proper sum of the products of the outer and inner terms, $11x$, we can write the answer.

Factors of 12	Possible Factors of Trinomial	Sum of the Products of the Outer and Inner Terms
1(12)	$(2x + 1)(x + 12)$	$25x$
2(6)	$(2x + 2)(x + 6)$	$14x$
3(4)	$(2x + 3)(x + 4)$	$11x$
4(3)	$(2x + 4)(x + 3)$	$10x$
6(2)	$(2x + 6)(x + 2)$	$10x$
12(1)	$(2x + 12)(x + 1)$	$14x$

Since the product of $(2x + 3)$ and $(x + 4)$ yields the correct bx term, $11x$, they are the correct factors.

$$2x^2 + 11x + 12 = (2x + 3)(x + 4)$$

In Example 1, our first factor could have been written with an x and the second with a $2x$. Had we done this we still would have obtained the correct answer:

$(x + 4)(2x + 3)$. In Example 1, we could have stopped once we found the pair of factors that yielded the $11x$. Instead, we listed all the factors so that you could study them.

EXAMPLE 2 Factor $5x^2 - 7x - 6$.

Solution: One factor must contain a $5x$ and the other an x. We now list the factors of -6 and look for the pair of factors that yields $-7x$.

Factors of −6	Possible Factors	Sum of the Products of the Outer and Inner Terms
$-1(6)$	$(5x - 1)(x + 6)$	$29x$
$-2(3)$	$(5x - 2)(x + 3)$	$13x$
$-3(2)$	$(5x - 3)(x + 2)$	$7x$
$-6(1)$	$(5x - 6)(x + 1)$	$-x$

Since we did not obtain the desired quantity, $-7x$, by writing the negative factor with the $5x$, we will now try listing the negative factor with the x.

Factors of −6	Possible Factors	Sum of the Products of the Outer and Inner Terms
$1(-6)$	$(5x + 1)(x - 6)$	$-29x$
$2(-3)$	$(5x + 2)(x - 3)$	$-13x$
$3(-2)$	$(5x + 3)(x - 2)$	$-7x$
$6(-1)$	$(5x + 6)(x - 1)$	x

We see that $(5x + 3)(x - 2)$ gives the $-7x$ we are looking for. Thus,

$$5x^2 - 7x - 6 = (5x + 3)(x - 2)$$

Again we listed all the possible combinations for you to study.

Helpful Hint

In Example 2, we were asked to factor $5x^2 - 7x - 6$. When we considered the product of $-3(2)$ in the first set of possible factors, we obtained

Factors of −6	Possible Factors	Sum of the Products of the Outer and Inner Terms
$-3(2)$	$(5x - 3)(x + 2)$	$7x$

Later in the problem we tried the factors $3(-2)$ and obtained the correct answer.

$3(-2)$	$(5x + 3)(x - 2)$	$-7x$

Helpful Hint continued on top of page 279.

When factoring a trinomial with a *negative constant,* if you obtain the bx term whose sign is the opposite of the one you are seeking, *reverse the signs on the constants* in the factors. This should give you the set of factors you are seeking.

EXAMPLE 3 Factor $8x^2 + 33x + 4$.

Solution: There are no factors common to all three terms. Since the first term is $8x^2$, there are a number of possible combinations for the first terms in the factors. Since $8 = 8 \cdot 1$ and $8 = 4 \cdot 2$, the possible factors may be of the form $(8x \quad)(x \quad)$ or $(4x \quad)(2x \quad)$. When this situation occurs, we generally start with the middle-size pair of factors. Thus, we begin with $(4x \quad)(2x \quad)$. If this pair does not lead to the solution, we will then try $(8x \quad)(x \quad)$. We now list the factors of the constant, 4. Since all signs are positive, we list only the positive factors of 4.

Factors of 4	Possible Factors	Sum of the Products of the Outer and Inner Terms
1(4)	$(4x + 1)(2x + 4)$	$18x$
2(2)	$(4x + 2)(2x + 2)$	$12x$
4(1)	$(4x + 4)(2x + 1)$	$12x$

Since we did not obtain the factors with $(4x \quad)(2x \quad)$, we now try $(8x \quad)(x \quad)$.

Factors of 4	Possible Factors	Sum of the Products of the Outer and Inner Terms
1(4)	$(8x + 1)(x + 4)$	$33x$
2(2)	$(8x + 2)(x + 2)$	$18x$
4(1)	$(8x + 4)(x + 1)$	$12x$

Since the product of $(8x + 1)$ and $(x + 4)$ yields the correct x term, $33x$, they are the correct factors.

$$8x^2 + 33x + 4 = (8x + 1)(x + 4)$$

EXAMPLE 4 Factor $4x^2 - 4x + 1$.

Solution: The factors must be of the form $(4x \quad)(x \quad)$ or $(2x \quad)(2x \quad)$. We start with the middle-size factors $(2x \quad)(2x \quad)$. Since the constant is positive and the coefficient of the x term is negative, both factors must be negative.

Factors of 1	Possible Factors	Sum of the Products of the Outer and Inner Terms
$(-1)(-1)$	$(2x - 1)(2x - 1)$	$-4x$

Since we found the correct factors, we can stop.

$$4x^2 - 4x + 1 = (2x - 1)(2x - 1) = (2x - 1)^2$$

EXAMPLE 5 Factor $2x^2 + 3x + 5$.

Solution: The factors will be of the form $(2x \quad)(x \quad)$. We need only consider the positive factors of 5. Can you explain why?

Factors of 5	Possible Factors	Sum of the Products of the Outer and Inner Terms
1(5)	$(2x + 1)(x + 5)$	$11x$
5(1)	$(2x + 5)(x + 1)$	$7x$

Since we have tried all possible combinations and we have not obtained the x term, $3x$, this trinomial cannot be factored.

If you come across a trinomial that cannot be factored, as in Example 5, do not leave the answer blank. Instead, write "cannot be factored." However, before you write the answer "cannot be factored," recheck your work and make sure you have tried every possible combination.

EXAMPLE 6 Factor $4x^2 + 7xy + 3y^2$.

Solution: This trinomial is different from the other trinomials in that the last term is not a constant but contains y^2. Don't let this scare you. The factoring process is the same, except that the second term of both factors will contain y. We begin by considering factors of the form $(2x \quad)(2x \quad)$. If we cannot find the factors, then we try factors of the form $(4x \quad)(x \quad)$.

Factors of 3	Possible Factors	Sum of the Products of the Outer and Inner Terms
1(3)	$(2x + y)(2x + 3y)$	$8xy$
3(1)	$(2x + 3y)(2x + y)$	$8xy$
1(3)	$(4x + y)(x + 3y)$	$13xy$
3(1)	$(4x + 3y)(x + y)$	$7xy$

$$4x^2 + 7xy + 3y^2 = (4x + 3y)(x + y)$$

Check: $(4x + 3y)(x + y) = 4x^2 + 4xy + 3xy + 3y^2 = 4x^2 + 7xy + 3y^2$

EXAMPLE 7 Factor $6x^2 - 13xy - 8y^2$.

Solution: We begin with factors of the form $(3x \quad)(2x \quad)$. If we cannot find the solution from these, we will try $(6x \quad)(x \quad)$. Since the last term, $-8y^2$, is negative, one factor will contain a plus sign and the other will contain a minus sign.

Factors of -8	Possible Factors	Sum of the Products of the Outer and Inner Terms
$1(-8)$	$(3x + y)(2x - 8y)$	$-22xy$
$2(-4)$	$(3x + 2y)(2x - 4y)$	$-8xy$
$4(-2)$	$(3x + 4y)(2x - 2y)$	$2xy$
$8(-1)$	$(3x + 8y)(2x - y)$	$13xy$

We are looking for $-13xy$. When we considered $8(-1)$, we obtained $13xy$. As explained in the Helpful Hint on page 278, if we reverse the signs of the numbers in the factors, we will obtain the factors we are seeking.

$$(3x + 8y)(2x - y) \qquad \text{Gives } 13xy$$

$$(3x - 8y)(2x + y) \qquad \text{Gives } -13xy$$

Therefore, $6x^2 - 13xy - 8y^2 = (3x - 8y)(2x + y)$.

Now we will look at an example in which all the terms of the trinomial have a common factor.

EXAMPLE 8 Factor $4x^3 + 10x^2 + 6x$.

Solution: *The first step in any factoring problem is to determine if all the terms contain a common factor. If so, factor out that common factor first.* In this example, $2x$ is common to all three terms. We begin by factoring out the $2x$. Then we continue factoring by trial and error.

$$\begin{aligned} 4x^3 + 10x^2 + 6x &= 2x(2x^2 + 5x + 3) \\ &= 2x(2x + 3)(x + 1) \end{aligned}$$

Method 2: Factoring by Grouping

2 We will now discuss the use of grouping. The steps in the box that follow give the procedure for factoring trinomials by grouping.

To Factor Trinomials of the Form $ax^2 + bx + c, a \neq 1$, by Grouping

1. Determine if there is a factor common to all three terms. If so, factor it out.
2. Find two numbers whose product is equal to the product of a times c, and whose sum is equal to b.
3. Rewrite the middle term, bx, as the sum or difference of two terms using the numbers found in step 2.
4. Factor by grouping as explained in Section 5.2.

This process will be made clear in Example 9. We will rework Examples 1 through 8 here using factoring by grouping. Example 9, which follows, is the same trinomial given in Example 1. After you study this method and try some exercises, you will gain a feel for which method you prefer using.

EXAMPLE 9 Factor $2x^2 + 11x + 12$.

Solution: First determine if there is a common factor to all the terms of the polynomial. There are no common factors (other than 1) to the three terms.

$$a = 2, \qquad b = 11, \qquad c = 12$$

1. We must find two numbers whose product is $a \cdot c$ and whose sum is b. We must therefore find two numbers whose product equals $2 \cdot 12$ or 24 and whose sum equals 11. Only the positive factors of 24 need be considered since all signs of the trinomial are positive.

Factors of 24	*Sum of Factors*
(1)(24)	1 + 24 = 25
(2)(12)	2 + 12 = 14
(3)(8)	3 + 8 = 11
(4)(6)	4 + 6 = 10

The desired factors are 3 and 8.

2. Rewrite the $11x$ term as the sum or difference of two terms using the values found in step 1. Therefore, we rewrite $11x$ as $3x + 8x$.

$$2x^2 + 11x + 12$$
$$= 2x^2 + 3x + 8x + 12$$

3. Now factor by grouping. Start by factoring out a common factor from the first two terms and a common factor from the last two terms. This procedure was discussed in Section 5.2.

x is common factor. 4 is common factor.

$$2x^2 + 3x + 8x + 12$$
$$= x(2x + 3) + 4(2x + 3)$$
$$= (x + 4)(2x + 3)$$

Note that in step 2 of Example 9 we rewrote $11x$ as $3x + 8x$. Would it have made a difference if we had written $11x$ as $8x + 3x$? Let us work it out and see.

$$2x^2 + 11x + 12$$
$$= 2x^2 + 8x + 3x + 12$$

$2x$ is common factor. 3 is common factor.

$$2x^2 + 8x + 3x + 12$$
$$= 2x(x + 4) + 3(x + 4)$$
$$= (2x + 3)(x + 4)$$

Since $(2x + 3)(x + 4) = (x + 4)(2x + 3)$, the factors are the same. We obtained the same answer by writing the $11x$ as either $3x + 8x$ or $8x + 3x$. *In general, when rewriting the middle term of the trinomial using the specific factors found, the terms may be listed in either order.* You should, however, check after you list the two terms to make sure that the sum of the terms you listed equals the middle term.

EXAMPLE 10 Factor $5x^2 - 7x - 6$.

Solution: There are no common factors other than 1.

$$a = 5, \qquad b = -7, \qquad c = -6$$

The product of a times c is $5(-6) = -30$. We must find two numbers whose product is -30 and whose sum is -7.

Factors of −30	*Sum of Factors*
$(-1)(30)$	$-1 + 30 = 29$
$(-2)(15)$	$-2 + 15 = 13$
$(-3)(10)$	$-3 + 10 = 7$
$(-5)(6)$	$-5 + 6 = 1$
$(-6)(5)$	$-6 + 5 = -1$
$(-10)(3)$	$-10 + 3 = -7$
$(-15)(2)$	$-15 + 2 = -13$
$(-30)(1)$	$-30 + 1 = -29$

Rewrite the middle term of the trinomial, $-7x$, as $-10x + 3x$.

$$\begin{aligned} & \overbrace{5x^2 - 7x - 6} \\ &= 5x^2 - 10x + 3x - 6 \quad \text{Now factor by grouping.} \\ &= 5x(x - 2) + 3(x - 2) \\ &= (5x + 3)(x - 2) \end{aligned}$$

In Example 10, we could have expressed the $-7x$ as $3x - 10x$ and obtained the same answer. Try working Example 10 by rewriting $-7x$ as $3x - 10x$.

Helpful Hint

Notice in Example 10 that we were looking for two factors of -30 whose sum was -7. When we considered the factors $(-3)(10)$ we obtained a sum of 7. The factors we eventually obtained that gave a sum of -7 were $(3)(-10)$. Note that when the *constant of the trinomial is negative,* if we switch the signs of the constant in the factors, the sign of the sum of the factors changes. Thus, when trying pairs of factors to obtain the middle term, if you obtain the opposite of the coefficient you are seeking, reverse the signs in the factors. This should give you the coefficient you are seeking.

EXAMPLE 11 Factor $8x^2 + 33x + 4$.

Solution: There are no common factors other than 1. We must find two numbers whose product is $8 \cdot 4$ or 32 and whose sum is 33. The numbers are 1 and 32.

Factors of 32	*Sum of Factors*
(1)(32)	$1 + 32 = 33$

Rewrite $33x$ as $32x + x$. Then factor by grouping.

$$\begin{aligned} &8x^2 + 33x + 4 \\ &= 8x^2 + 32x + x + 4 \\ &= 8x(x + 4) + 1(x + 4) \\ &= (8x + 1)(x + 4) \end{aligned}$$

Notice in Example 11 that we rewrote $33x$ as $32x + x$ rather than $x + 32x$. We did this to reinforce factoring out 1 from the last two terms of an expression. You should obtain the same answer if you rewrite $33x$ as $x + 32x$. Try this now.

EXAMPLE 12 Factor $4x^2 - 4x + 1$.

Solution: There are no common factors other than 1. We must find two numbers whose product is $4 \cdot 1$ or 4 and whose sum is -4. Since the product of a times c is positive and the coefficient of the x term is negative, both numerical factors must be negative.

Factors of 4	*Sum of Factors*
$(-1)(-4)$	$-1 + (-4) = -5$
$(-2)(-2)$	$-2 + (-2) = -4$

The desired factors are -2 and -2.

$$\begin{aligned} &4x^2 - 4x + 1 \\ &= 4x^2 - 2x - 2x + 1 && \text{Rewrite } -4x \text{ as } -2x - 2x. \\ &= 2x(2x - 1) - 2x + 1 \\ &= 2x(2x - 1) - 1(2x - 1) && \text{Rewrite } -2x + 1 \text{ as } -1(2x - 1). \\ &= (2x - 1)(2x - 1) \text{ or } (2x - 1)^2 \end{aligned}$$

When attempting to factor a trinomial, if there are no two integers whose product equals $a \cdot c$ and whose sum equals b, the trinomial cannot be factored.

EXAMPLE 13 Factor $2x^2 + 3x + 5$.

Solution: There are no common factors other than 1. We must find two numbers whose product is 10 and whose sum is 3. We need consider only positive factors of 10. Why?

Factors of 10	*Sum of Factors*
(1)(10)	$1 + 10 = 11$
(2)(5)	$2 + 5 = 7$

Since there are no factors of 10 whose sum is 3, we conclude that this trinomial cannot be factored.

EXAMPLE 14 Factor $4x^2 + 7xy + 3y^2$.

Solution: There are no common factors other than 1. This trinomial contains two variables. It is factored in basically the same manner as the previous examples. Find two numbers whose product is $4 \cdot 3$ or 12 and whose sum is 7. The two numbers are 4 and 3.

$$\begin{aligned} &4x^2 + 7xy + 3y^2 \\ &= 4x^2 + 4xy + 3xy + 3y^2 \\ &= 4x(x + y) + 3y(x + y) \\ &= (4x + 3y)(x + y) \end{aligned}$$

EXAMPLE 15 Factor $6x^2 - 13xy - 8y^2$.

Solution: There are no common factors other than 1. Find two numbers whose product is $6(-8)$ or -48 and whose sum is -13. Since the product is negative, one factor must be positive and the other negative. Some factors are given below.

Product of Factors	*Sum of Factors*
$(1)(-48)$	$1 + (-48) = -47$
$(2)(-24)$	$2 + (-24) = -22$
$(3)(-16)$	$3 + (-16) = -13$

There are many other factors, but we have found the pair we were looking for. The two numbers whose product is -48 and whose sum is -13 are 3 and -16.

$$\begin{aligned} &6x^2 - 13xy - 8y^2 \\ &= 6x^2 + 3xy - 16xy - 8y^2 \\ &= 3x(2x + y) - 8y(2x + y) \\ &= (3x - 8y)(2x + y) \end{aligned}$$

Check: $(3x - 8y)(2x + y)$

$$\begin{array}{ccccccc} \text{F} & & \text{O} & & \text{I} & & \text{L} \\ (3x)(2x) & + & (3x)(y) & + & (-8y)(2x) & + & (-8y)(y) \\ 6x^2 & + & 3xy & - & 16xy & - & 8y^2 \end{array}$$

$$6x^2 - 13xy - 8y^2$$

If you rework Example 15 by writing $-13xy$ as $-16xy + 3xy$, what answer would you obtain? Try it now and see. Remember that in any factoring problem our first step is to determine if all terms in the polynomial have a common factor other than 1. If so, we use the distributive property to factor the GCF from each term. We then continue to factor by one of the methods discussed in this chapter, if possible.

EXAMPLE 16 Factor $4x^3 + 10x^2 + 6x$.

Solution: The factor $2x$ is common to all three terms. Factor the $2x$ from each term of the polynomial.

$$4x^3 + 10x^2 + 6x = 2x(2x^2 + 5x + 3)$$

Now continue by factoring $2x^2 + 5x + 3$. The two numbers whose product is $2 \cdot 3$ or 6 and whose sum is 5 are 2 and 3.

$$\begin{aligned} &2x[2x^2 + \overbrace{5x}\, + 3] \\ &= 2x[2x^2 + 2x + 3x + 3] \\ &= 2x[2x(x + 1) + 3(x + 1)] \\ &= 2x(2x + 3)(x + 1) \end{aligned}$$

Helpful Hint

Which Method Should You Use to Factor a Trinomial?

If your instructor asks you to use a specific method, you should use that method. If your instructor does not require a specific method, you should use the method you feel most comfortable with. You may wish to start with the trial and error method if there are only a few possible factors to try. If you cannot find the factors by trial and error or if there are many possible factors to consider, you may wish to use the grouping procedure. With time and practice you will learn which method you feel most comfortable with and which method gives you greater success.

Exercise Set 5.4

Factor completely. If an expression cannot be factored, so state.

1. $2x^2 + 7x + 6$
2. $2x^2 + 5x + 3$
3. $6x^2 + 13x + 6$
4. $5x^2 + 13x + 6$
5. $3x^2 + 4x + 1$
6. $3x^2 + 5x + 2$
7. $2x^2 + 11x + 15$
8. $3x^2 - 2x - 8$
9. $4x^2 + 4x - 3$
10. $4x^2 - 11x + 7$
11. $5y^2 - 8y + 3$
12. $5m^2 - 16m + 3$
13. $5a^2 - 12a + 6$
14. $2x^2 - x - 1$
15. $4x^2 + 13x + 3$
16. $6y^2 - 19y + 15$
17. $5x^2 + 11x + 4$
18. $3x^2 - 2x - 5$
19. $5y^2 - 16y + 3$
20. $5x^2 + 2x + 7$
21. $4x^2 + 4x - 15$
22. $7x^2 + 43x + 6$
23. $7x^2 - 16x + 4$
24. $15x^2 - 19x + 6$
25. $3x^2 - 10x + 7$
26. $3y^2 - 22y + 7$
27. $5z^2 - 33z - 14$
28. $3z^2 - 11z - 6$
29. $8x^2 + 2x - 3$
30. $8x^2 + 6x - 9$
31. $10x^2 - 27x + 5$
32. $6x^2 + 7x - 10$
33. $8x^2 - 2x - 15$
34. $8x^2 + 13x - 6$

35. $6x^2 + 33x + 15$
36. $18x^2 - 3x - 10$
37. $6x^2 + 4x - 10$
38. $12z^2 + 32z + 20$
39. $6x^3 + 5x^2 - 4x$
40. $8x^3 + 8x^2 - 6x$
41. $4x^3 + 2x^2 - 6x$
42. $18x^3 - 21x^2 - 9x$
43. $6x^3 + 4x^2 - 10x$
44. $300x^2 - 400x - 400$
45. $60x^2 + 40x + 5$
46. $36x^2 - 36x + 9$
47. $2x^2 + 5xy + 2y^2$
48. $8x^2 - 8xy - 6y^2$
49. $2x^2 - 7xy + 3y^2$
50. $15x^2 - xy - 6y^2$
51. $18x^2 + 18xy - 8y^2$
52. $12a^2 - 34ab + 24b^2$
53. $6x^2 - 15xy - 36y^2$
54. $60x^2 - 125xy + 60y^2$

Write the polynomial whose factors are listed. Explain how you determined your answer.

55. $3x - 4$, $2x + 3$
56. $4x - 2$, $5x - 7$
57. 3, $2x + 5$, $x + 1$
58. $4x$, $2x - 5$, $3x + 2$
59. x^2, $x + 1$, $2x - 3$
60. $5x^2$, $3x - 7$, $2x + 3$

61. **(a)** If you know one binomial factor of a trinomial, explain how you can use division to find the second binomial factor of the trinomial (see Section 4.6).

(b) One factor of $18x^2 + 93x + 110$ is $3x + 10$. Use division to find the second factor.

62. What is the first step in factoring any trinomial?

63. How may any trinomial factoring problem be checked?

64. Explain in your own words the procedure used to factor a trinomial of the form $ax^2 + bx + c$, $a \neq 1$.

CUMULATIVE REVIEW EXERCISES

[2.5] **65.** Solve the equation $3x + 4 = -(x - 6)$.

[3.4] **66.** The perimeter of a rectangle is 22 feet. Find the dimensions of the rectangle if the length is two more than twice the width.

[5.1] **67.** Factor $36x^4y^3 - 12xy^2 + 24x^5y^6$.

[5.3] **68.** Factor $x^2 - 15x + 54$.

Group Activity/Challenge Problems

Factor each trinomial.

1. $18x^2 + 9x - 20$.
2. $8x^2 - 99x + 36$.
3. $15x^2 - 124x + 160$
4. $16x^2 - 62x - 45$
5. $72x^2 - 180x - 200$
6. $72x^2 + 417x - 420$

7. If one factor of $6x^2 + 235x + 2250$ is $3x + 50$, find the other factor. Explain how you determined your answer.

8. Two factors of the polynomial $2x^3 + 11x^2 + 3x - 36$ are $x + 3$ and $2x - 3$. Find the third factor. Explain how you determined your answer.

5.5 Special Factoring Formulas and a General Review of Factoring

Tape 8

1 Factor the difference of two squares.
2 Factor the sum and difference of two cubes.
3 Learn the general procedure for factoring a polynomial.

There are special formulas for certain types of factoring problems that are used very often. The special formulas we focus on in this section are the *difference of two squares* and the *sum and difference of two cubes.* There is no special formulas for the sum of two squares; this is because the sum of two squares cannot be factored using the set of real numbers. You will need to memorize the three formulas in this section so that you can use them whenever you need them.

Difference of Two Squares

1 Let us begin with the difference of two squares. Consider the binomial $x^2 - 9$. Note that each term of the binomial can be expressed as the square of some expression.

$$x^2 - 9 = x^2 - 3^2$$

This is an example of a difference of two squares. To factor the difference of two squares, it is convenient to use the difference of two squares formula which was introduced in Section 4.5.

Difference of Two Squares

$$a^2 - b^2 = (a + b)(a - b)$$

EXAMPLE 1 Factor $x^2 - 9$.

Solution: If we write $x^2 - 9$ as a difference of two squares, we have $x^2 - 3^2$. Using the difference of two squares formula, where a is replaced by x and b is replaced by 3, we obtain the following:

$$a^2 - b^2 = (a + b)(a - b)$$
$$x^2 - 3^2 = (x + 3)(x - 3)$$

EXAMPLE 2 Factor using the difference of two squares formula.

(a) $x^2 - 16$ **(b)** $4x^2 - 9$ **(c)** $16x^2 - 9y^2$

Solution:

(a) $x^2 - 16 = (x)^2 - (4)^2$
$= (x + 4)(x - 4)$

(b) $4x^2 - 9 = (2x)^2 - (3)^2$
$= (2x + 3)(2x - 3)$

(c) $16x^2 - 9y^2 = (4x)^2 - (3y)^2$
$= (4x + 3y)(4x - 3y)$

EXAMPLE 3 Factor the differences of squares.

(a) $16x^4 - 9y^4$ **(b)** $x^6 - y^4$

Solution: **(a)** Rewrite $16x^4$ as $(4x^2)^2$ and $9y^4$ as $(3y^2)^2$, then use the difference of two squares formula.

$$16x^4 - 9y^4 = (4x^2)^2 - (3y^2)^2$$
$$= (4x^2 + 3y^2)(4x^2 - 3y^2)$$

(b) Rewrite x^6 as $(x^3)^2$ and y^4 as $(y^2)^2$, then use the difference of two squares formula.

$$\begin{aligned} x^6 - y^4 &= (x^3)^2 - (y^2)^2 \\ &= (x^3 + y^2)(x^3 - y^2) \end{aligned}$$

EXAMPLE 4 Factor $4x^2 - 16y^2$ using the difference of two squares formula.

Solution: First remove the common factor, 4.

$$4x^2 - 16y^2 = 4(x^2 - 4y^2)$$

Now use the formula for the difference of two squares.

$$\begin{aligned} 4(x^2 - 4y^2) &= 4[(x)^2 - (2y)^2] \\ &= 4(x + 2y)(x - 2y) \end{aligned}$$

Notice in Example 4 that $4x^2 - 16y^2$ is the difference of two squares, $(2x)^2 - (4y)^2$. If you factor this difference of squares without first removing the common factor 4, the factoring may be more difficult. After you factor this difference of squares you will need to factor out the common factor 2 from each binomial factor, as illustrated below.

$$\begin{aligned} 4x^2 - 16y^2 &= (2x)^2 - (4y)^2 \\ &= (2x + 4y)(2x - 4y) \\ &= 2(x + 2y)2(x - 2y) \\ &= 4(x + 2y)(x - 2y) \end{aligned}$$

We obtain the same answer as we did in Example 4. However, since we did not factor out the common factor 4 first, we had to work a little harder to obtain the answer.

COMMON STUDENT ERROR The difference of two squares can be factored. However, it is not possible to factor the sum of two squares using real numbers.

Correct	Incorrect
$a^2 - b^2 = (a + b)(a - b)$	~~$a^2 + b^2 = (a + b)(a + b)$~~

Sum and Difference of Two Cubes

2 We begin our discussion of the sum and difference of two cubes with a multiplication of polynomials problem. Consider the product of $(a + b)(a^2 - ab + b^2)$.

$$\begin{array}{rrrr} & a^2 & - ab & + b^2 \\ & & a & + b \\ \hline & a^2b & - ab^2 & + b^3 \\ a^3 & - a^2b & + ab^2 & \\ \hline a^3 & & & + b^3 \end{array}$$

Thus, $(a + b)(a^2 - ab + b^2) = a^3 + b^3$. Since factoring is the opposite of multiplying, we may factor $a^3 + b^3$ as follows:

$$a^3 + b^3 = (a + b)(a^2 - ab + b^2)$$

We see, using the same procedure, that $a^3 - b^3 = (a - b)(a^2 + ab + b^2)$. The expression $a^3 + b^3$ is a sum of two cubes and the expression $a^3 - b^3$ is a difference of two cubes. The formulas for factoring the sum and the difference of two cubes follow.

Sum of Two Cubes

$$a^3 + b^3 = (a + b)(a^2 - ab + b^2)$$

Difference of Two Cubes

$$a^3 - b^3 = (a - b)(a^2 + ab + b^2)$$

Note that the trinomials $a^2 - ab + b^2$ and $a^2 + ab + b^2$ cannot be factored further. Now let us do some factoring problems using the sum and the difference of two cubes.

EXAMPLE 5 Factor $x^3 + 8$.

Solution: Rewrite $x^3 + 8$ as a sum of two cubes: $x^3 + 8 = (x)^3 + (2)^3$

Using the sum of cubes formula, if we let a correspond to x and b correspond to 2, we get

$$a^3 + b^3 = (a + b)(a^2 - a\,b + b^2)$$

$$\downarrow \quad \downarrow \quad \downarrow \quad \downarrow \quad \downarrow \quad \downarrow \quad \downarrow \quad \downarrow$$

$$x^3 + 2^3 = (x)^3 + (2)^3 = (x + 2)[x^2 - x(2) + 2^2]$$

$$= (x + 2)(x^2 - 2x + 4)$$

You can check the factoring by multiplying $(x + 2)(x^2 - 2x + 4)$. If factored correctly, the product of the factors will equal the original expression, $x^3 + 8$. Try it and see.

EXAMPLE 6 Factor $y^3 - 27$.

Solution: Rewrite $y^3 - 27$ as a difference of two cubes: $(y)^3 - (3)^3$. Using the difference of cubes formula, if we let a correspond to y and b correspond to 3, we get

$$a^3 - b^3 = (a - b)(a^2 + a\,b + b^2)$$

$$\downarrow \quad \downarrow \quad \downarrow \quad \downarrow \quad \downarrow \quad \downarrow \quad \downarrow \quad \downarrow$$

$$y^3 - 27 = (y)^3 - (3)^3 = (y - 3)[y^2 + y(3) + 3^2]$$

$$= (y - 3)(y^2 + 3y + 9)$$

EXAMPLE 7 Factor $8p^3 - k^3$.

Solution: Rewrite $8p^3 - k^3$ as a difference of two cubes. Because $(2p)^3 = 8p^3$, we can write

$$\begin{aligned} 8p^3 - k^3 &= (2p)^3 - (k)^3 \\ &= (2p - k)[(2p)^2 + (2p)(k) + k^2] \\ &= (2p - k)(4p^2 + 2pk + k^2) \end{aligned}$$

EXAMPLE 8 Factor $8r^3 + 27s^3$.

Solution: Rewrite $8r^3 + 27s^3$ as a sum of two cubes. Since $8r^3 = (2r)^3$ and $27s^3 = (3s)^3$, we write

$$\begin{aligned} 8r^3 + 27s^3 &= (2r)^3 + (3s)^3 \\ &= (2r + 3s)[(2r)^2 - (2r)(3s) + (3s)^2] \\ &= (2r + 3s)(4r^2 - 6rs + 9s^2) \end{aligned}$$

A General Review of Factoring

3 In this chapter we have presented several methods of factoring. We now combine techniques from this and previous sections.

Here is a general procedure for factoring any polynomial:

General Procedure to Factor a Polynomial

1. If all the terms of the polynomial have a greatest common factor other than 1, factor it out.
2. If the polynomial has two terms (or is a binomial), determine if it is a difference of two squares or a sum or a difference of two cubes. If so, factor using the appropriate formula.
3. If the polynomial has three terms, factor the trinomial using the methods discussed in Sections 5.3 and 5.4.
4. If the polynomial has more than three terms, try factoring by grouping.
5. As a final step, examine your factored polynomial to see if the terms in any factors have a common factor. If you find a common factor, factor it out at this point.

EXAMPLE 9 Factor $3x^4 - 27x^2$.

Solution: First see if the terms have a greatest common factor other than 1. Since $3x^2$ is common to both terms, factor it out.

$$\begin{aligned} 3x^4 - 27x^2 &= 3x^2(x^2 - 9) \\ &= 3x^2(x + 3)(x - 3) \end{aligned}$$

Note that $x^2 - 9$ is a difference of two squares.

EXAMPLE 10 Factor $3x^2y^2 - 6xy^2 - 24y^2$.

Solution: Begin by factoring the GCF, $3y^2$, from each term. Then factor the remaining trinomial.

$$\begin{aligned} 3x^2y^2 - 6xy^2 - 24y^2 &= 3y^2(x^2 - 2x - 8) \\ &= 3y^2(x - 4)(x + 2) \end{aligned}$$

EXAMPLE 11 Factor $10a^2b - 15ab + 20b$.

Solution: $10a^2b - 15ab + 20b = 5b(2a^2 - 3a + 4)$

Since $2a^2 - 3a + 4$ cannot be factored we stop here.

EXAMPLE 12 Factor $2xy + 4x + 2y + 4$.

Solution: Always begin by determining if all the terms in the polynomial have a common factor. In this example, 2 is the GCF. Factor 2 from each term.

$$2xy + 4x + 2y + 4 = 2(xy + 2x + y + 2)$$

Now factor by grouping.

$$\begin{aligned} &= 2[x(y + 2) + 1(y + 2)] \\ &= 2(x + 1)(y + 2) \end{aligned}$$

In Example 12, what would happen if we forgot to factor out the common factor 2? Let's rework the problem without first factoring out the 2, and see what happens. Factor $2x$ from the first two terms, and 2 from the last two terms.

$$\begin{aligned} 2xy + 4x + 2y + 4 &= 2x(y + 2) + 2(y + 2) \\ &= (2x + 2)(y + 2) \end{aligned}$$

In step 5 of the general factoring procedure on page 291, we are told to examine the factored polynomial to see if the terms in any factor have a common factor. If we study the factors, we see that the factor $2x + 2$ has a common factor of 2. If we factor out the 2 from $2x + 2$, we will obtain the same answer obtained in Example 12.

$$(2x + 2)(y + 2) = 2(x + 1)(y + 2)$$

EXAMPLE 13 Factor $12x^2 + 12x - 9$.

Solution: First factor out the common factor 3. Then factor the remaining trinomial by one of the methods discussed in Section 5.4 (either by grouping or trial and error).

$$\begin{aligned} 12x^2 + 12x - 9 &= 3(4x^2 + 4x - 3) \\ &= 3(2x + 3)(2x - 1) \end{aligned}$$

EXAMPLE 14 Factor $2x^4y + 54xy$.

Solution: First factor out the common factor $2xy$.

$$2x^4y + 54xy = 2xy(x^3 + 27)$$
$$= 2xy(x + 3)(x^2 - 3x + 9)$$

Note that $x^3 + 27$ is a sum of two cubes.

Exercise Set 5.5

Factor the difference of two squares.

1. $x^2 - 4$
2. $x^2 - 9$
3. $y^2 - 25$
4. $z^2 - 64$
5. $x^2 - 49$
6. $x^2 - a^2$
7. $x^2 - y^2$
8. $4x^2 - 9$
9. $9y^2 - 16$
10. $16x^2 - 9y^2$
11. $64a^2 - 36b^2$
12. $100x^2 - 81y^2$
13. $25x^2 - 16$
14. $y^4 - 4x^2$
15. $z^4 - 81x^2$
16. $9x^4 - 16y^4$
17. $9x^4 - 81y^2$
18. $4x^4 - 25y^4$
19. $49m^4 - 16n^2$
20. $2x^4 - 50y^2$
21. $20x^2 - 180$
22. $4x^3 - xy^2$
23. $16x^2 - 100y^4$
24. $27x^4 - 3y^2$

Factor the sum or difference of two cubes.

25. $x^3 + y^3$
26. $x^3 - y^3$
27. $a^3 - b^3$
28. $a^3 + b^3$
29. $x^3 + 8$
30. $x^3 - 8$
31. $x^3 - 27$
32. $a^3 + 27$
33. $a^3 + 1$
34. $a^3 - 1$
35. $8x^3 + 27$
36. $27y^3 - 8$
37. $27a^3 - 64$
38. $64 - x^3$
39. $27 - 8y^3$
40. $1 + 27y^3$
41. $8x^3 - 27y^3$
42. $64x^3 - 27y^3$

Factor completely.

43. $2x^2 - 2x - 12$
44. $3x^2 - 9x - 12$
45. $x^2y - 16y$
46. $2x^2 - 8$
47. $3x^2 + 6x + 3$
48. $3x^2 - 9x - 12x + 36$
49. $5x^2 + 10x - 15$
50. $4x^2 - 100$
51. $3xy - 6x + 9y - 18$
52. $x^2y + 2xy - 6xy - 12y$
53. $2x^2 - 72$
54. $4ya^2 - 36y$
55. $3x^2y - 27y$
56. $2x^3 - 50x$
57. $3x^3y^2 + 3y^2$
58. $x^4 - 8x$
59. $2x^3 - 16$
60. $x^3 - 64x$
61. $6x^2 - 4x + 24x - 16$
62. $4x^2y - 6xy - 20xy + 30y$
63. $3x^3 - 10x^2 - 8x$
64. $4x^3 - 22x^2 + 30x$
65. $4x^2 + 5x - 6$
66. $12a^2 - 36a + 27$
67. $25b^2 - 100$
68. $3b^2 - 75c^2$
69. $a^5b^2 - 4a^3b^4$
70. $12x^2 + 8x - 18x - 12$
71. $3x^4 - 18x^3 + 27x^2$
72. $a^6 + 4a^4b^2$
73. $x^3 + 25x$
74. $8y^2 + 23y - 3$
75. $y^4 - 16$
76. $16m^3 + 250$
77. $10a^2 + 25ab - 60b^2$
78. $ac + 2a + bc + 2b$
79. $2ab + 4a - 3b - 6$
80. $x^3 - 25x$
81. $9 - 9y^4$

82. **(a)** Write the formula for factoring the difference of two squares.
(b) In your own words, explain how to factor the difference of two squares.

83. **(a)** Write the formula for factoring the sum of two cubes.
(b) In your own words, explain how to factor the sum of two cubes.

84. **(a)** Write the formula for factoring the difference of two cubes.
(b) In your own words, explain how to factor the difference of two cubes.

CUMULATIVE REVIEW EXERCISES

[2.7] **85.** Solve the inequality and graph the solution on the number line: $6(x - 2) < 4x - 3 + 2x$.

[3.1] **86.** Solve the equation $2x - 5y = 6$ for y.

[4.1] **87.** Simplify $\left(\frac{4x^4y}{6xy^5}\right)^3$.

[4.2] **88.** Simplify $x^{-2}x^{-3}$.

Group Activity/ Challenge Problems

1. Explain why the sum of two squares, $a^2 + b^2$, cannot be factored using real numbers.

2. Factor $x^6 + 1$.

3. Factor $x^6 - 27y^9$.

4. Factor $x^2 - 6x + 9 - 4y^2$. *Hint:* write the first three terms as the square of a binomial.

5. Have you ever seen the proof that 1 is equal to 2? Here it is.

Let $a = b$, then square both sides of the equation:

$$a^2 = b^2$$

$$a^2 = b \cdot b$$

$$a^2 = ab \qquad \text{Substitute } a = b.$$

$$a^2 - b^2 = ab - b^2 \qquad \text{Subtract } b^2 \text{ from both sides of the equation.}$$

$$(a + b)(a - b) = b(a - b) \qquad \text{Factor both sides of the equation.}$$

$$\frac{(a + b)\cancel{(a - b)}}{\cancel{(a - b)}} = \frac{b\cancel{(a - b)}}{\cancel{(a - b)}} \qquad \text{Divide both sides of the equation by } (a - b) \text{ and divide out common fact}$$

$$a + b = b$$

$$b + b = b \qquad \text{Substitute } a = b.$$

$$2b = b$$

$$\frac{2\overset{1}{\cancel{b}}}{\underset{1}{\cancel{b}}} = \frac{\overset{1}{\cancel{b}}}{\underset{1}{\cancel{b}}} \qquad \text{Divide both sides of the equation by } b.$$

$$2 = 1$$

Obviously, $2 \neq 1$. Therefore, we must have made an error somewhere. Can you find it?

5.6 Solving Quadratic Equations Using Factoring

Tape 9

1. Recognize quadratic equations.
2. Solve quadratic equations using factoring.
3. Solve application problems by factoring quadratic equations.

Quadratic Equations

1 In this section we introduce **quadratic equations,** which are equations that contain a second-degree term and no term of a higher degree.

> **Quadratic Equation**
>
> Quadratic equations have the form
>
> $$ax^2 + bx + c = 0$$
>
> where a, b, and c are real numbers, $a \neq 0$.

Examples of Quadratic Equations

$$x^2 + 2x - 3 = 0$$
$$3x^2 - 4x = 0$$
$$2x^2 - 3 = 0$$

Quadratic equations like these, in which one side of the equation is written in descending order of the variable and the other side of the equation is zero, are said to be in **standard form.**

Some quadratic equations can be solved by factoring. Two methods for solving quadratic equations that cannot be solved by factoring are given in Chapter 10. To solve a quadratic equation by factoring, we use the zero-factor property.

You know that if you multiply by 0, the product is zero. That is, if $a = 0$ or $b = 0$, then $ab = 0$. The reverse is also true. If a product equals zero, at least one of its factors must be zero.

Zero-Factor Property

If $ab = 0$, then $a = 0$ or $b = 0$.

We now illustrate how the zero-factor property is used in solving equations.

EXAMPLE 1 Solve the equation $(x + 3)(x + 4) = 0$.

Solution: Since the product of the factors equals 0, according to the preceding rule, one or both factors must equal zero. Set each factor equal to 0, and solve each resulting equation.

$$\begin{aligned} x + 3 &= 0 & \text{or} \qquad x + 4 &= 0 \\ x + 3 - 3 &= 0 - 3 & x + 4 - 4 &= 0 - 4 \\ x &= -3 & \text{or} \qquad x &= -4 \end{aligned}$$

Thus, if x is either -3 or -4, the product of the factors is 0. The solutions to the equation are -3 and -4.

Check:

$$\begin{aligned} x &= -3 & x &= -4 \\ (x + 3)(x + 4) &= 0 & (x + 3)(x + 4) &= 0 \\ (-3 + 3)(-3 + 4) &= 0 & (-4 + 3)(-4 + 4) &= 0 \\ 0(1) &= 0 & -1(0) &= 0 \\ 0 &= 0 \quad \text{true} & 0 &= 0 \quad \text{true} \end{aligned}$$

EXAMPLE 2 Solve the equation $(4x - 3)(2x + 4) = 0$.

Solution: Set each factor equal to 0 and solve for x.

$$\begin{aligned} 4x - 3 &= 0 & \text{or} \qquad 2x + 4 &= 0 \\ 4x &= 3 & 2x &= -4 \\ x &= \frac{3}{4} & \text{or} \qquad x &= -2 \end{aligned}$$

The solutions to the equation are $\frac{3}{4}$ and -2.

Solving Quadratic Equations by Factoring

2 Now we give a general procedure for solving quadratic equations by factoring.

To Solve a Quadratic Equation by Factoring

1. Write the equation in standard form with the squared term positive. This will result in one side of the equation being 0.
2. Factor the side of the equation that is not 0.
3. Set each factor *containing a variable* equal to zero and solve each equation.
4. Check each solution found in step 3 in the original equation.

EXAMPLE 3 Solve the equation $2x^2 = 12x$.

Solution: To make the right side of the equation equal to 0, we subtract $12x$ from both sides of the equation. Then we factor out $2x$ from both terms.

$$\begin{aligned} 2x^2 &= 12x \\ 2x^2 - 12x &= 12x - 12x \\ 2x^2 - 12x &= 0 \\ 2x(x - 6) &= 0 \end{aligned}$$

Now set each factor equal to zero.

$$2x = 0 \quad \text{or} \quad x - 6 = 0$$
$$x = \frac{0}{2} \qquad\qquad x = 6$$
$$x = 0$$

The solutions to the quadratic equation are 0 and 6. Check by substituting $x = 0$, then $x = 6$ in $2x^2 = 12x$ yourself.

EXAMPLE 4 Solve the equation $x^2 + 10x + 28 = 4$.

Solution: To make the right side of the equation equal to 0, we subtract 4 from both sides of the equation. Then factor and solve.

$$x^2 + 10x + 24 = 0$$
$$(x + 4)(x + 6) = 0$$
$$x + 4 = 0 \quad \text{or} \quad x + 6 = 0$$
$$x = -4 \qquad\qquad x = -6$$

The solutions are -4 and -6. A check will show that both values satisfy the equation $x^2 + 10x + 28 = 4$.

EXAMPLE 5 Solve the equation $3x^2 + 2x - 12 = -7x$.

Solution: Since all terms are not on the same side of the equation, add $7x$ to both sides of the equation.

$$3x^2 + 9x - 12 = 0$$

Factor out the common factor.

$$3(x^2 + 3x - 4) = 0$$

Factor the remaining trinomial.

$$3(x + 4)(x - 1) = 0$$

Now solve for x.

$$x + 4 = 0 \quad \text{or} \quad x - 1 = 0$$
$$x = -4 \qquad\qquad x = 1$$

Since the 3 that was factored does not contain a variable, we do not set it equal to zero. The solutions to the quadratic equation are -4 and 1.

EXAMPLE 6 Solve the equation $-x^2 + 5x + 6 = 0$.

Solution: When the squared term is negative, we generally make it positive by multiplying both sides of the equation by -1.

$$-1(-x^2 + 5x + 6) = -1 \cdot 0$$
$$x^2 - 5x - 6 = 0$$

Note that the sign of each term on the left side of the equation changed and that the right side of the equation remained zero. Why? Now proceed as before.

$$x^2 - 5x - 6 = 0$$
$$(x - 6)(x + 1) = 0$$
$$x - 6 = 0 \quad \text{or} \quad x + 1 = 0$$
$$x = 6 \qquad\qquad x = -1$$

A check using the original equation will show that the solutions are 6 and -1.

COMMON STUDENT ERROR Be careful not to confuse factoring a polynomial with solving an equation.

Correct	*Incorrect*
Factor: $x^2 + 3x + 2$	Factor: $x^2 + 3x + 2$
$(x + 2)(x + 1)$	$(x + 2)(x + 1)$
	~~$x + 2 = 0$ or $x + 1 = 0$~~
	~~$x = -2$ or $x = -1$~~

Do you know what is wrong with the example on the right? The expression $x^2 + 3x + 2$ is a polynomial (a trinomial), and not an equation. Since it is not an equation, it cannot be solved. When you are given a polynomial, you cannot just add "= 0" to change it to an equation.

Correct

Solve: $x^2 + 3x + 2 = 0$

$$(x + 2)(x + 1) = 0$$
$$x + 2 = 0 \quad \text{or} \quad x + 1 = 0$$
$$x = -2 \quad \text{or} \quad x = -1$$

EXAMPLE 7 Solve the equation $x^2 = 9$.

Solution: Subtract 9 from both sides of the equation; then factor using the difference of two squares formula.

$$x^2 - 9 = 0$$
$$(x + 3)(x - 3) = 0$$
$$x + 3 = 0 \quad \text{or} \quad x - 3 = 0$$
$$x = -3 \qquad\qquad x = 3$$

Applications

3 Now let's solve some applied problems using quadratic equations.

EXAMPLE 8 The product of two numbers is 66. Find the two numbers if one number is 5 more than the other.

Solution: Let x = smaller number

$x + 5$ = larger number

$$x(x + 5) = 66$$
$$x^2 + 5x = 66$$
$$x^2 + 5x - 66 = 0$$
$$(x - 6)(x + 11) = 0$$
$$x - 6 = 0 \quad \text{or} \quad x + 11 = 0$$
$$x = 6 \qquad\qquad x = -11$$

Remember that x represents the smaller of the two numbers. This problem has two possible solutions.

	Solution 1	*Solution 2*
smaller number	6	-11
larger number	$x + 5 = 6 + 5 = 11$	$x + 5 = -11 + 5 = -6$

One solution is: smaller number 6, larger number 11. A second solution is: smaller number -11, larger number -6. You must give both solutions.

Check:

Product of the two numbers is 66.	$6 \cdot 11 = 66$	$(-11)(-6) = 66$
One number is 5 more than the other number.	11 is 5 more than 6	-6 is 5 more than -11

If the question had stated "the product of two *positive* numbers is 66," the only solution would be 6 and 11.

EXAMPLE 9 The marketing department of a large publishing house is planning to make a large rectangular sign to advertise a new book at a convention. They want the length of the sign to be 3 feet longer than the width (Fig. 5.3). Signs at the convention may have a maximum area of 54 square feet. Find the length and width of the sign if the area is to be 54 square feet.

FIGURE 5.3

Solution: Let x = width

$x + 3$ = length

$$\text{area} = \text{length} \cdot \text{width}$$
$$54 = (x + 3)x$$
$$54 = x^2 + 3x$$
$$0 = x^2 + 3x - 54$$
$$\text{or} \quad x^2 + 3x - 54 = 0$$
$$(x - 6)(x + 9) = 0$$
$$x - 6 = 0 \quad \text{or} \quad x + 9 = 0$$
$$x = 6 \qquad\qquad x = -9$$

Since the width of the sign cannot be a negative number, the only solution is

$$\text{width} = x = 6 \text{ feet} \qquad \text{length} = x + 3 = 6 + 3 = 9 \text{ feet}$$

The area, length · width, is 54 square feet, and the length is 3 feet more than the width, so the answer checks.

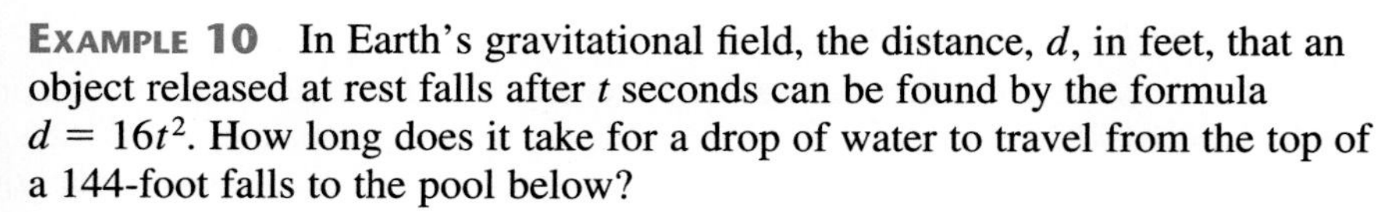

EXAMPLE 10 In Earth's gravitational field, the distance, d, in feet, that an object released at rest falls after t seconds can be found by the formula $d = 16t^2$. How long does it take for a drop of water to travel from the top of a 144-foot falls to the pool below?

Solution: Substitute 144 for d in the formula and then solve for t.

$$d = 16t^2$$
$$144 = 16t^2$$
$$\frac{144}{16} = t^2$$
$$9 = t^2$$

Now subtract 9 from both sides of the equation and write the equation with 0 on the right side to put the quadratic equation in standard form.

$$9 - 9 = t^2 - 9$$
$$0 = t^2 - 9$$
$$\text{or} \quad t^2 - 9 = 0$$
$$(t + 3)(t - 3) = 0$$
$$t + 3 = 0 \quad \text{or} \quad t - 3 = 0$$
$$t = -3 \qquad\qquad t = 3$$

Since t represents the number of seconds, it must be a positive number. Thus the only possible answer is 3 seconds. It takes 3 seconds for a drop of water (or any other object falling under the influence of gravity) to fall 144 feet.

Exercise Set 5.6

Solve.

1. $x(x + 2) = 0$
2. $3x(x - 4) = 0$
3. $4x(x - 9) = 0$
4. $(x + 3)(x - 5) = 0$
5. $(2x + 5)(x - 3) = 0$
6. $(2x + 3)(x - 5) = 0$
7. $x^2 - 16 = 0$
8. $x^2 - 25 = 0$
9. $x^2 - 12x = 0$
10. $x^2 + 8x = 0$
11. $9x^2 + 18x = 0$
12. $x^2 + 6x + 5 = 0$
13. $x^2 + x - 12 = 0$
14. $x^2 + 6x + 9 = 0$
15. $x^2 - 12x = -20$
16. $3y^2 - 2 = -y$
17. $z^2 + 3z = 18$
18. $3x^2 = -21x - 18$
19. $3x^2 - 6x - 72 = 0$
20. $x^2 = 3x + 18$
21. $x^2 + 19x = 42$
22. $3x^2 - 9x - 30 = 0$
23. $2y^2 + 22y + 60 = 0$
24. $w^2 + 45 + 18w = 0$
25. $-2x - 8 = -x^2$
26. $-9x + 20 = -x^2$
27. $-x^2 + 30x + 64 = 0$
28. $-y^2 + 12y - 11 = 0$
29. $x^2 - 3x - 18 = 0$
30. $z^2 + 16z = -64$
31. $3p^2 = 22p - 7$
32. $5w^2 - 16w = -3$
33. $3r^2 + r = 2$
34. $3x^2 = 7x + 20$
35. $4x^2 + 4x - 48 = 0$
36. $6x^2 + 13x + 6 = 0$
37. $6x^2 - 5x = 4$
38. $2x^2 - 4x - 6 = 0$
39. $2x^2 - 10x = -12$
40. $x^2 - 49 = 0$
41. $2x^2 = 32x$
42. $4x^2 - 9 = 0$
43. $x^2 = 36$
44. $2x^2 - 32 = 0$
45. $x^2 = 9$
46. $x^2 = 4$
47. $-t^2 = -81$
48. $-6r^2 = 5r - 6$

Express each problem as an equation, and solve.

49. The product of two consecutive positive even integers is 120. Find the two integers.
50. The product of two positive integers is 108. Find the two numbers if one is 3 more than the other.
51. The product of two positive integers is 64. Find the two integers if one number is four times the other.
52. The area of a rectangle is 36 square feet. Find the length and width if the length is four times the width.
53. The area of a rectangle is 54 square inches. Find the length and width if the length is 3 inches less than twice the width.
54. If each side of a square is increased by 6 meters, the area becomes 64 square meters. Find the length of a side of the original square.
55. If the length of the sign in Example 9 is to be 2 feet longer than the width and the area is to be 35 square feet, find the dimensions of the sign.
56. How long would it take for an object dropped from a helicopter to fall 256 feet to the ground? See Example 10.
57. Find the time required for a professional diver diving from a cliff 400 feet above the water to hit the water.

58. A shoe store owner finds that her daily profit, P, is approximated by the formula $P = n^2 - 15n - 10$, where n is the number of pairs of shoes she sells. How many pairs of shoes must she sell in a day for her profit to be $240?

59. The cost, C, for manufacturing x water sprinklers is given by the formula $C = x^2 - 27x - 20$. Find the number of water sprinklers manufactured at a cost of $70.

60. The sum, s, of the first n even numbers is given by the formula $s = n^2 + n$. Find n for the given sums: **(a)** $s = 12$; **(b)** $s = 30$.

61. In a league of x teams, the number of different ways, N, of selecting the league champion and the second-place team can be found by the formula $N = x(x - 1)$.

(a) The National Basketball Association currently has 26 teams. How many different first- and second-place finishers are possible?

(b) In a children's soccer league there are 90 different ways for the first- and second-place finishers to be selected. How many teams are in the league?

62. The area of a triangle may be found by the formula: area $= \frac{1}{2}$ base $\times$ height. A triangle has an area of 30 square feet. Find its base and height if its height is 2 feet less than twice the length of its base.

63. For a switchboard that handles n telephone lines, the maximum number of telephone connections, C, that it can make simultaneously is given by the formula $C = \dfrac{n(n - 1)}{2}$.

(a) How many telephone connections can a switchboard make simultaneously if it handles 12 lines?

(b) How many lines does a switchboard have if it can make 55 telephone connections simultaneously?

64. **(a)** When solving the equation $(x + 1)(x - 2) = 4$, explain why we **cannot** solve the equation by writing $x + 1 = 4$ or $x - 2 = 4$, and then solving for x.

(b) Solve the equation $(x + 1)(x - 2) = 4$.

65. When solving an equation like $3(x - 4)(x + 5) = 0$, we set factors $x - 4$ and $x + 5$ equal to 0, but we do not set the 3 equal to 0. Can you explain why?

CUMULATIVE REVIEW EXERCISES

[1.9] **66.** Evaluate $16 - 2^2 \cdot 12 \div 3 - 3^2$.

[4.4] **67.** Subtract $x^2 - 4x + 6$ from $3x + 2$.

[4.5] **68.** Multiply $(3x^2 + 2x - 4)(2x - 1)$

[4.6] **69.** Divide $\dfrac{6x^2 - 19x + 15}{3x - 5}$ by dividing the numerator by the denominator.

[5.4] **70.** Divide $\dfrac{6x^2 - 19x + 15}{3x - 5}$ by factoring the numerator and dividing out common factors.

Group Activity/ Challenge Problems

Create an equation whose solution is indicated. Explain how you determined your equation.

1. 6 and 3 **2.** 4 and -2 **3.** -5 and -9

4. Create a quadratic equation whose only solution is 5. Explain how you determined your answer.

5. Create an equation whose solutions are 0, 3, and 5. Explain how you determined your answer.

6. Create a quadratic equation whose solutions are $\frac{1}{2}$ and $\frac{2}{3}$. Explain how you determined your answer.

7. The break-even point for a manufacturer occurs when its cost of production, C, is equal to its revenue, R. The cost equation for a company is $C = 2x^2 - 20x + 600$ and its revenue equation is $R = x^2 + 50x - 400$, where x is the number of units produced and sold. How many units must be produced and sold for the manufacturer to break even? There are two values.

8. When a certain cannon is fired, the height in feet, of the cannonball at time t can be found by using the formula $h = -16t^2 + 128t$.

(a) Find the height of the cannonball 2 seconds after being fired.

(b) Find the time it takes for the cannonball to hit the ground.

Hint: What is the value of h at impact?

9. Solve the equation $x^3 + 3x^2 - 10x = 0$.

10. The product of two numbers is -40. Find the numbers if their sum is 3.

Summary

GLOSSARY

Composite number (253): A positive integer, other than 1, that is not prime.

Factor an expression (252): To factor an expression means to write the expression as a product of its factors.

Factors (252): If $a \cdot b = c$, then a and b are factors of c.

Greatest common factor (GCF) (253): The greatest factor that divides into each of the terms in an expression.

Prime number (253): An integer greater than 1 that has exactly two factors, itself and 1.

Quadratic equation (294): An equation of the form $ax^2 + bx + c = 0, a \neq 0$.

IMPORTANT FACTS

Difference of two squares: $a^2 - b^2 = (a + b)(a - b)$

Sum of two cubes: $a^3 + b^3 = (a + b)(a^2 - ab + b^2)$

Difference of two cubes: $a^3 - b^3 = (a - b)(a^2 + ab + b^2)$

Note: The sum of two squares, $a^2 + b^2$, cannot be factored using real numbers.

Zero-factor property: If $a \cdot b = 0$, then $a = 0$ or $b = 0$.

General Procedure to Factor a Polynomial

1. If all the terms of the polynomial have a greatest common factor other than 1, factor it out.
2. If the polynomial has two terms, determine if it is a difference of two squares or a sum or difference of two cubes. If so, factor using the appropriate formula.
3. If the polynomial has three terms, factor the trinomial using the methods discussed in Sections 5.3 and 5.4.
4. If the polynomial has more than three terms, try factoring by grouping.
5. As a final step, examine your factored polynomial to see if the terms in any factor have a common factor. If you find a common factor, factor it out at this point.

Review Exercises

[5.1] *Find the greatest common factor for each set of terms.*

1. $3x^2, x^4, x^5$
2. $3p, 6p^2, 9p^3$
3. $30x, 24, 36y^2$
4. $40x^2y^3, 36x^3y^4, 16x^5y^2z$
5. $9xyz, 12xz, 36, x^2y$
6. $-32x^5, 16x^2, 24x^2y$
7. $3(2x - 7), x(2x - 7)$
8. $x(x + 5), x + 5$

Factor each expression. If an expression cannot be factored, so state.

9. $4x - 16$
10. $10x + 5$
11. $24y^2 - 4y$
12. $55p^3 - 20p^2$
13. $24x^2y + 18x^3y^2$
14. $6xy - 12x^2y$
15. $2x^2 + 4x - 8$
16. $60x^4y^4 + 6x^9y^3 - 18x^5y^2$
17. $24x^2 - 13y^2 + 6xy$
18. $x(5x + 3) - 2(5x + 3)$
19. $3x(x - 1) - 2(x - 1)$
20. $2x(4x - 3) + 4x - 3$

2] *Factor by grouping.*

21. $x^2 + 4x + 2x + 8$
22. $x^2 - 3x + 4x - 12$
23. $x^2 - 7x + 7x - 49$
24. $2a^2 - 2ab - a + b$
25. $3xy + 3x + 2y + 2$
26. $x^2 + 3x - 2xy - 6y$
27. $5x^2 + 20x - x - 4$
28. $5x^2 - xy + 20xy - 4y^2$
29. $12x^2 - 8xy + 15xy - 10y^2$
30. $12x^2 + 15xy - 8xy - 10y^2$
31. $ab - a + b - 1$
32. $3x^2 - 9xy + 2xy - 6y^2$
33. $20x^2 - 12x + 15x - 9$
34. $6x^2 + 9x - 2x - 3$

3] *Factor completely. If an expression cannot be factored, so state.*

35. $x^2 + 7x + 10$
36. $x^2 - 8x + 15$
37. $x^2 - x - 20$
38. $x^2 + x - 20$
39. $x^2 - 11x + 30$
40. $x^2 - 15x + 56$
41. $x^2 - 12x - 44$
42. $x^2 + 11x + 24$
43. $x^3 + 5x^2 + 4x$
44. $x^3 - 3x^2 - 40x$
45. $x^2 - 2xy - 15y^2$
46. $4x^3 + 32x^2y + 60xy^2$

] *Factor completely. If an expression cannot be factored, so state.*

47. $2x^2 + 5x - 3$
48. $3x^2 + 13x + 4$
49. $4x^2 - 9x + 5$
50. $5x^2 - 13x - 6$
51. $4x^2 + 4x - 15$
52. $5x^2 - 32x + 12$
53. $3x^2 + 13x + 8$
54. $6x^2 + 31x + 5$
55. $4x^2 + 8x - 21$
56. $6x^2 + 11x - 10$
57. $8x^2 - 18x - 35$
58. $4x^2 + 20x + 25$
59. $9x^3 - 12x^2 + 4x$
60. $18x^3 - 24x^2 - 10x$
61. $4x^2 - 16xy + 15y^2$
62. $16x^2 - 22xy - 3y^2$

] *Factor completely.*

63. $x^2 - 16$
64. $x^2 - 64$
65. $4x^2 - 16$
66. $81x^2 - 9y^2$
67. $64x^4 - 81y^4$
68. $16 - 25y^2$
69. $4x^4 - 9y^4$
70. $100x^4 - 121y^4$
71. $x^3 - y^3$
72. $x^3 + y^3$
73. $a^3 + 8$
74. $a^3 - 1$
75. $a^3 + 27$
76. $x^3 - 8$
77. $8x^3 - y^3$
78. $27 - 8y^3$
79. $27x^4 - 75y^2$
80. $2x^3 - 128y^3$

-5.5] *Factor completely.*

81. $x^2 - 15x + 50$
82. $2x^2 - 16x + 32$
83. $4x^2 - 36$
84. $4y^2 - 64$
85. $8x^2 + 16x - 24$
86. $x^2 - 6x - 27$
87. $4x^2 - 4x - 15$
88. $6x^2 - 33x + 36$
89. $8x^3 - 8$
90. $x^3y - 27y$
91. $x^2y - xy + 4xy - 4y$
92. $6x^3 + 30x^2 + 9x^2 + 45x$
93. $x^2 + 5xy + 6y^2$
94. $2x^2 - xy - 10y^2$
95. $4x^2 - 20xy + 25y^2$
96. $16y^2 - 49z^2$
97. $ab + 7a + 6b + 42$
98. $16y^5 - 25y^7$
99. $2x^3 + 12x^2y + 16xy^2$
100. $6x^2 + 5xy - 21y^2$
101. $32x^3 + 32x^2 + 6x$
102. $y^4 - 1$

[5.6] *Solve.*

103. $x(x - 5) = 0$
104. $(x + 3)(x + 4) = 0$
105. $(x - 5)(3x + 2) = 0$
106. $x^2 - 3x = 0$
107. $5x^2 + 20x = 0$
108. $x^2 - 2x - 24 = 0$
109. $x^2 + 8x + 15 = 0$
110. $x^2 = -2x + 8$
111. $x^2 - 12 = -x$
112. $3x^2 + 21x + 30 = 0$
113. $x^2 - 6x + 8 = 0$
114. $6x^2 + 6x - 12 = 0$
115. $8x^2 - 3 = -10x$
116. $2x^2 + 15x = 8$
117. $4x^2 - 16 = 0$
118. $36x^2 - 49 = 0$
119. $26x - 48 = -4x^2$
120. $-48x = -12x^2 - 45$

[5.6] *Express each problem as an equation and solve.*

121. The product of two consecutive positive integers is 110. Find the two integers.
122. The product of two consecutive positive even integers is 48. Find the two integers.
123. The product of two positive integers is 40. Find the integers if the larger is 2 less than twice the smaller.
124. The area of a rectangle is 63 square feet. Find the length and width of the rectangle if the length is 2 feet greater than the width.
125. The side of one square is 4 inches longer than the side of another square. If the area of the larger square is 81 square inches, find the length of the side of each square.
126. The cost, C, for manufacturing x boxes of toothpicks is given by the formula $C = x^2 - 18x + 10$. Find the number of boxes that can be manufactured for $50.

Practice Test

1. Find the greatest common factor of $4x^4$, $12x^5$ and $10x^2$.
2. Find the greatest common factor of $6x^2y^3$, $9xy^2$, and $12xy^5$.

Factor completely.

3. $4x^2y - 8xy$
4. $24x^2y - 6xy + 9x$
5. $x^2 - 3x + 2x - 6$
6. $3x^2 - 12x + x - 4$
7. $5x^2 - 15xy - 3xy + 9y^2$
8. $x^2 + 12x + 32$
9. $x^2 + 5x - 24$
10. $x^2 - 9xy + 20y^2$
11. $2x^2 - 22x + 60$
12. $2x^3 - 3x^2 + x$
13. $12x^2 - xy - 6y^2$
14. $x^2 - 9y^2$
15. $x^3 + 27$

Solve.

16. $(x - 2)(2x - 5) = 0$
17. $x^2 + 6 = -5x$
18. $x^2 + 4x - 5 = 0$

Solve.

19. The product of two positive integers is 36. Find the two integers if the larger is 1 more than twice the smaller.
20. The area of a rectangle is 24 square meters. Find the length and width of the rectangle if its length is 2 meters greater than its width.

Chapter 6

Rational Expressions and Equations

See Section 6.3, Exercise 88.

Preview and Perspective

You worked with rational numbers when you discussed fractions in arithmetic. Now you will expand your knowledge to include fractions that contain variables. Fractions that contain variables are often referred to as rational expressions. The same basic procedures you used to reduce, add, subtract, multiply, and divide arithmetic fractions will be used with rational expressions in Sections 6.1 through 6.5. In Section 6.1 we define rational expressions and explain that rational expressions cannot have a denominator of zero. In Section 6.2, multiplication and division of rational expressions are discussed. In Sections 6.3 and 6.4 we explain how to add and subtract rational expressions. Complex fractions are discussed in Section 6.5.

Many real-life problems involve equations that contain rational expressions. Such equations are called rational equations. We discuss solving rational equations in Section 6.6 and give various real-life applications of rational equations in Sections 6.6 and 6.7. Since many formulas contain rational expressions, you will use the material you learn in this chapter in many other courses. In later mathematics courses you may graph rational equations that contain two variables. The material presented here will prepare you for that work.

To be successful in this chapter you need to have a complete understanding of factoring, which was presented in Chapter 5, especially Sections 5.3 and 5.4. You might wish to review reducing, adding, subtracting, multiplying, and dividing numerical fractions discussed in Section 1.2. The material presented in this chapter builds upon the procedures discussed in Section 1.2.

6.1 Reducing Rational Expressions

Tape 9

1. Determine the values for which a rational expression is defined.
2. Recognize the three signs of a fraction.
3. Reduce rational expressions.
4. Factor a negative 1 from a polynomial.

Rational Expressions

1 A **rational expression** (also called an **algebraic fraction**) is an algebraic expression of the form p/q, where p and q are polynomials and $q \neq 0$. Examples of rational expressions are

$$\frac{4}{5}, \quad \frac{x-6}{x}, \quad \frac{x^2+2x}{x-3}, \quad \frac{x}{x^2-4}$$

The denominator of a rational expression cannot equal 0 since division by 0 is not defined. In the expression $(x + 3)/x$, the value of x cannot be 0 because the denominator would then equal 0. We say that the expression $(x + 3)/x$ is *defined* for all real numbers except 0. It is *undefined* when x is 0. In $(x^2 + 4x)/(x - 3)$, the value of x cannot be 3 because the denominator would then be 0. What values of x cannot be used in the expression $x/(x^2 - 4)$? If you answered 2 and -2, you answered correctly. **Whenever we have a rational expression containing a variable in the denominator, we always assume that the value or values of the variable that make the denominator 0 are excluded.**

One method that can be used to determine the value or values of the variable that are excluded is to set the denominator equal to 0 and then solve the resulting equation for the variable.

EXAMPLE 1 Determine the value or values of the variable for which the rational expression is defined.

(a) $\dfrac{x+3}{2x-5}$ **(b)** $\dfrac{x+3}{x^2+3x-10}$

Solution: **(a)** We need to determine the value or values of x that make $2x - 5$ equal to 0 and exclude these. We can do this by setting $2x - 5$ equal to 0 and solving the equation for x.

$$2x - 5 = 0$$
$$2x = 5$$
$$x = \frac{5}{2}$$

Thus, we do not consider $x = \frac{5}{2}$ when we consider the rational expression $(x + 3)/(2x - 5)$. This expression is defined for all real numbers except $x = \frac{5}{2}$. We will sometimes shorten our answer and write $x \neq \frac{5}{2}$.

(b) To determine the value or values that are excluded, we set the denominator equal to zero and solve the equation for the variable.

$$x^2 + 3x - 10 = 0$$
$$(x + 5)(x - 2) = 0$$
$$x + 5 = 0 \quad \text{or} \quad x - 2 = 0$$
$$x = -5 \quad \text{or} \quad x = 2$$

Therefore, we do not consider the values $x = -5$ or $x = 2$ when we consider the rational expression $(x + 3)/(x^2 + 3x - 10)$, for both $x = -5$ and $x = 2$ make the denominator zero. This expression is defined for all real numbers except $x = -5$ and $x = 2$. Thus, $x \neq -5$ and $x \neq 2$.

Signs of a Fraction

2 Three signs are associated with any fraction: the sign of the numerator, the sign of the denominator, and the sign of the fraction.

sign of numerator

sign of fraction ⟶ $+\dfrac{-a}{+b}$

sign of denominator

Whenever any of the three signs is omitted, we assume it to be positive. For example,

$$\frac{a}{b} \quad \text{means} \quad +\frac{+a}{+b}$$

$$\frac{-a}{b} \quad \text{means} \quad +\frac{-a}{+b}$$

$$-\frac{a}{b} \quad \text{means} \quad -\frac{+a}{+b}$$

Changing any two of the three signs of a fraction does not change the value of a fraction. Thus,

$$\frac{-a}{b} = -\frac{a}{b} = \frac{a}{-b}$$

Generally, we do not write a fraction with a negative denominator. For example, the expression $\frac{2}{-5}$ would be written as either $\frac{-2}{5}$ or $-\frac{2}{5}$. The expression $\frac{x}{-(4-x)}$ can be written $\frac{x}{x-4}$ since $-(4-x) = -4 + x$ or $x - 4$.

Reducing Rational Expressions

3 A rational expression is **reduced to its lowest terms** when the numerator and denominator have no common factors other than 1. The fraction $\frac{9}{12}$ is not in reduced form because the 9 and 12 both contain the common factor 3. When the 3 is factored out, the reduced fraction is $\frac{3}{4}$.

$$\frac{9}{12} = \frac{\overset{1}{\cancel{3}} \cdot 3}{\underset{1}{\cancel{3}} \cdot 4} = \frac{3}{4}$$

The rational expression $\frac{ab-b^2}{2b}$ is not in reduced form because both the numerator and denominator have a common factor, b. To reduce this expression, factor b from each term in the numerator, then divide it out.

$$\frac{ab - b^2}{2b} = \frac{\cancel{b}(a-b)}{2\cancel{b}} = \frac{a-b}{2}$$

Thus, $\frac{ab-b^2}{2b}$ becomes $\frac{a-b}{2}$ when reduced to its lowest terms.

To Reduce Rational Expressions

1. Factor both the numerator and denominator as completely as possible.
2. Divide out any factors common to both the numerator and denominator.

EXAMPLE 2 Reduce $\frac{5x^3 + 10x^2 - 25x}{10x^2}$ to lowest terms.

Solution: Factor the greatest common factor, $5x$, from each term in the numerator. Since $5x$ is a common factor to both the numerator and denominator, divide it out.

$$\frac{5x^3 + 10x^2 - 25x}{10x^2} = \frac{\cancel{5x}(x^2 + 2x - 5)}{\cancel{5x} \cdot 2x}$$

$$= \frac{x^2 + 2x - 5}{2x}$$

EXAMPLE 3 Reduce $\frac{x^2 + 2x - 3}{x + 3}$ to lowest terms.

Solution: Factor the numerator; then divide out the common factor.

$$\frac{x^2 + 2x - 3}{x + 3} = \frac{\cancel{(x + 3)}(x - 1)}{\cancel{x + 3}} = x - 1$$

EXAMPLE 4 Reduce $\frac{x^2 - 16}{x - 4}$ to lowest terms.

Solution: Factor the numerator; then divide out common factors.

$$\frac{x^2 - 16}{x - 4} = \frac{(x + 4)\cancel{(x - 4)}}{\cancel{x - 4}} = x + 4$$

EXAMPLE 5 Reduce $\frac{2x^2 + 7x + 6}{x^2 - x - 6}$ to lowest terms.

Solution: Factor both the numerator and denominator, then divide out the common factor.

$$\frac{2x^2 + 7x + 6}{x^2 - x - 6} = \frac{(2x + 3)\cancel{(x + 2)}}{(x - 3)\cancel{(x + 2)}} = \frac{2x + 3}{x - 3}.$$

Note that $\frac{2x + 3}{x - 3}$ cannot be simplified any further.

COMMON STUDENT ERROR: Remember: Only common *factors* can be divided out when *multiplying* expressions.

Correct	*Incorrect*
$\frac{\overset{5}{\cancel{20}}\overset{x}{\cancel{x^2}}}{\underset{1}{\cancel{4}}\underset{1}{\cancel{x}}} = 5x$	$\frac{\overset{x}{\cancel{x^2}} - \overset{5}{\cancel{20}}}{\underset{1}{\cancel{x}} - \underset{1}{\cancel{4}}}$ (crossed out)

In the denominator of the example on the left, $4x$, the 4 and x are factors since they are *multiplied* together. The 4 and the x are also both factors of the numerator $20x^2$, since $20x^2$ can be written $4 \cdot x \cdot 5x$.

Some students incorrectly divide out *terms*. In the expression $\frac{x^2 - 20}{x - 4}$, the x and -4 are *terms* of the denominator, not factors, and therefore cannot be divided out.

Factoring Out

4 Recall that when -1 is factored from a polynomial, the sign of each term in the polynomial changes.

Examples

$$-3x + 5 = -1(3x - 5) = -(3x - 5)$$
$$6 - 2x = -1(-6 + 2x) = -(2x - 6)$$
$$-2x^2 + 3x - 4 = -1(2x^2 - 3x + 4) = -(2x^2 - 3x + 4)$$

Whenever the terms in a numerator and denominator differ only in their signs (one is the opposite or additive inverse of the other), we can factor out -1 from either the numerator or denominator and then divide out the common factor. This procedure is illustrated in Examples 6 and 7.

EXAMPLE 6 Reduce $\dfrac{3x-7}{7-3x}$ to lowest terms.

Solution: Since each term in the numerator differs only in sign from its like terms in the denominator, we will factor -1 from each term in the denominator.

$$\frac{3x-7}{7-3x} = \frac{3x-7}{-1(-7+3x)}$$

$$= \frac{\cancel{3x-7}}{-(\cancel{3x-7})} = -1$$

EXAMPLE 7 Reduce $\dfrac{4x^2-23x-6}{6-x}$ to lowest terms.

Solution: $$\frac{4x^2-23x-6}{6-x} = \frac{(4x+1)(x-6)}{6-x}$$

$$= \frac{(4x+1)\cancel{(x-6)}}{-1\cancel{(x-6)}}$$

$$= \frac{4x+1}{-1} = -(4x+1)$$

Note that $-4x-1$ is also an acceptable answer.

Exercise Set 6.1

Determine the value or values of the variable where the expression is defined. (See Example 1.)

1. $\dfrac{x+2}{x}$
2. $\dfrac{3}{x+5}$
3. $\dfrac{5}{x-6}$
4. $\dfrac{5}{2x-3}$
5. $\dfrac{x+4}{x^2-4}$
6. $\dfrac{7}{x^2-7x+12}$
7. $\dfrac{x-3}{x^2+6x-16}$
8. $\dfrac{x^2+3}{2x^2-13x+15}$

Reduce to lowest terms.

9. $\dfrac{x}{x+xy}$
10. $\dfrac{3x}{6x+9}$
11. $\dfrac{4x+12}{x+3}$
12. $\dfrac{3x^2+6x}{3x^2+9x}$
13. $\dfrac{x^3+6x^2+3x}{2x}$
14. $\dfrac{x^2y^2-xy+3y}{y}$
15. $\dfrac{x^2+2x+1}{x+1}$
16. $\dfrac{x-1}{x^2+2x-3}$
17. $\dfrac{x^2-2x}{x^2-4x+4}$
18. $\dfrac{x^2+3x-18}{2x-6}$
19. $\dfrac{x^2-x-6}{x^2-4}$
20. $\dfrac{x^2+6x+9}{x^2-9}$
21. $\dfrac{2x^2-4x-6}{x-3}$
22. $\dfrac{4x^2-12x-40}{2x^2-16x+30}$
23. $\dfrac{2x-3}{3-2x}$

24. $\dfrac{4x-6}{3-2x}$

25. $\dfrac{x^2-2x-8}{4-x}$

26. $\dfrac{5-x}{x^2-2x-15}$

27. $\dfrac{x^2+3x-18}{-2x^2+6x}$

28. $\dfrac{2x^2+5x-12}{2x-3}$

29. $\dfrac{2x^2+5x-3}{1-2x}$

30. $\dfrac{x^2-9}{x^2-2x-15}$

31. $\dfrac{6x^2+x-2}{2x-1}$

32. $\dfrac{4x^2-13x+10}{(x-2)^2}$

33. $\dfrac{6x^2+7x-20}{2x+5}$

34. $\dfrac{16x^2+24x+9}{4x+3}$

35. $\dfrac{6x^2-13x+6}{3x-2}$

36. $\dfrac{x^2-25}{(x-5)^2}$

37. $\dfrac{x^2-3x+4x-12}{x-3}$

38. $\dfrac{x^2-2x+4x-8}{2x^2+3x+8x+12}$

39. $\dfrac{2x^2-8x+3x-12}{2x^2+8x+3x+12}$

40. $\dfrac{x^3+1}{x^2-x+1}$

41. $\dfrac{x^3-8}{x-2}$

42. In any rational expression with a variable in the denominator, what do we always assume about the variable?

Explain why x can represent any real number in the expression.

43. $\dfrac{x-5}{x^2+1}$ **44.** $\dfrac{x+3}{x^2+4}$

Determine what values, if any, x cannot represent. Explain.

45. $\dfrac{x+3}{x-3}$ **46.** $\dfrac{x}{(x-4)^2}$

47. Is $\dfrac{x+4}{4-x}$ equal to -1? Explain.

48. Is $\dfrac{3x+2}{-3x-2}$ equal to 1? Explain.

49. What does it mean when a rational expression is reduced to its lowest terms?

50. In your own words, explain how to reduce a rational expression to lowest terms.

Explain why the expression cannot be reduced.

51. $\dfrac{2+3x}{6}$ **52.** $\dfrac{3x+4y}{12xy}$

Determine the denominator that will make the statement true. Explain how you obtained your answer.

53. $\dfrac{x^2-x-6}{\boxed{}}=x-3$ **54.** $\dfrac{2x^2+11x+12}{\boxed{}}=x+4$

Determine the numerator that will make the statement true. Explain how you obtained your answer.

55. $\dfrac{\boxed{}}{x+4}=x+3$ **56.** $\dfrac{\boxed{}}{x-5}=2x-1$

CUMULATIVE REVIEW EXERCISES

57. Solve the formula $z=\dfrac{x-y}{2}$ for y.

58. Find the measures of the three angles of a triangle if one angle is 30° greater than the smallest angle, and the third angle is 10° greater than three times the smallest angle.

[4.1] **59.** Simplify $\left(\dfrac{3x^6y^2}{9x^4y^3}\right)^2$.

[4.4] **60.** Subtract $6x^2-4x-8-(-3x^2+6x+9)$.

Group Activity/ Challenge Problems

*In Exercises 1–3 **(a)** determine the value or values that x cannot represent. **(b)** Reduce the expression to lowest terms.*

1. $\dfrac{x+3}{x^2-2x+3x-6}$

2. $\dfrac{x-4}{2x^2-5x-8x+20}$

3. $\dfrac{x+5}{2x^3+7x^2-15x}$

Reduce. Explain how you determined your answer.

4. $\dfrac{\frac{1}{5}x^5-\frac{2}{3}x^4}{x^4}$

5. $\dfrac{\frac{1}{5}x^5-\frac{2}{3}x^4}{\frac{1}{5}x^5-\frac{2}{3}x^4}$

6. $\dfrac{\frac{1}{5}x^5-\frac{2}{3}x^4}{\frac{2}{3}x^4-\frac{1}{5}x^5}$

6.2 Multiplication and Division of Rational Expressions

Tape 9

1 Multiply rational expressions.
2 Divide rational expressions.

Multiply Rational Expressions

1 In Section 1.2 we reviewed multiplication of numerical fractions. Recall that to multiply two fractions we multiply their numerators together and multiply their denominators together.

Multiplication

$$\frac{a}{b} \cdot \frac{c}{d} = \frac{a \cdot c}{b \cdot d}, \qquad b \neq 0 \quad \text{and} \quad d \neq 0$$

EXAMPLE 1 Multiply $\left(\frac{3}{5}\right)\left(\frac{-2}{9}\right)$.

Solution: First divide out common factors; then multiply.

$$\frac{\overset{1}{\cancel{3}}}{5} \cdot \frac{-2}{\underset{3}{\cancel{9}}} = \frac{1 \cdot (-2)}{5 \cdot 3} = -\frac{2}{15}$$

The same principles apply when multiplying rational expressions containing variables. Before multiplying, you should first divide out any factors common to both a numerator and a denominator.

To Multiply Rational Expressions

1. Factor all numerators and denominators completely.
2. Divide out common factors.
3. Multiply numerators together and multiply denominators together.

EXAMPLE 2 Multiply $\frac{3x^2}{2y} \cdot \frac{4y^3}{3x}$.

Solution: This problem can be represented as

$$\frac{3xx}{2y} \cdot \frac{4yyy}{3x}$$

$$\frac{\overset{1}{\cancel{3}}\overset{1}{\cancel{x}}x}{2y} \cdot \frac{4yyy}{\underset{1}{\cancel{3}}\underset{1}{\cancel{x}}} \qquad \text{Divide out the 3's and } x\text{'s.}$$

$$\frac{\overset{1}{\cancel{3}}\overset{1}{\cancel{x}}x}{\underset{1}{\cancel{2}}\underset{1}{\cancel{y}}} \cdot \frac{\overset{2}{\cancel{4}}\overset{1}{\cancel{y}}yy}{\underset{1}{\cancel{3}}\underset{1}{\cancel{x}}} \qquad \text{Divide both the 4 and the 2 by 2, and divide out the } y\text{'s.}$$

Now multiply the remaining numerators together and the remaining denominators together.

$$\frac{2xy^2}{1} \quad \text{or} \quad 2xy^2$$

Rather than illustrating this entire process when multiplying rational expressions, we will often proceed as follows:

$$\frac{3x^2}{2y} \cdot \frac{4y^3}{3x}$$

$$= \frac{\overset{1\ x}{\cancel{3}\cancel{x^2}}}{\underset{1\ 1}{\cancel{2}\cancel{y}}} \cdot \frac{\overset{2\ y^2}{\cancel{4}\cancel{y^3}}}{\underset{1\ 1}{\cancel{3}\cancel{x}}} = 2xy^2$$

EXAMPLE 3 Multiply $\dfrac{-2a^3b^2}{3x^3y} \cdot \dfrac{4a^2x}{5b^2y^3}$.

Solution: $\dfrac{-2a^3\cancel{b^2}}{3\underset{x^2}{\cancel{x^3}}y} \cdot \dfrac{4a^2\cancel{x}}{5\cancel{b^2}y^3} = \dfrac{-8a^5}{15x^2y^4}$.

In Example 3 when the b^2 was divided out from both the numerator and denominator we did not place a 1 above and below the b^2 factors. When an identical factor that appears in a numerator and denominator is factored out, we will generally not show the 1's.

EXAMPLE 4 Multiply $(x - 2) \cdot \dfrac{3}{x^2 - 2x}$.

Solution: $(x - 2) \cdot \dfrac{3}{x^2 - 2x} = \dfrac{\cancel{(x - 2)}}{1} \cdot \dfrac{3}{x\cancel{(x - 2)}} = \dfrac{3}{x}$.

EXAMPLE 5 Multiply $\dfrac{(x + 2)^2}{6x^2} \cdot \dfrac{3x}{x^2 - 4}$.

Solution:

$$\frac{(x + 2)^2}{6x^2} \cdot \frac{3x}{x^2 - 4} = \frac{(x + 2)(x + 2)}{6x^2} \cdot \frac{3x}{(x + 2)(x - 2)}$$

$$= \frac{\cancel{(x + 2)}(x + 2)}{\underset{2\ x}{\cancel{6}\cancel{x^2}}} \cdot \frac{\overset{1\ 1}{\cancel{3}\cancel{x}}}{\cancel{(x + 2)}(x - 2)} = \frac{x + 2}{2x(x - 2)}.$$

In Example 5 we could have multiplied the factors in the denominator to get $(x+2)/(2x^2-4x)$. This is also a correct answer. In this section we will leave rational answers with the numerator as a polynomial (in unfactored form) and the denominators in factored form, as was given in Example 5. This is consistent with how we will leave rational answers when we add and subtract rational expressions in later sections.

EXAMPLE 6 Multiply $\frac{x-3}{2x} \cdot \frac{4x}{3-x}$.

Solution: $\frac{x-3}{\underset{1}{\cancel{2x}}} \cdot \frac{\overset{2}{\cancel{4x}}}{3-x} = \frac{2(x-3)}{3-x}$.

This problem is still not complete. In Section 6.1 we showed that $3 - x$ is $-1(-3 + x)$ or $-1(x - 3)$. Thus,

$$\frac{2(x-3)}{3-x} = \frac{2\cancel{(x-3)}}{-1\cancel{(x-3)}} = -2$$

Helpful Hint

When only the signs differ in a numerator and denominator in a multiplication problem, factor out -1 *from either the numerator or denominator*; then divide out the common factor.

$$\frac{a-b}{x} \cdot \frac{y}{b-a} = \frac{\cancel{a-b}}{x} \cdot \frac{y}{-1\cancel{(a-b)}} = -\frac{y}{x}$$

EXAMPLE 7 Multiply $\frac{3x+2}{2x-1} \cdot \frac{4-8x}{3x+2}$.

Solution:

$$\frac{3x+2}{2x-1} \cdot \frac{4-8x}{3x+2} = \frac{3x+2}{2x-1} \cdot \frac{4(1-2x)}{3x+2}$$

$$= \frac{\cancel{(3x+2)}}{2x-1} \cdot \frac{4(1-2x)}{\cancel{(3x+2)}}.$$

Note that the factor $(1 - 2x)$ in the numerator of the second fraction differs only in sign from $(2x - 1)$, the denominator of the first fraction. We will therefore factor -1 from the numerator.

$$= \frac{\cancel{(3x+2)}}{2x-1} \cdot \frac{4(-1)(2x-1)}{\cancel{(3x+2)}}$$

$$= \frac{\cancel{(3x+2)}}{\cancel{(2x-1)}} \cdot \frac{-4\cancel{(2x-1)}}{\cancel{(3x+2)}} = \frac{-4}{1} = -4$$

EXAMPLE 8 Multiply $\frac{2x^2+7x-15}{4x^2-8x+3} \cdot \frac{2x^2+x-1}{x^2+6x+5}$.

Solution:

$$\frac{2x^2+7x-15}{4x^2-8x+3} \cdot \frac{2x^2+x-1}{x^2+6x+5} = \frac{(2x-3)(x+5)}{(2x-3)(2x-1)} \cdot \frac{(2x-1)(x+1)}{(x+1)(x+5)}$$

$$= \frac{\cancel{(2x-3)}\cancel{(x+5)}}{\cancel{(2x-3)}\cancel{(2x-1)}} \cdot \frac{\cancel{(2x-1)}\cancel{(x+1)}}{\cancel{(x+1)}\cancel{(x+5)}} = 1$$

EXAMPLE 9 Multiply $\frac{2x^3 - 14x^2 + 12x}{6y^2} \cdot \frac{-2y}{3x^2 - 3x}$.

Solution:

$$\frac{2x^3 - 14x^2 + 12x}{6y^2} \cdot \frac{-2y}{3x^2 - 3x} = \frac{2x(x^2 - 7x + 6)}{6y^2} \cdot \frac{-2y}{3x(x-1)}$$

$$= \frac{2x(x-6)(x-1)}{6y^2} \cdot \frac{-2y}{3x(x-1)}$$

$$= \frac{\overset{}{\cancel{2}}\cancel{x}(x-6)\cancel{(x-1)}}{\underset{3}{\cancel{6}}\underset{y}{\cancel{y^2}}} \cdot \frac{-2\cancel{y}}{3\cancel{x}\cancel{(x-1)}}$$

$$= \frac{-2(x-6)}{9y} = \frac{-2x + 12}{9y}$$

EXAMPLE 10 Multiply $\frac{x^2 - y^2}{x + y} \cdot \frac{x + 2y}{2x^2 - xy - y^2}$.

Solution:

$$\frac{x^2 - y^2}{x + y} \cdot \frac{x + 2y}{2x^2 - xy - y^2} = \frac{(x+y)(x-y)}{x+y} \cdot \frac{x+2y}{(2x+y)(x-y)}$$

$$= \frac{\cancel{(x+y)}\cancel{(x-y)}}{\cancel{(x+y)}} \cdot \frac{x+2y}{(2x+y)\cancel{(x-y)}}$$

$$= \frac{x+2y}{2x+y}$$

Divide Rational Expressions

2 In Chapter 1 we learned that to divide one fraction by a second we invert the divisor and multiply.

Division

$$\frac{a}{b} \div \frac{c}{d} = \frac{a}{b} \cdot \frac{d}{c} = \frac{ad}{bc}, \qquad b \neq 0, \quad d \neq 0, \quad \text{and} \quad c \neq 0$$

EXAMPLE 11 Divide.

(a) $\frac{3}{5} \div \frac{4}{5}$ **(b)** $\frac{2}{3} \div \frac{5}{6}$

Solution:

(a) $\frac{3}{5} \div \frac{4}{5} = \frac{3}{\underset{1}{\cancel{5}}} \cdot \frac{\overset{1}{\cancel{5}}}{4} = \frac{3 \cdot 1}{1 \cdot 4} = \frac{3}{4}$ **(b)** $\frac{2}{3} \div \frac{5}{6} = \frac{2}{\underset{1}{\cancel{3}}} \cdot \frac{\overset{2}{\cancel{6}}}{5} = \frac{2 \cdot 2}{1 \cdot 5} = \frac{4}{5}$

The same principles are used to divide rational expressions.

To Divide Rational Expressions

Invert the divisor (the second fraction) and multiply.

EXAMPLE 12 $\dfrac{5x^2}{z} \div \dfrac{4z^3}{3}$.

Solution: $\dfrac{5x^2}{z} \div \dfrac{4z^3}{3} = \dfrac{5x^2}{z} \cdot \dfrac{3}{4z^3} = \dfrac{15x^2}{4z^4}$.

EXAMPLE 13 $\dfrac{x^2-9}{x+4} \div \dfrac{x-3}{x+4}$.

Solution:

$$\frac{x^2-9}{x+4} \div \frac{x-3}{x+4} = \frac{x^2-9}{x+4} \cdot \frac{x+4}{x-3}$$

$$= \frac{(x+3)\cancel{(x-3)}}{\cancel{(x+4)}} \cdot \frac{\cancel{(x+4)}}{\cancel{(x-3)}} = x+3.$$

EXAMPLE 14 $\dfrac{-1}{2x-3} \div \dfrac{3}{3-2x}$.

Solution:

$$\frac{-1}{2x-3} \div \frac{3}{3-2x} = \frac{-1}{2x-3} \cdot \frac{3-2x}{3}$$

$$= \frac{-1}{\cancel{(2x-3)}} \cdot \frac{-1\cancel{(2x-3)}}{3}$$

$$= \frac{(-1)(-1)}{(1)(3)} = \frac{1}{3}.$$

EXAMPLE 15 $\dfrac{x^2+8x+15}{x^2} \div (x+3)^2$.

Solution:

$$\frac{x^2+8x+15}{x^2} \div (x+3)^2 = \frac{x^2+8x+15}{x^2} \cdot \frac{1}{(x+3)^2}$$

$$= \frac{(x+5)\cancel{(x+3)}}{x^2} \cdot \frac{1}{\cancel{(x+3)}(x+3)}$$

$$= \frac{x+5}{x^2(x+3)}.$$

EXAMPLE 16 $\dfrac{12x^2-22x+8}{3x} \div \dfrac{3x^2+2x-8}{2x^2+4x}$.

Solution:

$$\frac{12x^2-22x+8}{3x} \div \frac{3x^2+2x-8}{2x^2+4x} = \frac{12x^2-22x+8}{3x} \cdot \frac{2x^2+4x}{3x^2+2x-8}$$

$$= \frac{2(6x^2-11x+4)}{3x} \cdot \frac{2x(x+2)}{(3x-4)(x+2)}$$

$$= \frac{2\cancel{(3x-4)}(2x-1)}{3\cancel{x}} \cdot \frac{2\cancel{x}\cancel{(x+2)}}{\cancel{(3x-4)}\cancel{(x+2)}}$$

$$= \frac{4(2x-1)}{3} = \frac{8x-4}{3}.$$

Exercise Set 6.2

Multiply.

1. $\frac{3x}{2y} \cdot \frac{y^2}{6}$

2. $\frac{15x^3y^2}{z} \cdot \frac{z}{5xy^3}$

3. $\frac{16x^2}{y^4} \cdot \frac{5x^2}{y^2}$

4. $\frac{32m}{5n^3} \cdot \frac{-15m^2n^3}{4}$

5. $\frac{6x^5y^3}{5z^3} \cdot \frac{6x^4}{5yz^4}$

6. $\frac{x^2 - 4}{x^2 - 9} \cdot \frac{x + 3}{x - 2}$

7. $\frac{3x - 2}{3x + 2} \cdot \frac{4x - 1}{1 - 4x}$

8. $\frac{x - 6}{2x + 5} \cdot \frac{2x}{-x + 6}$

9. $\frac{x^2 + 7x + 12}{x + 4} \cdot \frac{1}{x + 3}$

10. $\frac{x^2 + 3x - 10}{2x} \cdot \frac{x^2 - 3x}{x^2 - 5x + 6}$

11. $\frac{a^2 - b^2}{a} \cdot \frac{a^2 + ab}{a + b}$

12. $\frac{x^2 - 25}{x^2 - 3x - 10} \cdot \frac{x + 2}{x}$

13. $\frac{6x^2 - 14x - 12}{6x + 4} \cdot \frac{x + 3}{2x^2 - 2x - 12}$

14. $\frac{2x^2 - 9x + 9}{8x - 12} \cdot \frac{2x}{x^2 - 3x}$

15. $\frac{x + 3}{x - 3} \cdot \frac{x^3 - 27}{x^2 + 3x + 9}$

16. $\frac{x^3 + 8}{x^2 - x - 6} \cdot \frac{x + 3}{x^2 - 2x + 4}$

Divide.

17. $\frac{6x^3}{y} \div \frac{2x}{y^2}$

18. $\frac{9x^3}{4} \div \frac{1}{16y^2}$

19. $\frac{25xy^2}{7z} \div \frac{5x^2y^2}{14z^2}$

20. $\frac{36y}{7z^2} \div \frac{3xy}{2z}$

21. $\frac{7a^2b}{xy} \div \frac{7}{6xy}$

22. $2xz \div \frac{4xy}{z}$

23. $\frac{3x^2 + 6x}{x} \div \frac{2x + 4}{x^2}$

24. $\frac{x - 3}{4y^2} \div \frac{x^2 - 9}{2xy}$

25. $(x - 3) \div \frac{x^2 + 3x - 18}{x}$

26. $\frac{1}{x^2 - 17x + 30} \div \frac{1}{x^2 + 7x - 18}$

27. $\frac{x^2 - 12x + 32}{x^2 - 6x - 16} \div \frac{x^2 - x - 12}{x^2 - 5x - 24}$

28. $\frac{a - b}{9a + 9b} \div \frac{a^2 - b^2}{a^2 + 2a + 1}$

29. $\frac{2x^2 + 9x + 4}{x^2 + 7x + 12} \div \frac{2x^2 - x - 1}{(x + 3)^2}$

30. $\frac{a^2 - b^2}{9} \div \frac{3a - 3b}{27x^2}$

31. $\frac{x^2 - y^2}{x^2 - 2xy + y^2} \div \frac{x + y}{x - y}$

32. $\frac{9x^2 - 9y^2}{6x^2y^2} \div \frac{3x + 3y}{12x^2y^5}$

Perform the operation indicated.

33. $\frac{12x^2}{6y^2} \cdot \frac{36xy^5}{12}$

34. $\frac{y^3}{8} \cdot \frac{9x^2}{y^3}$

35. $\frac{45a^2b^3}{12c^3} \cdot \frac{4c}{9a^3b^5}$

36. $\frac{-2xw}{y^5} \div \frac{6x^2}{y^6}$

37. $\frac{-xy}{a} \div \frac{-2ax}{6y}$

38. $\frac{27x}{5y^2} \div 3x^2y^2$

39. $\frac{80m^4}{49x^5y^7} \cdot \frac{14x^{12}y^5}{25m^5}$

40. $\frac{-18x^2y}{11z^2} \cdot \frac{22z^3}{x^2y^5}$

41. $(2x + 5) \cdot \frac{1}{4x + 10}$

42. $\frac{1}{4x - 3} \cdot (20x - 15)$

43. $\frac{1}{7x^2y} \div \frac{1}{21x^3y}$

44. $\frac{x^2y^5}{3z} \div \frac{3z}{2x}$

45. $\frac{12a^2}{4bc} \div \frac{3a^2}{bc}$

46. $\frac{4-x}{x-4} \cdot \frac{x-3}{3-x}$

47. $\frac{5-2x}{x+8} \cdot \frac{-x-8}{2x-5}$

48. $\frac{6x+6y}{a} \div \frac{12x+12y}{a^2}$

49. $\frac{2a+2b}{3} \div \frac{a^2-b^2}{a-b}$

50. $\frac{a^2b^2}{6x+6y} \div \frac{ab}{x^2-y^2}$

51. $\frac{1}{-x-4} \div \frac{x^2-7x}{x^2-3x-28}$

52. $\frac{x^2-5x-24}{x^2-x-12} \cdot \frac{x^2+x-6}{x^2-10x+16}$

53. $\frac{4x+4y}{xy^2} \cdot \frac{x^2y}{3x+3y}$

54. $\frac{x^2}{x^2-4} \cdot \frac{x^2-5x+6}{x^2-3x}$

55. $\frac{x^2+10x+21}{x+7} \div (x+3)$

56. $\frac{x^2-9x+14}{x^2-5x+6} \div \frac{x^2-5x-14}{x+2}$

57. $\frac{3x^2-x-2}{x+7} \div \frac{x-1}{4x^2+25x-21}$

58. $\frac{9x^2+6x-8}{x-3} \cdot \frac{(x-3)^2}{3x+4}$

59. $\frac{5x^2+17x+6}{x+3} \cdot \frac{x-1}{5x^2+7x+2}$

60. $\frac{2x+4y}{x^2+4xy+4y^2} \cdot \frac{x+2y}{2}$

61. $\frac{x^2-y^2}{x^2+xy} \cdot \frac{3x^2+6x}{3x^2-2xy-y^2}$

62. $\frac{x^2-4}{2y} \div \frac{2-x}{6xy}$

63. $\frac{2x^2+7x-15}{1-x} \div \frac{3x^2+13x-10}{x-1}$

64. $\frac{x^2-y^2}{8x^2-16xy+8y^2} \cdot \frac{4x-4y}{x+y}$

65. $\frac{x^2-4y^2}{x^2+3xy+2y^2} \cdot \frac{x+y}{x^2-4xy+4y^2}$

66. $\frac{x^3-64}{x+4} \div \frac{x^2+4x+16}{x^2+8x+16}$

What polynomial should be in the shaded area of the second fraction to make the statement true? Explain how you determined your answer.

67. $\frac{x+3}{x-4} \cdot \frac{\blacksquare}{x+3} = x+2$

68. $\frac{x-5}{x+2} \cdot \frac{\blacksquare}{x-5} = 2x-3$

69. $\frac{x-4}{x+5} \cdot \frac{x+5}{\blacksquare} = \frac{1}{x+3}$

70. $\frac{2x-1}{x-3} \cdot \frac{x-3}{\blacksquare} = \frac{1}{x-6}$

71. In your own words, explain how to multiply rational expressions.

72. In your own words, explain how to divide rational expressions.

CUMULATIVE REVIEW EXERCISES

[4.5] **73.** Multiply $(4x^3y^2z^4)(5xy^3z^7)$.

[4.6] **74.** Divide $\frac{4x^3-5x}{2x-1}$.

[5.4] **75.** Factor $3x^2-9x-30$.

[5.6] **76.** Solve $3x^2-9x-30=0$.

Group Activity/ Challenge Problems

Simplify.

1. $\left(\frac{x+2}{x^2-4x-12} \cdot \frac{x^2-9x+18}{x-2}\right) \div \frac{x^2+5x+6}{x^2-4}$

2. $\left(\frac{x^2-x-6}{2x^2-9x+9} \div \frac{x^2+x-12}{x^2+3x-4}\right) \cdot \frac{2x^2-5x+3}{x^2+x-2}$

Determine the polynomials that when placed in the shaded areas make the statement true. Explain how you determined your answer.

3. $\frac{\blacksquare}{\blacksquare} \cdot \frac{x^2+3x-4}{x^2-4x+3} = \frac{x-2}{x-5}$

4. $\frac{\blacksquare}{x^2+x-2} \cdot \frac{x^2+6x+8}{\blacksquare} = \frac{x+3}{x+5}$

6.3 Addition and Subtraction of Rational Expressions with a Common Denominator and Finding the Least Common Denominator

Tape 9

1. Add and subtract rational expressions with a common denominator.
2. Find the least common denominator.

Adding Expressions with a Common Denominator

1 Recall that when adding (or subtracting) two arithmetic fractions with a common denominator we add (or subtract) the numerators while keeping the common denominator.

Addition and Subtraction

$$\frac{a}{c} + \frac{b}{c} = \frac{a+b}{c},\ c \neq 0 \qquad \frac{a}{c} - \frac{b}{c} = \frac{a-b}{c},\ c \neq 0$$

EXAMPLE 1 Add: **(a)** $\frac{3}{8} + \frac{2}{8}$ **(b)** $\frac{5}{7} - \frac{1}{7}$

Solution: **(a)** $\frac{3}{8} + \frac{2}{8} = \frac{3+2}{8} = \frac{5}{8}$ **(b)** $\frac{5}{7} - \frac{1}{7} = \frac{5-1}{7} = \frac{4}{7}$

Note in Example 1(a) that we did not reduce $\frac{2}{8}$ to $\frac{1}{4}$. The fractions are given with a common denominator, 8. If $\frac{2}{8}$ was reduced to $\frac{1}{4}$, you would lose the common denominator that is needed to add or subtract fractions.

To Add or Subtract Rational Expressions with a Common Denominator

1. Add or subtract the numerators.
2. Place the sum or difference of the numerators found in step 1 over the common denominator.
3. Reduce the fraction if possible.

EXAMPLE 2 $\frac{3}{x+2} + \frac{x-4}{x+2}$.

Solution:

$$\frac{3}{x+2} + \frac{x-4}{x+2} = \frac{3+(x-4)}{x+2} = \frac{x-1}{x+2}.$$

EXAMPLE 3 $\dfrac{2x^2+5}{x+3}+\dfrac{6x-5}{x+3}$.

Solution: $\dfrac{2x^2+5}{x+3}+\dfrac{6x-5}{x+3}=\dfrac{2x^2+5+(6x-5)}{x+3}=\dfrac{2x^2+6x}{x+3}$.

Now factor $2x$ from each term in the numerator and reduce.

$$\frac{2x\cancel{(x+3)}}{\cancel{(x+3)}}=2x$$

EXAMPLE 4 $\dfrac{x^2+3x-2}{(x+5)(x-2)}+\dfrac{4x+12}{(x+5)(x-2)}$.

Solution:

$$\begin{aligned}\frac{x^2+3x-2}{(x+5)(x-2)}+\frac{4x+12}{(x+5)(x-2)}&=\frac{x^2+3x-2+(4x+12)}{(x+5)(x-2)}\\&=\frac{x^2+7x+10}{(x+5)(x-2)}\\&=\frac{\cancel{(x+5)}(x+2)}{\cancel{(x+5)}(x-2)}\\&=\frac{x+2}{x-2}\end{aligned}$$

When subtracting rational expressions, be sure to subtract the entire numerator of the fraction being subtracted. Study the following common student error very carefully.

COMMON STUDENT ERROR

Consider the subtraction

$$\frac{4x}{x-2}-\frac{2x+1}{x-2}$$

Many students begin problems of this type incorrectly. Here are the correct and incorrect ways of working this problem.

Correct

$$\begin{aligned}\frac{4x}{x-2}-\frac{2x+1}{x-2}&=\frac{4x-(2x+1)}{x-2}\\&=\frac{4x-2x-1}{x-2}\\&=\frac{2x-1}{x-2}\end{aligned}$$

Incorrect

$$\frac{4x}{x-2}-\frac{2x+1}{x-2}=\frac{4x-2x+1}{x-2}$$

Note that the entire numerator of the second fraction (not just the first term) **must be subtracted.** Also note that the sign of *each* term of the numerator being subtracted will change when the parentheses are removed.

EXAMPLE 5 $\dfrac{x^2-2x+3}{x^2+7x+12}-\dfrac{x^2-4x-5}{x^2+7x+12}$.

Solution:

$$\frac{x^2 - 2x + 3}{x^2 + 7x + 12} - \frac{x^2 - 4x - 5}{x^2 + 7x + 12} = \frac{x^2 - 2x + 3 - (x^2 - 4x - 5)}{x^2 + 7x + 12}$$

$$= \frac{x^2 - 2x + 3 - x^2 + 4x + 5}{x^2 + 7x + 12}$$

$$= \frac{2x + 8}{x^2 + 7x + 12}$$

$$= \frac{2\cancel{(x + 4)}}{(x + 3)\cancel{(x + 4)}}$$

$$= \frac{2}{x + 3}$$

EXAMPLE 6 $\dfrac{5x}{x - 6} - \dfrac{4x^2 - 16x - 18}{x - 6}$

Solution:

$$\frac{5x}{x - 6} - \frac{4x^2 - 16x - 18}{x - 6} = \frac{5x - (4x^2 - 16x - 18)}{x - 6}$$

$$= \frac{5x - 4x^2 + 16x + 18}{x - 6}$$

$$= \frac{-4x^2 + 21x + 18}{x - 6}$$

$$= \frac{-(4x^2 - 21x - 18)}{x - 6}$$

$$= \frac{-(4x + 3)\cancel{(x - 6)}}{\cancel{(x - 6)}}$$

$$= -(4x + 3) \text{ or } -4x - 3$$

Finding the Least Common Denominator

2 To add two fractions with unlike denominators, we must first obtain a common denominator. Now we explain how to find common denominators for rational expressions. We will use this information in Section 6.4 when we add and subtract rational expressions.

EXAMPLE 7 Add $\dfrac{3}{5} + \dfrac{4}{7}$.

Solution: The least common denominator (LCD) of the fractions $\frac{3}{5}$ and $\frac{4}{7}$ is 35. Thirty-five is the smallest number that is divisible by both denominators, 5 and 7. Rewrite each fraction so that its denominator is 35.

$$\frac{3}{5} + \frac{4}{7} = \frac{3}{5} \cdot \frac{7}{7} + \frac{4}{7} \cdot \frac{5}{5}$$

$$= \frac{21}{35} + \frac{20}{35} = \frac{41}{35} \text{ or } 1\frac{6}{35}$$

To add or subtract rational expressions, we must write each expression with a common denominator.

To Find the Least Common Denominator of Rational Expressions

1. Factor each denominator completely. Any factors that occur more than once should be expressed as powers. For example, $(x + 5)(x + 5)$ should be expressed as $(x + 5)^2$.
2. List all different factors (other than 1) that appear in any of the denominators. When the same factor appears in more than one denominator, write that factor with the highest power that appears.
3. The least common denominator is the product of all the factors listed in step 2.

EXAMPLE 8 Find the least common denominator.

$$\frac{1}{3} + \frac{1}{x}$$

Solution: The only factor (other than 1) of the first denominator is 3. The only factor (other than 1) of the second denominator is x. The LCD is therefore $3 \cdot x = 3x$.

EXAMPLE 9 Find the LCD.

$$\frac{3}{5x} - \frac{2}{x^2}$$

Solution: The factors that appear in the denominators are 5 and x. List each factor with its highest power. The LCD is the product of these factors.

highest power of x

$$\text{LCD} = 5 \cdot x^2 = 5x^2$$

EXAMPLE 10 Find the LCD.

$$\frac{1}{18x^3y} + \frac{5}{27x^2y^3}$$

Solution: Write both 18 and 27 as products of prime factors. $18 = 2 \cdot 3^2$ and $27 = 3^3$. *If you have forgotten how to write a number as a product of prime factors, read Section 5.1 or Appendix B now.*

$$\frac{1}{18x^3y} + \frac{5}{27x^2y^3} = \frac{1}{2 \cdot 3^2x^3y} + \frac{5}{3^3x^2y^3}$$

The factors that appear are 2, 3, x, and y. List the highest powers of each of these factors.

$$\text{LCD} = 2 \cdot 3^3 \cdot x^3 \cdot y^3 = 54x^3y^3$$

EXAMPLE 11 Find the LCD.

$$\frac{3}{x} - \frac{2y}{x + 5}$$

Solution: The factors in the denominators are x and $(x + 5)$. *Note that the x in the second denominator, $x + 5$, is a term, not a factor, since the operation is addition rather than multiplication.*

$$\text{LCD} = x(x + 5)$$

EXAMPLE 12 Find the LCD.

$$\frac{3}{2x^2 - 4x} + \frac{x^2}{x^2 - 4x + 4}$$

Solution: Factor both denominators.

$$\frac{3}{2x^2 - 4x} + \frac{x^2}{x^2 - 4x + 4} = \frac{3}{2x(x - 2)} + \frac{x^2}{(x - 2)(x - 2)}$$
$$= \frac{3}{2x(x - 2)} + \frac{x^2}{(x - 2)^2}$$

The factors in the denominators are 2, x, and $x - 2$. List the highest powers of each of these factors.

$$\text{LCD} = 2 \cdot x \cdot (x - 2)^2 = 2x(x - 2)^2$$

EXAMPLE 13 Find the LCD.

$$\frac{5x}{x^2 - x - 12} - \frac{6x^2}{x^2 - 7x + 12}$$

Solution: Factor both denominators.

$$\frac{5x}{x^2 - x - 12} - \frac{6x^2}{x^2 - 7x + 12} = \frac{5x}{(x + 3)(x - 4)} - \frac{6x^2}{(x - 3)(x - 4)}$$

The factors in the denominators are $x + 3$, $x - 4$, and $x - 3$.

$$\text{LCD} = (x + 3)(x - 4)(x - 3)$$

Although $(x - 4)$ is a common factor of each denominator, the highest power of that factor that appears in each denominator is 1.

EXAMPLE 14 Find the LCD.

$$\frac{3x}{x^2 - 14x + 48} + x + 9$$

Solution: Factor the denominator of the first term.

$$\frac{3x}{x^2 - 14x + 48} + x + 9 = \frac{3x}{(x - 6)(x - 8)} + x + 9$$

Since the denominator of $x + 9$ is 1, the expression can be rewritten as

$$\frac{3x}{(x - 6)(x - 8)} + \frac{x + 9}{1}$$

The LCD is therefore $1(x - 6)(x - 8)$ or simply $(x - 6)(x - 8)$.

Exercise Set 6.3

Add or subtract.

1. $\dfrac{x-1}{6}+\dfrac{x}{6}$

2. $\dfrac{x-7}{3}-\dfrac{4}{3}$

3. $\dfrac{2x+3}{5}-\dfrac{x}{5}$

4. $\dfrac{3x+6}{2}-\dfrac{x}{2}$

5. $\dfrac{1}{x}+\dfrac{x+2}{x}$

6. $\dfrac{3x+4}{x+1}+\dfrac{6x+5}{x+1}$

7. $\dfrac{x-3}{x}+\dfrac{x+3}{x}$

8. $\dfrac{x-4}{x}-\dfrac{x+4}{x}$

9. $\dfrac{x}{x-2}+\dfrac{2x+3}{x-2}$

10. $\dfrac{4x-3}{x-7}-\dfrac{2x+8}{x-7}$

11. $\dfrac{9x+7}{6x^2}-\dfrac{3x+4}{6x^2}$

12. $\dfrac{-2x-4}{x^2+2x+1}+\dfrac{3x+5}{x^2+2x+1}$

13. $\dfrac{-2x+6}{x^2+x-6}+\dfrac{3x-3}{x^2+x-6}$

14. $\dfrac{-x-4}{x^2-16}+\dfrac{2(x+4)}{x^2-16}$

15. $\dfrac{x+4}{3x+2}-\dfrac{x+4}{3x+2}$

16. $\dfrac{2x+4}{(x+2)(x-3)}-\dfrac{x+7}{(x+2)(x-3)}$

17. $\dfrac{2x+4}{x-7}-\dfrac{6x+5}{x-7}$

18. $\dfrac{x^2+2x}{3x}-\dfrac{x^2+5x+6}{3x}$

19. $\dfrac{x^2+4x+3}{x+2}-\dfrac{5x+9}{x+2}$

20. $\dfrac{4}{2x+3}+\dfrac{6x+5}{2x+3}$

21. $\dfrac{-2x+5}{5x-10}+\dfrac{2(x-5)}{5x-10}$

22. $\dfrac{x^2}{x+3}+\dfrac{9}{x+3}$

23. $\dfrac{x^2-2x-3}{x^2-x-6}+\dfrac{x-3}{x^2-x-6}$

24. $\dfrac{4x+12}{3-x}-\dfrac{3x+15}{3-x}$

25. $\dfrac{-x-7}{2x-9}-\dfrac{-3x-16}{2x-9}$

26. $\dfrac{x^2-2}{x^2+6x-7}-\dfrac{-4x+19}{x^2+6x-7}$

27. $\dfrac{x^2+6x}{(x+9)(x+5)}-\dfrac{27}{(x+9)(x+5)}$

28. $\dfrac{x^2-13}{x+4}-\dfrac{3}{x+4}$

29. $\dfrac{3x^2-7x}{4x^2-8x}+\dfrac{x}{4x^2-8x}$

30. $\dfrac{3x^2+15x}{x^3+2x^2-8x}+\dfrac{2x^2+5x}{x^3+2x^2-8x}$

31. $\dfrac{2x^2-6x+5}{2x^2+18x+16}-\dfrac{8x+21}{2x^2+18x+16}$

32. $\dfrac{x^3-10x^2+35x}{x(x-6)}-\dfrac{x^2+5x}{x(x-6)}$

33. $\dfrac{x^2+3x-6}{x^2-5x+4}-\dfrac{-2x^2+4x-4}{x^2-5x+4}$

34. $\dfrac{4x^2+5}{9x^2-64}-\dfrac{x^2-x+29}{9x^2-64}$

35. $\dfrac{5x^2+40x+8}{x^2-64}+\dfrac{x^2+9x}{x^2-64}$

36. $\dfrac{20x^2+5x+1}{6x^2+x-2}-\dfrac{8x^2-12x-5}{6x^2+x-2}$

*In Exercises 37–40 **(a)** Explain why the expression on the left side of the equal sign is not equal to the expression on the right side of the equal sign. **(b)** Show what the expression on the right side should be for it to be equal to the one on the left.*

37. $\dfrac{4x-3}{5x+4}-\dfrac{2x-7}{5x+4}\neq\dfrac{4x-3-2x-7}{5x+4}$

38. $\dfrac{6x-2}{x^2-4x+3}-\dfrac{3x^2-4x+5}{x^2-4x+3}\neq\dfrac{6x-2-3x^2-4x-}{x^2-4x+3}$

39. $\dfrac{4x+5}{x^2-6x}-\dfrac{-x^2+3x+6}{x^2-6x}\neq\dfrac{4x+5+x^2+3x+6}{x^2-6x}$

40. $\dfrac{3x^2-4}{x^2+2x}-\dfrac{2x-7}{x^2+2x}\neq\dfrac{3x^2-4-2x-7}{x^2+2x}$

Find the least common denominator.

41. $\dfrac{x}{3}+\dfrac{x-1}{3}$

42. $\dfrac{4-x}{5}-\dfrac{12}{5}$

43. $\dfrac{1}{2x}+\dfrac{1}{3}$

44. $\dfrac{1}{x+2}-\dfrac{3}{5}$

45. $\dfrac{3}{5x}+\dfrac{7}{2}$

46. $\dfrac{2}{3x}+1$

47. $\dfrac{2}{x^2}+\dfrac{3}{x}$

48. $\dfrac{5x}{x+1}+\dfrac{6}{x+2}$

49. $\dfrac{x+4}{2x+3}+x$

50. $\dfrac{x+4}{2x}+\dfrac{3}{7x}$

51. $\dfrac{x}{x+1}+\dfrac{4}{x^2}$

52. $\dfrac{x}{3x^2}+\dfrac{9}{15x^4}$

53. $\frac{x+3}{16x^2y} - \frac{5}{9x^3}$

54. $\frac{-4}{8x^2y^2} + \frac{7}{5x^4y^5}$

55. $\frac{x^2+3}{18x} - \frac{x-7}{12(x+5)}$

56. $\frac{x-7}{3x+5} - \frac{6}{x+5}$

57. $\frac{2x-7}{x^2+x} - \frac{x^2}{x+1}$

58. $\frac{9}{(x-4)(x+2)} - \frac{x+8}{x+2}$

59. $\frac{15}{36x^2y} + \frac{x+3}{15xy^3}$

60. $\frac{x^2-4}{x^2-16} + \frac{3}{x+4}$

61. $\frac{6}{2x+8} + \frac{6x+3}{3x-9}$

62. $6x^2 + \frac{9x}{x-3}$

63. $\frac{9x+4}{x+6} - \frac{3x-6}{x+5}$

64. $\frac{x+1}{x^2+11x+18} - \frac{x^2-4}{x^2-3x-10}$

65. $\frac{x-2}{x^2-5x-24} + \frac{3}{x^2+11x+24}$

66. $\frac{6x+5}{x^2-4} - \frac{3x}{x^2-5x-14}$

67. $\frac{6}{x+3} - \frac{x+5}{x^2-4x+3}$

68. $\frac{3x-8}{x^2-1} + \frac{x^2+5}{x+1}$

69. $\frac{2x}{x^2-x-2} - \frac{3}{x^2+4x+3}$

70. $\frac{6x+5}{x+2} + \frac{4x}{(x+2)^2}$

71. $\frac{3x-5}{x^2+4x+4} + \frac{3}{x+2}$

72. $\frac{9x+7}{(x+3)(x+2)} - \frac{4x}{(x-3)(x+2)}$

73. $\frac{4x^2}{x^2-9x+20} + x + 5$

74. $\frac{x-1}{x^2-9} + x - 4$

75. $\frac{x}{3x^2+16x-12} + \frac{6}{3x^2+17x-6}$

76. $\frac{2x-7}{2x^2+5x+2} - \frac{x^2}{3x^2+4x-4}$

77. $\frac{2x-3}{4x^2+4x+1} + \frac{x^2-4}{8x^2+10x+3}$

78. In your own words, explain how to add or subtract rational expressions with a common denominator.

79. When subtracting rational expressions, what must happen to the sign of each term of the numerator being subtracted?

80. In your own words, explain how to find the least common denominator of two rational expressions.

81. What is the least common denominator in the following subtraction? Explain how you determined your answer. $\frac{3x}{x^2+x-12} - \frac{2x}{-x^2-x+12}$

List the polynomial to be placed in the shaded area to make a true statement. Explain how you determined your answer.

82. $\frac{x^2-6x+3}{x+3} + \frac{\square}{x+3} = \frac{2x^2-5x-6}{x+3}$

83. $\frac{-x^2-4x+3}{2x+5} + \frac{\square}{2x+5} = \frac{5x-7}{2x+5}$

84. $\frac{4x^2-6x-7}{x^2-4} - \frac{\square}{x^2-4} = \frac{2x^2+x-3}{x^2-4}$

85. $\frac{-3x^2-9}{(x+4)(x-2)} - \frac{\square}{(x+4)(x-2)} = \frac{x^2+3x}{(x+4)(x-2)}$

CUMULATIVE REVIEW EXERCISES

86. Subtract $4\frac{3}{5} - 2\frac{5}{9}$.

87. Solve $6x + 4 = -(x + 2) - 3x + 4$.

88. The instructions on a bottle of concentrated hummingbird food indicate that 6 ounces of the concentrate should be mixed with 1 gallon (128 ounces) of water. If you wish to mix the concentrate with only 24 ounces of water, how much concentrate should be used?

89. The American Health Racquet Club has two payment plans. Plan 1 is a yearly membership fee of \$100 plus \$2 per hour for use of the court. Plan 2 is an annual membership fee of \$250 with no charge for court time.

(a) How many hours would Shamo have to play in a year to make the cost of plan 1 equal to the cost of plan 2?

(b) If Shamo plans to play an average of 4 hours per week for the year, which plan should he use?

[6.2] **90.** Divide $\frac{x^2+x-6}{2x^2+7x+3} \div \frac{x^2+5x+6}{x^2-4}$.

Group Activity/ Challenge Problems

Add or subtract.

1. $\dfrac{3x-2}{x^2-9} - \dfrac{4x^2-6}{x^2-9} + \dfrac{5x-1}{x^2-9}$

2. $\dfrac{x^2-6x+3}{x+2} + \dfrac{x^2-2x}{x+2} - \dfrac{2x^2-3x+5}{x+2}$

Find the least common denominator.

3. $\dfrac{3}{2x^3y^6} - \dfrac{5}{6x^5y^9} + \dfrac{1}{5x^{12}y^2}$

4. $\dfrac{x}{x-2} - \dfrac{4}{x^2-4} + \dfrac{3}{x+2}$

5. $\dfrac{4}{x^2-x-12} + \dfrac{3}{x^2-6x+8} + \dfrac{5}{x^2+x-6}$

6. $\dfrac{3}{2x^2-4x+3x-6} - \dfrac{4}{x^2-4} + \dfrac{5}{2x^2+7x+6}$

7. $\dfrac{7}{2x^2+x-10} + \dfrac{5}{2x^2+7x+5} - \dfrac{4}{-x^2+x+2}$

6.4 Addition and Subtraction of Rational Expressions

1 Add and subtract rational expressions.

Tape 10

In Section 6.3 we discussed how to add or subtract rational expressions with a common denominator. Now we discuss adding and subtracting rational expressions that are not given with a common denominator.

Add and Subtract Rational Expressions

1 The method used to add or subtract rational expressions with unlike denominators is outlined in Example 1.

EXAMPLE 1 Add $\dfrac{3}{x} + \dfrac{5}{y}$.

Solution: First, determine the LCD as outlined in Section 6.3.

$$\text{LCD} = xy$$

Now write each fraction with the LCD. We do this by multiplying **both** the numerator and denominator of each fraction by any factors needed to obtain the LCD.

In this problem the fraction on the left must be multiplied by y/y and the fraction on the right must be multiplied by x/x.

$$\frac{3}{x} + \frac{5}{y} = \frac{y}{y} \cdot \frac{3}{x} + \frac{5}{y} \cdot \frac{x}{x} = \frac{3y}{xy} + \frac{5x}{xy}$$

By multiplying both the numerator and denominator by the same factor, we are in effect multiplying by 1, which does not change the value of the fraction, only its appearance. Thus, the new fraction is equivalent to the original fraction.

Now add the numerators, while leaving the LCD alone.

$$\frac{3y}{xy} + \frac{5x}{xy} = \frac{3y+5x}{xy} \quad \text{or} \quad \frac{5x+3y}{xy}$$

To Add or Subtract Two Rational Expressions with Unlike Denominators

1. Determine the LCD.
2. Rewrite each fraction as an equivalent fraction with the LCD. This is done by multiplying both the numerator and denominator of each fraction by any factors needed to obtain the LCD.
3. Add or subtract the numerators while maintaining the LCD.
4. When possible, factor the remaining numerator and reduce the fraction.

EXAMPLE 2 Add $\frac{5}{4x^2y} + \frac{3}{14xy^3}$.

Solution: The LCD is $28x^2y^3$. We must write each fraction with the denominator $28x^2y^3$. To do this, multiply the left fraction by $7y^2/7y^2$ and the right fraction by $2x/2x$.

$$\frac{5}{4x^2y} + \frac{3}{14xy^3} = \frac{7y^2}{7y^2} \cdot \frac{5}{4x^2y} + \frac{3}{14xy^3} \cdot \frac{2x}{2x}$$

$$= \frac{35y^2}{28x^2y^3} + \frac{6x}{28x^2y^3}$$

$$= \frac{35y^2 + 6x}{28x^2y^3} \text{ or } \frac{6x + 35y^2}{28x^2y^3}$$

Helpful Hint

In Example 2 we multiplied the first fraction by $\frac{7y^2}{7y^2}$ and the second fraction by $\frac{2x}{2x}$ to get two fractions with a common denominator. How did we know what to multiply each fraction by? Many of you can determine this by observing the LCD and then determining what each denominator needs to be multiplied by to get the LCD. If this is not obvious, you can divide the LCD by the given denominator to determine what each fraction needs to be multiplied by. In Example 2, the LCD is $28x^2y^3$. If we divide $28x^2y^3$ by each given denominator, $4x^2y$ and $14xy^3$, we can determine what each respective fraction needs to be multiplied by.

$$\frac{28x^2y^3}{4x^2y} = 7y^2 \qquad \frac{28x^2y^3}{14xy^3} = 2x$$

Thus, $\frac{5}{4x^2y}$ needs to be multiplied by $\frac{7y^2}{7y^2}$ and $\frac{3}{14xy^3}$ needs to be multiplied by $\frac{2x}{2x}$ to obtain the LCD $28x^2y^3$.

EXAMPLE 3 Add $\dfrac{2}{x+2} + \dfrac{4}{x}$.

Solution: We must write each fraction with the LCD which is $x(x+2)$. To do this, multiply the fraction on the left by x/x and the fraction on the right by $(x+2)/(x+2)$.

$$\begin{aligned}\frac{2}{x+2} + \frac{4}{x} &= \frac{x}{x}\cdot\frac{2}{(x+2)} + \frac{4}{x}\cdot\frac{(x+2)}{(x+2)}\\ &= \frac{2x}{x(x+2)} + \frac{4(x+2)}{x(x+2)}\\ &= \frac{2x+4(x+2)}{x(x+2)}\\ &= \frac{2x+4x+8}{x(x+2)} = \frac{6x+8}{x(x+2)}\end{aligned}$$

Helpful Hint

Look at the answer to Example 3, $\dfrac{6x+8}{x(x+2)}$. Notice that the numerator could have been factored to obtain $\dfrac{2(3x+4)}{x(x+2)}$. Also notice that the denominator could have been multiplied to get $\dfrac{6x+8}{x^2+2x}$. All three of these answers are equivalent and each is correct. In this section, when writing answers, unless there is a common factor in the numerator and denominator we will leave the numerator in unfactored form and the denominator in factored form. If both the numerator and denominator have a common factor, we will factor the numerator and reduce the fraction.

EXAMPLE 4 Subtract $\dfrac{x}{x+5} - \dfrac{2}{x-3}$.

Solution: The LCD is $(x+5)(x-3)$. The fraction on the left must be multiplied by $(x-3)/(x-3)$ to obtain the LCD. The fraction on the right must be multiplied by $(x+5)/(x+5)$ to obtain the LCD.

$$\begin{aligned}\frac{x}{x+5} - \frac{2}{x-3} &= \frac{(x-3)}{(x-3)}\cdot\frac{x}{(x+5)} - \frac{2}{(x-3)}\cdot\frac{(x+5)}{(x+5)}\\ &= \frac{x(x-3)}{(x-3)(x+5)} - \frac{2(x+5)}{(x-3)(x+5)}\\ &= \frac{x^2-3x}{(x-3)(x+5)} - \frac{2x+10}{(x-3)(x+5)}\\ &= \frac{x^2-3x-(2x+10)}{(x-3)(x+5)}\\ &= \frac{x^2-3x-2x-10}{(x-3)(x+5)}\\ &= \frac{x^2-5x-10}{(x-3)(x+5)}\end{aligned}$$

EXAMPLE 5 Subtract $\frac{x+2}{x-4} - \frac{x+3}{x+4}$.

Solution: The LCD is $(x-4)(x+4)$.

$$\frac{x+2}{x-4} - \frac{x+3}{x+4} = \frac{(x+4)}{(x+4)} \cdot \frac{(x+2)}{(x-4)} - \frac{(x+3)}{(x+4)} \cdot \frac{(x-4)}{(x-4)}$$

$$= \frac{(x+4)(x+2)}{(x+4)(x-4)} - \frac{(x+3)(x-4)}{(x+4)(x-4)}$$

Use the FOIL method to multiply each numerator.

$$= \frac{x^2+6x+8}{(x+4)(x-4)} - \frac{x^2-x-12}{(x+4)(x-4)}$$

$$= \frac{x^2+6x+8-(x^2-x-12)}{(x+4)(x-4)}$$

$$= \frac{x^2+6x+8-x^2+x+12}{(x+4)(x-4)}$$

$$= \frac{7x+20}{(x+4)(x-4)}$$

Consider the problem

$$\frac{4}{x-3} + \frac{x+5}{3-x}$$

How do we add these rational expressions? We could write each fraction with the denominator $(x-3)(3-x)$. However, there is an easier way. Study the following Helpful Hint.

Helpful Hint

When adding or subtracting fractions whose denominators are opposites (and therefore differ only in signs), multiply both the numerator *and* the denominator of *either* of the fractions by -1. Then both fractions will have the same denominator.

$$\frac{x}{a-b} + \frac{y}{b-a} = \frac{x}{a-b} + \frac{-1}{-1} \cdot \frac{y}{(b-a)}$$

$$= \frac{x}{a-b} + \frac{-y}{a-b}$$

$$= \frac{x-y}{a-b}$$

EXAMPLE 6 Add $\dfrac{4}{x-3}+\dfrac{x+5}{3-x}$.

Solution: Since the denominators differ only in sign, we may multiply both the numerator and the denominator of either fraction by -1. Here we will multiply the numerator and denominator of the second fraction by -1 to obtain the common denominator $x - 3$.

$$\begin{aligned}\frac{4}{x-3}+\frac{x+5}{3-x} &= \frac{4}{x-3}+\frac{-1}{-1}\cdot\frac{(x+5)}{(3-x)}\\ &= \frac{4}{x-3}+\frac{-x-5}{x-3}\\ &= \frac{4-x-5}{x-3}\\ &= \frac{-x-1}{x-3}\end{aligned}$$

Recall from Section 6.1 that we can change any *two* signs of a fraction without changing its value. In Example 7 we change two signs of the fraction to simplify our work.

EXAMPLE 7 Subtract $\dfrac{x+2}{2x-5}-\dfrac{3x+5}{5-2x}$.

Solution: The denominators of the two fractions differ only in sign. If we change the signs of one of the denominators, we will have a common denominator. Here we change *two* of the fraction signs in the second fraction to obtain a common denominator.

$$\begin{aligned}\frac{x+2}{2x-5}-\frac{3x+5}{5-2x} &= \frac{x+2}{2x-5}+\frac{3x+5}{-(5-2x)}\\ &= \frac{x+2}{2x-5}+\frac{3x+5}{2x-5}\\ &= \frac{x+2+3x+5}{2x-5}\\ &= \frac{4x+7}{2x-5}\end{aligned}$$

In Example 7 we elected to change two signs of the second fraction. The same results could be obtained by multiplying both the numerator and denominator of either the first or second fraction by -1 as was done in Example 6. Try this now.

EXAMPLE 8 Add $\dfrac{3}{x^2+5x+6}+\dfrac{1}{3x^2+8x-3}$.

Solution:

$$\frac{3}{x^2+5x+6}+\frac{1}{3x^2+8x-3}=\frac{3}{(x+2)(x+3)}+\frac{1}{(3x-1)(x+3)}$$

The LCD is $(x+2)(x+3)(3x-1)$.

$$=\frac{(3x-1)}{(3x-1)}\cdot\frac{3}{(x+2)(x+3)}+\frac{1}{(3x-1)(x+3)}\cdot\frac{(x+2)}{(x+2)}$$

$$=\frac{9x-3}{(3x-1)(x+2)(x+3)}+\frac{x+2}{(3x-1)(x+2)(x+3)}$$

$$=\frac{10x-1}{(3x-1)(x+2)(x+3)}$$

COMMON STUDENT ERROR A common error in an addition or subtraction problem is to add or subtract the numerators and the denominators. Here is one such example.

Correct	*Incorrect*
$\frac{1}{x}+\frac{x}{1}=\frac{1}{x}+\frac{x^2}{x}$	~~$\frac{1}{x}+\frac{x}{1}=\frac{1+x}{x+1}$~~
$=\frac{1+x^2}{x}$ or $\frac{x^2+1}{x}$	~~$\frac{1}{x}-\frac{x}{1}=\frac{1-x}{x-1}$~~

Remember that to add or subtract fractions you must first have a common denominator. Then you add or subtract the numerators while maintaining the common denominator.

Another common mistake is to treat an addition or subtraction problem as a multiplication problem. You can divide out common factors only when *multiplying* expressions, not when adding or subtracting them.

Correct	*Incorrect*
$\frac{1}{x}\cdot\frac{x}{1}=\frac{1}{\not{x}_1}\cdot\frac{\not{x}^1}{1}$	~~$\frac{1}{x}+\frac{x}{1}=\frac{1}{\not{x}_1}+\frac{\not{x}^1}{1}$~~
$=1\cdot1=1$	~~$=1+1=2$~~

Exercise Set 6.4

Add or subtract.

1. $\frac{3}{x}+\frac{1}{2x}$
2. $\frac{1}{2x}+\frac{1}{2}$
3. $\frac{4}{x^2}+\frac{3}{2x}$
4. $3+\frac{5}{x}$
5. $2-\frac{1}{x^2}$
6. $\frac{5}{6y}+\frac{3}{4y^2}$
7. $\frac{2}{x^2}+\frac{3}{5x}$
8. $\frac{3}{x}-\frac{5}{x^2}$
9. $\frac{3}{4x^2y}+\frac{7}{5xy^2}$
10. $\frac{5}{12x^4y}-\frac{1}{5x^2y^3}$
11. $x+\frac{x}{y}$
12. $\frac{3x}{4y}+\frac{5}{6xy}$

13. $\dfrac{3x-1}{2x}+\dfrac{2}{3x}$

14. $\dfrac{3}{x}+4$

15. $\dfrac{5x}{y}+\dfrac{y}{x}$

16. $\dfrac{3}{5p}-\dfrac{5}{2p^2}$

17. $\dfrac{4}{5x^2}-\dfrac{6}{y}$

18. $\dfrac{x-3}{x}-\dfrac{x}{4x^2}$

19. $\dfrac{5}{x}+\dfrac{3}{x-2}$

20. $6-\dfrac{3}{x-3}$

21. $\dfrac{9}{a+3}+\dfrac{2}{a}$

22. $\dfrac{b}{a-b}+\dfrac{a+b}{b}$

23. $\dfrac{4}{3x}-\dfrac{2x}{3x+6}$

24. $\dfrac{2}{x-3}-\dfrac{4}{x-1}$

25. $\dfrac{4}{x-3}+\dfrac{2}{3-x}$

26. $\dfrac{3}{x-2}-\dfrac{1}{2-x}$

27. $\dfrac{7}{x+5}-\dfrac{4}{-x-5}$

28. $\dfrac{5}{2x-5}-\dfrac{3}{5-2x}$

29. $\dfrac{3}{x+1}+\dfrac{4}{x-1}$

30. $\dfrac{x}{2x-4}+\dfrac{3}{x-2}$

31. $\dfrac{x+5}{x-5}-\dfrac{x-5}{x+5}$

32. $\dfrac{x+7}{x+3}-\dfrac{x-3}{x+7}$

33. $\dfrac{x}{x^2-9}+\dfrac{4}{x+3}$

34. $\dfrac{5}{(x+4)^2}+\dfrac{2}{x+4}$

35. $\dfrac{x+3}{x^2-9}-\dfrac{3}{x+3}$

36. $\dfrac{3}{(x-2)(x+3)}+\dfrac{5}{(x+2)(x+3)}$

37. $\dfrac{2x+3}{x^2-7x+12}-\dfrac{2}{x-3}$

38. $\dfrac{x+3}{x^2-3x-10}-\dfrac{2}{x-5}$

39. $\dfrac{x^2}{x^2+2x-8}-\dfrac{x-4}{x+4}$

40. $\dfrac{x+1}{x^2-2x+1}-\dfrac{x+1}{x-1}$

41. $\dfrac{x-1}{x^2+4x+4}+\dfrac{x-1}{x+2}$

42. $\dfrac{y}{xy-x^2}-\dfrac{x}{y^2-xy}$

43. $\dfrac{1}{x^2+3x-10}+\dfrac{3}{x^2+x-6}$

44. $\dfrac{5}{x^2-9x+8}-\dfrac{3}{x^2-6x-16}$

45. $\dfrac{1}{x^2-4}+\dfrac{3}{x^2+5x+6}$

46. $\dfrac{x}{2x^2+7x-4}+\dfrac{2}{x^2-x-20}$

47. $\dfrac{x}{3x^2+5x-2}-\dfrac{4}{2x^2+7x+6}$

48. $\dfrac{x}{6x^2+7x+2}+\dfrac{5}{2x^2-3x-2}$

49. $\dfrac{x}{4x^2+11x+6}-\dfrac{2}{8x^2+2x-3}$

50. $\dfrac{x}{4x^2+13x+3}-\dfrac{4}{2x^2+x-15}$

51. **(a)** Explain in your own words a step-by-step procedure to add or subtract two rational expressions that have unlike denominators.

(b) Using the procedure outlined in part (a), add $\dfrac{x}{x^2-x-6}+\dfrac{3}{x^2-4}$.

CUMULATIVE REVIEW EXERCISES

[2.6] **52.** A videocassette recorder counter will go from 0 to 18 in 2 minutes. There are two movies on a VCR tape. If Phong Nguyen wishes to watch the second movie and the first movie is $1\frac{1}{2}$ hours long, what will be the number on the counter at the end of the first movie, where the second movie starts?

[2.7] **53.** Solve the inequality $3(x-2)+2<4(x+1)$ and graph the solution on the number line.

[4.6] **54.** Divide $(8x^2+6x-13)\div(2x+3)$.

[6.2] **55.** Multiply $\dfrac{x^2+xy-6y^2}{x^2-xy-2y^2}\cdot\dfrac{x^2-y^2}{x^2+2xy-3y^2}$

Group Activity/Challenge Problems

Add or subtract.

1. $\dfrac{x}{x-2}+\dfrac{3}{x+2}+\dfrac{4}{x^2-4}$

2. $\dfrac{4}{x^2+x-6}+\dfrac{x}{x+3}-\dfrac{5}{x-2}$

3. $\dfrac{x+6}{4-x^2}-\dfrac{x+3}{x+2}+\dfrac{x-3}{2-x}$

4. $\dfrac{3x-1}{x+2}+\dfrac{x}{x-3}-\dfrac{4}{2x+3}$

5. $\dfrac{3x}{x^2-4}+\dfrac{4}{x^3+8}$

6. $\dfrac{2}{x^2-x-6}+\dfrac{3}{x^2-2x-3}+\dfrac{1}{x^2+3x+2}$

6.5 Complex Fractions

Tape 10

1. Simplify complex fractions by combining terms.
2. Simplify complex fractions using multiplication first to clear fractions.

Simplify Complex Fractions

1 A **complex fraction** is one that has a fraction in its numerator or its denominator or in both its numerator and denominator. Examples of complex fractions are

$$\frac{\frac{2}{3}}{5} \qquad \frac{\frac{x+1}{x}}{3x} \qquad \frac{\frac{x}{y}}{x+1} \qquad \frac{\frac{a+b}{a}}{\frac{a-b}{b}}$$

$$\begin{array}{l} \text{Numerator of complex fraction} \\ \\ \text{Denominator of complex fraction} \end{array} \left\{ \frac{\dfrac{a+b}{a}}{\dfrac{a-b}{b}} \right. \longleftarrow \text{Main fraction line}$$

The expression above the main fraction line is the numerator, and the expression below the main fraction line is the denominator of the complex fraction.

There are two methods to simplify complex fractions. The first reinforces many of the concepts used in this chapter because we may need to add, subtract, multiply, and divide simpler fractions as we simplify the complex fraction. Many students prefer to use the second because the answer may be obtained more quickly. We will give three examples using the first method and then work the same three examples using the second method.

Combining Terms in the Numerator and Denominator

To Simplify a Complex Fraction by Combining Terms

1. Add or subtract the fractions in both the numerator and denominator of the complex fractions to obtain single fractions in both the numerator and the denominator.
2. Invert the denominator of the complex fraction and multiply the numerator by it.
3. Simplify further if possible.

EXAMPLE 1 Simplify $\dfrac{\frac{2}{3}+\frac{3}{4}}{\frac{3}{4}-\frac{2}{3}}$.

Solution: First, add the terms in the numerator and subtract the terms in the denominator to obtain single fractions in both the numerator and denominator. The LCD of both the numerator and the denominator is 12.

$$\frac{\frac{2}{3}+\frac{3}{4}}{\frac{3}{4}-\frac{2}{3}} = \frac{\frac{4}{4}\cdot\frac{2}{3}+\frac{3}{4}\cdot\frac{3}{3}}{\frac{3}{3}\cdot\frac{3}{4}-\frac{2}{3}\cdot\frac{4}{4}} = \frac{\frac{8}{12}+\frac{9}{12}}{\frac{9}{12}-\frac{8}{12}} = \frac{\frac{17}{12}}{\frac{1}{12}}$$

Next, invert the denominator and multiply the numerator by it.

$$\frac{\frac{17}{12}}{\frac{1}{12}} = \frac{17}{\cancel{12}_{1}}\cdot\frac{\cancel{12}^{1}}{1} = \frac{17}{1} = 17$$

EXAMPLE 2 Simplify $\dfrac{a+\frac{1}{x}}{x+\frac{1}{a}}$.

Solution: Express both the numerator and denominator of the complex fraction as single fractions. The LCD of the numerator is x and the LCD of the denominator is a.

$$\frac{a+\frac{1}{x}}{x+\frac{1}{a}} = \frac{\frac{x}{x}\cdot a+\frac{1}{x}}{\frac{a}{a}\cdot x+\frac{1}{a}} = \frac{\frac{ax}{x}+\frac{1}{x}}{\frac{ax}{a}+\frac{1}{a}} = \frac{\frac{ax+1}{x}}{\frac{ax+1}{a}}$$

Now invert the denominator and multiply the numerator by it.

$$= \frac{\cancel{(ax+1)}}{x}\cdot\frac{a}{\cancel{(ax+1)}} = \frac{a}{x}$$

EXAMPLE 3 Simplify $\dfrac{a}{\dfrac{1}{a}+\dfrac{1}{b}}$.

Solution:

$$\frac{a}{\frac{1}{a}+\frac{1}{b}} = \frac{a}{\frac{b}{b}\cdot\frac{1}{a}+\frac{1}{b}\cdot\frac{a}{a}} = \frac{a}{\frac{b}{ab}+\frac{a}{ab}} = \frac{a}{\frac{b+a}{ab}} = \frac{a}{1} \div \frac{b+a}{ab}$$

Now invert the divisor and multiply.

$$= \frac{a}{1}\cdot\frac{ab}{b+a} = \frac{a^2b}{b+a} \quad \text{or} \quad \frac{a^2b}{a+b}$$

Using Multiplication First to Clear Fractions

2 Here is the second method for simplifying complex fractions.

To Simplify a Complex Fraction Using Multiplication First

1. Find the least common denominator of *all* the denominators appearing in the complex fraction.
2. Multiply both the numerator and denominator of the complex fraction by the LCD found in step 1.
3. Simplify when possible.

We will now rework Examples 1, 2, and 3 using the above procedure.

EXAMPLE 4 Simplify $\dfrac{\dfrac{2}{3}+\dfrac{3}{4}}{\dfrac{3}{4}-\dfrac{2}{3}}$.

Solution: The denominators in the complex fraction are 3 and 4. The LCD of 3 and 4 is 12. Thus 12 is the LCD of the complex fraction. Multiply both the numerator and denominator of the complex fraction by 12.

$$\frac{\frac{2}{3}+\frac{3}{4}}{\frac{3}{4}-\frac{2}{3}} = \frac{12}{12}\cdot\frac{\left(\frac{2}{3}+\frac{3}{4}\right)}{\left(\frac{3}{4}-\frac{2}{3}\right)} = \frac{12\left(\frac{2}{3}\right)+12\left(\frac{3}{4}\right)}{12\left(\frac{3}{4}\right)-12\left(\frac{2}{3}\right)}$$

Now simplify.

$$= \frac{8+9}{9-8} = \frac{17}{1} = 17$$

Note that the answers to Examples 1 and 4 are the same.

EXAMPLE 5 Simplify $\dfrac{a + \dfrac{1}{x}}{x + \dfrac{1}{a}}$.

Solution: The denominators in the complex fraction are x and a. Therefore, the LCD of the complex fraction is ax. Multiply both the numerator and denominator of the complex fraction by ax.

$$\frac{a + \frac{1}{x}}{x + \frac{1}{a}} = \frac{ax}{ax} \cdot \frac{\left(a + \frac{1}{x}\right)}{\left(x + \frac{1}{a}\right)} = \frac{a^2x + a}{ax^2 + x}$$

$$= \frac{a\cancel{(ax + 1)}}{x\cancel{(ax + 1)}} = \frac{a}{x}$$

Note that the answers to Examples 2 and 5 are the same.

EXAMPLE 6 Simplify $\dfrac{a}{\dfrac{1}{a} + \dfrac{1}{b}}$.

Solution: The denominators in the complex fraction are 1 (from a in the numerator), a, and b. Therefore, the LCD of the complex fraction is ab. Multiply both the numerator and denominator of the complex fraction by ab.

$$\frac{a}{\frac{1}{a} + \frac{1}{b}} = \frac{ab}{ab} \cdot \frac{a}{\left(\frac{1}{a} + \frac{1}{b}\right)}$$

$$= \frac{a^2b}{ab\left(\frac{1}{a}\right) + ab\left(\frac{1}{b}\right)}$$

$$= \frac{a^2b}{b + a}$$

Note that the answers to Examples 3 and 6 are the same. When asked to simplify a complex fraction, you may use either method unless you are told by your instructor to use a specific method.

Exercise Set 6.5

Simplify.

1. $\dfrac{1 + \dfrac{3}{5}}{2 + \dfrac{1}{5}}$

2. $\dfrac{1 - \dfrac{9}{16}}{3 + \dfrac{4}{5}}$

3. $\dfrac{2 + \dfrac{3}{8}}{1 + \dfrac{1}{3}}$

4. $\dfrac{\dfrac{3}{5} + \dfrac{2}{7}}{\dfrac{1}{5} + \dfrac{5}{6}}$

5. $\dfrac{\dfrac{4}{9} - \dfrac{3}{8}}{4 - \dfrac{3}{5}}$

6. $\dfrac{1 - \dfrac{x}{y}}{x}$

7. $\dfrac{\dfrac{x^2y}{4}}{\dfrac{2}{x}}$

8. $\dfrac{\dfrac{15a}{b^2}}{\dfrac{b^3}{5}}$

9. $\dfrac{\dfrac{8x^2y}{3z^3}}{\dfrac{4xy}{9z^5}}$

10. $\dfrac{\dfrac{36x^4}{5y^4z^5}}{\dfrac{9xy^2}{15z^5}}$

11. $\dfrac{x + \dfrac{1}{y}}{\dfrac{x}{y}}$

12. $\dfrac{x - \dfrac{x}{y}}{\dfrac{1 + x}{y}}$

13. $\dfrac{\dfrac{9}{x} + \dfrac{3}{x^2}}{3 + \dfrac{1}{x}}$

14. $\dfrac{\dfrac{2}{a} + \dfrac{1}{2a}}{a + \dfrac{a}{2}}$

15. $\dfrac{3 - \dfrac{1}{y}}{2 - \dfrac{1}{y}}$

16. $\dfrac{\dfrac{x}{x - y}}{\dfrac{x^2}{y}}$

17. $\dfrac{\dfrac{x}{y} - \dfrac{y}{x}}{\dfrac{x + y}{x}}$

18. $\dfrac{1}{\dfrac{1}{x} + y}$

19. $\dfrac{\dfrac{a^2}{b} - b}{\dfrac{b^2}{a} - a}$

20. $\dfrac{\dfrac{1}{x} + \dfrac{2}{x^2}}{2 + \dfrac{1}{x^2}}$

21. $\dfrac{\dfrac{a}{b} - 2}{\dfrac{-a}{b} + 2}$

22. $\dfrac{\dfrac{x^2 - y^2}{x}}{\dfrac{x + y}{x^3}}$

23. $\dfrac{\dfrac{4x + 8}{3x^2}}{\dfrac{4x}{6}}$

24. $\dfrac{\dfrac{1}{a} + \dfrac{1}{b}}{ab}$

25. $\dfrac{\dfrac{1}{a} + \dfrac{1}{b}}{\dfrac{1}{ab}}$

26. $\dfrac{\dfrac{1}{a} + 1}{\dfrac{1}{b} - 1}$

27. $\dfrac{\dfrac{a}{b} + \dfrac{1}{a}}{\dfrac{b}{a} + \dfrac{1}{a}}$

28. $\dfrac{\dfrac{1}{a} + \dfrac{1}{b}}{\dfrac{1}{a}}$

29. $\dfrac{\dfrac{1}{x} - \dfrac{1}{y}}{\dfrac{1}{x} + \dfrac{1}{y}}$

30. $\dfrac{\dfrac{1}{x^2} + \dfrac{1}{x}}{\dfrac{1}{x} + \dfrac{1}{x^2}}$

31. $\dfrac{\dfrac{1}{x^2} + \dfrac{1}{x}}{\dfrac{1}{y} + \dfrac{1}{y^2}}$

32. $\dfrac{\dfrac{x}{y} - \dfrac{1}{x}}{\dfrac{y}{x} + 1}$

33. What is a complex fraction?

34. (a) Select the method you prefer to use to simplify complex fractions. Then write in your own words a step-by-step procedure for simplifying complex fractions using that method.

(b) Using the procedure given in part (a) simplify

$$\dfrac{\dfrac{2}{x} - \dfrac{3}{y}}{x + \dfrac{1}{y}}$$

CUMULATIVE REVIEW EXERCISES

35. Solve the equation $4x + 3(x - 2) = 7x - 6$.

36. What is a polynomial?

[5.1] **37.** Simplify $(4x^2y^3)^2(3xy^4)$.

[6.4] **38.** Subtract $\dfrac{x}{3x^2 + 17x - 6} - \dfrac{2}{x^2 + 3x - 18}$.

Group Activity/ Challenge Problems

1. The efficiency of a jack, E, is expressed by the formula $E = \dfrac{\frac{1}{2}h}{h + \frac{1}{2}}$, where h is determined by the pitch of the jack's thread. Determine the efficiency of a jack if h is
(a) $\dfrac{2}{3}$ **(b)** $\dfrac{4}{5}$

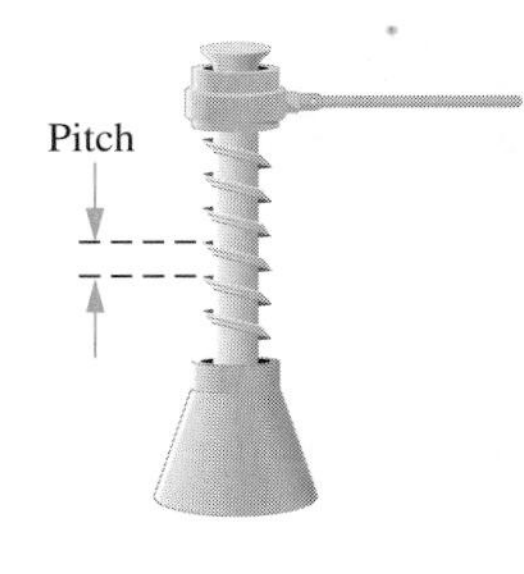

Simplify.

2. $\dfrac{\frac{x}{y} + \frac{y}{x} + \frac{1}{x}}{\frac{x}{y} + y}$ **3.** $\dfrac{\frac{a}{b} + b - \frac{1}{a}}{\frac{a}{b^2} - \frac{b}{a} + \frac{1}{a^2}}$ **4.** $\dfrac{x}{1 + \frac{x}{1 + x}}$

6.6 Solving Rational Equations

1 Solve rational equations.

Tape 10

Solve Rational Equations

1 In Sections 6.1 through 6.5 we focused on how to add, subtract, multiply, and divide rational expressions. Now we are ready to solve rational equations. A **rational equation** is one that contains one or more rational expressions.

To Solve Rational Equations

1. Determine the LCD of all fractions in the equation.
2. Multiply **both** sides of the equation by the LCD. **This will result in every term in the equation being multiplied by the LCD.**
3. Remove any parentheses and combine like terms on each side of the equation.
4. Solve the equation using the properties discussed in earlier chapters.
5. Check your solution in the original equation.

The purpose of multiplying both sides of the equation by the LCD (step 2) is to eliminate all fractions from the equation. After both sides of the equation are multiplied by the LCD, the resulting equation should contain no fractions. We will omit some of the checks to save space.

EXAMPLE 1 Solve $\dfrac{x}{3} + 2x = 7$.

Solution:

$$3\left(\frac{x}{3} + 2x\right) = 7 \cdot 3 \qquad \text{Multiply both sides of the equation by the LCD, 3.}$$

$$\cancel{3}\left(\frac{x}{\cancel{3}}\right) + 3 \cdot 2x = 7 \cdot 3 \qquad \text{Distributive property.}$$

$$x + 6x = 21$$

$$7x = 21$$

$$x = 3$$

Check:

$$\frac{x}{3} + 2x = 7$$

$$\frac{3}{3} + 2(3) = 7$$

$$1 + 6 = 7$$

$$7 = 7 \quad \text{true}$$

EXAMPLE 2 Solve the equation $\frac{3}{4} + \frac{5x}{9} = \frac{x}{6}$.

Solution: Multiply both sides of the equation by the LCD, 36.

$$36\left(\frac{3}{4} + \frac{5x}{9}\right) = \frac{x}{6} \cdot 36$$

$$\overset{9}{\cancel{36}}\left(\frac{3}{\cancel{4}}\right) + \overset{4}{\cancel{36}}\left(\frac{5x}{\cancel{9}}\right) = \frac{x}{\cancel{6}} \cdot \overset{6}{\cancel{36}}$$

$$27 + 20x = 6x$$

$$27 = -14x$$

$$\frac{-27}{14} = x$$

A check will show that $\frac{-27}{14}$ is the solution.

EXAMPLE 3 Solve the equation $\frac{x}{4} + 3 = 2(x - 2)$.

Solution: Multiply both sides of the equation by the LCD, 4.

$$\frac{x}{4} + 3 = 2(x - 2)$$

$$4\left(\frac{x}{4} + 3\right) = 4\,[2(x - 2)]$$

$$4\left(\frac{x}{4}\right) + 4(3) = 4[2(x - 2)]$$

$$\cancel{4}\left(\frac{x}{\cancel{4}}\right) + 4(3) = 8(x - 2)$$

$$x + 12 = 8(x - 2)$$

$$x + 12 = 8x - 16$$

$$12 = 7x - 16$$

$$28 = 7x$$

$$4 = x$$

Warning: **Whenever a variable appears in any denominator of a rational equation, it is necessary to check your answer in the original equation. If the answer obtained makes any denominator equal to zero, that value is not a solution to the equation.** Such values are called **extraneous roots** or **extraneous solutions.**

EXAMPLE 4 Solve the equation $3 - \frac{4}{x} = \frac{5}{2}$.

Solution: Multiply both sides of the equation by the LCD, $2x$.

$$2x\left(3 - \frac{4}{x}\right) = \left(\frac{5}{2}\right) \cdot 2x$$

$$2x(3) - 2x\left(\frac{4}{x}\right) = \left(\frac{5}{2}\right) \cdot 2x$$

$$6x - 8 = 5x$$

$$x - 8 = 0$$

$$x = 8$$

Check:

$$3 - \frac{4}{x} = \frac{5}{2}$$

$$3 - \frac{4}{8} = \frac{5}{2}$$

$$3 - \frac{1}{2} = \frac{5}{2}$$

$$\frac{5}{2} = \frac{5}{2} \quad \text{true}$$

Since 8 does check, it is the solution to the equation.

EXAMPLE 5 Solve the equation $\frac{x - 7}{x + 2} = \frac{1}{4}$.

Solution: The LCD is $4(x + 2)$. Multiply both sides of the equation by the LCD.

$$4(x + 2) \cdot \frac{(x - 7)}{(x + 2)} = \frac{1}{4} \cdot 4(x + 2)$$

$$4(x - 7) = 1(x + 2)$$

$$4x - 28 = x + 2$$

$$3x - 28 = 2$$

$$3x = 30$$

$$x = 10$$

A check will show that 10 is the solution.

In Section 2.6 we illustrated that proportions of the form

$$\frac{a}{b} = \frac{c}{d}$$

can be cross-multiplied to obtain $a \cdot d = b \cdot c$. Example 5 is a proportion and can also be solved by cross-multiplying, as is done in Example 6.

EXAMPLE 6 Use cross-multiplication to solve the equation $\frac{3}{x+4} = \frac{4}{x-1}$.

Solution:

$$\frac{3}{x+4} = \frac{4}{x-1}$$
$$3(x-1) = 4(x+4)$$
$$3x - 3 = 4x + 16$$
$$-x - 3 = 16$$
$$-x = 19$$
$$x = -19$$

A check will show that -19 is the solution to the equation.

Now let us examine some examples that involve quadratic equations. Recall from Section 5.6 that quadratic equations have the form $ax^2 + bx + c = 0$, where $a \neq 0$.

EXAMPLE 7 Solve the equation $x + \frac{12}{x} = -7$.

Solution:

$$x \cdot \left(x + \frac{12}{x}\right) = -7 \cdot x \qquad \text{Multiply both sides of the equation by } x.$$
$$x(x) + x\left(\frac{12}{x}\right) = -7x$$
$$x^2 + 12 = -7x$$
$$x^2 + 7x + 12 = 0$$
$$(x+3)(x+4) = 0$$
$$x + 3 = 0 \quad \text{or} \quad x + 4 = 0$$
$$x = -3 \qquad\qquad x = -4$$

Check:

$x = -3$		$x = -4$	
$x + \frac{12}{x} = -7$		$x + \frac{12}{x} = -7$	
$-3 + \frac{12}{-3} = -7$		$-4 + \frac{12}{-4} = -7$	
$-3 + (-4) = -7$		$-4 + (-3) = -7$	
$-7 = -7$	true	$-7 = -7$	true

The solutions are -3 and -4.

EXAMPLE 8 Solve the equation $\frac{2x-5}{x-6} = \frac{7}{x-6}$.

Solution: Using cross-multiplication we get

$$\frac{2x-5}{x-6} = \frac{7}{x-6}$$

$$(2x-5)(x-6) = 7(x-6)$$

$$2x^2 - 17x + 30 = 7x - 42$$

$$2x^2 - 24x + 72 = 0$$

$$2(x^2 - 12x + 36) = 0$$

$$2(x-6)(x-6) = 0$$

$$x - 6 = 0 \quad \text{or} \quad x - 6 = 0$$

$$x = 6 \qquad\qquad x = 6$$

Check: $x = 6$

$$\frac{2x-5}{x-6} = \frac{7}{x-6}$$

$$\frac{2(6)-5}{6-6} = \frac{7}{6-6}$$

$$\frac{7}{0} = \frac{7}{0}, \quad \frac{7}{0} \text{ is not a real number}$$

Since 7/0 is not a real number, 6 is an extraneous solution. Thus, this equation has *no solution.*

EXAMPLE 9 Solve the equation $\dfrac{x^2}{x-4} = \dfrac{16}{x-4}$.

Solution: If we try to solve this proportion using cross-multiplication we will get $x^2(x-4) = 16(x-4)$, which simplifies to $x^3 - 4x^2 = 16x - 64$. This is an example of a *cubic equation* since the greatest exponent on the variable x is 3. Since solving cubic equations is beyond the scope of this course, we will need to try another procedure to solve the original equation. If we multiply both sides of the original equation by the least common denominator, $x - 4$, we can solve the equation, as follows.

$$(x-4) \cdot \frac{x^2}{(x-4)} = \frac{16}{(x-4)} \cdot (x-4)$$

$$x^2 = 16$$

$$x^2 - 16 = 0 \quad \text{This is a difference of two squares.}$$

$$(x+4)(x-4) = 0$$

$$x + 4 = 0 \quad \text{or} \quad x - 4 = 0$$

$$x = -4 \qquad\qquad x = 4$$

Check: $x = -4$

$$\frac{x^2}{x-4} = \frac{16}{x-4}$$

$$\frac{(-4)^2}{-4-4} = \frac{16}{-4-4}$$

$$\frac{16}{-8} = \frac{16}{-8}$$

$$-2 = -2 \quad \text{true}$$

$x = 4$

$$\frac{x^2}{x-4} = \frac{16}{x-4}$$

$$\frac{(4)^2}{4-4} = \frac{16}{4-4}$$

$$\frac{16}{0} = \frac{16}{0} \quad \text{not a solution}$$

Since 4 results in a denominator of 0, $x = 4$ is *not* a solution to the equation. The 4 is an *extraneous root.* The only solution to the equation is $x = -4$.

EXAMPLE 10 Solve the equation $\dfrac{2x}{x^2-4} + \dfrac{1}{x-2} = \dfrac{2}{x+2}$.

Solution: First factor $x^2 - 4$.

$$\frac{2x}{(x+2)(x-2)} + \frac{1}{x-2} = \frac{2}{x+2}$$

Multiply both sides of the equation by the LCD, $(x + 2)(x - 2)$.

$$(x+2)(x-2) \cdot \left[\frac{2x}{(x+2)(x-2)} + \frac{1}{x-2}\right] = \frac{2}{x+2} \cdot (x+2)(x-2)$$

$$(x+2)(x-2) \cdot \frac{2x}{(x+2)(x-2)} + (x+2)(x-2) \cdot \frac{1}{(x-2)} = \frac{2}{(x+2)} \cdot (x+2)(x-2)$$

$$\cancel{(x+2)}\cancel{(x-2)} \cdot \frac{2x}{\cancel{(x+2)}\cancel{(x-2)}} + (x+2)\cancel{(x-2)} \cdot \frac{1}{\cancel{(x-2)}} = \frac{2}{\cancel{(x+2)}} \cdot \cancel{(x+2)}(x-2)$$

$$2x + (x+2) = 2(x-2)$$

$$2x + x + 2 = 2x - 4$$

$$3x + 2 = 2x - 4$$

$$x + 2 = -4$$

$$x = -6$$

A check will show that -6 is the solution to the equation.

Helpful Hint

Some students confuse adding and subtracting rational expressions with solving rational equations. When adding or subtracting rational expressions, we must rewrite each expression with a common denominator. When solving a rational equation, we multiply both sides of the equation by the LCD to eliminate fractions from the equation. Consider the following two problems. Note that the one on the right is an equation because it contains an equal sign. We will work both problems. The LCD for both problems is $x(x + 4)$.

Adding Rational Expressions

$$\frac{x+2}{x+4} + \frac{3}{x}$$

We rewrite each fraction with the LCD, $x(x + 4)$.

$$= \frac{x}{x} \cdot \frac{x+2}{x+4} + \frac{3}{x} \cdot \frac{x+4}{x+4}$$

$$= \frac{x(x+2)}{x(x+4)} + \frac{3(x+4)}{x(x+4)}$$

$$= \frac{x^2+2x}{x(x+4)} + \frac{3x+12}{x(x+4)}$$

$$= \frac{x^2+2x+3x+12}{x(x+4)}$$

$$= \frac{x^2+5x+12}{x(x+4)}$$

Solving Rational Equations

$$\frac{x+2}{x+4} = \frac{3}{x}$$

We eliminate fractions by multiplying both sides of the equation by the LCD, $x(x + 4)$.

$$(x)(x+4)\left(\frac{x+2}{x+4}\right) = \frac{3}{x}(x)(x+4)$$

$$x(x+2) = 3(x+4)$$

$$x^2 + 2x = 3x + 12$$

$$x^2 - x - 12 = 0$$

$$(x-4)(x+3) = 0$$

$$x - 4 = 0 \quad \text{or} \quad x + 3 = 0$$

$$x = 4 \quad \text{or} \quad x = -3$$

The numbers 4 and -3 on the right will both check and are thus solutions to the equation.

Note that when adding and subtracting rational expressions we usually end up with an algebraic expression. When solving rational equations, the solution will be a numerical value or values. The equation on the right could also be solved using cross-multiplication.

Exercise Set 6.6

Solve each equation, and check your solution

1. $\frac{3}{5} = \frac{x}{10}$ **2.** $\frac{3}{k} = \frac{9}{6}$ **3.** $\frac{5}{12} = \frac{20}{x}$ **4.** $\frac{x}{4} = \frac{-10}{8}$

5. $\frac{a}{25} = \frac{12}{10}$ **6.** $\frac{9c}{10} = \frac{9}{5}$ **7.** $\frac{9}{3b} = \frac{-6}{2}$ **8.** $\frac{5}{8} = \frac{2b}{80}$

9. $\frac{x+4}{9} = \frac{5}{9}$ **10.** $\frac{1}{4} = \frac{z+1}{8}$ **11.** $\frac{4x+5}{6} = \frac{7}{2}$ **12.** $\frac{a}{5} = \frac{a-3}{2}$

13. $\frac{6x+7}{10} = \frac{2x+9}{6}$ **14.** $\frac{n}{10} = 9 - \frac{n}{5}$ **15.** $\frac{x}{3} - \frac{3x}{4} = \frac{1}{12}$

16. $\frac{2}{8} + \frac{3}{4} = \frac{w}{5}$ **17.** $\frac{3}{4} - x = 2x$ **18.** $\frac{2}{y} + \frac{1}{2} = \frac{5}{2y}$

19. $\frac{5}{3x} + \frac{3}{x} = 1$ **20.** $\frac{x}{4} - \frac{x}{6} = \frac{1}{4}$ **21.** $\frac{x-1}{x-5} = \frac{4}{x-5}$

22. $\dfrac{2x+3}{x+1}=\dfrac{3}{2}$

23. $\dfrac{5y-3}{7}=\dfrac{15y-2}{28}$

24. $\dfrac{2}{x+1}=\dfrac{1}{x-2}$

25. $\dfrac{5}{-x-6}=\dfrac{2}{x}$

26. $\dfrac{4}{y-3}=\dfrac{6}{y+3}$

27. $\dfrac{2x-3}{x-4}=\dfrac{5}{x-4}$

28. $\dfrac{3}{x}+4=\dfrac{3}{x}$

29. $\dfrac{x-2}{x+4}=\dfrac{x+1}{x+10}$

30. $\dfrac{x-3}{x+1}=\dfrac{x-6}{x+5}$

31. $\dfrac{2x-1}{3}-\dfrac{3x}{4}=\dfrac{5}{6}$

32. $x+\dfrac{3}{x}=\dfrac{12}{x}$

33. $x+\dfrac{6}{x}=-5$

34. $\dfrac{15}{x}+\dfrac{9x-7}{x+2}=9$

35. $\dfrac{3y-2}{y+1}=4-\dfrac{y+2}{y-1}$

36. $\dfrac{2b}{b+1}=2-\dfrac{5}{2b}$

37. $\dfrac{1}{x+3}+\dfrac{1}{x-3}=\dfrac{-5}{x^2-9}$

38. $c-\dfrac{c}{3}+\dfrac{c}{5}=26$

39. $\dfrac{a}{a-3}+\dfrac{3}{2}=\dfrac{3}{a-3}$

40. $\dfrac{3x}{x^2-9}+\dfrac{1}{x-3}=\dfrac{3}{x+3}$

41. $\dfrac{2}{x-3}-\dfrac{4}{x+3}=\dfrac{8}{x^2-9}$

42. $\dfrac{x+1}{x+3}+\dfrac{x-3}{x-2}=\dfrac{2x^2-15}{x^2+x-6}$

43. $\dfrac{y}{2y+2}+\dfrac{2y-16}{4y+4}=\dfrac{y-3}{y+1}$

44. $\dfrac{3}{x+3}+\dfrac{5}{x+4}=\dfrac{12x+19}{x^2+7x+12}$

45. $\dfrac{1}{2}+\dfrac{1}{x-1}=\dfrac{2}{x^2-1}$

46. $\dfrac{2y}{y+2}=\dfrac{y}{y+3}-\dfrac{3}{y^2+5y+6}$

47. $\dfrac{x+2}{x^2-x}=\dfrac{6}{x^2-1}$

48. $\dfrac{2}{x-2}-\dfrac{1}{x+1}=\dfrac{5}{x^2-x-2}$

49. **(a)** Explain in your own words the steps to use to solve rational equations.
(b) Using the procedure given in part (a), solve the equation $\dfrac{1}{x-1}-\dfrac{1}{x+1}=\dfrac{3x}{x^2-1}$.

50. Consider the equation $\dfrac{x^2}{x+3}=\dfrac{25}{x+3}$.
(a) Explain why you may have difficulty trying to solve this equation by cross multiplication.
(b) Find the solution to the equation.

51. Consider the following problems:

Simplify: $\dfrac{x}{3}-\dfrac{x}{4}+\dfrac{1}{x-1}$, Solve: $\dfrac{x}{3}-\dfrac{x}{4}=\dfrac{1}{x-1}$

(a) Explain the difference between the two types of problems.
(b) Explain how you would work each problem to obtain the correct answer.
(c) Find the correct answer to each problem.

CUMULATIVE REVIEW EXERCISES

[3] 52. With a monthly bus pass that costs \$36 per month, each bus ride costs 40 cents. Without the monthly bus pass, each bus ride costs \$1.60. How many bus rides would Steve have to take per month so that the total cost with the bus pass is the same as the total cost without the bus pass?

[4] 53. Two angles are supplementary angles if their sum measures 180°. Find the two supplementary angles if the larger angle is 30° greater than twice the smaller angle.

[4.7] 54. How long will it take to fill a 600-gallon Jacuzzi if water is flowing into the Jacuzzi at a rate of 8 gallons a minute?

[2.2] [5.6] 55. Explain the difference between a linear equation and a quadratic equation, and give an example of each.

Group Activity/ Challenge Problems

1. A formula frequently used in optics is

$$\frac{1}{p} + \frac{1}{q} = \frac{1}{f}$$

where p represents the distance of the object from a mirror (or lens), q represents the distance of the image from the mirror (or lens), and f represents the focal length of the mirror (or lens). If a mirror has a focal length of 10 centimeters, how far from the mirror will the image appear when the object is 30 centimeters from the mirror?

2. In electronics the total resistance R_T, of resistors wired in a parallel circuit is determined by the formula

$$\frac{1}{R_T} = \frac{1}{R_1} + \frac{1}{R_2} + \frac{1}{R_3} + \cdots + \frac{1}{R_n}$$

where $R_1, R_2, R_3, \ldots, R_n$ are the resistances of the individual resistors (measured in ohms) in the circuit.

(a) Find the total resistance if two resistors, one of 200 ohms and the other of 300 ohms, are wired in a parallel circuit.

(b) If three identical resistors are to be wired in parallel, what should be the resistance of each resistor if the total resistance of the circuit is to be 300 ohms?

3. Can an equation of the form $\frac{a}{x} + 1 = \frac{a}{x}$ have a real number solution for any real number a? Explain your answer.

4. (a) Solve the equation $\frac{1}{x} + \frac{1}{3} = \frac{2}{x}$ and check your answer.

(b) If we take the reciprocal of each term in the equation we get the equation $\frac{x}{1} + \frac{3}{1} = \frac{x}{2}$ or $x + 3 = \frac{x}{2}$. Do you believe that the reciprocal of the answer you found in part (a) will be the solution to this equation? Explain your answer.

(c) Solve the equation $x + 3 = \frac{x}{2}$ and check your answer.

6.7 Applications of Rational Equations

Tape 11

1. Set up and solve problems containing rational expressions.
2. Set up and solve motion problems.
3. Set up and solve work problems.

Applications Containing Rational Expressions

1 Many applications of algebra involve rational equations. After we represent the application as an equation, we solve the rational equation as we did in Section 6.6

EXAMPLE 1 The area of a triangle is 27 square feet. Find the base and height if its height is 3 feet less than twice its base.

Solution: Let x = base

then $2x - 3$ = height (See Fig. 6.1.)

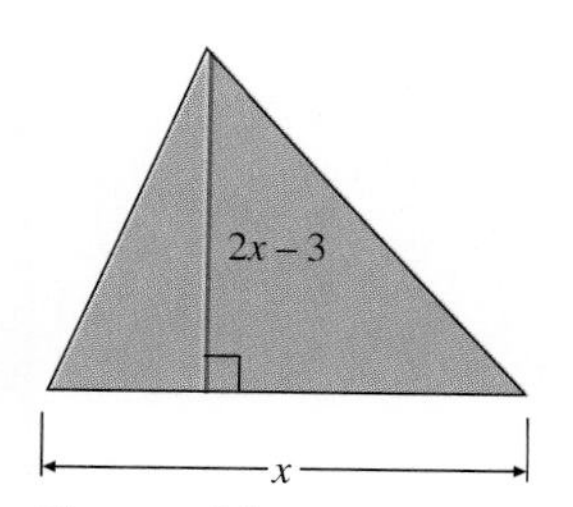

FIGURE 6.1

$$\text{area} = \frac{1}{2} \cdot \text{base} \cdot \text{height}$$

$$27 = \frac{1}{2}(x)(2x - 3)$$

$$2(27) = \cancel{2}\left[\frac{1}{\cancel{2}}(x)(2x - 3)\right]$$ Multiply both sides of the equation by 2.

$$54 = x(2x - 3)$$

$$54 = 2x^2 - 3x$$

$$0 = 2x^2 - 3x - 54$$

$$\text{or} \quad 2x^2 - 3x - 54 = 0$$

$$(2x + 9)(x - 6) = 0$$

$$2x + 9 = 0 \quad \text{or} \quad x - 6 = 0$$

$$2x = -9 \quad \text{or} \quad x = 6$$

$$x = -\frac{9}{2}$$

Since the base of a triangle cannot be negative, we can eliminate $-\frac{9}{2}$ as an answer to our problem.

$$\text{Base} = x = 6 \text{ feet}$$

$$\text{Height} = 2x - 3 = 2(6) - 3 = 9 \text{ feet}$$

Check:

$$a = \frac{1}{2}bh$$

$$27 = \frac{1}{2}(6)(9)$$

$$27 = 27 \quad \text{true}$$

EXAMPLE 2 One number is 3 times another number. The sum of their reciprocals is 4. Find the numbers.

Solution: Let x = smaller number
then $1/(3x)$ = larger number

The reciprocal of the smaller number is $1/x$ and the reciprocal of the larger number is $1/(3x)$. The sum of their reciprocals is 4; thus

$$\frac{1}{x} + \frac{1}{3x} = 4$$

$$3x\left(\frac{1}{x} + \frac{1}{3x}\right) = 3x(4)$$

$$3\cancel{x}\left(\frac{1}{\cancel{x}}\right) + \cancel{3x}\left(\frac{1}{\cancel{3x}}\right) = 12x$$

$$3 + 1 = 12x$$

$$4 = 12x$$

$$\frac{4}{12} = x$$

$$\frac{1}{3} = x$$

The smaller number is $\frac{1}{3}$; the larger number is $3x = 3(\frac{1}{3}) = 1$.

Check:

$$\frac{1}{x} + \frac{1}{3x} = 4$$

$$\frac{1}{\frac{1}{3}} + \frac{1}{3(\frac{1}{3})} = 4$$

$$3 + 1 = 4$$

$$4 = 4 \quad \text{true}$$

Motion Problems

2 In Chapter 4 we discussed motion problems. Recall that

$$\text{distance} = \text{rate} \cdot \text{time}$$

If we solve this equation for time, we obtain

$$\text{time} = \frac{\text{distance}}{\text{rate}} \quad \text{or} \quad t = \frac{d}{r}$$

This equation is useful in solving motion problems when the total time of travel for two objects or the time of travel between two points is known.

EXAMPLE 3 A river has a current of 3 miles per hour. If it takes Jack's motorboat the same time to go 10 miles downstream as 6 miles upstream, find the speed of his boat in still water.

Solution: Let r = speed (or rate) of boat in still water
then $r + 3$ = speed of boat downstream (with current)
and $r - 3$ = speed of boat upstream (against current)

	Distance	Rate	Time
Downstream	10	$r + 3$	$\frac{10}{r+3}$
Upstream	6	$r - 3$	$\frac{6}{r-3}$

Since the time it takes to travel 10 miles downstream is the same as the time to travel 6 miles upstream, we set the times equal to each other and then solve the resulting equation.

$$\text{time downstream} = \text{time upstream}$$

$$\frac{10}{r+3} = \frac{6}{r-3} \qquad \text{Now cross-multiply.}$$

$$10(r - 3) = 6(r + 3)$$

$$10r - 30 = 6r + 18$$

$$4r - 30 = 18$$

$$4r = 48$$

$$r = 12$$

Since this rational equation contains a variable in a denominator the solution must be checked. A check will show that 12 satisfies the equation. Thus, the speed of the boat in still water is 12 miles per hour.

EXAMPLE 4 Mr. Blake rides his bike every Saturday morning in the county park. During the first part of the ride he is peddling mostly uphill and his average speed is 12 miles an hour. After a certain point, he is traveling mostly downhill and averages 18 miles per hour. If the total distance he travels is 30 miles and the total time he rides is 2 hours, how long did he ride at each speed?

Solution: Let d = distance traveled at 12 miles per hour
then $30 - d$ = distance traveled at 18 miles per hour

	Distance	Rate	Time
Uphill	d	12	$\frac{d}{12}$
Downhill	$30 - d$	18	$\frac{30-d}{18}$

Since the total time spent riding is 2 hours, we write

$$\text{time going uphill} + \text{time going downhill} = 2 \text{ hours}$$

$$\frac{d}{12} + \frac{30-d}{18} = 2$$

$$36\left(\frac{d}{12} + \frac{30-d}{18}\right) = 36 \cdot 2$$

$$\overset{3}{\cancel{36}}\left(\frac{d}{\cancel{12}}\right) + \overset{2}{\cancel{36}}\left(\frac{30-d}{\cancel{18}}\right) = 72$$

$$3d + 60 - 2d = 72$$

$$d + 60 = 72$$

$$d = 12$$

The answer to the problem is not 12. Remember that the question asked us to *find the time spent* traveling at each speed. The variable d does not represent time, but represents the distance traveled at 12 miles per hour. To find the time traveled and to answer the question asked, we need to evaluate $\frac{d}{12}$ and $\frac{30-d}{18}$ for $d = 12$ miles.

Time at 12 mph

$$\frac{d}{12} = \frac{12}{12} = 1$$

Time at 18 mph

$$\frac{30-d}{18} = \frac{30-12}{18} = \frac{18}{18} = 1$$

Thus, 1 hour was spent traveling at each rate. We leave the check for you to do.

EXAMPLE 5 A car and train take parallel routes from Los Angeles to the California State Fair in Sacramento. The train averages 70 miles per hour and the car averages 50 miles per hour. If the train arrives at the fair 2.2 hours before the car, find the distance from Los Angeles to the fair.

Solution: Let d = distance from Los Angeles to the fair.

	Distance	Rate	Time
Train	d	70	$\frac{d}{70}$
Car	d	50	$\frac{d}{50}$

We are given that the car ride is 2.2 hours longer than the train ride. Therefore, to make the two travel times equal, we need to add 2.2 hours to the travel time of the train. Using this information, we set up the following equation.

$$\text{time for car ride} = \text{time for train ride} + 2.2 \text{ hours}$$

$$\frac{d}{50} = \frac{d}{70} + 2.2$$

Now multiply both sides of the equation by the LCD, 350.

$$350\left(\frac{d}{50}\right) = 350\left(\frac{d}{70} + 2.2\right)$$

$$7d = 350\left(\frac{d}{70}\right) + 350(2.2)$$

$$7d = 5d + 770$$

$$2d = 770$$

$$d = 385$$

Therefore, the distance from Los Angeles to the State Fair in Sacramento is 385 miles.

In Example 5 we added 2.2 hours to the time of the train to obtain an equation. We could have subtracted 2.2 hours from the time of the car to obtain an equivalent equation. Rework Example 5 now by subtracting 2.2 hours from the car's time.

Work Problems

3 Problems in which two or more machines or people work together to complete a certain task are sometimes referred to as **work problems.** Work problems often involve equations containing fractions. Generally, work problems are based on the fact that the fractional part of the work done by person 1 (or machine 1) plus the fractional part of the work done by person 2 (or machine 2) is equal to the total amount of work done by both people (or both machines). *We represent the total amount of work done by the number 1, which represents one whole job completed.*

Part of task done by first person or machine	+	Part of task done by second person or machine	=	1 (one whole task completed)

To determine the part of the task completed by each person or machine, we use the formula

part of task completed = rate · time

This formula is very similar to the formula

amount = rate · time

that was discussed in Section 4.7. To determine the part of the task completed, we need to determine the rate. Suppose that Paul can do a particular task in 6 hours. Then he would complete 1/6 of the task per hour. Thus, his rate is 1/6 of the task per hour. If Audrey can do a particular task in 5 minutes, her rate is 1/5 of the task per minute. In general, if a person or machine can complete a task in t units of time, the rate is $1/t$.

EXAMPLE 6 Mr. Donaldson can paint a house by himself in 20 hours. Mr. Cronkite can paint the same house by himself in 30 hours. How long will it take them to paint the house if they work together?

Solution: Let t = the time, in hours, for both men to paint the house together. We will construct a table to help us in finding the part of the task completed by Mr. Donaldson and by Mr. Cronkite in t hours.

	Rate of Work	Time Worked	Part of Task
Mr. Donaldson	$\frac{1}{20}$	t	$\frac{t}{20}$
Mr. Cronkite	$\frac{1}{30}$	t	$\frac{t}{30}$

$$\left(\begin{array}{c}\text{part of house painted}\\ \text{by Mr. Donaldson in } t \text{ hours}\end{array}\right) + \left(\begin{array}{c}\text{part of house painted}\\ \text{by Mr. Cronkite in } t \text{ hours}\end{array}\right) = 1 \text{ (whole house painted)}$$

$$\frac{t}{20} \qquad + \qquad \frac{t}{30} \qquad = 1$$

Now multiply both sides of the equation by the LCD, 60.

$$\frac{t}{20} + \frac{t}{30} = 1$$

$$60\left(\frac{t}{20} + \frac{t}{30}\right) = 60 \cdot 1$$

$$60\left(\frac{t}{20}\right) + 60\left(\frac{t}{30}\right) = 60$$

$$3t + 2t = 60$$

$$5t = 60$$

$$t = 12$$

Thus, the two men working together can paint the house in 12 hours.

EXAMPLE 7 One pipe can fill a tank in 4 hours and another pipe can empty it in 6 hours. If the valves to both pipes are open, how long will it take to fill the tank?

Solution: Let t = amount of time to fill the tank with both valves open.

	Rate of Work	Time	Part of Task
Pipe Filling Tank	$\frac{1}{4}$	t	$\frac{t}{4}$
Pipe Emptying Tank	$\frac{1}{6}$	t	$\frac{t}{6}$

As one pipe is filling, the other is emptying the tank. Thus, the pipes are working against each other. Therefore, instead of adding the parts of the task, as was done in Example 6, where the people worked together, we will subtract the parts of the task.

$$\begin{pmatrix}\text{part of tank}\\ \text{filled in } t \text{ hours}\end{pmatrix} - \begin{pmatrix}\text{part of tank}\\ \text{emptied in } t \text{ hours}\end{pmatrix} = 1 \text{ (total tank filled)}$$

$$\frac{t}{4} - \frac{t}{6} = 1$$

$$12\left(\frac{t}{4} - \frac{t}{6}\right) = 12 \cdot 1$$

$$12\left(\frac{t}{4}\right) - 12\left(\frac{t}{6}\right) = 12$$

$$3t - 2t = 12$$

$$t = 12$$

The tank will be filled in 12 hours.

EXAMPLE 8 Dolores and Maryann are auto mechanics at Simpson's garage. When Dolores removes and rebuilds a car's transmission by herself, it takes her 10 hours. When Dolores and Maryann work together to remove and rebuild the transmission, it takes them 6 hours. How long does it take Maryann by herself to remove and rebuild the transmission?

Solution: Let t = time for Maryann to remove and rebuild the transmission by herself. Then Maryann's rate of work is $1/t$. Let us make a table to help analyze the problem. In the table we use the fact that together they can remove and rebuild the transmission in 6 hours.

	Rate of Work	Time	Part of Task
Dolores	$\frac{1}{10}$	6	$\frac{6}{10}$ or $\frac{3}{5}$
Maryann	$\frac{1}{t}$	6	$\frac{6}{t}$

$$\begin{pmatrix}\text{part of task completed}\\ \text{by Dolores}\end{pmatrix} + \begin{pmatrix}\text{part of task completed}\\ \text{by Maryann}\end{pmatrix} = 1$$

$$\frac{3}{5} \qquad + \qquad \frac{6}{t} \qquad = 1$$

Now multiply both sides of the equation by the LCD, $5t$.

$$5t\left(\frac{3}{5} + \frac{6}{t}\right) = 5t \cdot 1$$

$$5t\left(\frac{3}{5}\right) + 5t\left(\frac{6}{t}\right) = 5t$$

$$3t + 30 = 5t$$

$$30 = 2t$$

$$15 = t$$

A check will show that 15 is the solution to the equation. Thus, it takes Maryann 15 hours by herself to remove and rebuild the transmission.

EXAMPLE 9 Mr. and Mrs. O'Connor are handwriting thank-you notes to guests who attended their wedding. Mrs. O'Connor by herself could write all the notes in 8 hours and Mr. O'Connor could write all the notes by himself in 7 hours. After Mrs. O'Connor has been writing thank-you notes for 5 hours by herself, she must leave town on business. Mr. O'Connor then continues the task of writing the thank-you notes. How long will it take Mr. O'Connor to finish writing the remaining notes?

Solution: Let t = time it will take Mr. O'Connor to finish writing the notes.

	Rate	Time	Part of Task
Mrs. O'Connor	$\frac{1}{8}$	5	$\frac{5}{8}$
Mr. O'Connor	$\frac{1}{7}$	t	$\frac{t}{7}$

$$\begin{pmatrix}\text{part of cards written}\\ \text{by Mrs. O'Connor}\end{pmatrix} + \begin{pmatrix}\text{part of cards written}\\ \text{by Mr. O'Connor}\end{pmatrix} = 1$$

$$\frac{5}{8} \qquad + \qquad \frac{t}{7} \qquad = 1$$

$$\begin{aligned} 56\left(\frac{5}{8} + \frac{t}{7}\right) &= 56 \cdot 1 \\ 56\left(\frac{5}{8}\right) + 56\left(\frac{t}{7}\right) &= 56 \\ 35 + 8t &= 56 \\ 8t &= 21 \\ t &= \frac{21}{8} \text{ or } 2\frac{5}{8} \end{aligned}$$

Thus, it will take Mr. O'Connor $2\frac{5}{8}$ hours to complete the cards.

Exercise Set 6.7

*For exercises 1–32, **(a)** write an equation that can be used to solve the problem and **(b)** solve the problem.*

Geometry Problems

1. The base of a triangle is 6 centimeters greater than its height. Find the base and height if the area is 80 square centimeters.
2. The height of a triangle is 1 centimeter less than twice its base. Find the base and height if the triangle's area is 33 square centimeters.

Number Problems

3. One number is three times as large as another. The sum of their reciprocals is $\frac{4}{3}$. Find the two numbers.
4. The numerator of the fraction $\frac{3}{4}$ is increased by an amount so that the value of the resulting fraction is $\frac{5}{2}$. Find the amount that the numerator was increased.

5. The reciprocal of 3 plus the reciprocal of 5 is the reciprocal of what number?

6. One number is 4 times as large as another. The sum of their reciprocals is $\frac{5}{8}$. Find the two numbers.

Motion Problems

7. Jim can row 4 miles per hour in still water. It takes him as long to row 6 miles upstream as 10 miles downstream. How fast is the current?

8. In the Pixie River a boat travels 9 miles upstream in the same amount of time it travels 11 miles downstream. If the current of the river is 2 miles per hour, find the speed of the boat in still water.

9. Ms. Duncan took her two sons water skiing in still water. She drove the motor boat one way on the water pulling the younger son at 30 miles per hour. Then she turned around and pulled her older son in the opposite direction the same distance at 30 miles per hour. If the total time spent skiing was $\frac{1}{2}$ hour, how far did each son travel?

See Exercise 9

10. A business executive traveled 1800 miles by jet and then traveled an additional 300 miles on a private propeller plane. If the speed of the jet is four times the speed of the prop plane and the entire trip took 5 hours, find the speed of each plane.

11. One car travels 30 kilometers per hour faster than another. In the time it takes the slower car to travel 250 kilometers the faster car travels 400 kilometers. Find the speed of both cars.

12. Maria walked at a speed of 2 miles per hour and jogged at a speed of 4 miles per hour. If she jogged 3 miles farther than she walked and the total time for the trip was 3 hours, how far did she walk and how far did she jog?

13. On a treadmill, Mario walks a distance of 2 miles. He then doubles the speed of the treadmill and jogs for another 2 miles. If the total time he spent on the treadmill was 1 hour, find the speeds at which he walks and jogs.

14. In her daily morning activity, Dawn jogs a specific distance at 8 miles per hour and then walks another distance at 4 miles per hour. If the total distance she travels is 6 miles and the total time spent in her outing is 1.2 hours, find the distance she jogs and the distance she walks.

15. Apostolos jogs and then walks in alternating intervals. When he jogs he averages 5 miles per hour and when he walks he averages 2 miles per hour. If he walks and jogs a total of 3 miles in a total of 0.9 hour, what time is spent jogging and what time is spent walking?

16. A Boeing 747 flew from San Francisco to Honolulu, a distance of 2800 miles. Flying with the wind, it averaged 600 miles per hour. When the wind changed from a tailwind to a headwind, the plane's speed dropped to 500 miles per hour. If the total time of the trip was 5 hours, determine the length of time it flew at each speed.

17. Sean and his father Scott begin skiing the same cross-country ski trail at the same time. If Sean, who averages 9 miles per hour, finishes the trail 0.5 hour quicker than his father, who averages 6 miles per hour, find the length of the trail.

See Exercise 17

18. Jan swims freestyle at an average speed of 40 meters per minute. He swims using the breast stroke at an average speed of 30 meters per minute. Jan decides to swim freestyle across Echo Lake. After resting he swims back across the lake using the breast stroke. If his return trip took 20 minutes longer than his trip going, what is the width of the lake where he crossed?

Work Problems

19. One truck carrying fresh water takes 6 hours to dispense enough water to meet the needs of residents of a small community that was recently flooded. A newer

model truck requires only 3 hours to dispense the same quantity of water. How long would it take them working together to dispense enough water to meet the water needs of the community?

20. Mr. Dell fertilizes the farm in 6 hours. Mrs. Dell fertilizes the farm in 7 hours. How long will it take them to fertilize their farm if they work together?

21. A $\frac{1}{2}$-inch-diameter hose can fill a swimming pool in 8 hours. A $\frac{4}{5}$-inch-diameter hose can fill the same pool in 5 hours. How long will it take to fill the pool when both hoses are used?

22. A conveyor belt operating at full speed can fill a tank with topsoil in 3 hours. When a valve at the bottom of the tank is opened, the tank will empty in 4 hours. If the conveyor belt is operating at full speed and the valve at the bottom of the tank is open, how long will it take to fill the tank?

23. If the stopper is in the basin, the water from the rinse cycle of a washing machine will fill the basin in 4 minutes. If the stopper from the full basin is removed and no additional water is coming into the basin, the basin will empty in 5 minutes. If there is no stopper in the empty basin and water from the rinse cycle starts filling the basin, how long will it take for the basin to be filled?

24. One bottling machine can meet the company's daily production of filled and capped bottles in 8 hours. When a second bottling machine is also running, the daily production of bottles can be completed in 3 hours. How long would it take the second bottling machine to meet the daily production if it were working alone?

25. At the NCNB Savings Bank it takes a computer 4 hours to process and print payroll checks. When a second computer is used and the two computers work together, the checks can be processed and printed in 3 hours. How long would it take the second computer by itself to process and print the payroll checks?

26. The Wilsons own a large farm where they grow wheat. With the large tractor Mrs. Wilson can plow the entire farm in 6 days. With a smaller tractor Mr. Wilson can plow it in 10 days. Mrs. Wilson begins plowing the farm but after 4 days has problems with the large tractor and must stop. Mr. Wilson then begins plowing with the smaller tractor. How much longer will it take him to finish plowing the farm?

27. A construction company with two backhoes has contracted to dig a long trench for drainage pipes. The larger backhoe can dig the entire trench by itself in 12 days. The smaller backhoe can dig the entire trench by itself in 15 days. The large backhoe begins working on the trench by itself, but after 5 days it is transferred to a different job and the smaller backhoe begins working on the trench. How long will it take for the smaller backhoe to complete the job?

28. A large farm owns three hay balers. The oldest can pick up and bale an acre of hay in 3 hours while each of the two newer balers can work an acre in 2 hours.

(a) How long would it take the three balers working together to pick up and bale one acre?

(b) If the area of the farm is 500 acres, how long would it take the three balers working together to pick up and bale all the hay on the farm?

29. A boat designed to skim oil off the surface of the water has two skimmers. One skimmer can fill the boat's holding tank in 60 hours while the second skimmer can fill the boat's holding tank in 50 hours. There is also a valve in the holding tank that is used to transfer the oil to a larger vessel. If no new oil is coming into the holding tank, a full holding tank of skimmed oil can be transferred to a larger tank in 30 hours. If both skimmers begin skimming and the valve on the holding tank is opened, how long will it take for the empty holding tank on the skimmer to fill?

30. When only the cold-water valve is opened, a washtub will fill in 8 minutes. When only the hot-water valve is opened, the washtub will fill in 12 minutes. When the drain of the washtub is open, it will drain completely in 7 minutes. If both the hot- and cold-water valves are open and the drain is open, how long will it take for the washtub to fill?

31. Assume that the residents of continent A can deplete all the world's fresh water supply in 300 years; the residents of continent B can deplete the world's fresh water supply in 100 years; and the residents of continent C can deplete all the world's fresh water supply in 200 years. How long would it take for the three populations together to deplete all the world's fresh water? (Some resources report that fresh water is still abundant. However, because the supply is limited and water pollution has increased, it can support at most one more doubling of demand, which is expected to occur in 20 to 30 years.)

32. Susan can lay a walkway in 20 hours, and Patty can lay the same walkway in 25 hours. After Susan has been working by herself for 11 hours, Patty decides to join her; how long will it take them working together to complete the walkway?

CUMULATIVE REVIEW EXERCISES

[1.9] **33.** Evaluate $6 - [(3 - 5^2) \div 11]^2 + 18 \div 3$.

[2.1] **34.** Simplify $\frac{1}{2}(x + 3) - (2x + 6)$.

[6.2] **35.** Divide $\dfrac{x^2 - 14x + 48}{x^2 - 5x - 24} \div \dfrac{2x^2 - 13x + 6}{2x^2 + 5x - 3}$.

[6.4] **36.** Subtract $\dfrac{x}{6x^2 - x - 15} - \dfrac{5}{9x^2 - 12x - 5}$.

Group Activity/ Challenge Problems

1. The reciprocal of the difference of a certain number and 3 is twice the reciprocal of the difference of twice the number and 6. Find the number(s).
2. If three times a number is added to twice the reciprocal of the number, the answer is 5. Find the number(s).
3. Donald and Juniper McDonald, whose parents own a strawberry farm, must each pick the same number of buckets of strawberries a day. Donald, who is older, picks an average of 6 buckets of berries per hour, while Juniper picks an average of 3 buckets of berries per hour. If Donald and Juniper begin picking strawberries at the same time, and Donald finishes 1.5 hours before Juniper, how many buckets of strawberries must each pick?
4. A mail processing machine can sort a large bin of mail in 1 hour. A newer model can sort the same quantity of mail in 40 minutes. If they operate together, how long will it take them to sort the bin of mail?
5. An old photocopying machine can make 10 copies per minute, while the newer one can make 40 copies a minute.
   (a) How long will it take the old machine to make 400 copies?
   (b) How long will it take the new machine to make 400 copies?
   (c) How long will it take both machines working together to make 400 copies?
6. One tape-duplicating machine can make 100 copies in 10 minutes, while another can make 50 copies in 10 minutes. How long will it take both machines working together to make 100 copies?

Summary

GLOSSARY

Complex fraction (333): A fraction that has a fraction in its numerator or its denominator, or in both its numerator and denominator.

Extraneous root or extraneous solution (339): A value obtained when solving an equation that is not a solution to the original equation.

Rational expression (306): An expression of the form $\frac{p}{q}$, where p and q are polynomials and $q \neq 0$.

Reduced to lowest terms (308): An algebraic fraction is reduced to its lowest terms when the numerator and denominator have no common factors other than 1.

IMPORTANT FACTS

For any fraction: $-\frac{a}{b} = \frac{-a}{b} = \frac{a}{-b}, \; b \neq 0$

To add fractions: $\frac{a}{c} + \frac{b}{c} = \frac{a+b}{c}, \; c \neq 0$

To subtract fractions: $\frac{a}{c} - \frac{b}{c} = \frac{a-b}{c}, \; c \neq 0$

To multiply fractions: $\frac{a}{b} \cdot \frac{c}{d} = \frac{ac}{bd}, \; b \neq 0, \; d \neq 0$

To divide fractions: $\frac{a}{b} \div \frac{c}{d} = \frac{a}{b} \cdot \frac{d}{c} = \frac{ad}{bc}, \; b \neq 0, \; c \neq 0, \; d \neq 0$

$\text{Time} = \frac{\text{distance}}{\text{rate}}$

Review Exercises

[6.1] *Determine the values of the variable for which the following expressions are defined.*

1. $\dfrac{6}{2x - 8}$
2. $\dfrac{5}{x^2 - 7x + 12}$
3. $\dfrac{2}{2x^2 - 13x + 15}$

Reduce to lowest terms.

4. $\dfrac{x}{x - xy}$
5. $\dfrac{x^3 + 4x^2 + 12x}{x}$
6. $\dfrac{9x^2 + 6xy}{3x}$
7. $\dfrac{x^2 + x - 12}{x - 3}$
8. $\dfrac{x^2 - 4}{x - 2}$
9. $\dfrac{2x^2 - 7x + 3}{3 - x}$
10. $\dfrac{x^2 - 2x - 24}{x^2 + 6x + 8}$
11. $\dfrac{3x^2 - 8x - 16}{x^2 - 8x + 16}$
12. $\dfrac{2x^2 - 21x + 40}{4x^2 - 4x - 15}$

[6.2] *Multiply.*

13. $\dfrac{4y}{3x} \cdot \dfrac{4x^2y}{2}$
14. $\dfrac{15x^2y^3}{3z} \cdot \dfrac{6z^3}{5xy^3}$
15. $\dfrac{40a^3b^4}{7c^3} \cdot \dfrac{14c^5}{5a^5b}$
16. $\dfrac{1}{x - 2} \cdot \dfrac{2 - x}{2}$
17. $\dfrac{-x + 2}{3} \cdot \dfrac{6x}{x - 2}$
18. $\dfrac{a - 2}{a + 3} \cdot \dfrac{a^2 + 4a + 3}{a^2 - a - 2}$

Divide.

19. $\dfrac{6y^3}{x} \div \dfrac{y^3}{6x}$
20. $\dfrac{8xy^2}{z} \div \dfrac{x^4y^2}{4z^2}$
21. $\dfrac{3x + 3y}{x^2} \div \dfrac{x^2 - y^2}{x^2}$
22. $\dfrac{1}{a^2 + 8a + 15} \div \dfrac{3}{a + 5}$
23. $(x + 3) \div \dfrac{x^2 - 4x - 21}{x - 7}$
24. $\dfrac{x^2 - 3xy - 10y^2}{6x} \div \dfrac{x + 2y}{12x^2}$

[6.3] *Add or subtract.*

25. $\dfrac{x}{x + 2} - \dfrac{2}{x + 2}$
26. $\dfrac{4x}{x + 2} + \dfrac{8}{x + 2}$
27. $\dfrac{9x - 4}{x + 8} + \dfrac{76}{x + 8}$
28. $\dfrac{7x - 3}{x^2 + 7x - 30} - \dfrac{3x + 9}{x^2 + 7x - 30}$
29. $\dfrac{4x^2 - 11x + 4}{x - 3} - \dfrac{x^2 - 4x + 10}{x - 3}$
30. $\dfrac{6x^2 - 4x}{2x - 3} - \dfrac{(-3x + 12)}{2x - 3}$

Find the least common denominator.

31. $\dfrac{x}{3} + \dfrac{5x}{8}$
32. $\dfrac{4}{3x} + \dfrac{8}{5x^2}$
33. $\dfrac{2}{3x^3y^4} - \dfrac{3}{10x^6y^2}$
34. $\dfrac{6}{x + 1} - \dfrac{3x}{x}$
35. $\dfrac{6x + 3}{x + 2} + \dfrac{4}{x - 3}$
36. $\dfrac{7x - 12}{x^2 + x} - \dfrac{4}{x + 1}$
37. $\dfrac{9x - 3}{x + y} - \dfrac{4x + 7}{x^2 - y^2}$
38. $\dfrac{4x^2}{x - 7} + 8x^2$
39. $\dfrac{19x - 5}{x^2 + 2x - 35} + \dfrac{3x - 2}{x^2 + 9x + 14}$

[6.4] *Add or subtract.*

40. $\dfrac{4}{2x} + \dfrac{x}{x^2}$
41. $\dfrac{1}{4x} + \dfrac{6x}{xy}$
42. $\dfrac{5x}{3xy} - \dfrac{4}{x^2}$
43. $5 - \dfrac{3}{x + 3}$
44. $\dfrac{a + c}{c} - \dfrac{a - c}{a}$
45. $\dfrac{3}{x + 3} + \dfrac{4}{x}$
46. $\dfrac{2}{3x} - \dfrac{3}{3x - 6}$
47. $\dfrac{4}{x + 5} + \dfrac{6}{(x + 5)^2}$
48. $\dfrac{x + 2}{x^2 - x - 6} + \dfrac{x - 3}{x^2 - 8x + 15}$

.2–6.4] *Perform the operation indicated.*

49. $\frac{x+4}{x+3} - \frac{x-3}{x+4}$

50. $6 + \frac{x}{x+2}$

51. $\frac{4x}{a+2} \div \frac{8x^2}{a-2}$

52. $\frac{x+3}{x^2-9} + \frac{2}{x+3}$

53. $\frac{4x+4y}{x^2y} \cdot \frac{y^3}{8x}$

54. $\frac{4}{(x+2)(x-3)} - \frac{4}{(x-2)(x+2)}$

55. $\frac{x+5}{x^2-15x+50} - \frac{x-2}{x^2-25}$

56. $\frac{x^2-y^2}{x-y} \cdot \frac{x+y}{xy+x^2}$

57. $\frac{4x^2-16y^2}{9} \div \frac{(x+2y)^2}{12}$

58. $\frac{a^2-9a+20}{a-4} \cdot \frac{a^2-8a+15}{a^2-10a+25}$

59. $\frac{4x^2-16}{x^2+7x+10} \div \frac{x^2-2x}{2x^2+6x-20}$

60. $\frac{x}{x^2-1} - \frac{2}{3x^2-2x-5}$

5] *Simplify each complex fraction.*

61. $\dfrac{1+\frac{5}{12}}{\frac{3}{8}}$

62. $\dfrac{4-\frac{9}{16}}{1+\frac{5}{8}}$

63. $\dfrac{\frac{15xy}{6z}}{\frac{3x}{z^2}}$

64. $\dfrac{36x^4y^2}{\frac{9xy^5}{4z^2}}$

65. $\dfrac{x+\frac{1}{y}}{y^2}$

66. $\dfrac{x-\frac{x}{y}}{\frac{1+x}{y}}$

67. $\dfrac{\frac{4}{x}+\frac{2}{x^2}}{6-\frac{1}{x}}$

68. $\dfrac{\frac{x}{x+y}}{\frac{x^2}{2x+2y}}$

69. $\dfrac{\frac{1}{a}}{\frac{1}{a^2}}$

70. $\dfrac{\frac{1}{a}+2}{\frac{1}{a}+\frac{1}{a}}$

71. $\dfrac{\frac{1}{x^2}+\frac{1}{x}}{\frac{1}{x^2}-\frac{1}{x}}$

72. $\dfrac{\frac{3x}{y}-x}{\frac{y}{x}-1}$

] *Solve.*

73. $\frac{3}{x} = \frac{8}{24}$

74. $\frac{4}{a} = \frac{16}{4}$

75. $\frac{x+3}{5} = \frac{9}{5}$

76. $\frac{x}{6} = \frac{x-4}{2}$

77. $\frac{3x+4}{5} = \frac{2x-8}{3}$

78. $\frac{x}{5} + \frac{x}{2} = -14$

79. $4 - \frac{5}{x+5} = \frac{x}{x+5}$

80. $\frac{4}{x} - \frac{1}{6} = \frac{1}{x}$

81. $\frac{1}{x-2} + \frac{1}{x+2} = \frac{1}{x^2-4}$

82. $\frac{x-3}{x-2} + \frac{x+1}{x+3} = \frac{2x^2+x+1}{x^2+x-6}$

83. $\frac{x}{x^2-9} + \frac{2}{x+3} = \frac{4}{x-3}$

Solve.

84. It takes Lee 5 hours to mow Mr. Young's lawn. It takes Pat 4 hours to mow the same lawn. How long will it take them working together to mow Mr. Young's lawn?

85. A $\frac{3}{4}$-inch-diameter hose can fill a swimming pool in 7 hours. A $\frac{5}{16}$-inch-diameter hose can siphon all the water out of a full pool in 12 hours. How long will it take to fill the pool if while one hose is filling the pool the other hose is siphoning water from the pool?

86. One number is four times as large as another. The sum of their reciprocals is $\frac{1}{2}$. Find the numbers.

87. A Greyhound bus can travel 400 kilometers in the same time that an Amtrak train can travel 600 kilometers. If the speed of the train is 40 kilometers per hour greater than that of the bus, find the speeds of the bus and the train.

Practice Test

Perform the indicated operations.

1. $\dfrac{3x^2y}{4z^2} \cdot \dfrac{8xz^3}{9y^4}$
2. $\dfrac{a^2 - 9a + 14}{a - 2} \cdot \dfrac{a^2 - 4a - 21}{(a - 7)^2}$
3. $\dfrac{x^2 - 9y^2}{3x + 6y} \div \dfrac{x + 3y}{x + 2y}$
4. $\dfrac{16}{y^2 + 2y - 15} \div \dfrac{4y}{y - 3}$
5. $\dfrac{6x + 3}{2y} + \dfrac{x - 5}{2y}$
6. $\dfrac{7x^2 - 4}{x + 3} - \dfrac{6x + 7}{x + 3}$
7. $\dfrac{5}{x} + \dfrac{3}{2x^2}$
8. $5 - \dfrac{6x}{x + 2}$
9. $\dfrac{x - 5}{x^2 - 16} - \dfrac{x - 2}{x^2 + 2x - 8}$

Simplify.

10. $\dfrac{3 + \dfrac{5}{8}}{2 - \dfrac{3}{4}}$
11. $\dfrac{x + \dfrac{x}{y}}{\dfrac{1}{x}}$

Solve.

12. $\dfrac{x}{3} - \dfrac{x}{4} = 5$
13. $\dfrac{x}{x - 8} + \dfrac{6}{x - 2} = \dfrac{x^2}{x^2 - 10x + 16}$

Solve.

14. Mr. Johnson, on his tractor, can level a 1-acre field in 8 hours. Mr. Hackett, on his tractor, can level a 1-acre field in 5 hours. If they work together, how long will it take them to level a 1-acre field?

Cumulative Review Test

1. Evaluate $3x^2 - 2xy - 7$ when $x = -3$ and $y = 5$.
2. Evaluate $-4 - [2(-6 \div 3)^2] \div 2$.
3. Solve the equation $4y + 3 = -2(y + 6)$.
4. Simplify $\left(\dfrac{6x^2y^3}{2x^5y}\right)^3$.
5. Solve the formula $P = 2E + 3R$ for R.
6. Simplify $(6x^2 - 3x - 5) - (3x^2 + 8x - 9)$.
7. Multiply $(4x^2 - 6x + 3)(3x - 5)$.
8. Factor $6a^2 - 6a - 5a + 5$.
9. Factor $10x^2 - 5x + 5$.
10. Factor $x^2 - 10x + 24$.
11. Factor $6x^2 - 11x - 10$.
12. Solve $2x^2 = 11x - 12$.
13. Multiply $\dfrac{x^2 - 9}{x^2 - x - 6} \cdot \dfrac{x^2 - 2x - 8}{2x^2 - 7x - 4}$.
14. Subtract $\dfrac{x}{x + 4} - \dfrac{3}{x - 5}$.
15. Add $\dfrac{4}{x^2 - 3x - 10} + \dfrac{2}{x^2 + 5x + 6}$
16. Solve the equation $\dfrac{x}{6} - \dfrac{x}{4} = \dfrac{1}{8}$.
17. Solve the equation $\dfrac{1}{x - 4} + \dfrac{2}{x - 3} = \dfrac{4}{x^2 - 7x + 12}$.
18. A school district allows its employees to choose from two medical plans. With plan 1, the employee pays 10% of all medical bills (the school district pays the balance). With plan 2, the employee pays the school district a one-time payment of $100, then the employee pays 5% of all medical bills. What total medical bills would result in the employee paying the same amount with the two plans?
19. A grocer wishes to mix 6 pounds of Chippy dog food worth $3 per pound with Hippy dog food, worth $4 per pound. How much of the Hippy dog food should be mixed if the dog food mixture is to sell for $3.20 per pound?
20. Chiquita rides her bike up a hill at an average speed of 4 miles per hour. Then she rides down the hill and rides to her friend's house at an average speed of 12 miles per hour. If the total distance traveled is 6 miles and the total time spent riding is 1 hour, find the distance she rides at 4 miles per hour and the distance she rides at 12 miles per hour.

Chapter 7

Graphing Linear Equations

See Section 7.1, Group Activities 1.

Preview and Perspective

In this chapter we explain how to graph linear equations. The graphs of linear equations are straight lines. We will graph linear equations further in Chapter 8, when we graph systems of equations. The material presented in this chapter will be used again in Chapter 10 when we graph other types of equations, whose graphs are not straight lines.

Graphing is one of the most important topics in mathematics, and each year its importance increases. If you take additional mathematics courses you will graph many different types of equations. The material presented in this chapter should give you a good background for graphing in later courses. Graphs are also used in many professions and industries. They are used to display information and to make projections about the future.

In Section 7.1 we introduce and discuss pie, bar, and line graphs. You see graphs of this type daily in newspapers and magazines, yet many students do not understand how to interpret them. Once you are able to interpret these graphs you will be in a better position to make important personal decisions.

In Section 7.2 we introduce the Cartesian coordinate system and explain how to plot points. In Section 7.3 we discuss two methods for graphing linear equations: by plotting points and by using the x and y intercepts. The slope of a line is discussed in Section 7.4. In Section 7.5 we discuss a third procedure, using slope, for graphing a linear equation.

Functions are discussed in Section 7.6. Functions are a unifying concept in mathematics. In this chapter we give a brief and somewhat informal introduction to functions. Functions will be discussed in much more depth in later mathematics courses.

We solved inequalities in one variable in Section 2.7. In Section 7.7 we will solve and graph linear inequalities in two variables. Graphing linear inequalities is an extension of graphing linear equations. In later mathematics courses, you may graph other types of inequalities in two variables.

This is an important chapter. If you plan on taking another mathematics course, graphs and functions will probably be a significant part of that course.

7.1 Pie, Bar, and Line Graphs

Tape 11

1. Learn to interpret pie graphs.
2. Learn to interpret bar and line graphs.

In this section we discuss pie graphs, bar graphs, and line graphs. You see such graphs daily in newspapers and magazines.

Pie Graphs

1 A *pie graph*, also called a *circle graph*, displays information using a circle. The circle is divided into pieces called *sectors*. Figure 7.1 illustrates a pie graph showing sources of business start-up capital.

In Figure 7.1, notice that 73% of the \$82,300 used to start the average new business in the survey came from the individuals themselves and their family and friends. The sector of the circle that represents the individuals, their family, and friends should be 73% of the area of the entire circle. Outside investors should be 13% of the area of the entire circle, and so on. The total percent indicated in the circle should be 100%. Note that 73% + 13% + 8% + 6% = 100%. How much

Founders of businesses risked an average of $82,300 to start their companies. The money came from:

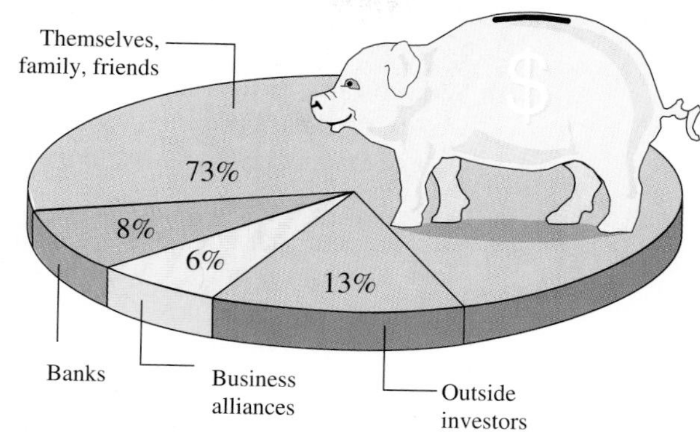

FIGURE 7.1

money of the $82,300 came from the individuals, their family, and friends? To determine this amount, we find 73% of $82,300 as follows:

$$(0.73)(82{,}300) = 60{,}079$$

Thus, $60,079 came from the individuals, their family, and friends. The amount from outside investors, banks, and business alliances is determined as follows:

Outside investors:	$(0.13)(82{,}300) = 10{,}699$
Banks:	$(0.08)(82{,}300) = 6584$
Business alliances:	$(0.06)(82{,}300) = 4938$

Notice that the sum of the four amounts is $82,300. That is, $60,079 + $10,699 + $6584 + $4938 = $82,300.

EXAMPLE 1 Consider the pie graph in Figure 7.2, which shows federal health care spending in 1993.

(a) What should the sum of the amounts in the four sectors equal? Explain.

(b) What percent of the total budget went to Medicaid?

(c) What percent of the area of the circle should be designated as Medicaid?

(d) Redraw the pie graph so that it lists percents rather than dollar amounts.

Where federal health-care spending goes ('93, in billions)

Total $254.2

Medicare
$134.1
$80.3
$24.9
Veterans affairs $14.9
Medicaid
Other

FIGURE 7.2

Solution:

(a) Since the total budget is \$254.2 billion, the sum of the four sectors should be \$254.2 billion. Note that \$134.1 billion + \$80.3 billion + \$24.9 billion + \$14.9 billion = \$254.2 billion.

(b) \$80.3 billion of the \$254.2 billion goes to Medicaid. To find the percent, we divide the amount that goes to Medicaid by the total budget, then change the decimal answer to a percent. The procedure follows.

$$\text{Percent that goes to Medicaid} = \frac{\text{amount for Medicaid}}{\text{total budget}}$$

$$= \frac{80.3}{254.2} \approx 0.316 \approx 31.6\%$$

(c) Since about 31.6% of the budget is for Medicaid, the portion of the circle illustrating Medicaid should be about 31.6% of the area of the circle.

(d) To redraw this circle with percents, we need to determine the percent of the total for each sector. We follow the same procedure we used to find the percent that goes for Medicaid to find the other percents.

$$\text{Percent that goes to Medicare} = \frac{134.1}{254.2} \approx 0.528 \approx 52.8\%$$

$$\text{Percent that goes to veterans' affairs} = \frac{14.9}{254.2} \approx 0.059 \approx 5.9\%$$

$$\text{Percent that goes to other} = \frac{24.9}{254.2} \approx 0.098 \approx 9.8\%$$

Notice that 31.6% + 52.8% + 5.9% + 9.8% = 100.1%. We get slightly more than 100% because of roundoff error. The circle graph is illustrated with percents in Figure 7.3.

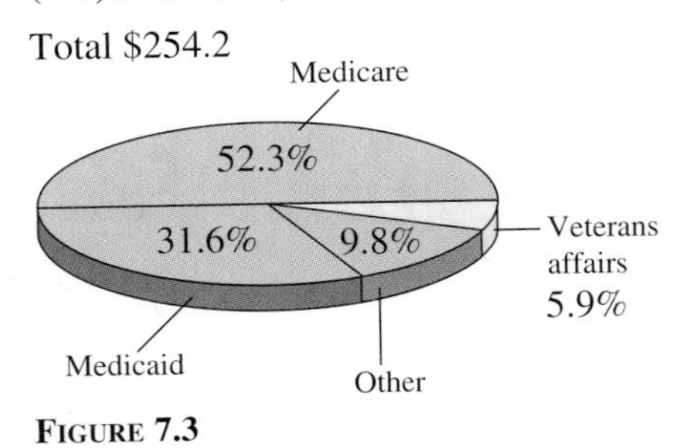

FIGURE 7.3

Bar and Line Graphs

2 Now we discuss bar and line graphs. A typical bar graph is shown in Figure 7.4. This bar graph shows the net income of Blockbuster Entertainment. In a bar graph, one of the axes indicates amounts. Bars are used to show the amounts of each item. The amounts may be listed on either the vertical or horizontal axis. The other axis will give additional information, such as the years the amounts were recorded.

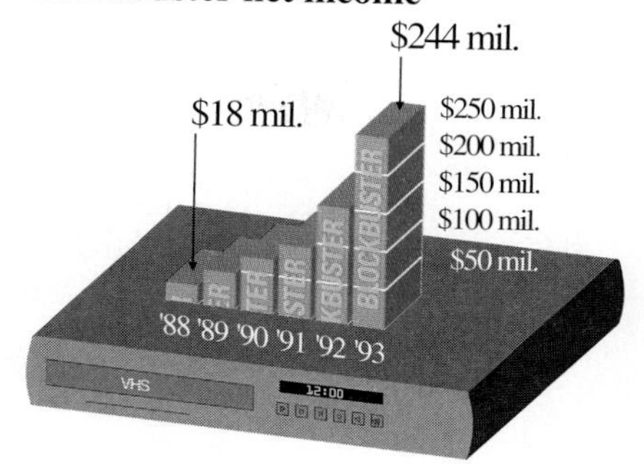

FIGURE 7.4

Consider the bar graph in Figure 7.4. To determine the annual income of Blockbuster Entertainment in any year, find the bar for that year and estimate where the top of the bar would intersect the vertical axis if the top of the bar were extended. For example, in 1989 we see that the income was slightly under $50 million, but in 1992 the income had grown to about $150 million. From this graph we see that the annual income of Blockbuster has grown each year since 1988. This is due primarily to the increase in movie rentals. Can you estimate the total income of Blockbuster from 1988 up to and including 1993? Since each bar represents the annual income for a specific year, the total income can be found by adding the income for each of the six years. Estimate the total income of Blockbuster from 1988 to 1993 now. Your total should be about $620 million. Answers will differ somewhat based on individual estimates. For example, one student might estimate 1990 income to be $60 million, while another might estimate it to be $70 million.

Bar graphs can be used to illustrate trends. However, you need to be aware that trends change. For example, by looking at this graph you might be led to believe that Blockbuster Entertainment will continue to increase its annual income year after year. However, you might form a different opinion after looking at the *line graph* indicated in Figure 7.5.

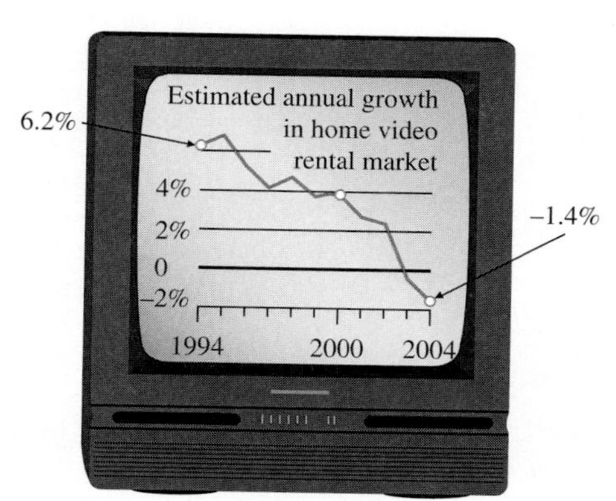

FIGURE 7.5

A line graph generally has amounts indicated on the vertical axis, and some measure of time, such as years or months, listed on its horizontal axis. To construct a line graph, mark the corresponding amounts above the respective time periods. Then connect the marks consecutively with straight-line segments. Notice that the amount listed on the vertical axis in Figure 7.5 is percent. This graph indicates that the estimated annual growth in the home video market rental is expected to increase by 6.2% in 1994 and peak in 1995. By the year 2000 the increase in growth is expected to slow to about 4%. Between 2002 and 2003 the rental home video market is expected to begin decreasing in size. The decrease in the growth from 1995 may be due to the increase in the use of cable television and the number of television channels available.

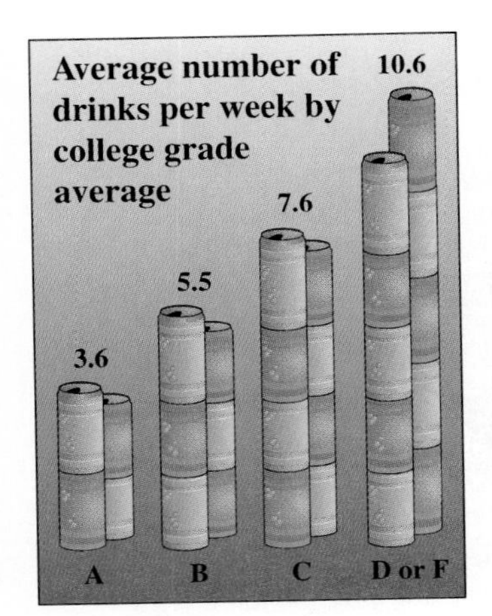

FIGURE 7.6

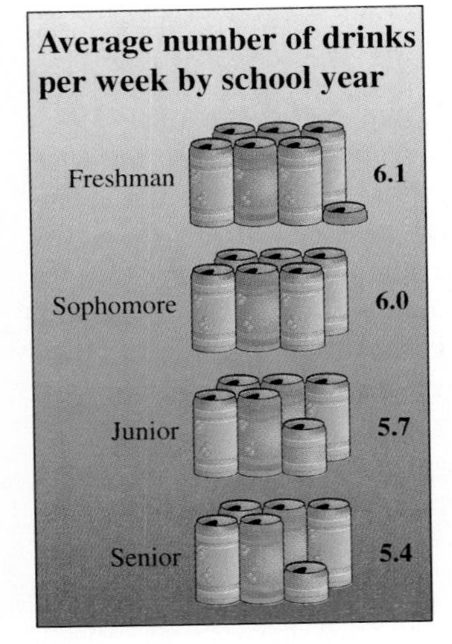

FIGURE 7.7

EXAMPLE 2 Consider the graphs indicated in Figures 7.6 and 7.7. The information for these graphs was provided by the Southern Illinois University Core Alcohol and Drug Survey.

(a) Explain why both of these figures may be considered bar graphs.

(b) Redraw Figure 7.6 as a bar graph with a vertical axis indicating the number of drinks on the right side of the graph.

(c) Explain what Figure 7.6 shows.

(d) Redraw Figure 7.7 as a bar graph with the horizontal axis indicating the number of drinks.

(e) Explain what Figure 7.7 shows.

Solution:

(a) In Figures 7.6 and 7.7, pictures of cans are used instead of bars. Pictures are often used in place of bars to make the graph more interesting or appealing. However, the same information is indicated whether bars or cans are used.

(b) Figure 7.8 shows the same information as illustrated in Figure 7.6.

(c) Figure 7.6 indicates that, in general, college students who achieve better grades drink less.

(d) Figure 7.9 illustrates the same information as illustrated in Figure 7.7.

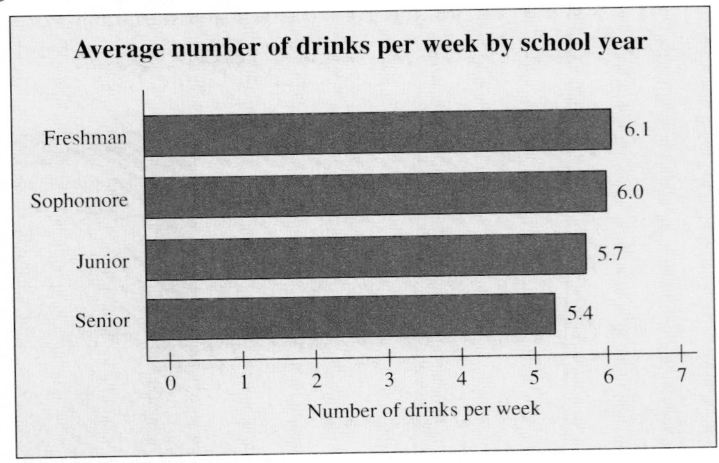

FIGURE 7.9

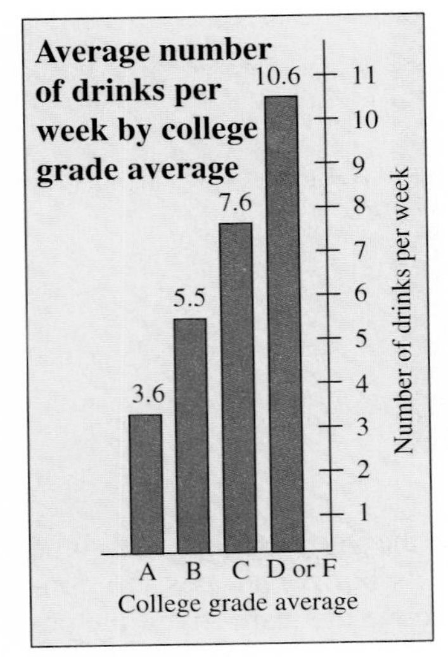

FIGURE 7.8

(e) Figure 7.7 illustrates that the average number of drinks consumed decreases from the freshmen year through the senior year.

Figure 7.10 illustrates a line graph showing the increase in global temperature from 1960 to 1990. The information for this graph was provided by the U.S. Department of Agriculture. Notice the break above the zero in the

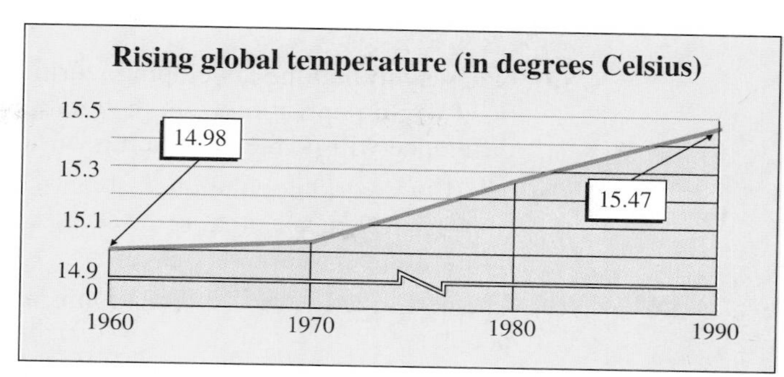

FIGURE 7.10

vertical axis. If all the values from 0 to 15.5 were to be shown, this line graph would be very high and require more space. To save space we "break the axis," as illustrated in this figure. The break indicates that a part of the axis has been omitted.

In Figure 7.11 we have redrawn the bar graph in Figure 7.9 with a break in the horizontal axis. The break gives the graph a different appearance and makes the difference between the classes appear greater.

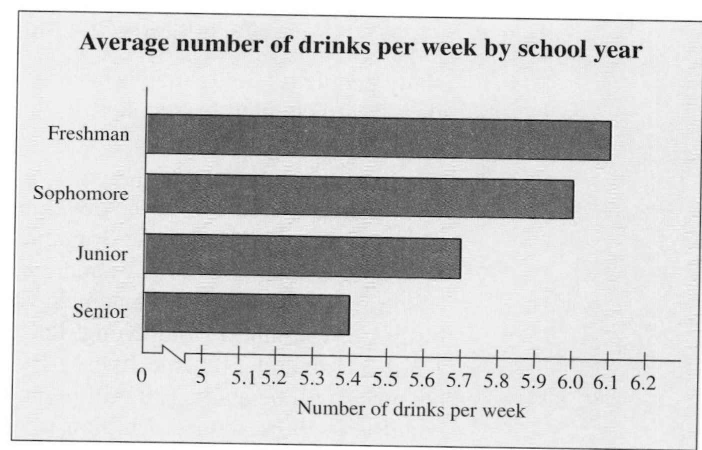

FIGURE 7.11

Some bar and line graphs are formed by placing individual parts on top of one another as illustrated in Example 3.

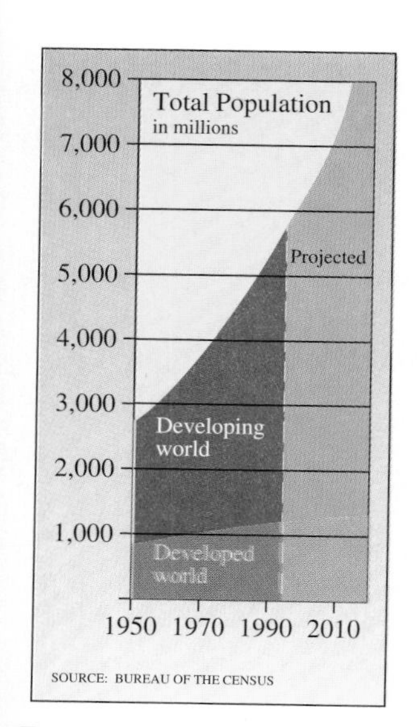

FIGURE 7.12

EXAMPLE 3 The line graph in Figure 7.12 shows Bureau of the Census data for past, present, and projected population for both the developed and developing world.

(a) Explain how to find the developing world population for any given year.
Estimate:

(b) the world's population in 1994 (indicated by the vertical dashed line),

(c) the developed world's population in 1994,

(d) the developing world's population in 1994, and,

(e) the increase in the world's population from 1950 to 2010.

(f) Explain what this graph shows.

Solution:

(a) One way to find the developing world population is to subtract the developed world population from the total population for the given year. This difference will be the developing world population.

(b) The 1994 world population is about 5700 million or 5.7 billion people (the article indicates that in 1994 the world's population is increasing by 94 million per year).

(c) The developed world's population is about 1300 million or 1.3 billion people.

(d) The developing world population for 1994 can be found by subtracting the developed world population from the total population:

developing world population in 1994 = 5.7 billion − 1.3 billion = 4.4 billion

(e) The increase in the world's population can be found by subtracting the world's population in 1950 from the projected world's population in 2010. The world's population in 1950 was about 2.7 billion and the projected world population in 2010 is about 7.9 billion:

increase in the world's population from 1950 to 2010
= 7.9 billion − 2.7 billion = 5.2 billion

Many people do not realize how large 5.2 billion is. To give you some idea, if you counted by seconds, it would take about 165 years to count to 5.2 billion.

(f) The line graph shows that the world's population is increasing rapidly. The greatest increase is in the developing countries and regions (such as the Gaza Strip, Ethiopia, and Somalia, where the average number of births per woman are 7.9, 7.5, and 6.8, respectively). The developed countries are growing at a much slower rate (the average number of births per woman in Hong Kong, Italy, Japan, and the United States are 1.2, 1.3, 1.6, and 2.0 respectively). By 2010 the increase in the developed countries will be about 120 million and the increase in the developing countries will be about 1.7 billion.

Exercise Set 7.1

1. Use the circle graph below to answer the following questions.

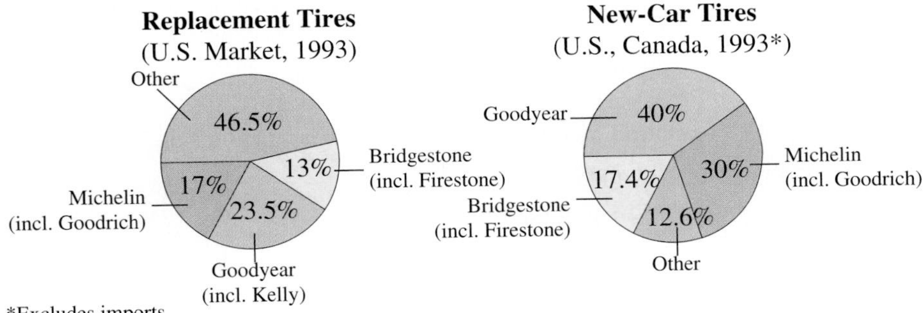

*Excludes imports.

(a) In 1993, what percent of replacement tires were manufactured by Goodyear, Michelin, or Bridgestone?

(b) By observing the graph, is it possible to determine whether Goodyear sold more replacement tires or more new car tires in 1993? Explain.

(c) By observing the graph, is it possible to determine the number of new tires sold by Michelin in 1993? Explain.

(d) By observing the graphs, is it possible to determine whether Goodyear, Michelin, or Bridgestone made the greatest profit in 1993? Explain.

2. Use the circle graph below to answer the following questions.

Distribution of Federal Spending (1993)

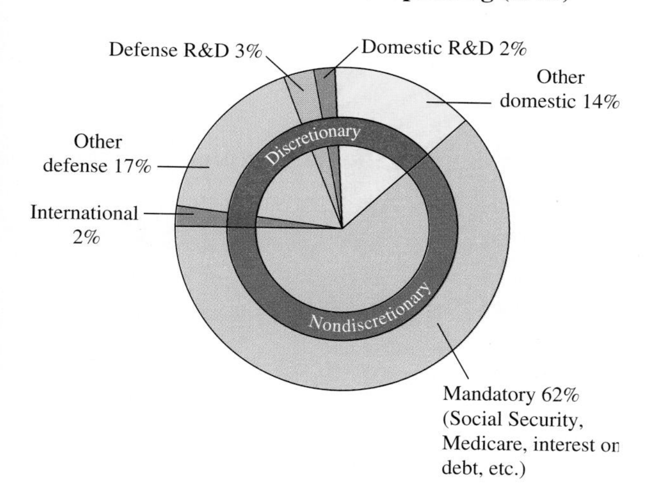

What percent of the federal budget was for

(a) discretionary spending?

(b) nondiscretionary spending?

(c) research and development (total)?

(d) If the total 1993 federal spending was \$1.47 trillion, what was the total spent on research and development?

(e) How much was spent on mandatory items?

3. The following graph shows a breakdown of revenue from the 1993 Virginia State Fair held at Strawberry Hill in Richmond.

How much money was collected from

(a) ticket sales?

(b) fees from concessionaires and vendors?

State Fair 1993 Revenue: \$3.6 Million

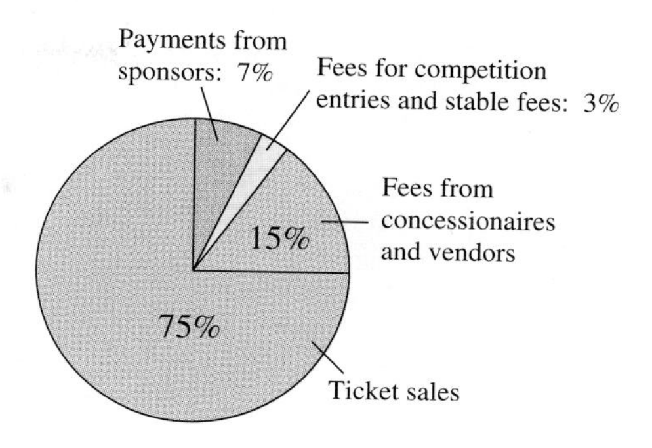

(c) Redraw this pie graph using dollar amounts collected rather than percents.

4. Consider the bar graph illustrated below (computers). Information for this graph was obtained by Motorola.

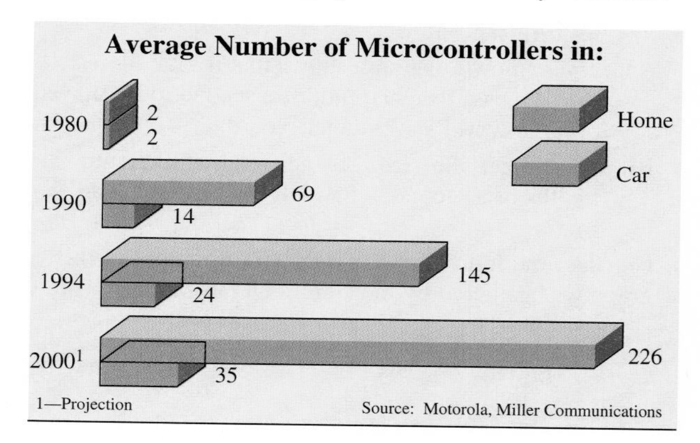

(a) What was the increase in the number of microcomputers in the typical house from 1990 to 1994?

(b) What was the increase in the number of microcomputers in the typical car from 1990 to 1994?

(c) List at least 10 places where microcomputers may be found in a home.

(d) Do you believe that this explosive growth in the use of microcomputers will continue in the typical house? Explain.

5. The following bar graph shows changes in the record time, for the world's men 100 meter race.

Fastest Record Slow to Change

It took more than 26 years to shave one-tenth of a second from the world men's 100-meter record.

From	To	Difference	Length of time
9.95	9.93	.02	15 years
9.93	9.92	.01	5 years
9.92	9.90	.02	3 years
9.90	9.86	.04	2.5 months
9.86	9.85	.01	3 years

Note: Canada's Ben Johnson ran times of 9.83 and 9.79, but both were invalidated because of drug use.

(a) How long was the men's 100-meter record at 9.95 seconds?

(b) How long did it take for the record to be reduced from 9.95 seconds to 9.90 seconds, and by what part of a second had the record been reduced during this time period?

(c) Explain how this graph shows that it took more than 26 years for one-tenth of a second to be shaved from the world's 100-meter record.

(d) If next year, the record is reduced by three hundredths of a second, what will be the new record time?

6. Consider the following bar graph. Information for this graph was gathered by the Gallup Organization.

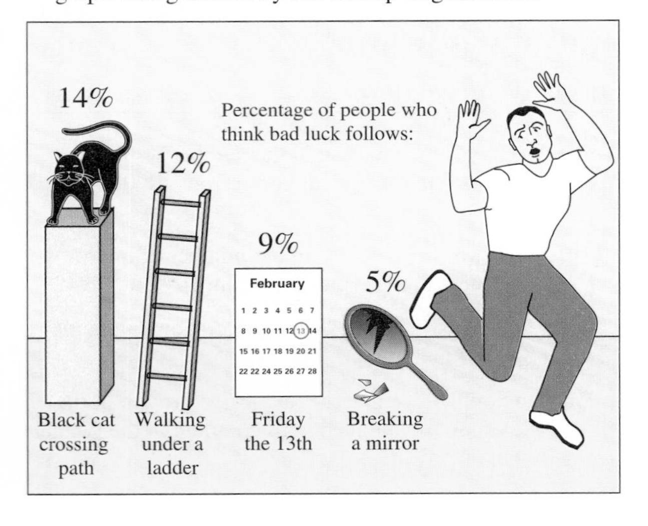

(a) Do more people believe that walking under a ladder is more likely to bring bad luck than breaking a mirror? Explain.

(b) If we add the four percents, we get a total of 40%. Does this mean that 40% of the population believe that bad luck will follow the occurrence of one these items? Explain.

See Exercise 5.

(c) Redraw this bar graph listing percent on the horizontal.

7. The bar graph in the following figure provides information on the number of cinema screens worldwide. Source: Screen Digest.

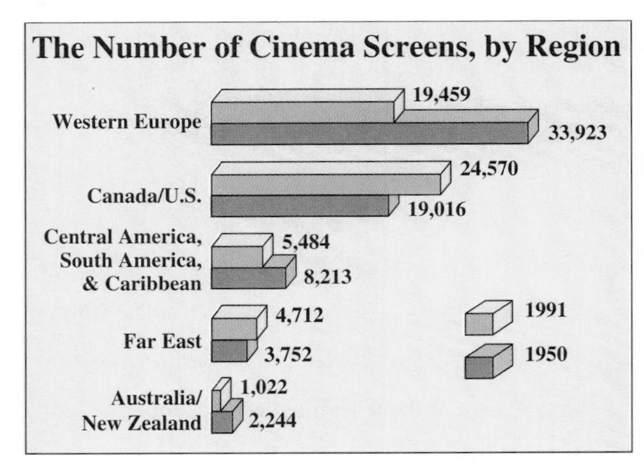

(a) Has the number of cinema screens increased or decreased in western Europe from 1950 to 1991, and by how much?

(b) Has the number of cinema screens increased or decreased in Canada/U.S. from 1950 to 1991, and by how much?

(c) What was the total number of cinema screens in 1950? What was the total number in 1991?

See Exercise 7.

8. The bar graph below shows 1993 military expenditures as a percentage of gross domestic product in selected countries. *Source:* United Nations.

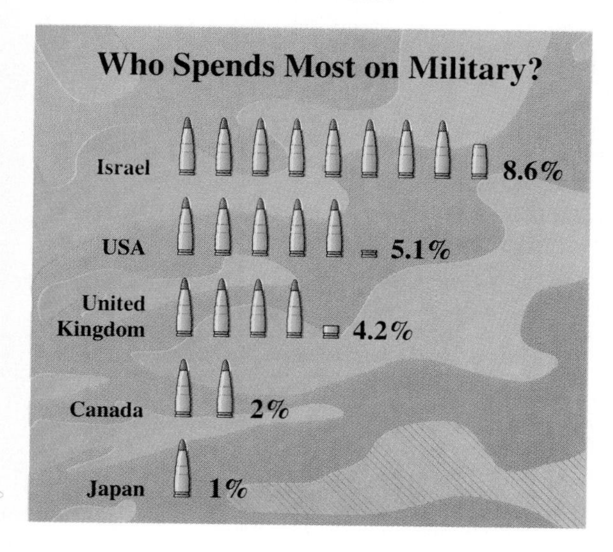

(a) By observing the graph, is it possible to determine how many dollars were spent by the United States on military expenditures in 1993? Explain.

(b) In 1993 the U.S. gross domestic product was $6377.9 billion. Determine the amount the United States spent on military expenditures in 1993.

(c) In 1993, Israel's gross domestic product was $74 billion. Determine the amount Israel spent on military expenditures in 1993.

9. The following graph illustrates new AIDS cases reported worldwide.

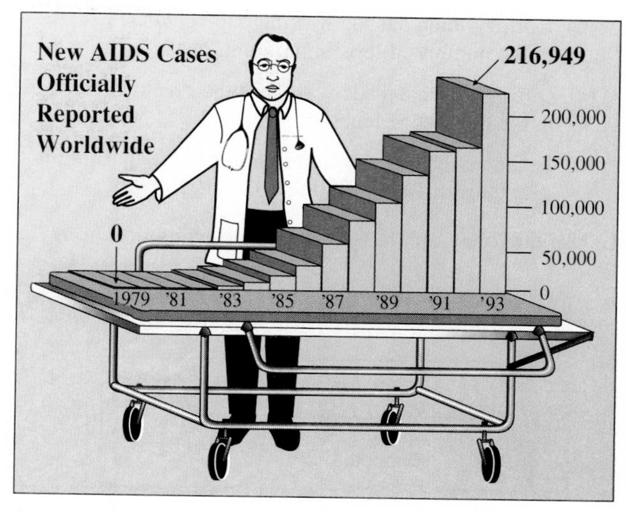

Estimate the number of new AIDS cases reported in

(a) 1986

(b) 1987

(c) 1992.

(d) Estimate the number of AIDS cases reported to the World Health Organization from 1979 through 1993. (The World Health Organization suspects that in 1993 the number of AIDS cases was much higher than that reported: about 4 million people with AIDS, plus 17 million people infected with HIV.)

10. The following line graph shows legal U.S. immigration by decades, beginning 1901 to 1910 and ending 1981 to 1990. For example, from the graph we can see that during the decade 1901-1910 there were 8,795,386 immigrants to legally enter the United States.

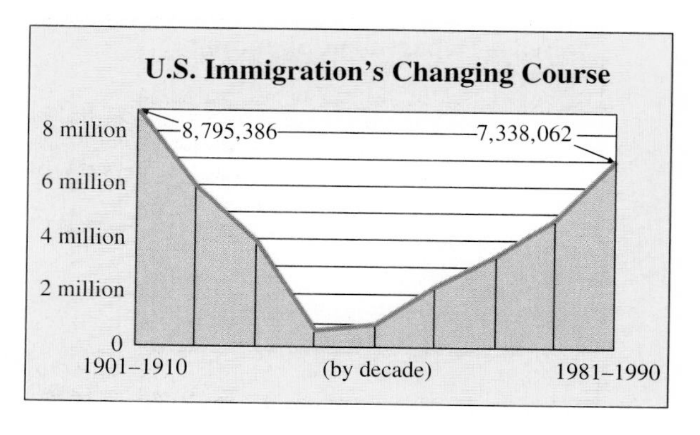

(a) Describe the trend in U.S. immigration from 1901–1910 to 1981–1990.

(b) During which decade was there the greatest decrease in the number of legal immigrants?

(c) During which decade was there the greatest increase in the number of legal immigrants?

(d) Estimate the increase in the number of legal immigrants from 1971–1980 to 1981–1990.

11. Use the line graph to answer the following questions.

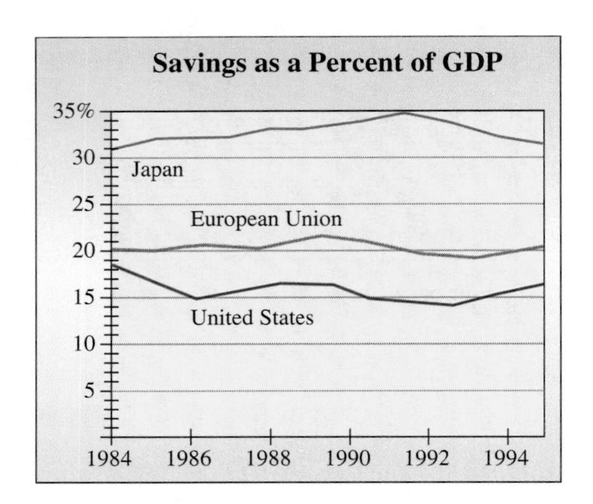

(a) Estimate savings as a percent of gross domestic product for Japan, for the European Union, and for the United States in 1984.

(b) Describe the savings trend in the United States from 1984 to 1994.

(c) Describe the savings trend in Japan from 1984 to 1994.

12. Consider the following graph, with information supplied by the Aerospace Industry Association.

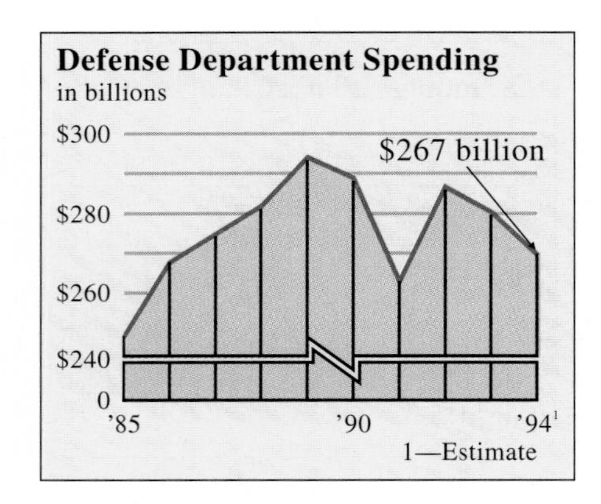

Estimate the amount the United States spent on defense in

(a) 1989

(b) 1991

(c) 1992.

(d) Estimate the difference in the amount spent in 1991 and 1994.

(e) Estimate the total spent on defense from 1985 through 1994.

13. Consider the line graph that follows.

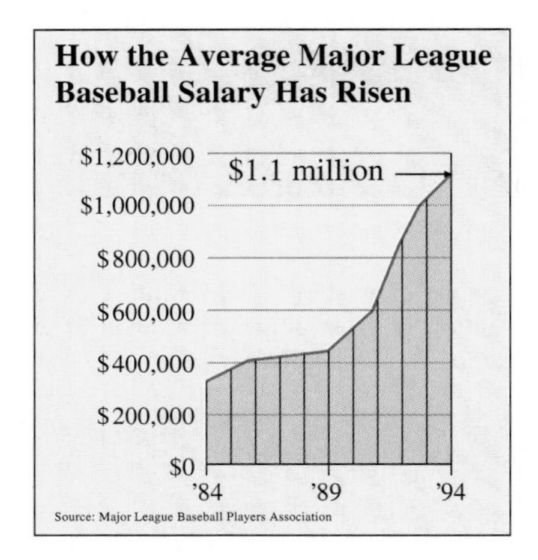

(a) Estimate the average baseball player's salary in 1984.

(b) When did baseball players' salaries start to increase rapidly?

(c) Estimate the average baseball player's salary in 1989.

(d) Estimate the increase in salary from 1989 to 1994.

(e) Estimate the percent increase in salary from 1989 to 1994 by dividing the increase in salary from 1989 to 1994 [part (d)] by the 1989 salary.

(f) Estimate the average yearly percent increase from 1989 to 1994 by dividing the percent increase [part (e)] by the number of years from 1989 to 1994, which is 5 years.

14. Consider the following text and line graphs.

Population and the Environment

The world's demand for wood — half of which is used as the primary source of energy in developing countries — has continued to climb. The use of wood is threatening the world's forests and raising the amount of carbon dioxide in the air. Some scientists say carbon dioxide is the prime contributor to global warming, the theoretical "greenhouse effect." Even though developed nations are working to cut carbon dioxide emissions, levels are expected to continue upward because of activities in developing countries.

Wood production (millions of cubic meters)

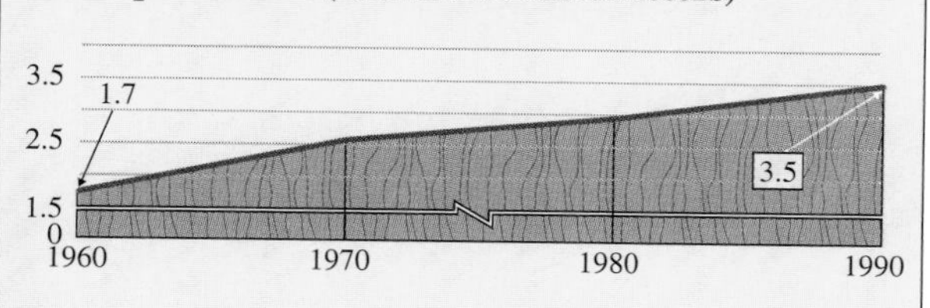

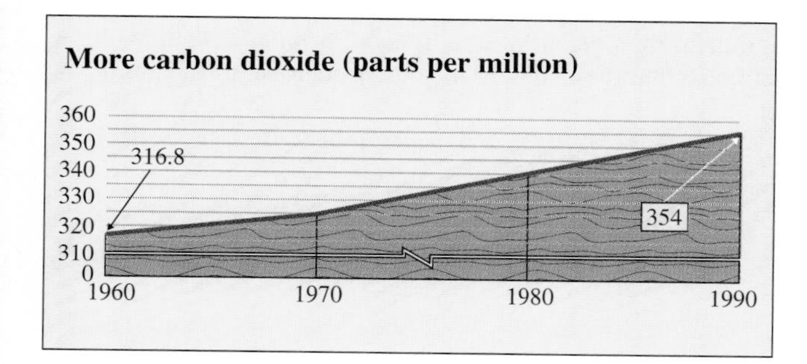

(a) Find the difference in wood production from 1960 to 1990.

(b) Estimate the percent increase in wood production by dividing the increase from 1960 to 1990 found in part (a) by the amount of wood produced in 1960.

(c) Estimate the annual percent increase in wood production from 1960 to 1990 by dividing the percent increase found in part (b) by the number of years from 1960 to 1990, which is 30.

(d) Using information provided in parts (a)–(c), determine the annual percent increase in carbon dioxide from 1960 to 1990.

15. The information for the following graph is provided by the World Bank.

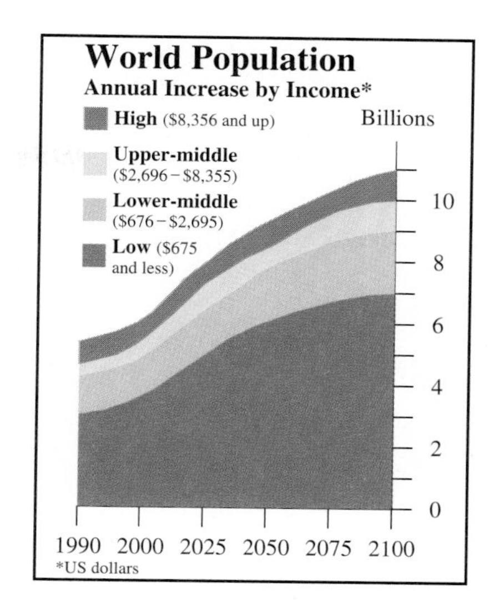

(a) In 1990, estimate the world population classified as having a low income.

(b) In 2100, estimate the projected world population classified as having a low income.

(c) In 1990 estimate the world population classified as having a high income.

(d) In 2100 estimate the world population classified as having a high income.

(e) In 2100 estimate the world population classified as having a lower–middle income.

(f) Explain what this graph shows.

See Exercise 15.

16. Describe a pie graph.

17. Describe a bar graph.

18. Describe a line graph.

19. Find two pie graphs in newspapers or magazines, cut them out and paste them in your notebook, and explain what they show.

20. Find two bar graphs in newspapers or magazines, cut them out and paste them in your notebook, and explain what they show.

21. Find two line graphs in newspapers or magazines, cut them out and paste them in your notebook, and explain what they show.

CUMULATIVE REVIEW EXERCISES

[4.4] **22.** Simplify $3x - \frac{2}{3} + \frac{3}{5}x - \frac{1}{2}$

[4.6] **23.** Divide $3x^2 - 4x + 6$ by $x - 2$.

[5.5] **24.** Factor $4x^2 - 16$.

[6.6] **25.** Solve $x - \frac{2}{3} = \frac{3}{5}x - \frac{1}{2}$

Group Activity/ Challenge Problems

1. The Disney Corporation receives its income from theme parks and resorts, films, and consumer products. The percent income from each source in 1992 is indicated below. Construct a pie graph which indicates the percent of total income from each source. Explain how you drew the sectors so that they contain the correct percent of the area. (*Hint:* A circle contains 360°.)

Theme parks and resorts	45%
Films	35%
Consumer products	20%

2. The average cost of making a movie, in millions of dollars, in a variety of countries in 1991, is indicated below.

U.S.	7.63
U.K.	5.32
France	3.86
Belgium	3.44
Australia	3.34
Italy	2.70
Germany	1.86
India	0.12

Illustrate this information using a bar graph with the cost on the horizontal axis. (The average cost of making a movie in only U.S. major studios exceeds $24 million.)

3. The following chart indicates the number of TV households and the number of cable households for various countries.

(a) For each country determine the cable households as a percentage of total TV households. Round answers to the nearest percent.

(b) Illustrate the percent of cable households relative to total TV households for the countries indicated using a bar graph with percent on the vertical axis.

Country	TV Households	Cable Households
Belgium	3,700,000	3,400,000
Netherlands	6,100,000	4,900,000
Canada	9,756,000	7,537,000
United States	92,740,000	54,890,000
Germany	33,281,000	9,900,000
Japan	40,000,000	6,170,000
United Kingdom	21,600,000	490,000

4. The number of cellular telephone subscribers from 1984 to 1993 is indicated below. Draw a line graph indicating the information.

Year	84	85	86	87	88	89	90	91	92	93
Subscribers (millions)	0.1	0.3	0.7	1.2	2.1	3.5	5.3	7.6	11.0	16.0

7.2 The Cartesian Coordinate System and Linear Equations in Two Variables

Tape 11

1. Plot points in the Cartesian coordinate system.
2. Determine whether an ordered pair is a solution to a linear equation.

The Cartesian Coordinate System

1 In this chapter we discuss several procedures that can be used to draw graphs. A **graph** shows the relationship between two variables in an equation. Many algebraic relationships are easier to understand if we can see a picture of them. We draw graphs using the **Cartesian (or rectangular) coordinate system.** The Cartesian coordinate system is named for its developer, the French mathematician and philosopher René Descartes (1596–1650).

The Cartesian coordinate system provides a means of locating and identifying points just as the coordinates on a map help us find cities and other locations. Consider the map of the Great Smoky Mountains (see Fig. 7.13 on page 376). Can you find Cades Cove on the map? If I tell you that it is in grid A3, you can probably find it much quicker and easier.

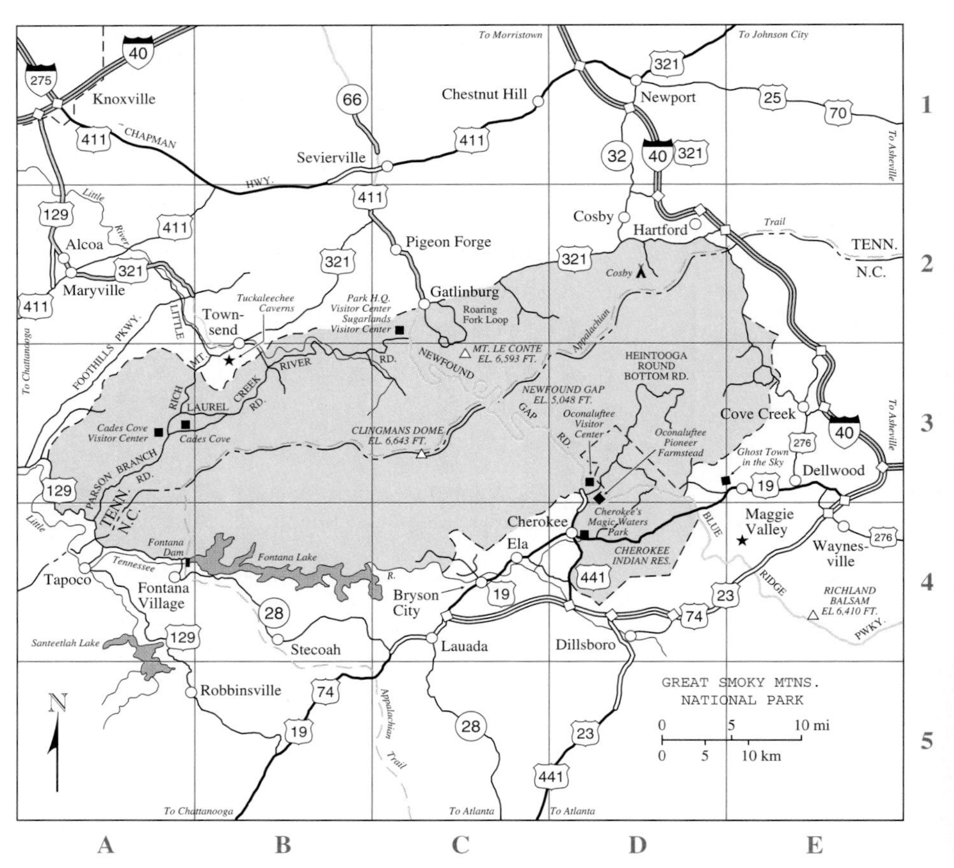

FIGURE 7.13

y axis

II I

Origin

x axis

III IV

FIGURE 7.14

The Cartesian coordinate system is a grid system, like that of a map, except that it is formed by two axes (or number lines) drawn perpendicular to each other. The two intersecting axes form four **quadrants** (Fig. 7.14).

The horizontal axis is called the ***x* axis.** The vertical axis is called the ***y* axis.** The point of intersection of the two axes is called the **origin.** At the origin the value of x is 0 and the value of y is 0. Starting from the origin and moving to the right along the x axis, the numbers increase. Starting from the origin and moving to the left, the numbers decrease (Fig. 7.15). Starting from the origin and moving up the y axis, the numbers increase. Starting from the origin and moving down, the numbers decrease.

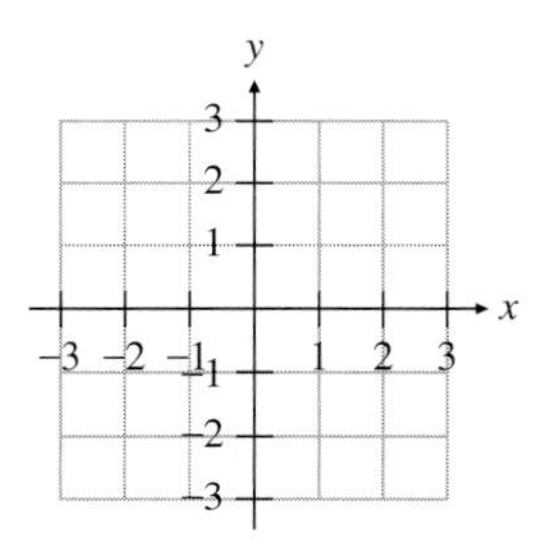

FIGURE 7.15

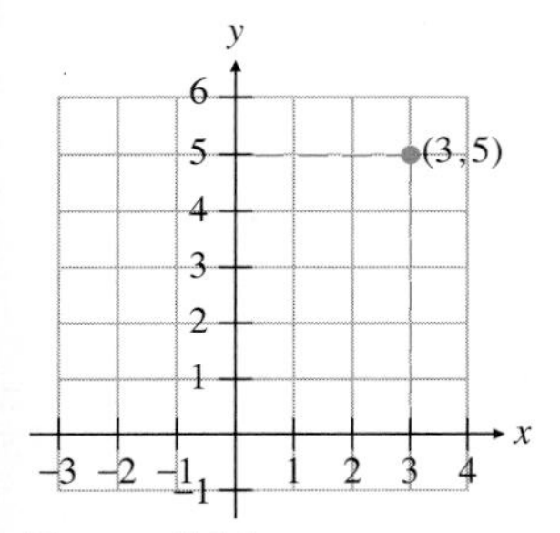

FIGURE 7.16

To locate a point, it is necessary to know both the value of x and the value of y, or the **coordinates,** of the point. When the x and y coordinates of a point are placed in parentheses, *with the x coordinate listed first,* we have an **ordered pair.** In the ordered pair (3, 5) the x coordinate is 3 and the y coordinate is 5. The point corresponding to the ordered pair (3, 5) is plotted in Figure 7.16. The phrase "the point corresponding to the ordered pair (3, 5)" is often abbreviated "the point (3, 5)". For example, if we write "the point $(-1, 2)$" it means "the point corresponding to the ordered pair $(-1, 2)$."

EXAMPLE 1 Plot each point on the same axes.

(a) $A(4, 2)$ **(b)** $B(2, 4)$ **(c)** $C(-3, 1)$
(d) $D(4, 0)$ **(e)** $E(-2, -5)$ **(f)** $F(0, -3)$
(g) $G(0, 3)$ **(h)** $H(6, -\frac{7}{2})$ **(i)** $I(-\frac{3}{2}, -\frac{5}{2})$

Solution: The first number in each ordered pair is the x coordinate and the second number is the y coordinate. The points are plotted in Figure 7.17.

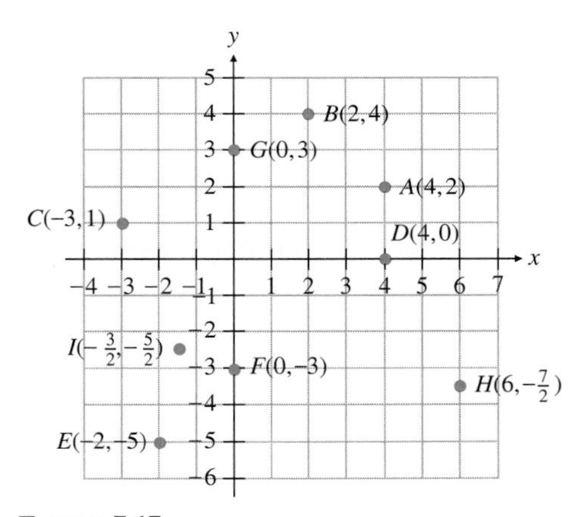

FIGURE 7.17

Note that when the x coordinate is 0, as in Example 1(f) and (g), the point is on the y axis. When the y coordinate is 0, as in Example 1(d), the point is on the x axis.

EXAMPLE 2 List the ordered pairs for each point shown in Figure 7.18.

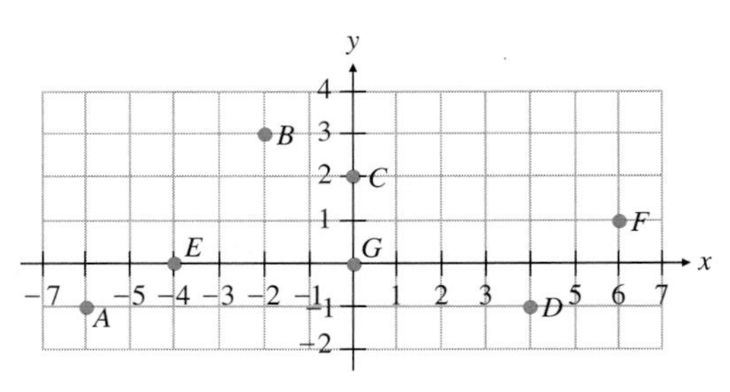

FIGURE 7.18

Solution: Remember to give the x value first in the ordered pair.

Point	Ordered Pair
A	$(-6, -1)$
B	$(-2, 3)$
C	$(0, 2)$
D	$(4, -1)$
E	$(-4, 0)$
F	$(6, 1)$
G	$(0, 0)$

Linear Equations

2 In Section 7.3 we will learn to graph linear equations in two variables. Below we explain how to identify a linear equation in two variables.

A **linear equation in two variables** is an equation that can be put in the form

$$ax + by = c$$

where a, b, and c are real numbers.

The graphs of equations of the form $ax + by = c$ *are straight lines. For this reason such equations are called linear.* Linear equations may be written in various forms, as we will show later. A linear equation in the form $ax + by = c$ is said to be in **standard form.**

Examples of Linear Equations

$$3x - 2y = 4$$
$$y = 5x + 3$$
$$x - 3y + 4 = 0$$

Note in the examples that only the equation $3x - 2y = 4$ is in standard form. However, the bottom two equations can be written in standard form, as follows:

$$y = 5x + 3 \qquad x - 3y + 4 = 0$$
$$-5x + y = 3 \qquad x - 3y = -4$$

Most of the equations we have discussed thus far have contained only one variable. Exceptions to this include formulas used in application sections. Consider the linear equation in *one* variable, $2x + 3 = 5$. What is its solution?

$$2x + 3 = 5$$
$$2x = 2$$
$$x = 1$$

This equation has only one solution, 1.

Check: $2x + 3 = 5$

$$2(1) + 3 = 5$$
$$5 = 5 \quad \text{true}$$

Now consider the linear equation in *two* variables, $y = x + 1$. What is the solution? Since the equation contains two variables, its solutions must contain two numbers, one for each variable. One pair of numbers that satisfies this equation is $x = 1$ and $y = 2$. To see that this is true, we substitute both values into the equation and see that the equation checks.

Check: $y = x + 1$

$2 = 1 + 1$

$2 = 2$ true

We write this answer as an ordered pair by writing the x and y values within parentheses separated by a comma. Remember the x value is always listed first since the form of an ordered pair is (x, y). Therefore, one possible solution to this equation is the ordered pair $(1, 2)$. The equation $y = x + 1$ has other possible solutions, as follows.

	Solution	*Solution*	*Solution*
	$x = 2, y = 3$	$x = -1, y = 0$	$x = -3, y = -2$
Check:	$y = x + 1$	$y = x + 1$	$y = x + 1$
	$3 = 2 + 1$	$0 = -1 + 1$	$-2 = -3 + 1$
	$3 = 3$ true	$0 = 0$ true	$-2 = -2$ true

Solution Written as an Ordered Pair

$(2, 3)$ $(-1, 0)$ $(-3, -2)$

How many possible solutions does the equation $y = x + 1$ have? The equation $y = x + 1$ has an unlimited or *infinite number* of possible solutions. Since it is not possible to list all the specific solutions, the solutions are illustrated with a graph. **A graph of an equation is an illustration of a set of points whose coordinates satisfy the equation.** Figure 7.19a shows the points $(2, 3)$, $(-1, 0)$, and $(-3, -2)$ plotted in the Cartesian coordinate system. Figure 7.19b shows a straight line drawn through the three points. Every point on this line will satisfy the equation $y = x + 1$, so this graph illustrates all the solutions of $y = x + 1$. Check to see if the ordered pair $(1, 2)$, which is on the line, satisfies the equation.

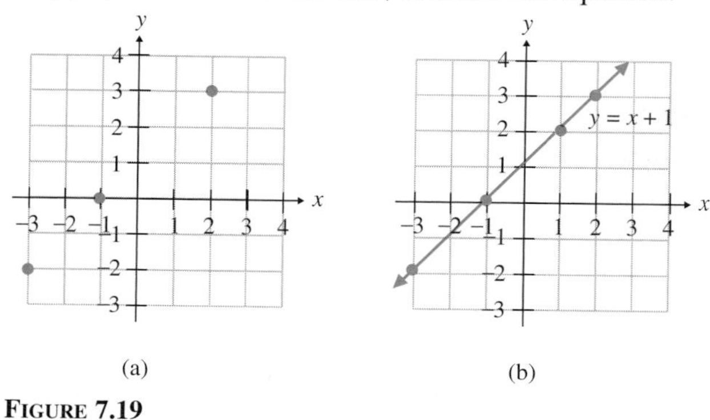

FIGURE 7.19

EXAMPLE 3 Determine which of the following ordered pairs satisfy the equation $2x + 3y = 12$.

(a) $(2, 3)$ **(b)** $(3, 2)$ **(c)** $(8, -\frac{4}{3})$

Solution: To determine if the ordered pairs are solutions, we substitute them into the equation.

(a)
$$2x + 3y = 12$$
$$2(2) + 3(3) = 12$$
$$4 + 9 = 12$$
$$13 = 12, \text{ false}$$

(2, 3) is *not* a solution

(b)
$$2x + 3y = 12$$
$$2(3) + 3(2) = 12$$
$$6 + 6 = 12$$
$$12 = 12, \text{ true}$$

(3, 2) is a solution

(c)
$$2x + 3y = 12$$
$$2(8) + 3(-\tfrac{4}{3}) = 12$$
$$16 - 4 = 12$$
$$12 = 12, \text{ true}$$

$(8, -\frac{4}{3})$ is a solution.

In Example 3, if we plotted the two solutions (3, 2) and $(8, -\frac{4}{3})$ and connected the two points with a straight line, we would get the graph of the equation $2x + 3y = 12$. The coordinates of every point on this line would satisfy the equation.

In Figure 7.19b, what do you notice about the points (2, 3), (1, 2), (−1, 0), and (−3, −2)? You probably said that they are in a straight line. A set of points that are in a straight line are said to be **collinear.** *In Section 7.3 when you graph linear equations by plotting points, the points you plot should all be collinear.*

EXAMPLE 4 Determine if the three points given are collinear.

(a) (2, 7), (0, 3), and (−2, −1)

(b) (0, 5), $(\frac{5}{2}, 0)$, and (5, −5)

(c) (−2, −5), (0, 1), and (5, 8)

Solution: We plot the points to determine if they are collinear. The solution is shown in Figure 7.20.

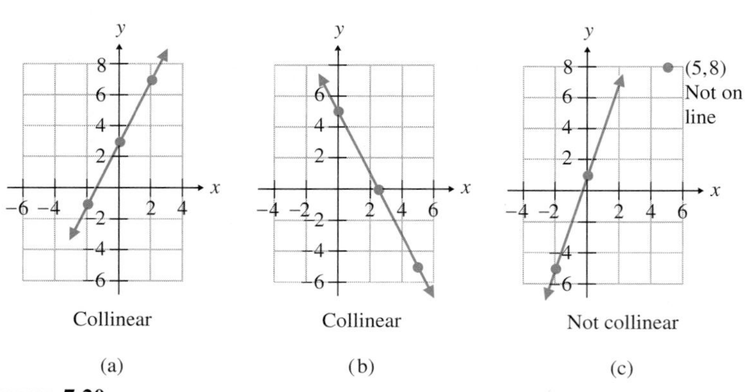

FIGURE 7.20

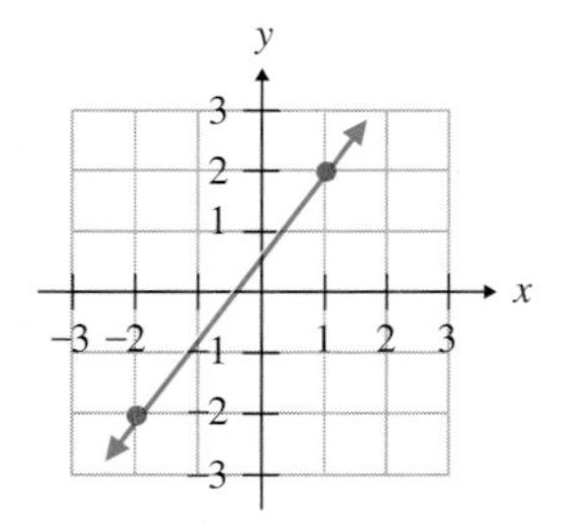

FIGURE 7.21

How many points do you need to graph a linear equation? As mentioned earlier, **the graph of every linear equation of the form $ax + by = c$ will be a straight line.** Since only two points are needed to draw a straight line (Fig. 7.21), only two points are needed to graph a linear equation. However, if you graph a linear equation using only two points and you have made an error in determining or plotting one of those points, your graph will be wrong and you will not know it. If you use at least three points to plot your graph, as in Figure 7.19b and they are collinear, you probably have not made a mistake.

EXAMPLE 5

(a) Determine which of the following ordered pairs satisfy the equation $2x + y = 4$.

$$(2, 0),\ (0, 4),\ (3, -1),\ (-1, 6)$$

(b) Plot all the points on the same axes and draw a straight line through the three points that are collinear.

(c) Does the graph support your answer in part (a)? Explain.

(d) What does this straight line represent?

Solution:

(a) We substitute values for x and y into the equation $2x + y = 4$ and determine if they check.

Check:

$(2, 0)$

$$2x + y = 4$$
$$2(2) + 0 = 4$$
$$4 = 4 \quad \text{true}$$

$(0, 4)$

$$2x + y = 4$$
$$2(0) + 4 = 4$$
$$4 = 4 \quad \text{true}$$

$(3, -1)$

$$2x + y = 4$$
$$2(3) - 1 = 4$$
$$6 - 1 = 4$$
$$5 = 4 \quad \text{false}$$

$(-1, 6)$

$$2x + y = 4$$
$$2(-1) + 6 = 4$$
$$-2 + 6 = 4$$
$$4 = 4 \quad \text{true}$$

All the points except $(3, -1)$ check and are solutions to the equation.

(b) See Figure 7.22

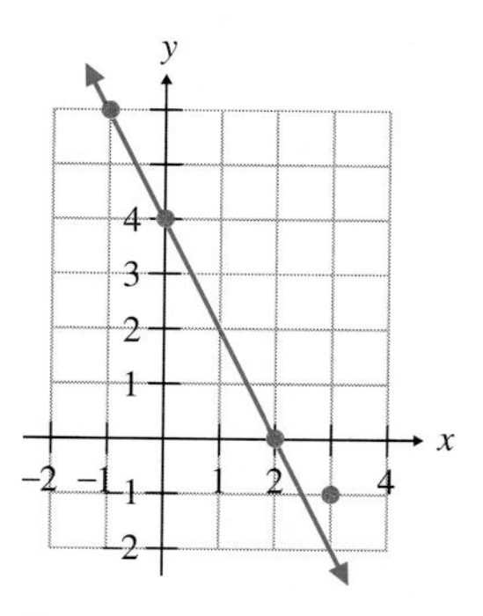

FIGURE 7.22

(c) Yes, the three points that are solutions are collinear. The point $(3, -1)$, which is not a solution, is not collinear with the other points.

(d) The straight line represents all solutions of $2x + y = 4$. The coordinates of every point on this line satisfy the equation $2x + y = 4$.

Exercise Set 7.2

Indicate the quadrant in which each of the points belong.

1. $(2, 5)$ **2.** $(-3, 6)$ **3.** $(-4, -2)$

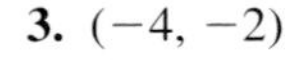

4. $(5, -3)$ **5.** $(-8, 5)$ **6.** $(-6, 30)$

7. $(7, 93)$ **8.** $(83, -57)$ **9.** $(-124, -132)$

10. $(75, -200)$ **11.** $(-8, 42)$ **12.** $(-46, -192)$

13. List the ordered pairs corresponding to each point.

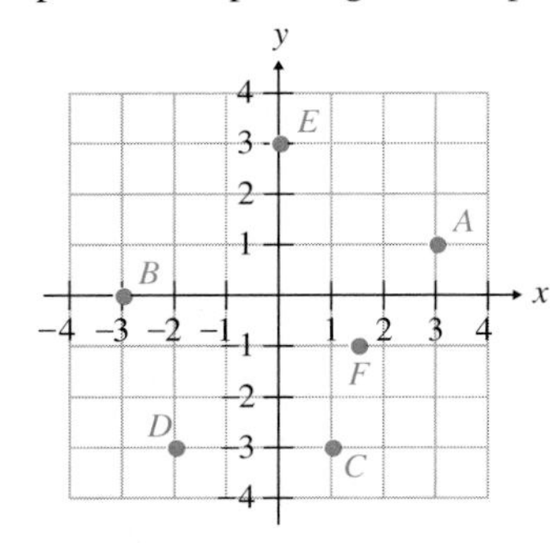

14. List the ordered pairs corresponding to each point.

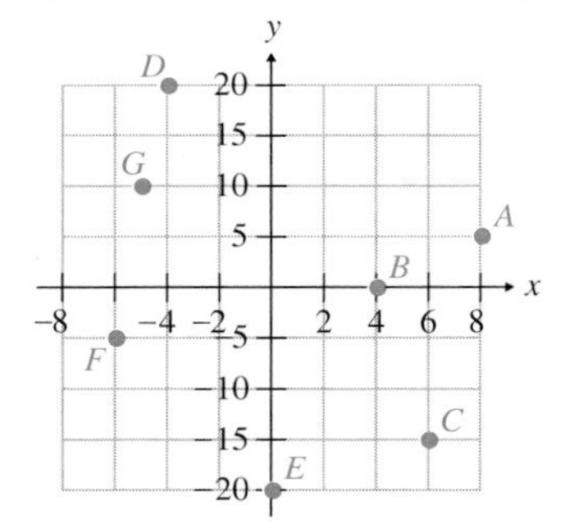

Plot each point on the same set of axes.

15. **(a)** $(4, 2)$ **(b)** $(-3, 2)$ **(c)** $(0, -3)$ **(d)** $(-2, 0)$ **(e)** $(-3, -4)$ **(f)** $(-4, -2)$

16. **(a)** $(-3, -1)$ **(b)** $(2, 0)$ **(c)** $(-3, 2)$ **(d)** $(\frac{1}{2}, -4)$ **(e)** $(-4, 2)$ **(f)** $(0, 4)$

17. **(a)** $(4, 0)$ **(b)** $(-1, 3)$ **(c)** $(2, 4)$ **(d)** $(0, -2)$ **(e)** $(-3, -3)$ **(f)** $(2, -3)$

Plot the following points. Then determine if they are collinear.

18. **(a)** $(1, -1)$ **(b)** $(5, 3)$ **(c)** $(-3, -5)$ **(d)** $(0, -2)$ **(e)** $(2, 0)$

19. **(a)** $(1, -2)$ **(b)** $(0, -5)$ **(c)** $(3, 1)$ **(d)** $(-1, -8)$ **(e)** $(\frac{1}{2}, -\frac{7}{2})$

20. **(a)** $(0, 2)$ **(b)** $(-1, 3)$ **(c)** $(-2, 4)$ **(d)** $(3, 0)$ **(e)** $(4, -2)$

*In Exercises 21–26, **(a)** determine which three of the four ordered pairs satisfy the given equation. **(b)** Plot all the points on the same axes and draw a straight line through the three points* that are collinear. ***(c)** Does the graph support your answer in part (a)? Explain. **(d)** What does this straight line represent? (See Example 5.)*

21. $y = x + 1$, **(a)** $(0, 1)$ **(b)** $(-1, 0)$ **(c)** $(2, 3)$ **(d)** $(1, 1)$

22. $2x + y = -4$, **(a)** $(-2, 0)$ **(b)** $(-2, 1)$ **(c)** $(0, -4)$ **(d)** $(-1, -2)$

23. $3x - 2y = 6$, **(a)** $(4, 0)$ **(b)** $(2, 0)$ **(c)** $(\frac{2}{3}, -2)$ **(d)** $(\frac{4}{3}, -1)$

24. $2x - 4y = 6$, **(a)** $(3, 0)$ **(b)** $(2, -\frac{1}{2})$ **(c)** $(0, -\frac{3}{2})$ **(d)** $(-1, -1)$

25. $2x + 4y = 4$, **(a)** $(2, 3)$ **(b)** $(1, \frac{1}{2})$ **(c)** $(0, 1)$ **(d)** $(-2, 2)$

26. $y = \frac{1}{2}x + 2$, **(a)** $(0, 2)$ **(b)** $(2, 0)$ **(c)** $(-2, 1)$ **(d)** $(4, 4)$

27. In an ordered pair, which coordinate is always listed first?

28. What is another name for the Cartesian coordinate system?

29. **(a)** Is the *horizontal axis* the x or y axis in the Cartesian coordinate system?
(b) Is the *vertical axis* the x or y axis?

30. What is the *origin* in the Cartesian coordinate system?

31. We can refer to the *x axis* and we can refer to the *y axis*. We can also refer to the *x* and *y axes*. Explain when we use the word *axis* and when we use the word *axes*.

32. Explain how to plot the point $(-3, 5)$ in the Cartesian coordinate system.

33. What does the graph of a linear equation illustrate?

34. Why are arrowheads added to the ends of graphs of linear equations?

35. What will the graph of a linear equation look like?

36. (a) How many points are needed to graph a linear equation? **(b)** Why is it always a good idea to use three or more points when graphing a linear equation?

37. What is the standard form of a linear equation?

38. When graphing linear equations the points that are plotted should all be *collinear*. Explain what this means.

CUMULATIVE REVIEW EXERCISES

] **39.** Subtract $6x^2 - 4x + 5$ from $-2x^2 - 5x + 9$.

] **40.** Multiply $(3x^2 - 4x + 5)(2x - 3)$.

[5.2] **41.** Factor by grouping $x^2 - 2x + 3xy - 6y$.

[6.4] **42.** Subtract $\frac{3}{x+2} - \frac{4}{x+1}$.

Group Activity/ Challenge Problems

In Section 7.3 we discuss how to find ordered pairs to plot when graphing linear equations. Let us see if you can draw some graphs now. For the following ***(a)*** *select any three values for x and find the corresponding values of y;* ***(b)*** *plot the points (they should be collinear);* ***(c)*** *draw the graph.*

1. $y = x$ **2.** $y = 2x$ **3.** $y = x + 1$ **4.** $2x + y = 4$

5. Another type of coordinate system that is used to identify a location or position on the Earth's surface involves *latitude and longitude*. Do research and write a report explaining how a specific location on the surface of the Earth can be represented using latitude and longitude. In your report approximate the location of your college using latitude and longitude.

7.3 Graphing Linear Equations

Tape 12

1. Graph linear equations by plotting points.
2. Graph linear equations of the form $ax + by = 0$.
3. Graph linear equations using the x and y intercepts.
4. Graph horizontal and vertical lines.

In Section 7.2 we explained the Cartesian Coordinate System, how to plot points, and how to recognize linear equations in two variables. Now we are ready to graph linear equations. *In this section we discuss two methods that can be used to graph linear equations: (1) graphing by plotting points, and (2) graphing using the x and y intercepts.* In Section 7.4 we discuss graphing using the slope and the y intercept. We begin by discussing graphing by plotting points.

Graphing by Plotting Points

1 Graphing by plotting points is the most versatile method of graphing because we can also use it to graph second- and higher-degree equations. We will graph quadratic equations, which are second-degree equations, by plotting points in Chapter 10.

Graphing Linear Equations by Plotting Points

1. Solve the linear equation for the variable y. That is, get the variable y by itself on the left side of the equal sign.
2. Select a value for the variable x. Substitute this value in the equation for x and find the corresponding value of y. Record the ordered pair (x, y).
3. Repeat step 2 with two different values of x. This will give you two additional ordered pairs.
4. Plot the three ordered pairs. The three points should be collinear. If they are not collinear, recheck your work for mistakes.
5. *With a straightedge,* draw a straight line through the three points. Draw arrowheads on each end of the line to show that the line continues indefinitely in both directions.

In step 1, you need to solve the equation for y. If you have forgotten how to do this, review Section 3.1. In steps 2 and 3, you need to select values for x. The values you choose to select are up to you. However, you should choose values small enough so that the ordered pairs obtained can be plotted on the axes. Since y is often easy to find when $x = 0$, 0 is always a good value to select for x.

EXAMPLE 1 Graph the equation $y = 3x + 6$.

Solution: First we determine that this is a linear equation. Its graph must therefore be a straight line. The equation is already solved for y. Select three values for x, substitute them in the equation, and find the corresponding values for y. We will arbitrarily select the values 0, 2, and -3 for x. The calculations that follow show that when $x = 0$, $y = 6$, when $x = 2$, $y = 12$, and when $x = -3$, $y = -3$.

x	$y = 3x + 6$	*Ordered Pair*
0	$y = 3(0) + 6 = 6$	$(0, 6)$
2	$y = 3(2) + 6 = 12$	$(2, 12)$
-3	$y = 3(-3) + 6 = -3$	$(-3, -3)$

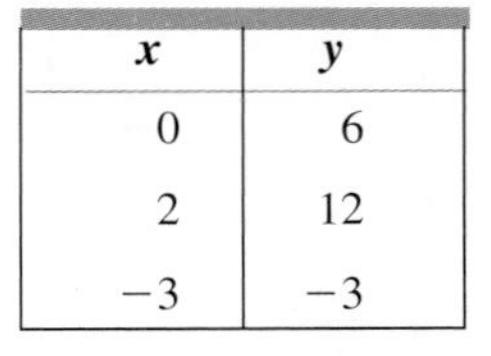

x	y
0	6
2	12
-3	-3

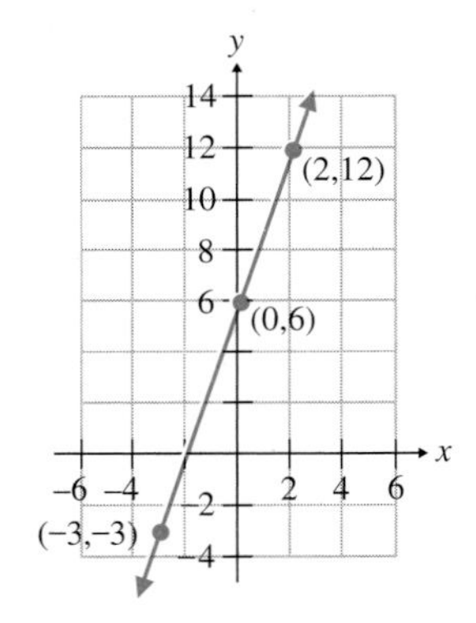

FIGURE 7.23

It is sometimes convenient to list the x and y values in a table. Then plot the three ordered pairs on the same axes (Fig. 7.23).

Since the three points are collinear, the graph appears correct. Connect the three points with a straight line. Place arrowheads at the ends of the line to show that the line continues infinitely in both directions.

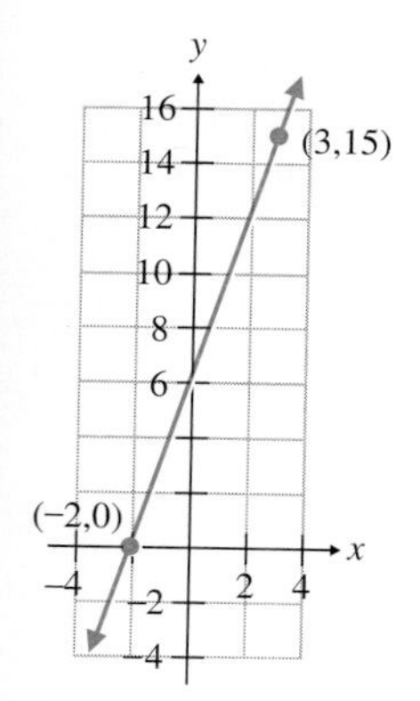

FIGURE 7.24

To graph the equation $y = 3x + 6$, we arbitrarily used the three values $x = 0$, $x = 2$, and $x = -3$. We could have selected three entirely different values and obtained exactly the same graph. When selecting values to substitute for x, use values that make the equation easy to evaluate.

The graph drawn in Example 1 represents the set of *all* ordered pairs that satisfy the equation $y = 3x + 6$. If we select any point on this line, the ordered pair represented by that point will be a solution to the equation $y = 3x + 6$. Similarly, any solution to the equation will be represented by a point on the line. Let us select some points on the graph, say, (3, 15) and (−2, 0), and verify that they are solutions to the equation (Fig. 7.24).

Check (3, 15):
$$\begin{aligned} y &= 3x + 6 \\ 15 &= 3(3) + 6 \\ 15 &= 9 + 6 \\ 15 &= 15 \quad \text{true} \end{aligned}$$

Check (−2, 0):
$$\begin{aligned} y &= 3x + 6 \\ 0 &= 3(-2) + 6 \\ 0 &= -6 + 6 \\ 0 &= 0 \quad \text{true} \end{aligned}$$

Remember, a graph of an equation is an illustration of the set of points whose coordinates satisfy the equation.

EXAMPLE 2 Graph the equation $2y = 4x - 12$.

Solution: We begin by solving the equation for y. This will make it easier to determine ordered pairs that satisfy the equation. To solve the equation for y, divide both sides of the equation by 2.

$$2y = 4x - 12$$
$$y = \frac{4x - 12}{2} = \frac{4x}{2} - \frac{12}{2} = 2x - 6$$

Now select three values for x and find the corresponding values for y using the equation $y = 2x - 6$.

$$y = 2x - 6$$
Let $x = 0$, $y = 2(0) - 6 = -6$
Let $x = 2$, $y = 2(2) - 6 = -2$
Let $x = 3$, $y = 2(3) - 6 = 0$

x	y
0	−6
2	−2
3	0

Plot the points and draw the straight line (Fig. 7.25).

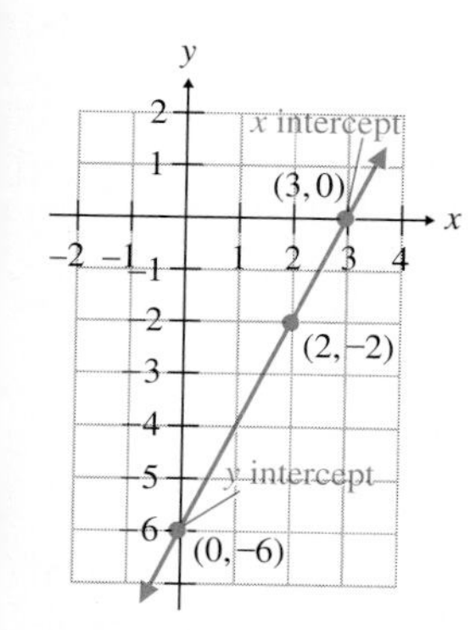

FIGURE 7.25

Graphing Linear Equations of the Form $ax + by = 0$

2 In Example 3 we graph an equation of the form $ax + by = 0$, which is a linear equation whose constant is 0.

EXAMPLE 3 Graph the equation $2x + 5y = 0$.

Solution: We begin by solving the equation for y.

$$2x + 5y = 0$$
$$5y = -2x$$
$$y = -\frac{2x}{5} \quad \text{or} \quad y = -\frac{2}{5}x$$

Now we select values for x and find the corresponding values of y. Which values shall we select for x? Notice that the coefficient of the x term is a fraction, with the denominator 5. If we select values for x that are multiples of the denominator, such as . . . $-15, -10, -5, 0, 5, 10, 15, \ldots$, the 5 in the denominator will divide out. This will give us integer values for y. We will arbitrarily select the values $x = -5$, $x = 0$, and $x = 5$.

$$y = -\frac{2}{5}x$$

Let $x = -5$, $\quad y = \left(-\frac{2}{5}\right)(-5) = 2$

Let $x = 0$, $\quad y = \left(-\frac{2}{5}\right)(0) = 0$

Let $x = 5$, $\quad y = -\frac{2}{5}(5) = -2$

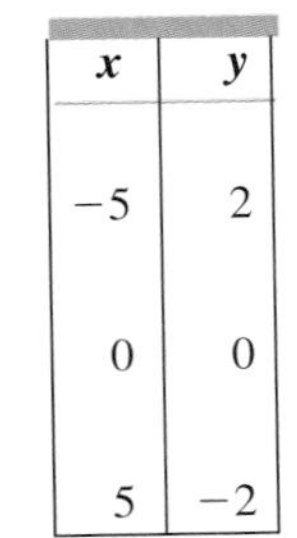

x	y
-5	2
0	0
5	-2

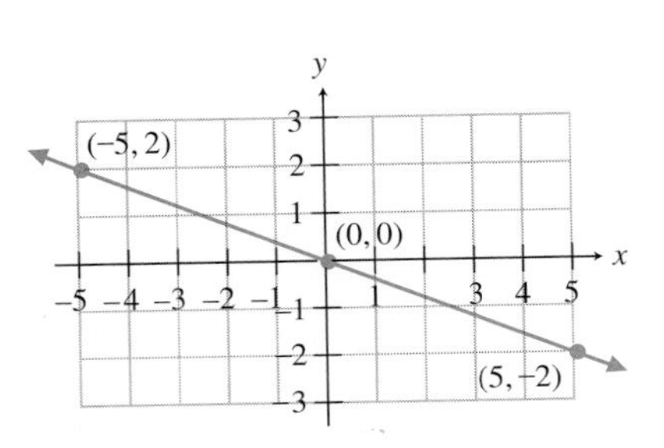

FIGURE 7.26

Now plot the points and draw the graph (Fig. 7.26).

The graph in Example 3 passes through the origin. The graph of every linear equation with a constant of 0 (equations of the form $ax + by = 0$) will pass through the origin.

Graphing Using Intercepts

3 Now we discuss graphing linear equations using the x and y intercepts. Let us examine two points on the graph in Figure 7.25. Note that the graph crosses the x axis at 3. Therefore, (3, 0) is called the **x intercept**. Note that the x intercept has an x coordinate of 3 and a y coordinate of 0. In general, an x intercept will be of the form $(x, 0)$. The x in the ordered pair is the value where the graph crosses the x axis.

The graph in Figure 7.25 crosses the y axis at -6. Therefore, $(0, -6)$ is called the **y intercept**. Note that the y intercept has an x coordinate of 0 and a y coordinate of -6. In general, a y intercept will be of the form $(0, y)$. The y in the ordered pair represents the value where the graph crosses the y axis.

Note that the graph in Figure 7.26 crosses both the x and y axes at the origin. Thus, both the x and y intercepts of this graph are (0, 0).

It is often convenient to graph linear equations by finding their x and y intercepts. To graph an equation using the x and y intercepts, use the procedure that follows.

Graphing Linear Equations Using the x and y Intercepts

1. **Find the y intercept by setting x in the given equation equal to 0 and finding the corresponding value of y.**
2. **Find the x intercept by setting y in the given equation equal to 0 and finding the corresponding value of x.**

3. Determine a check point by selecting a nonzero value for x and finding the corresponding value of y.
4. Plot the y intercept (where the graph crosses the y axis), the x intercept (where the graph crosses the x axis), and the check point. The three points should be collinear. If not, recheck your work.
5. *Using a straightedge,* draw a straight line through the three points. Draw an arrowhead at both ends of the line to show that the line continues indefinitely in both directions.

Helpful Hint

Since only two points are needed to determine a straight line, it is not absolutely necessary to determine and plot the check point in step 3. However, *if you use only the x and y intercepts to draw your graph and one of those points is wrong, your graph will be incorrect and you will not know it.* It is always a good idea to use three points when graphing a linear equation.

EXAMPLE 4 Graph the equation $3y = 6x + 12$ by plotting the x and y intercepts.

Solution: To find the y intercept (where the graph crosses the y axis), set $x = 0$ and find the corresponding value of y.

$$3y = 6x + 12$$
$$3y = 6(0) + 12$$
$$3y = 0 + 12$$
$$3y = 12$$
$$y = \frac{12}{3} = 4$$

The graph crosses the y axis at 4. The ordered pair representing the y intercept is (0, 4). To find the x intercept (where the graph crosses the x axis), set $y = 0$ and find the corresponding value of x.

$$3y = 6x + 12$$
$$3(0) = 6x + 12$$
$$0 = 6x + 12$$
$$-12 = 6x$$
$$\frac{-12}{6} = x$$
$$-2 = x$$

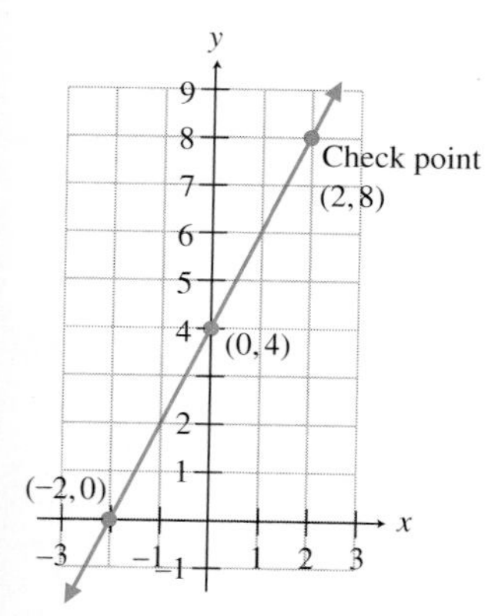

FIGURE 7.27

The graph crosses the x axis at -2. The ordered pair representing the x intercept is $(-2, 0)$. Now plot the intercepts (Fig. 7.27).

Before we graph the equation, we will select a nonzero value for x, find the corresponding value of y, and make sure that it is collinear with the x and y intercepts. This third point is our check point.

$$\text{Let } x = 2$$
$$3y = 6x + 12$$
$$3y = 6(2) + 12$$
$$3y = 12 + 12$$
$$3y = 24$$
$$y = \frac{24}{3} = 8$$

Plot the check point (2, 8). Since the three points are collinear, draw the straight line through all three points.

EXAMPLE 5 Graph the equation $2x + 3y = 9$ by finding the x and y intercepts.

Solution:

Find y Intercept	*Find x Intercept*	*Check Point*
Let $x = 0$	Let $y = 0$	Let $x = 2$
$2x + 3y = 9$	$2x + 3y = 9$	$2x + 3y = 9$
$2(0) + 3y = 9$	$2x + 3(0) = 9$	$2(2) + 3y = 9$
$0 + 3y = 9$	$2x + 0 = 9$	$4 + 3y = 9$
$3y = 9$	$2x = 9$	$3y = 5$
$y = 3$	$x = \frac{9}{2}$	$y = \frac{5}{3}$

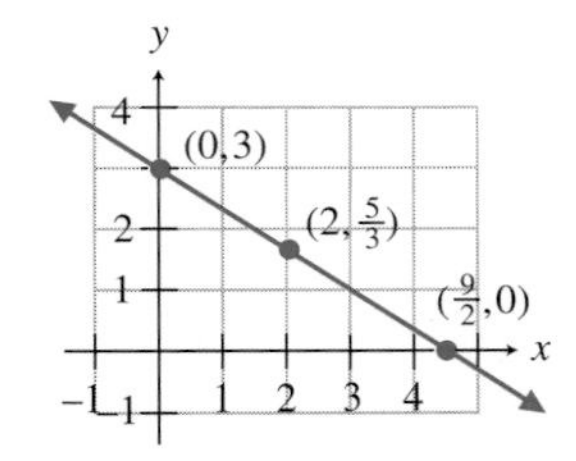

FIGURE 7.28

The three ordered pairs are (0, 3,), $\left(\frac{9}{2}, 0\right)$, and $\left(2, \frac{5}{3}\right)$.

The three points appear to be collinear. Draw a straight line through all three points (Fig. 7.28).

EXAMPLE 6 Graph the equation $y = 20x + 60$.

Solution:

Find y Intercept	*Find x Intercept*	*Check Point*
Let $x = 0$	Let $y = 0$	Let $x = 3$
$y = 20x + 60$	$y = 20x + 60$	$y = 20x + 60$
$y = 20(0) + 60$	$0 = 20x + 60$	$y = 20(3) + 60$
$y = 60$	$-60 = 20x$	$y = 60 + 60$
	$-3 = x$	$y = 120$

The three ordered pairs are (0, 60), (−3, 0), and (3, 120). Since the values of y are large, we let each interval on the y axis be 15 units rather than 1 (Fig. 7.29). In addition, the length of the intervals on the y axis will be made smaller than those on the x axis. Sometimes you will have have to use different scales on the x and y axes, as illustrated, to accommodate the graph. Now plot the points and draw the graph.

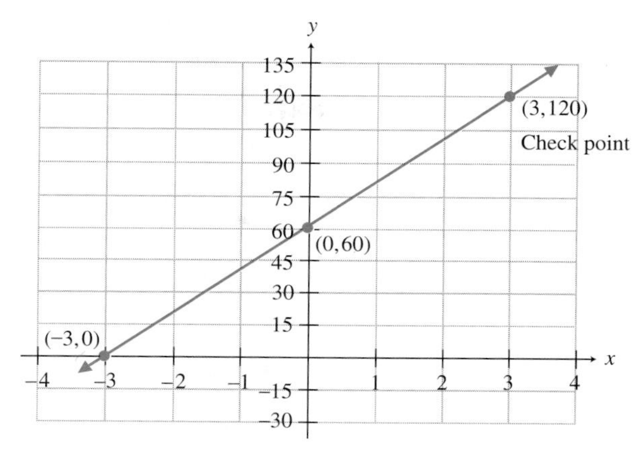

FIGURE 7.29

When selecting the scales for your axes, you should realize that different scales will result in the same equation having a different appearance. Consider the graphs shown in Figure 7.30. Both graphs represent the same equation, $y = x$. In Figure 7.30a, both the x and y axes have the same scale. In Figure 7.30b, the x and y axes do not have the same scale. Both graphs are correct in that each represents the graph of $y = x$. The difference in appearance is due to the difference in scales on the x axis. When possible, keep the scales on the x and y axis the same, as in Figure 7.30a.

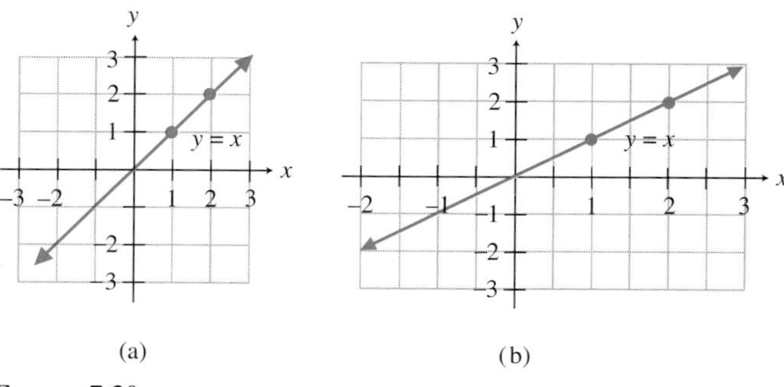

FIGURE 7.30

EXAMPLE 7 Graph the equation $3y + 2x = -4$.

(a) By solving the equation for y, selecting three values of x, and finding the corresponding values of y.

(b) By using the x and y intercepts.

Solution:

(a) $3y + 2x = -4$

$$3y = -2x - 4$$

$$y = \frac{-2x - 4}{3} = -\frac{2}{3}x - \frac{4}{3}$$

When selecting values for x, we select those that are multiples of 3 so that the arithmetic will be easier. Let us select 0, 3, and -3.

$$y = -\frac{2}{3}x - \frac{4}{3}$$

Let $x = 0$, $\quad y = -\frac{2}{3}(0) - \frac{4}{3} = 0 - \frac{4}{3} = -\frac{4}{3}$

Let $x = 3$, $\quad y = -\frac{2}{3}(3) - \frac{4}{3} = -\frac{6}{3} - \frac{4}{3} = -\frac{10}{3}$

Let $x = -3$, $\quad y = -\frac{2}{3}(-3) - \frac{4}{3} = \frac{6}{3} - \frac{4}{3} = \frac{2}{3}$

x	y
0	$-\frac{4}{3}$
3	$-\frac{10}{3}$
-3	$\frac{2}{3}$

The graph is shown in Figure 7.31. Note that the coordinates are not always integers.

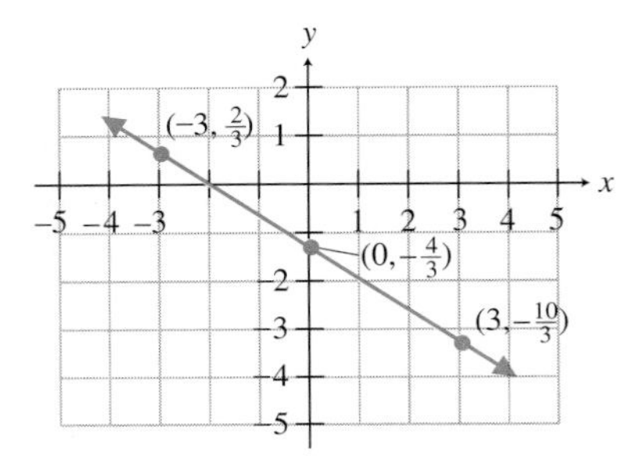

FIGURE 7.31

(b) $3y + 2x = -4$

Find y Intercept	*Find x Intercept*	*Check Point*
Let $x = 0$	Let $y = 0$	Let $x = 3$, then
$3y + 2x = -4$	$3y + 2x = -4$	$y = -\frac{10}{3}$
$3y + 2(0) = -4$	$3(0) + 2x = -4$	from part (a).
$3y = -4$	$2x = -4$	$\left(3, -\frac{10}{3}\right)$
$y = -\frac{4}{3}$	$x = -2$	

The graph is shown in Figure 7.32. Note that the graphs in Figures 7.31 and 7.32 are the same.

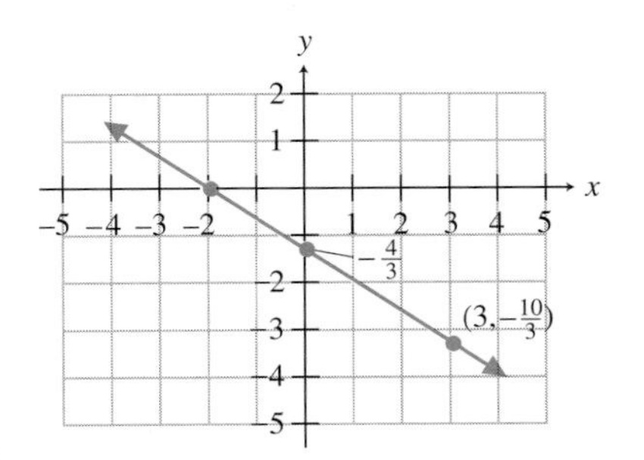

FIGURE 7.32

Horizontal and Vertical Lines

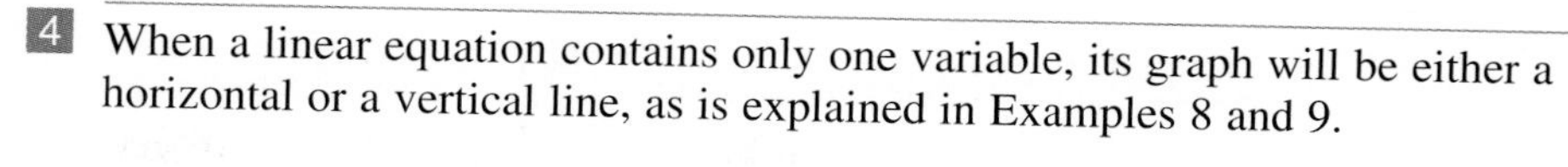

4 When a linear equation contains only one variable, its graph will be either a horizontal or a vertical line, as is explained in Examples 8 and 9.

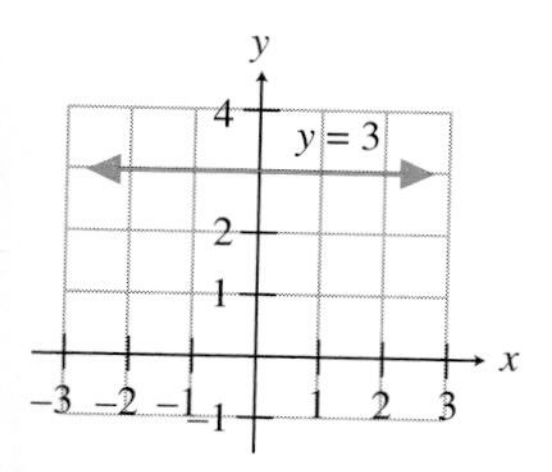

FIGURE 7.33

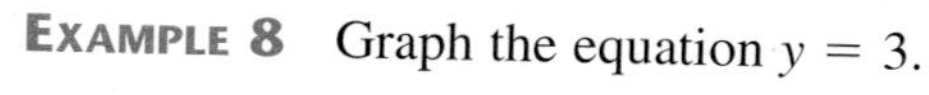

EXAMPLE 8 Graph the equation $y = 3$.

Solution: This equation can be written as $y = 3 + 0x$. Thus, for any value of x selected, y will be 3. The graph of $y = 3$ is illustrated in Figure 7.33.

The graph of an equation of the form $y = b$ is a horizontal line whose y intercept is $(0, b)$

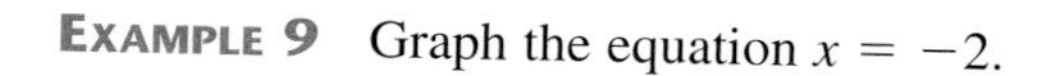

EXAMPLE 9 Graph the equation $x = -2$.

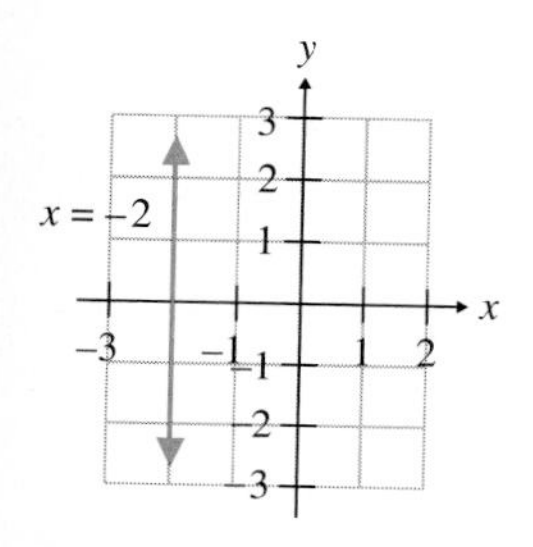

FIGURE 7.34

Solution: This equation can be written as $x = -2 + 0y$. Thus, for any value of y selected, x will have a value of -2. The graph of $x = -2$ is illustrated in Figure 7.34.

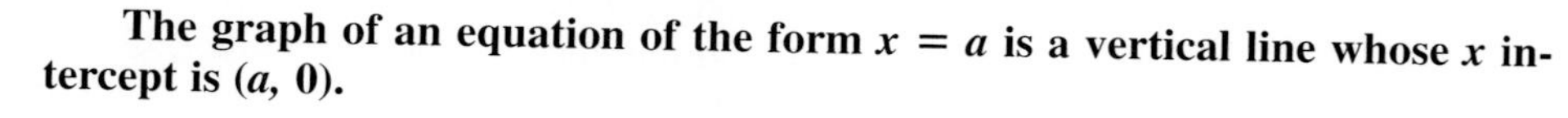

The graph of an equation of the form $x = a$ is a vertical line whose x intercept is $(a, 0)$.

Before we leave this section, let us look at an application of graphing. We will see additional applications of graphing linear equations in Sections 7.6 and 8.4.

EXAMPLE 10 The salary, R, received by a salesperson is \$200 per week plus a 7% commission on all sales, s.

(a) Write an equation for the salary, R, in terms of the sales, s.

(b) Graph the salary for sales of \$0 up to and including \$20,000.

(c) From the graph estimate the salary if the salesperson's weekly sales are \$15,000.

(d) By observing the graph, estimate the sales needed for the salesperson to earn a salary of \$800.

Solution:

(a) Since s is the amount of sales, a 7% commission on s dollars in sales is $0.07s$.

$$\text{Salary received} = \$200 + \text{commission}$$
$$R = 200 + 0.07s$$

(b) We select three values for s and find the corresponding values of R.

$$R = 200 + 0.07s$$

Let $s = 0$ $\qquad R = 200 + 0.07(0) = 200$

Let $s = 10{,}000$ $\qquad R = 200 + 0.07(10{,}000) = 900$

Let $s = 20{,}000$ $\qquad R = 200 + 0.07(20{,}000) = 1600$

s	R
0	200
10,000	900
20,000	1600

The graph is illustrated in Figure 7.35. Notice that since we only graph the equation for values of s from 0 to 20,000, we do not place arrowheads on the ends of the graph.

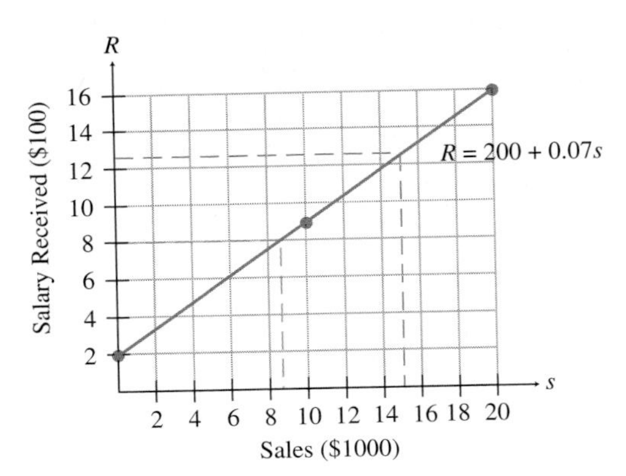

FIGURE 7.35

(c) To determine the weekly salary earned on sales of \$15,000, locate \$15,000 on the sales axis. Then draw a vertical line up to where it intersects the graph, the *red* line in Figure 7.35. Now draw a horizontal line across to the salary axis. Since the horizontal line crosses the salary axis at about \$1250, weekly sales of \$15,000 would result in a weekly salary of about \$1250. You can find the exact salary by substituting 15,000 for s in the equation $R = 200 + 0.07s$ and finding the value of R. Do this now.

(d) To find the sales needed for the salesperson to earn a weekly salary of \$800, we find \$800 on the salary axis. We then draw a horizontal line from the point to the graph, as shown with the *green* line in Figure 7.35. We then draw a vertical line from the point of intersection of the graph to the sales axis. This value on the sales axis represents the sales needed for the salesperson to earn \$800. Thus, sales of about \$8600 per week would result in a salary of \$800. An exact answer can be found by substituting 800 for R in the equation $R = 200 + 0.07s$ and solving the equation for s. Do this now.

Helpful Hint

In Chapter 2 we solved linear equations in one variable. Consider the solution to the equation $3x + 5 = x + 9$.

$$\begin{aligned} 3x + 5 &= x + 9 \\ 2x + 5 &= 9 \\ 2x &= 4 \\ x &= 2 \end{aligned}$$

A check will show that 2 makes the statement true, so 2 is the solution to the equation.

Consider the equation $3x + 5 = x + 9$ again. This time let us set one side of the equation equal to zero, as follows.

$$\begin{aligned} 3x + 5 &= x + 9 \\ 2x + 5 &= 9 \\ 2x - 4 &= 0 \end{aligned}$$

Now if we replace the 0 with y, we get $y = 2x - 4$. The graph of $y = 2x - 4$ follows.

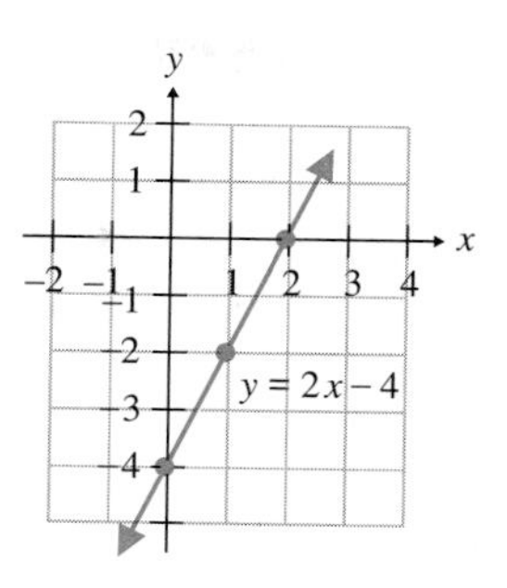

Notice the x intercept of this graph is at 2, the value obtained for the solution of the equation $3x + 5 = x + 9$. Is this just a coincidence, or will this process always result in the x intercept of the graph being the solution to the linear equation in one variable? This process will always result in the x intercept being the solution to the linear equation in one variable. Let's discuss why.

Consider the two equations $3x + 5 = x + 9$ and $2x - 4 = 0$. These equations are equivalent since they have the same solution. The solution to the equation $2x - 4 = 0$ is the value of x that makes the expression $2x - 4$ have a value of 0. If we replace the 0 in the equation with y, we get $y = 2x - 4$. The value of y is 0 where the graph crosses the x axis, or at the x intercept. Thus, the value of x where the graph crosses the x axis is the solution to both $2x - 4 = 0$ and $3x + 5 = x + 9$.

Calculator Corner

Calculators are available that will graph equations on the calculator display. Such calculators are called graphing calculators or graphers. Some graphing calculators available at this time include:

Casio	*Hewlett-Packard*	*Sharp*	*Texas Instruments*
FX7000	HP 48S/Sx	EL-9200	TI 81
FX7700	HP 48G/Gx	EL-9300	TI 82
FX 8700			TI 85

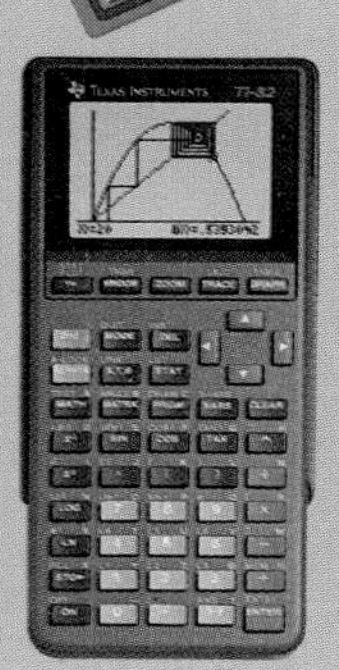

The TI 82 is pictured on the left.

The cost of graphing calculators start at about $65 and increase, depending on the model and features you select. Using graphing calculators in mathematics courses is becoming more and more common. You may find yourself using a graphing calculator if you take additional mathematics courses.

Since each calculator functions a little differently, you should consult with your mathematics instructor before you purchase one. A particular graphing calculator may be required in a later course.

Exercise Set 7.3

Find the missing coordinate in the solutions for $2x + y = 6$.

1. (2, ?)
2. (−1, ?)
3. (?, −5)
4. (?, −3)
5. (?, 0)
6. $(\frac{1}{2}, ?)$

Find the missing coordinate in the solutions for $3x - 2y = 8$.

7. (2, ?)
8. (0, ?)
9. (?, 0)
10. $(?, -\frac{1}{2})$
11. (−3, ?)
12. (?, − 5)

Graph each equation.

13. $y = 4$
14. $x = -2$
15. $x = 3$
16. $y = 2$

Graph by plotting points. Plot at least three points for each graph.

17. $y = 4x - 2$
18. $y = -x + 3$
19. $y = 6x + 2$
20. $y = x - 4$

21. $y = -\frac{1}{2}x + 3$
22. $2y = 2x + 4$
23. $6x - 2y = 4$
24. $4x - y = 5$

25. $5x - 2y = 8$
26. $3x + 2y = 0$
27. $6x + 5y = 30$
28. $-2x - 3y = 6$

29. $-4x + 5y = 0$
30. $8y - 16x = 24$
31. $y = 20x + 40$
32. $2y - 50 = 100x$

33. $y = \frac{2}{3}x$

34. $y = -\frac{3}{5}x$

35. $y = \frac{1}{2}x + 4$

36. $y = -\frac{2}{5}x + 2$

37. $2y = 3x + 6$

38. $-4x - y = -2$

39. $-4x + 8y = 16$

40. $4x - 6y = 10$

Graph using the x and y intercepts.

41. $y = 2x + 4$

42. $y = -2x + 6$

43. $y = 4x - 3$

44. $y = -3x + 8$

45. $y = -6x + 5$

46. $y = 4x + 16$

47. $2y + 3x = 12$

48. $-2x + 3y = 10$

49. $4x = 3y - 9$

50. $7x + 14y = 21$

51. $\frac{1}{2}x + y = 4$

52. $30x + 25y = 50$

53. $6x - 12y = 24$

54. $25x + 50y = 100$

55. $8y = 6x - 12$

56. $-3y - 2x = -6$

57. $30y + 10x = 45$

58. $120x - 360y = 720$

59. $40x + 6y = 40$

60. $20x - 240 = -60y$

61. $\frac{1}{3}x + \frac{1}{4}y = 12$

62. $\frac{1}{5}x - \frac{2}{3}y = 60$

63. $\frac{1}{2}x = \frac{2}{5}y - 80$

64. $\frac{2}{3}y = \frac{5}{4}x + 120$

Write the equation represented by the given graph. (See Examples 8 and 9.)

65.

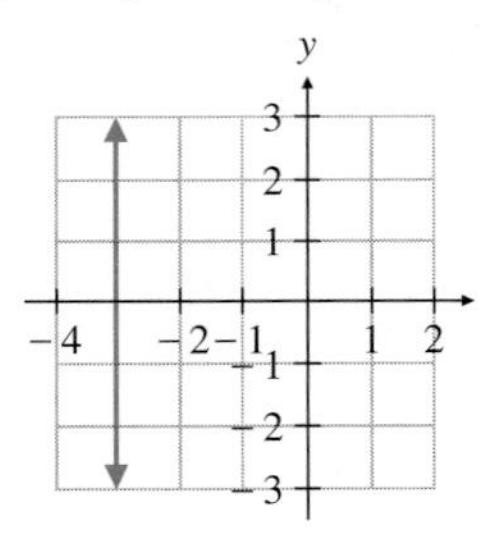

66.

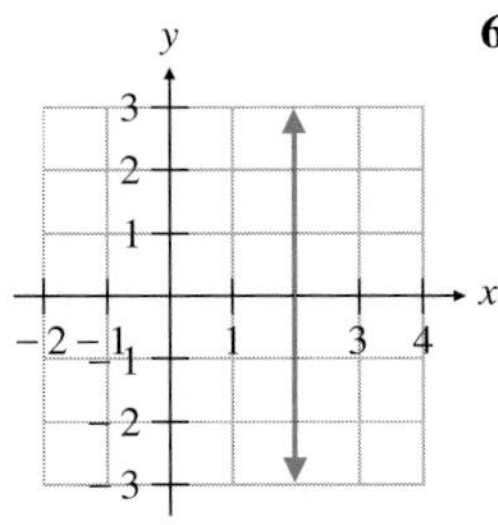

67.

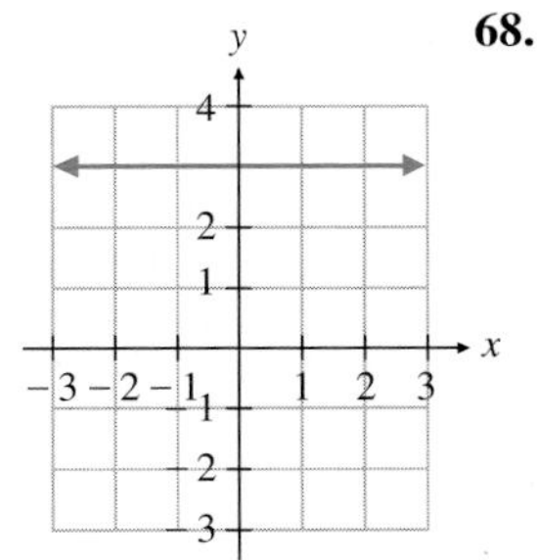

68.

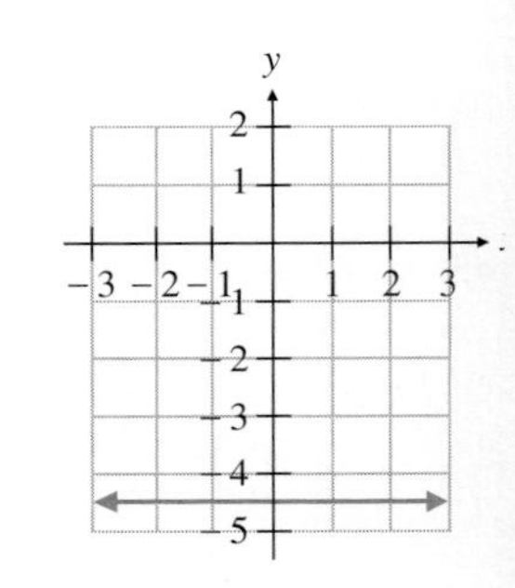

Study Example 10 before working Exercises 69–72.

69. The cost, C, of renting a large truck is \$25 per day plus \$1 per mile, m.

(a) Write an equation for the cost in terms of the miles driven.

(b) Graph the equation for values up to and including 100 miles.

(c) Estimate the cost of driving 50 miles in one day.

(d) Estimate the miles driven if the cost for one day is \$60.

70. Simple interest is calculated by the simple interest formula, interest = principal x rate x time or $I = prt$. If the principal is \$10,000 and the rate is 5%

(a) Write an equation for simple interest in terms of time.

(b) Graph the equation for times of 0 to 20 years inclusive.

(c) What is the simple interest for 10 years?

(d) If the simple interest is \$500, find the length of time.

71. The weekly profit, P, of a video rental store can be approximated by the formula $P = 1.5n - 200$, where n is the number of tapes rented weekly.

(a) Draw a graph of profit in terms of tape rentals for up to and including 1000 tapes.

(b) Estimate the weekly profit if 500 tapes are rented.

(c) Estimate the number of tapes rented if the week's profit is \$1000.

72. The cost, C, of playing tennis in the Downtown Tennis Club includes an annual \$200 membership fee plus \$10 per hour, h, of court time.

(a) Write an equation for the annual cost of playing tennis at the Downtown Tennis Club in terms of hours played.

(b) Graph the equation for up to and including 300 hours.

(c) Estimate the cost for playing 200 hours in a year.

(d) If the annual cost for playing tennis was \$1200, estimate how many hours of tennis were played.

73. Explain how to find the x and y intercepts of a line.

74. How many points are needed to graph a straight line? How many points should be used? Why?

75. What will the graph of $y = b$ look like for any real number b?

76. What will the graph of $x = a$ look like for any real number a?

77. In Example 10 (c) and (d), we made an estimate. Why is it sometimes not possible to obtain an exact answer from a graph?

78. In Example 10 does the salary, R, depend on the sales, s, or do the sales depend on the salary? Explain.

Determine the coefficients to be placed in the shaded areas so that the graph of the equation will be a line with the x and y intercepts specified. Explain how you determined your answer.

79. $\square x + \square y = 20$; x intercept of 4, y intercept of 5

80. $\square x + \square y = 18$; x intercept of -3, y intercept of 6

81. $\square x - \square y = -12$; x intercept of -2, y intercept of 3

82. $\square x - \square y = 30$; x intercept of -5, y intercept of -15

CUMULATIVE REVIEW EXERCISES

83. Solve the equation $4(x - 2) - (3 - x) = 2x + 4$.

84. Two cyclists are 18 miles apart headed toward each other. One is traveling at a speed of 3 miles per hour faster than the other. If they meet in 1.5 hours, find the speed of each.

[5.6] **85.** Solve the equation $2x^2 = -23x + 12$.

[6.6] **86.** Solve the equation $x - 14 = \dfrac{-48}{x}$.

Group Activity/ Challenge Problems

1. (a) Graph each of the following equations on the same axes:
$y = 2x + 4, \quad y = 2x + 2, \quad y = 2x - 2.$

(b) What do you notice about the graphs?

(c) Explain why you think this happens. We discuss this material further in Section 7.5.

2. **(a)** Carefully graph the following equations on the same axes: $y = 2x - 1$, $y = -x + 5$.
(b) Determine the point of intersection of the two graphs.
(c) Substitute the values for x and y at the point of intersection into each of the two equations and determine if the point of intersection satisfies each equation.
(d) Do you believe there are any other ordered pairs that satisfy both equations? Explain your answer. We will study equations like these, called systems of equations, in Chapter 8.

3. A straight line has an x intercept of (2, 0) and a y intercept of (0, 4).
(a) Draw the line.
(b) Determine the equation of the line. Explain how you determined your answer. We study problems like this in Section 7.5.

In chapter 10 we will be graphing quadratic equations. The graphs of quadratic equations are *not* straight lines. Graph each of the following quadratic equations by selecting values for x and find the corresponding values of y, then plot the points. Make sure you plot a sufficient number of points to get an accurate graph.

4. $y = x^2 - 4$ **5.** $y = x^2 - 2x - 8$

7.4 Slope of a Line

Tape 12

1 Find the slope of a line.
2 Recognize positive and negative slopes.
3 Examine the slopes of horizontal and vertical lines.

Slope

1 The slope of a line is an important concept in many areas of mathematics. A knowledge of slope is helpful in understanding linear equations.

The slope of a line is a measure of the *steepness* of the line. The **slope of a line** is a ratio of the vertical change to the horizontal change between any two selected points on the line. As an example, consider the two points (3, 6) and (1, 2), (Fig. 7.36a).

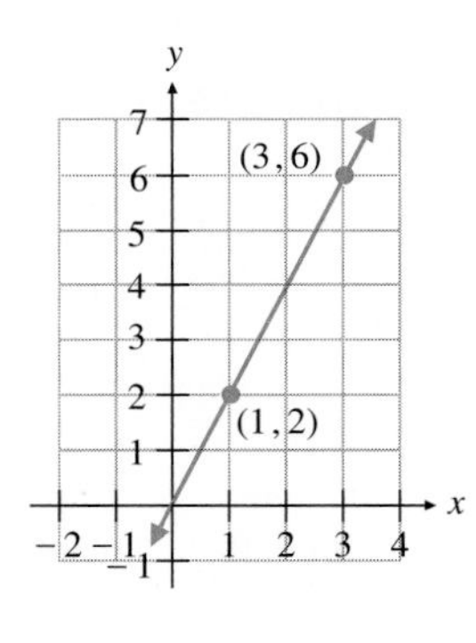

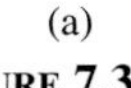

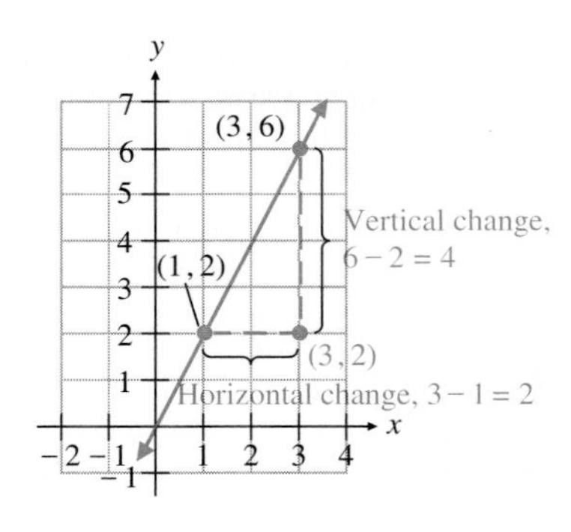

(b)

FIGURE 7.36

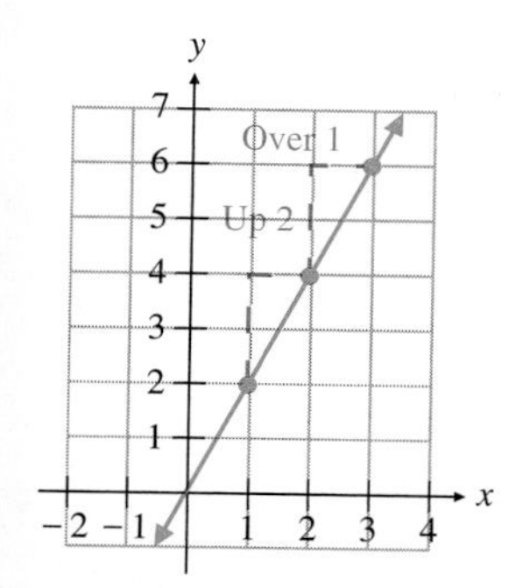

FIGURE 7.37

If we draw a line parallel to the x axis through the point (1, 2) and a line parallel to the y axis through the point (3, 6), the two lines intersect at (3, 2), (Fig. 7.36b). From the figure we can determine the slope of the line. The vertical change (along the y axis) is $6 - 2$, or 4 units. The horizontal change (along the x axis) is $3 - 1$, or 2 units.

$$\text{Slope} = \frac{\text{vertical change}}{\text{horizontal change}} = \frac{4}{2} = 2$$

Thus, the slope of the line through these two points is 2. By examining the line connecting these two points, we can see that as the graph moves up 2 units on the y axis it moves to the right 1 unit on the x axis (Fig. 7.37).

Now we present the procedure to find the slope of a line between any two points (x_1, y_1) and (x_2, y_2). Consider Figure 7.38.

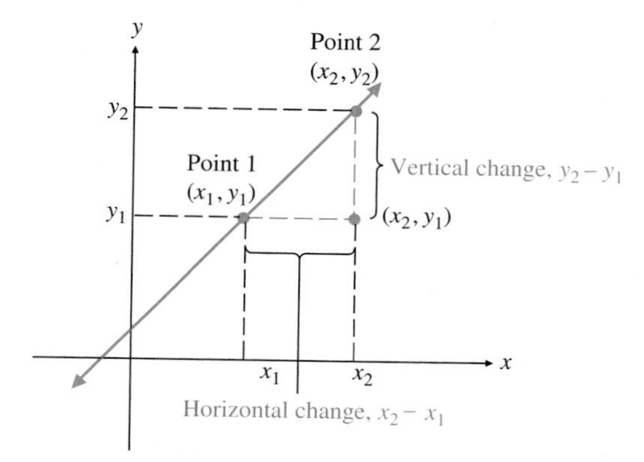

FIGURE 7.38

The vertical change can be found by subtracting y_1 from y_2. The horizontal change can be found by subtracting x_1 from x_2.

Slope of a Line Through the Points (x_1, y_1) and (x_2, y_2)

$$\text{Slope} = \frac{\text{change in } y \text{ (vertical change)}}{\text{change in } x \text{ (horizontal change)}} = \frac{y_2 - y_1}{x_2 - x_1}$$

It makes no difference which two points are selected when finding the slope of a line. It also makes no difference which point you label (x_1, y_1) or (x_2, y_2). The Greek capital letter delta, Δ, is often used to represent the words "the change in." Thus, the slope, which is symbolized by the letter m, is indicated as

$$m = \frac{\Delta y}{\Delta x} = \frac{y_2 - y_1}{x_2 - x_1}$$

EXAMPLE 1 Find the slope of the line through the points $(-6, -1)$ and $(3, 5)$.

Solution: We will designate $(-6, -1)$ as (x_1, y_1) and $(3, 5)$ as (x_2, y_2).

$$m = \frac{y_2 - y_1}{x_2 - x_1} = \frac{5 - (-1)}{3 - (-6)} = \frac{5 + 1}{3 + 6} = \frac{6}{9} = \frac{2}{3}$$

Thus, the slope is $\frac{2}{3}$.

If we had designated (3, 5) as (x_1, y_1) and (−6, −1) as (x_2, y_2), we would have obtained the same results.

$$m = \frac{y_2 - y_1}{x_2 - x_1} = \frac{-1 - 5}{-6 - 3} = \frac{-6}{-9} = \frac{2}{3}$$

Positive and Negative Slopes

2 A straight line for which the value of y increases as x increases has a **positive slope** (Fig. 7.39a). A line with a positive slope rises as it moves from left to right. A straight line for which the value of y decreases as x increases has a **negative slope** (Fig. 7.39b). A line with a negative slope falls as it moves from left to right.

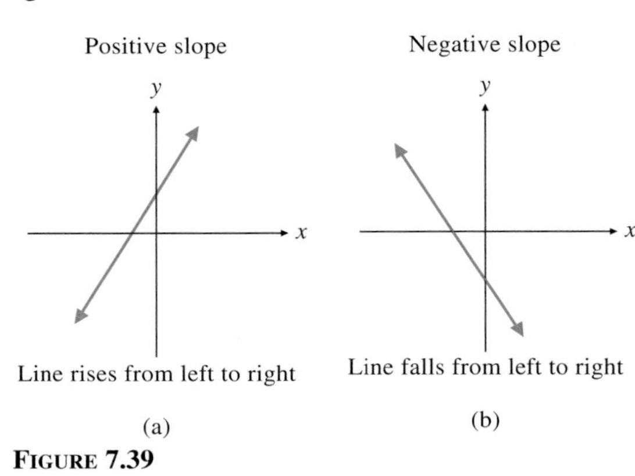

FIGURE 7.39

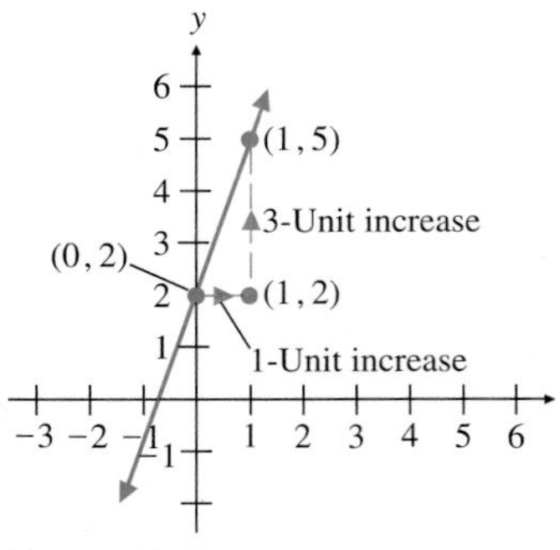

FIGURE 7.40

EXAMPLE 2 Consider the line in Figure 7.40.

(a) Determine its slope by observing the vertical change and horizontal change between the points (1, 5) and (0, 2).

(b) Calculate the slope of the line using the two given points.

Solution:

(a) The first thing you should notice is that the slope is positive since the line rises from left to right. Now determine the vertical change between the two points. The vertical change is +3 units. Next determine the horizontal change between the two points. The horizontal change is +1 unit. Since the slope is the ratio of the vertical change to the horizontal change between any two points, and since the slope is positive, the slope of the line is $\frac{3}{1}$ or 3.

(b) We can use any two points on the line to determine its slope. Since we are given the ordered pairs (1, 5) and (0, 2) we will use them.

Let (x_2, y_2) be (1, 5) Let (x_1, y_1) be (0, 2)

$$m = \frac{y_2 - y_1}{x_2 - x_1} = \frac{5 - 2}{1 - 0} = \frac{3}{1} = 3$$

Note that the slope obtained in part **(b)** agrees with the slope obtained in part **(a)**. If we had designated (1, 5) as (x_1, y_1) and (0, 2) as (x_2, y_2), the slope would not have changed. Try reversing (x_1, y_1) and (x_2, y_2) and see that you will still obtain a slope of 3.

EXAMPLE 3 Find the slope of the line in Figure 7.41 by observing the vertical change and horizontal change between the two points shown.

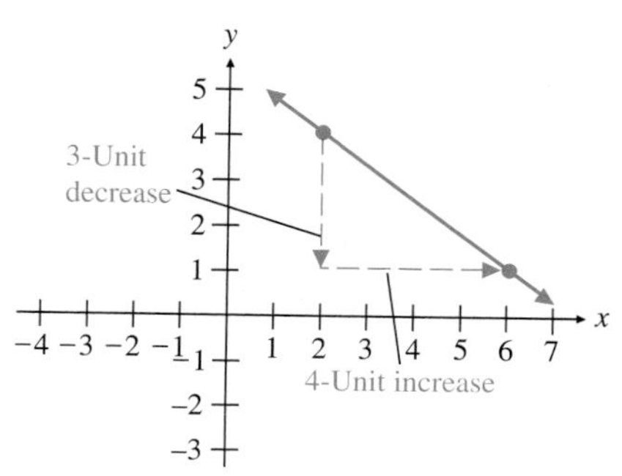

FIGURE 7.41

Solution: Since the graph falls from left to right we should realize that the line has a negative slope. The vertical change between the two given points is -3 units since it is decreasing. The horizontal change between the two given points is 4 units since it is increasing. Since the ratio of the vertical change to the horizontal change is -3 units to 4 units, the slope of this line is

$$\frac{-3}{4} \text{ or } -\frac{3}{4}.$$

Using the two points shown in Fig. 7.41 and the definition of slope, calculate the slope of the line. You should obtain the same answer as we did in Example 3.

Slope of Horizontal and Vertical Lines

3 Now we consider the slope of horizontal and vertical lines.

Consider the graph of $y = 3$ (Fig. 7.42). What is its slope?

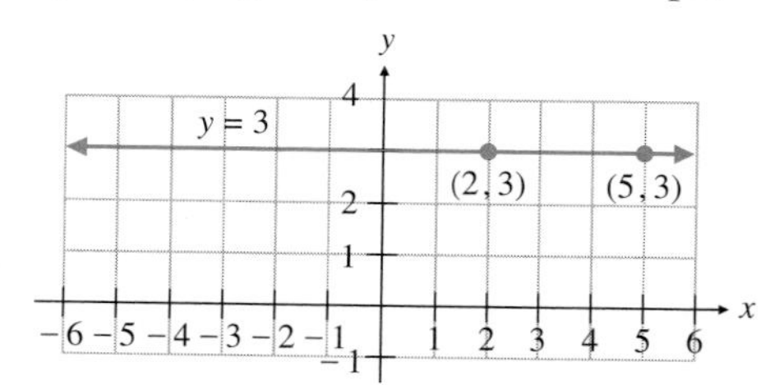

FIGURE 7.42

The graph is parallel to the x axis and goes through the points (2, 3) and (5, 3). Select (5, 3) as (x_2, y_2) and (2, 3) as (x_1, y_1). Then the slope of the line is

$$m = \frac{y_2 - y_1}{x_2 - x_1} = \frac{3 - 3}{5 - 2} = \frac{0}{3} = 0$$

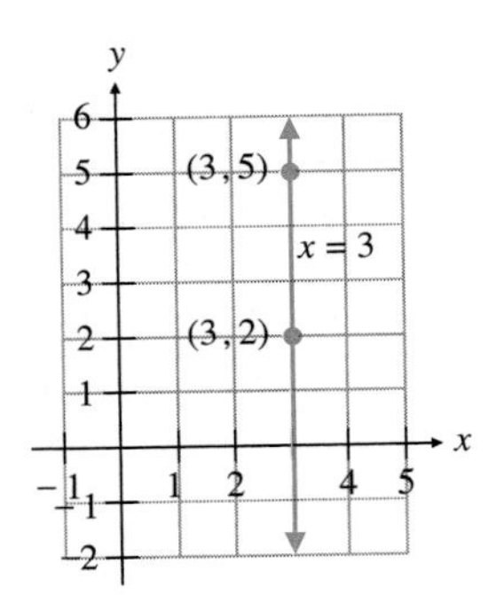

FIGURE 7.43

Since there is no change in y, this line has a slope of 0. **Every horizontal line has a slope of 0.**

Now we discuss vertical lines. Consider the graph of $x = 3$ (Fig. 7.43). What is its slope?

The graph is parallel to the y axis and goes through the points (3, 2) and (3, 5). Select (3, 5) as (x_2, y_2) and (3, 2) as (x_1, y_1). Then the slope of the line is

$$m = \frac{y_2 - y_1}{x_2 - x_1} = \frac{5 - 2}{3 - 3} = \frac{3}{0}$$

We learned in Section 1.7 that $\frac{3}{0}$ is undefined. Thus, we say that the slope of this line is undefined. **The slope of any vertical line is undefined.**

Exercise Set 7.4

Find the slope of the line through the given points.

1. (4, 1) and (5, 6)
2. (8, −2) and (6, −4)
3. (9, 0) and (5, −2)
4. (5, −6) and (6, −5)
5. (3, 8) and (−3, 8)
6. (−4, 2) and (6, 5)
7. (−4, 6) and (−2, 6)
8. (9, 3) and (5, −6)
9. (3, 4) and (3, −2)
10. (−7, 5) and (3, −4)
11. (−4, 2) and (5, −3)
12. (−9, −6) and (−3, −1)
13. (−1, 7) and (4, −3)
14. (0, 4) and (6, −2)

By observing the vertical and horizontal change of the line between the two points indicated, determine the slope of the line.

15.

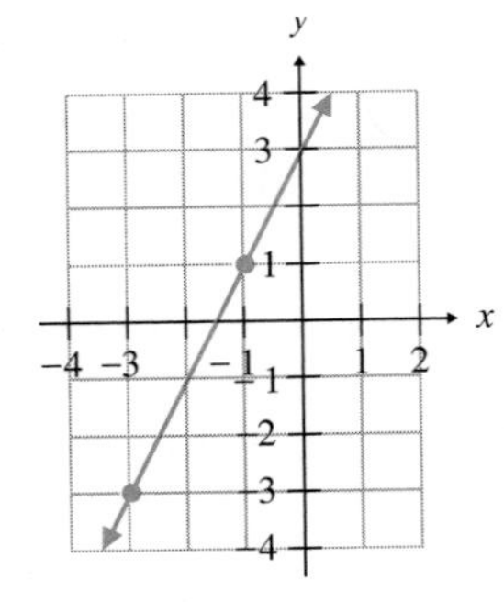

16.

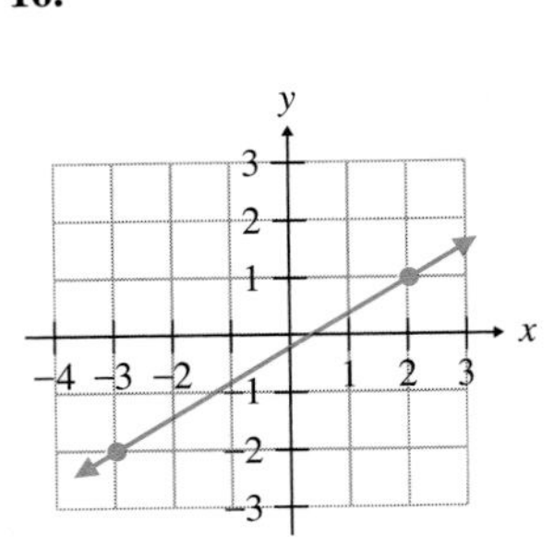

17.

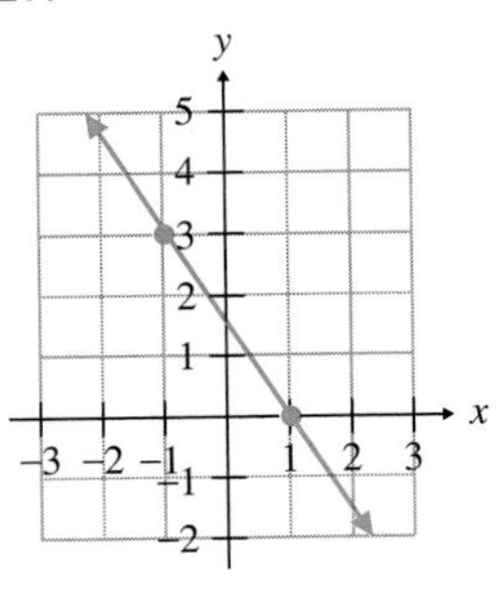

18.

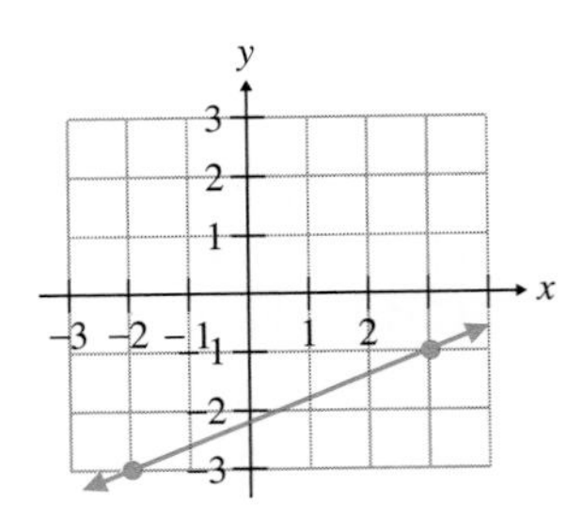

19.

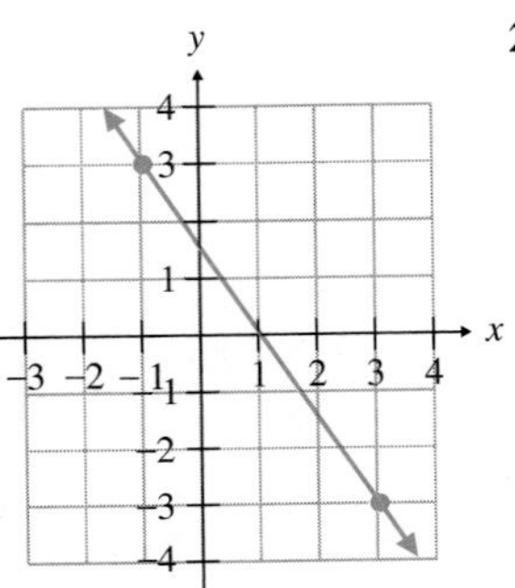

20.

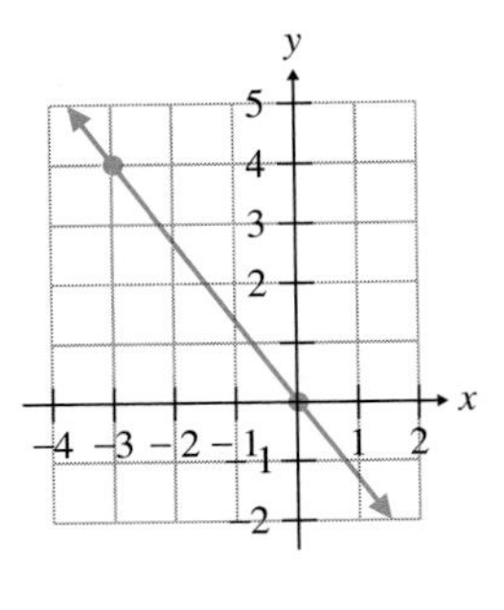

21.

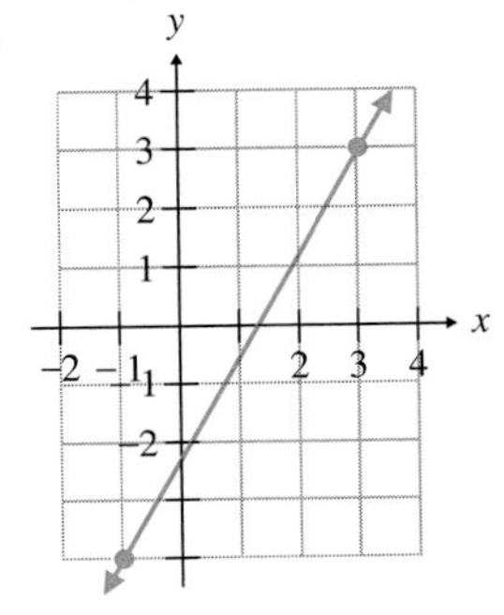

22.

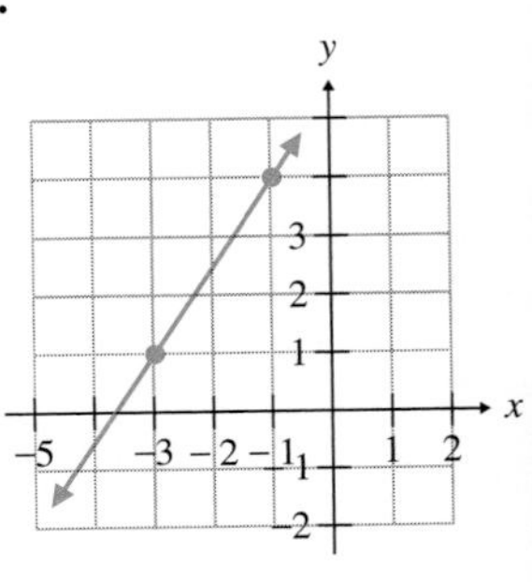

23.

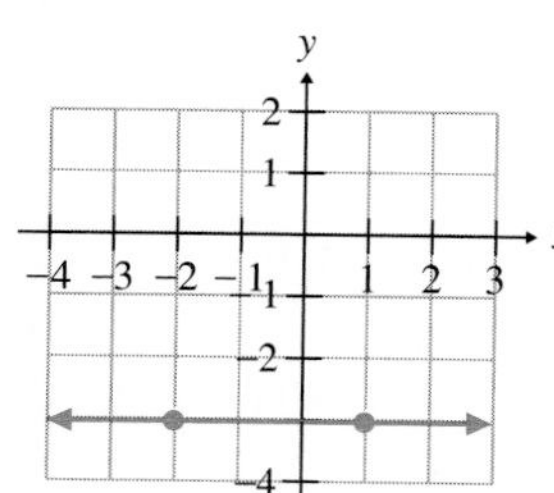

24.

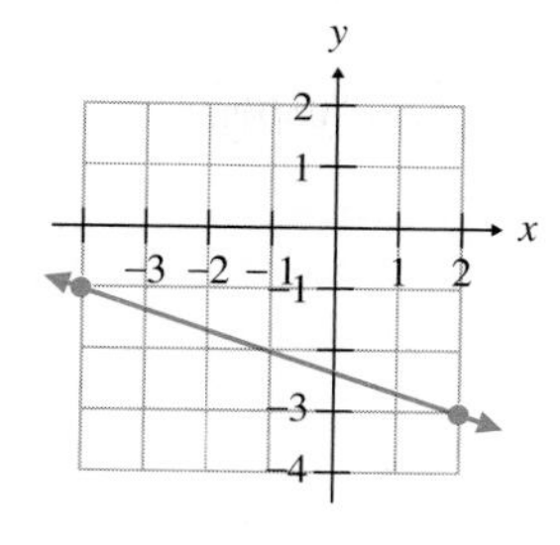

25.

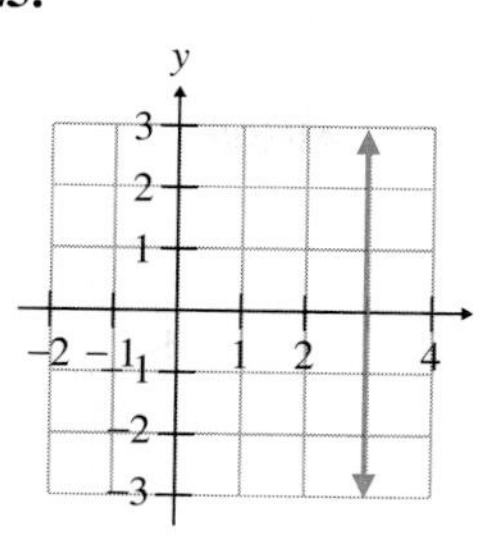

26.

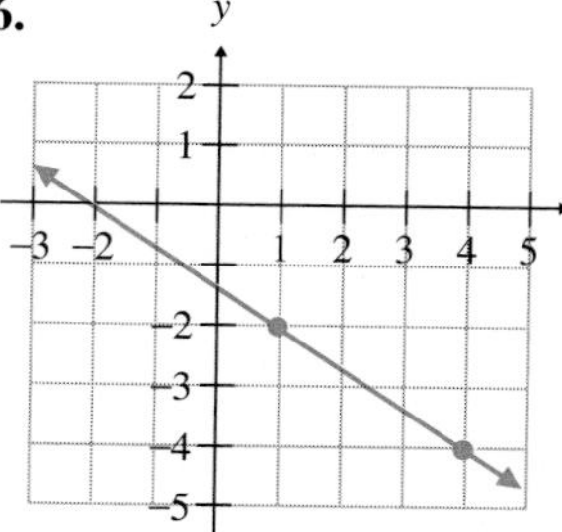

27. Explain what is meant by the slope of a line.

28. Explain how to find the slope of a line.

29. What does it mean when a line has a positive slope?

30. What does it mean when a line has a negative slope?

31. Explain how to tell by observation if the line has a positive slope or negative slope.

32. What is the slope of any horizontal line? Explain your answer.

33. Do vertical lines have a slope? Explain.

CUMULATIVE REVIEW EXERCISES

We have spent a great deal of time discussing and solving equations. For each type of equation that follows **(a)** *give a general description of the equation, and* **(b)** *give a specific example of the equation (answers will vary).*

2] **34.** A linear equation in one variable.

5] **35.** A quadratic equation in one variable.

[6.6] **36.** A rational equation in one variable.

[7.2] **37.** A linear equation in two variables.

Group Activity/ Challenge Problems

1. Find the slope of the line through the points $(\frac{1}{2}, -\frac{3}{8})$ and $(-\frac{4}{9}, -\frac{7}{2})$.

2. If one point on a line is $(6, -4)$ and the slope of the line is $-\frac{5}{3}$, identify another point on the line.

3. One point on a line is $(-5, 2)$ and the slope of the line is $-\frac{3}{4}$. A second point on the line has a y coordinate of -7. Find the x coordinate of the point.

4. The slope of a hill and the slope of a line both measure steepness. However, there are several important differences.

(a) Explain how you think the slope of a hill is determined.

(b) Is the slope of a line, graphed in the Cartesian coordinate system, measured in any specific unit? Is the slope of a hill measured in any specific unit?

5. A quadrilateral (a four-sided figure) has four vertices (the points where the sides meet). Vertex A is at $(0, 1)$, vertex B is at $(6, 2)$, vertex C is at $(5, 4)$, and vertex D is at $(1, -1)$.

(a) Graph the quadrilateral in the Cartesian coordinate system.

(b) Find the slope of sides AC, CB, DB, and AD.

(c) Do you believe this figure is a parallelogram? Explain.

6. The following graph shows the world's population estimated to the year 2016.

(a) Find the slope of the line segment between each pair of points, that is, *ab*, *bc*, and so on. Remember, the second coordinate is in billions. Thus, for example, 0.5 billion is actually 500,000,000.

(b) Would you say that this graph represents a linear equation? Explain.

World Population Growth

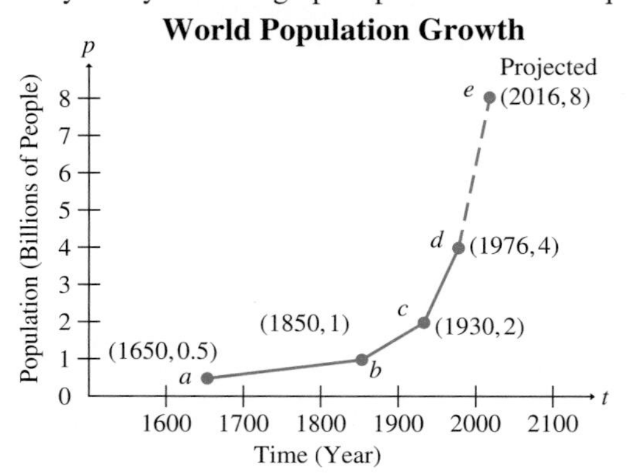

7. Consider the graph on the left.

(a) Do you believe that the slope of the dashed line from a to d will be the same as the average of the slopes of the three solid lines?

(b) Find the slope of the dashed line from a to d.

(c) Find the slope of each of the three solid lines.

(d) Find the average of the three slopes found in part (c)

(e) Determine if your answer in part (a) appears correct.

(f) What do you think this means?

7.5 Slope–Intercept and Point–Slope Forms of a Linear Equation

Tape 12

1. Write a linear equation in slope–intercept form.
2. Graph a linear equation using the slope and y intercept.
3. Use the slope and y intercept to determine the equation of a line.
4. Determine if two lines are parallel.
5. Use the slope and a point on the line or two points on a line to determine the equation of the line.
6. Compare the three methods of graphing linear equations.

In Section 7.2 we introduced the **standard form** of a linear equation, $ax + by = c$. In this section we introduce two more forms, the slope–intercept form and the point–slope form. We begin our discussion with the slope–intercept form.

Slope–Intercept Form

1 A very important form of a linear equation is the **slope–intercept form,** $\boldsymbol{y = mx + b}$**.** The graph of an equation of the form $y = mx + b$ will always be a straight line with a **slope of** $\boldsymbol{m}$ and a $\boldsymbol{y}$ **intercept of** $\boldsymbol{b}$**.** For example, the graph of the equation $y = 3x - 4$ will be a straight line with a slope of 3 and a y intercept of -4. The graph of $y = -2x + 5$ will be a straight line with a slope of -2 and a y intercept of 5.

Slope–Intercept Form of a Linear Equation

$$y = mx + b$$

where m is the slope, and b* is the y intercept of the line.

slope (→ m) **y intercept** (→ b)

$$y = mx + b$$

Equations in Slope–Intercept Form	*Slope*	*y Intercept*
$y = 3x - 6$	3	-6
$y = \frac{1}{2}x + \frac{3}{2}$	$\frac{1}{2}$	$\frac{3}{2}$
$y = -5x + 3$	-5	3
$y = -\frac{2}{3}x - \frac{3}{5}$	$-\frac{2}{3}$	$-\frac{3}{5}$

To write a linear equation in slope–intercept form, solve the equation for y.

Once the equation is solved for y, the numerical coefficient of the x term will be the slope, and the constant term will be the y intercept.

EXAMPLE 1 Write the equation $-3x + 4y = 8$ in slope–intercept form. State the slope and y intercept.

Solution: To write this equation in slope–intercept form, we solve the equation for y.

$$-3x + 4y = 8$$

$$4y = 3x + 8$$

$$y = \frac{3x + 8}{4}$$

$$y = \frac{3}{4}x + \frac{8}{4}$$

$$y = \frac{3}{4}x + 2$$

The slope is $\frac{3}{4}$, and the y intercept is 2.

*In this section, when we refer to b as the y intercept, it means that the y intercept is $(0, b)$.

Graphing Linear Equations Using the Slope and y Intercept

 In Section 7.4 we discussed two methods of graphing a linear equation. They were (1) by plotting points and (2) using the x and y intercepts. Now we present a third method. This method makes use of the slope and the y intercept. Remember that when we solve an equation for y we put the equation in slope–intercept form. Once in this form we can determine the slope and the y intercept of the graph from the equation. The procedure to use to graph by this method follows.

To Graph Linear Equations Using the Slope and y Intercept

1. Solve the linear equation for y. That is, get the equation in slope–intercept form, $y = mx + b$.
2. Note the slope, m, and y intercept, b.
3. Plot the y intercept on the y axis.
4. Use the slope to find a second point on the graph.
   (a) If the slope is **positive,** a second point can be found by moving **up and to the right.** Thus, if the slope is of the form $\frac{p}{q}$, we can find a second point by moving *up* p units and to the *right* q units.
   (b) If the slope is **negative,** a second point can be found by moving **down and to the right** (or **up and to the left**). Thus, if the slope is of the form $-\frac{p}{q}\left(\text{or } \frac{-p}{q} \text{ or } \frac{p}{-q}\right)$, we can find a second point by moving *down* p units and to the *right* q units (or *up* p units and to the *left* q units).
5. With a straightedge, draw a straight line through the two points. Draw arrowheads at the ends of the line to show that the line continues indefinitely in both directions.

EXAMPLE 2 Write the equation $-3x + 4y = 8$ in slope–intercept form; then use the slope and y intercept to graph $-3x + 4y = 8$.

Solution: In Example 1 we solved $-3x + 4y = 8$ for y. We found that

$$y = \frac{3}{4}x + 2$$

The slope of the line is 3/4 and the y intercept is 2. Mark the first point, the y intercept, at 2 on the y axis (Fig. 7.44). Now we use the slope 3/4 , to find a second point. Since the slope is positive, we move up 3 units and to the right 4 units to find the second point. A second point will be at (4, 5). We can continue this process to obtain a third point at (8, 8). Now draw a straight line through the three points. Notice that the line has a positive slope, which is what we expected.

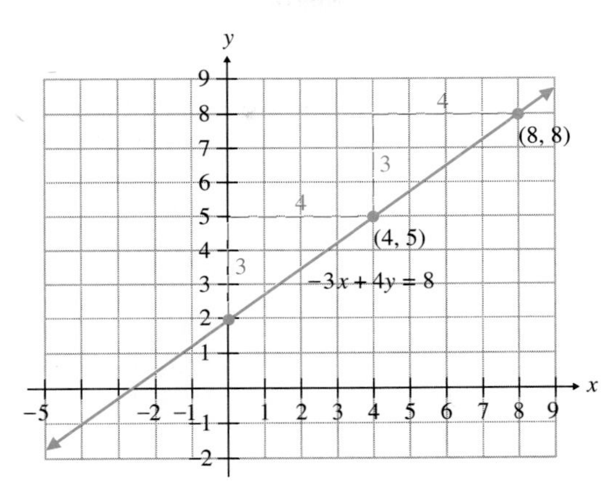

FIGURE 7.44

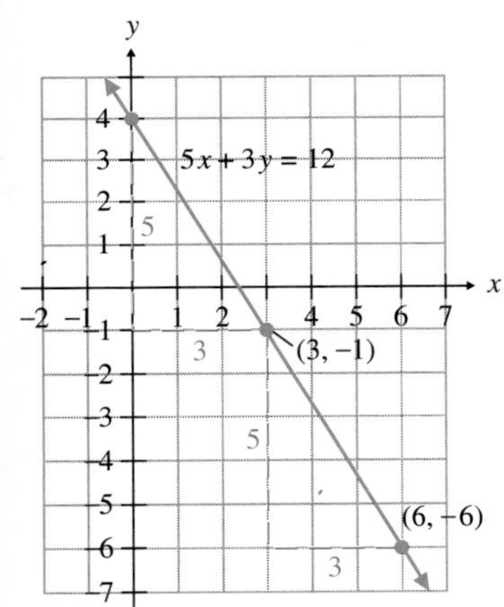

FIGURE 7.45

EXAMPLE 3 Graph the equation $5x + 3y = 12$ by using the slope and y intercept.

Solution: Solve the equation for y.

$$5x + 3y = 12$$

$$3y = -5x + 12$$

$$y = \frac{-5x + 12}{3}$$

$$y = -\frac{5}{3}x + 4$$

Thus, the slope is $-5/3$ and the y intercept is 4. Begin by marking a point at 4 on the y axis (Fig. 7.45). Then move down 5 units and to the right 3 units to determine the next point. We move down and to the right (or up and to the left) because the slope is negative and a line with a negative slope must fall as it goes from left to right. Finally, draw the straight line between the plotted points.

Determine the Equation of a Line

3 Now that we know how to use the slope–intercept form of a line, we can use it to write the equation of a given line. To do so, we need to determine the slope, m, and y intercept, b, of the line. Once we determine these values we can write the equation in slope–intercept form, $y = mx + b$. For example, if we determine the slope of a line is -4 and the y intercept is 6, the equation of the line is $y = -4x + 6$.

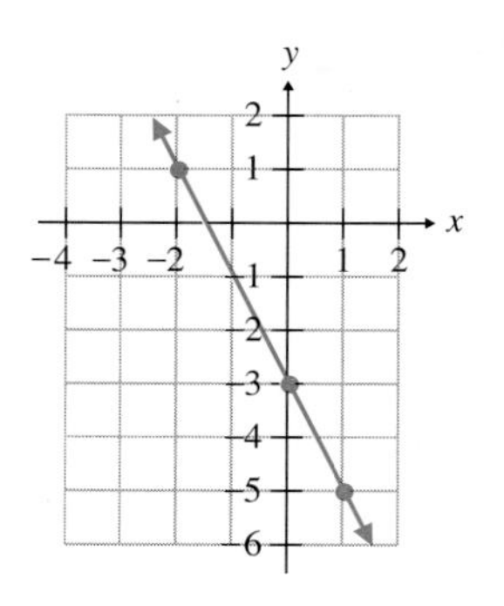

FIGURE 7.46

EXAMPLE 4 Determine the equation of the line shown in Figure 7.46.

Solution: The graph shows that the y intercept is -3. Now we need to determine the slope of the line. Since the graph falls from left to right, it has a negative slope. We can see that the vertical change is 2 units for each horizontal change of 1 unit. Thus, the slope of the line is -2. The slope can also be determined by selecting any two points on the line and calculating the slope. Let us use the point $(-2, 1)$ to represent (x_2, y_2) and the point $(0, -3)$ to represent (x_1, y_1).

$$m = \frac{\Delta y}{\Delta x} = \frac{y_2 - y_1}{x_2 - x_1} = \frac{1 - (-3)}{-2 - 0} = \frac{1 + 3}{-2} = \frac{4}{-2} = -2$$

Again we obtain a slope of -2. The slope–intercept form of a line is $y = mx + b$, where m is the slope and b is the y intercept. Substituting -2 for m and -3 for b gives us the equation of the line in Figure 7.46, which is $y = -2x - 3$.

Parallel Lines

4 We will discuss the meaning of parallel lines shortly, but before we do we will work Example 5.

EXAMPLE 5 Determine if both equations represent lines that have the same slope.

$$6x + 3y = 8$$
$$-4x - 2y = -3$$

Solution: Solve each equation for y to get the equations in slope–intercept form.

$$6x + 3y = 8 \qquad\qquad -4x - 2y = -3$$
$$3y = -6x + 8 \qquad\qquad -2y = 4x - 3$$
$$y = \frac{-6x + 8}{3} \qquad\qquad y = \frac{4x - 3}{-2}$$
$$y = -2x + \frac{8}{3} \qquad\qquad y = -2x + \frac{3}{2}$$

Both lines have the same slope of -2. Notice, however, that their y intercepts are different.

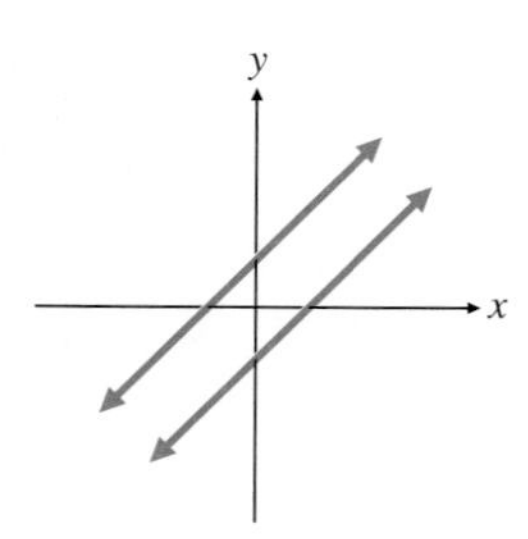

FIGURE 7.47

Two lines are **parallel** when they do not intersect no matter how far they are extended. Figure 7.47 illustrates two parallel lines. **Lines with the same slope are parallel** (or identical) **lines.** The graphs of the equations in Example 5 are parallel lines since they both have the same slope, -2. Note that the two equations represent different lines since their y intercepts are different.

To Determine If Two Lines Are Parallel

Write both equations in slope–intercept form and compare the slopes of the two lines. If both lines have the same slope, but different y intercepts, then the lines are parallel. If the slopes are not the same, the lines are not parallel. If both equations have the same slope and the same y intercept then both equations represent the same line.

EXAMPLE 6

(a) Determine whether or not the following equations represent parallel lines.
(b) Graph both equations on the same set of axes.

$$y = 2x + 4$$
$$-4x + 2y = -2$$

Solution:

(a) Write each equation in slope–intercept form and compare their slopes. The equation $y = 2x + 4$ is already in slope–intercept form.

$$-4x + 2y = -2$$
$$2y = 4x - 2$$
$$y = \frac{4x - 2}{2} = 2x - 1$$

The two equations are now

$$y = 2x + 4$$
$$y = 2x - 1$$

Since both equations have the same slope, 2, but different y intercepts, the equations represent parallel lines.

(b) We now graph $y = 2x + 4$ and $y = 2x - 1$ on the same axes (Fig. 7.48). Remember that $y = 2x - 1$ is the equation $-4x + 2y = -2$ in slope–intercept form.

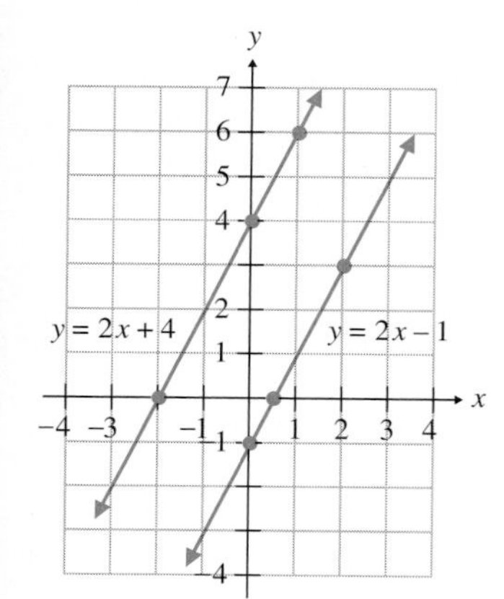

FIGURE 7.48

Point–Slope Form of a Linear Equation

5 Thus far we have discussed the standard form of a linear equation, $ax + by = c$, and the slope–intercept form of a linear equation, $y = mx + b$. Now we will discuss another form, called the *point–slope form.*

When the slope of a line and a point on the line are known, we can use the point–slope form to determine the equation of the line. The point–slope form can be obtained by beginning with the slope between any selected point (x, y) and a fixed point (x_1, y_1) on a line.

$$m = \frac{y - y_1}{x - x_1} \quad \text{or} \quad \frac{m}{1} = \frac{y - y_1}{x - x_1}$$

Now cross-multiply to obtain

$$m(x - x_1) = y - y_1 \quad \text{or} \quad y - y_1 = m(x - x_1)$$

Point–Slope Form of a Linear Equation

$$y - y_1 = m(x - x_1)$$

where m is the slope of the line and (x_1, y_1) is a point on the line.

EXAMPLE 7 Write an equation of the line that goes through the point (2, 3) and has a slope of 4.

Solution: The slope m is 4. The point on the line is (2, 3); use this point for (x_1, y_1) in the formula. Substitute 4 for m, 2 for x_1, and 3 for y_1 in the point–slope form of a linear equation.

$$\begin{aligned} y - y_1 &= m(x - x_1) \\ y - 3 &= 4(x - 2) \\ y - 3 &= 4x - 8 \\ y &= 4x - 5 \end{aligned}$$

The graph of $y = 4x - 5$ has a slope of 4 and passes through the point (2, 3).

The answer to Example 7 was given in slope–intercept form. The answer could have also been given in standard form. Therefore, two other acceptable answers are $-4x + y = -5$ and $4x - y = 5$. Your instructor may specify the form in which the equation is to be given.

Helpful Hint

We have discussed three forms of a linear equation. We summarize the three forms below:

Standard Form	Examples
$ax + by = c$	$2x - 3y = 8$ $-5x + y = -2$
Slope–Intercept Form	**Examples**
$y = mx + b$ (m is the slope, b is the y intercept)	$y = 2x - 5$ $y = -\frac{3}{2}x + 2$
Point–Slope Form	**Examples**
$y - y_1 = m(x - x_1)$ (m is the slope, (x_1, y_1) is a point on the line)	$y - 3 = 2(x + 4)$ $y + 5 = -4(x - 1)$

We now discuss how to use the point–slope form to determine the equation of a line when two points on the line are known.

EXAMPLE 8 Find an equation of the line through the points $(-1, -3)$ and (4, 2). Write the equation in slope–intercept form.

Solution: To use the point–slope form, we must first find the slope of the line through the two points. To determine the slope, let us designate $(-1, -3)$ as (x_1, y_1) and (4, 2) as (x_2, y_2).

$$m = \frac{y_2 - y_1}{x_2 - x_1} = \frac{2 - (-3)}{4 - (-1)} = \frac{2 + 3}{4 + 1} = \frac{5}{5} = 1$$

We can use either point (one at a time) in determining the equation of the line. This example will be worked out using both points to show that the solutions obtained are identical.

Using the point $(-1, -3)$ as (x_1, y_1),

$$\begin{aligned} y - y_1 &= m(x - x_1) \\ y - (-3) &= 1[x - (-1)] \\ y + 3 &= x + 1 \\ y &= x - 2 \end{aligned}$$

Using the point $(4, 2)$ as (x_1, y_1),

$$\begin{aligned} y - y_1 &= m(x - x_1) \\ y - 2 &= 1(x - 4) \\ y - 2 &= x - 4 \\ y &= x - 2 \end{aligned}$$

Note that the equations for the line are identical.

Helpful Hint

In the exercise set at the end of this section, you will be asked to write a linear equation in slope–intercept form. Even though you will eventually write the equation in slope–intercept form, you may need to start your work with the point–slope form. Below we indicate the initial form to use to solve the problem.

Begin with the **slope–intercept form** if you know:

The slope of the line and the y intercept

Begin with the **point–slope form** if you know:

(a) The slope of the line and a point on the line, or

(b) Two points on the line (first find the slope, then use the point–slope form)

Summary of Three Methods of Graphing Linear Equations

6 We have discussed three methods to graph a linear equation: (1) plotting points, (2) using the x and y intercepts, and (3) using the slope and y intercept. In Example 9 we graph an equation using all three methods. No single method is always the easiest to use. If the equation is given in slope–intercept form, $y = mx + b$, then graphing by plotting points or by using the slope and y intercept might be easier. If the equation is given in standard form, $ax + by = c$, then graphing using the intercepts might be easier. Unless your teacher specifies that you should graph by a specific method, you may use the method that you feel most comfortable with. Graphing by plotting is the most versatile method since it can also be used to graph equations that are not straight lines.

EXAMPLE 9 Graph $3x - 2y = 8$ **(a)** by plotting points; **(b)** using the x and y intercepts; and **(c)** using the slope and y intercept.

Solution: For parts **(a)** and **(c)** we must write the equation in slope–intercept form.

$$3x - 2y = 8$$

$$-2y = -3x + 8$$

$$y = \frac{-3x + 8}{-2} = \frac{3}{2}x - 4$$

(a) *Plotting points:* Substitute values for x and find the corresponding values of y. Then plot the ordered pairs and draw the graph (Fig. 7.49).

$$y = \frac{3}{2}x - 4$$

x	y
0	−4
2	−1
4	2

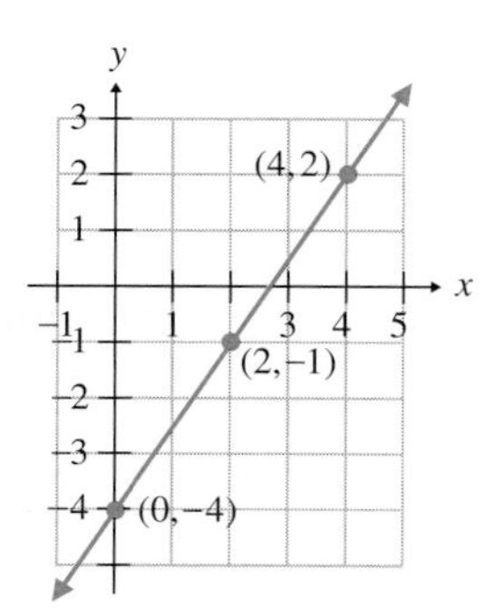

FIGURE 7.49

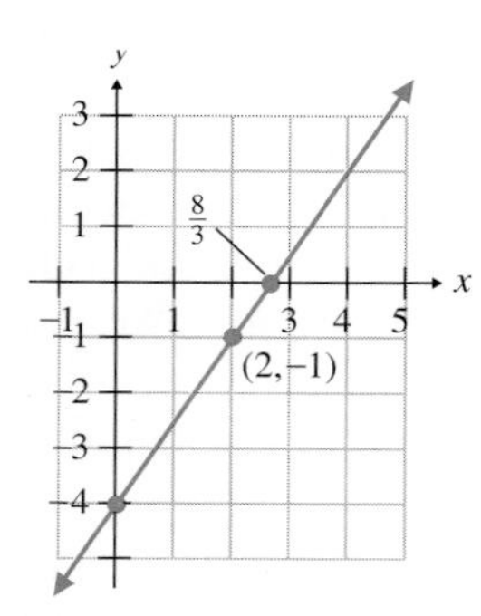

FIGURE 7.50

(b) *Intercepts:* Find the x and y intercepts and a check point. Then plot the points and draw the graph (Fig. 7.50).

$$3x - 2y = 8$$

x Intercept

Let $y = 0$

$$3x - 2y = 8$$
$$3x - 2(0) = 8$$
$$3x = 8$$
$$x = \frac{8}{3}$$

y Intercept

Let $x = 0$

$$3x - 2y = 8$$
$$3(0) - 2y = 8$$
$$-2y = 8$$
$$y = -4$$

Check Point

Let $x = 2$

$$3x - 2y = 8$$
$$3(2) - 2y = 8$$
$$6 - 2y = 8$$
$$-2y = 2$$
$$y = -1$$

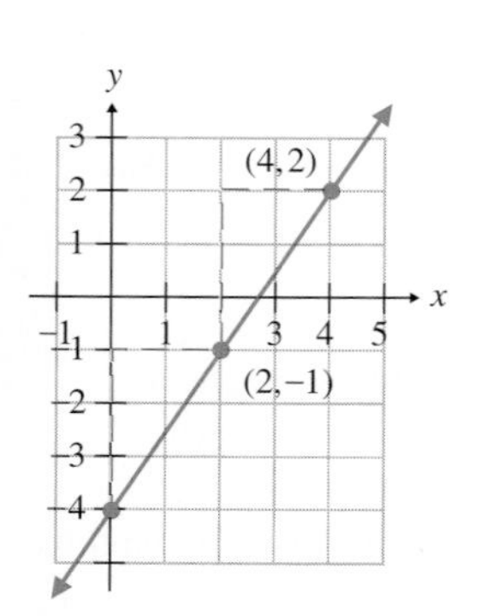

FIGURE 7.51

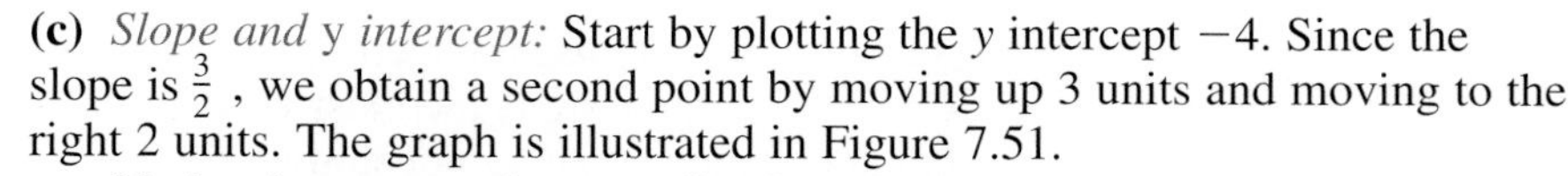

(c) *Slope and y intercept:* Start by plotting the y intercept -4. Since the slope is $\frac{3}{2}$, we obtain a second point by moving up 3 units and moving to the right 2 units. The graph is illustrated in Figure 7.51.

Notice that we get the same line by all three methods.

Exercise Set 7.5

Determine the slope and y intercept of the line represented by each equation. Graph the line using the slope and y intercept. (See Examples 2 and 3.)

1. $y = 2x - 1$

2. $y = 3x + 2$

3. $y = -x + 5$

4. $y = 2x$

5. $y = -4x$

6. $2x + y = 5$

7. $-2x + y = -3$

8. $3x - y = -2$

9. $3x + 3y = 9$

10. $5x - 2y = 10$

11. $-x + 2y = 8$

12. $5x + 10y = 15$

13. $4x = 6y + 9$

14. $4y = 5x - 12$

15. $-6x = -2y + 8$

16. $6y = 5x - 9$

17. $-3x + 8y = -8$

18. $16y = 8x + 32$

19. $3x = 2y - 4$

20. $20x = 80y + 40$

Determine the equation of each line. (See example 4.)

21.

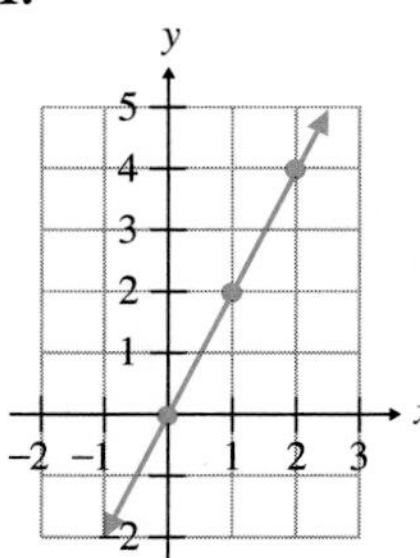

22.

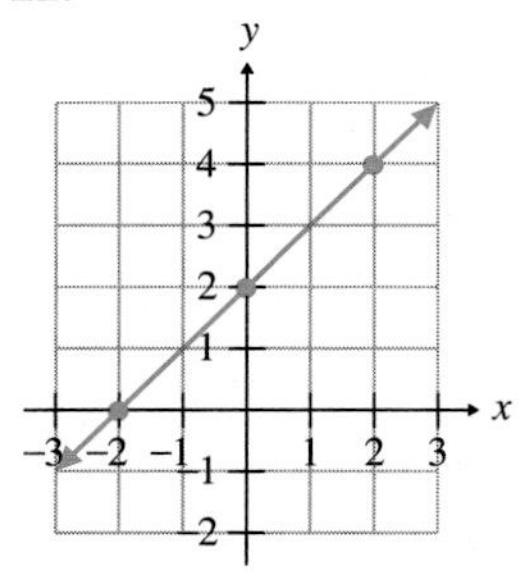

23.

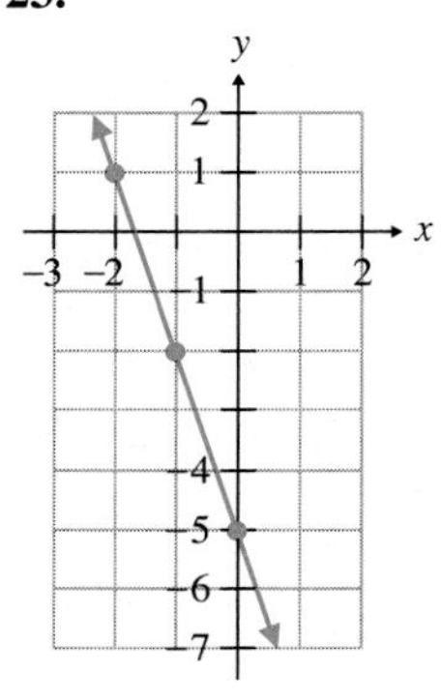

24.

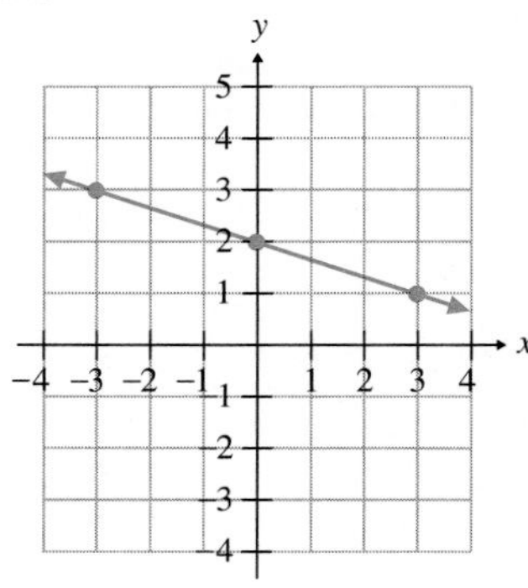

25.

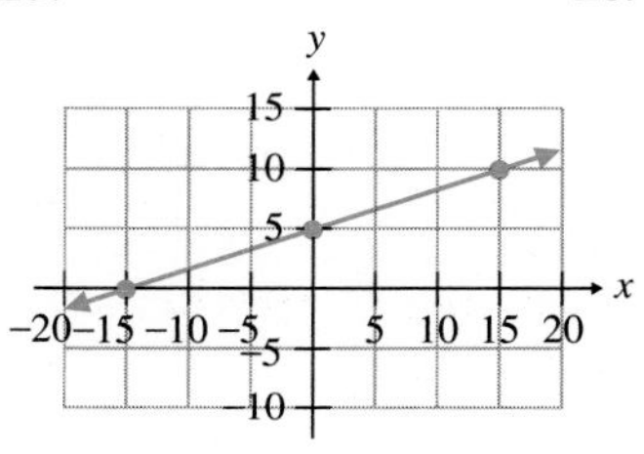

26.

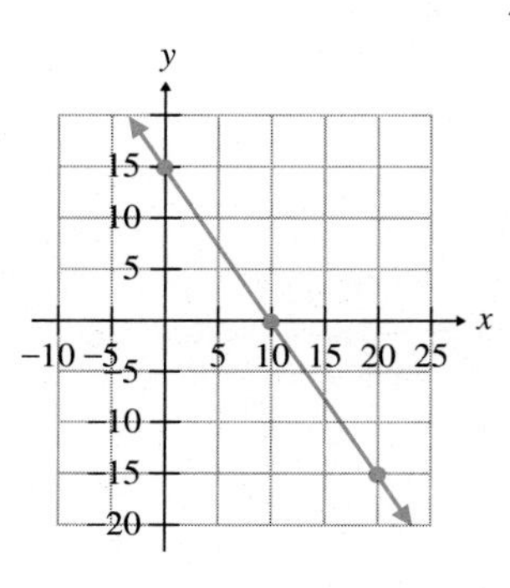

27.

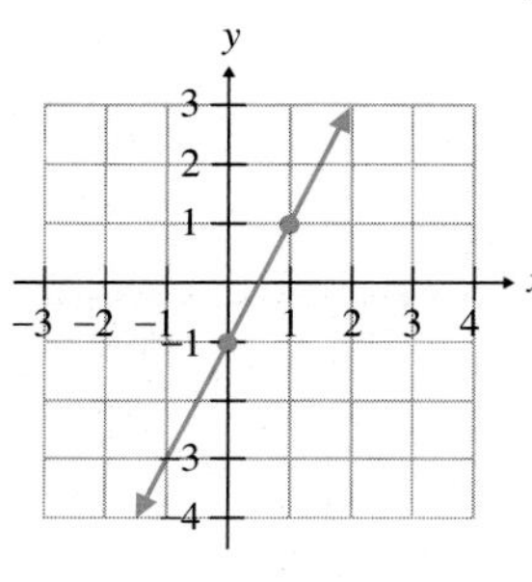

28.

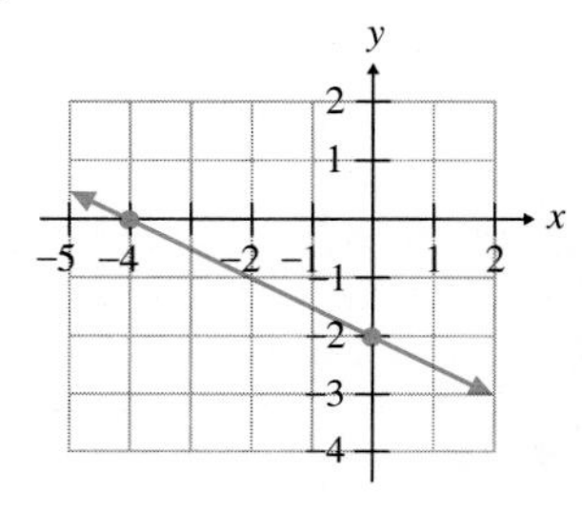

Determine if the lines are parallel. (See Examples 5 and 6.)

29. $y = 2x - 4$
$y = 2x + 3$

30. $2x + 3y = 8$
$y = -\dfrac{2}{3}x + 5$

31. $4x + 2y = 9$
$8x = 4 - 4y$

32. $3x - 5y = 7$
$5y + 3x = 2$

33. $2x + 5y = 9$
$-x + 3y = 9$

34. $6x + 2y = 8$
$4x - 9 = -y$

35. $y = \dfrac{1}{2}x - 6$
$3y = 6x + 9$

36. $2y - 6 = -5x$
$y = -\dfrac{5}{2}x - 2$

Write the equation of the line, with the given properties, in slope–intercept form. (See the Helpful Hint on page 410 and Examples 7 and 8.)

37. Slope = 5, through (0, 4)

38. Slope = 4, through (2, 3)

39. Slope = −2, through (−4, 5)

40. Slope = −1, through (6, 0)

41. Slope = $\frac{1}{2}$, through (−1, −5)

42. Slope = $-\frac{2}{3}$, through (−1, −2)

43. Slope = $\frac{3}{5}$, y intercept = 7

44. Slope = $\frac{1}{2}$, y intercept = −3

45 Through (−4, −2) and (−2, 4)

46. Through (6, 3) and (5, 2)

47. Through (−4, 6) and (4, −6)

48. Through (1, 0) and (−2, 4)

49. Through (10, 3) and (0, −2)

50. Through (−6, −2) and (5, −3)

51. Slope = 5.2, y intercept = −1.6

52. Slope = $-\frac{5}{8}$, y intercept = $-\frac{7}{10}$

Graph the equation by **(a)** *plotting points,* **(b)** *using the x and y intercepts, and* **(c)** *using the slope and y intercept. (See Example 9.)*

53. $3x - 2y = 4$

54. $4x + 3y = 6$

55. $2x - 3y = -6$

56. When you are given an equation in a form other than slope–intercept form, how can you change it to slope–intercept form?

57. Explain how you can determine if two lines are parallel without actually graphing them.

58. Consider the two equations $40x - 60y = 100$ and $-40x + 60y = 80$.

(a) When these equations are graphed, will the two lines have the same slope? Explain how you determined your answer.

(b) Will these two equations when graphed be parallel lines?

59. Write **(a)** the standard form of a linear equation, **(b)** the slope–intercept form of a linear equation, and **(c)** the point–slope form of a linear equation.

60. Suppose that you were asked to write the equation of a line with the properties given below. Which form of a linear equation—standard form, slope–intercept form, or point–slope form—would you start with? Explain your answer.

(a) The slope of the line and the y intercept of the line

(b) The slope and a point on the line

(c) Two points on the line

61. (a) Explain in your own words, in a step-by-step manner, how to graph a linear equation using each of the three methods discussed: plotting points, using the x and y intercepts, and using the slope and y intercept.

(b) Graph $-3x + 2y = 4$ using each of the three methods.

Cumulative Review Exercises

[6.5] **62.** Divide $\dfrac{9x^3 - 3x^2 - 9x + 4}{3x + 2}$.

[6.7] **63.** Consuella has 1 liter of a 5% saltwater solution. How much pure water, without salt, will Consuella need to add to the 5% solution to obtain a 2% saltwater solution?

[6.2] **64.** Divide $\dfrac{x^2 + 2x - 8}{x^2 - 16} \div \dfrac{2x^2 - 5x - 3}{x^2 - 7x + 12}$.

[6.4] **65.** Add $\dfrac{3}{x - 2} + \dfrac{5}{x + 2}$.

[6.6] **66.** Solve the equation $x + \dfrac{30}{x} = 11$.

[6.7] **67.** The area of a triangle is 36 square feet. Find the base and height if its height is 7 feet less than twice its base.

Group Activity/Challenge Problems

1. Determine the equation of the line with y intercept at 4 that is parallel to the line $2x + y = 6$. Explain how you determined your answer.

2. Determine if the line with x intercept at 5 and y intercept at -2 is parallel to the line $4x + 10y = 20$. Explain how you determined your answer.

3. Will a line through the points (60, 30) and (20, 90) be parallel to the line with x intercept at 2 and y intercept at 3? Explain how you determined your answer.

4. Write an equation of the line parallel to $3x - 4y = 6$ that passes through the point $(-4, -1)$

5. Two lines are **perpendicular** and cross at right angles when their slopes are negative reciprocals of each other. The negative reciprocal of any number a is $-1/a$. For example, the negative reciprocal of 2 is $-\frac{1}{2}$ and the negative reciprocal of $-\frac{3}{4}$ is $\frac{4}{3}$. Write an equation of the line perpendicular to $-5x + 2y = -4$ that passes through the point $(2, \frac{1}{2})$.

6. Write the equation $3x = 4y + 6$ in point–slope form. Explain how you obtained your answer. There is more than one acceptable answer.

7. Determine the equation of the straight line that intersects the greatest number of shaded points on the following graph.

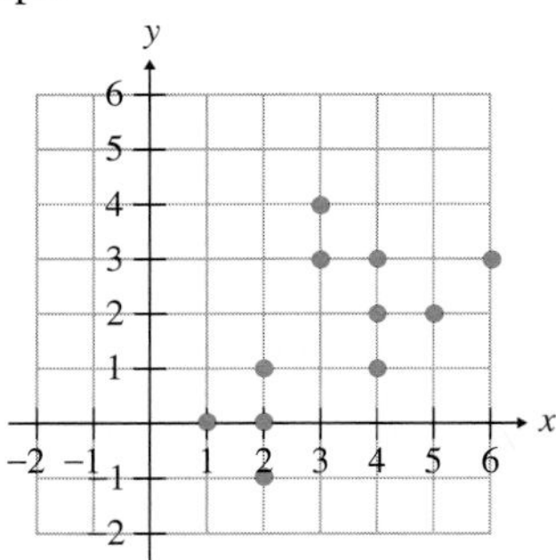

7.6 Functions

Tape 13

1. Identify relations.
2. Find the domain and range of a relation.
3. Use the vertical line test to determine if a relation is a function.
4. Graph linear functions.

In this section we introduce relations and functions. As you will learn shortly, a function is a special type of relation. Functions are a common thread in mathematics courses from algebra through calculus. In this section we give an informal introduction to relations and functions. The information given here will prove very valuable to you if you plan on taking additional mathematics courses. We will also be discussing functions in Chapter 10, when we graph quadratic equations.

Recall from Section 1.3 that a set is a collection of **elements.** In this section the elements will be numbers or ordered pairs. Remember that sets are indicated by using braces, { }. For example, in the set {1, 2, 3, 4, 5} the elements are 1, 2, 3, 4, and 5. Three dots at the end of a set indicate that the set continues in the same manner. For example, {1, 2, 3, 4, 5, . . . } represents the set of all natural numbers.

Relations

1 Now we will discuss relations. A **relation** is any set of ordered pairs. A relation may be indicated by (1) an equation in two variables, (2) a set of ordered pairs, or (3) a graph. Consider the equation $y = x + 2$, where x is an integer between 1 and 4 inclusive. This equation and every other equation in two variables is a relation. We can obtain some ordered pairs that satisfy the equation by selecting values for x and finding the corresponding values of y.

	$y = x+2$	*Ordered Pair*
Let $x = 1$	$y = 1 + 2 = 3$	(1, 3)
Let $x = 2$	$y = 2 + 2 = 4$	(2, 4)
Let $x = 3$	$y = 3 + 2 = 5$	(3, 5)
Let $x = 4$	$y = 4 + 2 = 6$	(4, 6)

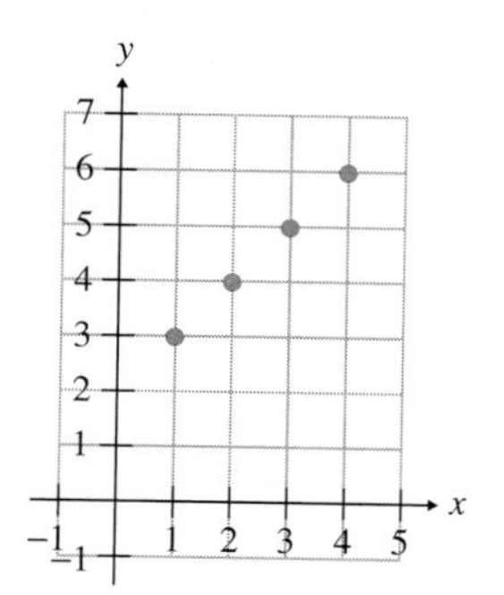

FIGURE 7.52

The set of ordered pairs {(1, 3), (2, 4), (3, 5), (4, 6)} is a relation. In fact, every set of ordered pairs is a relation. If we plot the set of ordered pairs, we get the graph in Figure 7.52. The graph in Figure 7.52 is a relation. In fact, every graph drawn in the Cartesian coordinate system is a relation. As you can see, the word *relation* is a general term that can be used to describe any relationship between two variables.

Suppose that you are buying navel oranges at the supermarket. The cost of each orange is 20¢. If you purchase 1 orange, your cost will be 20¢. If you purchase 2 oranges, your cost will be 2(20¢), or 40¢. Three oranges cost 3(20¢), or 60¢, and so on (see Table 7.1). In general, if you purchase n oranges your cost will be 20¢ times the number of oranges. We can express this relationship with the equation $c = 0.20n$, where c is the cost, in dollars, of n oranges. In the equation $c = 0.20n$ the cost, c, depends on the number of oranges, n. Thus, we call c the **dependent variable** and n the **independent variable.**

TABLE 7.1

Number of Oranges, n	Cost in Dollars, c
0	0.00
1	0.20
2	0.40
3	0.60
.	.
.	.
.	.
n	$0.20n$

Domain and Range

2 In a relation the set of values that can be used for the independent variable is called the **domain.** The set of values that represent the dependent variable is called the **range**. Since the values in the domain are substituted into the equation, the values in the domain determine the values in the range. Consider the equation for the cost of n oranges, $c = 0.20n$. What is the domain and what is the range? The domain is the set of "input values" that can be used for n, the number of oranges. Since we cannot purchase a fractional part of an orange, or a negative number of oranges, the domain is the set of whole numbers {0, 1, 2, 3, 4, . . . }. We can purchase from 0 oranges to any fixed number of oranges. The three dots at the end of the set indicate that the set continues indefinitely. Notice that the numbers in the left column of Table 7.1 are the numbers that made up the domain.

When the values in the domain 0, 1, 2, 3, . . . are substituted for n in the formula $c = 0.20n$, the values we get are 0.00, 0.20, 0.40, 0.60, ... The set of these "output values" form the range. Thus the range is {0.00, 0.20, 0.40, 0.60, 0.80, . . . }. Notice that the values in the right column of Table 7.1 are the numbers that make up the range.

If we list the values in Table 7.1 as a set of ordered pairs, we get {(0, 0.00), (1, 0.20), (2, 0.40), (3, 0.60), . . . }. Note that the *domain is the set of first coordinates in the set of ordered pairs,* and the *range is the set of second coordinates in the set of ordered pairs.*

If we refer back to the equation $y = x + 2$, where x is an integer between 1 and 4 inclusive, we see that the domain, the set of values of x, is $\{1, 2, 3, 4\}$. The range, the set of values of y, is $\{3, 4, 5, 6\}$.

When a graph is given, its domain and range may be determined by observation. The domain is the set of values of x, the first coordinate in each ordered pair, and the range is the set of values of y, the second coordinate in each ordered pair.

EXAMPLE 1 State the domain and the range of the relation shown in Figure 7.53.

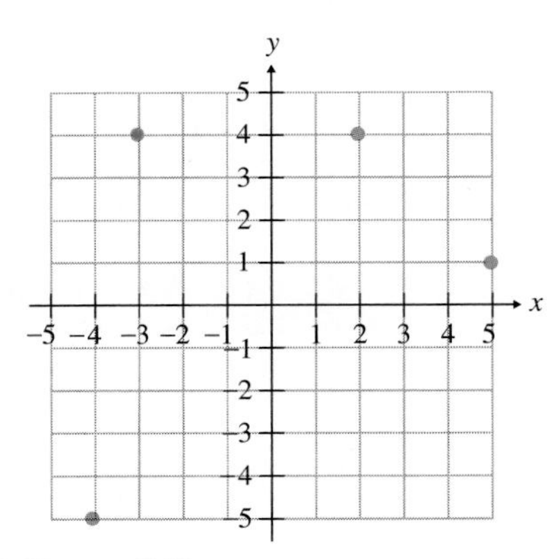

FIGURE 7.53

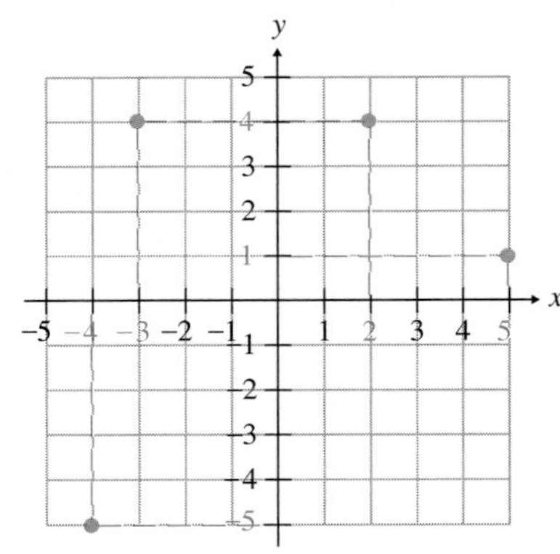

FIGURE 7.54

Solution: The domain is the set of values of x, which are indicated in red on the x axis in Figure 7.54. The domain is $\{-4, -3, 2, 5\}$. The range is the set of values of y, which are indicated in blue on the y axis in Figure 7.54. The range is $\{-5, 1, 4\}$.

In our discussions thus far the domains have been limited to integer values. In Example 2 we present a relation whose domain is not limited to integer values.

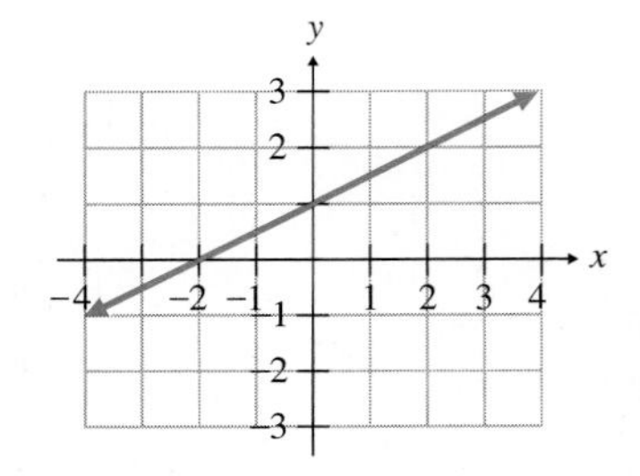

FIGURE 7.55

EXAMPLE 2 Determine the domain and range of the relation shown in Figure 7.55.

Solution: The arrowheads on the line indicate that the graph continues indefinitely in both directions. From the graph we can see that any real value of x can be used in the domain. Therefore, we say that the domain is *the set of real numbers*. Since the symbol $\mathbb{R}$ is used to represent the set of real numbers, we can write, domain: $\mathbb{R}$. Similarly, every value of y will be included in the range, and the range is also *the set of real numbers*. Therefore, we may write, range: $\mathbb{R}$.

Functions

3 Now we discuss functions. A **function** is a special type of relation. For a relation to be a function, each first coordinate in the set of ordered pairs must have a *unique* second coordinate. Is the set of ordered pairs $\{(4, 5), (3, 2), (-2, -3), (2, 5), (1, 6)\}$ a function? Do any of the ordered pairs have the same first coordinates and a different second coordinate? If so, this relation would not be a function. *Since no two ordered pairs have the same first coordinate and a different second coordinate, this set of ordered pairs is a function.* Now consider the set of ordered pairs $\{(4, 5), (3, 2), (-2, -3), (4, 1), (5, -2)\}$. Is this set of ordered pairs a function? Since the two ordered pairs (4, 5) and (4, 1) have the same first coordinate and a different second

coordinate, this set of ordered pairs is *not a function.* Note that the second coordinates in the set of ordered pairs may repeat. However, the first coordinates cannot repeat if the set of ordered pairs is to be a function.

Function

A function is a relation in which no two ordered pairs have the same first coordinate and a different second coordinate.

A function may also be defined as a relation in which each element of the domain corresponds to exactly one element in the range. In other words, each value of x must correspond to a unique value of y. This may sound a bit confusing but the following graph can help in your understanding. Let us graph the two sets of ordered pairs $\{(4, 5), (3, 2), (-2, -3), (2, 5), (1, 6)\}$ and $\{(4, 5), (3, 2), (-2, -3), (4, 1), (5, -2)\}$ on different axes (Fig. 7.56 (a) and (b) respectively).

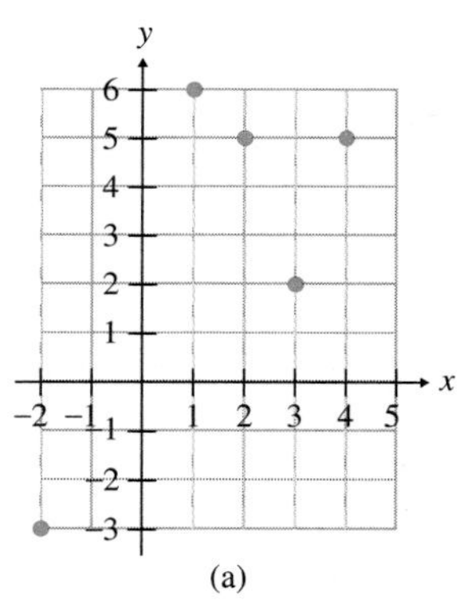

(a)

First set of ordered pairs
Function

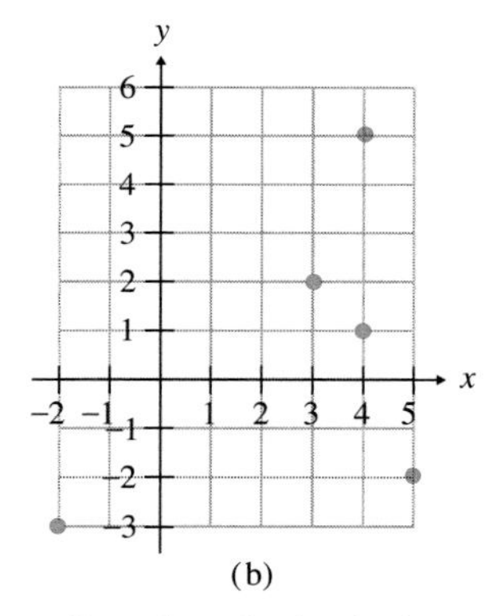

(b)

Second set of ordered pairs
Not a Function

FIGURE 7.56

Notice that in Figure 7.56a, if a vertical line is drawn through each point, no vertical line intersects more than one point. This indicates that no two ordered pairs have the same first (or x) coordinate. This also indicates that for each element in the domain (the values of x) there is a unique element in the range (the values of y). Thus, each value of x corresponds to a unique value of y. Therefore, this set of points is a function.

Now look at Figure 7.56b. If a vertical line is drawn through each point, one vertical line passes through two points. The red vertical line intersects both (4, 5) and (4, 1). This indicates that there are two ordered pairs that have the same first (or x) coordinate and a different second coordinate. Therefore, this set of ordered pairs is *not* a function. Also notice that each element in the domain *does not* correspond to a unique element in the range. The number 4 in the domain corresponds to two numbers, 5 and 1, in the range.

To determine if a graph is a function, we can use the **vertical line test** which we just introduced. **If a vertical line can be drawn through any part of a graph and the vertical line intersects another part of the graph, then each value of x does not have a unique value of y and the graph is not a function. If a vertical line cannot be drawn to intersect the graph at more than one point, each value of x has a unique value of y, and the graph is a function.**

EXAMPLE 3

Using the vertical line test, determine whether or not the following graphs are functions.

(a) **(b)** **(c)**

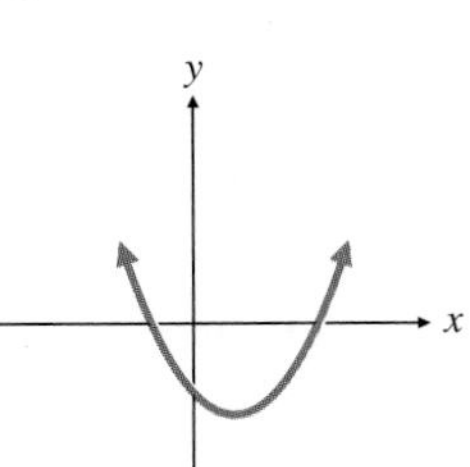
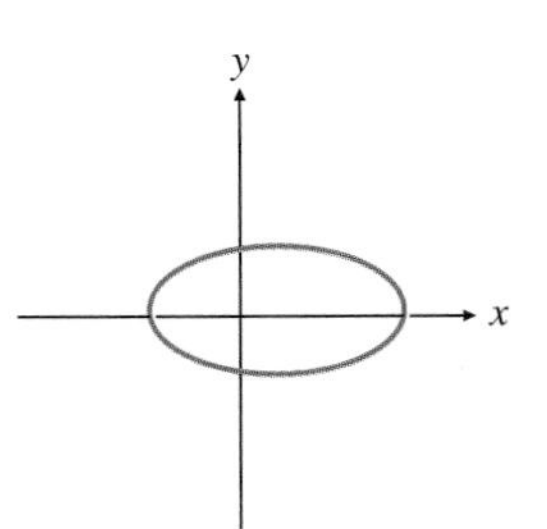

Solution:

(a) **(b)** **(c)**

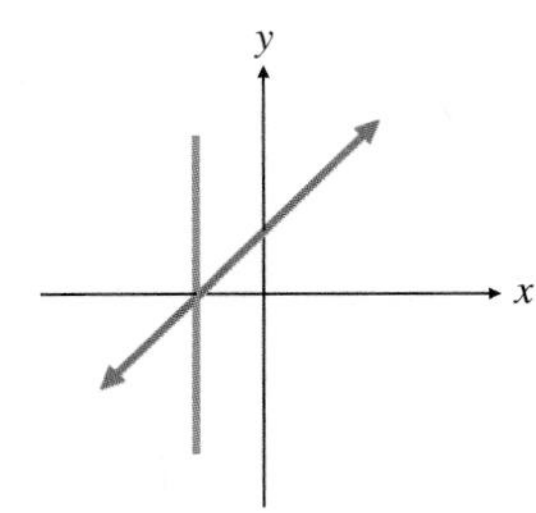
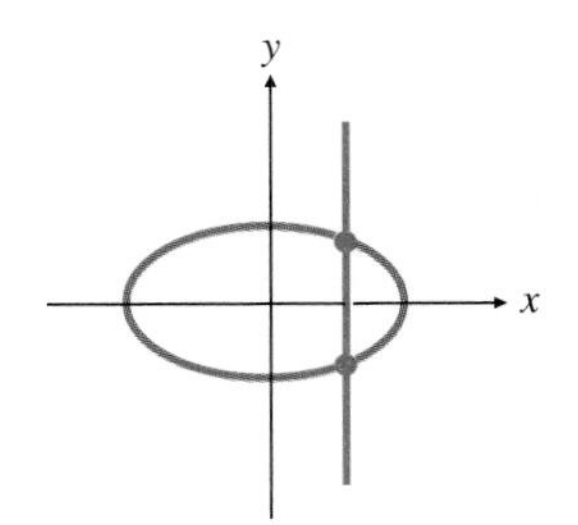

The graphs in parts **(a)** and **(b)** are functions since it is not possible to draw a vertical line that intersects the graph at more than one point. The graph in part **(c)** is not a function since a vertical line can be drawn to intersect the graph at more than one point.

Let us return to the equation $y = x + 2$, where x is an integer in the set $\{1, 2, 3, 4\}$. Study its graph (Fig. 7.52). This relation is a function since its graph passes the vertical line test. Since the graph is a function the equation $y = x + 2$ for $x = 1, 2, 3, 4$ is also a function. Since the value of y in the equation or function depends on the value of x, we say that *y is a function of x*. The notation $y = f(x)$ is used to show that y is a function of the variable x. For this function we can write

$$y = f(x) = x + 2$$

The notation $f(x)$ is read "f of x" and *does not mean f* times x.

To evaluate a function for a specific value of x, substitute that value for x everywhere the x appears in the function. For example, to evaluate the function $f(x) = x + 2$ at $x = 1$, we do the following:

$$y = f(x) = x + 2$$
$$y = f(1) = 1 + 2 = 3$$

Thus, when x is 1, y is 3. When $x = 4$, $y = 6$, as illustrated below.

$$y = f(x) = x + 2$$
$$y = f(4) = 4 + 2 = 6$$

The notation $f(1)$ is read "f of 1" and $f(4)$ is read "f of 4."

Recall that the domain of the function $y = x + 2$ was the integers 1, 2, 3, and 4. Figure 7.57 displays how the function assigns each value of x in the domain to exactly one value of y in the range.

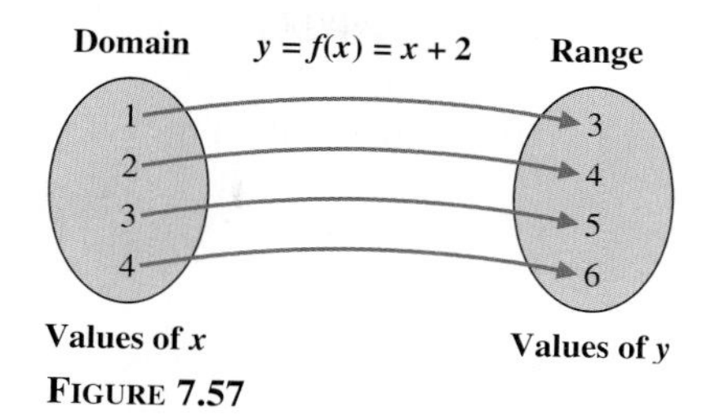

FIGURE 7.57

Sometimes we may use the notation $f(x)$ in place of y. For example, the function $y = x + 2$ may sometimes be written as $f(x) = x + 2$. Always keep in mind that $f(x)$ is the same as y.

EXAMPLE 4 For the function $f(x) = x^2 + 2x - 3$, find **(a)** $f(4)$ and **(b)** $f(-5)$. **(c)** If $x = -1$, determine the value of y.

Solution:

(a) Substitute 4 for each x in the function, and then evaluate.

$$f(x) = x^2 + 2x - 3$$
$$f(4) = 4^2 + 2(4) - 3 = 16 + 8 - 3 = 21$$

(b) $f(x) = x^2 + 2x - 3$

$$f(-5) = (-5)^2 + 2(-5) - 3 = 25 - 10 - 3 = 12$$

(c) Since $y = f(x)$ we evaluate $f(x)$ at -1.

$$f(x) = x^2 + 2x - 3$$
$$f(-1) = (-1)^2 + 2(-1) - 3 = 1 - 2 - 3 = -4$$

Thus, when $x = -1$, $y = -4$.

Graph Linear Functions

4 The graphs of all equations of the form $y = ax + b$ will be straight lines that are functions. Therefore, we may refer to equations of the form $y = ax + b$ as **linear functions.** Equations of the form $f(x) = ax + b$ are also linear functions since $f(x)$ is the same as y. We may graph linear functions as shown in Example 5.

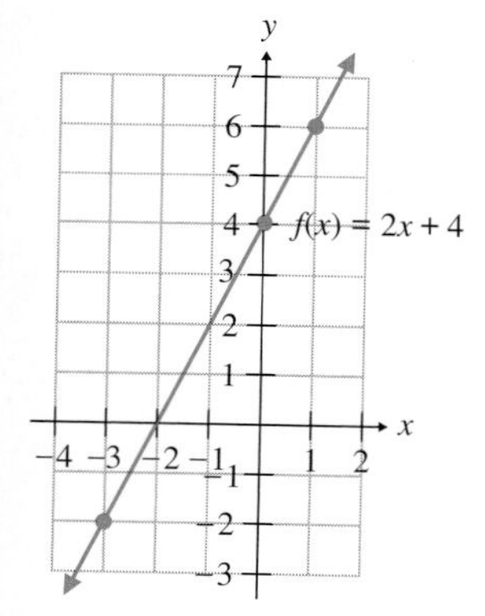

FIGURE 7.58

EXAMPLE 5 Graph $f(x) = 2x + 4$.

Solution: Since $f(x)$ is the same as y, write $y = f(x) = 2x + 4$. Select values for x and find the corresponding values for y or $f(x)$.

$$y = f(x) = 2x + 4$$

$x = -3 \qquad y = f(-3) = 2(-3) + 4 = -2$

$x = 0 \qquad y = f(0) = 2(0) + 4 = 4$

$x = 1 \qquad y = f(1) = 2(1) + 4 = 6$

x	y
−3	−2
0	4
1	6

Now plot the points and draw the graph of the function (Fig. 7.58).

EXAMPLE 6 The weekly profit, p, of an ice skating rink is a function of the number of skaters per week, n. The function approximating the profit is $p = f(n) = 8n - 600$, where $0 \leq n \leq 400$.

(a) Construct a graph showing the relationship between the number of skaters and the weekly profit.
(b) Estimate the profit if there are 200 skaters in a given week.

FIGURE 7.59

Solution:

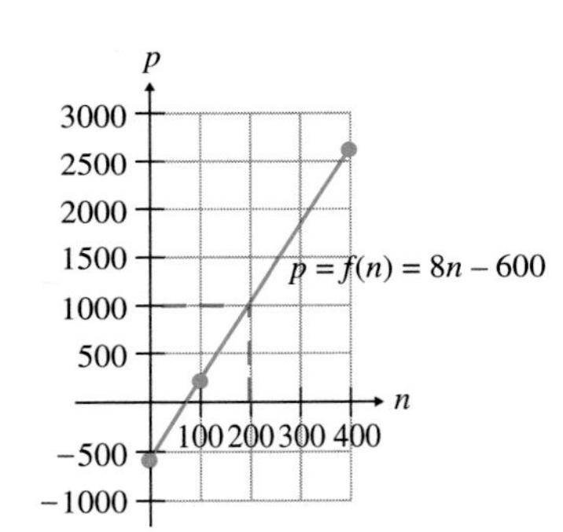

FIGURE 7.59

(a) Select values for n, and find the corresponding values for p. Then draw the graph (Fig. 7.59). Notice there are no arrowheads on the line because the function is defined only for values of n between 0 and 400 inclusive.

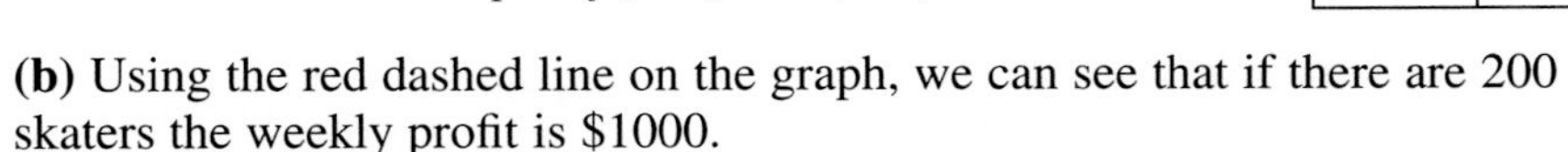

$$p = f(n) = 8n - 600$$

Let $n = 0$ $\quad p = f(0) = 8(0) - 600 = -600$

Let $n = 100$ $\quad p = f(100) = 8(100) - 600 = 200$

Let $n = 400$ $\quad p = f(400) = 8(400) - 600 = 2600$

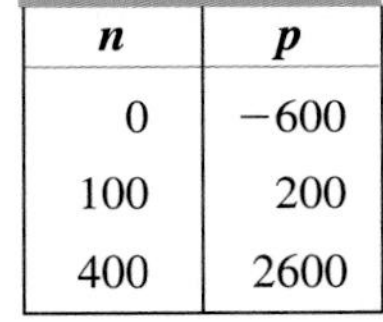

n	p
0	−600
100	200
400	2600

(b) Using the red dashed line on the graph, we can see that if there are 200 skaters the weekly profit is \$1000.

We will discuss functions further in Section 10.4, when graphing quadratic equations.

Exercise Set 7.6

Determine which of the relations are also functions. Give the range and domain of each relation or function.

1. $\{(4, 4), (2, 2), (3, 5), (1, 3), (5, 1)\}$

2. $\{(2, 1), (4, 0), (3, 5), (2, 2), (5, 1)\}$

3. $\{(5, -2), (3, 0), (1, 2), (1, 4), (2, 4), (7, 5)\}$

4. $\{(-2, 1), (1, -3), (3, 4), (4, 5), (-2, 0)\}$

5. $\{(5, 0), (3, -4), (0, -1), (3, 2), (1, 1)\}$

6. $\{(-1, 3), (-3, 4), (0, 3), (5, 2), (3, 5), (2, 5)\}$

7. $\{(4, 0), (0, -3), (1, 5), (1, 0), (1, 2)\}$

8. $\{(4, -3), (3, -7), (4, -9), (3, 5)\}$

9. $\{(0, 3), (1, 3), (2, 3), (3, 3), (4, 3)\}$

10. $\{(3, 5), (2, 4), (1, 0), (0, 1), (-1, 5)\}$

The domain and range of a relation are illustrated. (a) Construct a set of ordered pairs that represent the relation. (b) Determine if the relation is a function. Explain your answer.

11.

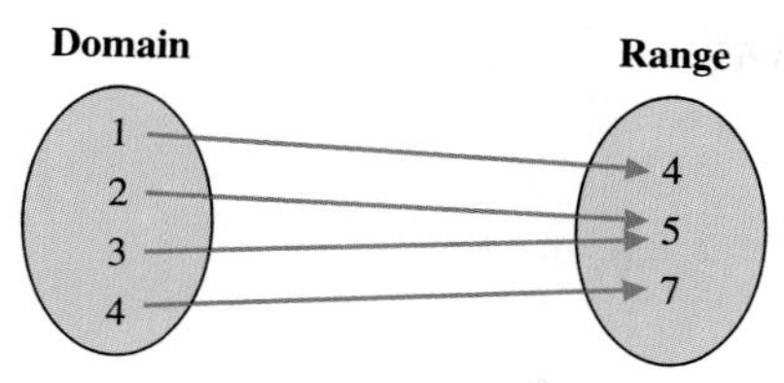

12.

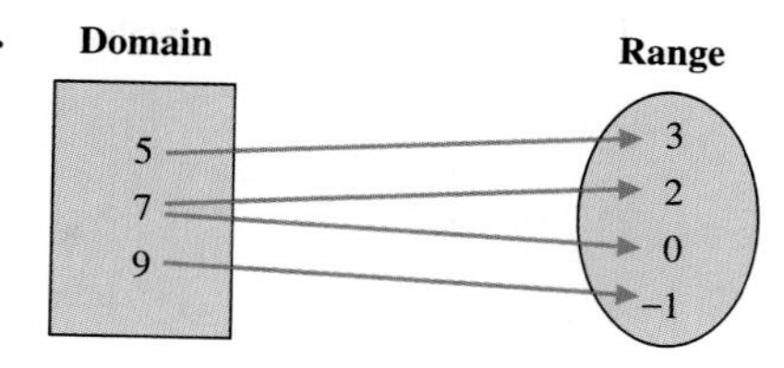

13.

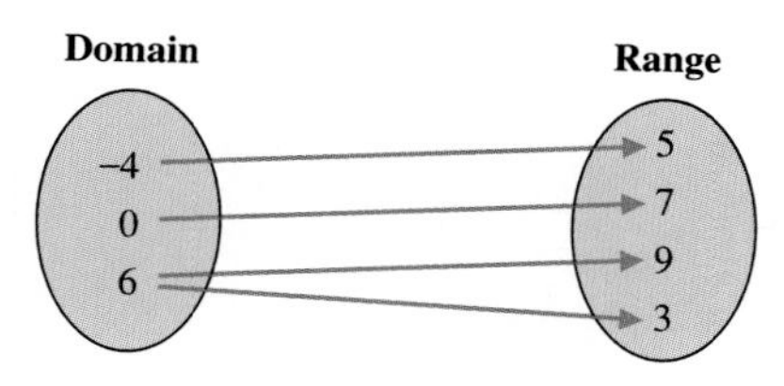

14.

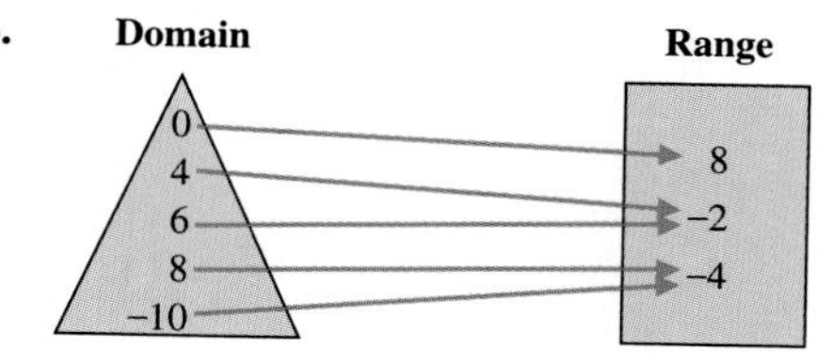

Use the vertical line test to determine if the relation is also a function.

15.

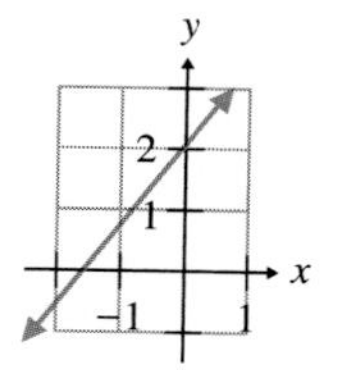

16.

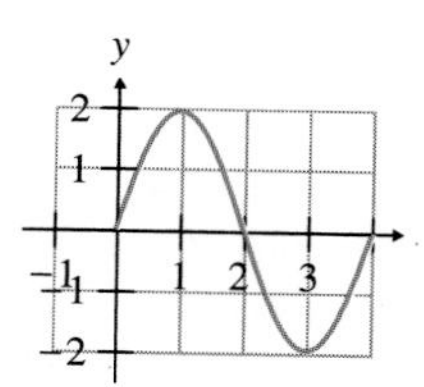

17.

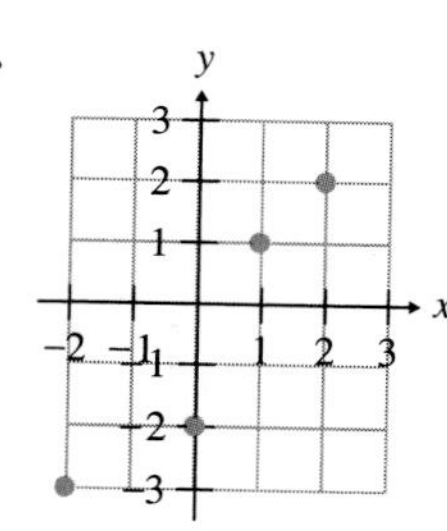

18.

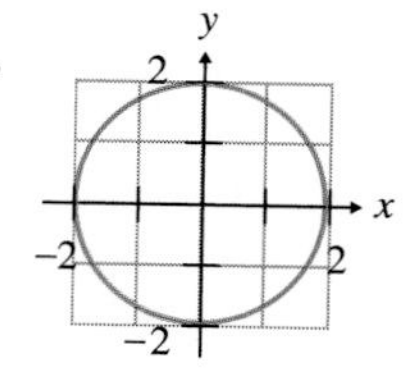

19.

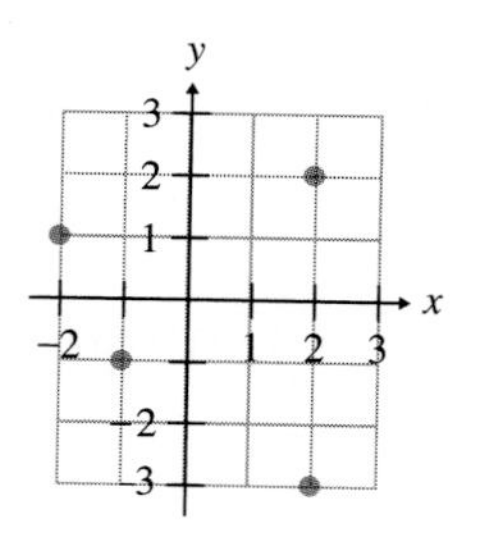

20.

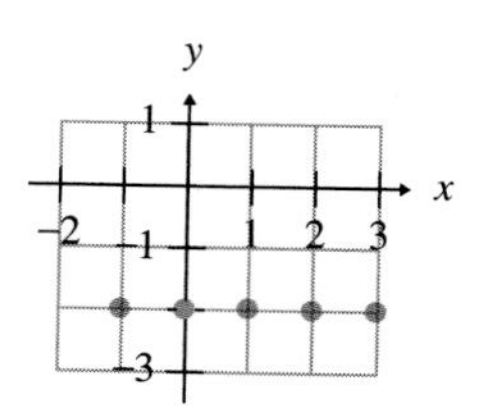

21.

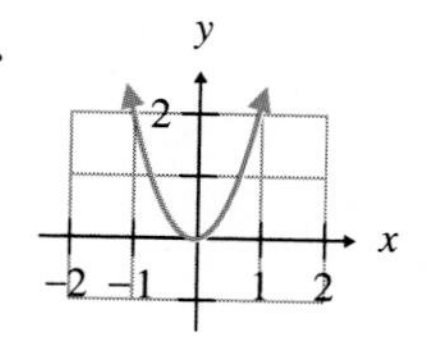

22.

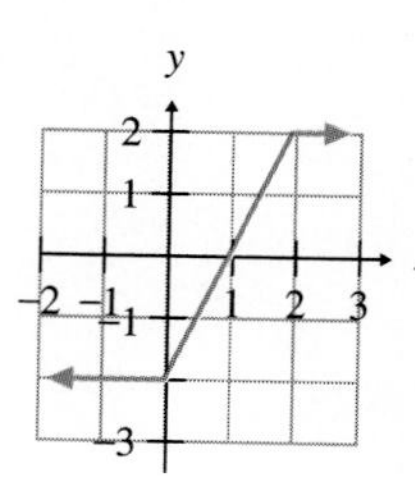

23.

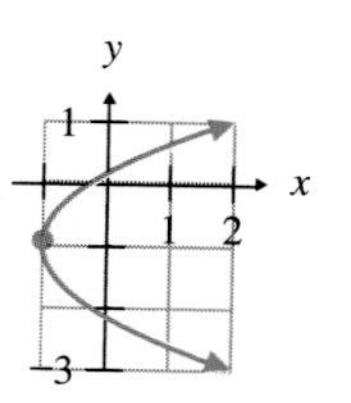

24.

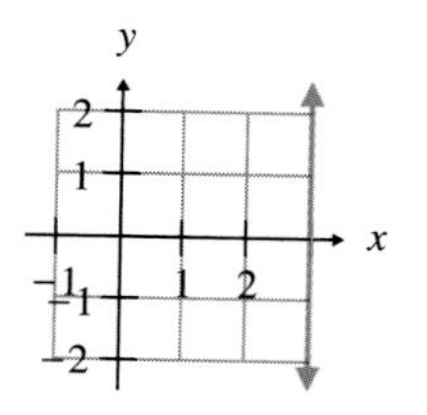

25.

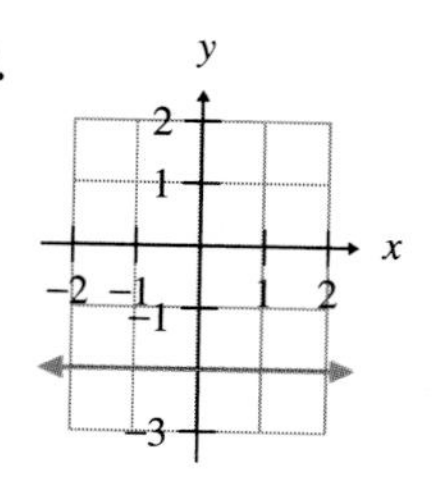

26.

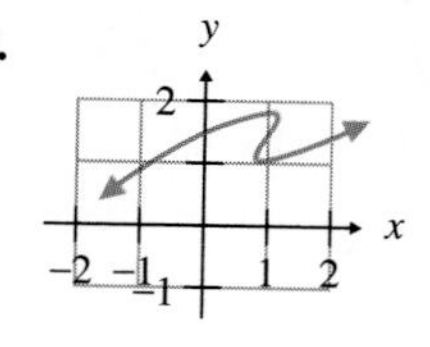

Evaluate the functions at the values indicated.

27. $f(x) = 2x + 3$; find **(a)** $f(2)$ **(b)** $f(-2)$

28. $f(x) = -3x + 5$; find **(a)** $f(0)$ **(b)** $f(1)$

29. $f(x) = x^2 - 4$; find **(a)** $f(2)$ **(b)** $f(3)$

30. $f(x) = 2x^2 + 3x - 4$ **(a)** $f(1)$ **(b)** $f(-3)$

31. $f(x) = 3x^2 - x + 5$ **(a)** $f(0)$ **(b)** $f(2)$

32. $f(x) = \frac{1}{2}x - 4$; find **(a)** $f(10)$ **(b)** $f(-4)$

33. $f(x) = \frac{x + 4}{2}$; find **(a)** $f(2)$ **(b)** $f(6)$

34. $f(x) = \frac{1}{2}x^2 + 6$; find **(a)** $f(2)$ **(b)** $f(-2)$

Graph each function.

35. $f(x) = 2x + 1$

36. $f(x) = -x + 4$

37. $f(x) = 3x - 1$

38. $f(x) = 4x + 2$

39. $f(x) = -2x + 4$

40. $f(x) = -x + 5$

41. $f(x) = -3x - 3$

42. $f(x) = -4x$

43. The cost, c, in dollars, of repairing a highway can be estimated by the formula $c = 2000 + 6000m$, where m is the number of miles to be repaired.

(a) Draw a graph of the function for up to and including 6 miles.

(b) Estimate the cost of repairing 2 miles of road.

See Exercise 43.

44. The cost, c, in dollars, of a cross-country train trip can be estimated by the function $c = 50 + 0.15m$, where m is the distance traveled in miles.

(a) Draw a graph of the function for up to and including 3000 miles traveled.

(b) Estimate the cost of a 1000-mile trip.

45. A discount stock broker commission, c, on stock trades is \$25 plus 2% of the sales value, s. Therefore, the broker's commission is a function of the sales, $c = 25 + 0.02s$.

(a) Draw a graph illustrating the broker's commission on sales up to and including \$10,000.

(b) If the sales value of a trade is \$8000, estimate the broker's commission.

46. A state's auto registration fee, f, is \$20 plus \$15 per 1000 pounds of the vehicle's gross weight. The registration fee is a function of the vehicle's weight, $f = 20 + 0.015w$, where w is the weight of the vehicle in pounds.

(a) Draw a graph of the function for vehicle weights up to and including 10,000 pounds.

(b) Estimate the registration fee of a vehicle whose gross weight is 4000 pounds.

47. A new singing group, The Three Bugs, sign a recording contract with the Squash Record label. Their contract provides them with a signing bonus of \$10,000, plus an 8% royalty on the sales, s, of their new record, Hey Jud! Their income, i, is a function of their sales, $i = 10{,}000 + 0.08s$.

(a) Draw a graph of the function for sales of up to and including \$100,000.

(b) Estimate their income if their sales are \$20,000.

48. A monthly electric bill, m, in dollars, consists of a \$20 monthly fee plus \$0.07 per kilowatthour, k, of electricity used. The amount of the bill is a function of the kilowatthours used, $m = 20 + 0.07k$.

(a) Draw a graph for up to and including 3,000 kilowatthours of electricity used in a month.

(b) Estimate the bill if 1500 kilowatthours of electricity are used.

49. The 1994 Internal Revenue Tax Rate Schedule indicates that the income tax, t, for a single person earning no more than \$22,100 is 15% of their taxable income, i. The tax is a function of their income, $t = 0.15i$.

(a) Draw a graph of the function for taxable incomes up to and including \$22,100.

(b) Estimate the tax if a single person's taxable income is \$15,000.

50. What is a relation?

51. What is a function?

52. What is the domain of a relation or function?

53. What is the range of a relation or function?

54. **(a)** Is every relation a function?

(b) Is every function a relation? Explain your answer.

55. If two distinct ordered pairs in a relation have the same first coordinate, can the relation be a function? Explain.

56. If a relation consists of six ordered pairs and the domain of the relation consists of five values of x, can the relation be a function? Explain.

57. If a relation consists of six ordered pairs and the range of the relation consists of five values of y, can the relation be a function? Explain.

58. In a function is it necessary for each value of y in the range to have a unique value of x in the domain? Explain.

Consider the following graphs. Recall from Section 2.7 that an open circle at the end of a line segment means that the endpoint is not included in the answer. A solid circle at the end of a line segment indicates that the endpoint is included in the answer. Determine whether the following graphs are functions. Explain your answer.

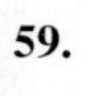

59.

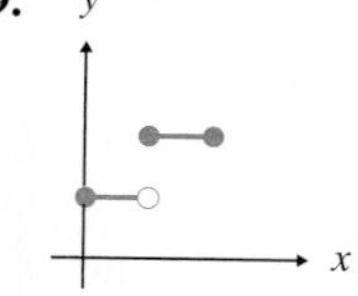

60.

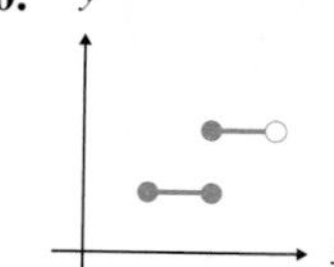

61.

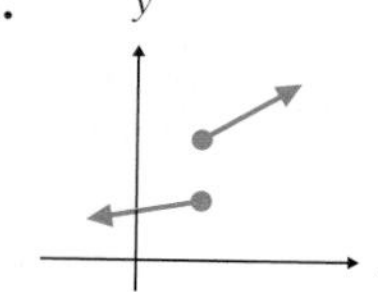

62.

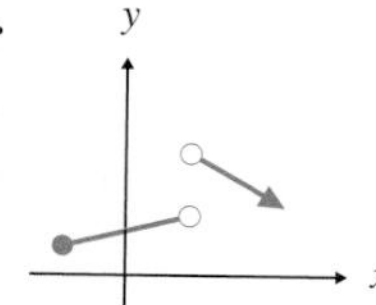

CUMULATIVE REVIEW EXERCISES

63. Solve the equation $3x - 4 = 5(x - 2) - 1$.

64. If it requires \$15 worth of gas to travel 280 miles, how much will it cost to travel 1000 miles?

[2.7] **65.** Solve the inequality $3x + 9 < -x + 5$ and graph the solution on the number line.

[3.1] **66.** Solve the formula $A = 2l + 2w$, for w.

Group Activity/ Challenge Problems

1. Give three real-life examples (different from those already given) of a quantity that is a function of another. Write each as a function, and indicate what each variable represents.
2. $f(x) = \frac{1}{2}x^2 - 3x + 5$; find **(a)** $f\left(\frac{1}{2}\right)$ **(b)** $f\left(\frac{2}{3}\right)$ **(c)** $f(0.2)$
3. $f(x) = x^2 + 2x - 3$; find **(a)** $f(1)$ **(b)** $f(2)$ **(c)** $f(a)$. Explain how you determined your answer to part **(c)**.
4. The cost of mailing a first-class letter at the time of this writing is 32 cents for the first ounce and 23 cents for each additional ounce. A graph showing the cost of mailing a letter first class is pictured on the right.
   (a) Does this graph represent a function? Explain your answer.
   (b) What is the cost of mailing a 4-ounce package first class?
   (c) What is the cost of mailing a 3.6-ounce package first class?

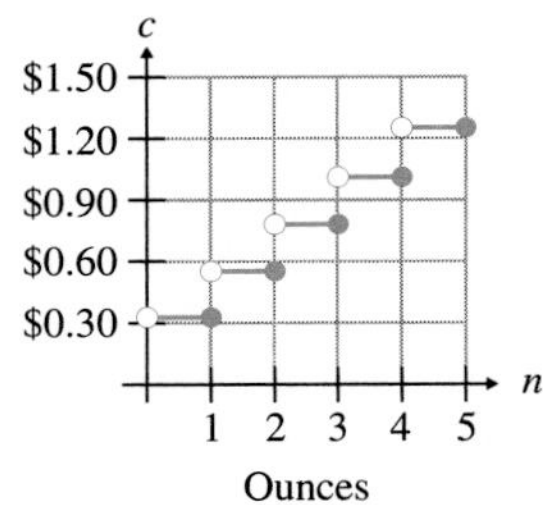

7.7 Graphing Linear Inequalities

Tape 13

1 Graph linear inequalities in two variables.

Graphing Inequalities in Two Variables

1 A linear inequality results when the equal sign in a linear equation is replaced with an inequality sign. Examples of linear inequalities in two variables are

$$3x + 2y > 4 \qquad -x + 3y < -2$$
$$-x + 4y \geq 3 \qquad 4x - y \leq 4$$

To Graph a Linear Inequality

1. Replace the inequality symbol with an equal sign.
2. Draw the graph of the equation in step 1. If the original inequality contained the symbol $\geq$ or $\leq$, draw the graph using a solid line. If the original inequality contained the symbol $>$ or $<$, draw the graph using a dashed line.
3. Select any point not on the line and determine if this point is a solution to the original inequality. If the selected point is a solution, shade the region on the side of the line containing this point. If the selected point does not satisfy the inequality, shade the region on the side of the line not containing this point.

EXAMPLE 1 Graph the inequality $y < 2x - 4$.

Solution: First graph the equation $y = 2x - 4$ (Fig. 7.60). Since the original inequality contains a less than sign, $<$, use a dashed line when drawing the graph. The dashed line indicates that the points on this line are not solutions to the inequality $y < 2x - 4$.

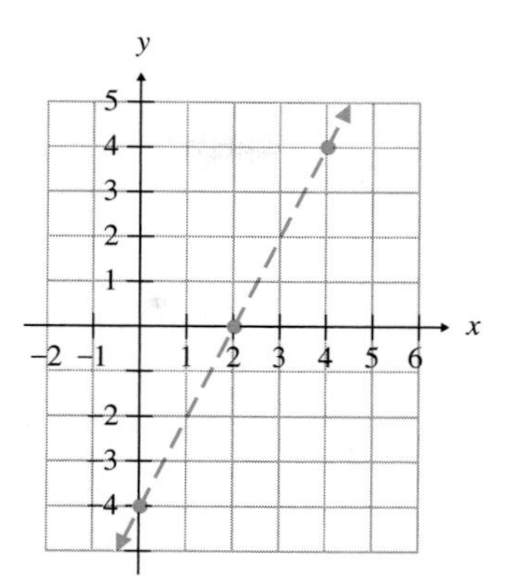

FIGURE 7.60

Next select a point not on the line and determine if this point satisfies the inequality. Often the easiest point to use is the origin, (0, 0).

Check:

$$y < 2x - 4$$
$$0 < 2(0) - 4$$
$$0 < 0 - 4$$
$$0 < -4 \quad \text{false}$$

Since 0 is not less than -4, the point (0, 0) does not satisfy the inequality. The solution will therefore be all the points on the opposite side of the line from the point (0, 0). Shade in this region (Fig. 7.61).

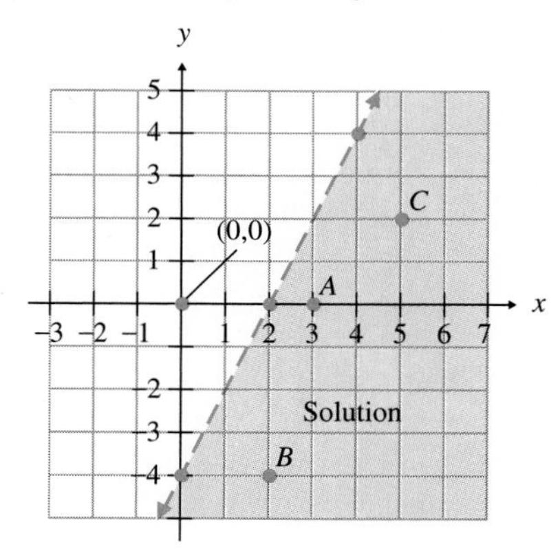

FIGURE 7.61

Every point in the shaded area satisfies the given inequality. Let us check a few selected points A, B, and C.

Point A	*Point B*	*Point C*
(3, 0)	(2, −4)	(5, 2)
$y < 2x - 4$	$y < 2x - 4$	$y < 2x - 4$
$0 < 2(3) - 4$	$-4 < 2(2) - 4$	$2 < 2(5) - 4$
$0 < 2$ true	$-4 < 0$ true	$2 < 6$ true

All points in the shaded area in Figure 7.61 satisfy the inequality $y < 2x - 4$. The points in the unshaded area to the left of the dashed line would satisfy the inequality $y > 2x - 4$.

EXAMPLE 2 Graph the inequality $y \geq -\frac{1}{2}x$.

Solution: Graph the equation $y = -\frac{1}{2}x$. Since the inequality symbol is $\geq$, we use a solid line to indicate that the points on the line are solutions to the inequality (Fig. 7.62).

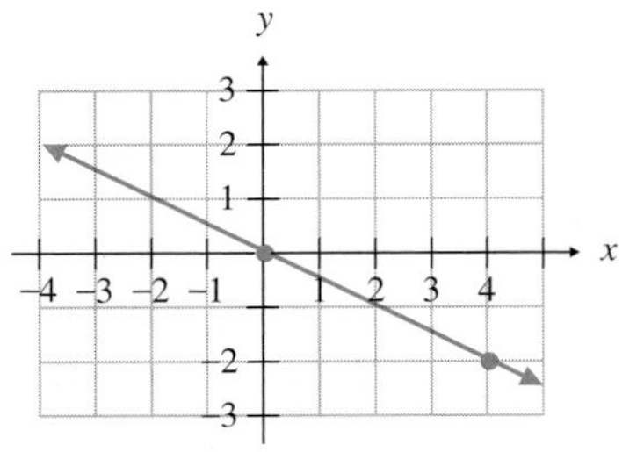

FIGURE 7.62

Since the point (0, 0) is on the line, we cannot select it as our test point. Let us select the point (3, 1).

$$y \geq -\frac{1}{2}x$$

$$1 \geq -\frac{1}{2}(3)$$

$$1 \geq -\frac{3}{2} \qquad \text{true}$$

Since the ordered pair (3, 1) satisfies the inequality, every point on the same side of the line as (3, 1) will also satisfy the inequality $y \geq -\frac{1}{2}x$. Shade this region (Fig. 7.63). Every point in the shaded region as well as every point on the line satisfies the inequality.

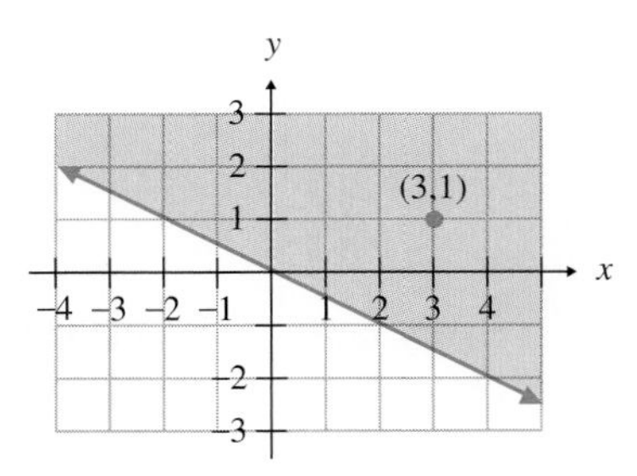

FIGURE 7.63

In some of the exercises you may need to solve the inequality for y before graphing. For example, to graph $-2x + y < -4$ you would solve the inequality for y to obtain $y < 2x - 4$. Then you would graph the inequality $y < 2x - 4$. Note that $y < 2x - 4$ was graphed in Figure 7.61.

Exercise Set 7.7

Graph each inequality.

1. $x > 3$

2. $y < -2$

3. $x \geq \frac{5}{2}$

4. $y < x$

5. $y \geq 2x$

6. $y > -2x$

7. $y < x - 4$

8. $y < 2x + 1$

9. $y < -3x + 4$

10. $y \geq 2x + 4$

11. $y \geq \frac{1}{2}x - 4$

12. $y < 3x + 4$

13. $y \leq \frac{1}{3}x + 3$

14. $y > 4x + 1$

15. $3x + y \leq 5$

16. $3x - 2 < y$

17. $2x + y \leq 3$

18. $3y > 2x - 3$

19. $y - 4 \leq -x$

20. $4x - 2y \leq 6$

21. $2y - 5x < -6$

22. $-2x \leq -3y + 9$

23. $y \geq -4x + 3$

24. $y > -\frac{x}{2} + 2$

25. When graphing inequalities that contain either $\leq$ or $\geq$, explain why the points on the line will be solutions to the inequality.

26. When graphing inequalities that contain either $<$ or $>$, explain why the points on the line will not be solutions to the inequality.

CUMULATIVE REVIEW EXERCISES

[...7] **27.** Solve the inequality $2(x - 3) < 4(x - 2) - 4$ and graph the solution on the number line.

[...] **28.** Use the simple interest formula $i = prt$ to find the principal if the simple interest Manuel gained over a 3-year period at a rate of 8% is $300.

[6.6] **29.** Solve the formula $C = \frac{5}{9}(F - 32)$ for F.

[7.5] **30.** The equation of a line is $6x - 5y = 9$. Find the slope and y intercept of the line.

Group Activity/ Challenge Problems

1. Indicate whether the phrase given means less than, less than or equal to, greater than, or greater than or equal to.
   (a) no more than **(b)** no less than **(c)** at most **(d)** at least
2. Which of the following inequalities have the same graphs? Explain how you determined your answer.
   (a) $2x - y > 4$ **(b)** $-2x + y < -4$ **(c)** $y < 2x - 4$ **(d)** $-2y + 4x < -8$
3. How do the graphs of $2x + 3y > 6$ and $2x + 3y < 6$ differ?
4. Consider the two inequalities $2x + 1 > 5$ and $2x + y > 5$.
   (a) How many variables does the inequality $2x + 1 > 5$ contain?
   (b) How many variables does the inequality $2x + y > 5$ contain?
   (c) What is the solution to $2x + 1 > 5$? Indicate the solution on the number line.
   (d) Graph $2x + y > 5$.

Graph the inequality in the first quadrant only, that is, where $x \geq 0$ *and* $y \geq 0$*.*

5. A toy company must ship x toy cars to one outlet and y toy cars to a second outlet. The maximum number of toy cars that the manufacturer can produce and ship is 200. We can represent this situation with the inequality $x + y \leq 200$. Graph the inequality.
6. An auto dealer wishes to sell x cars and y trucks this year and he needs to sell a total of at least 100 vehicles.
   (a) Represent this situation as an inequality.
   (b) Graph the inequality.
7. Each newly released videotape cost \$2 to rent and each older videotape cost \$1 to rent. James wishes to rent x new and y older tapes, but he can spend no more than \$10.
   (a) Represent this situation with an inequality.
   (b) Graph the inequality.

Summary

GLOSSARY

Bar graph (364): A graph that displays information using either horizontal or vertical bars.

Cartesian (or rectangular) coordinate system (375): Two axes intersecting at right angles that are used when drawing graphs.

Collinear (380): A set of points in a straight line is collinear.

Domain (417): The set of values that can be used for the independent variable in a relation or function.

Function (418): A relation in which no two ordered pairs have the same first coordinate and a different second coordinate.

Graph (379): An illustration of the set of points whose coordinates satisfy an equation or an inequality.

Line graph (365): A graph that displays information using connected line segments. The horizontal axis is usually some measure of time.

Linear equation in two variables (378): An equation that can be written in the form $ax + by = c$.

Negative slope (400): A line has a negative slope if the values of y decrease as the values of x increase.

Origin (376): The point of intersection of the x and y axes.

Parallel lines (408): Lines that never intersect.

Pie (or circle) graph (362): A graph that displays information in the form of a circle.

Positive slope (400): A line has a positive slope if the values of y increase as the values of x increase.

Range (417): The set of values that represent the dependent variable in a relation or function.

Relation (416): Any set of ordered pairs.

Slope of a line (398): The ratio of the vertical change to the horizontal change between any two points on the line.

***x* axis** (376): The horizontal axis in the Cartesian coordinate system.

***x* intercept** (386): The value of x at the point where the graph crosses the x axis.

***y* axis** (376): The vertical axis in the Cartesian coordinate system.

***y* intercept** (386): The value of y at the point where the graph crosses the y axis.

IMPORTANT FACTS

To find the *x* intercept: Set $y = 0$ and find the corresponding value of x.

To find the *y* intercept: Set $x = 0$ and find the corresponding value of y.

Slope of line, *m*, through points (x_1, y_1) and (x_2, y_2):

$$m = \frac{y_2 - y_1}{x_2 - x_1}$$

Methods of Graphing

$$y = 3x - 4$$

By Plotting Points

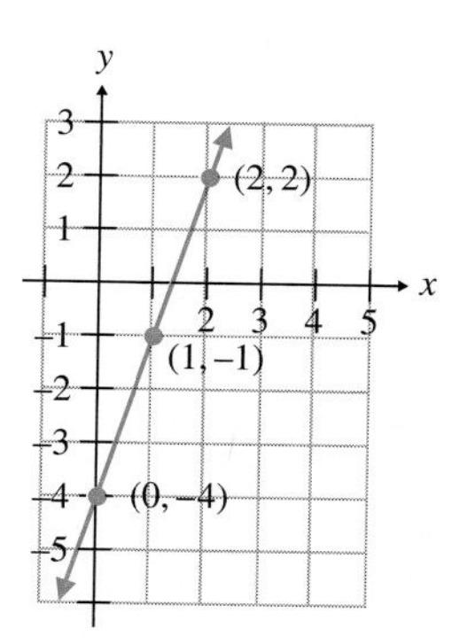

Using Intercepts

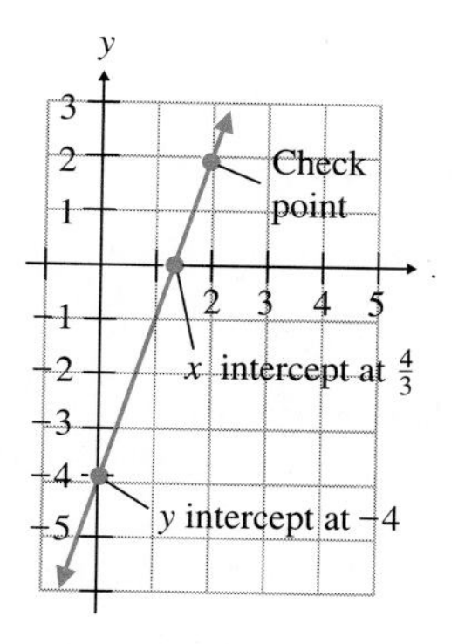

Using Slope and y Intercept

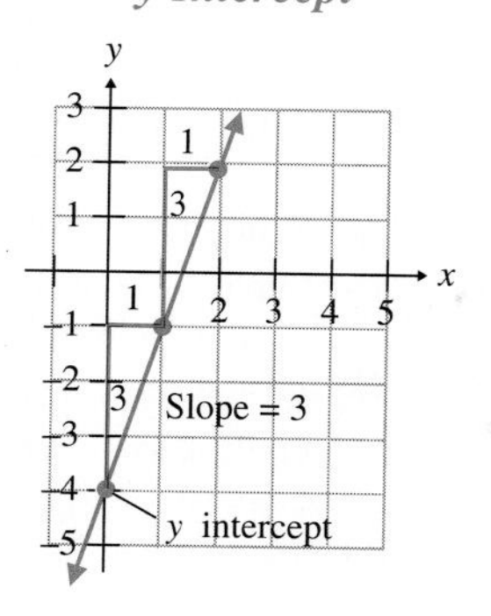

Standard form of a linear equation: $ax + by = c$.

Slope-intercept form of a linear equation: $y = mx + b$.

Point-slope form of a linear equation: $y - y_1 = m(x - x_1)$.

Review of slope

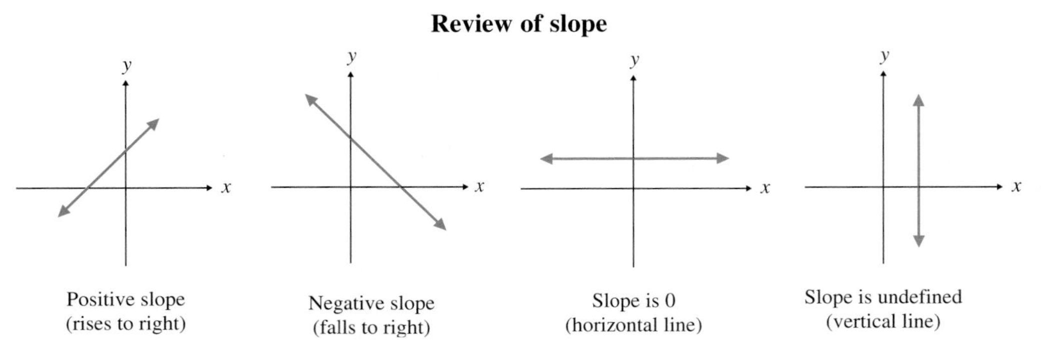

Vertical line test: If a vertical line can be drawn through any part of a graph and the vertical line intersects another part of the graph, the graph is not a function.

Review Exercises

[7.1] **1.** The following pie graph illustrates the percent of dollars charged to credit cards in the United States in 1994.

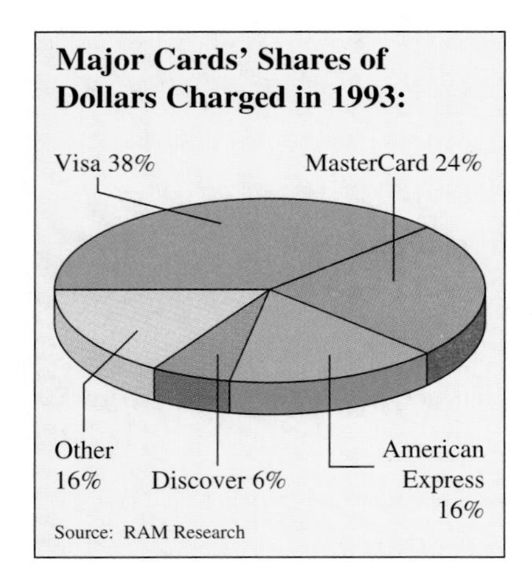

(a) What percent of all charges were charged to Visa?

(b) What percent of all charges were charged to either Visa or MasterCard?

(c) In 1994, $635 billion dollars were charged to credit cards. How much was charged to American Express cards?

(d) Redraw the pie graph using dollar figures rather than percents in the sectors.

2. The following chart shows how consumer shopping habits have changed over the past five years.

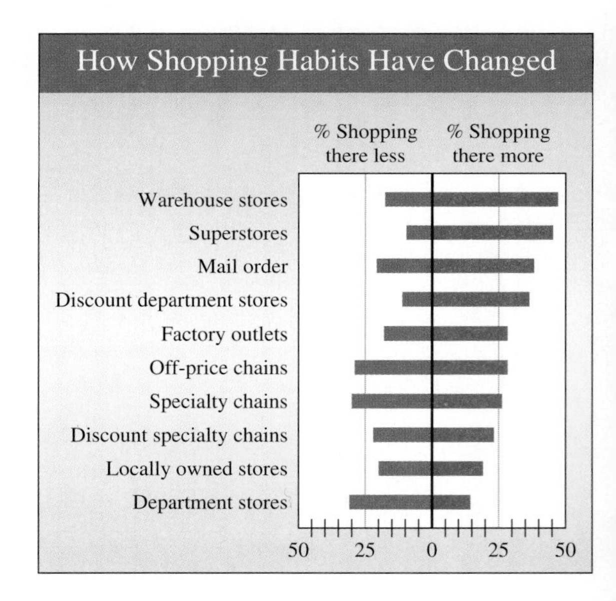

(a) Estimate the percent who said they shopped more at a warehouse store.

(b) Estimate the percent who said they shopped less at a warehouse store.

(c) Overall, is there a greater or smaller percent of people shopping at a warehouse store? List the percent increase or decrease.

(d) Which category had the biggest percent gain overall? Explain your answer.

(e) Describe the general change in our shopping habits from 1987 to 1994.

3. The following line graph indicates the worldwide market share, in percent, of DOS (disk operating system) software, Windows software, and Macintosh software from 1990 to 1994.

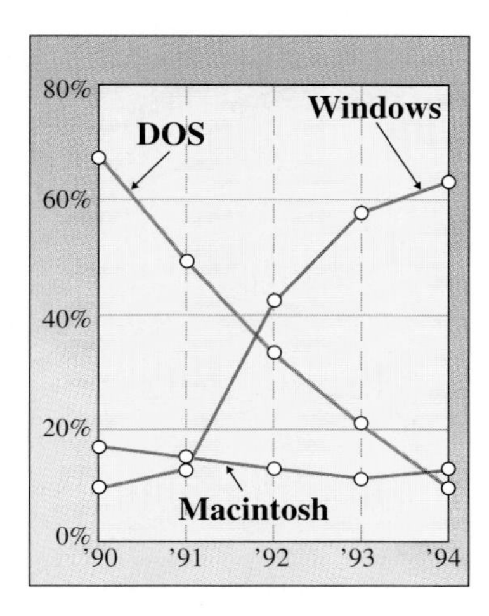

(a) Estimate the market share of software for DOS in 1990 and 1994.

(b) Estimate when the market share of Windows software equaled the market share for DOS.

(c) Estimate the total market share of all software for Windows, DOS, or Macintosh in 1990.

(d) Estimate the total market share of all software for Windows, DOS, or Macintosh in 1994.

4. The following two graphs show changes in the US prison population and the crimes committed since 1979.

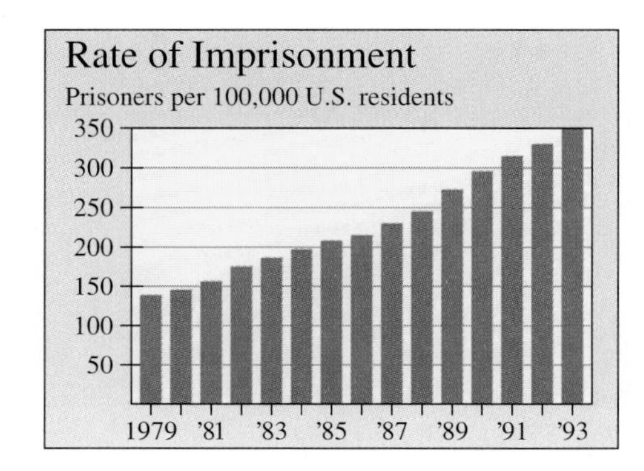

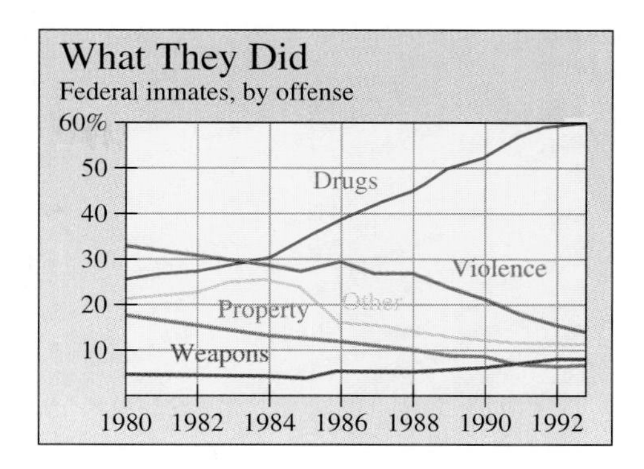

(a) Estimate the rate of imprisonment in 1979 and in 1993, and estimate the increase.

(b) Estimate the percent of inmates imprisoned because of drug charges in 1980 and 1992.

(c) Estimate the sum of the percents for the five categories in the line graph in 1980. Does your answer seem reasonable? Explain.

(d) Repeat part **(c)** for 1992.

(e) Describe the general change in the prison population from 1980 to 1992.

[7.2] **5.** Plot each ordered pair on the same axes.

(a) $A(5, 3)$

(b) $B(0, 6)$

(c) $C(5, \frac{1}{2})$

(d) $D(-4, 3)$

(e) $E(-6, -1)$

(f) $F(-2, 0)$

6. Determine if the points given are collinear: $(0, -4)$, $(6, 8)$, $(-2, 0)$, $(4, 4)$

7. Which of the following ordered pairs satisfy the equation $2x + 3y = 9$?

(a) $(4, 3)$ **(b)** $(0, 3)$

(c) $(-1, 4)$ **(d)** $(2, \frac{5}{3})$

[7.3] **8.** Find the missing coordinate in the following solutions for $3x - 2y = 8$ **(a)** $(2, ?)$ **(b)** $(0, ?)$ **(c)** $(?, 4)$ **(d)** $(?, 0)$

Graph each equation using the method of your choice.

9. $y = 2$

10. $x = -3$

11. $y = 3x$

12. $y = 2x - 1$

13. $y = -3x + 4$

14. $y = -\frac{1}{2}x + 4$

15. $2x + 3y = 6$

16. $3x - 2y = 12$

17. $2y = 3x - 6$

18. $4x - y = 8$

19. $-5x - 2y = 10$

20. $3x = 6y + 9$

21. $25x + 50y = 100$

22. $3x - 2y = 270$

23. $\frac{2}{3}x = \frac{1}{4}y + 20$

[7.4] *Find the slope of the line through the given points.*

24. $(3, -7)$ and $(-2, 5)$

25. $(-4, -2)$ and $(8, -3)$

26. $(-2, -1)$ and $(-4, 3)$

27. What is the slope of a horizontal line?

28. What is the slope of a vertical line?

29. Define the slope of a straight line.

Find the slope of the lines graphed below.

30.

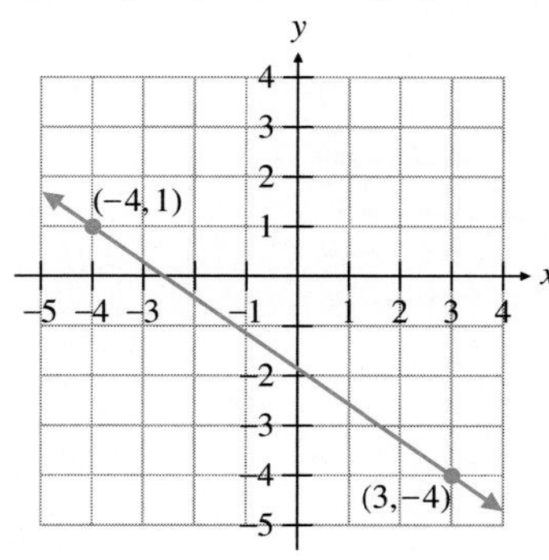

31.

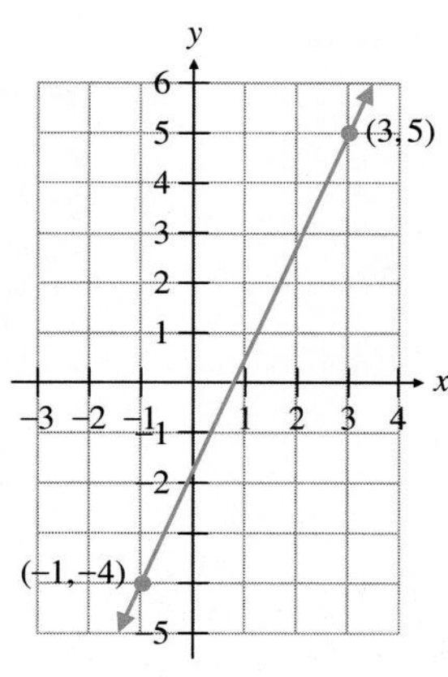

32.

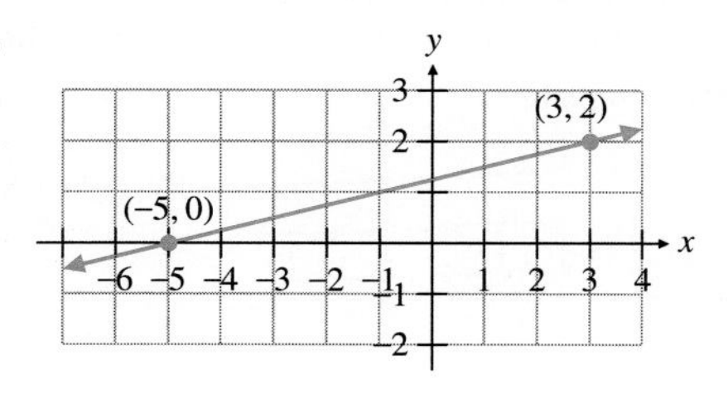

Determine the slope and y intercept of the graph of each equation.

33. $y = -x + 4$

34. $y = -4x + \frac{1}{2}$

35. $2x + 3y = 8$

36. $3x + 6y = 9$

37. $4y = 6x + 12$

38. $3x + 5y = 12$

39. $9x + 7y = 15$

40. $4x - 8 = 0$

41. $3y + 9 = 0$

Write the equation of each line.

42.

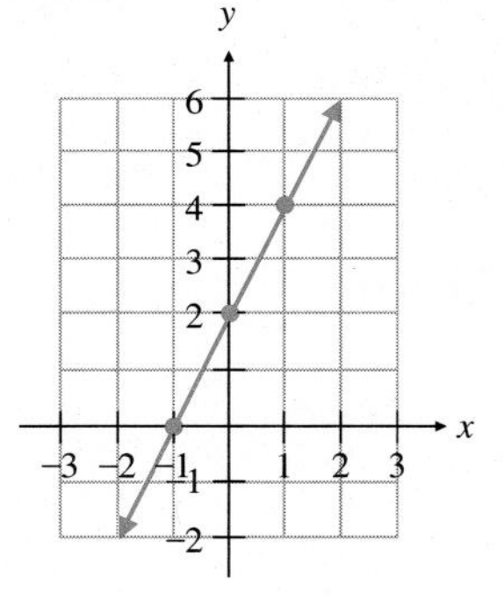

43.

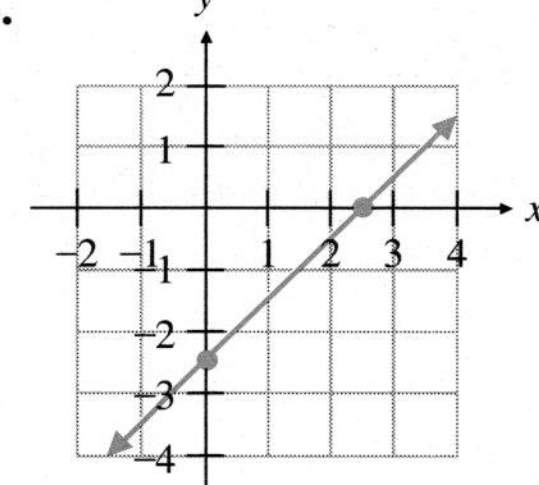

44.

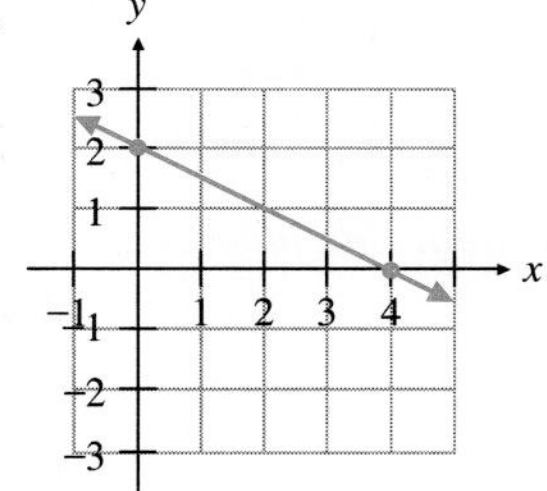

Determine if the two lines are parallel.

45. $y = 3x - 6$
$6y = 18x + 6$

46. $2x - 3y = 9$
$3x - 2y = 6$

47. $y = \frac{4}{9}x + 5$
$4x = 9y + 2$

48. $4x = 6y + 3$
$-2x = -3y + 10$

Find the equation of the line with the properties given.

49. Slope = 2, through (3, 4)

50. Slope = −3, through (−1, 5)

51. Slope = $-\frac{2}{3}$, through (3, 2)

52. Slope = 0, through (4, 2)

53. Slope is undefined, through (3, 5)

54. Slope = −2, y-intercept = −4

55. Through (−2, 3) and (0, −4)

56. Through (−4, −2) and (−4, 3)

Determine which of these relations are also functions. Give the domain and range of each.

57. {(0, 2), (4, −3), (1, 5), (2, −1), (6, 4)}

58. {(3, 1), (4, 2), (4, 5), (6, 1), (7, 0)}

59. {(3, 1), (4, 1), (5, 1), (6, 2), (3, −3)}

60. {(5, −2), (3, −2), (4, −2), (9, −2), (−2, −2)}

The domain and range of a relation are illustrated. ***(a)*** *Construct a set of ordered pairs that represent the relation.* ***(b)*** *Determine if the relation is a function. Explain your answer.*

61. **Domain** **Range**

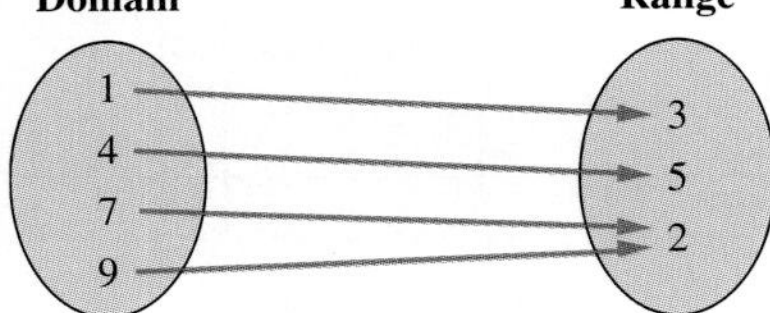

62. **Domain** **Range**

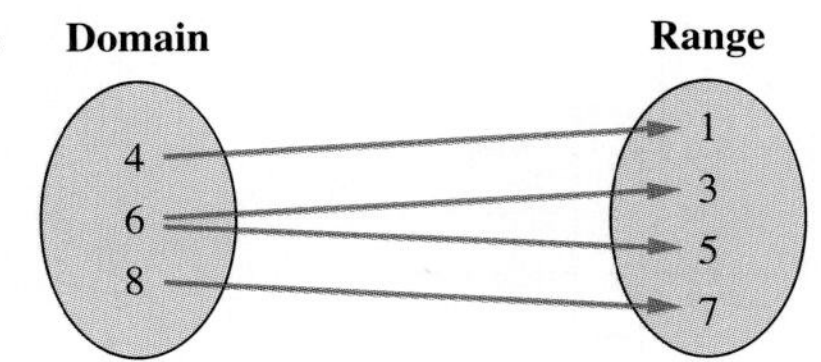

Use the vertical line test to determine if the relation is also a function.

63.

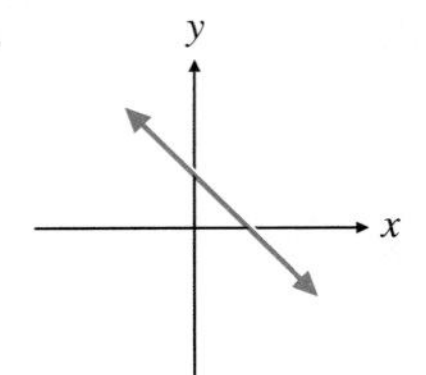

64.

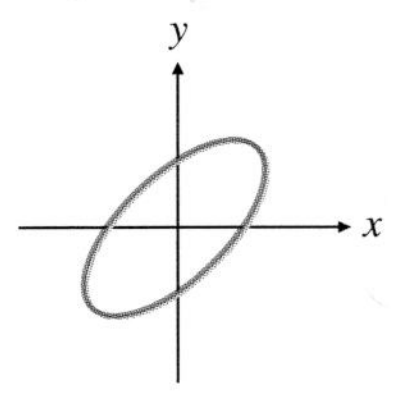

65.

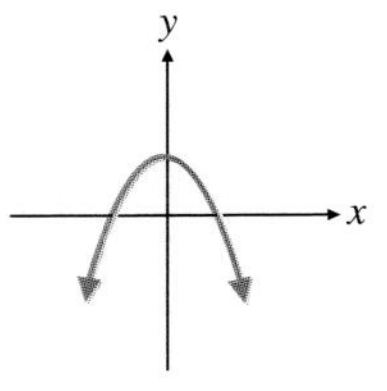

66.

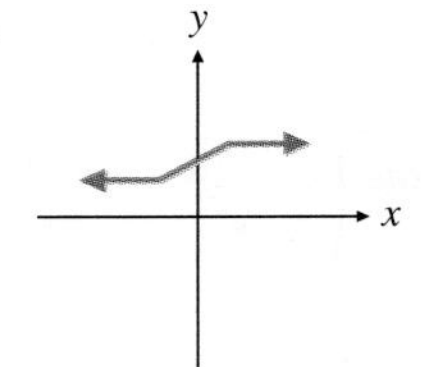

Evaluate the functions at the values indicated.

67. $f(x) = 3x - 4$; find **(a)** $f(2)$ **(b)** $f(-5)$

68. $f(x) = -4x - 5$; find **(a)** $f(-4)$ **(b)** $f(8)$

69. $f(x) = \frac{1}{3}x - 5$; find **(a)** $f(3)$ **(b)** $f(-9)$

70. $f(x) = 2x^2 - 4x + 6$; find **(a)** $f(3)$ **(b)** $f(-5)$

Graph the following functions.

71. $f(x) = 3x - 5$

72. $f(x) = -2x + 3$

73. $f(x) = -4x$

74. $f(x) = -3x - 1$

75. A discount stockbroker charges $25 plus 3 cents per share of stock bought or sold. The broker's commission, c, in dollars, is a function of the number of shares, n, bought or sold, $c = 25 + 0.03n$.

(a) Draw a graph illustrating the broker's commission for up to and including 10,000 shares of stock.

(b) Estimate the commission if 1000 shares of a stock are purchased.

76. The monthly profit, p, of an Everything For A Dollar store can be estimated by the function $p = 4x - 1600$, where x represents the number of items sold.

(a) Draw a graph of the function for up to and including 1000 items sold.

(b) Estimate the profit if 400 items are sold.

[7.7] *Graph each inequality.*

77. $y \geq -3$

78. $x < 4$

79. $y < 3x$

80. $y > 2x + 1$

Graph each inequality.

81. $y \leq 4x - 3$ **82.** $-6x + y \geq 5$ **83.** $y < -x + 4$ **84.** $3y + 6 \leq x$

Practice Test

1. The following bar graph shows the 1994 market shares, in percent, of over-the-counter pain-reliever drugs.

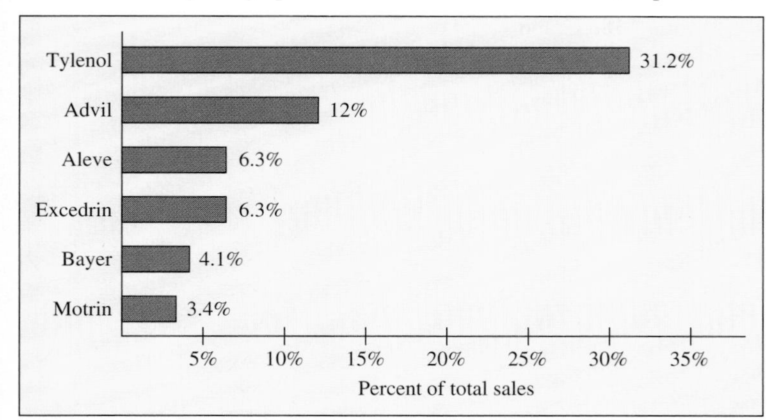

(a) If a bar was to be added for "other," what percent would correspond to it? Explain how you determined your answer.

(b) By what percent does sales of Advil exceed sales of Excedrin?

(c) In 1994 the annual sales for pain relievers was \$2.6 billion. What was the annual dollar sales of Tylenol?

2. Which of the following ordered pairs satisfy the equation $3y = 5x - 9$?

(a) $(3, 2)$ **(b)** $(\frac{9}{5}, 0)$

(c) $(-2, -6)$ **(d)** $(0, 3)$

3. Find the slope of the line through the points $(-4, 3)$ and $(2, -5)$.

4. Find the slope and y intercept of $4x - 9y = 15$.

5. Write an equation of the graph in the accompanying figure.

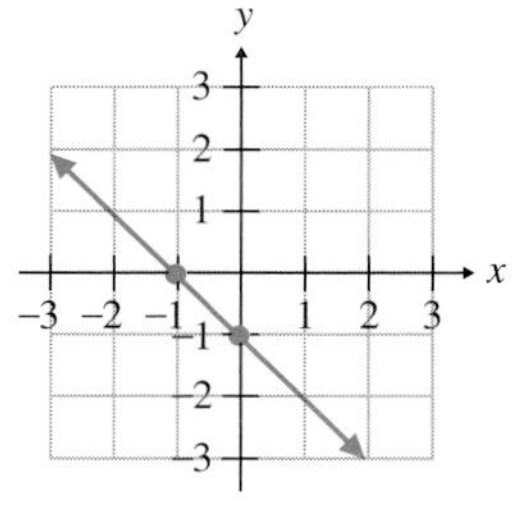

6. Write, in slope–intercept form, an equation of the line with a slope of 3 passing through the point $(1, 3)$.

7. Write, in slope–intercept form, an equation of the line passing through the points $(3, -1)$ and $(-4, 2)$.

8. Determine if the following equations represent parallel lines. Explain how you determined your answer.

$$2y = 3x - 6 \text{ and } y - \frac{3}{2}x = -5.$$

9. Graph $x = -5$. **10.** Graph $y = 3x - 2$.

11. Graph $3x + 5y = 15$. **12.** Graph $3x - 2y = 8$.

13. Define a function.

14. (a) Determine whether the relation that follows is a function. Explain your answer.

$$\{(1, 2), (3, -4), (5, 3), (1, 0), (6, 5)\}$$

(b) Give the domain and range of the relation or function.

15. Determine if the following graphs are functions. Explain how you determined your answer.

(a)

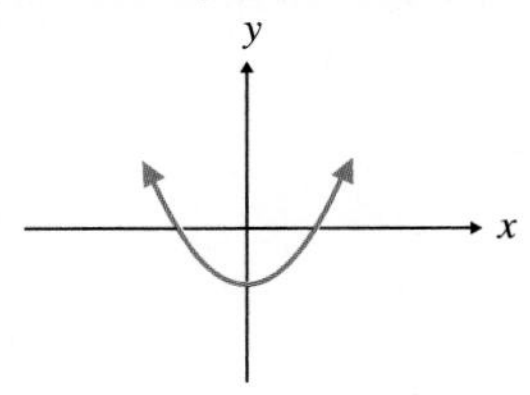

(b)

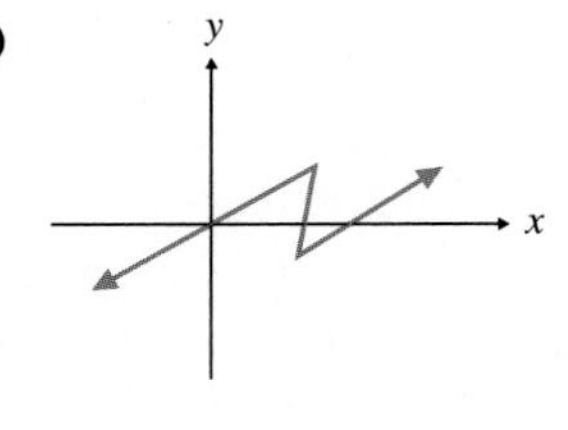

16. Graph the function $f(x) = 2x - 4$.

17. Graph $y \geq -3x + 5$.

18. Graph $y < 4x - 2$.

Chapter 8

Systems of Linear Equations

See Section 8.2, Exercise 23.

Preview and Perspective

In this chapter we learn how to express application problems as systems of linear equations and how to solve systems of linear equations. People in business work with many variables and unknown quantities. For example, a company considers overhead cost, cost of material, labor cost, maximum possible production, selling price of the item, and a host of other items when seeking to maximize their profit. The business may express the relationships between the variables in several equations or inequalities. These equations or inequalities form a *system* of equations or inequalities. The solution of the system of equations or inequalities gives the value of the variables (such as production, price, labor cost) for which the company can maximize profits. Such problems are sometimes called linear programming problems. Computers may be used to solve complex systems of equations but first someone must develop the system of equations.

In this chapter we explain three procedures for solving systems of equations. In Section 8.1 we solve systems using graphs. In Section 8.2 we solve systems using substitution. In Section 8.3 we solve systems using the addition (or elimination) method. In Section 8.4 we explain how to express real life application problems as systems of linear equations. Lastly, in Section 8.5 we build on the graphical solution presented in Section 8.1, and solve systems of linear *inequalities* graphically.

Systems of linear equations may also be solved using matrices, determinants, or graphing calculators. You may solve systems of linear equations using one or more of these techniques in intermediate algebra. In this course we deal only with linear systems of equations in two variables. The graphs of such equations are straight lines. In more advanced courses, you may solve systems of nonlinear equations. In a nonlinear system the graph of at least one of the equations is not a straight line.

To be successful with Section 8.1, solving systems of equations graphically, you need to understand the procedure for graphing straight lines presented in Sections 7.3 through 7.5. To be successful with Section 8.5, systems of inequalities, you need to understand how to graph linear inequalities, which was presented in Section 7.7.

8.1 Solving Systems of Equations Graphically

Tape 13

1. Determine whether or not an ordered pair is a solution to a system of equations.
2. Determine if a system of equations is consistent, inconsistent, or dependent.
3. Solve a system of equations graphically.

Solutions to a System of Linear Equations

1 When we seek a common solution to two or more linear equations, the equations are called **simultaneous linear equations** or a **system of linear equations.** An example of a system of linear equations follows:

$$\left.\begin{aligned} (1)\ y &= x + 5 \\ (2)\ y &= 2x + 4 \end{aligned}\right\} \quad \text{system of linear equations}$$

The **solution to a system of equations** is the ordered pair or pairs that satisfy all equations in the system. The solution to the system above is (1, 6).

Check:	*In Equation (1)*	*In Equation (2)*
	(1, 6)	(1, 6)
	$y = x + 5$	$y = 2x + 4$
	$6 = 1 + 5$	$6 = 2(1) + 4$
	$6 = 6$ true	$6 = 6$ true

Because the ordered pair (1, 6) satisfies *both* equations, it is a solution to the system of equations. Notice that the ordered pair (2, 7) satisfies the first equation but does not satisfy the second equation.

Check:	*In Equation (1)*	*In Equation (2)*
	(2, 7)	(2, 7)
	$y = x + 5$	$y = 2x + 4$
	$7 = 2 + 5$	$7 = 2(2) + 4$
	$7 = 7$ true	$7 = 8$ false

Since the ordered pair (2, 7) does not satisfy both equations, it is *not* a solution to the system of equations.

EXAMPLE 1 Determine which of the following ordered pairs satisfy the system of equations.

$$y = 2x - 8$$
$$2x + y = 4$$

(a) $(2, -4)$ **(b)** $(4, -4)$ **(c)** $(3, -2)$

Solution:

(a) Substitute 2 for x and -4 for y in each equation.

$y = 2x - 8$	$2x + y = 4$
$-4 = 2(2) - 8$	$2(2) + (-4) = 4$
$-4 = 4 - 8$	$4 - 4 = 4$
$-4 = -4$ true	$0 = 4$ false

Since $(2, -4)$ does not satisfy both equations, it is not a solution to the system of equations.

(b) Substitute 4 for x and -4 for y in each equation.

$y = 2x - 8$	$2x + y = 4$
$-4 = 2(4) - 8$	$2(4) + (-4) = 4$
$-4 = 8 - 8$	$8 - 4 = 4$
$-4 = 0$ false	$4 = 4$ true

Since $(4, -4)$ does not satisfy both equations it is not a solution to the system of equations.

(c) Substitute 3 for x and -2 for y in each equation.

$$\begin{aligned} y &= 2x - 8 \\ -2 &= 2(3) - 8 \\ -2 &= 6 - 8 \\ -2 &= -2 \quad \text{true} \end{aligned} \qquad \begin{aligned} 2x + y &= 4 \\ 2(3) + (-2) &= 4 \\ 6 - 2 &= 4 \\ 4 &= 4 \quad \text{true} \end{aligned}$$

Since $(3, -2)$ satisfies both equations, it is a solution to the system of linear equations.

In this chapter we discuss three methods for finding the solution to a system of equations: the graphical method, the substitution method, and the addition method. In this section we discuss the graphical method.

Consistent, Inconsistent and Dependent Systems

2 The **solution to a system of linear equations** is the ordered pair (or pairs) common to all lines in the system when the lines are graphed. When two lines are graphed, three situations are possible, as illustrated in Figure 8.1.

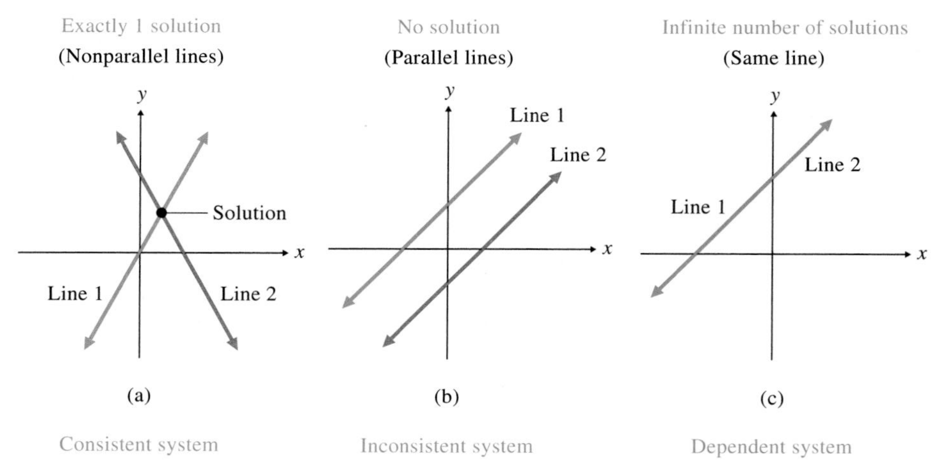

FIGURE 8.1

In Figure 8.1a, lines 1 and 2 are not parallel lines. They intersect at exactly one point. This system of equations has *exactly one solution.* This is an example of a **consistent system of equations.** A consistent system of equations is a system of equations that has a solution.

In Figure 8.1b, lines 1 and 2 are two different parallel lines. The lines do not intersect, and this system of equations has *no solution.* This is an example of an **inconsistent system of equations.** An inconsistent system of equations is a system of equations that has no solution.

In Figure 8.1c, lines 1 and 2 are actually the same line. In this case, every point on the line satisfies both equations and is a solution to the system of equations. This system has *an infinite number of solutions.* This is an example of a **dependent system of equations.** A dependent system of linear equations is a system of equations that has an infinite number of solutions. If a system of two linear equations is dependent, then both equations represent the same line. *Note that a dependent system is also a consistent system since it has a solution.*

We can determine if a system of linear equations is consistent, inconsistent, or dependent by writing each equation in slope–intercept form and comparing the slopes and y intercepts. If the slopes of the lines are different (Fig. 8.1a), the system is consistent. If the slopes are the same but the y intercepts are different (Fig. 8.1b), the system is inconsistent. If both the slopes and the y intercepts are the same (Fig. 8.1c), the system is dependent.

EXAMPLE 2 Determine whether the system is consistent, inconsistent, or dependent.

$$3x + 4y = 8$$
$$6x + 8y = 4$$

Solution: Write each equation in slope–intercept form and then compare the slopes and the y intercepts.

$$\begin{aligned} 3x + 4y &= 8 \\ 4y &= -3x + 8 \\ y &= \frac{-3x + 8}{4} \\ y &= \frac{-3}{4}x + 2 \end{aligned}$$

$$\begin{aligned} 6x + 8y &= 4 \\ 8y &= -6x + 4 \\ y &= \frac{-6x + 4}{8} \\ y &= \frac{-6}{8}x + \frac{4}{8} \\ y &= \frac{-3}{4}x + \frac{1}{2} \end{aligned}$$

Since the equations have the same slope, $-\frac{3}{4}$, and different y intercepts, the lines are parallel. This system of equations is therefore inconsistent and has no solution.

Solve Systems Graphically

3

To obtain the solution to a system of equations graphically, graph each equation and determine the point or points of intersection.

EXAMPLE 3 Solve the following system of equations graphically.

$$2x + y = 11$$
$$x + 3y = 18$$

Solution: Find the x and y intercepts of each graph; then draw the graphs.

$2x + y = 11$	*Ordered Pair*	$x + 3y = 18$	*Ordered Pair*
Let $x = 0$; then $y = 11$	$(0, 11)$	Let $x = 0$; then $y = 6$	$(0, 6)$
Let $y = 0$; then $x = \frac{11}{2}$	$\left(\frac{11}{2}, 0\right)$	Let $y = 0$; then $x = 18$	$(18, 0)$

The two graphs (Fig. 8.2) appear to intersect at the point (3, 5). The point (3, 5) may be the solution to the system of equations. To be sure, however, we must check to see that (3, 5) satisfies *both* equations.

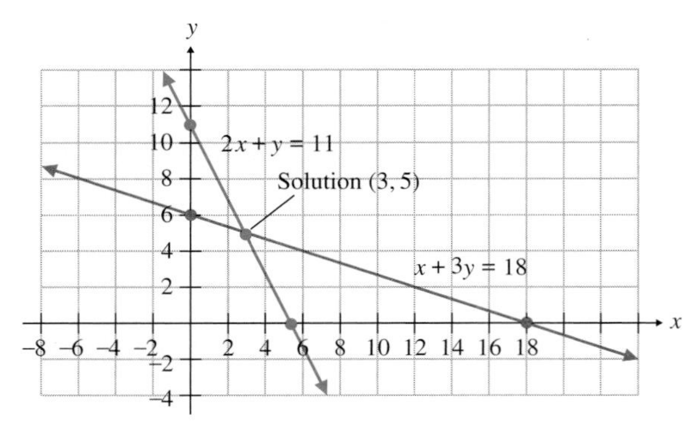

FIGURE 8.2

Check:

$$\begin{aligned} 2x + y &= 11 \\ 2(3) + 5 &= 11 \\ 11 &= 11 \quad \text{true} \end{aligned} \qquad \begin{aligned} x + 3y &= 18 \\ 3 + 3(5) &= 18 \\ 18 &= 18 \quad \text{true} \end{aligned}$$

Since the ordered pair (3, 5) checks in both equations, it is the solution to the system of equations. This system of equations is consistent.

EXAMPLE 4 Solve the following system of equations graphically.

$$2x + y = 3$$
$$4x + 2y = 12$$

Solution:

$2x + y = 3$	*Ordered Pair*	$4x + 2y = 12$	*Ordered Pair*
Let $x = 0$; then $y = 3$	(0, 3)	Let $x = 0$; then $y = 6$	(0, 6)
Let $y = 0$; then $x = \frac{3}{2}$	$\left(\frac{3}{2}, 0\right)$	Let $y = 0$; then $x = 3$	(3, 0)

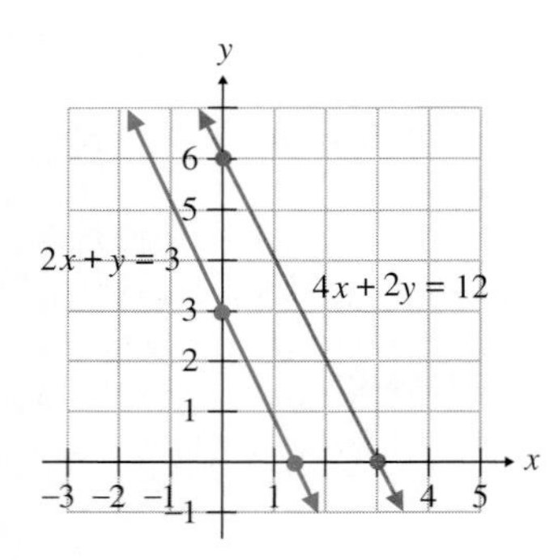

FIGURE 8.3

The two lines (Fig. 8.3) appear to be parallel.

To show that the two lines are indeed parallel, write each equation in slope–intercept form.

$$\begin{aligned} 2x + y &= 3 \\ y &= -2x + 3 \end{aligned} \qquad \begin{aligned} 4x + 2y &= 12 \\ 2y &= -4x + 12 \\ y &= -2x + 6 \end{aligned}$$

Both equations have the same slope, -2, and different y intercepts; thus the lines must be parallel. Since parallel lines do not intersect, this system of equations has no solution. This system of equations is inconsistent.

EXAMPLE 5 Solve the following system of equations graphically.

$$x - \frac{1}{2}y = 2$$
$$y = 2x - 4$$

Solution:

$x - \frac{1}{2}y = 2$	*Ordered Pair*	$y = 2x - 4$	*Ordered Pair*
Let $x = 0$; then $y = -4$	$(0, -4)$	Let $x = 0$; then $y = -4$	$(0, -4)$
Let $y = 0$; then $x = 2$	$(2, 0)$	Let $y = 0$; then $x = 2$	$(2, 0)$

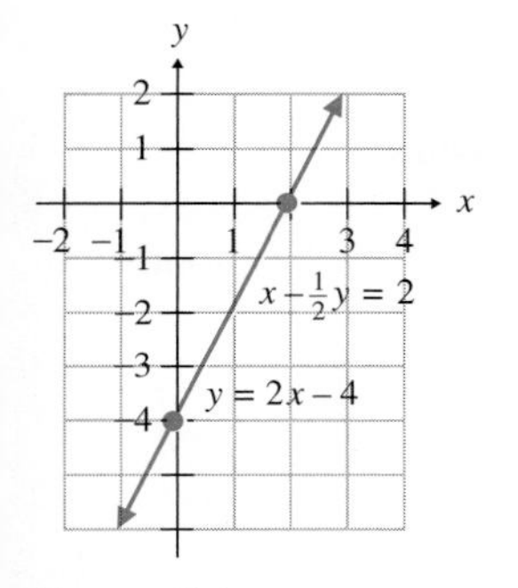

FIGURE 8.4

Because the lines have the same x and y intercepts, both equations represent the same line (Fig. 8.4). When the equations are changed to slope–intercept form, it becomes clear that the equations are identical and the system is dependent.

$$x - \frac{1}{2}y = 2 \qquad\qquad y = 2x - 4$$
$$2\left(x - \frac{1}{2}y\right) = 2(2)$$
$$2x - y = 4$$
$$-y = -2x + 4$$
$$y = 2x - 4$$

The solution to this system of equations is all the points on the line.

When graphing a system of equations, the intersection of the lines is not always easy to read on the graph. For example, the solution to a system of equations may be the ordered pair $(\frac{5}{9}, -\frac{4}{11})$. In cases like this, it is not easy to find the exact value of the solution by observation, but you should be able to give an approximate answer. An approximate answer to this system might be $(\frac{1}{2}, -\frac{1}{3})$ or $(0.6, -0.3)$. The accuracy of your answer will depend on how carefully you draw the graphs and on the scale of the graph paper used.

In Section 3.3 we solved a problem involving security systems using only one variable. In the example that follows, we will work that same problem using two variables and illustrate the solution in the form of a graph. Although an answer may sometimes be easier to obtain using only one variable, a graph of the situation may help you to visualize the total picture better. Before you read Example 6, you may wish to review the solution to Example 11 in Section 3.3.

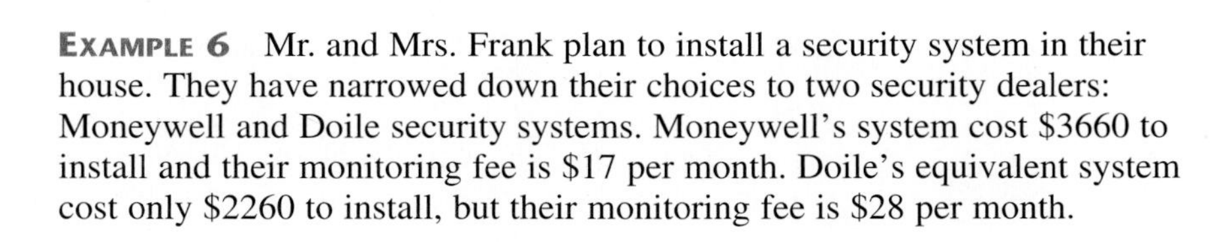

EXAMPLE 6 Mr. and Mrs. Frank plan to install a security system in their house. They have narrowed down their choices to two security dealers: Moneywell and Doile security systems. Moneywell's system cost \$3660 to install and their monitoring fee is \$17 per month. Doile's equivalent system cost only \$2260 to install, but their monitoring fee is \$28 per month.

(a) Assuming that their monthly monitoring fees do not change, in how many months would the total cost of Moneywell's and Doile's system be the same?

(b) If both dealers guarantee not to raise monthly fees for 10 years, and if you plan to use the system for 10 years, which system would be the least expensive?

Solution:

(a) We need to determine the number of months when both systems will have the same total cost.

Let n = number of months
c = total cost of the security system over n months

Now write an equation to represent the cost of each system using the two variables c and n.

Moneywell

$$\text{Total cost} = \begin{pmatrix}\text{initial}\\ \text{cost}\end{pmatrix} + \begin{pmatrix}\text{fees over}\\ n \text{ months}\end{pmatrix}$$

$$c = 3360 + 17n$$

Doile

$$\text{Total cost} = \begin{pmatrix}\text{initial}\\ \text{cost}\end{pmatrix} + \begin{pmatrix}\text{fees over}\\ n \text{ months}\end{pmatrix}$$

$$c = 2260 + 28n$$

Thus, our system of equations is

$$c = 3360 + 17n$$
$$c = 2260 + 28n$$

Now graph each equation.

$$c = 3360 + 17n$$

Let $n = 0$ $\quad c = 3360 + 17(0) = 3360$

Let $n = 100$ $\quad c = 3360 + 17(100) = 5060$

Let $n = 150$ $\quad c = 3360 + 17(150) = 5910$

n	c
0	3360
100	5060
150	5910

$$c = 2260 + 28n$$

Let $n = 0$ $\quad c = 2260 + 28(0) = 2260$

Let $n = 100$ $\quad c = 2260 + 28(100) = 5060$

Let $n = 150$ $\quad c = 2260 + 28(150) = 6460$

n	c
0	2260
100	5060
150	6460

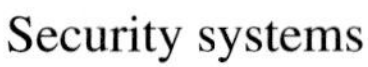

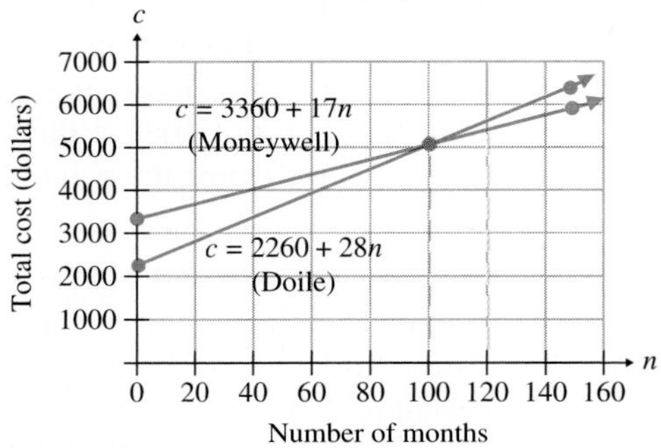

FIGURE 8.5

The graph (Fig. 8.5) shows that the total cost of the two security systems would be the same in 100 months. This is the same answer obtained in Example 11 in Section 3.3.

(b) Since 10 years is 120 months, draw a dashed vertical line at $n = 120$ months and see where it intersects the two lines. Since at 120 months the Doile line is higher than the Moneywell line, the cost for the Doile system for 120 months is more than the cost of the Moneywell system. Therefore, the cost of the Moneywell system would be less expensive for 10 years.

Applications of systems of equations are discussed further in Section 8.5

Helpful Hint

In the Helpful Hint in Section 7.3 we gave one graphical interpretation of the solution of a linear equation in one variable. Now we will give another. Consider the equation $3x - 1 = x + 1$. Its solution is 1, as is illustrated below.

$$3x - 1 = x + 1$$
$$2x - 1 = 1$$
$$2x = 2$$
$$x = 1$$

Let us set each side of the equation $3x - 1 = x + 1$ equal to y to obtain the following system of equations:

$$y = 3x - 1$$
$$y = x + 1$$

The graphical solution of this system of equations is illustrated below.

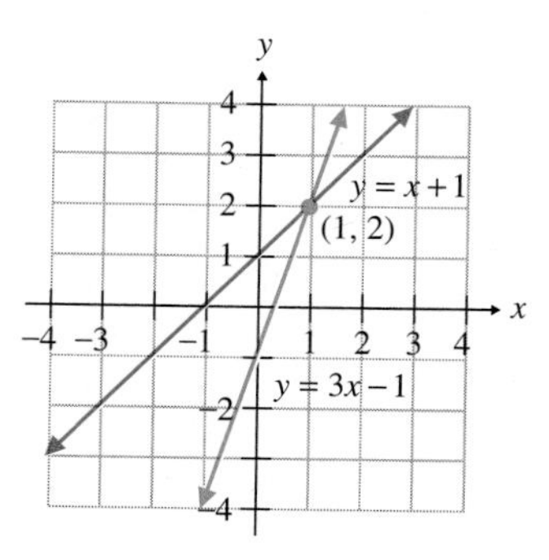

A check will show that the solution to the system is (1, 2). Notice that the x coordinate of the solution of the system, 1, is the solution to the linear equation in one variable, $3x - 1 = x + 1$. Thus, you can solve a linear equation in one variable by changing it to a system of linear equation in two variables. The x coordinate of the solution to the system will be the solution to the linear equation in one variable.

Exercise Set 8.1

Determine which, if any, of the following ordered pairs satisfy each system of linear equations.

1. $y = 3x - 4$
$y = -x + 4$
(a) $(-2, 2)$ **(b)** $(-4, -8)$ **(c)** $(2, 2)$

2. $y = -4x$
$y = -2x + 8$
(a) $(0, 0)$ **(b)** $(-4, 16)$ **(c)** $(2, -8)$

3. $y = 2x - 3$
$y = x + 5$
(a) $(8, 13)$ **(b)** $(4, 5)$ **(c)** $(4, 9)$

4. $x + 2y = 4$
$y = 3x + 3$
(a) $(0, 2)$ **(b)** $(-2, 3)$ **(c)** $(4, 15)$

5. $3x - y = 6$
$2x + y = 9$
(a) $(3, 3)$ **(b)** $(4, -2)$ **(c)** $(-6, 3)$

6. $y = 2x + 4$
$y = 2x - 1$
(a) $(0, 4)$ **(b)** $(3, 10)$ **(c)** $(-2, 0)$

7. $2x - 3y = 6$
$y = \frac{2}{3}x - 2$
(a) $(3, 0)$ **(b)** $(3, -2)$ **(c)** $(6, 2)$

8. $y = -x + 4$
$2y = -2x + 8$
(a) $(2, 5)$ **(b)** $(0, 4)$ **(c)** $(5, -1)$

9. $3x - 4y = 8$
$2y = \frac{2}{3}x - 4$
(a) $(0, -2)$ **(b)** $(1, -6)$ **(c)** $\left(-\frac{1}{3}, -\frac{9}{4}\right)$

10. $2x + 3y = 6$
$-2x + 5 = y$
(a) $\left(\frac{1}{2}, \frac{5}{3}\right)$ **(b)** $(2, 1)$ **(c)** $\left(\frac{9}{4}, \frac{1}{2}\right)$

Identify each system of linear equations (lines are labeled 1 and 2) as consistent, inconsistent, or dependent. State whether the system has exactly one solution, no solution, or an infinite number of solutions.

11.

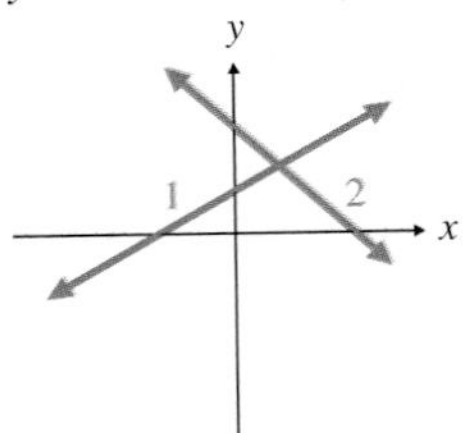

12.

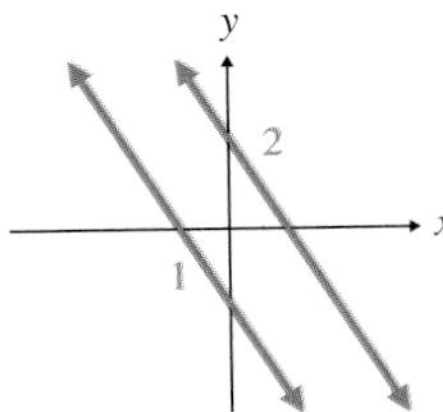

13.

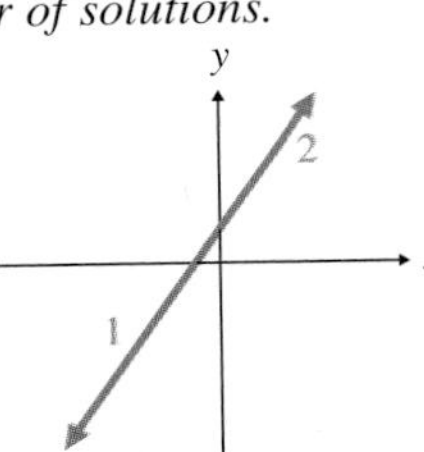

14.

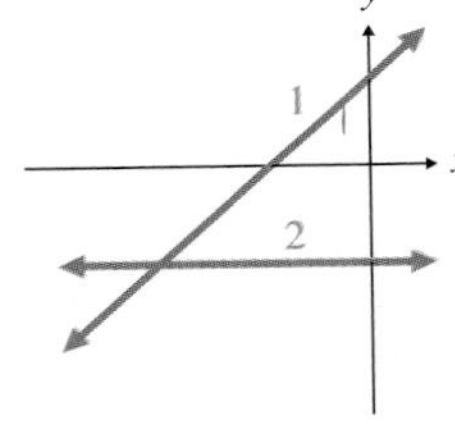

15.

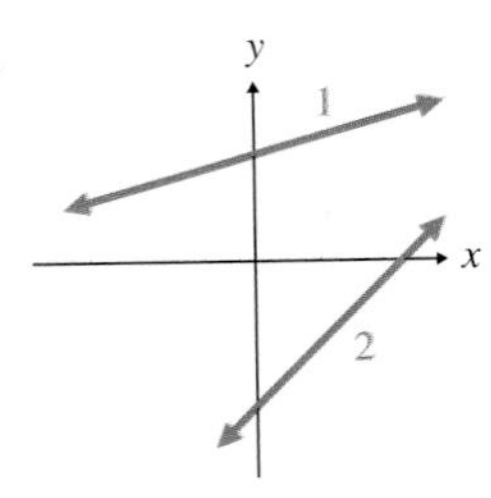

16.

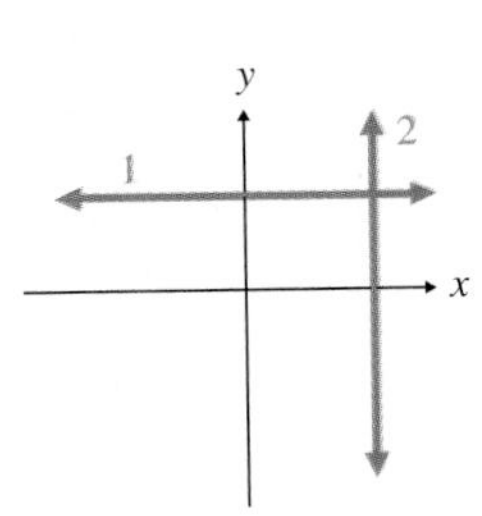

17.

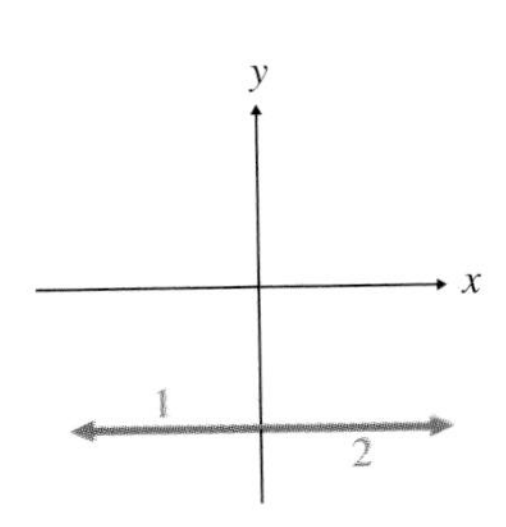

18.

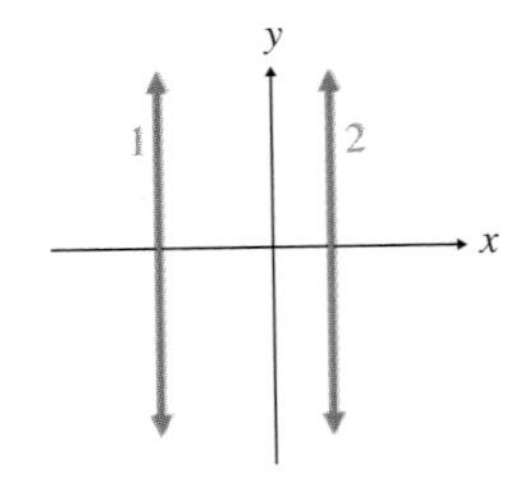

Express each equation in slope–intercept form. Without graphing the equations, state whether the system of equations has exactly one solution, no solution, or an infinite number of solutions.

19. $y = 3x - 2$
$2y = 4x - 6$

20. $x + y = 6$
$x - y = 6$

21. $3y = 2x + 3$
$y = \frac{2}{3}x - 2$

22. $y = \frac{1}{2}x + 4$
$2y = x + 8$

23. $4x = y - 6$
$3x = 4y + 5$

24. $x + 2y = 6$
$2x + y = 4$

25. $2x = 3y + 4$
$6x - 9y = 12$

26. $x - y = 2$
$2x - 2y = -2$

27. $y = \frac{3}{2}x + \frac{1}{2}$
$3x - 2y = -\frac{1}{2}$

28. $x - y = 3$
$\frac{1}{2}x - 2y = -6$

29. $3x + 5y = -7$
$-3x - 5y = -7$

30. $3x = 7y - 9$
$3x = -7y + 9$

Determine the solution to each system of equations graphically. If the system is dependent or inconsistent, so state.

31. $y = x + 2$
$y = -x + 2$

32. $y = 2x + 4$
$y = -3x - 6$

33. $y = 3x - 6$
$y = -x + 6$

34. $y = 2x - 1$
$2y = 4x + 6$

35. $2x = 4$
$y = -3$

36. $x + y = 5$
$2y = x - 2$

37. $y = x + 2$
$x + y = 4$

38. $2x + y = 6$
$2x - y = -2$

39. $y = -\frac{1}{2}x + 4$
$x + 2y = 6$

40. $x + 2y = -4$
$2x - y = -3$

41. $x + 2y = 8$
$2x - 3y = 2$

42. $4x - y = 5$
$2y = 8x - 10$

43. $2x + 3y = 6$
$4x = -6y + 12$

44. $2x + 3y = 6$
$2x + y = -2$

45. $y = 3$
$y = 2x - 3$

46. $x = 3$
$y = 2x - 2$

47. $x - 2y = 4$
$2x - 4y = 8$

48. $3x + y = -6$
$2x = 1 + y$

49. $2x + y = -2$
$6x + 3y = 6$

50. $y = 2x - 3$
$y = -x$

51. $4x - 3y = 6$
$2x + 4y = 14$

52. $2x + 6y = 6$
$y = -\frac{1}{3}x + 1$

53. $2x - 3y = 0$
$x + 2y = 0$

54. $2x = 4y - 12$
$-4x + 8y = 8$

In Exercises 55–57, find the solution by graphing the system of equations.

55. In Example 6, if Moneywell's system cost \$4400 plus \$15 per month and Doile's system cost \$3400 plus \$25 per month, after how many months would the total cost of the two systems be the same?

56. A college is considering purchasing one of two computer systems. In system A the host computer costs \$3600 and the terminals cost \$400 each. In system B the host computer costs \$2400 but the terminals cost \$600 each. We can represent this situation with the system of equations

$$c = 3600 + 400n$$
$$c = 2400 + 600n$$

where c is the total cost and n is the number of terminals. Find the number of terminals for which the systems cost the same.

57. The Evergreen Landscape Service charges a consultation fee of \$200 plus \$60 per hour for labor. The Out of Sight Landscape Service charges a consultation fee of \$300 plus \$40 per hour for labor. We can represent this situation with the system of equations

$$c = 200 + 60h$$
$$c = 300 + 40h$$

where c is the total cost and h is the number of hours of labor. Find the number of hours of labor for the two services to have the same total cost.

SEE EXERCISE 57.

58. What does the solution to a system of equations represent?

59. Given the system of equations $5x - 4y = 10$ and $12y = 15x - 20$, determine without graphing if the graphs of the two equations will be parallel lines. Explain how you determined your answer.

60. (a) What is a consistent system of equations? (b) What is an inconsistent system of equations? (c) What is a dependent system of equations?

61. Explain how to determine without graphing if a system of linear equations has exactly one solution, no solution, or an infinite number of solutions.

62. When a dependent system of two linear equations is graphed, what will be the results?

CUMULATIVE REVIEW EXERCISES

[1.3] 63. Consider this set of numbers

$$\{6, -4, 0, \sqrt{3}, 2\tfrac{1}{2}, -\tfrac{9}{5}, 4.22, -\sqrt{7}\}$$

List the numbers that are:

(a) Natural numbers
(b) Whole numbers
(c) Integers
(d) Rational numbers
(e) Irrational numbers
(f) Real numbers

[5.2] 64. Factor $xy - 4x + 3y - 12$.

[5.4] 65. Factor $8x^2 + 2x - 15$.

[6.4] 66. Subtract $\dfrac{x}{x+3} - 2$.

[6.6] 67. Solve $\dfrac{x}{x+3} - 2 = 0$.

Group Activity/ Challenge Problems

Suppose that a system of three linear equations is graphed on the same axes. Find the maximum number of points where two or more of the lines can intersect if:

1. The three lines have the same slope but different y intercepts
2. The three lines have the same slope and the same y intercept
3. Two lines have the same slope but different y intercepts and the third line has a different slope
4. The three lines have different slopes but the same y intercept
5. The three lines have different slopes but two have the same y intercept
6. The three lines have different slopes and different y intercepts

Tape 13

8.2 Solving Systems of Equations by Substitution

1 Solve systems of equations by substitution.

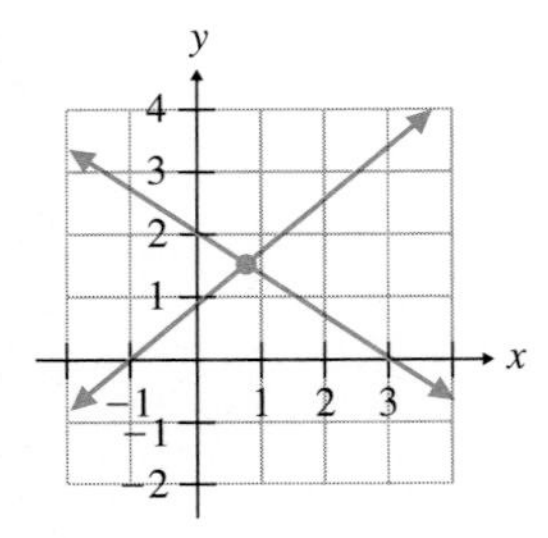

FIGURE 8.6

Often, a graphic solution to a system of equations may be inaccurate since you must estimate the coordinates of the point of intersection. For example, can you determine the solution to the system of equations shown in Figure 8.6? You may estimate the solution to be $\left(\frac{7}{10}, \frac{3}{2}\right)$ when it may actually be $\left(\frac{4}{5}, \frac{8}{5}\right)$. When an exact solution is necessary, the system should be solved algebraically, either by substitution or by addition of equations.

Solving Systems by Substitution

1 The procedure for solving a system of equations by substitution is illustrated in Example 1. The procedure for solving by addition is presented in Section 8.3. Regardless of which of the two algebraic techniques is used to solve a system of equations, our immediate goal remains the same, that is, to obtain one equation containing only one unknown.

EXAMPLE 1 Solve the following system of equations by substitution.

$$2x + y = 11$$
$$x + 3y = 18$$

Solution: Begin by solving for one of the variables in either of the equations. You may solve for any of the variables; however, if you solve for a variable with a numerical coefficient of 1, you may avoid working with fractions. In this system the y term in $2x + y = 11$ and the x term in $x + 3y = 18$ both have a numerical coefficient of 1.

Let us solve for y in $2x + y = 11$.

$$2x + y = 11$$
$$y = -2x + 11$$

Next, substitute $-2x + 11$ for y in the *other equation*, $x + 3y = 18$, and solve for the remaining variable, x.

$$x + 3y = 18$$
$$x + 3(-2x + 11) = 18$$
$$x - 6x + 33 = 18$$
$$-5x + 33 = 18$$
$$-5x = -15$$
$$x = 3$$

Finally, substitute $x = 3$ in the equation that is solved for y and find the value of y.

$$y = -2x + 11$$
$$y = -2(3) + 11$$
$$y = -6 + 11$$
$$y = 5$$

The solution is the ordered pair (3, 5).

Note that this solution is identical to the graphical solution obtained in Example 3 of Section 8.1.

To Solve a System of Equations by Substitution

1. Solve for a variable in either equation. (If possible, solve for a variable with a numerical coefficient of 1 to avoid working with fractions.)

continued on top of next page

2. Substitute the expression found for the variable in step 1 into the other equation.
3. Solve the equation determined in step 2 to find the value of one variable.
4. Substitute the value found in step 3 into the equation obtained in step 1 to find the other variable.

EXAMPLE 2 Solve the following system of equations by substitution.

$$2x + y = 3$$
$$4x + 2y = 12$$

Solution: Solve for y in $2x + y = 3$.

$$2x + y = 3$$
$$y = -2x + 3$$

Now substitute the expression $-2x + 3$ for y in the *other equation,* $4x + 2y = 12$, and solve for x.

$$4x + 2y = 12$$
$$4x + 2(-2x + 3) = 12$$
$$4x - 4x + 6 = 12$$
$$6 = 12 \quad \text{false}$$

Since the statement $6 = 12$ is false, the system has no solution. (Therefore, the graphs of the equations will be parallel lines and the system is inconsistent because it has no solution.)

Note that the solution in Example 2 is identical to the graphical solution obtained in Example 4 of Section 8.1. Figure 8.3 on page 444 shows the parallel lines.

EXAMPLE 3 Solve the following system of equations by substitution.

$$x - \frac{1}{2}y = 2$$
$$y = 2x - 4$$

Solution: The equation $y = 2x - 4$ is already solved for y. Substitute $2x - 4$ for y in the other equation, $x - \frac{1}{2}y = 2$, and solve for x.

$$x - \frac{1}{2}y = 2$$
$$x - \frac{1}{2}(2x - 4) = 2$$
$$x - x + 2 = 2$$
$$2 = 2 \quad \text{true}$$

Notice that the sum of the x terms is 0, and when simplified x is no longer part of the equation. *Since the statement $2 = 2$ is true, this system has an infinite number of solutions. Therefore, the graphs of the equations represent the same line and the system is dependent.*

Note that the solution in Example 3 is identical to the solution obtained graphically in Example 5 of Section 8.1. Figure 8.4 on page 445 shows the graphs of both equations are the same line.

EXAMPLE 4 Solve the following system of equations by substitution.

$$\begin{aligned} 2x + 4y &= 6 \\ 4x - 2y &= -8 \end{aligned}$$

Solution: None of the variables in either equation has a numerical coefficient of 1. However, since the numbers 4 and 6 are both divisible by 2, if we solve the first equation for x, we will avoid having to work with fractions.

$$\begin{aligned} 2x + 4y &= 6 \\ 2x &= -4y + 6 \\ \frac{2x}{2} &= \frac{-4y + 6}{2} \\ x &= -2y + 3 \end{aligned}$$

Now substitute $-2y + 3$ for x in the other equation, $4x - 2y = -8$, and solve for the remaining variable y.

$$\begin{aligned} 4x - 2y &= -8 \\ 4\overbrace{(-2y + 3)} - 2y &= -8 \\ -8y + 12 - 2y &= -8 \\ -10y + 12 &= -8 \\ -10y &= -20 \\ y &= 2 \end{aligned}$$

Finally, solve for x by substituting $y = 2$ in the equation previously solved for x.

$$\begin{aligned} x &= -2y + 3 \\ x &= -2(2) + 3 = -4 + 3 = -1 \end{aligned}$$

The solution is $(-1, 2)$.

Helpful Hint

Remember that a solution to a system of linear equations must contain both an x and a y value. Don't solve the system for one of the variables and forget to solve for the other.

EXAMPLE 5 Solve the following system of equations by substitution.

$$4x + 4y = 3$$
$$2x = 2y + 5$$

Solution: We will elect to solve for x in the second equation.

$$2x = 2y + 5$$
$$x = \frac{2y + 5}{2}$$
$$x = y + \frac{5}{2}$$

Now substitute $y + \frac{5}{2}$ for x in the other equation.

$$4x + 4y = 3$$
$$4\left(y + \frac{5}{2}\right) + 4y = 3$$
$$4y + 10 + 4y = 3$$
$$8y + 10 = 3$$
$$8y = -7$$
$$y = -\frac{7}{8}$$

Finally, find the value of x.

$$x = y + \frac{5}{2}$$
$$x = -\frac{7}{8} + \frac{5}{2} = -\frac{7}{8} + \frac{20}{8} = \frac{13}{8}$$

The solution is the ordered pair $\left(\frac{13}{8}, -\frac{7}{8}\right)$.

Exercise Set 8.2

Find the solution to each system of equations using substitution.

1. $x + 2y = 4$
$2x - 3y = 1$

2. $y = x + 3$
$y = -x - 5$

3. $x + y = -2$
$x - y = 0$

4. $2x + y = 3$
$2y = 6 - 4x$

5. $2x + y = 3$
$2x + y + 5 = 0$

6. $y = 2x + 4$
$y = -2$

7. $x = 4$
$x + y + 5 = 0$

8. $y = 2x - 13$
$-4x - 7 = 9y$

9. $x - \frac{1}{2}y = 2$
$y = 2x - 4$

10. $2x + 3y = 7$
$6x - y = 1$

11. $3x + y = -1$
$y = 3x + 5$

12. $y = -2x + 5$
$x + 3y = 0$

13. $y = \frac{1}{3}x - 2$
$x - 3y = 6$

14. $x = y + 4$
$3x + 7y = -18$

15. $2x + 3y = 7$
$6x - 2y = 10$

16. $4x - 3y = 6$
$2x + 4y = 5$

17. $3x - y = 14$
$6x - 2y = 10$

18. $5x - 2y = -7$
$5 = y - 3x$

19. $4x - 5y = -4$
$3x = 2y - 3$

20. $3x + 4y = 10$
$4x + 5y = 14$

21. $5x + 4y = -7$
$x - \frac{5}{3}y = -2$

22. $\frac{1}{2}x + y = 4$
$3x + \frac{1}{4}y = 6$

23. In Seattle, Washington, the temperature is 82°F, but it is decreasing by 2 degrees an hour. The temperature, T, at time t, in hours, is represented by $T = 82 - 2t$. In Spokane, Washington, the temperature is 64°F, but it is increasing by 2.5 degrees per hour. The temperature, T, can be represented by $T = 64 + 2.5t$. **(a)** If the temperature continues decreasing and increasing at the same rate in these cities, how long it will be before both cities have the same temperature? **(b)** When both cities have the same temperature, what will that temperature be?

24. John, a salesperson for an entertainment company, earns a weekly salary of $300 plus $3 for each video tape he sells. His weekly salary can be represented by $S = 300 + 3n$. He is being offered a chance to change salary plans. Under the new plan he will earn a weekly salary of $400 plus $2 for each video tape he sells. His weekly salary under this new plan can be represented by $S = 400 + 2n$. How many tapes would John need to sell in a week for his salary to be the same under both plans?

25. Jim's car is at the 100 mile marker on the interstate highway 15 miles behind Kathy's car. Jim's car is traveling at 65 miles per hour and Kathy's car is traveling at 60 miles per hour. The mile marker that Jim will be at in t hours can be found by $m = 100 + 65t$ and the mile marker that Kathy will be at in t hours can be found by $m = 115 + 60t$. **(a)** Determine the time it will take for Jim to catch Kathy. **(b)** At what mile marker will they be when they meet?

26. When solving the system of equations

$$3x + 6y = 9$$
$$4x + 3y = 5$$

by substitution, which variable, in which equation, would you choose to solve for to make the solution easier? Explain your answer.

27. When solving a system of linear equations by substitution, how will you know if the system is inconsistent?

28. When solving a system of linear equations by substitution, how will you know if the system is dependent?

CUMULATIVE REVIEW EXERCISES

Evaluate.

29. -3^3 **30.** $(-3)^3$

31. -3^4 **32.** $(-3)^4$

33. If the directions on a box of Quik-Bake Brownie Mix state that 2 eggs must be added to every 3 cups of brownie mix, how many eggs must be added to 9 cups of brownie mix?

34. If triangles ABC and $A'B'C'$ are similar triangles, find the length of side x.

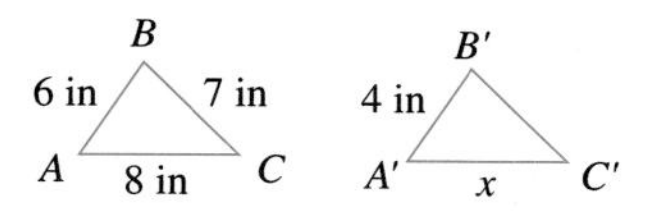

[4.1] *Simplify.*

35. $\left(\frac{3x^2y^4}{x^3y^2}\right)^2$

36. $(4x^2y^3)^3(3x^4y^5)^2$

[6.6] **37.** Solve the equation $\frac{3}{x-12} + \frac{5}{x-5} = \frac{5}{x^2 - 17x + 60}$.

Group Activity/ Challenge Problems

1. For the system of equations

$$y = ax + b$$
$$y = cx + d$$

where $a \neq c$, **(a)** find the x coordinate of the solution. **(b)** Use your answer to part **(a)** to find the x coordinate of the system of equations

$$y = 3x + 2$$
$$y = x + 6$$

(c) Find the solution to the system of equations in part **(b)**.

2. **(a)** Solve the system of equations

$$2y = 4x + 600$$
$$2y = 2x + 1200$$

(b) Suppose that both sides of each equation in the system are divided by 2, to obtain the following system.

$$y = 2x + 300$$
$$y = x + 600$$

Do you believe the solution to this system will be the same as the solution in part **(a)**? If not, do you believe the solution to this system will be related in any way to the solution obtained in part **(a)**? Explain your answer without solving the system.

(c) Solve the system in part **(b)** and compare the solution with the answer you gave in part **(b)**.

3. **(a)** Solve the system of equations

$$y = 4x + 600$$
$$y = 2x + 1200$$

(b) Suppose that the right side of each equation in the system is divided by 2, to obtain the following system.

$$y = 2x + 300$$
$$y = x + 600$$

Do you believe the solution to this system will be the same as the solution in part **(a)**? If not, do you believe the solution to this system will be related in any way to the solution obtained in part **(a)**? Explain your answer without solving the system.

(c) Solve the system in part **(b)** and compare your solution to the answer you gave in part **(b)**.

4. In a laboratory a large metal ball is heated to a temperature of 180° F. This metal ball is then placed in a gallon of oil at a temperature of 20° F. When placed in the oil the ball loses temperature at the rate of 10° per minute while the oil's temperature rises at a rate of 6° per minute.

(a) How long will it take for the ball and oil to reach the same temperature?

(b) When the ball and oil reach the same temperature, what will that temperature be?

5. In intermediate algebra you may solve systems containing three equations with three variables. Solve the following system of equations.

$$x = 4$$
$$2x - y = 6$$
$$-x + y + z = -3$$

8.3 Solving Systems of Equations by the Addition Method

Tape 14

1 Solve systems of equations by addition.

The Addition Method

1 A third, and often the easiest, method of solving a system of equations is by the addition (or elimination) method. *The object of this process is to obtain two equations whose sum will be an equation containing only one variable.* Always keep in mind that our immediate goal is to obtain one equation containing only one unknown.

EXAMPLE 1 Solve the following system of equations using the addition method.

$$x + y = 6$$
$$2x - y = 3$$

Solution: Note that one equation contains $+y$ and the other contains $-y$. By adding the equations, we can eliminate the variable y and obtain one equation containing only one variable, x. When added, $+y$ and $-y$ sum to 0, and so the variable y is eliminated.

$$\begin{aligned} x + y &= 6 \\ \underline{2x - y} &\underline{= 3} \\ 3x \quad\; &= 9 \end{aligned}$$

Now solve for the remaining variable, x.

$$\frac{3x}{3} = \frac{9}{3}$$
$$x = 3$$

Finally, solve for y by substituting $x = 3$ in either of the original equations.

$$x + y = 6$$
$$3 + y = 6$$
$$y = 3$$

The solution is (3, 3).

Check the answer in *both* equations.

Check:

$$x + y = 6 \qquad\qquad 2x - y = 3$$
$$3 + 3 = 6 \qquad\qquad 2(3) - 3 = 3$$
$$6 = 6 \quad \text{true} \qquad\qquad 6 - 3 = 3$$
$$3 = 3 \quad \text{true}$$

EXAMPLE 2 Solve the following system of equations using the addition method.

$$-x + 3y = 8$$
$$x + 2y = -13$$

Solution: By adding the equations we can eliminate the variable x.

$$\begin{aligned} -x + 3y &= 8 \\ x + 2y &= -13 \\ \hline 5y &= -5 \\ \frac{5y}{5} &= \frac{-5}{5} \\ y &= -1 \end{aligned}$$

Now solve for x by substituting $y = -1$ in either of the original equations.

$$\begin{aligned} x + 2y &= -13 \\ x + 2(-1) &= -13 \\ x - 2 &= -13 \\ x &= -11 \end{aligned}$$

The solution is $(-11, -1)$.

Now we state the procedure for solving a system of equations by the addition method.

To Solve a System of Equations by the Addition (or Elimination) Method

1. If necessary, rewrite each equation so that the terms containing variables appear on the left side of the equal sign and any constants appear on the right side of the equal sign.
2. If necessary, multiply one or both equations by a constant(s) so that when the equations are added the resulting sum will contain only one variable.
3. Add the equations. This will result in a single equation containing only one variable.
4. Solve for the variable in the equation in step 3.
5. Substitute the value found in step 4 into either of the original equations. Solve that equation to find the value of the remaining variable.

In step 2 we indicate it may be necessary to multiply one or both equations by a constant. In this text we will use brackets [ ], to indicate that both sides of the equation within the brackets are to be multiplied by some constant. Thus, for example, $2[x + y = 1]$ means that both sides of the equation $x + y = 1$ are to be multiplied by 2. We write

$$2[x + y = 1] \qquad \text{gives} \qquad 2x + 2y = 2$$

Similarly, $-3[4x - 2y = 5]$ means both sides of the equation $4x - 2y = 5$ are to be multiplied by -3. We write

$$-3[4x - 2y = 5] \qquad \text{gives} \qquad -12x + 6y = -15$$

The use of this notation may make it easier for you to follow the procedure used to solve the problem.

EXAMPLE 3 Solve the following system of equations using the addition method.

$$2x + y = 6$$
$$3x + y = 5$$

Solution: The object of the addition process is to obtain two equations whose sum will be an equation containing only one variable. If we add these two equations, none of the variables will be eliminated. However, if we multiply either equation by -1 and then add, the terms containing y will sum to zero, and we will accomplish our goal. We will multiply the top equation by -1.

$$-1[2x + y = 6] \quad \text{gives} \quad -2x - y = -6$$
$$3x + y = 5 \qquad\qquad 3x + y = 5$$

Remember that both sides of the equation must be multiplied by -1. This process changes the sign of each term in the equation being multiplied without changing the solution to the system of equations. Now add the two equations on the right.

$$\begin{aligned} -2x - y &= -6 \\ 3x + y &= 5 \\ \hline x \qquad &= -1 \end{aligned}$$

Solve for y in either of the original equations.

$$\begin{aligned} 2x + y &= 6 \\ 2(-1) + y &= 6 \\ -2 + y &= 6 \\ y &= 8 \end{aligned}$$

The solution is $(-1, 8)$.

EXAMPLE 4 Solve the following system of equations using the addition method.

$$2x + y = 11$$
$$x + 3y = 18$$

Solution: To eliminate the variable x, we multiply the second equation by -2 and add the two equations.

$$2x + y = 11 \quad \text{gives} \quad 2x + y = 11$$
$$-2[x + 3y = 18] \qquad\qquad -2x - 6y = -36$$

Now add:

$$\begin{aligned} 2x + y &= 11 \\ -2x - 6y &= -36 \\ \hline -5y &= -25 \\ y &= 5 \end{aligned}$$

Solve for x.

$$\begin{aligned} 2x + y &= 11 \\ 2x + 5 &= 11 \\ 2x &= 6 \\ x &= 3 \end{aligned}$$

The solution (3, 5) is the same as the solution obtained graphically in Example 3 of Section 8.1 and by substitution in Example 1 of Section 8.2.

In Example 4, we could have multiplied the first equation by -3 to eliminate the variable y.

$$\begin{aligned} -3[2x + y &= 11] \qquad \text{gives} \qquad & -6x - 3y &= -33 \\ x + 3y &= 18 & x + 3y &= 18 \end{aligned}$$

Now add:

$$\begin{aligned} -6x - 3y &= -33 \\ x + 3y &= 18 \\ \hline -5x &= -15 \\ x &= 3 \end{aligned}$$

Solve for y.

$$\begin{aligned} 2x + y &= 11 \\ 2(3) + y &= 11 \\ 6 + y &= 11 \\ y &= 5 \end{aligned}$$

The solution remains the same, (3, 5).

EXAMPLE 5 Solve the following system of equations using the addition method.

$$\begin{aligned} 4x + 2y &= -18 \\ -2x - 5y &= 10 \end{aligned}$$

Solution: To eliminate the variable x, we can multiply the second equation by 2 and then add.

$$\begin{aligned} 4x + 2y &= -18 \qquad \text{gives} \qquad & 4x + 2y &= -18 \\ 2[-2x - 5y &= 10] & -4x - 10y &= 20 \end{aligned}$$

$$\begin{aligned} 4x + 2y &= -18 \\ -4x - 10y &= 20 \\ \hline -8y &= 2 \\ y &= -\frac{1}{4} \end{aligned}$$

Solve for x:

$$4x + 2y = -18$$
$$4x + 2\left(-\frac{1}{4}\right) = -18$$
$$4x - \frac{1}{2} = -18$$
$$2\left(4x - \frac{1}{2}\right) = 2(-18)$$ **Multiply both sides of the equation by 2 to remove fractions.**
$$8x - 1 = -36$$
$$8x = -35$$
$$x = -\frac{35}{8}$$

The solution is $\left(-\frac{35}{8}, -\frac{1}{4}\right)$.

Check the solution $\left(-\frac{35}{8}, -\frac{1}{4}\right)$ in both equations.

Check:

$$4x + 2y = -18 \qquad\qquad -2x - 5y = 10$$
$$4\left(-\frac{35}{8}\right) + 2\left(-\frac{1}{4}\right) = -18 \qquad\qquad -2\left(-\frac{35}{8}\right) - 5\left(-\frac{1}{4}\right) = 10$$
$$-\frac{35}{2} - \frac{1}{2} = -18 \qquad\qquad \frac{35}{4} + \frac{5}{4} = 10$$
$$-\frac{36}{2} = -18 \qquad\qquad \frac{40}{4} = 10$$
$$-18 = -18 \quad \text{true} \qquad\qquad 10 = 10 \quad \text{true}$$

Note that the solution to Example 5 contains fractions. You should not always expect to get integers as answers.

Example 6 Solve the following system of equations using the addition method.

$$2x + 3y = 6$$
$$5x - 4y = -8$$

Solution: The variable x can be eliminated by multiplying the first equation by -5 and the second by 2 and then adding the equations.

$$-5[2x + 3y = 6] \quad \text{gives} \quad -10x - 15y = -30$$
$$2[5x - 4y = -8] \qquad\qquad 10x - 8y = -16$$

$$\begin{aligned} -10x - 15y &= -30 \\ 10x - 8y &= -16 \\ \hline -23y &= -46 \\ y &= 2 \end{aligned}$$

The same value could have been obtained for y by multiplying the first equation by 5 and the second by -2 and then adding. Try it now and see.

Solve for x.

$$
\begin{aligned}
2x + 3y &= 6 \\
2x + 3(2) &= 6 \\
2x + 6 &= 6 \\
2x &= 0 \\
x &= 0
\end{aligned}
$$

The solution to this system is (0, 2).

EXAMPLE 7 Solve the following system of equations using the addition method.

$$
\begin{aligned}
2x + y &= 3 \\
4x + 2y &= 12
\end{aligned}
$$

Solution: The variable y can be eliminated by multiplying the first equation by -2 and then adding the two equations.

$$
\begin{aligned}
-2[2x + y = 3] \qquad &\text{gives} \qquad -4x - 2y = -6 \\
4x + 2y = 12 \qquad & \qquad\qquad\quad 4x + 2y = 12
\end{aligned}
$$

$$
\begin{array}{r}
-4x - 2y = -6 \\
4x + 2y = 12 \\
\hline
0 = 6 \quad \text{false}
\end{array}
$$

Since $0 = 6$ *is a false statement, this system has no solution.* The system is inconsistent. The graphs of the equations will be parallel lines.

This solution is identical to the solutions obtained by graphing in Example 4 of Section 8.1 and by substitution in Example 2 of Section 8.2.

EXAMPLE 8 Solve the following system of equations using the addition method.

$$
\begin{aligned}
x - \frac{1}{2}y &= 2 \\
y &= 2x - 4
\end{aligned}
$$

Solution: First align the x and y terms on the left side of the equation.

$$
\begin{aligned}
x - \frac{1}{2}y &= 2 \\
-2x + y &= -4
\end{aligned}
$$

Now proceed as in the previous examples.

$$2\left[x - \frac{1}{2}y = 2\right] \quad \text{gives} \quad 2x - y = 4$$
$$-2x + y = -4 \qquad\qquad -2x + y = -4$$

$$\begin{array}{r} 2x - y = 4 \\ -2x + y = -4 \\ \hline 0 = 0 \end{array} \quad \text{true}$$

Since $0 = 0$ is a true statement, the system is dependent and has an infinite number of solutions. When graphed, both equations will be the same line. This solution is the same as the solutions obtained by graphing in Example 5 of Section 8.1 and by substitution in Example 3 of Section 8.2.

EXAMPLE 9 Solve the following system of equations using the addition method.

$$2x + 3y = 7$$
$$5x - 7y = -3$$

Solution: We can eliminate the variable x by multiplying the first equation by -5 and the second by 2.

$$-5[2x + 3y = 7] \quad \text{gives} \quad -10x - 15y = -35$$
$$2[5x - 7y = -3] \qquad\qquad 10x - 14y = -6$$

$$\begin{array}{r} -10x - 15y = -35 \\ 10x - 14y = -6 \\ \hline -29y = -41 \\ y = \dfrac{41}{29} \end{array}$$

We can now find x by substituting $y = \frac{41}{29}$ into one of the original equations and solving for x. If you try this, you will see that although it can be done, the calculations are messy. An easier method of solving for x is to go back to the original equations and eliminate the variable y.

$$7[2x + 3y = 7] \quad \text{gives} \quad 14x + 21y = 49$$
$$3[5x - 7y = -3] \qquad\qquad 15x - 21y = -9$$

$$\begin{array}{r} 14x + 21y = 49 \\ 15x - 21y = -9 \\ \hline 29x = 40 \\ x = \dfrac{40}{29} \end{array}$$

Thus, the solution is $\left(\frac{40}{29}, \frac{41}{29}\right)$.

Helpful Hint

We have illustrated three methods for solving a system of linear equations: graphing, substitution, and addition. When you are given a system of equations, which method should you use to solve the system? When you need an exact solution, graphing should not be used. Of the two algebraic methods, the addition method may be easier to use if there are no numerical coefficients of 1 in the system. If one or more of the variables has a coefficient of 1, you may wish to use either method.

Exercise Set 8.3

Solve each system of equations using the addition method.

1. $x + y = 8$
$x - y = 4$

2. $x - y = 6$
$x + y = 4$

3. $-x + y = 5$
$x + y = 1$

4. $x + y = 10$
$-x + y = -2$

5. $x + 2y = 15$
$x - 2y = -7$

6. $3x + y = 10$
$4x - y = 4$

7. $4x + y = 6$
$-8x - 2y = 20$

8. $5x + 3y = 30$
$3x + 3y = 18$

9. $-5x + y = 14$
$-3x + y = -2$

10. $2x - y = 7$
$3x + 2y = 0$

11. $3x + y = 10$
$3x - 2y = 16$

12. $-4x + 3y = 0$
$5x - 6y = 9$

13. $4x - 3y = 8$
$2x + y = 14$

14. $2x - 3y = 4$
$2x + y = -4$

15. $5x + 3y = 6$
$2x - 4y = 5$

16. $6x - 4y = 9$
$2x - 8y = 3$

17. $4x - 2y = 6$
$y = 2x - 3$

18. $5x - 2y = -4$
$-3x - 4y = -34$

19. $3x - 2y = -2$
$3y = 2x + 4$

20. $5x + 4y = 10$
$-3x - 5y = 7$

21. $3x - 4y = 6$
$-6x + 8y = -4$

22. $2x - 3y = 11$
$-3x = -5y - 17$

23. $3x - 5y = 0$
$2x + 3y = 0$

24. $5x - 2y = 8$
$4y = 10x - 16$

25. $5x - 4y = 20$
$-3x + 2y = -15$

26. $5x = 2y - 4$
$3x - 5y = 6$

27. $6x + 2y = 5$
$3y = 5x - 8$

28. $4x - 3y = -4$
$3x - 5y = 10$

29. $4x + 5y = 0$
$3x = 6y + 4$

30. $4x - 3y = 8$
$-3x + 4y = 9$

31. $x - \frac{1}{2}y = 4$
$3x + y = 6$

32. $2x - \frac{1}{3}y = 6$
$5x - y = 4$

33. When solving a system of linear equations by the addition method, how will you know if the system is inconsistent?

34. When solving a system of linear equations by the addition method, how will you know if the system is dependent?

35. **(a)** In your own words explain the procedure to follow to solve a system of linear equations by the addition method.

(b) Solve the system that follows by the procedure given in part **(a)**.

$$3x - 2y = 10$$
$$2x + 5y = 13$$

CUMULATIVE REVIEW EXERCISES

[4.1] **36.** Simplify $\left(\dfrac{16x^4y^6z^3}{4x^6y^5z^7}\right)^3$.

[4.7] **37.** A highway crew is paving a new highway. On average, they pave 110 feet of highway each day. How long will it take the crew to pave 2420 feet?

[6.2] **38.** Simplify $\dfrac{2x^2 - x - 6}{x^2 - 7x + 10} \div \dfrac{2x^2 - 7x - 4}{x^2 - 9x + 20}$.

[6.4] **39.** Simplify $\dfrac{x}{x^2 - 1} - \dfrac{3}{x^2 - 16x + 15}$.

Group Activity/ Challenge Problems

Create a system of linear equations in two variables that has the listed solution. Explain how you obtained your system.

1. (2, 3)

2. (4, −2)

Solve each system of equations using the addition method. (Hint: *First remove all fractions.*)

3. $\dfrac{x+2}{2} - \dfrac{y+4}{3} = 4.$

$\dfrac{x+y}{2} = \dfrac{1}{2} + \dfrac{x-y}{3}$

4. $\dfrac{5x}{2} + 3y = \dfrac{9}{2} + y$

$\dfrac{1}{4}x - \dfrac{1}{2}y = 6x + 12$

5. (a) Solve the system of equations

$$4x + 2y = 1000$$
$$2x + 4y = 800$$

(b) If we divide all the terms in the top equation by 2 we get the following system.

$$2x + y = 500$$
$$2x + 4y = 800$$

How will the solution to this system compare to the solution in part **(a)**? Explain and then check your explanation by solving this system.

(c) If we divided all the terms in both equations given in part **(a)** by 2 we obtain the following system.

$$2x + y = 500$$
$$x + 2y = 400$$

How will the solution to this system compare to the solution in part **(a)**? Explain and then check your explanation by solving the system.

6. In intermediate algebra you may solve systems of three equations with three unknowns. Solve the following system.

$$x + 2y - z = 2$$
$$2x - y + z = 3$$
$$2x + y + z = 7$$

Hint: work with one pair of equations to get two equations in two unknowns. Then work with a different pair of equations to get two equations in the same two unknowns. Then solve the system of two equations in two unknowns.

8.4 Applications of Systems of Equations

Tape 14

1 Use systems of equations to solve application problems.

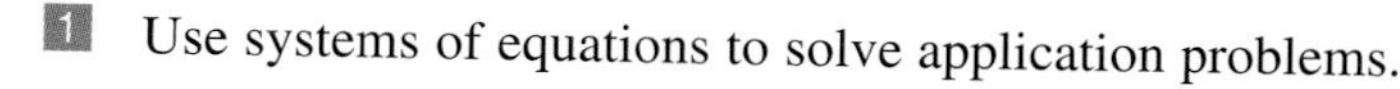

Applications

1 The method you use to solve a system of equations may depend on whether you wish to see "the entire picture" or are interested in finding the exact solution. If you are interested in the trend as the variable changes, you might decide to graph the equations. If you want only the solution, that is, the ordered pair common to both equations, you might use one of the two algebraic methods to find the common solution. Many of the application problems solved in earlier chapters using only one variable can also be solved using two variables.

EXAMPLE 1 Two angles are **complementary angles** when the sum of their measures is 90°. Angles x and y are complementary and angle x is 24° more than angle y. Find angles x and y.

Solution: Since the angles are complementary, the sum of their measures must be 90°. Thus one equation in the system is $x + y = 90$. Since angle x is 24° greater than angle y, the second equation is $x = y + 24$.

system of equations
$$\begin{cases} x + y = 90 \\ x = y + 24 \end{cases}$$

Subtract y from each side of the second equation. Then use the addition method to solve.

$$\begin{array}{rcr} x + y &=& 90 \\ x - y &=& 24 \\ \hline 2x \quad &=& 114 \\ x &=& 57 \end{array}$$

Now substitute 57 for x in the first equation and solve for y.

$$\begin{aligned} x + y &= 90 \\ 57 + y &= 90 \\ y &= 33 \end{aligned}$$

Angle x is 57° and angle y is 33°. Note that their sum is 90° and angle x is 24° greater than angle y.

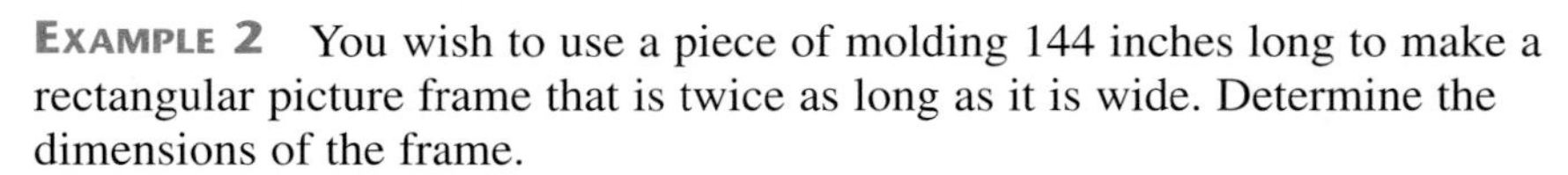

EXAMPLE 2 You wish to use a piece of molding 144 inches long to make a rectangular picture frame that is twice as long as it is wide. Determine the dimensions of the frame.

Solution: The formula for the perimeter of a rectangle is $P = 2l + 2w$.

Let w = width of the frame
l = length of the frame

Since the perimeter is to be 144 inches, one equation in the system is

$$144 = 2l + 2w$$

Since the length is to be twice the width, $l = 2w$.

system of equations
$$\begin{cases} 144 = 2l + 2w \\ l = 2w \end{cases}$$

We will solve this system by substitution. Substitute $2w$ for l in the equation $144 = 2l + 2w$ to obtain

$$\begin{aligned} 144 &= 2l + 2w \\ 144 &= 2(2w) + 2w \\ 144 &= 4w + 2w \\ 144 &= 6w \\ 24 &= w \end{aligned}$$

Therefore, the width is 24 inches. Since the length is twice the width, the length is 2(24) or 48 inches.

EXAMPLE 3 A plane can fly 600 miles per hour with the wind and 450 miles per hour against the wind. Find the speed of the wind and the speed of the plane in still air.

Solution: Let p = speed of the plane in still air
w = speed of the wind

If p equals the speed of the plane in still air and w equals the speed of the wind, then $p + w$ equals the speed of the plane flying with the wind, and $p - w$ equals the speed of the plane flying against the wind. We make use of this information when writing the system of equations.

Speed of plane flying with wind: $p + w = 600$
Speed of plane flying against wind: $p - w = 450$
(system of equations)

We will use the addition method because the sum of the terms containing w is zero.

$$\begin{aligned} p + w &= 600 \\ p - w &= 450 \\ \hline 2p \quad &= 1050 \\ p &= 525 \end{aligned}$$

The plane's speed is 525 miles per hour in still air. Now substitute 525 for p in the first equation and solve for w.

$$\begin{aligned} p + w &= 600 \\ 525 + w &= 600 \\ w &= 75 \end{aligned}$$

The wind's speed is 75 miles per hour.

EXAMPLE 4 Only two-axle vehicles are permitted to cross a bridge that leads to a state park on an island. The toll for the bridge is 50 cents for motorcycles and $1.00 for cars and trucks. On Saturday, the toll booth attendant collected a total of $150, and the vehicle counter recorded 170 vehicles crossing the bridge. How many motorcycles and how many cars and trucks crossed the bridge that day?

Solution: Let x = number of motorcycles
y = number of cars and trucks

Since a total of 170 vehicles crossed the bridge, one equation is $x + y = 170$. The second equation comes from the tolls collected.

Tolls from motorcycles + tolls from cars and trucks = 150
$$0.50x + 1.00y = 150$$

system of equations
$$\begin{cases} x + y = 170 \\ 0.50x + 1.00y = 150 \end{cases}$$

Since the first equation can be easily solved for y, we will solve this system by substitution. Solving for y in $x + y = 170$ gives $y = 170 - x$. Substitute $170 - x$ for y in the second equation and solve for x.

$$\begin{aligned} 0.50x + 1.00y &= 150 \\ 0.50x + 1.00(170 - x) &= 150 \\ 0.50x + 170 - 1.00x &= 150 \\ 170 - 0.5x &= 150 \\ -0.5x &= -20 \\ \frac{-0.5x}{-0.5} &= \frac{-20}{-0.5} \\ x &= 40 \end{aligned}$$

Forty motorcycles crossed the bridge. The total number of vehicles that crossed the bridge is 170. Therefore, $170 - 40$ or 130 cars and trucks crossed the bridge that Saturday.

EXAMPLE 5 Mrs. Beal needs to purchase a new engine for her car and have it installed by a mechanic. She is considering two garages: Sally's garage and Scotty's garage. At Sally's garage, the parts cost $800 and the labor cost is $25 per hour. At Scotty's garage, the parts cost $575 and labor cost is $50 per hour.

(a) How many hours would the repairs need to take for the total cost at each garage to be the same?

(b) If both garages estimate that the repair will take 8 hours, which garage would be the least expensive?

Solution: **(a)** Let n = number of hours of labor
c = total cost of repairs

Sally's	*Scotty's*
total cost = parts + labor	total cost = parts + labor
$c = 800 + 25n$	$c = 575 + 50n$

system of equations $\begin{cases} c = 800 + 25n \\ c = 575 + 50n \end{cases}$

$$\begin{aligned} \text{total cost at Sally's} &= \text{total cost at Scotty's} \\ 800 + 25n &= 575 + 50n \\ 225 + 25n &= 50n \\ 225 &= 25n \\ 9 &= n \end{aligned}$$

If 9 hours of labor is required, the cost from both garages would be equal.

(b) If repairs take 8 hours, Scotty's would be less expensive as shown below.

Sally's	*Scotty's*
$c = 800 + 25n$	$c = 575 + 50n$
$c = 800 + 25(8)$	$c = 575 + 50(8)$
$c = 1000$	$c = 975$

For 8 hours labor, the cost of repairs at Sally's is $1000 and the cost of repair at Scotty's is $975.

Simple Interest Problems

We introduced the simple interest formula, interest = principal × rate × time or $i = prt$ in Section 3.1. Now we will use this formula to solve a simple interest problem using a system of equations. In Example 6 and the examples that follow, we organize the information in a table.

EXAMPLE 6 Emil has invested a total of $12,000 in two savings accounts. One account earns 5% simple interest and the other earns 8% simple interest. Find the amount invested in each account if he receives a total of $840 interest after 1 year.

Solution: Let x = principal invested at 5%
y = principal invested at 8%

	Principal	Rate	Time	Interest
5% account	x	0.05	1	$0.05x$
8% account	y	0.08	1	$0.08y$

Since the total interest is $840, one of our equations is

$$0.05x + 0.08y = 840$$

Because the total principal invested is $12,000, our second equation is

$$x + y = 12{,}000$$

$$\left.\begin{aligned} 0.05x + 0.08y &= 840 \\ x + y &= 12{,}000 \end{aligned}\right\} \text{system of equations}$$

We will multiply our first equation by 100 to eliminate the decimal numbers. This gives the system

$$\begin{aligned} 5x + 8y &= 84{,}000 \\ x + y &= 12{,}000 \end{aligned}$$

To eliminate the x, we will multiply the second equation by -5 and then add the results to the first equation.

$$\begin{aligned} 5x + 8y &= 84{,}000 \\ -5[x + y &= 12{,}000] \end{aligned} \quad \text{gives} \quad \begin{aligned} 5x + 8y &= 84{,}000 \\ -5x - 5y &= -60{,}000 \\ \hline 3y &= 24{,}000 \\ y &= 8000 \end{aligned}$$

Now solve for x.

$$\begin{aligned} x + y &= 12{,}000 \\ x + 8000 &= 12{,}000 \\ x &= 4000 \end{aligned}$$

Thus, $4000 is invested at 5% and $8000 is invested at 8%.

Motion Problems with Two Rates

We introduced the distance formula, distance = rate · time or $d = rt$, in Section 4.7 and worked additional motion problems in Section 6.7. Now we introduce a method, using two variables and a system of equations, to solve motion problems that involve two rates. When working problems using this method, we use the information given to write two equations in two variables. Then we proceed to solve the system of equations using one of the three methods introduced in this chapter. Often when working motion problems with two different rates it is helpful to construct a table indicating the information given. We do this in Examples 7 and 8.

EXAMPLE 7 It takes Malcolm 3 hours in his motorboat to make a 48-mile trip downstream with the current. The return trip against the current takes him 4 hours. Find **(a)** the speed of the motorboat in still water, and **(b)** the speed of the current.

Solution: **(a)** Let us make a sketch of the situation (Fig. 8.7).

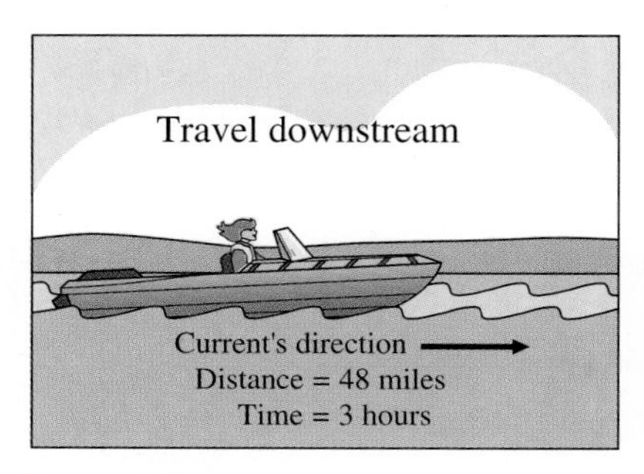

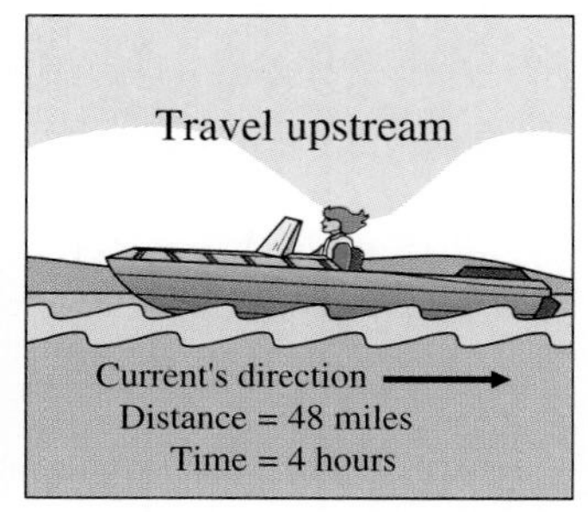

FIGURE 8.7

Let m = speed of the motorboat in still water
c = speed of the current

Boat's Direction	Rate	Time	Distance
With current	$m + c$	3	$3(m + c)$
Against current	$m - c$	4	$4(m - c)$

Since the distances traveled downstream and upstream are both 48 miles, our system of equations is

$$3(m + c) = 48$$
$$4(m - c) = 48$$

If we divide both sides of the top equation by 3 and both sides of the bottom equation by 4, we obtain a simplified system of equations.

$$\frac{3(m + c)}{3} = \frac{48}{3} \qquad \frac{4(m - c)}{4} = \frac{48}{4}$$
$$m + c = 16 \qquad m - c = 12$$

Now we solve the simplified system of equations.

$$\begin{aligned} m + c &= 16 \\ m - c &= 12 \\ \hline 2m \quad &= 28 \\ m &= 14 \end{aligned}$$

Therefore, the speed of the boat in still water is 14 miles per hour.

(b) The speed of the current may be found by substituting 14 for m in either of the simplified equations. We will use $m + c = 16$.

$$m + c = 16$$
$$14 + c = 16$$
$$c = 2$$

Thus, the speed of the current is 2 miles per hour.

EXAMPLE 8 Dawn and Chris go hiking down the Grand Canyon. Chris begins hiking 0.5 hour before Dawn. Chris travels at 2 miles per hour and Dawn travels at 1.5 miles per hour. How long will it take, after Chris begins hiking, for Chris and Dawn to be 2 miles apart?

Solution: We need to find the time it takes for Chris and Dawn to become separated by 2 miles.

Let x = time Chris is hiking
y = time Dawn is hiking

Hiker	Rate	Time	Distance
Chris	2	x	$2x$
Dawn	1.5	y	$1.5y$

Since Chris begins hiking 0.5 hour before Dawn, our first equation is

$$x = y + 0.5$$

Note that if we add 0.5 hour to Dawn's time we get the time Chris has been hiking.

Our second equation is obtained from the fact that the distance between the two hikers must be 2 miles. Since the hikers are traveling in the same direction, we must subtract their distances to obtain a difference of 2 miles. Since Chris is traveling at a faster speed and started first, we subtract Dawn's distance from Chris's distance.

$$\text{Chris's distance} - \text{Dawn's distance} = 2 \text{ miles}$$
$$2x - 1.5y = 2$$

The system of equations is

$$x = y + 0.5$$
$$2x - 1.5y = 2$$

The first equation is already solved for x. Substituting $y + 0.5$ for x in the second equation, we get

$$2(y + 0.5) - 1.5y = 2$$
$$2y + 1 - 1.5y = 2$$
$$0.5y + 1 = 2$$
$$0.5y = 1$$
$$y = \frac{1}{0.5} = 2$$

Thus, the time Dawn has been hiking is 2 hours. Since Chris has been hiking 0.5 hour longer than Dawn, Chris has been hiking for 2.5 hours when the distance between them is 2 miles.

Mixture Problems

Mixture problems were solved with one variable in Section 4.7. Now we will solve mixture problems using two variables and systems of equations. Recall that any problem in which two or more quantities are combined to produce a different quantity, or a single quantity is separated into two or more quantities, may be considered a mixture problem.

EXAMPLE 9 Deborah, who owns a coffee shop, wishes to mix Amaretto coffee that sells for \$6 per pound with 12 pounds of Kona coffee that sells for \$7.50 per pound.

(a) How many pounds of Amaretto coffee must be mixed with the 12 pounds of Kona coffee to obtain a mixture worth \$6.50 per pound?

(b) How much of the mixture will be produced?

Solution:

(a) We are asked to find the number of pounds of Amaretto coffee.

Let x = number of pounds of Amaretto coffee

y = number of pounds of mixture

Often it is helpful to make a sketch of the situation. After we draw a sketch we will construct a table. In our sketch we will use barrels to mix the coffee in (Fig. 8.8).

FIGURE 8.8

The value of the coffee is found by multiplying the number of pounds by the price per pound.

Coffee	Price	Number of Pounds	Value of Coffee
Amaretto	6	x	$6x$
Kona	7.50	12	7.50(12)
Mixture	6.50	y	$6.50y$

Our two equations come from the following information:

$$\begin{pmatrix}\text{number of pounds}\\ \text{of Amaretto coffee}\end{pmatrix} + \begin{pmatrix}\text{number of pounds}\\ \text{of Kona coffee}\end{pmatrix} = \begin{pmatrix}\text{number of pounds}\\ \text{in mixture}\end{pmatrix}$$

$$x + 12 = y$$

value of Amaretto coffee + value of Kona coffee = value of mixture

$$6x + 7.50(12) = 6.50y$$

system of equations $\begin{cases} x + 12 = y \\ 6x + 7.50(12) = 6.50y \end{cases}$

Now substitute $x + 12$ for y in the bottom equation and solve for x.

$$\begin{aligned} 6x + 7.50(12) &= 6.50y \\ 6x + 7.50(12) &= 6.50(x + 12) \\ 6x + 90 &= 6.50x + 78 \\ 90 &= 0.50x + 78 \\ 12 &= 0.50x \\ 24 &= x \end{aligned}$$

Thus, 24 pounds of Amaretto coffee must be mixed with 12 pounds of Kona coffee.

(b) The total mixture will be 24 + 12 or 36 pounds.

EXAMPLE 10 A 50% sulfuric acid solution is to be mixed with a 75% sulfuric acid solution to get 60 liters of a 60% sulfuric acid solution. How many liters of the 50% solution and the 75% solution should be mixed?

Solution: Let x = number of liters of 50% solution
y = number of liters of 75% solution

The problem is displayed in Figure 8.9.

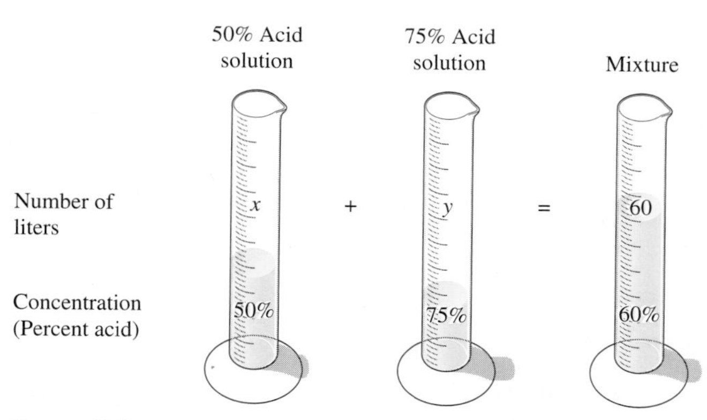

FIGURE 8.9

The acid content of a solution is found by multiplying the volume of the solution by the concentration of the solution.

	Number of Liters	Concentration	Acid Content
50% Solution	x	0.50	$0.50x$
75% Solution	y	0.75	$0.75y$
Mixture	60	0.60	0.60(60)

Because the total volume of the combination is 60 liters, we have

$$x + y = 60$$

From the table we see that

$$\begin{pmatrix}\text{acid content of}\\ \text{50\% solution}\end{pmatrix} + \begin{pmatrix}\text{acid content of}\\ \text{75\% solution}\end{pmatrix} = \begin{pmatrix}\text{acid content}\\ \text{of mixture}\end{pmatrix}$$

$$0.5x + 0.75y = 0.6(60)$$

system of equations
$$\begin{cases} x + y = 60 \\ 0.5x + 0.75y = 0.6(60) \end{cases}$$

We will solve this system by substitution. Solve for y in the first equation.

$$x + y = 60$$
$$y = 60 - x$$

Substitute $60 - x$ for y in the second equation.

$$0.5x + 0.75y = 0.6(60)$$
$$0.5x + 0.75(60 - x) = 36$$
$$0.5x + 45 - 0.75x = 36$$
$$-0.25x + 45 = 36$$
$$-0.25x = -9$$
$$x = \frac{-9}{-0.25} = 36$$

Now solve for y.

$$y = 60 - x$$
$$y = 60 - 36$$
$$y = 24$$

Thus, 36 liters of a 50% acid solution should be mixed with 24 liters of a 75% acid solution to obtain 60 liters of a 60% acid solution.

Exercise Set 8.4

Express each exercise as a system of linear equations, and then find the solution. Use a calculator where appropriate.

1. The sum of two integers is 26. Find the numbers if one number is 2 greater than twice the other.
2. The difference of two integers is 25. Find the two numbers if the larger is 1 less than three times the smaller.
3. Angles A and B are complementary angles. If angle B is 18° greater than angle A, find the measure of each angle. (See Example 1.)

4. Two angles are **supplementary angles** when the sum of their measures is 180°. If angles A and B are supplementary angles, and angle A is four times as large as angle B, find the measure of each angle.

5. If angles A and B are supplementary angles and angle A is 52° greater than angle B, find the measure of each angle.

6. A rectangular picture frame will be made from a piece of molding 60 inches long. What dimensions will the frame have if the length is to be 6 inches greater than the width? (See Example 2.)

7. The perimeter of a rectangular plot of land is 800 feet. If the length is 100 feet greater than the width, find the dimensions of the piece of land.

8. Paul has 25 rare currency bills in his safe deposit box. If their face values total \$101 and Paul has only \$2 and \$5 bills, find the number of each type of currency in his collection.

9. Robin collects 1-ounce gold dollar coins. Her collection consists of 14 coins, which are either gold United States Eagles or gold Canadian Maple Leafs. The total value of her collection is \$6560. If the value of the Eagles is \$480 each and the value of the Maple Leafs is \$460 each, find the number of Eagles and Maple Leafs that Robin owns.

10. Max Cisneros plants corn and wheat on his 62-acre farm in Albuquerque. He estimates that his income before deducting expenses is \$3000 per acre of corn and \$2200 per acre of wheat. Find the number of acres of corn and wheat planted if his total income before expenses is \$158,800.

11. Karla bought five times as many shares of BancOne stock as she did of Microsoft stock. The BancOne stock cost \$37 a share and the Microsoft stock cost \$75 a share. If her total cost for all the stock was \$7,800, how many shares of each did she purchase?

12. A plane can fly 540 miles per hour with the wind and 490 miles per hour against the wind. Find the speed of the plane in still air and the speed of the wind. (See Example 3.)

13. Carlos can paddle a kayak 4.5 miles per hour with the current and 3.2 miles per hour against the current. Find the speed of the kayak in still water and the speed of the current.

14. The population of Alpine Mountain is 40,000 and it is growing by 800 per year. The population of Beautiful Valley is 66,000 and it is decreasing by 500 per year. How long will it take for both areas to have the same population?

15. Sol's Club Discount Warehouse has two membership plans. Under plan A the customer pays a \$50 annual membership and 85% of the manufacturer's recommended list price. Under plan B the annual membership fee is \$100 and the customer pays 80% of the manufacturer's recommended list price. How much merchandise, in dollars, would one have to purchase to pay the same amount under both plans?

16. Condita, a salesperson, is considering two job offers. She would be selling the same product at each company. At the AMEXI Company, Condita's salary would be \$300 per week plus a 5% commission on sales. At the ROMAX company, her salary would be \$200 per week plus an 8% commission of sales.
    (a) What weekly dollar volume of sales would Condita need to make for the total income from both companies to be the same?
    (b) If she expects to make sales of \$4000, which company would give the greater salary?

17. Jerry is a Financial Planner for Gnocci & Co. His salary is a flat 40% commission of sales. As an employee he has no overhead. He is considering starting his own company. Then 100% of sales would be income to him. However, he estimates his monthly overhead for office rent, secretary, utilities, and so on, would be about \$1500 per month.
    (a) How much in sales would Jerry's own company need to make in a month for him to make the same income he did as an employee of Gnocci & Co.?
    (b) Suppose when Jerry opens his own office that in addition to the \$1500 per month overhead, he has a one-time cost of \$6000 for the purchase of office equipment. If he estimates his monthly sales at \$3000, how long would it take to recover the initial \$6000 cost?

Review Example 6 before working Exercises 18–20.

18. Mr. and Mrs. McAdams invest a total of \$8000 in two savings accounts. One account gives 10% simple interest and the other 8% simple interest. Find the amount placed in each account if they receive a total of \$750 in interest after 1 year.

19. The Websters wish to invest a total of \$12,500 in two savings accounts. One account pays 10% simple inter-

est and the other, $5\frac{1}{4}\%$ simple interest. The Websters wish their interest from the two accounts to be at least \$1200 at the end of the year. Find the minimum amount that can be placed in the account giving 10% interest.

20. The Cohens invested a total of \$10,000. Part of the money was placed in a savings account paying 5% simple interest. The rest was placed in a fixed annuity paying 6% simple interest. If the total interest received for the year was \$540, how much had been invested in each account?

Review Examples 7 and 8 before working Exercises 21–26.

21. During a race, Teresa's speed boat travels 4 miles per hour faster than Jill's speed boat. If Teresa's boat finishes the race in 3 hours and Jill finishes the race in 3.2 hours, find the speed of each boat.

22. Bob started to drive from Columbus, Ohio, toward Chicago, Illinois, a distance of 903 miles. At the same time Bob started, Mickey starts driving to Columbus, Ohio, from Chicago, Illinois. If the two meet after 7 hours, and Mickey's speed averages 15 miles per hour greater than Bob's speed, find the speed of each car.

23. A United Airlines jet leaves New York's Kennedy Airport headed for Los Angeles International Airport, a distance of 2700 miles. At the same time, a Delta Airlines jet leaves the Los Angeles Airport headed for Kennedy Airport. Due to headwinds and tailwinds, the United jet's average speed is 100 miles per hour greater than that of the Delta jet. If the two jets pass one another after 3 hours, find the speed of each plane.

24. Two cars start at the same time 240 miles apart and travel toward each other. One car travels 6 miles per hour faster than the other. If they meet after 2 hours, what was the average speed of each car?

25. Micki and Petra go jogging along the same trail. Micki starts 0.3 hour before Petra. If Micki jogs at a rate of 5 miles per hour and Petra jogs at 8 miles per hour, how long after Petra starts would it take for Petra to catch up to Micki?

26. Simon trots his horse Chipmunk east at 8 miles per hour. One half-hour later, Michelle starts at the same point and canters her horse Buttermilk west at 16 miles per hour. How long will it take after Michelle starts riding for Michelle and Simon to be separated by 10 miles?

Review Examples 9 and 10 before working Exercises 27–32.

27. Marie, a chemist, has a 25% hydrochloric acid solution and a 50% hydrochloric acid solution. How many liters of each should she mix to get 10 liters of a hydrochloric acid solution with a 40% acid concentration?

28. Moura Williams, a druggist, needs 1000 milliliters of a 10% phenobarbital solution. She has only 5% and 25% phenobarbital solutions available. How many milliliters of each solution should she mix to obtain the desired solution?

29. Janet wishes to mix 30 pounds of coffee that will sell for \$160. To obtain the mixture, she will mix coffee that sells for \$5 per pound with coffee that sells for \$7 per pound. How many pounds of each type coffee should she use?

30. Jason has milk that is 5% butterfat and skim milk without butterfat. How much 5% milk and how much skim milk should he mix to make 100 gallons of milk that is 3.5% butterfat?

31. The All Natural Juice Company sells apple juice for 12 cents an ounce and apple drink for 6 cents an ounce. They wish to market and sell for 10 cents an ounce cans of juice drink that are part juice and part drink. How many ounces of each will be used if the juice drink is to be sold in 8-ounce cans?

32. Pierre's recipe for Quiche Lorraine calls for 16 ounces (or 2 cups) of light cream, which is 20% milk fat. It is often difficult to find light cream with 20% milk fat at the supermarket. What is commonly found is heavy cream, which is 36% milk fat, and half and half, which is 10.5% milk fat. How many ounces of the heavy cream and how much of the half and half should be mixed to obtain 16 ounces of light cream that is 20% milkfat?

CUMULATIVE REVIEW EXERCISES

[1.9] **33.** Evaluate $3(4x - 3)^2 - 2y^2 - 1$ when $x = 3$ and $y = -2$.

[4.2] **34.** Simplify $(3x^4)^{-2}$.

[4.4] **35.** Indicate whether or not each of the following is a polynomial. If it is not a polynomial explain why. Give the degree of each polynomial.

(a) $3x^3 + 2x - 6$ **(b)** $\frac{1}{2}x^4 - 3x^2 - 2$

(c) $x^3 - 2x^2 - x^{-1}$

[4.6] **36.** Divide $(2x^2 + 5x - 10) \div (x + 4)$.

Group Activity/ Challenge Problems

1. Two pressurized tanks are connected by a controlled pressure valve as shown in the figure.

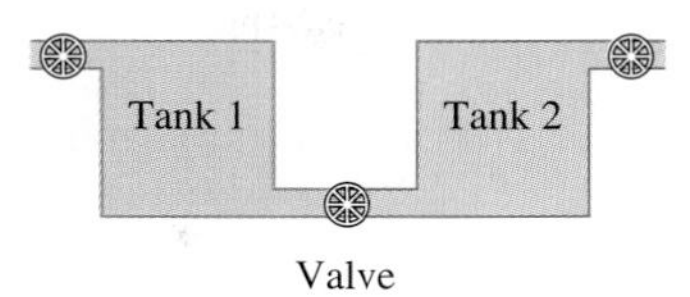

Initially, the internal pressure in tank 1 is 200 pounds per square inch, and the internal pressure in tank 2 is 20 pounds per square inch. The pressure valve is opened slightly to reduce the pressure in tank 1 by 2 pounds per square inch per minute. This increases the pressure in tank 2 by 2 pounds per square inch per minute. At this rate, how long will it take for the pressure to be equal in both tanks?

2. Mrs. O'Neil is considering cars for purchase. Car A has a list price of \$10,500 and gets an average of 40 miles per gallon. Car B has a list price of \$9500 and gets an average of 20 miles per gallon. Being a conservationist, Mrs. O'Neil wishes to purchase car A but is concerned about its greater initial cost. She plans to keep the car for many years. If she purchases car A, how many miles would she need to drive for the total cost of car A to equal the total cost of car B? Assume gasoline costs of \$1.25 per gallon.

3. Two brothers jog to school daily. The older jogs at 9 miles per hour, the younger, at 5 miles per hour. When the older brother reaches the school, the younger brother is $\frac{1}{2}$ mile away. How far is the school from the boys' house?

4. By weight, an alloy of brass is 70% copper and 30% zinc. Another alloy of brass is 40% copper and 60% zinc. How many grams of each of these alloys must be melted and combined to obtain 300 grams of a brass alloy that is 60% copper and 40% zinc?

8.5 Systems of Linear Inequalities

Tape 14  

1 Solve systems of linear inequalities graphically.

Solve Graphically

1 In Section 7.7, we learned how to graph linear inequalities in two variables. In Section 8.1, we learned how to solve systems of equations graphically. In this section, we discuss how to solve systems of linear inequalities graphically. The **solution to a system of linear inequalities** is the set of points that satisfies all inequalities in the system. Although a system of linear inequalities may contain more than two inequalities, in this book, except in the Group Activity Exercises, we will consider systems with only two inequalities.

To Solve a System of Linear Inequalities

Graph each inequality on the same axes. The solution is the set of points that satisfies all the inequalities in the system.

EXAMPLE 1 Determine the solution to the system of inequalities.

$$x + 2y \le 6$$
$$y > 2x - 4$$

Solution: First graph the inequality $x + 2y \le 6$ (Fig. 8.10).

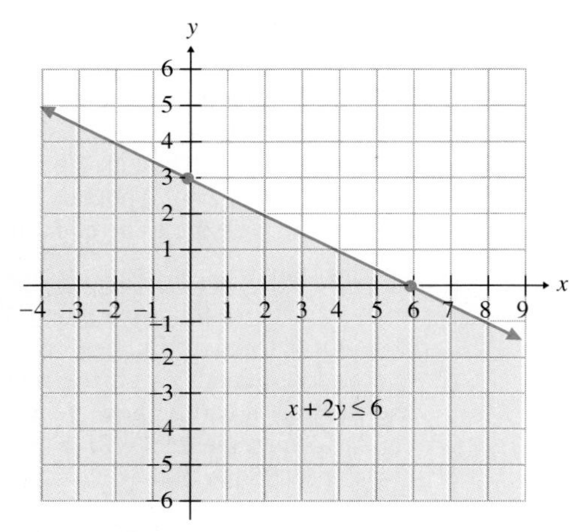

FIGURE 8.10

Now, on the same axes, graph the inequality $y > 2x - 4$ (Fig. 8.11). Note that the line is dashed. Why?

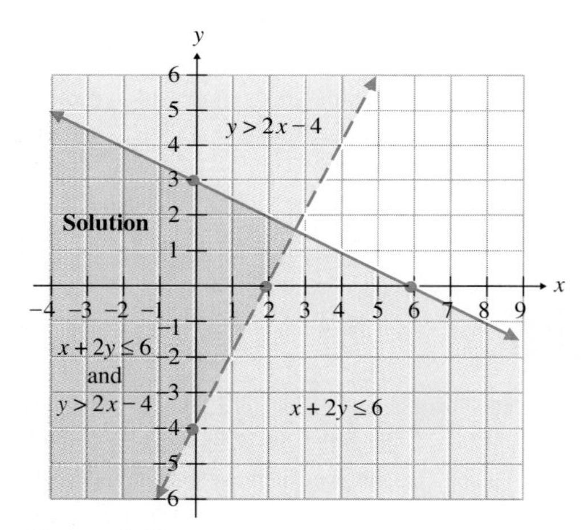

FIGURE 8.11

The solution is the set of points common to both inequalities. It is the part of the graph that contains both shadings. The dashed line is not part of the solution. However, the part of the solid line that satisfies both inequalities is part of the solution.

EXAMPLE 2 Determine the solution to the system of inequalities.

$$2x + 3y \ge 4$$
$$2x - y > -6$$

Solution: Graph $2x + 3y \geq 4$ (Fig. 8.12). Graph $2x - y > -6$ on the same axes (Fig. 8.13). The solution is the part of the graph with both shadings and the part of the solid line that satisfies both inequalities.

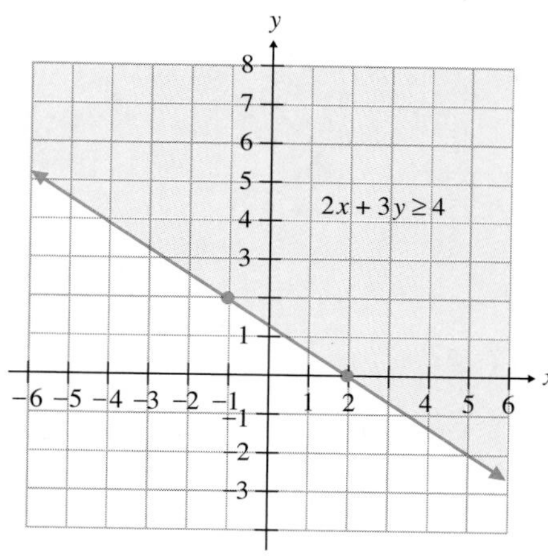

FIGURE 8.12

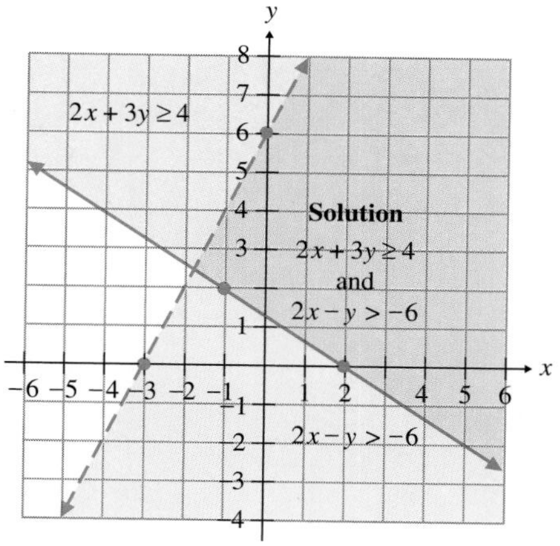

2x + 3y ≥ 4
Solution
2x + 3y ≥ 4
and
2x − y > −6
2x − y > −6

FIGURE 8.13

EXAMPLE 3 Determine the solution to the system of inequalities.

$$y < 2$$
$$x > -3$$

Solution: Graph both inequalities on the same axes (Fig. 8.14).

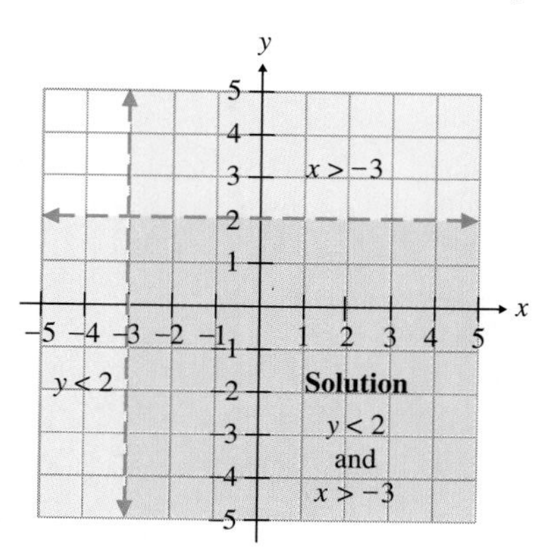

FIGURE 8.14

Exercise Set 8.5

Determine the solution to each system of inequalities.

1. $x + y > 2$
   $x - y < 2$

2. $y \leq 3x - 2$
   $y > -4x$

3. $y \leq x$
   $y < -2x + 4$

4. $2x + 3y < 6$
$4x - 2y \geq 8$

5. $y > x + 1$
$y \geq 3x + 2$

6. $x + 3y \geq 6$
$2x - y > 4$

7. $x - 2y < 6$
$y \leq -x + 4$

8. $y \leq 3x + 4$
$y < 2$

9. $4x + 5y < 20$
$x \geq -3$

10. $3x - 4y \leq 12$
$y > -x + 4$

11. $x \leq 4$
$y \geq -2$

12. $x \geq 0$
$y \leq 0$

13. $x > -3$
$y > 1$

14. $4x + 2y > 8$
$y \leq 2$

15. $-2x + 3y \geq 6$
$x + 4y \geq 4$

16. Can a system of linear inequalities have no solution? Explain your answer with the use of your own example.

17. Is it possible to construct a system of two nonparallel linear inequalities that has no solution? Explain.

CUMULATIVE REVIEW EXERCISES

[4.5] **18.** Multiply $(x^2 - 2x + 7)(3x - 4)$.

[5.2] **19.** Factor $xy + x - 3y - 3$.

[5.3] **20.** Factor $x^2 - 13x + 42$.

[5.4] **21.** Factor $6x^2 - x - 2$.

Group Activity/ Challenge Problems

Determine the solution to the following system of inequalities.

1. $x + 2y \leq 6$
$2x - y < 2$
$y > 2$

2. $x \geq 0$
$y \geq 0$
$y \leq 2x + 4$
$y \leq -x + 6$

3. $x \geq 0$
$y \geq 0$
$5x + y \leq 40$
$10x + y \leq 60$
$x \leq 4$

Summary

GLOSSARY

Consistent system of linear equations (442): A system of linear equations that has a solution.

Dependent system of linear equations (442): A system of linear equations that has an infinite number of solutions.

Inconsistent system of linear equations (442): A system of linear equations that has no solution.

Simultaneous linear equations or a system of linear equations (440): Two or more linear equations considered together.

Solution to a system of equations (441): The ordered pair or pairs that satisfy all equations in the system.

Solution to a system of inequalities (477): The set of ordered pairs that satisfies all inequalities in the system.

IMPORTANT FACTS

Consistent, exactly 1 solution

Inconsistent, no solution

Dependent, infinite number of solutions

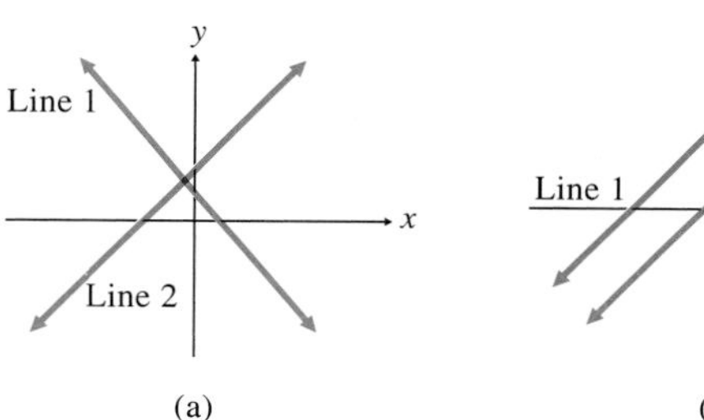

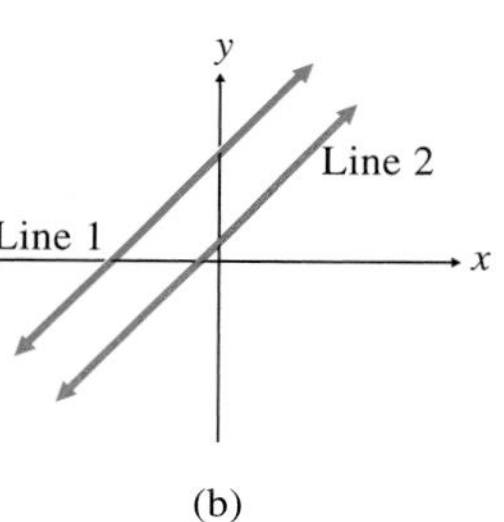

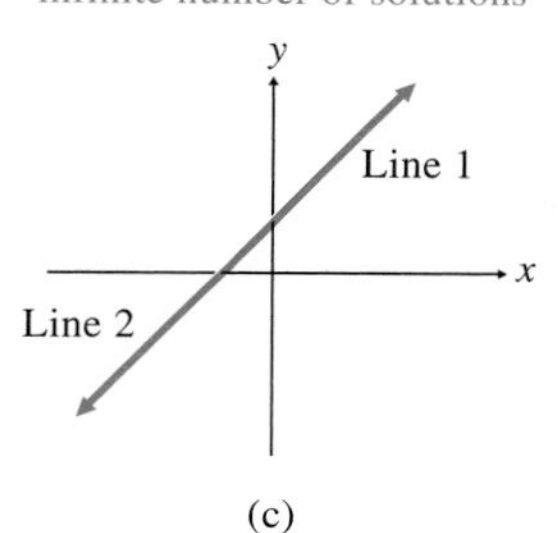

(a) (b) (c)

Three methods that can be used to solve a system of linear equations are the (1) graphical method, (2) substitution method, and (3) addition (or elimination) method.

Review Exercises

Determine which, if any, of the ordered pairs satisfy each system of equations.

1. $y = 3x - 2$
$2x + 3y = 5$
(a) $(0, -2)$ **(b)** $(2, 4)$ **(c)** $(1, 1)$

2. $y = -x + 4$
$3x + 5y = 15$
(a) $\left(\frac{5}{2}, \frac{3}{2}\right)$ **(b)** $(0, 4)$ **(c)** $\left(\frac{1}{2}, \frac{3}{5}\right)$

Identify each system of linear equations as consistent, inconsistent, or dependent. State whether the system has exactly one solution, no solution, or an infinite number of solutions.

3.

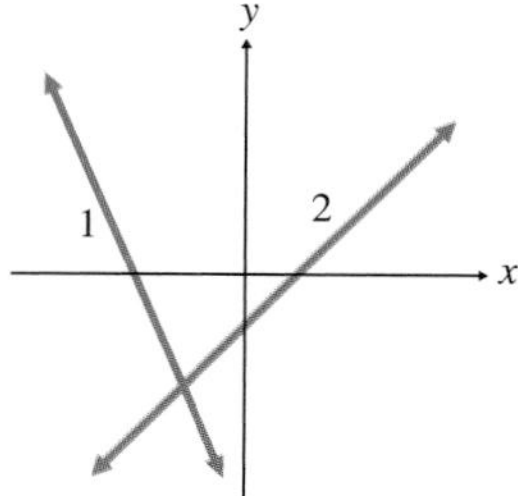

4.

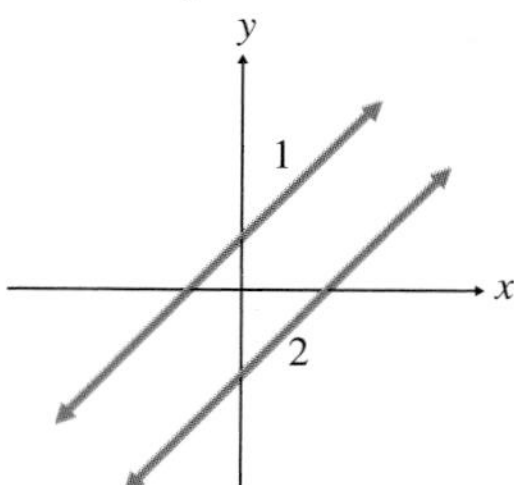

5.

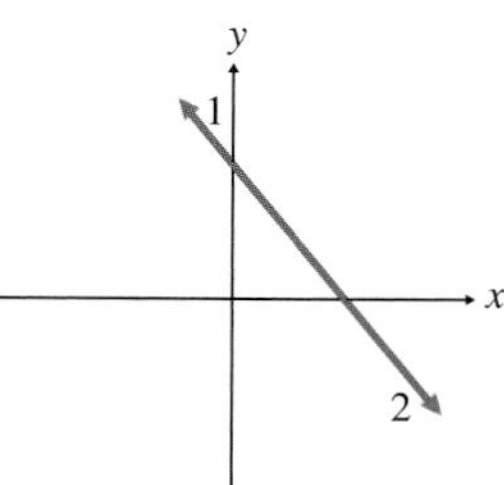

6.

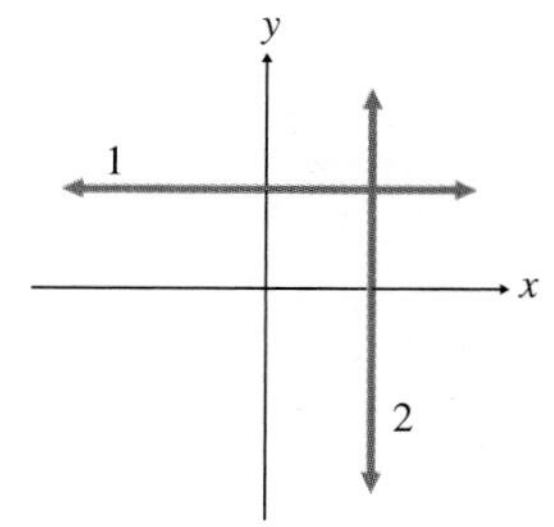

Write each equation in slope–intercept form. Without graphing or solving the system of equations, state whether the system of linear equations has exactly one solution, no solution, or an infinite number of solutions.

7. $x + 2y = 8$
$3x + 6y = 12$

8. $y = -3x - 6$
$2x + 3y = 8$

9. $y = \frac{1}{2}x - 4$
$x - 2y = 8$

10. $6x = 4y - 8$
$4x = 6y + 8$

Determine the solution to each system of equations graphically.

11. $y = x + 3$
$y = 2x + 5$

12. $x = -2$
$y = 3$

13. $y = 3$
$y = -2x + 5$

14. $x + 3y = 6$
$y = 2$

15. $x + 2y = 8$
$2x - y = -4$

16. $y = x - 3$
$2x - 2y = 6$

17. $2x + y = 0$
$4x - 3y = 10$

18. $x + 2y = 4$
$\frac{1}{2}x + y = -2$

[8.2] *Find the solution to each system of equations using substitution.*

19. $y = 2x - 8$
$2x - 5y = 0$

20. $x = 3y - 9$
$x + 2y = 1$

21. $2x + y = 5$
$3x + 2y = 8$

22. $2x - y = 6$
$x + 2y = 13$

23. $3x + y = 17$
$2x - 3y = 4$

24. $x = -3y$
$x + 4y = 6$

25. $4x - 2y = 10$
$y = 2x + 3$

26. $2x + 4y = 8$
$4x + 8y = 16$

27. $2x - 3y = 8$
$6x + 5y = 10$

28. $4x - y = 6$
$x + 2y = 8$

[8.3] *Find the solution to each system of equations using the addition method.*

29. $x + y = 6$
$x - y = 10$

30. $x + 2y = -3$
$2x - 2y = 6$

31. $2x + 3y = 4$
$x + 2y = -6$

32. $x + y = 12$
$2x + y = 5$

33. $4x - 3y = 8$
$2x + 5y = 8$

34. $-2x + 3y = 15$
$3x + 3y = 10$

35. $2x + y = 9$
$-4x - 2y = 4$

36. $2x + 2y = 8$
$y = 4x - 3$

37. $3x + 4y = 10$
$-6x - 8y = -20$

38. $2x - 5y = 12$
$3x - 4y = -6$

[8.4] *Express as a system of linear equations, and then find the solution.*

39. The sum of two integers is 48. Find the two numbers if the larger is 3 less than twice the smaller.

40. A plane flies 600 miles per hour with the wind and 530 miles per hour against the wind. Find the speed of the wind and the speed of the plane in still air.

41. ABC Truck Rental charges $20 per day plus 50 cents per mile, while Murtz Truck Rental charges $35 per day plus 40 cents per mile. How far would you have to travel in one day for the total cost of both rental companies to be the same?

42. The Hackets invested a total of $16,000. Part of the money was placed in a savings account paying 4% simple interest. The rest was placed in a savings account paying 6% simple interest. If the total interest received for the year was $760, how much had they invested in each account?

43. Ron drives from Charleston, South Carolina to Louisville, Kentucky, a distance of 600 miles. At the same time, Audra starts driving from Louisville to Charleston along the same route. If the two meet after driving 5 hours and Audra's average speed was 6 miles per hour greater than Ron's, find the average speed of each car.

44. Green Turf's grass seed costs 60 cents a pound and Agway's grass seed costs 45 cents a pound. How many pounds of each were used to make a 40-pound mixture that cost $20.25?

45. A chemist has a 30% acid solution and a 50% acid solution. How much of each must be mixed to get 6 liters of a 40% acid solution?

[8.5] *Determine the solution to each system of inequalities.*

46. $x + y > 2$
$2x - y \leq 4$

47. $2x - 3y \leq 6$
$x + 4y > 4$

48. $2x - 6y > 6$
$x > -2$

49. $x < 2$
$y \geq -3$

Practice Test

1. Determine which, if any, of the ordered pairs satisfy the system of equations.

$$x + 2y = -6$$
$$3x + 2y = -12$$

(a) $(0, -6)$ **(b)** $\left(-3, -\frac{3}{2}\right)$ **(c)** $(2, -4)$

Identify each system as consistent, inconsistent, or dependent. State whether the system has exactly one solution, no solution, or an infinite number of solutions.

2.

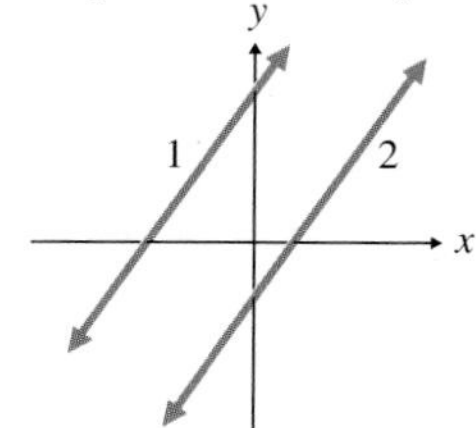

3.

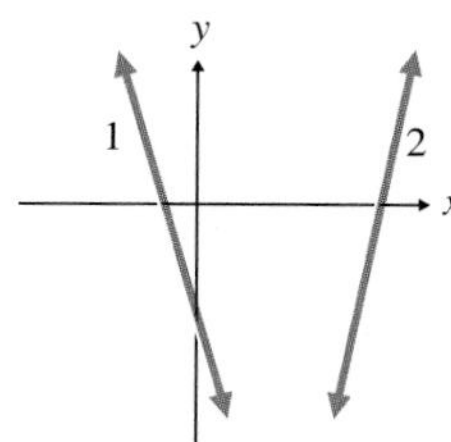

4.

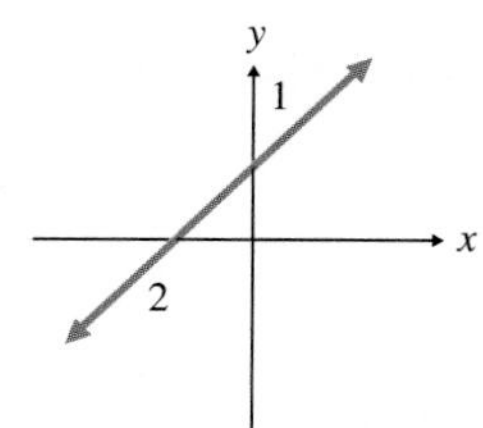

Write each equation in slope–intercept form. Then determine, without solving the system, whether the system of equations has exactly one solution, no solution, or an infinite number of solutions.

5. $3y = 6x - 9$
   $2x - y = 6$

6. $3x + 2y = 10$
   $3x - 2y = 10$

Solve each system of equations graphically.

7. $y = 3x - 2$
   $y = -2x + 8$

8. $3x - 2y = -3$
   $3x + y = 6$

Solve each system of equations using substitution.

9. $3x + y = 8$
   $x - y = 6$

10. $4x - 3y = 9$
    $2x + 4y = 10$

Solve each system of equations using the addition method.

11. $2x + y = 5$
    $x + 3y = -10$

12. $3x + 2y = 12$
    $-2x + 5y = 8$

Express the problem as a system of linear equations, and then find the solution.

13. Budget Rent a Car Agency charges \$40 per day plus 8 cents per mile to rent a certain model car. Hertz charges \$45 per day plus 3 cents per mile to rent the same car. How many miles will have to be driven in one day for the cost of Budget's car to equal the cost of Hertz's car?

14. Albert's Grocery sells cashews for \$6.00 a pound and peanuts for \$4.50 a pound. How much of each must Albert mix to get 20 pounds of a mixture that he can sell for \$5.00 per pound?

15. Determine the solution to the system of inequalities.

$$2x + 4y < 8$$
$$x - 3y \geq 6$$

Cumulative Review Test

1. Evaluate $\dfrac{|-4| + |-16| \div 2^2}{3 - [2 - (4 \div 2)]}$.
2. Solve the equation $4(x - 2) + 6(x - 3) = 2 - 4x$.
3. Solve the equation $3x^2 - 13x + 12 = 0$.
4. Solve the equation $\dfrac{1}{3}(x + 2) + \dfrac{1}{4} = 8$.
5. Solve the equation $\dfrac{1}{x - 3} + \dfrac{1}{x + 3} = \dfrac{1}{x^2 - 9}$.
6. Find the length of side x if the two figures are similar.

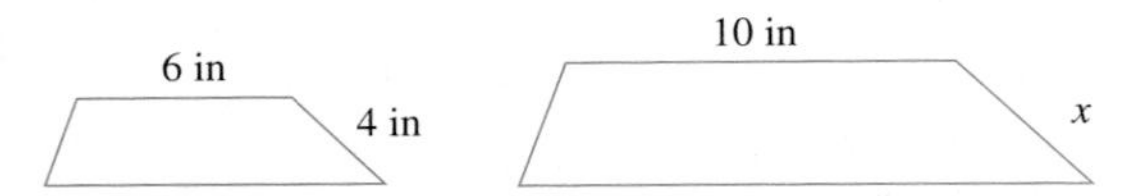

7. Simplify $(x^5y^3)^4(2x^3y^5)$.
8. Factor $6x^2 - 11x + 4$.
9. Subtract $\dfrac{4}{x^2 - 9} - \dfrac{3}{x^2 - 9x + 18}$.
10. Divide $\dfrac{x^2 - 7x + 12}{2x^2 - 11x + 12} \div \dfrac{x^2 - 9}{x^2 - 16}$.
11. Graph the equation $2x - 3y = 6$ by plotting points.
12. Graph the equation $3x + 2y = 9$ using the x and y intercepts.
13. Graph the inequality $2x - y < 6$.
14. Without graphing the equation, determine if the following system of equations has exactly one solution, no solutions, or an infinite number of solutions. Explain.

$$3x = 2y + 8$$
$$-4y = -6x + 12$$

15. Solve the system of equations graphically.

$$x + 2y = 2$$
$$2x - 3y = -3$$

16. Solve the system of equations using the addition method.

$$2x + 3y = 4$$
$$x - 4y = 6$$

17. A factory worker can inspect 40 units in 15 minutes. How long will it take her to inspect 160 units?
18. **(a)** An author is trying to decide between two publishing contract offers. The PCR Publishing Company offers an initial grant of \$20,000 plus a 10% royalty rate on sales. The ARA Publishing Company offers an initial grant of \$10,000 plus a 12% royalty rate on sales. How many dollars of sales are needed for the author's total income to be the same from both companies?

    (b) If the author expects total dollar sales of \$200,000, which company would result in the higher income?
19. How many liters of a 20% hydrochloric acid solution and how many liters of a 35% hydrochloric acid solution should be mixed to get 10 liters of a 25% hydrochloric acid solution?
20. Mr. and Mrs. Pontilo own a pizza shop. Mr. Pontilo can clean the pizza shop by himself in 50 minutes. Mrs. Pontilo can clean the pizza shop by herself in 60 minutes. How long will it take them to clean the pizza shop if they work together?

Chapter 9

Roots and Radicals

See Section 9.6, Exercise 37.

Preview and Perspective

In this chapter we study roots, radical expressions, and radical equations. In the great majority of this chapter we discuss only square roots. Square roots are one type of radical expression. In Sections 9.1 through 9.4 we learn how to evaluate a square root, how to simplify square root expressions, and how to add, subtract, multiply, and divide expressions that contain square roots. In Section 9.5 we discuss solving equations that contain square roots. Section 9.6, applications of radicals, is an extension of Section 9.5 and presents some real-life uses of square roots. In Section 9.7 we discuss cube roots and higher roots. We strongly suggest that you use a calculator with a square root key, $\sqrt{x}$, for this and the next chapter.

The material presented in this chapter will be used throughout Chapter 10. In particular, simplifying square root expressions containing only constants, like Examples 1 through 5 in Section 9.2, will be used in Sections 10.1 and 10.2.

Radical expressions and equations play an important part in mathematics and the sciences. Many mathematical and scientific formulas involve radicals. As you will see in Chapter 10, one of the most important formulas in mathematics, the quadratic formula, contains a square root.

The emphasis of this chapter is on square roots. Higher roots are introduced in Section 9.7. If you take intermediate algebra, you will study higher roots in more depth. Additionally, in intermediate algebra, or in later mathematics courses, you may find yourself graphing radical equations in two variables. The material presented here will help prepare you for that work also.

9.1 Evaluating Square Roots

Tape 14

1. Evaluate square roots of real numbers.
2. Recognize that not all square roots represent real numbers.
3. Determine whether a square root is a rational or an irrational number.
4. Write square roots as exponential expressions.

Evaluate Square Roots

1 In this section we introduce a number of important concepts related to radicals. We first discuss some terminology. Let us begin with square roots. Square roots are one type of radical expression that you will use in both mathematics and science.

$\sqrt{x}$ is read "the square root of x."

The $\sqrt{\ }$ is called the **radical sign.** The number or expression inside the radical sign is called the **radicand.**

radical sign

$$\sqrt{x}$$

radicand

The entire expression, including the radical sign and radicand, is called the **radical expression.**

Another part of a radical expression is its **index.** The index tells the "root" of the expression. Square roots have an index of 2. The index of square roots is generally not written.

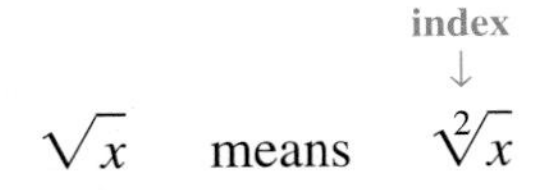

$$\sqrt{x} \quad \text{means} \quad \sqrt[2]{x}$$

Other types of radical expressions have different indexes. For example, $\sqrt[3]{x}$, which is read "the cube root of x," has an index of 3. Cube roots are discussed in Section 9.7.

Examples of Square Roots	*How Read*	*Radicand*
$\sqrt{8}$	the square root of 8	8
$\sqrt{5x}$	the square root of $5x$	$5x$
$\sqrt{\frac{x}{2y}}$	the square root of x over $2y$	$\frac{x}{2y}$

Every positive number has two square roots, a principal or positive square root and a negative square root.

> The **principal or positive square root** of a positive real number x, written $\sqrt{x}$, is that *positive* number whose square equals x.

Examples

$\sqrt{25} = 5$ since $5^2 = 5 \cdot 5 = 25$

$\sqrt{36} = 6$ since $6^2 = 6 \cdot 6 = 36$

$\sqrt{\frac{1}{4}} = \frac{1}{2}$ since $\left(\frac{1}{2}\right)^2 = \left(\frac{1}{2}\right)\left(\frac{1}{2}\right) = \frac{1}{4}$

$\sqrt{\frac{4}{9}} = \frac{2}{3}$ since $\left(\frac{2}{3}\right)^2 = \left(\frac{2}{3}\right)\left(\frac{2}{3}\right) = \frac{4}{9}$

The **negative square root** of a positive real number x, written $-\sqrt{x}$, is the additive inverse or opposite of the principal square root. For example, $-\sqrt{25} = -5$ and $-\sqrt{36} = -6$. **Whenever we use the term square root in this book, we mean the principal or positive square root.**

EXAMPLE 1 Evaluate: **(a)** $\sqrt{64}$ **(b)** $\sqrt{100}$

Solution: **(a)** $\sqrt{64} = 8$ since $8^2 = (8)(8) = 64$.

(b) $\sqrt{100} = 10$ since $(10)^2 = 100$.

EXAMPLE 2 Evaluate: **(a)** $-\sqrt{64}$ **(b)** $-\sqrt{100}$

Solution: **(a)** $\sqrt{64} = 8$. Now we take the opposite of both sides to get

$$-\sqrt{64} = -8.$$

(b) Similarly, $-\sqrt{100} = -10$.

Square Roots That Are Not Real Numbers

2 You must understand that **square roots of negative numbers are not real numbers.** Consider $\sqrt{-4}$; to what is $\sqrt{-4}$ equal? To evaluate $\sqrt{-4}$, we must find some number whose square equals -4. But we know that the square of any nonzero number must be a positive number. Therefore, no number squared is -4, so $\sqrt{-4}$ has no real value. Numbers like $\sqrt{-4}$, or the square root of any negative number, are called **imaginary numbers.** The study of imaginary numbers is beyond the scope of this book.

EXAMPLE 3 Indicate whether the radical expression is a real or imaginary number.

(a) $-\sqrt{9}$ (b) $\sqrt{-9}$ (c) $\sqrt{-37}$ (d) $-\sqrt{37}$

Solution:

(a) Real (equal to -3) (b) Imaginary (c) Imaginary (d) Real

Suppose we have an expression like $\sqrt{x}$, where x represents some number. For the radical $\sqrt{x}$ to be a real number, and not imaginary, we must assume that x is a nonnegative number.

In this chapter, unless stated otherwise, we will assume that all expressions that are radicands represent nonnegative numbers.

Rational and Irrational Numbers

3 To help in our discussion of rational and irrational numbers, we will define perfect squares. The numbers 1, 4, 9, 16, 25, 36, 49, . . . are called **perfect squares** because each number is *the square of a natural number.* When a perfect square is a factor of a radicand, we may refer to it as a **perfect square factor.**

$1, 2, 3, 4, 5, 6, 7, \ldots$ natural numbers

$1^2, 2^2, 3^2, 4^2, 5^2, 6^2, 7^2, \ldots$ the squares of the natural numbers

$1, 4, 9, 16, 25, 36, 49, \ldots$ perfect squares

What are the next two perfect squares? Note that the square root of a perfect square is an integer. That is, $\sqrt{1} = 1$, $\sqrt{4} = 2$, $\sqrt{9} = 3$, $\sqrt{16} = 4$, and so on.

Table 9.1 illustrates the 20 smallest perfect squares. You may wish to refer to this table when simplifying radical expressions.

Perfect Square	Square Root of Perfect Square		Value	Perfect Square	Square Root of Perfect Square		Value
1	$\sqrt{1}$	=	1	121	$\sqrt{121}$	=	11
4	$\sqrt{4}$	=	2	144	$\sqrt{144}$	=	12
9	$\sqrt{9}$	=	3	169	$\sqrt{169}$	=	13
16	$\sqrt{16}$	=	4	196	$\sqrt{196}$	=	14
25	$\sqrt{25}$	=	5	225	$\sqrt{225}$	=	15
36	$\sqrt{36}$	=	6	256	$\sqrt{256}$	=	16
49	$\sqrt{49}$	=	7	289	$\sqrt{289}$	=	17
64	$\sqrt{64}$	=	8	324	$\sqrt{324}$	=	18
81	$\sqrt{81}$	=	9	361	$\sqrt{361}$	=	19
100	$\sqrt{100}$	=	10	400	$\sqrt{400}$	=	20

A **rational number** is one that can be written in the form $\frac{a}{b}$, where a and b are integers, and $b \neq 0$. Examples of rational numbers are $\frac{1}{2}, \frac{3}{5}, -\frac{9}{2}$, 4, and 0. All integers are rational numbers since they can be expressed with a denominator of 1. For example, $4 = \frac{4}{1}$ and $0 = \frac{0}{1}$. The square roots of perfect squares are also rational numbers since each is an integer. When a rational number is written as a decimal, it will be either a terminating or repeating decimal.

Terminating Decimals	*Repeating Decimals*
$\frac{1}{2} = 0.5$	$\frac{1}{3} = 0.333\ldots$
$\frac{5}{8} = 0.625$	$\frac{4}{9} = 0.444\ldots$
$\sqrt{4} = 2.0$	$\frac{1}{6} = 0.1666\ldots$

Real numbers that are not rational numbers are called **irrational numbers.** Irrational numbers when written as decimals are nonterminating, nonrepeating decimals. The square root of every positive integer that is not a perfect square is an irrational number. For example, $\sqrt{2}$ and $\sqrt{3}$ are irrational numbers. The 20 square roots listed in Table 9.1 are rational numbers. All other square roots of integers between 1 and 400 are irrational numbers. For example, since $\sqrt{230}$ is not in Table 9.1, it is an irrational number. Furthermore, since $\sqrt{230}$ is between $\sqrt{225}$ and $\sqrt{256}$ in Table 9.1, the value of $\sqrt{230}$ is between 15 and 16.

Calculator Corner

Evaluating Square Roots on the Calculator

The square root key, $\sqrt{x}$, on calculators can be used to find square roots of nonnegative numbers. For example, to find the square root of 4, we press

$$4\ \sqrt{x}$$

The calculator then displays the answer 2. Since 2 is an integer, $\sqrt{4}$ is a rational number.

What would the calculator display if we evaluate $\sqrt{7}$?

$$7\ \sqrt{x}$$

The display shows 2.6457513. Note that $\sqrt{7}$ is an irrational number, or a nonrepeating, nonterminating decimal. The exact decimal value of $\sqrt{7}$, or any other irrational number, can never be given exactly. The answers given on a calculator display are only close approximations of their value.

Suppose we tried to evaluate $\sqrt{-4}$ on a calculator. What would the calculator give as an answer?

$$4\ {+/-}\ \sqrt{x}$$

The calculator would give an error message, because the square root of -4, or the square root of any other negative number, is not a real number.

The symbol $\approx$ means "is approximately equal to." From time to time we may use this symbol. For example, we may write $\sqrt{2} \approx 1.414$. This is read "the square root of 2 is approximately equal to 1.414." Recall that $\sqrt{2}$ is not a perfect square, so its square root cannot be evaluated exactly.

EXAMPLE 4 Use your calculator or Table 9.1 to determine whether the following square roots are rational or irrational numbers.

(a) $\sqrt{118}$ **(b)** $\sqrt{169}$ **(c)** $\sqrt{64}$ **(d)** $\sqrt{200}$

Solution:

(a) Irrational **(b)** Rational, equal to 13

(c) Rational, equal to 8 **(d)** Irrational

Changing Square Roots to Exponential Form

4 Radical expressions can be written in exponential form. Since we are discussing square roots, we will show how to write square roots in exponential form. Writing other radicals in exponential form will be discussed in Section 9.7. We introduce this information here because your instructor may wish to use exponential form to help explain certain concepts.

Recall that the index of square roots is 2. For example,

$$\sqrt{x} \text{ means } \sqrt[2]{x}$$

We use the index, 2, when writing square roots in exponential form. To change from an expression in square root form to an expression in exponential form, simply write the radicand of the square root to the 1/2 power, as follows:

Writing Square Root in Exponential Form

$$\sqrt{\blacksquare} = \blacksquare^{\frac{1}{2}} \leftarrow \text{index of square root}$$

↑ radicand ↑

For example, $\sqrt{8}$ in exponential form is $8^{1/2}$, and $\sqrt{2xy} = (2xy)^{1/2}$. Other examples are

Square Root Form		*Exponential Form*
$\sqrt{16}$	=	$(16)^{1/2}$
$\sqrt{3x}$	=	$(3x)^{1/2}$
$\sqrt{12x^2y}$	=	$(12x^2y)^{1/2}$

EXAMPLE 5 Write each radical expression in exponential form.

(a) $\sqrt{6}$ **(b)** $\sqrt{8x}$

Solution: **(a)** $6^{1/2}$ **(b)** $(8x)^{1/2}$

We can also convert an expression from exponential form to radical form. To do so, we reverse the process. For example, $(5x)^{1/2}$ can be written $\sqrt{5x}$ and $(20x^4)^{1/2}$ can be written $\sqrt{20x^4}$.

The rules of exponents presented in Sections 4.1 and 4.2 apply to rational (or fractional) exponents. For example,

$$(x^2)^{1/2} = x^{2 \cdot 1/2} = x^1 = x$$
$$(xy)^{1/2} = x^{1/2}y^{1/2}$$
and
$$x^{1/2} \cdot x^{3/2} = x^{(1/2)+(3/2)} = x^{4/2} = x^2$$

Exercise Set 9.1

Evaluate each square root.

1. $\sqrt{1}$
2. $\sqrt{4}$
3. $\sqrt{0}$
4. $\sqrt{64}$
5. $-\sqrt{81}$
6. $\sqrt{9}$
7. $\sqrt{121}$
8. $\sqrt{100}$
9. $-\sqrt{16}$
10. $-\sqrt{36}$
11. $\sqrt{144}$
12. $\sqrt{49}$
13. $\sqrt{169}$
14. $\sqrt{225}$
15. $-\sqrt{1}$
16. $-\sqrt{100}$
17. $\sqrt{81}$
18. $-\sqrt{25}$
19. $-\sqrt{121}$
20. $-\sqrt{169}$
21. $\sqrt{\frac{1}{4}}$
22. $\sqrt{\frac{4}{9}}$
23. $\sqrt{\frac{9}{16}}$
24. $\sqrt{\frac{25}{64}}$
25. $-\sqrt{\frac{4}{25}}$
26. $-\sqrt{\frac{100}{144}}$
27. $\sqrt{\frac{36}{49}}$
28. $\sqrt{\frac{81}{121}}$

Use your calculator to evaluate each square root.

29. $\sqrt{8}$
30. $\sqrt{2}$
31. $\sqrt{15}$
32. $\sqrt{30}$
33. $\sqrt{80}$
34. $\sqrt{79}$
35. $\sqrt{81}$
36. $\sqrt{52}$
37. $\sqrt{97}$
38. $\sqrt{5}$
39. $\sqrt{3}$
40. $\sqrt{40}$

Answer true or false.

41. $\sqrt{25}$ is a rational number.
42. $\sqrt{-4}$ is not a real number.
43. $\sqrt{-5}$ is not a real number.
44. $\sqrt{5}$ is an irrational number.
45. $\sqrt{9}$ is an irrational number.
46. $\sqrt{\frac{1}{4}}$ is a rational number.
47. $\sqrt{\frac{4}{9}}$ is a rational number.
48. $\sqrt{231}$ is a rational number.
49. $\sqrt{125}$ is a rational number.
50. $\sqrt{27}$ is an irrational number.
51. $\sqrt{(18)^2}$ is an integer.
52. $\sqrt{(12)^2}$ is an integer.

Write in exponential form.

53. $\sqrt{7}$
54. $\sqrt{24}$
55. $\sqrt{17}$
56. $\sqrt{16}$
57. $\sqrt{5x}$
58. $\sqrt{6y}$
59. $\sqrt{12x^2}$
60. $\sqrt{25x^2y}$
61. $\sqrt{19xy^2}$
62. $\sqrt{34x^3y}$
63. $\sqrt{40x^3}$
64. $\sqrt{36x^3y^3}$

65. Whenever we see an expression in a square root, what assumption do we make about the expression? Why do we make this assumption?

66. In your own words, explain how you would determine if the square root of a positive integer less than 400 is a rational or irrational number **(a)** by using a calculator, and **(b)** without the use of a calculator.

67. In your own words, explain why the square root of a negative number is not a real number.

CUMULATIVE REVIEW EXERCISES

Solve.

68. $\dfrac{2x}{x^2-4} + \dfrac{1}{x-2} = \dfrac{2}{x+2}$.

69. $\dfrac{4x}{x^2+6x+9} - \dfrac{2x}{x+3} = \dfrac{x+1}{x+3}$.

[6.7] **70.** A conveyor belt operating at full speed can fill a large silo with corn in 6 hours. A second conveyor belt operating at full speed can fill the same silo in 5 hours. How long will it take both conveyor belts operating together to fill the silo?

[7.3] **71.** Find the slope of the line through the points $(-5, 3)$ and $(6, 7)$.

[7.5] **72.** If $f(x) = x^2 - 4x - 5$, find $f(-3)$.

Group Activity/ Challenge Problems

1. Is $\sqrt{0}$ **(a)** a real number? **(b)** a positive number? **(c)** a negative number? **(d)** a rational number? **(e)** an irrational number? Explain your answer.

We discuss the following concepts in Sections 9.2 and 9.3.

2. **(a)** Is $\sqrt{4} \cdot \sqrt{9}$ equal to $\sqrt{4 \cdot 9}$?

(b) Is $\sqrt{9} \cdot \sqrt{25}$ equal to $\sqrt{9 \cdot 25}$?

(c) Using these two examples, can you guess what $\sqrt{a} \cdot \sqrt{b}$ is equal to (provided that $a \geq 0$, $b \geq 0$)?

(d) Create your own problem like those given in parts (a) and (b) and see if the answer you gave in part (c) works with your numbers.

3. **(a)** Is $\sqrt{2^2}$ equal to 2?

(b) Is $\sqrt{5^2}$ equal to 5?

(c) Using these two examples can you guess what $\sqrt{a^2}$, $a \geq 0$, is equal to?

(d) Create your own problem like those given in parts **(a)** and **(b)** and see if the answer you gave in part **(c)** works with your numbers.

4. **(a)** Is $\dfrac{\sqrt{16}}{\sqrt{4}}$ equal to $\sqrt{\dfrac{16}{4}}$?

(b) Is $\dfrac{\sqrt{36}}{\sqrt{9}}$ equal to $\sqrt{\dfrac{36}{9}}$?

(c) Using these two examples, can you guess what $\dfrac{\sqrt{a}}{\sqrt{b}}$ is equal to (provided that $a \geq 0$, $b > 0$)?

(d) Create your own problem like those given in parts **(a)** and **(b)** and see if the answer you gave in part **(c)** works with your numbers.

In Section 9.7 we will explain that a radical expression of the form $\sqrt[n]{a^m}$, $a \geq 0$ can be written in exponential form as $x^{m/n}$. Change each of the following to exponential form, then simplify if possible.

5. $\sqrt{x^3}$ **6.** $\sqrt[3]{y^6}$ **7.** $\sqrt[4]{(2x)^8}$ **8.** $\sqrt{(3x)^6}$

9.2 Multiplying and Simplifying Square Roots

Tape 15

1 Use the product rule to simplify radicals containing constants.

2 Use the product rule to simplify radicals containing variables.

Product Rule for Radicals

1 To simplify square roots in this section we will make use of the **product rule for radicals.**

Product Rule for Radicals

$$\sqrt{a} \cdot \sqrt{b} = \sqrt{a \cdot b}, \text{ provided } a \geq 0, b \geq 0 \qquad \text{Rule 1}$$

The product rule says that the product of two square roots is equal to the square root of the product. The product rule applies only when both a and b are nonnegative, since the square roots of negative numbers are not real numbers.

Example of the Product Rule

$$\left.\begin{array}{r}\sqrt{1}\cdot\sqrt{60}\\ \sqrt{2}\cdot\sqrt{30}\\ \sqrt{3}\cdot\sqrt{20}\\ \sqrt{4}\cdot\sqrt{15}\\ \sqrt{6}\cdot\sqrt{10}\end{array}\right\} = \sqrt{60}$$

Note that $\sqrt{60}$ can be factored into any of these forms.

To Simplify the Square Root of a Constant

1. Write the constant as a product of the largest perfect square factor and another factor.
2. Use the product rule to write the expression as a product of square roots, with each square root containing one of the factors.
3. Find the square root of the perfect square factor.

EXAMPLE 1 Simplify $\sqrt{60}$.

Solution: The only perfect square factor of 60 is 4.

$$\begin{aligned}\sqrt{60} &= \sqrt{4\cdot 15}\\ &= \sqrt{4}\cdot\sqrt{15}\\ &= 2\sqrt{15}\end{aligned}$$

Since 15 is not a perfect square and has no perfect square factors, this expression cannot be simplified further. The expression $2\sqrt{15}$ is read "two times the square root of fifteen" or "two radical fifteen."

EXAMPLE 2 Simplify $\sqrt{75}$.

Solution: $\sqrt{75} = \sqrt{25\cdot 3} = \sqrt{25}\cdot\sqrt{3}$
$= 5\sqrt{3}$

EXAMPLE 3 Simplify $\sqrt{80}$.

Solution: $\sqrt{80} = \sqrt{16\cdot 5} = \sqrt{16}\cdot\sqrt{5}$
$= 4\sqrt{5}$

EXAMPLE 4 Simplify $\sqrt{180}$.

Solution: $\sqrt{180} = \sqrt{36\cdot 5} = \sqrt{36}\cdot\sqrt{5}$
$= 6\sqrt{5}$

Helpful Hint

When simplifying a square root, it is not uncommon for students to use a perfect square factor that is not the largest perfect square factor of the radicand. Let us consider Example 3 again. Four is also a perfect square factor of 80.

$$\sqrt{80} = \sqrt{4 \cdot 20} = \sqrt{4} \cdot \sqrt{20} = 2\sqrt{20}$$

Since 20 itself contains a perfect square factor of 4, the problem is not complete. Rather than starting the entire problem again, you can continue the simplification process as follows.

$$\sqrt{80} = 2\sqrt{20} = 2\sqrt{4 \cdot 5} = 2\sqrt{4} \cdot \sqrt{5} = 2 \cdot 2 \cdot \sqrt{5} = 4\sqrt{5}$$

Now the result checks with the answer in Example 3.

EXAMPLE 5 Simplify $\sqrt{156}$.

Solution: $\sqrt{156} = \sqrt{4 \cdot 39} = \sqrt{4} \cdot \sqrt{39}$
$= 2\sqrt{39}$

Although 39 can be factored into $3 \cdot 13$, neither of these factors is a perfect square. Thus, the answer can be simplified no further.

The Product Rule

2 Now we will simplify radicals that contain variables in the radicand.

In Section 9.1 we noted that certain numbers in a radicand were **perfect squares.** We will also refer to certain expressions that contain a variable as perfect squares. When a radical contains a variable (or number) raised to an **even exponent,** that variable (or number) and exponent together also form a perfect square. For example, in the expression $\sqrt{x^4}$, the x^4 is a perfect square since the exponent, 4, is even. In the expression $\sqrt{x^5}$, the x^5 is not a perfect square since the exponent is odd. However, x^4 is a **perfect square factor** of x^5 because x^4 is a perfect square and x^4 is a factor of x^5. Note that $x^5 = x^4 \cdot x$.

To evaluate square roots when the radicand is a perfect square, we use the following rule.

$$\sqrt{a^{2 \cdot n}} = a^n \qquad a \geq 0 \qquad \text{Rule 2}$$

This rule states that **the square root of a variable raised to an even power equals the variable raised to one-half that power.** To explain this rule, we can write the square root expression $\sqrt{a^{2n}}$ in exponential form, and then simplify as follows.

$$\sqrt{a^{2n}} = (a^{2n})^{1/2} = a^n$$

Examples of rule 2 follow.

Examples

$$\sqrt{x^2} = x$$
$$\sqrt{y^4} = y^2$$
$$\sqrt{x^{12}} = x^6$$
$$\sqrt{x^{20}} = x^{10}$$

A special case of rule 2 (when $n = 1$) is

$$\boldsymbol{\sqrt{a^2} = a, \qquad a \geq 0}$$

EXAMPLE 6 Simplify: **(a)** $\sqrt{x^{54}}$ **(b)** $\sqrt{x^4y^6}$ **(c)** $\sqrt{x^8y^2}$ **(d)** $\sqrt{y^8z^{12}}$

Solution:

(a) $\sqrt{x^{54}} = x^{27}$ **(b)** $\sqrt{x^4y^6} = \sqrt{x^4}\sqrt{y^6} = x^2y^3$

(c) $\sqrt{x^8y^2} = x^4y$ **(d)** $\sqrt{y^8z^{12}} = y^4z^6$

To Simplify the Square Root of a Radicand Containing a Variable Raised to an Odd Power

1. Express the variable as the product of two factors, one of which has an exponent of 1 (the other will therefore be a perfect square factor).
2. Use the product rule to simplify.

Examples 7 and 8 illustrate this procedure.

EXAMPLE 7 Simplify.

(a) $\sqrt{x^3}$ **(b)** $\sqrt{y^{11}}$ **(c)** $\sqrt{x^{99}}$

Solution:

(a) $\sqrt{x^3} = \sqrt{x^2 \cdot x} = \sqrt{x^2} \cdot \sqrt{x}$ (Remember that x means x^1.)

$= x \cdot \sqrt{x}$ or $x\sqrt{x}$

(b) $\sqrt{y^{11}} = \sqrt{y^{10} \cdot y} = \sqrt{y^{10}} \cdot \sqrt{y}$

$= y^5\sqrt{y}$

(c) $\sqrt{x^{99}} = \sqrt{x^{98} \cdot x} = \sqrt{x^{98}} \cdot \sqrt{x}$

$= x^{49}\sqrt{x}$

More complex radicals can be simplified using the product rule for radicals and the principles discussed in this section.

EXAMPLE 8 Simplify.

(a) $\sqrt{16x^3}$ **(b)** $\sqrt{32x^2}$ **(c)** $\sqrt{32x^3}$

Solution: Write each expression as the product of square roots, one of which has a radicand that is a perfect square.

(a) $\sqrt{16x^3} = \sqrt{16x^2} \cdot \sqrt{x}$
$= 4x\sqrt{x}$

(b) $\sqrt{32x^2} = \sqrt{16x^2} \cdot \sqrt{2}$
$= 4x\sqrt{2}$

(c) $\sqrt{32x^3} = \sqrt{16x^2} \cdot \sqrt{2x}$
$= 4x\sqrt{2x}$

EXAMPLE 9 Simplify.

(a) $\sqrt{50x^2y}$ **(b)** $\sqrt{48x^3y^2}$ **(c)** $\sqrt{98x^9y^7}$

Solution:

(a) $\sqrt{50x^2y} = \sqrt{25x^2} \cdot \sqrt{2y}$
$= 5x\sqrt{2y}$

(b) $\sqrt{48x^3y^2} = \sqrt{16x^2y^2} \cdot \sqrt{3x}$
$= 4xy\sqrt{3x}$

(c) $\sqrt{98x^9y^7} = \sqrt{49x^8y^6} \cdot \sqrt{2xy}$
$= 7x^4y^3\sqrt{2xy}$

The radicand of your simplified answer should not contain any perfect square factors or any variables with an exponent greater than 1.

Now let us look at an example where we use the product rule to multiply two radicals before simplifying.

EXAMPLE 10 Multiply and then simplify.

(a) $\sqrt{2} \cdot \sqrt{8}$ **(b)** $\sqrt{2x} \cdot \sqrt{8}$ **(c)** $\sqrt{2x} \cdot \sqrt{8x}$

Solution:

(a) $\sqrt{2} \cdot \sqrt{8} = \sqrt{2 \cdot 8} = \sqrt{16} = 4$

(b) $\sqrt{2x} \cdot \sqrt{8} = \sqrt{16x} = \sqrt{16} \cdot \sqrt{x} = 4\sqrt{x}$

(c) $\sqrt{2x} \cdot \sqrt{8x} = \sqrt{16x^2} = 4x$

EXAMPLE 11 Multiply and then simplify.

(a) $\sqrt{8x^3y} \cdot \sqrt{4xy^5}$ **(b)** $\sqrt{5xy^6}\sqrt{6x^3y}$

Solution:

(a) $\sqrt{8x^3y} \cdot \sqrt{4xy^5} = \sqrt{32x^4y^6} = \sqrt{16x^4y^6} \cdot \sqrt{2}$
$= 4x^2y^3\sqrt{2}$

(b) $\sqrt{5xy^6} \cdot \sqrt{6x^3y} = \sqrt{30x^4y^7} = \sqrt{x^4y^6} \cdot \sqrt{30y}$
$= x^2y^3\sqrt{30y}$

In part (b), 30 can be factored in many ways. However, none of the factors are perfect squares, so we leave the answer as given.

Exercise Set 9.2

Simplify.

1. $\sqrt{16}$
2. $\sqrt{64}$
3. $\sqrt{8}$
4. $\sqrt{75}$
5. $\sqrt{96}$
6. $\sqrt{125}$
7. $\sqrt{32}$
8. $\sqrt{52}$
9. $\sqrt{160}$
10. $\sqrt{28}$
11. $\sqrt{48}$
12. $\sqrt{27}$
13. $\sqrt{108}$
14. $\sqrt{128}$
15. $\sqrt{156}$
16. $\sqrt{180}$
17. $\sqrt{256}$
18. $\sqrt{212}$
19. $\sqrt{900}$
20. $\sqrt{x^4}$
21. $\sqrt{y^6}$
22. $\sqrt{x^9}$
23. $\sqrt{x^2y^4}$
24. $\sqrt{x^2y}$
25. $\sqrt{x^9y^{12}}$
26. $\sqrt{x^4y^5z^6}$
27. $\sqrt{a^2b^4c}$
28. $\sqrt{a^3b^9c^{11}}$
29. $\sqrt{3x^3}$
30. $\sqrt{12x^4y^2}$
31. $\sqrt{50x^2y^3}$
32. $\sqrt{125x^3y^5}$
33. $\sqrt{200y^5z^{12}}$
34. $\sqrt{64xyz^5}$
35. $\sqrt{243q^2b^3c}$
36. $\sqrt{500ab^4c^3}$
37. $\sqrt{128x^3yz^5}$
38. $\sqrt{112x^6y^8}$
39. $\sqrt{250x^4yz}$
40. $\sqrt{98x^4y^4z}$

Simplify.

41. $\sqrt{8}\cdot\sqrt{3}$
42. $\sqrt{5}\cdot\sqrt{5}$
43. $\sqrt{18}\cdot\sqrt{3}$
44. $\sqrt{60}\cdot\sqrt{5}$
45. $\sqrt{75}\cdot\sqrt{6}$
46. $\sqrt{30}\cdot\sqrt{5}$
47. $\sqrt{3x}\sqrt{5x}$
48. $\sqrt{4x^3}\sqrt{4x}$
49. $\sqrt{5x^2}\sqrt{8x^3}$
50. $\sqrt{15x^2}\sqrt{6x^5}$
51. $\sqrt{12x^2y}\sqrt{6xy^3}$
52. $\sqrt{20xy^4}\sqrt{6x^5}$
53. $\sqrt{18xy^4}\sqrt{3x^2y}$
54. $\sqrt{40x^2y^4}\sqrt{6x^3y^5}$
55. $\sqrt{15xy^6}\sqrt{6xyz}$
56. $\sqrt{14xyz^5}\sqrt{3xy^2z^6}$
57. $\sqrt{9x^4y^6}\sqrt{4x^2y^4}$
58. $\sqrt{3x^3yz^6}\sqrt{6x^4y^5z^6}$
59. $(\sqrt{4x})^2$
60. $(\sqrt{6x^2})^2$
61. $(\sqrt{13x^4y^6})^2$
62. $\sqrt{36x^2y^7}\sqrt{2x^4y}$
63. $(\sqrt{4x})^2(\sqrt{5x})^2$
64. $(\sqrt{3x})^2(\sqrt{5x})^2$

Which coefficients and exponents should be placed in the shaded areas to make a true statement? Explain how you obtained your answer.

65. $\sqrt{16x^{\square}y^6} = 4x^2y^3$
66. $\sqrt{\square x^4y^{\square}} = 4x^2y^4$
67. $\sqrt{4x^{\square}y^{\square}} = 2x^3y^2\sqrt{y}$
68. $\sqrt{3x^4y^{\square}}\cdot\sqrt{3x^{\square}y^5} = 3x^5y^7\sqrt{xy}$
69. $\sqrt{2x^{\square}y^5}\cdot\sqrt{\square x^3y^{\square}} = 4x^7y^6\sqrt{x}$
70. $\sqrt{32x^4z^{\square}}\cdot\sqrt{\square x^{\square}z^{12}} = 8x^5z^9\sqrt{z}$

71. In your own words, state the product rule for radicals and explain what it means.
72. (a) In your own words, explain how to simplify a square root containing only a constant.
    (b) Simplify $\sqrt{20}$ using the procedure you gave in part (a).
73. Explain why the product rule cannot be used to simplify the problem $\sqrt{-4}\cdot\sqrt{-9}$.
74. We learned that for $a \geq 0$, $\sqrt{a^{2\cdot n}} = a^n$. Explain in your own words what this means.
75. (a) Explain how to simplify the square root of a radical containing a variable raised to an odd power.
    (b) Simplify $\sqrt{x^9}$ using the procedure stated in part (a).
76. (a) Explain why $\sqrt{32x^3}$ is not a simplified expression.
    (b) Simplify $\sqrt{32x^3}$.
77. (a) Explain why $\sqrt{75x^5}$ is not a simplified expression.
    (b) Simplify $\sqrt{75x^5}$.

CUMULATIVE REVIEW EXERCISES

[6.2] **78.** Divide $\dfrac{3x^2 - 16x - 12}{3x^2 - 10x - 8} \div \dfrac{x^2 - 7x + 6}{3x^2 - 11x - 4}$.

[7.4] **79.** Write the equation $3x + 6y = 9$ in slope–intercept form and indicate the slope and the y-intercept.

[7.6] **80.** Graph $6x - 5y \geq 30$.

[8.3] **81.** Solve the system of equations
$3x - 4y = 6,$
$5x - 3y = 5.$

Group Activity/ Challenge Problems

1. We know that $\sqrt{a} \cdot \sqrt{b} = \sqrt{a \cdot b}$ if $a \geq 0$ and $b \geq 0$. Does $\sqrt{\dfrac{a}{b}} = \dfrac{\sqrt{a}}{\sqrt{b}}$ if $a \geq 0$ and $b > 0$? Try several pairs of values for a and b and see.
2. Is $\sqrt{6.25}$ a rational or irrational number? Explain how you determined your answer.
3. **(a)** Will the product of two rational numbers always be a rational number? Explain and give an example to support your answer.
   (b) Will the product of two irrational numbers always be an irrational number? Explain and give an example to support your answer.
4. **(a)** In Section 9.4 we will be multiplying expressions like $(\sqrt{a} + \sqrt{b})(\sqrt{a} - \sqrt{b})$ using the FOIL method. Can you find this product now?
   (b) Multiply $(\sqrt{6} + \sqrt{3})(\sqrt{6} - \sqrt{3})$.
5. The area of a square is found by the formula $A = s^2$. We will learn later that we can rewrite this formula as $s = \sqrt{A}$.
   (a) If the area is 16 square feet, what is the length of a side?
   (b) If the area is doubled, is the length of a side doubled? Explain.
   (c) To double the length of a side of a square, how much must the area be increased? Explain.

9.3 Dividing and Simplifying Square Roots

Tape 15

1. Understand what it means for a radical to be simplified.
2. Use the quotient rule to simplify radicals.
3. Rationalize denominators.

Simplifying Square Roots

1 In this section we will use a new rule, the quotient rule, to simplify radicals containing fractions. However, before we do that, we need to discuss what it means for a square root to be simplified.

A Square Root Is Simplified When

1. No radicand has a factor that is a perfect square.
2. No radicand contains a fraction.
3. No denominator contains a square root.

All three criteria must be met for an expression to be simplified. Let us look at some radical expressions that *are not simplified.*

Radical	*Reason Not Simplified*	
$\sqrt{8}$	Contains perfect square factor, 4. ($\sqrt{8} = \sqrt{4} \cdot \sqrt{2} = 2\sqrt{2}$)	(Criteria 1)
$\sqrt{x^3}$	Contains perfect square factor, x^2. ($\sqrt{x^3} = \sqrt{x^2} \cdot \sqrt{x} = x\sqrt{x}$)	(Criteria 1)
$\sqrt{\dfrac{1}{2}}$	Radicand contains a fraction.	(Criteria 2)
$\dfrac{1}{\sqrt{2}}$	Square root in the denominator.	(Criteria 3)

The Quotient Rule for Radicals

2 The quotient rule for radicals states that the quotient of two square roots is equal to the square root of the quotient.

Quotient Rule for Radicals

$$\frac{\sqrt{a}}{\sqrt{b}} = \sqrt{\frac{a}{b}}, \quad \text{provided} \quad a \geq 0, b > 0 \qquad \text{Rule 3}$$

Examples 1 through 4 illustrate how the quotient rule is used to simplify square roots.

Example 1 Simplify.

(a) $\sqrt{\dfrac{8}{2}}$ **(b)** $\sqrt{\dfrac{25}{5}}$ **(c)** $\sqrt{\dfrac{9}{4}}$

Solution: When the square root contains a fraction, divide out any factor common to both the numerator and denominator. If the square root still contains a fraction, use the quotient rule for radicals to simplify.

(a) $\sqrt{\dfrac{8}{2}} = \sqrt{4} = 2$ **(b)** $\sqrt{\dfrac{25}{5}} = \sqrt{5}$ **(c)** $\sqrt{\dfrac{9}{4}} = \dfrac{\sqrt{9}}{\sqrt{4}} = \dfrac{3}{2}$

Example 2 Simplify.

(a) $\sqrt{\dfrac{16x^2}{8}}$ **(b)** $\sqrt{\dfrac{64x^4y}{2x^2y}}$ **(c)** $\sqrt{\dfrac{3x^2y^4}{27x^4}}$ **(d)** $\sqrt{\dfrac{15xy^5z^2}{3x^5yz}}$

Solution: First divide out any factors common to both the numerator and denominator; then use the quotient rule for radicals to simplify.

(a) $\sqrt{\dfrac{16x^2}{8}} = \sqrt{2x^2} = \sqrt{x^2}\sqrt{2} = x\sqrt{2}$

(b) $\sqrt{\dfrac{64x^4y}{2x^2y}} = \sqrt{32x^2} = \sqrt{16x^2}\sqrt{2} = 4x\sqrt{2}$

(c) $\sqrt{\dfrac{3x^2y^4}{27x^4}} = \sqrt{\dfrac{y^4}{9x^2}} = \dfrac{\sqrt{y^4}}{\sqrt{9x^2}} = \dfrac{y^2}{3x}$

(d) $\sqrt{\dfrac{15xy^5z^2}{3x^5yz}} = \sqrt{\dfrac{5y^4z}{x^4}} = \dfrac{\sqrt{5y^4z}}{\sqrt{x^4}} = \dfrac{\sqrt{y^4}\sqrt{5z}}{\sqrt{x^4}} = \dfrac{y^2\sqrt{5z}}{x^2}$

When you are given a fraction containing a radical expression in both the numerator and the denominator, use the quotient rule to simplify, as in Examples 3 and 4.

EXAMPLE 3 Simplify.

(a) $\dfrac{\sqrt{2}}{\sqrt{8}}$ **(b)** $\dfrac{\sqrt{75}}{\sqrt{3}}$

Solution: **(a)** $\dfrac{\sqrt{2}}{\sqrt{8}} = \sqrt{\dfrac{2}{8}} = \sqrt{\dfrac{1}{4}} = \dfrac{\sqrt{1}}{\sqrt{4}} = \dfrac{1}{2}$

(b) $\dfrac{\sqrt{75}}{\sqrt{3}} = \sqrt{\dfrac{75}{3}} = \sqrt{25} = 5$

EXAMPLE 4 Simplify.

(a) $\dfrac{\sqrt{32x^4y^3}}{\sqrt{8xy}}$ **(b)** $\dfrac{\sqrt{75x^8y^4}}{\sqrt{3x^5y^8}}$

Solution:

(a) $\dfrac{\sqrt{32x^4y^3}}{\sqrt{8xy}} = \sqrt{\dfrac{32x^4y^3}{8xy}} = \sqrt{4x^3y^2} = \sqrt{4x^2y^2} \cdot \sqrt{x} = 2xy\sqrt{x}$

(b) $\dfrac{\sqrt{75x^8y^4}}{\sqrt{3x^5y^8}} = \sqrt{\dfrac{75x^8y^4}{3x^5y^8}} = \sqrt{\dfrac{25x^3}{y^4}} = \dfrac{\sqrt{25x^3}}{\sqrt{y^4}} = \dfrac{\sqrt{25x^2}\sqrt{x}}{\sqrt{y^4}} = \dfrac{5x\sqrt{x}}{y^2}$

Rationalize Denominators

3 When the denominator of a fraction contains the square root of a number that is not a perfect square, we generally simplify the expression by **rationalizing the denominator. To rationalize a denominator means to remove all radicals from the denominator.** We rationalize the denominator because it is easier (without a calculator) to obtain the approximate value of a number like $\sqrt{2}/2$ than a number like $1/\sqrt{2}$.

To rationalize a denominator, multiply *both* the numerator and the denominator of the fraction by the square root that appears in the denominator or by the square root of a number that makes the denominator a perfect square.

EXAMPLE 5 Simplify $\dfrac{1}{\sqrt{2}}$.

Solution: Since $\sqrt{2} \cdot \sqrt{2} = \sqrt{4} = 2$, we multiply both the numerator and denominator by $\sqrt{2}$.

$$\frac{1}{\sqrt{2}} = \frac{1}{\sqrt{2}} \cdot \frac{\sqrt{2}}{\sqrt{2}} = \frac{\sqrt{2}}{\sqrt{4}} = \frac{\sqrt{2}}{2}$$

The answer $\frac{\sqrt{2}}{2}$ is simplified because it satisfies the three requirements stated earlier.

In Example 5, multiplying both the numerator and denominator by $\sqrt{2}$ is equivalent to multiplying the fraction by 1, which does not change its value.

COMMON STUDENT ERROR An expression under a square root *cannot* be divided by an expression not under a square root.

Correct	*Incorrect*
$\frac{\sqrt{2}}{2}$ cannot be simplified any further.	$\frac{\sqrt{\cancel{2}^{1}}}{\cancel{2}_{1}} = \sqrt{1} = 1$
$\frac{\sqrt{6}}{3}$ cannot be simplified any further.	$\frac{\sqrt{\cancel{6}^{2}}}{\cancel{3}_{1}} = \sqrt{2}$
$\frac{\sqrt{x^3}}{x} = \frac{\sqrt{x^2}\sqrt{x}}{x} = \frac{\cancel{x}\sqrt{x}}{\cancel{x}} = \sqrt{x}$	$\frac{\sqrt{x^{\cancel{3}2}}}{\cancel{x}_{1}} = \sqrt{x^2} = x$

Each of the following simplifications is correct because the constants divided out are not under square roots.

Correct	*Correct*
$\frac{\overset{2}{\cancel{6}}\sqrt{2}}{\underset{1}{\cancel{3}}} = 2\sqrt{2}$	$\frac{\cancel{x}\sqrt{2}}{\cancel{x}} = \sqrt{2}$
$\frac{\overset{1}{\cancel{4}}\sqrt{3}}{\underset{2}{\cancel{8}}} = \frac{\sqrt{3}}{2}$	$\frac{3x^{\cancel{2}}\sqrt{5}}{\cancel{x}} = 3x\sqrt{5}$

EXAMPLE 6 Simplify.

(a) $\sqrt{\frac{2}{3}}$ **(b)** $\sqrt{\frac{x^2}{18}}$

Solution:

(a) $\sqrt{\frac{2}{3}} = \frac{\sqrt{2}}{\sqrt{3}} = \frac{\sqrt{2}}{\sqrt{3}} \cdot \frac{\sqrt{3}}{\sqrt{3}} = \frac{\sqrt{6}}{3}$

(b) $\sqrt{\frac{x^2}{18}} = \frac{\sqrt{x^2}}{\sqrt{18}} = \frac{x}{\sqrt{9} \cdot \sqrt{2}} = \frac{x}{3\sqrt{2}}$

Now rationalize the denominator.

$$\frac{x}{3\sqrt{2}} \cdot \frac{\sqrt{2}}{\sqrt{2}} = \frac{x\sqrt{2}}{3\sqrt{4}} = \frac{x\sqrt{2}}{3 \cdot 2} = \frac{x\sqrt{2}}{6}$$

Part **(b)** can also be rationalized as follows:

$$\sqrt{\frac{x^2}{18}} = \frac{\sqrt{x^2}}{\sqrt{18}} = \frac{x}{\sqrt{18}} \cdot \frac{\sqrt{2}}{\sqrt{2}} = \frac{x\sqrt{2}}{\sqrt{36}} = \frac{x\sqrt{2}}{6}$$

Note that $\frac{x}{\sqrt{18}} \cdot \frac{\sqrt{18}}{\sqrt{18}}$ will also give us the same result when simplified.

Exercise Set 9.3

Simplify each radical.

1. $\sqrt{\frac{12}{3}}$
2. $\sqrt{\frac{8}{2}}$
3. $\sqrt{\frac{27}{3}}$
4. $\sqrt{\frac{16}{4}}$
5. $\frac{\sqrt{18}}{\sqrt{2}}$
6. $\frac{\sqrt{3}}{\sqrt{27}}$
7. $\sqrt{\frac{1}{25}}$
8. $\sqrt{\frac{16}{25}}$
9. $\sqrt{\frac{9}{49}}$
10. $\sqrt{\frac{4}{81}}$
11. $\frac{\sqrt{10}}{\sqrt{490}}$
12. $\sqrt{\frac{16x^3}{4x}}$
13. $\sqrt{\frac{40x^3}{2x}}$
14. $\sqrt{\frac{45x^2}{16x^2y^4}}$
15. $\sqrt{\frac{9xy^4}{3y^3}}$
16. $\sqrt{\frac{50x^3y^6}{10x^3y^8}}$
17. $\sqrt{\frac{25x^6y}{45x^6y^3}}$
18. $\sqrt{\frac{14xyz^5}{56x^3y^3z^4}}$
19. $\sqrt{\frac{72xy}{72x^3y^5}}$
20. $\frac{\sqrt{16x^4}}{\sqrt{8x}}$
21. $\frac{\sqrt{32x^5}}{\sqrt{8x}}$
22. $\frac{\sqrt{60x^2y^2}}{\sqrt{6x^2y^4}}$
23. $\frac{\sqrt{16x^4y}}{\sqrt{25x^6y^3}}$
24. $\frac{\sqrt{72}}{\sqrt{36x^2y^6}}$
25. $\frac{\sqrt{45xy^6}}{\sqrt{9xy^4z^2}}$
26. $\frac{\sqrt{24x^2y^6}}{\sqrt{8x^4z^4}}$
27. $\frac{\sqrt{72x^{12}y^{20}}}{\sqrt{2x^2y^4}}$
28. $\frac{\sqrt{144x^{60}y^{32}}}{\sqrt{12x^{40}y^{18}}}$

Simplify each radical.

29. $\frac{3}{\sqrt{2}}$
30. $\frac{2}{\sqrt{3}}$
31. $\frac{4}{\sqrt{8}}$
32. $\frac{6}{\sqrt{6}}$
33. $\frac{5}{\sqrt{10}}$
34. $\frac{9}{\sqrt{50}}$
35. $\sqrt{\frac{2}{5}}$
36. $\sqrt{\frac{7}{12}}$
37. $\sqrt{\frac{3}{15}}$
38. $\sqrt{\frac{3}{10}}$
39. $\sqrt{\frac{x^2}{2}}$
40. $\sqrt{\frac{x^2}{7}}$
41. $\sqrt{\frac{x^2}{8}}$
42. $\sqrt{\frac{x^3}{18}}$
43. $\sqrt{\frac{x^4}{5}}$
44. $\sqrt{\frac{x^3}{11}}$
45. $\sqrt{\frac{x^6}{15y}}$
46. $\sqrt{\frac{x^5y}{12y^2}}$
47. $\sqrt{\frac{8x^4y^2}{32x^2y^3}}$
48. $\sqrt{\frac{27xz^4}{6y^4}}$

49. $\sqrt{\dfrac{18yz}{75x^4y^5z^3}}$ **50.** $\dfrac{\sqrt{25x^5}}{\sqrt{100xy^5}}$ **51.** $\dfrac{\sqrt{90x^4y}}{\sqrt{2x^5y^5}}$ **52.** $\dfrac{\sqrt{120xyz^2}}{\sqrt{9xy^2}}$

53. What are the three requirements for a square root to be considered simplified?

Explain why each radical is not simplified. Simplify the radicals.

54. $\sqrt{32}$ **55.** $\sqrt{\dfrac{1}{3}}$ **56.** $\dfrac{3}{\sqrt{5}}$

Explain why the radical can or cannot be simplified. Simplify the radical if possible.

57. $\dfrac{\sqrt{3}}{3}$ **58.** $\dfrac{4\sqrt{3}}{2}$ **59.** $\dfrac{x^2\sqrt{2}}{x}$

60. $\dfrac{\sqrt{10}}{5}$ **61.** $\dfrac{\sqrt{x}}{x}$ **62.** $\dfrac{\sqrt{6}}{2}$

63. In your own words, state the quotient rule for radicals and explain what it means.

64. Explain why the quotient rule cannot be used to simplify the problem $\dfrac{\sqrt{-10}}{\sqrt{-2}}$.

65. (a) Explain, in your own words, how to rationalize the denominator of a fraction of the form $\dfrac{a}{\sqrt{b}}$.

(b) Rationalize $\dfrac{a}{\sqrt{b}}$.

CUMULATIVE REVIEW EXERCISES

[6.4] **66.** Factor $3x^2 - 12x - 96$.

[7.1] **67.** Reduce $\dfrac{x-1}{x^2-1}$ to lowest terms.

[7.6] **68.** Solve the equation $x + \dfrac{24}{x} = 10$.

[8.1] **69.** Solve the system of equations graphically.

$$y = 2x - 2$$
$$2x + 3y = 10$$

Group Activity/ Challenge Problems

1. (a) Will the quotient of two rational numbers (denominator not equal to 0) always be a rational number? Explain and give an example to support your answer.

(b) Will the quotient of two irrational numbers (denominator not 0) always be an irrational number? Explain and give an example to support your answer.

2. In Section 9.4 we will simplify expressions like $\dfrac{2}{\sqrt{6}+\sqrt{3}}$ by multiplying both the numerator and denominator of the fraction by $\sqrt{6} - \sqrt{3}$. Simplify this expression now.

3. (a) Is $\sqrt{10}$ twice as large as $\sqrt{5}$? Explain. **(b)** What is the quotient of $\dfrac{\sqrt{10}}{\sqrt{5}}$ equal to?

(c) What number is twice as large as $\sqrt{5}$? Explain how you determined your answer.

Fill in the shaded area to make the expression true. Explain how you determined your answer.

4. $\sqrt{\dfrac{\blacksquare}{4x^2}} = 4x^4$ **5.** $\dfrac{\sqrt{32x^5}}{\sqrt{\blacksquare}} = 2x^2$

6. $\dfrac{1}{\sqrt{\blacksquare}} = \dfrac{\sqrt{2}}{2}$ **7.** $\dfrac{3x}{\sqrt{\blacksquare}} = \dfrac{3\sqrt{2x}}{2}$

9.4 Adding and Subtracting Square Roots

Tape 15

1. Add and subtract like and unlike square roots.
2. Rationalize a denominator that contains a binomial.
3. Simplify a fraction containing two numerical terms in the numerator where one is a radical.

Adding and Subtracting Square Roots

1 **Like square roots** are square roots having the same radicands. Like square roots are added in much the same manner that like terms are added. **To add like square roots,** add their coefficients and then multiply that sum by the like square root.

Examples of Adding Like Terms

$$2x + 3x = (2 + 3)x = 5x$$

$$4x + x = 4x + 1x = (4 + 1)x = 5x$$

Examples of Adding Like Square Roots

$$2\sqrt{7} + 3\sqrt{7} = (2 + 3)\sqrt{7} = 5\sqrt{7}$$

$$4\sqrt{x} + \sqrt{x} = 4\sqrt{x} + 1\sqrt{x} = (4 + 1)\sqrt{x} = 5\sqrt{x}$$

Note that adding like square roots is an application of the distributive property.

$$2\sqrt{7} + 3\sqrt{7} = (2 + 3)\sqrt{7}$$

$$= 5\sqrt{7}$$

Other Examples of Adding Like Square Roots

$$2\sqrt{5} - 3\sqrt{5} = (2 - 3)\sqrt{5} = -1\sqrt{5} = -\sqrt{5}$$

$$\sqrt{x} + \sqrt{x} = 1\sqrt{x} + 1\sqrt{x} = (1 + 1)\sqrt{x} = 2\sqrt{x}$$

$$6\sqrt{2} + 3\sqrt{2} - \sqrt{2} = (6 + 3 - 1)\sqrt{2} = 8\sqrt{2}$$

$$\frac{2\sqrt{3}}{5} + \frac{1\sqrt{3}}{5} = \left(\frac{2}{5} + \frac{1}{5}\right)\sqrt{3} = \frac{3}{5}\sqrt{3} \text{ or } \frac{3\sqrt{3}}{5}$$

EXAMPLE 1 Simplify if possible.

(a) $4\sqrt{3} + 2\sqrt{3} - 2$ (b) $\sqrt{5} - 4\sqrt{5} + 5$

(c) $5 + 3\sqrt{2} - \sqrt{2} + 3$ (d) $2\sqrt{3} + 5\sqrt{2}$

Solution: (a) $4\sqrt{3} + 2\sqrt{3} - 2 = 6\sqrt{3} - 2$

(b) $\sqrt{5} - 4\sqrt{5} + 5 = -3\sqrt{5} + 5$

(c) $5 + 3\sqrt{2} - \sqrt{2} + 3 = 8 + 2\sqrt{2}$

(d) Cannot be simplified since the radicands are different.

EXAMPLE 2 Simplify.

(a) $2\sqrt{x} - 3\sqrt{x} + 4\sqrt{x}$ (b) $3\sqrt{x} + x + 4\sqrt{x}$

(c) $x + \sqrt{x} + 2\sqrt{x} + 3$ (d) $x\sqrt{x} + 3\sqrt{x} + x$

(e) $\sqrt{xy} + 2\sqrt{xy} - \sqrt{x}$

Solution:

(a) $2\sqrt{x} - 3\sqrt{x} + 4\sqrt{x} = 3\sqrt{x}$

(b) $3\sqrt{x} + x + 4\sqrt{x} = x + 7\sqrt{x}$ Only $3\sqrt{x}$ and $4\sqrt{x}$ can be combined.

(c) $x + \sqrt{x} + 2\sqrt{x} + 3 = x + 3\sqrt{x} + 3$ Only $\sqrt{x}$ and $2\sqrt{x}$ can be combined.

(d) $x\sqrt{x} + 3\sqrt{x} + x = (x + 3)\sqrt{x} + x$ Only $x\sqrt{x}$ and $3\sqrt{x}$ can be combined.

(e) $\sqrt{xy} + 2\sqrt{xy} - \sqrt{x} = 3\sqrt{xy} - \sqrt{x}$ Only $\sqrt{xy}$ and $2\sqrt{xy}$ can be combined.

Unlike square roots are square roots having different radicands. It is sometimes possible to change unlike square roots into like square roots, as in Examples 3, 4, and 5.

EXAMPLE 3 Simplify $\sqrt{2} + \sqrt{18}$.

Solution: Since 18 has a perfect square factor, 9, we write 18 as a product of the perfect square factor and another factor.

$$\begin{aligned}\sqrt{2} + \sqrt{18} &= \sqrt{2} + \sqrt{9 \cdot 2} \\ &= \sqrt{2} + \sqrt{9}\sqrt{2} \\ &= \sqrt{2} + 3\sqrt{2} \\ &= 4\sqrt{2}\end{aligned}$$

EXAMPLE 4 Simplify $\sqrt{24} - \sqrt{54}$.

Solution: Write each radicand as a product of a perfect square factor and another factor.

$$\begin{aligned}\sqrt{24} - \sqrt{54} &= \sqrt{4 \cdot 6} - \sqrt{9 \cdot 6} \\ &= \sqrt{4}\sqrt{6} - \sqrt{9}\sqrt{6} \\ &= 2\sqrt{6} - 3\sqrt{6} = -\sqrt{6}\end{aligned}$$

EXAMPLE 5 Simplify.

(a) $2\sqrt{8} - \sqrt{32}$ **(b)** $3\sqrt{12} + 5\sqrt{27} + 2$ **(c)** $\sqrt{120} - \sqrt{75}$

Solution: **(a)**
$$\begin{aligned}2\sqrt{8} - \sqrt{32} &= 2\sqrt{4 \cdot 2} - \sqrt{16 \cdot 2} \\ &= 2\sqrt{4}\sqrt{2} - \sqrt{16}\sqrt{2} \\ &= 2 \cdot 2\sqrt{2} - 4\sqrt{2} \\ &= 4\sqrt{2} - 4\sqrt{2} \\ &= 0\end{aligned}$$

(b)
$$\begin{aligned}3\sqrt{12} + 5\sqrt{27} + 2 &= 3\sqrt{4 \cdot 3} + 5\sqrt{9 \cdot 3} + 2 \\ &= 3\sqrt{4}\sqrt{3} + 5\sqrt{9}\sqrt{3} + 2 \\ &= 3 \cdot 2\sqrt{3} + 5 \cdot 3\sqrt{3} + 2 \\ &= 6\sqrt{3} + 15\sqrt{3} + 2 \\ &= 21\sqrt{3} + 2\end{aligned}$$

(c)
$$\begin{aligned}\sqrt{120} - \sqrt{75} &= \sqrt{4 \cdot 30} - \sqrt{25 \cdot 3} \\ &= \sqrt{4}\sqrt{30} - \sqrt{25}\sqrt{3} \\ &= 2\sqrt{30} - 5\sqrt{3}\end{aligned}$$

Since 30 has no perfect square factors and since the radicands are different, the expression $2\sqrt{30} - 5\sqrt{3}$ cannot be simplified any further.

COMMON STUDENT ERROR The product rule presented in Section 9.2 was $\sqrt{a} \cdot \sqrt{b} = \sqrt{a \cdot b}$. The same principle **does not apply to** addition.

Incorrect

~~$\sqrt{a} + \sqrt{b} = \sqrt{a + b}$~~

For example, to evaluate $\sqrt{9} + \sqrt{16}$,

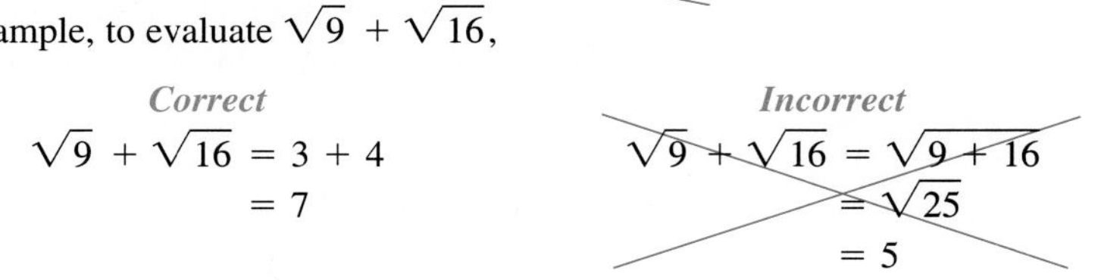

Rationalize Denominators

2 When the denominator of a rational expression is a binomial with a square root term, we again **rationalize the denominator.** We do this by multiplying both the numerator and the denominator of the fraction by the **conjugate** of the denominator. The conjugate of a binomial is a binomial having the same two terms with the sign of the second term changed.

Binomial	*Its Conjugate*
$3 + \sqrt{2}$	$3 - \sqrt{2}$
$\sqrt{5} - 3$	$\sqrt{5} + 3$
$2\sqrt{3} - \sqrt{5}$	$2\sqrt{3} + \sqrt{5}$
$-x + \sqrt{3}$	$-x - \sqrt{3}$

When a binomial is multiplied by its conjugate using the FOIL method, the outer and inner terms will add to zero.

EXAMPLE 6 Multiply $(2 + \sqrt{3})(2 - \sqrt{3})$ using the FOIL method.

Solution:

$$(2 + \sqrt{3})(2 - \sqrt{3}) \quad \overset{F}{2(2)} + \overset{O}{2(-\sqrt{3})} + \overset{I}{2(\sqrt{3})} + \overset{L}{\sqrt{3}(-\sqrt{3})}$$

$$= 4 - 2\sqrt{3} + 2\sqrt{3} - \sqrt{9}$$

$$= 4 - \sqrt{9}$$

$$= 4 - 3 = 1$$

EXAMPLE 7 Multiply $(\sqrt{3} - \sqrt{5})(\sqrt{3} + \sqrt{5})$ using the FOIL method.

Solution:

$$(\sqrt{3} - \sqrt{5})(\sqrt{3} + \sqrt{5})$$

F, O, I, L

$$\begin{aligned}
&\quad\ \text{F} \qquad\qquad \text{O} \qquad\qquad\quad \text{I} \qquad\qquad\quad \text{L}\\
&\sqrt{3}\cdot\sqrt{3} + \sqrt{3}\cdot\sqrt{5} + (-\sqrt{5})(\sqrt{3}) + (-\sqrt{5})(\sqrt{5})\\
&= \sqrt{9} + \sqrt{15} - \sqrt{15} - \sqrt{25}\\
&= \sqrt{9} - \sqrt{25}\\
&= 3 - 5 = -2
\end{aligned}$$

Now let's try some examples where we rationalize the denominator when the denominator is a binomial with one or more radical terms.

EXAMPLE 8 Simplify $\dfrac{5}{2 + \sqrt{3}}$.

Solution: To rationalize the denominator, multiply both the numerator and the denominator by $2 - \sqrt{3}$, which is the conjugate of $2 + \sqrt{3}$.

$$\begin{aligned}
\frac{5}{2 + \sqrt{3}} \cdot \frac{2 - \sqrt{3}}{2 - \sqrt{3}} &= \frac{5(2 - \sqrt{3})}{(2 + \sqrt{3})(2 - \sqrt{3})}\\
&= \frac{5(2 - \sqrt{3})}{4 - 3}\\
&= \frac{5(2 - \sqrt{3})}{1}\\
&= 5(2 - \sqrt{3}) = 10 - 5\sqrt{3}
\end{aligned}$$

Note that $-5\sqrt{3} + 10$ is also an acceptable answer.

EXAMPLE 9 Simplify $\dfrac{6}{\sqrt{5} - \sqrt{2}}$.

Solution: Multiply both the numerator and the denominator of the fraction by $\sqrt{5} + \sqrt{2}$, the conjugate of $\sqrt{5} - \sqrt{2}$.

$$\begin{aligned}
\frac{6}{\sqrt{5} - \sqrt{2}} \cdot \frac{\sqrt{5} + \sqrt{2}}{\sqrt{5} + \sqrt{2}} &= \frac{6(\sqrt{5} + \sqrt{2})}{5 - 2}\\
&= \frac{\overset{2}{\cancel{6}}(\sqrt{5} + \sqrt{2})}{\underset{1}{\cancel{3}}}\\
&= 2(\sqrt{5} + \sqrt{2}) = 2\sqrt{5} + 2\sqrt{2}
\end{aligned}$$

EXAMPLE 10 Simplify $\dfrac{\sqrt{3}}{2 - \sqrt{6}}$.

Solution: Multiply both the numerator and the denominator of the fraction by $2 + \sqrt{6}$, the conjugate of $2 - \sqrt{6}$.

$$\begin{aligned}\frac{\sqrt{3}}{2-\sqrt{6}} \cdot \frac{2+\sqrt{6}}{2+\sqrt{6}} &= \frac{\sqrt{3}(2+\sqrt{6})}{4-6} \\ &= \frac{2\sqrt{3}+\sqrt{3}\cdot\sqrt{6}}{-2} \\ &= \frac{2\sqrt{3}+\sqrt{18}}{-2} \\ &= \frac{2\sqrt{3}+\sqrt{9}\cdot\sqrt{2}}{-2} \\ &= \frac{2\sqrt{3}+3\sqrt{2}}{-2} \\ &= \frac{-2\sqrt{3}-3\sqrt{2}}{2}\end{aligned}$$

EXAMPLE 11 Simplify $\dfrac{x}{x+\sqrt{y}}$.

Solution: Multiply both the numerator and the denominator of the fraction by the conjugate of the denominator, $x - \sqrt{y}$.

$$\frac{x}{x+\sqrt{y}} \cdot \frac{x-\sqrt{y}}{x-\sqrt{y}} = \frac{x(x-\sqrt{y})}{x^2-y} = \frac{x^2-x\sqrt{y}}{x^2-y}$$

Remember, you cannot divide out the x^2 terms because they are not factors.

A Radical in the Numerator

3 In Chapter 10, when using the quadratic formula, you will need to simplify fractional expressions that contain a radical in the numerator. For example, you might have to simplify the following expressions.

$$\frac{4+8\sqrt{3}}{4}, \quad \frac{-2+2\sqrt{6}}{2}$$

To Simplify a Fraction Whose Numerator Contains a Constant Term and a Square Root Term

1. Simplify the radical expression as far as possible.
2. If possible, factor out the GCF from each term in the numerator.
3. If possible, divide out a common factor from the GCF that was factored out of the numerator in step 2 and the denominator.

Examples 12 and 13 will illustrate this process.

EXAMPLE 12 Simplify **(a)** $\dfrac{4+8\sqrt{3}}{4}$ **(b)** $\dfrac{-2+2\sqrt{6}}{8}$

Solution: **(a)** Factor 4 from both terms in the numerator. Then divide out the common factor, 4, from the numerator and denominator.

$$\frac{4 + 8\sqrt{3}}{4} = \frac{\not{4}(1 + 2\sqrt{3})}{\not{4}} = 1 + 2\sqrt{3}$$

(b) Factor 2 from both terms in the numerator. Then divide out the common factor, 2, from the numerator and denominator.

$$\frac{-2 + 2\sqrt{6}}{8} = \frac{\not{2}(-1 + \sqrt{6})}{\underset{4}{\not{8}}} = \frac{-1 + \sqrt{6}}{4}$$

Note that the factoring of the numerators in parts **(a)** and **(b)** of Example 12 may be checked by multiplying using the distributive property.

(a) $4(1 + 2\sqrt{3}) = 4 + 8\sqrt{3}$ **(b)** $2(-1 + \sqrt{6}) = -2 + 2\sqrt{6}$

You should always check your factoring by mentally multiplying back the factors.

COMMON STUDENT ERROR Remember only common *factors* can be divided out. Do not make the mistakes on the right.

Cannot Be Simplified	*Incorrect*
$\frac{3 + 2\sqrt{5}}{2}$	$\frac{3 + 2\sqrt{5}}{2} = \frac{3 + \overset{1}{\not{2}}\sqrt{5}}{\underset{1}{\not{2}}} = 3 + \sqrt{5}$
$\frac{4 + 3\sqrt{5}}{2}$	$\frac{\overset{2}{\not{4}} + 3\sqrt{5}}{\underset{1}{\not{2}}} = 2 + 3\sqrt{5}$
$\frac{3 + \sqrt{6}}{2}$	$\frac{3 + \sqrt{\not{6}^{3}}}{\underset{1}{\not{2}}} = 3 + \sqrt{3}$

EXAMPLE 13 Simplify if possible.

(a) $\frac{12 - \sqrt{27}}{3}$ **(b)** $\frac{10 + \sqrt{20}}{5}$ **(c)** $\frac{5 + 3\sqrt{10}}{5}$

Solution:

(a) First simplify $\sqrt{27}$.

$$\frac{12 - \sqrt{27}}{3} = \frac{12 - \sqrt{9}\sqrt{3}}{3} = \frac{12 - 3\sqrt{3}}{3} = \frac{\not{3}(4 - \sqrt{3})}{\not{3}} = 4 - \sqrt{3}$$

(b) First simplify $\sqrt{20}$.

$$\frac{10 + \sqrt{20}}{5} = \frac{10 + \sqrt{4}\sqrt{5}}{5} = \frac{10 + 2\sqrt{5}}{5} = \frac{2(5 + \sqrt{5})}{5}$$

Since the 2 in the numerator and the 5 in the denominator have no common factors, the expression $\frac{10 + \sqrt{20}}{5}$ can be simplified to $\frac{10 + 2\sqrt{5}}{5}$ or $\frac{2(5 + \sqrt{5})}{5}$, but it cannot be simplified further.

(c) First notice that $\sqrt{10}$ cannot be simplified. Since the two terms in the numerator have no common factor, this fraction cannot be simplified.

Exercise Set 9.4

Simplify each radical expression.

1. $5\sqrt{6} - 2\sqrt{6}$
2. $\sqrt{5} + 2\sqrt{5}$
3. $6\sqrt{7} - 8\sqrt{7}$
4. $4\sqrt{10} + 6\sqrt{10} - \sqrt{10} + 2$
5. $2\sqrt{3} - 2\sqrt{3} - 4\sqrt{3} + 5$
6. $12\sqrt{15} + 5\sqrt{15} - 8\sqrt{15}$
7. $4\sqrt{x} + \sqrt{x}$
8. $-2\sqrt{x} - 3\sqrt{x}$
9. $-\sqrt{x} + 6\sqrt{x} - 2\sqrt{x}$
10. $3\sqrt{y} - 6\sqrt{y}$
11. $3\sqrt{y} - \sqrt{y} + 3$
12. $3\sqrt{5} - \sqrt{x} + 4\sqrt{5} + 3\sqrt{x}$
13. $\sqrt{x} + \sqrt{y} + x + 3\sqrt{y}$
14. $2 + 3\sqrt{y} - 6\sqrt{y} + 5$
15. $3 + 4\sqrt{x} - 6\sqrt{x}$
16. $4\sqrt{x} + 6\sqrt{x} - 3\sqrt{x} + 2x$
17. $5 - 2\sqrt{z} - \sqrt{z} - 5\sqrt{z}$
18. $-3\sqrt{7} + \sqrt{7} - 2\sqrt{x} - 7\sqrt{x}$

Simplify each radical expression.

19. $\sqrt{8} - \sqrt{12}$
20. $\sqrt{27} + \sqrt{45}$
21. $\sqrt{200} - \sqrt{72}$
22. $\sqrt{75} + \sqrt{108}$
23. $\sqrt{125} + \sqrt{20}$
24. $\sqrt{60} - \sqrt{135}$
25. $4\sqrt{50} - \sqrt{72} + \sqrt{8}$
26. $-4\sqrt{90} + 3\sqrt{40} + 2\sqrt{10}$
27. $-6\sqrt{75} + 4\sqrt{125}$
28. $4\sqrt{80} - \sqrt{75}$
29. $5\sqrt{250} - 9\sqrt{80}$
30. $7\sqrt{108} - 6\sqrt{180}$
31. $8\sqrt{64} - \sqrt{96}$
32. $3\sqrt{250} + 5\sqrt{160}$

Multiply.

33. $(3 + \sqrt{2})(3 - \sqrt{2})$
34. $(\sqrt{6} + 3)(\sqrt{6} - 3)$
35. $(6 - \sqrt{5})(6 + \sqrt{5})$
36. $(\sqrt{8} - 3)(\sqrt{8} + 3)$
37. $(\sqrt{x} + 3)(\sqrt{x} - 3)$
38. $(\sqrt{x} + 5)(\sqrt{x} - 5)$
39. $(\sqrt{6} + x)(\sqrt{6} - x)$
40. $(\sqrt{y} - 3)(\sqrt{y} + 3)$
41. $(\sqrt{x} + y)(\sqrt{x} - y)$
42. $(\sqrt{5x} + \sqrt{y})(\sqrt{5x} - \sqrt{y})$
43. $(2\sqrt{x} + 3\sqrt{y})(2\sqrt{x} - 3\sqrt{y})$
44. $(4\sqrt{2x} + \sqrt{3y})(4\sqrt{2x} - \sqrt{3y})$

Simplify each radical expression.

45. $\frac{4}{2 + \sqrt{3}}$
46. $\frac{3}{\sqrt{6} - 5}$
47. $\frac{3}{\sqrt{5} + 2}$
48. $\frac{4}{\sqrt{2} - 7}$
49. $\frac{2}{\sqrt{2} + \sqrt{3}}$
50. $\frac{5}{\sqrt{6} + \sqrt{3}}$
51. $\frac{8}{\sqrt{5} - \sqrt{8}}$
52. $\frac{1}{\sqrt{17} - \sqrt{8}}$
53. $\frac{2}{6 + \sqrt{x}}$
54. $\frac{5}{\sqrt{x} - 3}$
55. $\frac{6}{4 - \sqrt{y}}$
56. $\frac{5}{3 + \sqrt{x}}$
57. $\frac{4}{\sqrt{x} - y}$
58. $\frac{9}{x + \sqrt{y}}$
59. $\frac{x}{\sqrt{x} + \sqrt{y}}$
60. $\frac{\sqrt{3}}{\sqrt{x} - \sqrt{3}}$
61. $\frac{\sqrt{x}}{\sqrt{5} + \sqrt{x}}$
62. $\frac{x}{\sqrt{x} - y}$

Simplify each radical expression if possible. (See Examples 12 and 13.)

63. $\dfrac{8 - 4\sqrt{3}}{2}$

64. $\dfrac{6 - 4\sqrt{2}}{2}$

65. $\dfrac{6 + 24\sqrt{5}}{3}$

66. $\dfrac{10 + 2\sqrt{5}}{5}$

67. $\dfrac{4 + 3\sqrt{6}}{2}$

68. $\dfrac{4 - \sqrt{20}}{2}$

69. $\dfrac{6 + 2\sqrt{75}}{3}$

70. $\dfrac{5 + 2\sqrt{50}}{5}$

71. $\dfrac{-2 + 4\sqrt{80}}{10}$

72. $\dfrac{12 - 6\sqrt{72}}{3}$

73. $\dfrac{60 - 40\sqrt{18}}{100}$

74. $\dfrac{-50 - \sqrt{200}}{60}$

Find the product of the given binomial and its conjugate.

75. $a - \sqrt{b}$

76. $\sqrt{a} + b$

77. $\sqrt{a} - \sqrt{b}$

78. $\sqrt{a} + \sqrt{b}$

79. Under what conditions can two square roots be added or subtracted?

80. **(a)** Explain in your own words how to rationalize the denominator in a fraction of the form $\dfrac{a}{b + \sqrt{c}}$.

(b) Rationalize $\dfrac{a}{b + \sqrt{c}}$.

CUMULATIVE REVIEW EXERCISES

81. Solve the equation $3(2x - 6) = 4(x - 9) + 3x$.

82. Solve the equation $2x^2 - x - 36 = 0$.

83. Subtract $\dfrac{1}{x^2 - 4} - \dfrac{2}{x - 2}$.

84. Mr. Moreno can stack a pile of wood in 20 minutes. With his wife's help, together they can stack the wood in 12 minutes. How long would it take Mrs. Moreno to stack the wood by herself?

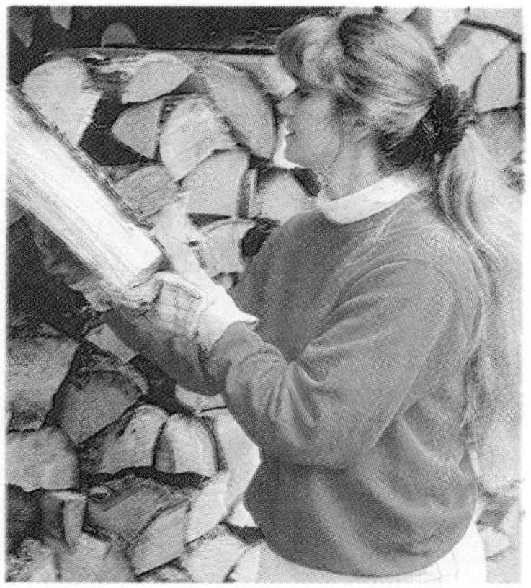

SEE EXERCISE 84.

Group Activity/ Challenge Problems

1. **(a)** Is the sum or difference of two rational expressions always a rational expression? Explain and give an example.

(b) Is the sum or difference of two irrational expressions always an irrational expression? Explain and give an example.

Find the perimeter and area of the following figures.

2.

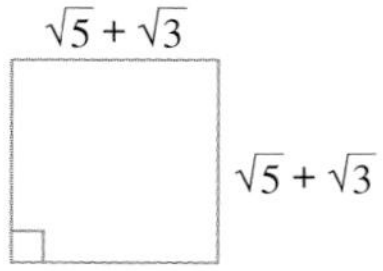

3.

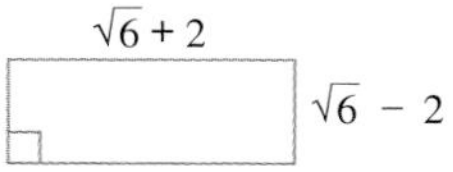

4.

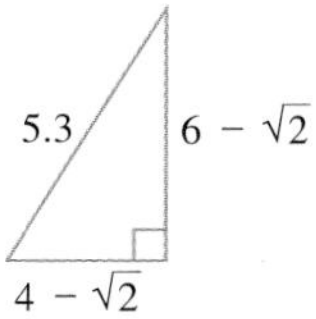

Fill in the shaded area to make the statement true. Explain how you determined your answer.

5. $-5\sqrt{} + 2\sqrt{3} + 3\sqrt{27} = -9\sqrt{3}$

6. $\dfrac{3}{5 + \sqrt{}} = \dfrac{3(5 - \sqrt{3})}{22}$

9.5 Solving Radical Equations

Tape 15  

1. Solve radical equations with only one square root term.
2. Solve radical equations with two square root terms.

Solve Radical Equations Containing One Square Root

1 A **radical equation** is an equation that contains a variable in a radicand. Some examples of radical equations are

$$\sqrt{x} = 3, \qquad \sqrt{x+4} = 6, \qquad \sqrt{x-2} = x - 6$$

To Solve a Radical Equation Containing Only One Square Root Term

1. Use the appropriate properties to rewrite the equation with the square root term by itself on one side of the equation. We call this *isolating* the radical.
2. Combine like terms.
3. Square both sides of the equation to remove the square root.
4. Solve the equation for the variable.
5. Check the solution in the original equation for extraneous roots.

The following examples illustrate this procedure.

EXAMPLE 1 Solve the equation $\sqrt{x} = 6$.

Solution: The square root containing the variable is already by itself on one side of the equation. Square both sides of the equation.

$$\begin{aligned} \sqrt{x} &= 6 \\ (\sqrt{x})^2 &= (6)^2 \\ x &= 36 \end{aligned}$$

Check:

$$\begin{aligned} \sqrt{x} &= 6 \\ \sqrt{36} &= 6 \\ 6 &= 6 \quad \text{true} \end{aligned}$$

EXAMPLE 2 Solve the equation $\sqrt{x+4} = 6$.

Solution: The square root containing the variable is already by itself on one side of the equation. Square both sides of the equation.

$$\begin{aligned} \sqrt{x+4} &= 6 \\ (\sqrt{x+4})^2 &= 6^2 \\ x + 4 &= 36 \\ x + 4 - 4 &= 36 - 4 \\ x &= 32 \end{aligned}$$

Check:
$$\sqrt{x+4} = 6$$
$$\sqrt{32+4} = 6$$
$$\sqrt{36} = 6$$
$$6 = 6 \quad \text{true}$$

EXAMPLE 3 Solve the equation $\sqrt{x} + 4 = 6$.

Solution: Since the 4 is outside the square root sign, we first subtract 4 from both sides of the equation to isolate the square root term.

$$\sqrt{x} + 4 = 6$$
$$\sqrt{x} + 4 - 4 = 6 - 4$$
$$\sqrt{x} = 2$$

Now square both sides of the equation.

$$(\sqrt{x})^2 = 2^2$$
$$x = 4$$

A check will show that 4 is the solution.

Helpful Hint

When you square both sides of an equation, you may introduce extraneous roots. An **extraneous root** is a number obtained when solving an equation that is not a solution to the original equation. Therefore, equations where both sides are squared in the process of finding their solutions should always be checked for extraneous roots by substituting the numbers found back in the **original** equation.

Consider the equation

$$x = 5$$

Now square both sides.

$$x^2 = 25$$

Note that the equation $x = 5$ is true only when x is 5. However the equation $x^2 = 25$ is true for both 5 and -5. When we squared $x = 5$, we introduced the extraneous root -5.

EXAMPLE 4 Solve the equation $\sqrt{x} = -5$

Solution:
$$\sqrt{x} = -5$$
$$(\sqrt{x})^2 = (-5)^2$$
$$x = 25$$

Check:
$$\sqrt{x} = -5$$
$$\sqrt{25} = -5$$
$$5 = -5 \quad \text{false}$$

Since the check results in a false statement, the number 25 is an extraneous root and is not a solution to the given equation. Thus, the equation $\sqrt{x} = -5$ has no real solutions.

In Example 4, you might have realized without working the problem that there is no solution. In the original equation, the left side is nonnegative and the right side is negative; thus they cannot possibly be equal.

EXAMPLE 5 Solve the equation $\sqrt{2x - 3} = x - 3$.

Solution: Square both sides of the equation.

$$(\sqrt{2x - 3})^2 = (x - 3)^2$$
$$2x - 3 = (x - 3)(x - 3)$$
$$2x - 3 = x^2 - 6x + 9$$

Now solve the quadratic equation as explained in Section 5.6. Move the $2x$ and -3 to the right side of the equation to obtain

$$0 = x^2 - 8x + 12 \quad \text{or} \quad x^2 - 8x + 12 = 0$$

Now solve for x by factoring.

$$x^2 - 8x + 12 = 0$$
$$(x - 6)(x - 2) = 0$$
$$x - 6 = 0 \quad \text{or} \quad x - 2 = 0$$
$$x = 6 \qquad\qquad x = 2$$

Check:

$x = 6$		$x = 2$	
$\sqrt{2x - 3} = x - 3$		$\sqrt{2x - 3} = x - 3$	
$\sqrt{2(6) - 3} = 6 - 3$		$\sqrt{2(2) - 3} = 2 - 3$	
$\sqrt{9} = 3$		$\sqrt{1} = -1$	
$3 = 3$	**true**	$1 = -1$	**false**

The solution is 6. Two is not a solution to the equation.

EXAMPLE 6 Solve the equation $2x - 5\sqrt{x} - 3 = 0$.

Solution: First rewrite the equation so that the square root containing the variable is by itself on one side of the equation.

$$2x - 5\sqrt{x} - 3 = 0$$
$$-5\sqrt{x} = -2x + 3$$
$$\text{or} \qquad 5\sqrt{x} = 2x - 3$$

Now square both sides of the equation.

$$(5\sqrt{x})^2 = (2x - 3)^2$$
$$5^2(\sqrt{x})^2 = (2x - 3)^2$$
$$25x = (2x - 3)(2x - 3)$$
$$25x = 4x^2 - 12x + 9$$
$$0 = 4x^2 - 37x + 9$$
$$\text{or} \qquad 4x^2 - 37x + 9 = 0$$
$$(4x - 1)(x - 9) = 0$$
$$4x - 1 = 0 \quad \text{or} \quad x - 9 = 0$$
$$4x = 1 \qquad\qquad x = 9$$
$$x = \frac{1}{4}$$

Check:

$$x = \frac{1}{4}$$
$$2x - 5\sqrt{x} - 3 = 0$$
$$2\left(\frac{1}{4}\right) - 5\sqrt{\frac{1}{4}} - 3 = 0$$
$$\frac{1}{2} - 5\left(\frac{1}{2}\right) - 3 = 0$$
$$\frac{1}{2} - \frac{5}{2} - 3 = 0$$
$$-\frac{4}{2} - 3 = 0$$
$$-2 - 3 = 0$$
$$-5 = 0 \quad \text{false}$$

$$x = 9$$
$$2x - 5\sqrt{x} - 3 = 0$$
$$2(9) - 5\sqrt{9} - 3 = 0$$
$$18 - 5(3) - 3 = 0$$
$$18 - 15 - 3 = 0$$
$$0 = 0 \quad \text{true}$$

The solution is 9; $\frac{1}{4}$ is not a solution.

Solve Radical Equations Containing Two Square Roots

2 Consider the radical equations

$$\sqrt{x+1} = \sqrt{x-3}, \qquad \sqrt{x+5} - \sqrt{2x+4} = 0$$

These equations are different from those previously discussed because they have two square root terms containing the variable x. To solve equations of this type, rewrite the equation, when necessary, so that there is only one square root on each side of the equation. Then square both sides of the equation. Examples 7 and 8 illustrate this procedure.

EXAMPLE 7 Solve the equation $\sqrt{2x+2} = \sqrt{3x-5}$

Solution: Since each side of the equation already contains one square root, it is not necessary to rewrite the equation. Square both sides of the equation, then solve for x.

$$(\sqrt{2x+2})^2 = (\sqrt{3x-5})^2$$
$$2x + 2 = 3x - 5$$
$$2 = x - 5$$
$$7 = x$$

Check:

$$\sqrt{2x+2} = \sqrt{3x-5}$$
$$\sqrt{2(7)+2} = \sqrt{3(7)-5}$$
$$\sqrt{16} = \sqrt{16}$$
$$4 = 4 \quad \text{true}$$

The solution is 7.

EXAMPLE 8 Solve the equation $3\sqrt{x-2} - \sqrt{7x+4} = 0$.

Solution: Add $\sqrt{7x+4}$ to both sides of the equation to get one square root on each side of the equation. Then square both sides of the equation.

$$
\begin{aligned}
3\sqrt{x-2} - \sqrt{7x+4} + \sqrt{7x+4} &= 0 + \sqrt{7x+4} \\
3\sqrt{x-2} &= \sqrt{7x+4} \\
(3\sqrt{x-2})^2 &= (\sqrt{7x+4})^2 \\
9(x-2) &= 7x+4 \\
9x - 18 &= 7x + 4 \\
2x - 18 &= 4 \\
2x &= 22 \\
x &= 11
\end{aligned}
$$

Check:

$$
\begin{aligned}
3\sqrt{x-2} - \sqrt{7x+4} &= 0 \\
3\sqrt{11-2} - \sqrt{7(11)+4} &= 0 \\
3\sqrt{9} - \sqrt{77+4} &= 0 \\
3(3) - \sqrt{81} &= 0 \\
9 - 9 &= 0 \quad \text{true}
\end{aligned}
$$

Helpful Hint

In Example 6 when we simplified $(5\sqrt{x})^2$ we obtained $25x$ and in Example 8 when we simplified $(3\sqrt{x-2})^2$ we obtained $9(x-2)$. Remember, by the power rule for exponents (discussed in Section 4.1) that when a product of factors is raised to a power, each of the factors is raised to that power. Thus, we see that

$$(5\sqrt{x})^2 = 5^2(\sqrt{x})^2 = 25x \quad \text{and} \quad (3\sqrt{x-2})^2 = 3^2(\sqrt{x-2})^2 = 9(x-2)$$

Exercise Set 9.5

Solve each equation. If the equation has no real solution, so state.

1. $\sqrt{x} = 8$
2. $\sqrt{x} = 5$
3. $\sqrt{x} = -3$
4. $\sqrt{x-3} = 6$
5. $\sqrt{x+5} = 3$
6. $\sqrt{2x-4} = 2$
7. $\sqrt{2x+4} = -6$
8. $\sqrt{x} - 4 = 8$
9. $\sqrt{x} + 3 = 5$
10. $4 + \sqrt{x} = 9$
11. $6 = 4 + \sqrt{x}$
12. $2 = 8 - \sqrt{x}$
13. $4 + \sqrt{x} = 2$
14. $\sqrt{3x+4} = x - 2$
15. $\sqrt{2x-5} = x - 4$
16. $\sqrt{x^2+8} = x + 2$
17. $\sqrt{2x-6} = \sqrt{5x-27}$
18. $2\sqrt{x+3} = 10$
19. $\sqrt{3x+3} = \sqrt{5x-1}$
20. $\sqrt{2x-5} = \sqrt{x+2}$
21. $\sqrt{3x+9} = 2\sqrt{x}$
22. $x - 6 = \sqrt{3x}$
23. $\sqrt{4x-5} = \sqrt{x+9}$
24. $\sqrt{x^2+3} = x + 1$
25. $3\sqrt{x} = \sqrt{x+8}$
26. $x - 5 = \sqrt{x^2-35}$
27. $4\sqrt{x} = x + 3$
28. $2x - 1 = -\sqrt{x}$
29. $\sqrt{2x-3} = 2\sqrt{3x-2}$
30. $6 - 2\sqrt{3x} = 0$
31. $\sqrt{x^2+5} = x + 5$
32. $2\sqrt{4x-3} = 10$
33. $5 + \sqrt{x-5} = x$
34. $\sqrt{4x+5} + 5 = 2x$
35. $\sqrt{8-7x} = x - 2$
36. $1 + \sqrt{x+1} = x$

37. What is a radical equation?

38. What is an extraneous root?

39. Why is it necessary to always check solutions to radical equations?

40. (a) Write in your own words a step-by-step procedure for solving equations containing a single square root term.
(b) Solve the equation $\sqrt{x+1} - 1 = 1$ using the procedure outlined in part (a).

CUMULATIVE REVIEW EXERCISES

1] 41. Solve the system of equations graphically.

$$3x - 2y = 6$$
$$y = 2x - 4$$

2] 42. Solve the system of equations by substitution.

$$3x - 2y = 6$$
$$y = 2x - 4$$

[8.3] 43. Solve the system of equations by the addition method.

$$3x - 2y = 6$$
$$y = 2x - 4$$

[8.4] 44. A boat can travel at a speed of 12 miles per hour with the current and 4 miles per hour against the current. Find the speed of the boat in still water and the speed of the current.

Group Activity/ Challenge Problems

Solve each equation. (Hint: *You will need to square both sides of the equation twice.)*

1. $\sqrt{x+2} = \sqrt{x+16}$
2. $\sqrt{x+1} = 2 - \sqrt{x}$
3. $\sqrt{x+7} = 5 - \sqrt{x-8}$

Radical equations in two variables can be graphed by selecting values for x *and finding the corresponding values of* y *as was done in Section 7.3.* **(a)** *Graph the equation using values of* x *that make the radicand greater than or equal to 0.* **(b)** *Is the graph linear? Explain.* **(c)** *Is the graph a function? (See Section 7.6). Explain.*

4. $y = \sqrt{x}$
5. $y = \sqrt{x+2}$
6. $y = \sqrt{x-2}$

9.6 Applications of Radicals

Tape 16

1. Use the Pythagorean theorem to solve application problems.
2. Use the distance formula to solve application problems.
3. Use radicals to solve science problems.

In this section we will focus on some of the many important applications of radicals. We will discuss the Pythagorean theorem and the distance formula, and then give a few additional applications of radicals.

Pythagorean Theorem

1 A **right triangle** is a triangle that contains a right, or 90°, angle (Fig. 9.1). The two shorter sides of a right triangle are called the **legs** and the side opposite the right angle is called the **hypotenuse.** The Pythagorean theorem expresses the relationship between the legs of a right triangle and its hypotenuse.

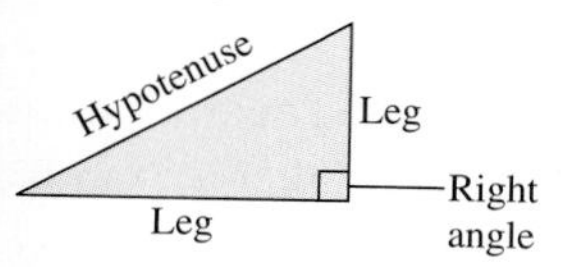

FIGURE 9.1

Pythagorean Theorem

The square of the hypotenuse of a right triangle is equal to the sum of the squares of the two legs.

If a and b represent the legs, and c represents the hypotenuse, then

$$a^2 + b^2 = c^2$$

In Section 9.5, when we solved equations containing square roots, we raised both sides of the equation to the second power to eliminate the square roots. When we solve problems using the Pythagorean theorem, we will raise both sides of the equation to the $\frac{1}{2}$ power to remove the square on one of the variables. We can do this because the rules of exponents presented in Sections 4.1 and 4.2 also apply to fractional exponents. Since lengths are positive, the values of a, b, and c in the Pythagorean theorem must represent positive values.

EXAMPLE 1 Find the hypotenuse of the right triangle whose legs are 3 feet and 4 feet.

Solution: Draw a picture of the problem (Fig. 9.2). When drawing the picture, it makes no difference which leg is called a and which leg is called b.

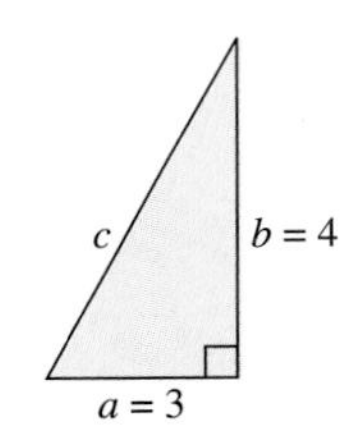

FIGURE 9.2

$$\begin{aligned} a^2 + b^2 &= c^2 \\ 3^2 + 4^2 &= c^2 \\ 9 + 16 &= c^2 \\ 25 &= c^2 \\ (25)^{1/2} &= (c^2)^{1/2} && \text{Raise both sides of equations to the } \tfrac{1}{2} \text{ power.} \\ \sqrt{25} &= c \\ 5 &= c \end{aligned}$$

The hypotenuse is 5 feet.

Check:

$$\begin{aligned} a^2 + b^2 &= c^2 \\ 3^2 + 4^2 &= 5^2 \\ 9 + 16 &= 25 \\ 25 &= 25 && \text{true} \end{aligned}$$

In Example 1, we could also solve $25 = c^2$ for c by taking the square root of each side of the equation. We will discuss the square root property that allows us to do this in Section 10.1.

EXAMPLE 2 The hypotenuse of a right triangle is 12 inches. Find the second leg if one leg is 8 inches.

$c = 12$ $b = ?$ $a = 8$

FIGURE 9.3

Solution: First, draw a sketch of the triangle (Fig. 9.3).

$$a^2 + b^2 = c^2$$
$$8^2 + b^2 = (12)^2$$
$$64 + b^2 = 144$$
$$b^2 = 80$$
$$(b^2)^{1/2} = (80)^{1/2}$$
$$b = \sqrt{80}, \quad \text{or} \quad \text{approximately } 8.94 \text{ inches}$$

EXAMPLE 3 A regulation baseball diamond is a square with 90 feet between bases. How far is second base from home plate?

Solution: Draw the baseball diamond (Fig. 9.4). We are asked to find the distance from second base to home plate. This distance is the hypotenuse of the triangle, c, shown in Figure 9.5.

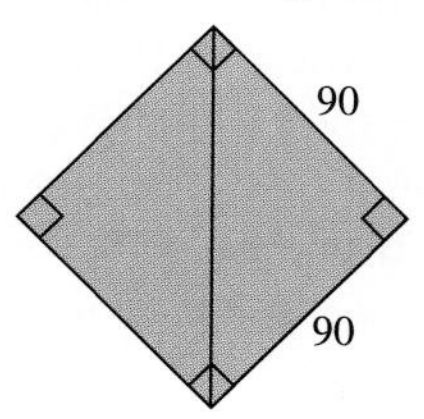

FIGURE 9.4

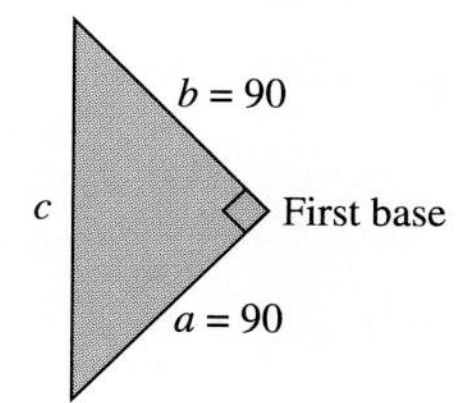

FIGURE 9.5

$$a^2 + b^2 = c^2$$
$$(90)^2 + (90)^2 = c^2$$
$$8100 + 8100 = c^2$$
$$16{,}200 = c^2$$
$$c = \sqrt{16{,}200}, \quad \text{or} \quad \text{approximately } 127.28 \text{ feet}$$

Distance Formula

2 The distance formula can be used to find the distance between two points, (x_1, y_1) and (x_2, y_2), in the Cartesian coordinate system.

Distance Formula

$$d = \sqrt{(x_2 - x_1)^2 + (y_2 - y_1)^2}$$

EXAMPLE 4 Find the length of the line segment between the points $(-1, -4)$ and $(5, -2)$.

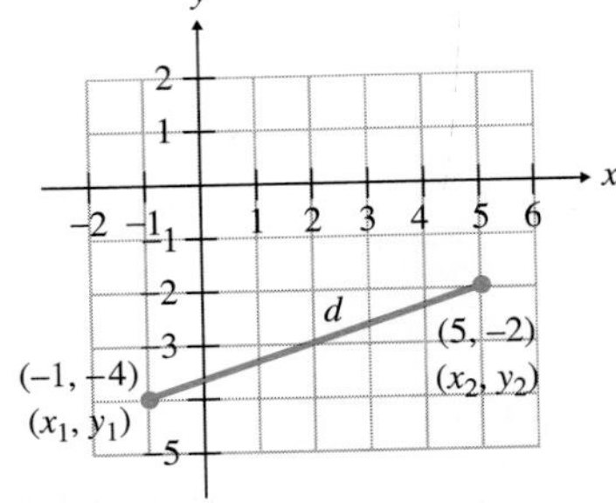

FIGURE 9.6

Solution: The two points are illustrated in Figure 9.6. It makes no difference which points are labeled (x_1, y_1), and (x_2, y_2). Let $(5, -2)$ be (x_2, y_2) and $(-1, -4)$ be (x_1, y_1). Thus $x_2 = 5$, $y_2 = -2$ and $x_1 = -1$, $y_1 = -4$.

$$\begin{aligned} d &= \sqrt{(x_2 - x_1)^2 + (y_2 - y_1)^2} \\ &= \sqrt{[5 - (-1)]^2 + [-2 - (-4)]^2} \\ &= \sqrt{(5 + 1)^2 + (-2 + 4)^2} \\ &= \sqrt{6^2 + 2^2} \\ &= \sqrt{36 + 4} = \sqrt{40}, \quad \text{or} \quad \text{approximately } 6.32 \end{aligned}$$

Thus, the distance between $(-1, -4)$ and $(5, -2)$ is approximately 6.32 units.

Science Applications

3 Radicals are often used in science and mathematics courses. Examples 5 through 7 illustrate some scientific applications of radicals.

EXAMPLE 5 During the sixteenth and seventeenth centuries, Galileo Galilei, using the (leaning) Tower of Pisa, performed many experiments with objects falling freely under the influence of gravity. He showed, for example, that a rock dropped from, say, 10 feet, hit the ground with a higher velocity than did the same rock dropped from 5 feet. A formula for the velocity of an object in feet per second (neglecting air resistance) after it has fallen a certain distance is

$$v = \sqrt{2gh}$$

where g is the acceleration of gravity and h is the height the object has fallen in feet. On Earth the acceleration of gravity, g, is approximately 32 feet per second squared.

(a) Find the velocity of a rock after it has fallen 5 feet.

(b) Find the velocity of a coffee mug after it has fallen 10 feet.

(c) Find the velocity of a tube of toothpaste after it has fallen 100 feet.

Solution: **(a)** Begin by substituting 32 for g in the given equation.

$$v = \sqrt{2gh} = \sqrt{2(32)h} = \sqrt{64h}$$

At $h = 5$ feet,

$$v = \sqrt{64(5)} = \sqrt{320} \approx 17.9 \text{ feet per second}$$

After a rock has fallen 5 feet, its velocity is approximately 17.9 feet per second.

(b) After falling 10 feet,

$$v = \sqrt{64(10)} = \sqrt{640} \approx 25.3 \text{ feet per second}$$

The velocity of a coffee mug after falling 10 feet is approximately 25.3 feet per second.

(c) After falling 100 feet.

$$v = \sqrt{64(100)} = \sqrt{6400} = 80 \text{ feet per second}$$

The velocity of a tube of toothpaste after falling 100 feet is 80 feet per second.

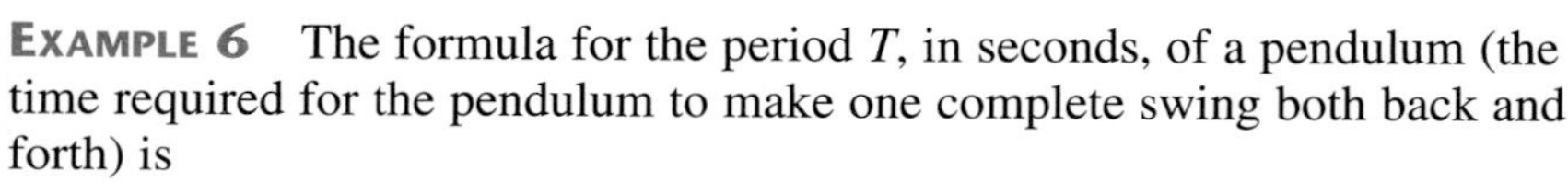

EXAMPLE 6 The formula for the period T, in seconds, of a pendulum (the time required for the pendulum to make one complete swing both back and forth) is

$$T = 2\pi\sqrt{L/32}$$

where L is the length of the pendulum in feet (Fig. 9.7). Find the period of the pendulum if its length is 8 feet. Use 3.14 as an approximation for π.

Solution:

$$T \approx 2(3.14)\sqrt{\frac{8}{32}}$$

$$\approx 6.28\sqrt{\frac{1}{4}}$$

$$\approx 6.28\left(\frac{1}{2}\right) = 3.14 \text{ seconds}$$

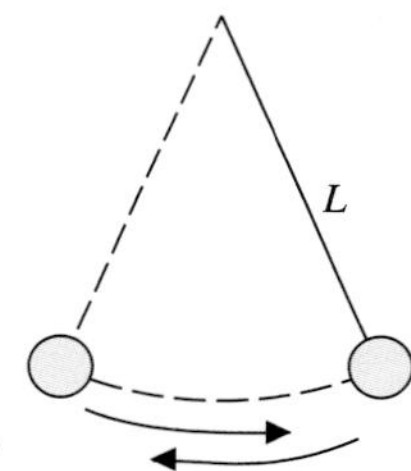

FIGURE 9.7

A pendulum 8 feet long takes about 3.14 seconds to make one complete swing.

EXAMPLE 7 For any planet, its "year" is the time it takes for the planet to revolve once around the sun. The number of "Earth days," N, in a given planet's year, is approximated by the formula

$$N = 0.2(\sqrt{R})^3$$

where R is the mean distance from the sun in millions of kilometers. Find the number of Earth days in the year of each planet illustrated in Figure 9.8.

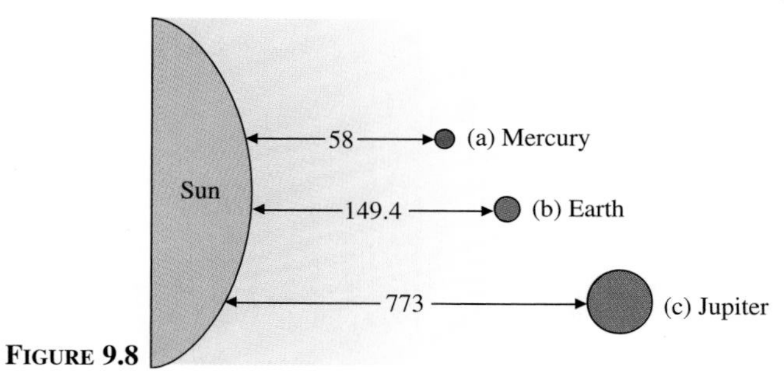

FIGURE 9.8

Solution:

(a) Mercury:

$$\begin{aligned} N &\approx 0.2(\sqrt{58})^3 \\ &\approx 0.2(7.6)^3 \\ &\approx 0.2(441.7) \\ &\approx 88.3 \end{aligned}$$

It takes Mercury about 88 Earth days to revolve once around the sun.

(b) Earth:

$$\begin{aligned} N &\approx 0.2(\sqrt{149.4})^3 \\ &\approx 0.2(12.2)^3 \\ &\approx 0.2(1826.1) \\ &\approx 365.2 \end{aligned}$$

It takes Earth about 365 Earth days to revolve once around the sun (an answer that should not be surprising).

(c) Jupiter:

$$\begin{aligned} N &\approx 0.2(\sqrt{773})^3 \\ &\approx 0.2(27.8)^3 \\ &\approx 0.2(21{,}491.6) \\ &\approx 4298 \end{aligned}$$

It takes Jupiter about 4298 Earth days (or about 11.8 Earth years) to revolve once about the sun.

Exercise Set 9.6

Use the Pythagorean theorem to find the quantity indicated. Round answers to the nearest hundredth.

1.

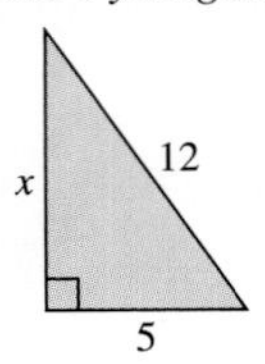

2.

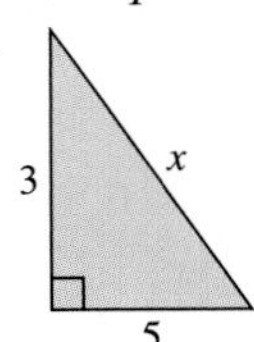

3.

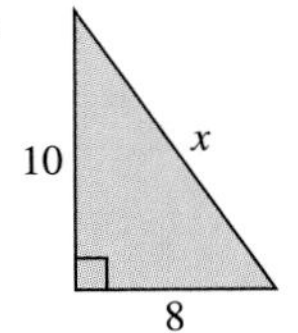

4.

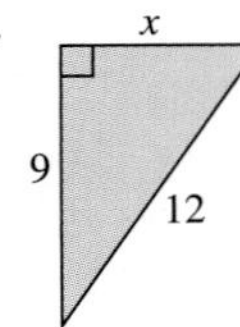

5.

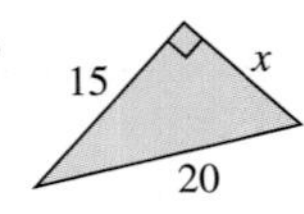

6.

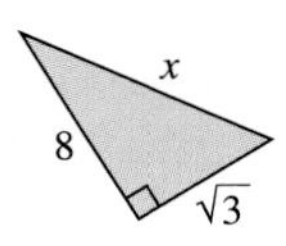

7.

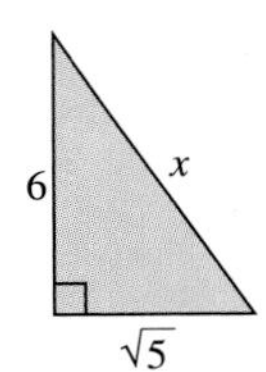

8.

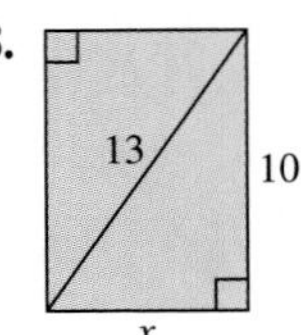

9.

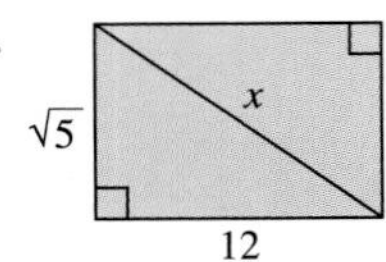

10.

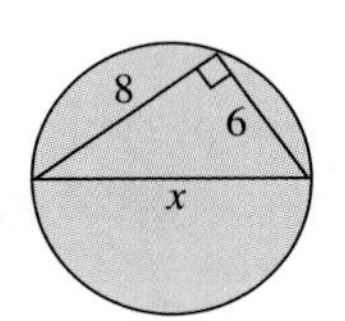

11.

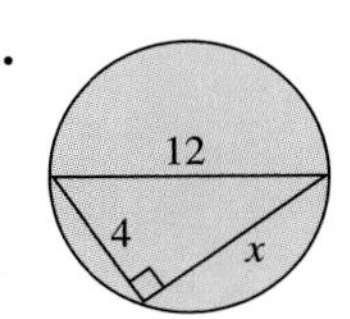

12.

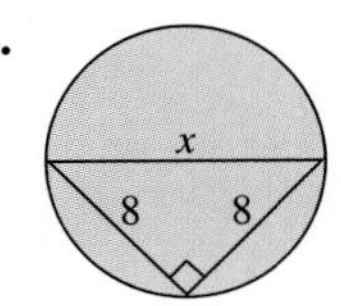

13. A football field is 120 yards long from end zone to end zone. Find the length of the diagonal from one end zone to the other if the width of the field is 53.3 yards.

14. A boxing ring is a square 16 feet by 16 feet (actual ring size will vary with country and state). Find the distance from one boxer's corner to the other boxer's corner.

15. How long a wire is needed to reach from the top of a 4-meter telephone pole to a point 1.5 meters from the base of the pole?

16. An 8-meter extension ladder leans against a house. The base of the ladder is 2 meters from the house. How high is the top of the ladder?

Use the distance formula to find the length of the line segments between each pair of points. (See Example 4.)

17. $(-4, 3)$ and $(-1, 4)$

18. $(4, -3)$ and $(6, 2)$

19. $(-8, 4)$ and $(4, -8)$

20. $(0, 5)$ and $(-6, -4)$

Solve.

21. Find the side of a square that has an area of 144 square inches. (Use $A = s^2$.)

22. A formula for the area of a circle is $A = \pi r^2$, where π is approximately 3.14 and r is the radius of the circle. Find the radius of a circle whose area is 20 square inches.

23. Find the radius of a circle whose area is 80 square feet.

24. Find the length of the diagonal of a rectangle with a length of 12 inches and width of 5 inches.

25. Find the length of the diagonal of a rectangle with a length of 9 inches and width of 12 inches.

26. Find the length of the diagonal of a rectangle with a length of 25 inches and width of 10 inches.

27. Find the velocity of a lamp after it has fallen 80 feet. (Use $v = \sqrt{2gh}$; refer to Example 5.)

28. Find the velocity of a plate after it has fallen 1000 feet.

29. An olive is dropped from the Empire State Building, height 1250 feet. What is its velocity when it strikes the ground?

30. With what velocity will a suitcase dropped from a plane 1 mile (5280 feet) high strike the ground?

31. Find the period of a 40-foot pendulum. (Use $T = 2\pi\sqrt{L/32}$, refer to Example 6.)

32. Find the period of a 60-foot pendulum.

33. Find the period of a 10-foot pendulum.

34. Find the number of Earth days in the year of the planet Mars, whose mean distance from the sun is 227 million kilometers. [Use $N = 0.2(\sqrt{R})^3$, and refer to Example 7.]

35. Find the number of Earth days in the year of the planet Saturn, whose mean distance from the sun is 1418 million kilometers.

36. When two forces, F_1 and F_2, pull at right angles to each other as illustrated, the resultant, or the effective force, R, can be found by the formula

$$R = \sqrt{F_1^2 + F_2^2}$$

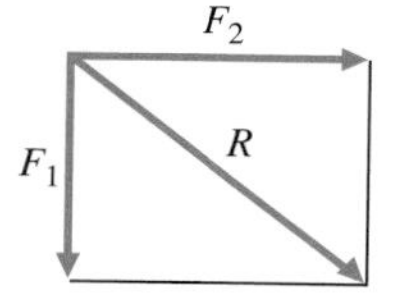

Two cars at a 90° angle to each other are trying to pull a third out of the mud, as shown. If car A exerts 600 pounds of force and car B exerts 800 pounds of force, find the resulting force on the car stuck in the mud.

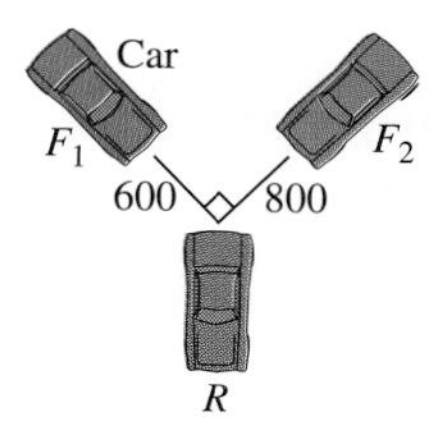

37. The escape velocity in meters per second, or the velocity needed for a spacecraft to escape a planet's gravitational field, is found by the formula

$$v_e = \sqrt{2gR}$$

where g is the force of gravity of the planet and R is the radius of the planet in meters. Find the escape velocity for Earth where $g = 9.75$ meters per second squared and $R = 6{,}370{,}000$ meters.

CUMULATIVE REVIEW EXERCISES

] 38. Solve the inequality $2(x + 3) < 4x - 6$.

] 39. Simplify $(4x^{-4}y^3)^{-1}$.

[4.4] 40. Simplify:

$$x^3 - 2x^2 - 6x + 4 - (3x^3 - 6x^2 + 8)$$

[4.6] 41. Divide $\dfrac{5x^4 - 9x^3 + 6x^2 - 4x - 3}{3x^2}$

Group Activity/ Challenge Problems

1. The length of a rectangle is 3 inches more than its width. If the length of the diagonal is 15 inches, find the dimensions of the rectangle.
2. The force of gravity on the moon is $\frac{1}{6}$ of that on Earth. If a camera falls from a rocket 100 feet above the surface of the moon, with what velocity will it strike the moon? (Use $v = \sqrt{2gh}$; see Example 5.)
3. Find the length of a pendulum if the period is 2 seconds. (Use $T = 2\pi\sqrt{L/32}$; refer to Example 6.)
4. The length of the diagonal of a rectangular solid (see the figure) is given by

$$d = \sqrt{a^2 + b^2 + c^2}$$

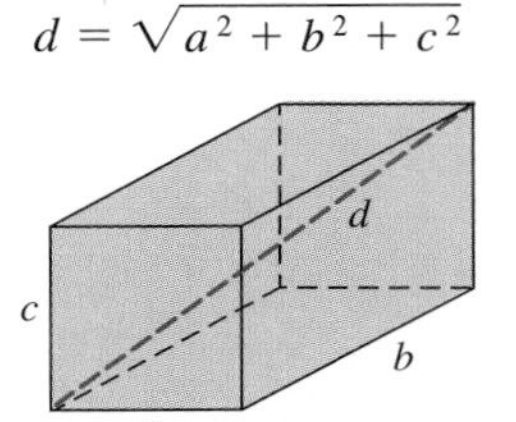

Find the length of the diagonal of a suitcase 37 inches long, 15 inches wide, and 9 inches deep.

9.7 Higher Roots and Rational Exponents

Tape 16

1 Evaluate cube and fourth roots.
2 Simplify cube and fourth roots.
3 Write radical expressions in exponential form.

Evaluating Cube and Fourth Roots

1 In this section we will use the same basic concepts used in Sections 9.1 through 9.4 to work with radicals with indexes of 3 and 4. Now we introduce cube and fourth roots.

$\sqrt[3]{a}$ is read "the cube root of a"
$\sqrt[4]{a}$ is read "the fourth root of a"

Note that

$$\sqrt[3]{a} = b \quad \text{if} \quad b^3 = a$$

and

$$\sqrt[4]{a} = b \quad \text{if} \quad b^4 = a, \qquad b > 0$$

Examples

$\sqrt[3]{8} = 2$ since $2^3 = 2 \cdot 2 \cdot 2 = 8$
$\sqrt[3]{-8} = -2$ since $(-2)^3 = (-2)(-2)(-2) = -8$
$\sqrt[3]{27} = 3$ since $3^3 = 3 \cdot 3 \cdot 3 = 27$
$\sqrt[4]{16} = 2$ since $2^4 = 2 \cdot 2 \cdot 2 \cdot 2 = 16$
$\sqrt[4]{81} = 3$ since $3^4 = 3 \cdot 3 \cdot 3 \cdot 3 = 81$

EXAMPLE 1 Evaluate: **(a)** $\sqrt[3]{-27}$ **(b)** $\sqrt[3]{125}$

Solution:

(a) To find $\sqrt[3]{-27}$, we must find the number that when cubed is -27.

$$\sqrt[3]{-27} = -3 \qquad \text{since } (-3)^3 = -27$$

(b) To find $\sqrt[3]{125}$, we must find the number that when cubed is 125.

$$\sqrt[3]{125} = 5 \qquad \text{since } 5^3 = 125$$

Note that the cube root of a positive number is a positive number and the cube root of a negative number is a negative number. The radicand of a fourth root (or any even root) must be a nonnegative number for the expression to be a real number. For example, $\sqrt[4]{-16}$ is not a real number because no real number raised to the fourth power can be a negative number.

Calculator Corner

EVALUATING CUBE AND HIGHER ROOTS ON A CALCULATOR

Scientific calculators contain a key that finds cube roots and higher roots. If your calculator contains a x^y or y^x key, then you can find these roots. To find cube and higher roots, you need to use both the inverse key, inv, and either the x^y or y^x key. (Some calculators use a 2nd key in place of the inv key.)

To find the value of $\sqrt[3]{216}$ using a calculator, press the following keys:

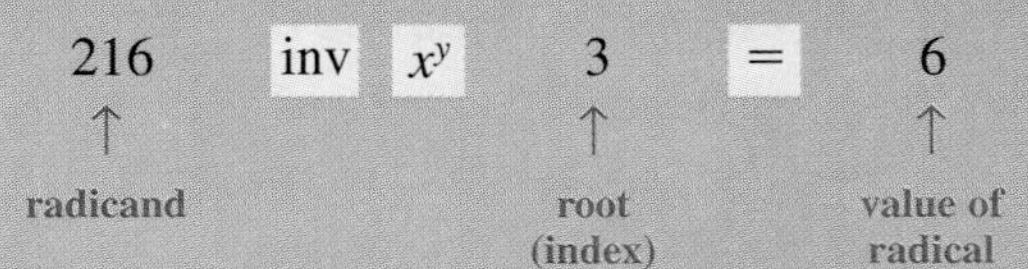

Thus, $\sqrt[3]{216} = 6$. If your calculator has a y^x key, use that key instead of the x^y key in the preceding sequence. To find the value of $\sqrt[4]{618}$ using a calculator, press

618 inv x^y 4 = 4.98594

Thus, $\sqrt[4]{618} \approx 4.98594$.

To find an odd root of a negative number, find the odd root of that positive number and then place a negative sign before the value. For example, to find $\sqrt[3]{-64}$, find $\sqrt[3]{64}$, which is 4; then place a negative sign before the value to get -4. Thus, $\sqrt[3]{-64} = -4$.

It will be helpful in the explanations that follow if we define perfect cubes. A **perfect cube** is a number that is the cube of a natural number.

1, 2, 3, 4, 5, 6, 7, 8, 9, 10, ...	Natural numbers
$1^3, 2^3, 3^3, 4^3, 5^3, 6^3, 7^3, 8^3, 9^3, 10^3, \ldots$	Cubes of natural numbers
1, 8, 27, 64, 125, 216, 343, 512, 729, 1000, ...	Perfect cubes

Note that $\sqrt[3]{1} = 1$, $\sqrt[3]{8} = 2$, $\sqrt[3]{27} = 3$, $\sqrt[3]{64} = 4$, and so on.

Perfect fourth powers can be expressed in a similar manner.

1, 2, 3, 4, 5, 6, ...	Natural numbers
$1^4, 2^4, 3^4, 4^4, 5^4, 6^4, \ldots$	Fourth powers of the natural numbers
1, 16, 81, 256, 625, 1296, ...	Perfect fourth powers

Note that $\sqrt[4]{1} = 1$, $\sqrt[4]{16} = 2$, $\sqrt[4]{81} = 3$, $\sqrt[4]{256} = 4$, and so on.

You may wish to refer to these numbers when evaluating cube and fourth roots.

Simplifying Cube and Fourth Roots

2 The product rule used in simplifying square roots can be expanded to indexes greater than 2.

Product Rule for Radicals

$$\sqrt[n]{a}\sqrt[n]{b} = \sqrt[n]{ab}, \qquad \text{for } a \geq 0, b \geq 0$$

To simplify a cube root whose radicand is a constant, write the radicand as the product of a perfect cube and another number. Then simplify, using the product rule.

EXAMPLE 2 Simplify: **(a)** $\sqrt[3]{32}$ **(b)** $\sqrt[3]{54}$ **(c)** $\sqrt[4]{32}$

Solution:

(a) Eight is a perfect cube that is a factor of the radicand, 32. Therefore, we simplify as follows:

$$\sqrt[3]{32} = \sqrt[3]{8 \cdot 4} = \sqrt[3]{8}\sqrt[3]{4} = 2\sqrt[3]{4}$$

(b) $\sqrt[3]{54} = \sqrt[3]{27 \cdot 2} = \sqrt[3]{27}\sqrt[3]{2} = 3\sqrt[3]{2}$

(c) Write $\sqrt[4]{32}$ as a product of a perfect fourth power and another number, then simplify. From the listing above, we see that 16 is a perfect fourth power. Since 16 is a factor of 32, we simplify as follows:

$$\sqrt[4]{32} = \sqrt[4]{16 \cdot 2} = \sqrt[4]{16}\sqrt[4]{2} = 2\sqrt[4]{2}$$

Writing Radicals in Exponential Form

3 A radical expression can be written in **exponential form** by using the following rule.

$$\sqrt[n]{a} = a^{1/n}, \qquad a \geq 0 \qquad \text{Rule 4}$$

Examples

$\sqrt{8} = 8^{1/2}$ $\qquad$ $\sqrt{x} = x^{1/2}$

$\sqrt[3]{4} = 4^{1/3}$ $\qquad$ $\sqrt[4]{9} = 9^{1/4}$

$\sqrt[3]{x} = x^{1/3}$ $\qquad$ $\sqrt[4]{y} = y^{1/4}$

$\sqrt[3]{5x^2} = (5x^2)^{1/3}$ $\qquad$ $\sqrt[4]{3y^2} = (3y^2)^{1/4}$

Notice $\sqrt{8} = 8^{1/2}$ and $\sqrt{x} = x^{1/2}$, which is consistent with what we learned in Section 9.1. This concept can be expanded as follows.

Power Index

$\sqrt[n]{a^m} = (\sqrt[n]{a})^m = a^{m/n}$, for $a \geq 0$ and m and n integers **Rule 5**

As long as the radicand is nonnegative, we can change from one form to another.

Examples

$\sqrt[3]{8^4} = (\sqrt[3]{8})^4 = 2^4 = 16$ $\qquad$ $\sqrt[3]{x^3} = x^{3/3} = x^1 = x$

$27^{2/3} = (\sqrt[3]{27})^2 = 3^2 = 9$ $\qquad$ $\sqrt[4]{y^{12}} = y^{12/4} = y^3$

EXAMPLE 3 Simplify: **(a)** $\sqrt[3]{y^{15}}$ **(b)** $\sqrt[4]{x^{24}}$

Solution: Write each radical expression in exponential form, then simplify.

(a) $\sqrt[3]{y^{15}} = y^{15/3} = y^5$

(b) $\sqrt[4]{x^{24}} = x^{24/4} = x^6$

The rules of exponents discussed in Section 4.1 and 4.2 also apply when the exponents are fractions. In Example 4(c) we use the negative exponent rule and in Example 6 we use the product and power rules with fractional exponents.

EXAMPLE 4 Evaluate: **(a)** $8^{5/3}$ **(b)** $16^{5/4}$ **(c)** $8^{-2/3}$

Solution: To evaluate we write each exponential expression in radical form.

(a) $8^{5/3} = (\sqrt[3]{8})^5 = 2^5 = 32$

(b) $16^{5/4} = (\sqrt[4]{16})^5 = 2^5 = 32$

(c) Recall from Section 4.2 that $x^{-m} = \frac{1}{x^m}$. Thus,

$$8^{-2/3} = \frac{1}{8^{2/3}} = \frac{1}{(\sqrt[3]{8})^2} = \frac{1}{2^2} = \frac{1}{4}$$

EXAMPLE 5 Write each of the following radicals in exponential form.

(a) $\sqrt[3]{x^5}$ **(b)** $\sqrt[4]{y^7}$ **(c)** $\sqrt[4]{z^{15}}$

Solution: **(a)** $\sqrt[3]{x^5} = x^{5/3}$ **(b)** $\sqrt[4]{y^7} = y^{7/4}$ **(c)** $\sqrt[4]{z^{15}} = z^{15/4}$

COMMON STUDENT ERROR Students may make mistakes simplifying expressions that contain negative exponents. Be careful when working such problems. The following is a common error.

Correct	*Incorrect*
$27^{-2/3} = \dfrac{1}{27^{2/3}}$	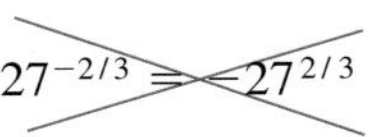

The expression $27^{-2/3}$ simplifies to $\frac{1}{9}$. Can you show how?

EXAMPLE 6 Simplify: **(a)** $\sqrt{x} \cdot \sqrt[4]{x}$ **(b)** $(\sqrt[4]{x^2})^8$

Solution: To simplify we change each radical expression to exponential form, then apply the rules of exponents.

(a)
$$\begin{aligned}\sqrt{x} \cdot \sqrt[4]{x} &= x^{1/2} \cdot x^{1/4} \\ &= x^{(1/2)+(1/4)} \\ &= x^{(2/4)+(1/4)} \\ &= x^{3/4} \\ &= \sqrt[4]{x^3}\end{aligned}$$

(b)
$$\begin{aligned}(\sqrt[4]{x^2})^8 &= (x^{2/4})^8 \\ &= (x^{1/2})^8 \\ &= x^4\end{aligned}$$

This section was meant to give you a brief introduction to roots other than square roots. If you take a course in intermediate algebra, you may study these concepts in more depth.

Exercise Set 9.7

Evaluate.

1. $\sqrt[3]{8}$ **2.** $\sqrt[3]{27}$ **3.** $\sqrt[3]{-8}$ **4.** $\sqrt[3]{-27}$
5. $\sqrt[4]{16}$ **6.** $\sqrt[3]{125}$ **7.** $\sqrt[4]{81}$ **8.** $\sqrt[4]{1}$
9. $\sqrt[3]{-1}$ **10.** $\sqrt[3]{-125}$ **11.** $\sqrt[3]{64}$ **12.** $\sqrt[3]{-64}$

Simplify.

13. $\sqrt[3]{54}$ **14.** $\sqrt[3]{32}$ **15.** $\sqrt[3]{16}$ **16.** $\sqrt[3]{24}$
17. $\sqrt[3]{81}$ **18.** $\sqrt[3]{128}$ **19.** $\sqrt[4]{32}$ **20.** $\sqrt[3]{250}$
21. $\sqrt[3]{40}$ **22.** $\sqrt[4]{48}$

Simplify.

23. $\sqrt[3]{x^3}$ **24.** $\sqrt[3]{y^6}$ **25.** $\sqrt[4]{y^{12}}$ **26.** $\sqrt[4]{y^{16}}$
27. $\sqrt[3]{x^{12}}$ **28.** $\sqrt[3]{x^9}$ **29.** $\sqrt[4]{y^4}$ **30.** $\sqrt[4]{y^{24}}$
31. $\sqrt[3]{x^{15}}$ **32.** $\sqrt[3]{x^{18}}$

Evaluate.

33. $8^{4/3}$ **34.** $27^{4/3}$ **35.** $16^{3/4}$ **36.** $81^{3/4}$
37. $1^{5/3}$ **38.** $16^{5/2}$ **39.** $9^{3/2}$ **40.** $64^{2/3}$
41. $81^{3/4}$ **42.** $25^{3/2}$ **43.** $125^{4/3}$ **44.** $8^{-1/3}$
45. $27^{-2/3}$ **46.** $16^{-3/4}$ **47.** $8^{-5/3}$ **48.** $64^{-2/3}$

Write each radical in exponential form.

49. $\sqrt[3]{x^7}$
50. $\sqrt[3]{x^6}$
51. $\sqrt[3]{x^4}$
52. $\sqrt[4]{x^7}$
53. $\sqrt[4]{y^{15}}$
54. $\sqrt[4]{x^9}$
55. $\sqrt[4]{y^{21}}$
56. $\sqrt[4]{x^5}$

Simplify and write the answer in exponential form.

57. $\sqrt[3]{x} \cdot \sqrt[3]{x}$
58. $\sqrt[4]{x} \cdot \sqrt[4]{x}$
59. $\sqrt[4]{x^2} \cdot \sqrt[4]{x^2}$
60. $\sqrt[3]{x} \cdot \sqrt[3]{x^5}$
61. $(\sqrt[3]{x^2})^6$
62. $(\sqrt[4]{x^3})^4$
63. $(\sqrt[4]{x^2})^4$
64. $(\sqrt[4]{x^6})^2$
65. Show that for $x = 8$, $(\sqrt[3]{x})^2 = \sqrt[3]{x^2}$
66. How do you read the following radicals?
    (a) $\sqrt{8}$ (b) $\sqrt[3]{8}$ (c) $\sqrt[4]{8}$
67. (a) In your own words, explain how to change an expression written in exponential form to radical form.
    (b) Using the procedure given in part (a), write $x^{5/8}$ in radical form.
68. (a) In your own words, explain how to change an expression written in radical form to exponential form.
    (b) Using the procedure given in part (a), write $\sqrt[4]{x^9}$ in exponential form.

Cumulative Review Exercises

69. Evaluate $-x^2 + 4xy - 6$ when $x = 2$ and $y = -4$.
[5.4] 70. Factor $3x^2 - 28x + 32$.
71. Graph $y = \frac{2}{3}x - 4$.
[9.3] 72. Simplify $\sqrt{\frac{64x^3y^7}{2x^4}}$.

Group Activity/ Challenge Problems

Simplify.

1. $\sqrt[3]{xy} \cdot \sqrt[3]{x^2y^2}$
2. $\sqrt[4]{3x^2y} \cdot \sqrt[4]{27x^6y^3}$
3. $\sqrt[4]{32} - \sqrt[4]{2}$
4. $\sqrt[3]{3x^3y} + \sqrt[3]{24x^3y}$
5. To rationalize the denominator in $\frac{1}{\sqrt[3]{2}}$ we can multiply both the numerator and denominator by $\sqrt[3]{2^2}$. **(a)** Explain why this procedure will give an integer in the denominator. **(b)** Rationalize the denominator. **(c)** Explain why we cannot rationalize the denominator by multiplying both numerator and denominator by $\sqrt[3]{2}$.

Read Group Activity Exercise 5 above, then rationalize the denominator.

6. $\frac{1}{\sqrt[3]{x}}$
7. $\frac{1}{\sqrt[3]{x^2}}$
8. $\frac{3}{\sqrt[3]{2x^2}}$
9. $\frac{4}{\sqrt[3]{5y^4}}$

Summary

Glossary

Conjugate (506): The conjugate of $a + b$ is $a - b$.

Hypotenuse (517): The side opposite the right angle in a right triangle.

Index of a radical (486): The root of a radical expression.

Irrational numbers (489): Real numbers that are not rational numbers.

Legs of a right triangle (517): The two shorter sides of a right triangle.

Like square roots (504): Square roots having the same radicand.

Perfect square (488): When a radicand contains a variable (or number) raised to an even exponent, that variable (or number) and exponent together form a perfect square.

Principal or positive square root (487): The principal or positive square root of a positive real number x, written $\sqrt{x}$, is that positive number whose square equals x.

Radical equation (512): An equation that contains a variable in a radicand.

Radical expression (486): A mathematical expression containing a radical.

Radical sign (486): $\sqrt{\ }$.

Radicand (486): The expression within the radical sign.

Rational number (488): A number that can be written in the form $\frac{a}{b}$, where a and b are both integers, $b \neq 0$.

Rationalize the denominator (500): Removing radical expressions from the denominator of a fraction.

Right triangle (517): A triangle with a 90° angle.

Unlike square roots (505): Square roots having different radicands.

IMPORTANT FACTS

Numbers that are perfect squares: 1, 4, 9, 16, 25, 36, 49, 64, . . .

Numbers that are perfect cubes: 1, 8, 27, 64, 125, 216, 343, 512, . . .

Product rule for radicals: $\sqrt{a} \cdot \sqrt{b} = \sqrt{ab},\ a \geq 0, b \geq 0$

$\sqrt{a^{2 \cdot n}} = a^n,\ a \geq 0$

$\sqrt{a^2} = a,\ a \geq 0$

Quotient rule for radicals: $\frac{\sqrt{a}}{\sqrt{b}} = \sqrt{\frac{a}{b}},\quad a \geq 0, b > 0$

$\sqrt[n]{a} = a^{1/n},\ a \geq 0$

$\sqrt[n]{a^m} = (\sqrt[n]{a})^m = a^{m/n},\ a \geq 0$

Pythagorean theorem: $a^2 + b^2 = c^2$

Distance formula: $d = \sqrt{(x_2 - x_1)^2 + (y_2 - y_1)^2}$

Review Exercises

[9.1] *Evaluate.*

1. $\sqrt{25}$ **2.** $\sqrt{36}$ **3.** $-\sqrt{81}$

Write in exponential form.

4. $\sqrt{8}$ **5.** $\sqrt{26x}$ **6.** $\sqrt{20xy^2}$

[9.2] *Simplify.*

7. $\sqrt{32}$ **8.** $\sqrt{44}$ **9.** $\sqrt{45x^5y^4}$

10. $\sqrt{125x^4y^6}$ **11.** $\sqrt{15x^5yz^3}$ **12.** $\sqrt{48ab^4c^5}$

Simplify.

13. $\sqrt{8} \cdot \sqrt{12}$ **14.** $\sqrt{5x} \cdot \sqrt{5x}$ **15.** $\sqrt{18x} \cdot \sqrt{2xy}$

16. $\sqrt{25x^2y} \cdot \sqrt{3y}$ **17.** $\sqrt{20xy^4} \cdot \sqrt{5xy^3}$ **18.** $\sqrt{8x^3y} \cdot \sqrt{3y^4}$

.3] *Simplify.*

19. $\dfrac{\sqrt{32}}{\sqrt{2}}$ **20.** $\sqrt{\dfrac{10}{250}}$ **21.** $\sqrt{\dfrac{7}{28}}$

22. $\dfrac{3}{\sqrt{5}}$ **23.** $\sqrt{\dfrac{5x}{12}}$ **24.** $\sqrt{\dfrac{x}{6}}$

25. $\sqrt{\dfrac{x^2}{2}}$ **26.** $\sqrt{\dfrac{x^5}{8}}$ **27.** $\sqrt{\dfrac{60xy^5}{4x^5y^3}}$

28. $\sqrt{\dfrac{30x^4y}{15x^2y^4}}$ **29.** $\dfrac{\sqrt{90}}{\sqrt{8x^3y^2}}$ **30.** $\dfrac{\sqrt{2x^4yz^4}}{\sqrt{7x^5yz^2}}$

31. $\dfrac{3}{1 + \sqrt{2}}$ **32.** $\dfrac{5}{3 - \sqrt{6}}$ **33.** $\dfrac{\sqrt{3}}{2 + \sqrt{x}}$

34. $\dfrac{2}{\sqrt{x} - 5}$ **35.** $\dfrac{\sqrt{5}}{\sqrt{x} + \sqrt{3}}$

4] *Simplify.*

36. $6\sqrt{3} - 2\sqrt{3}$ **37.** $6\sqrt{2} - 8\sqrt{2} + \sqrt{2}$ **38.** $3\sqrt{x} - 5\sqrt{x}$

39. $\sqrt{x} + 3\sqrt{x} - 4\sqrt{x}$ **40.** $\sqrt{8} - \sqrt{2}$ **41.** $7\sqrt{40} - 2\sqrt{10}$

42. $2\sqrt{98} - 4\sqrt{72}$ **43.** $3\sqrt{18} + 5\sqrt{50} - 2\sqrt{32}$ **44.** $4\sqrt{27} + 5\sqrt{80} + 2\sqrt{12}$

5] *Solve.*

45. $\sqrt{x} = 9$ **46.** $\sqrt{x} = -2$ **47.** $\sqrt{x - 3} = 6$

48. $\sqrt{3x + 1} = 5$ **49.** $\sqrt{2x + 4} = \sqrt{3x - 5}$ **50.** $4\sqrt{x} - x = 4$

51. $\sqrt{x^2 + 4} = x + 2$ **52.** $\sqrt{3x + 5} - \sqrt{5x - 9} = 0$ **53.** $3\sqrt{2x + 3} = 9$

5] *Find the length indicated.*

54.

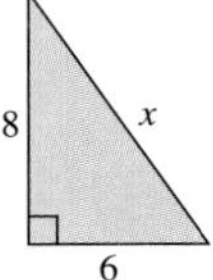

55.

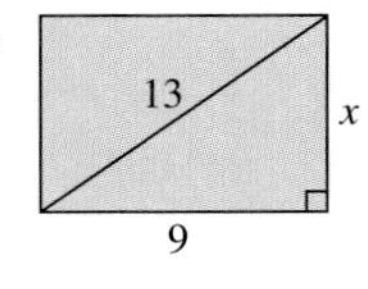

56.

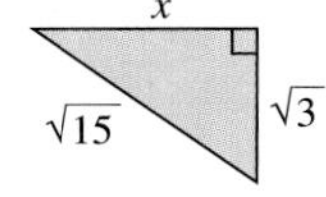

57.

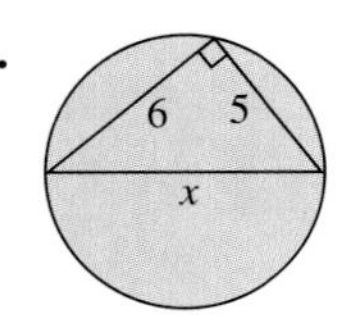

58. Jason leans a 12-foot ladder against a house. If the base of the ladder is 3 feet from the house, how high is the ladder on the house?

59. Find the diagonal of a rectangle of length 15 inches and width 6 inches.

60. Find the straight-line distance between the points $(-5, 4)$ and $(3, -2)$.

61. Find the length of the line segment between the points $(6, 5)$ and $(-6, 8)$.

62. Find the side of a square whose area is 121 square feet.

] *Evaluate.*

63. $\sqrt[3]{8}$ **64.** $\sqrt[3]{-27}$ **65.** $\sqrt[4]{16}$

Simplify.

66. $\sqrt[3]{16}$ **67.** $\sqrt[3]{24}$ **68.** $\sqrt[4]{32}$

69. $\sqrt[3]{48}$ **70.** $\sqrt[3]{54}$ **71.** $\sqrt[4]{96}$

72. $\sqrt[3]{x^{15}}$ **73.** $\sqrt[3]{x^{12}}$ **74.** $\sqrt[4]{y^{16}}$

Evaluate.

75. $8^{2/3}$ **76.** $16^{1/2}$ **77.** $27^{-2/3}$

78. $64^{2/3}$ **79.** $16^{-3/4}$ **80.** $25^{3/2}$

Write in exponential form.

81. $\sqrt[3]{x^5}$ **82.** $\sqrt[3]{x^{10}}$ **83.** $\sqrt[4]{y^9}$

84. $\sqrt{x^5}$ **85.** $\sqrt{y^{11}}$ **86.** $\sqrt[4]{x^7}$

Simplify.

87. $\sqrt{x} \cdot \sqrt{x}$ **88.** $\sqrt[3]{x} \cdot \sqrt[3]{x}$ **89.** $\sqrt[3]{x^2} \cdot \sqrt[3]{x^7}$

90. $\sqrt[4]{x^2} \cdot \sqrt[4]{x^6}$ **91.** $(\sqrt[3]{x^3})^2$ **92.** $(\sqrt[3]{x^2})^3$

93. $(\sqrt[4]{x^2})^6$ **94.** $(\sqrt[4]{x^3})^8$

Practice Test

1. Write $\sqrt{3xy}$ in exponential form.

Simplify.

2. $\sqrt{(x+3)^2}$ **3.** $\sqrt{96}$ **4.** $\sqrt{12x^2}$

5. $\sqrt{32x^4y^5}$ **6.** $\sqrt{8x^2y} \cdot \sqrt{6xy}$ **7.** $\sqrt{15xy^2} \cdot \sqrt{5x^3y^3}$

8. $\sqrt{\dfrac{5}{125}}$ **9.** $\dfrac{\sqrt{3xy^2}}{\sqrt{48x^3}}$ **10.** $\dfrac{1}{\sqrt{2}}$

11. $\sqrt{\dfrac{4x}{5}}$ **12.** $\sqrt{\dfrac{40x^2y^5}{6x^3y^7}}$ **13.** $\dfrac{3}{2+\sqrt{5}}$

14. $\dfrac{6}{\sqrt{x}-3}$ **15.** $\sqrt{48} + \sqrt{75} + 2\sqrt{3}$ **16.** $4\sqrt{x} - 6\sqrt{x} - \sqrt{x}$

Solve.

17. $\sqrt{x+5} = 9$ **18.** $2\sqrt{x-4} + 4 = x$

Solve.

19. Find the value of x in the triangle shown.

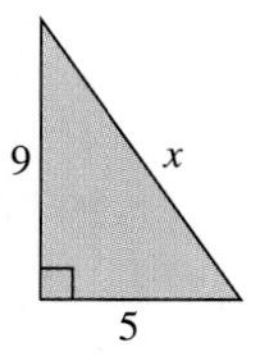

20. Find the length of the line segment between the points $(-2, -4)$ and $(5, 1)$.

21. Evaluate $27^{-4/3}$.

22. Simplify $\sqrt[3]{x^4} \cdot \sqrt[3]{x^{11}}$.

Chapter 10

Quadratic Equations

See Section 10.1, Example 6

Preview and Perspective

Much of this book has dealt with solving linear equations. Quadratic equations are another important category of equations. Quadratic equations were introduced and solved by factoring in Section 5.6. You may wish to review that section now. Recall that quadratic equations are of the form $ax^2 + bx + c = 0$, where $a \neq 0$. Not every quadratic equation can be solved by factoring. In this chapter we present two important procedures that can be used to solve quadratic equations: completing the square and the quadratic formula.

In Section 10.1 we present the square root property which is used when solving quadratic equations by completing the square. In Section 10.2 we introduce completing the square for solving quadratic equations. The quadratic formula is discussed in Section 10.3. The quadratic formula is most typically used when solving quadratic equations that cannot be factored. The quadratic formula is used in intermediate algebra and other mathematics courses. We graphed linear equations in Chapter 7. In Section 10.4 we graph quadratic equations. Section 10.4 is a very valuable section, for it lays the groundwork for graphing cubic and higher-degree equations in other mathematics courses.

One important reason for learning to solve quadratic equations in one variable ($ax^2 + bx + c = 0$) is that the real solutions will be the x intercepts of the graph of the quadratic equation in two variables ($y = ax^2 + bx + c$). Thus, solving quadratic equations in one variable will be an aid when graphing quadratic equations in two variables.

There are many real-life applications of quadratic equations. You will see some of them in this chapter.

10.1 The Square Root Property

Tape 16

1. Know that every positive real number has two square roots.
2. Solve quadratic equations using the square root property.

In Section 5.6 we solved quadratic equations by factoring. Recall that **quadratic equations** are equations of the form

$$ax^2 + bx + c = 0,$$

where a, b, and c are real numbers, $a \neq 0$. A quadratic equation in this form is said to be in **standard form.** Solving quadratic equations by factoring is the preferred technique when the factors can be found quickly. To refresh your memory, below we will solve the equation $x^2 - 2x - 8 = 0$ by factoring. If you need further examples, review Section 5.6.

$$x^2 - 2x - 8 = 0$$

$$(x - 4)(x + 2) = 0$$

$$x - 4 = 0 \quad \text{or} \quad x + 2 = 0$$

$$x = 4 \qquad\qquad x = -2$$

The solutions of this equation are 4 and -2.

Not every quadratic equation can be factored easily and many cannot be factored at all. In this chapter we give two techniques, completing the square and the quadratic formula, for solving quadratic equations that cannot be solved by factoring.

Positive and Negative Square Roots

1 In Section 9.1 we stated that every positive number has two square roots. Thus far we have been using only the positive or principal square root. In this section we use both the positive and negative square roots of a number.

The positive square root of 25 is 5.

$$\sqrt{25} = 5$$

The negative square root of 25 is -5.

$$-\sqrt{25} = -5$$

Notice that $5 \cdot 5 = 25$ and $(-5)(-5) = 25$. The two square roots of 25 are $+5$ and -5. A convenient way to indicate the two square roots of a number is to use the plus or minus symbol, $\pm$. For example, the square roots of 25 can be indicated ± 5, read "plus or minus 5."

Number	*Both Square Roots*
36	± 6
100	± 10
7	$\pm\sqrt{7}$

The value of a number like $-\sqrt{5}$ can be found by finding the value of $\sqrt{5}$ on your calculator and then taking its opposite or negative value.

$$\sqrt{5} = 2.24 \quad \text{(rounded to two decimal places)}$$
$$-\sqrt{5} = -2.24$$

Consider the equation

$$x^2 = 25$$

We can see by substitution that this equation has two solutions, 5 and -5.

Check:	$x = 5$	$x = -5$
	$x^2 = 25$	$x^2 = 25$
	$5^2 = 25$	$(-5)^2 = 25$
	$25 = 25$ true	$25 = 25$ true

Therefore, the solutions to the equation $x^2 = 25$ are 5 or -5 (or ± 5).

The Square Root Property

 In general, for any quadratic equation of the form $x^2 = a$, we can use the square root property to obtain the solution.

Square Root Property

If $x^2 = a$ then $x = \sqrt{a}$ or $x = -\sqrt{a}$ (abbreviated $x = \pm\sqrt{a}$).

For example, if $x^2 = 7$, then by the square root property, $x = \sqrt{7}$ or $x = -\sqrt{7}$. We may also write $x = \pm\sqrt{7}$.

EXAMPLE 1 Solve the equation $x^2 - 9 = 0$.

Solution: Add 9 to both sides of the equation to get the variable all by itself on one side of the equation.

$$x^2 = 9$$

Now use the square root property.

$$x = \pm\sqrt{9}$$
$$x = \pm 3$$

Check in the original equation.

Check:

$x = 3$		$x = -3$	
$x^2 - 9 = 0$		$x^2 - 9 = 0$	
$3^2 - 9 = 0$		$(-3)^2 - 9 = 0$	
$9 - 9 = 0$		$9 - 9 = 0$	
$0 = 0$	true	$0 = 0$	true

EXAMPLE 2 Solve the equation $x^2 + 7 = 71$.

Solution: Begin by subtracting 7 from both sides of the equation.

$$x^2 + 7 = 71$$
$$x^2 = 64$$
$$x = \pm\sqrt{64}$$
$$x = \pm 8$$

EXAMPLE 3 Solve the equation $x^2 - 7 = 0$.

Solution:

$$x^2 - 7 = 0$$
$$x^2 = 7$$
$$x = \pm\sqrt{7}$$

EXAMPLE 4 Solve the equation $(x - 3)^2 = 4$.

Solution: Begin by using the square root property.

$$(x - 3)^2 = 4$$
$$x - 3 = \pm\sqrt{4}$$
$$x - 3 = \pm 2$$
$$x - 3 + 3 = 3 \pm 2 \quad \text{Add 3 to both sides of the equation.}$$
$$x = 3 \pm 2$$
$$x = 3 + 2 \quad \text{or} \quad x = 3 - 2$$
$$x = 5 \quad \text{or} \quad x = 1$$

The solutions are 1 and 5.

EXAMPLE 5 Solve the equation $(3x + 4)^2 = 32$.

Solution:

$$(3x + 4)^2 = 32$$
$$3x + 4 = \pm\sqrt{32}$$
$$3x + 4 = \pm\sqrt{16}\sqrt{2}$$
$$3x + 4 = \pm 4\sqrt{2}$$
$$3x + 4 - 4 = -4 \pm 4\sqrt{2} \quad \text{Subtract 4 from both sides of the equation.}$$
$$3x = -4 \pm 4\sqrt{2}$$
$$x = \frac{-4 \pm 4\sqrt{2}}{3}$$

Thus, the solutions are $\frac{-4 + 4\sqrt{2}}{3}$ and $\frac{-4 - 4\sqrt{2}}{3}$.

EXAMPLE 6 Mrs. Albert wants a rectangular garden. To make her garden look most appealing, the length of the rectangle is to be 1.62 times its width (a rectangle with this length-to-width ratio is called a *golden rectangle*; Fig. 10.1). Find the dimensions of the rectangle if the rectangle is to have an area of 6000 square feet.

Solution: Let x = width of rectangle; then $1.62x$ = length of rectangle.

$$\text{Area} = \text{length} \cdot \text{width}$$
$$6000 = (1.62x)x$$
$$6000 = 1.62x^2$$
$$\text{or} \quad 1.62x^2 = 6000$$
$$x^2 = \frac{6000}{1.62} \approx 3703.7$$
$$x \approx \pm\sqrt{3703.7} \approx \pm 60.86 \text{ feet}$$

Since the width cannot be negative, the width, x, is 60.86 feet. The length is $1.62(60.86) = 98.59$ feet.

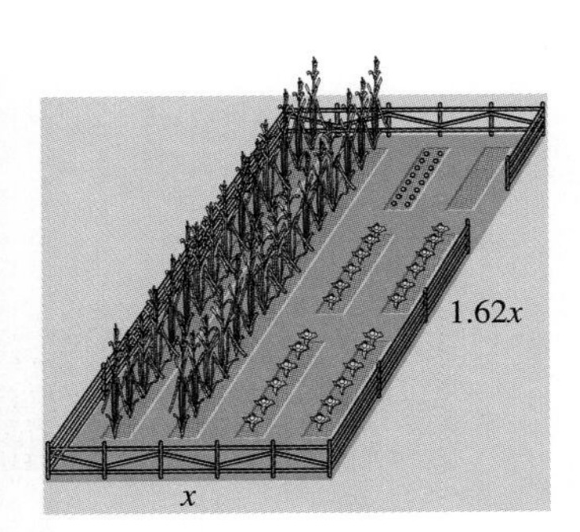

FIGURE 10.1

Check: area = length · width

$$6000 = (98.59)(60.86)$$

$$6000 \approx 6000.19$$ true (There is a slight round-off error due to rounding off decimal answers.)

Note that the answer did not come out to be a whole number. In many real-life situations this is the case. You should not feel uncomfortable when this occurs.

Exercise Set 10.1

Solve.

1. $x^2 = 16$
2. $x^2 = 25$
3. $x^2 = 100$
4. $x^2 = 49$
5. $y^2 = 36$
6. $z^2 = 9$
7. $x^2 = 10$
8. $a^2 = 15$
9. $x^2 = 8$
10. $w^2 = 24$
11. $3x^2 = 12$
12. $5y^2 = 45$
13. $2w^2 = 34$
14. $5x^2 = 90$
15. $2x^2 + 1 = 19$
16. $3x^2 - 4 = 8$
17. $4w^2 - 3 = 12$
18. $3y^2 + 8 = 36$
19. $5x^2 - 9 = 30$
20. $2x^2 + 3 = 51$

Solve.

21. $(x + 1)^2 = 4$
22. $(x - 2)^2 = 9$
23. $(x - 3)^2 = 16$
24. $(x + 5)^2 = 25$
25. $(x + 4)^2 = 36$
26. $(x - 4)^2 = 100$
27. $(x - 1)^2 = 12$
28. $(x + 3)^2 = 18$
29. $(x + 6)^2 = 20$
30. $(x - 4)^2 = 32$
31. $(x + 2)^2 = 25$
32. $(x + 6)^2 = 75$
33. $(x - 9)^2 = 100$
34. $(x - 3)^2 = 15$
35. $(2x + 3)^2 = 18$
36. $(3x - 2)^2 = 30$
37. $(4x + 1)^2 = 20$
38. $(5x - 6)^2 = 100$
39. $(2x - 6)^2 = 18$
40. $(3x - 5)^2 = 90$

41. The length of a rectangle is twice its width. Find the length and width if the area is 80 square feet.
42. The length of a rectangle is 3 times its width. Find the length and width if its area is 96 square feet.
43. Write an equation that has solutions of 6 and -6.
44. Write an equation that has solutions of $\sqrt{7}$ and $-\sqrt{7}$.
45. Fill in the shaded area to make a true statement. The equation $x^2 - \square = 27$ has solutions of 6 and -6. Explain how you determined your answer.
46. Fill in the shaded area to make a true statement. The equation $x^2 + \square = 45$ has solutions of 8 and -8. Explain how you determined your answer.
47. What is the standard form of a quadratic equation?
48. **(a)** If we have a quadratic equation of the form $ax^2 + bx + c = 0$ where $a < 0$, how can we change the equation to the form $ax^2 + bx + c = 0$, where $a > 0$?
    (b) Rewrite $-3x^2 + 5x - 6 = 0$ so that the coefficient of the x^2 term is positive.

CUMULATIVE REVIEW EXERCISES

[5.4] 49. Factor $4x^2 - 10x - 24$.

[6.6] 50. Solve $\dfrac{3x - 7}{x - 4} = \dfrac{5}{x - 4}$.

[6.7] 51. Collette invests $10,000 in two savings accounts. Part of the money is put into an account paying 6% simple interest. The rest is put into a savings account paying 8% simple interest. If the interest for the year from both accounts totals $760, how much money was in each account?

[7.4] 52. Determine the equation of the line illustrated.

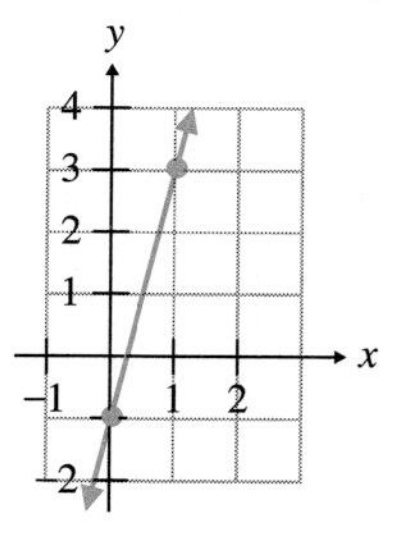

Group Activity/ Challenge Problems

Use the square root property to solve for the variable indicated. Assume that all variables represent positive numbers. You may wish to review Section 3.1 Formulas, before working these problems. List only the positive square root.

1. $A = s^2$ for s **2.** $I = p^2 r$ for p **3.** $A = \pi r^2$ for r

4. $a^2 + b^2 = c^2$ for b **5.** $I = \dfrac{k}{d^2}$ for d **6.** $A = p(1 + r)^2$ for r

10.2 Solving Quadratic Equations by Completing the Square

Tape 16

1. Write perfect square trinomials.
2. Solve quadratic equations by completing the square.

Quadratic equations that cannot be solved using factoring can be solved by completing the square or by the quadratic formula. In this section we focus on completing the square. In Section 10.3 we use the quadratic formula.

Perfect Square Trinomials

1 A **perfect square trinomial** is a trinomial that can be expressed as the square of a binomial. Some examples follow.

Perfect Square Trinomials	*Factors*	*Square of a Binomial*
$x^2 + 6x + 9$	$= (x + 3)(x + 3)$	$= (x + 3)^2$
$x^2 - 6x + 9$	$= (x - 3)(x - 3)$	$= (x - 3)^2$
$x^2 + 10x + 25$	$= (x + 5)(x + 5)$	$= (x + 5)^2$
$x^2 - 10x + 25$	$= (x - 5)(x - 5)$	$= (x - 5)^2$

Notice that each of the squared terms in the preceding perfect square trinomials has a numerical coefficient of 1. When the coefficient of the squared term is 1, there is an important relationship between the coefficient of the x term and the constant. In every perfect square trinomial of this type, *the constant term is the square of one-half the coefficient of the x term.*

Consider the perfect square trinomial $x^2 + 6x + 9$. The coefficient of the x term is 6 and the constant is 9. Note that the constant, 9, is the square of one-half the coefficient of the x term.

$$x^2 + 6x + 9$$

$$\left[\frac{1}{2}(6)\right]^2 = 3^2 = 9$$

Consider the perfect square trinomial $x^2 - 10x + 25$. The coefficient of the x term is -10 and the constant is 25. Note that

$$x^2 - 10x + 25$$

$$\left[\frac{1}{2}(-10)\right]^2 = (-5)^2 = 25$$

Consider the expression $x^2 + 8x + \square$. Can you determine what number must be placed in the colored box to make the trinomial a perfect square trinomial? If you answered 16, you answered correctly.

$$x^2 + 8x + \square$$

$$\left[\frac{1}{2}(8)\right]^2 = 4^2 = 16$$

The perfect square trinomial is $x^2 + 8x + 16$. Note that $x^2 + 8x + 16 = (x + 4)^2$. Let us examine perfect square trinomials a little further.

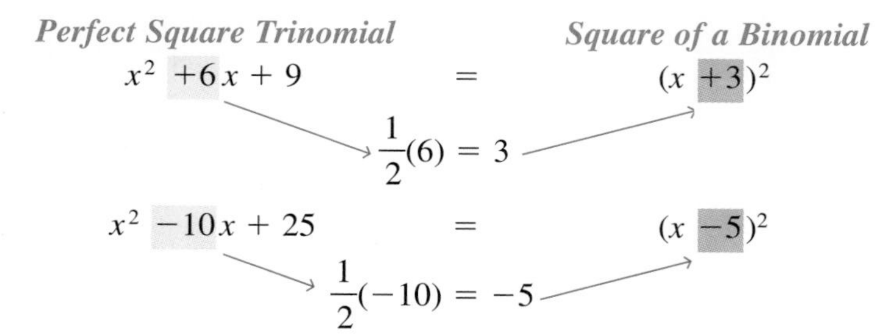

Note that when a perfect square trinomial is written as the square of a binomial *the constant in the binomial is one-half the value of the coefficient of the x term in the perfect square trinomial.*

Solving Quadratic Equations by Completing the Square

2 The procedure for solving a quadratic equation by completing the square is illustrated in the following example.

EXAMPLE 1 Solve the equation $x^2 + 6x + 5 = 0$ by completing the square.

Solution: First we make sure that the squared term has a coefficient of 1. In Example 5 we explain what to do if the coefficient is not 1. Next we wish to get the terms containing a variable by themselves on the left side of the equation. Therefore, we subtract 5 from both sides of the equation.

$$x^2 + 6x + 5 = 0$$
$$x^2 + 6x = -5$$

Determine one-half the numerical coefficient of the x term. In this example the x term is $6x$.

$$\frac{1}{2}(6) = 3$$

Square this number.

$$(3)^2 = (3)(3) = 9$$

Add this product to both sides of the equation.

$$x^2 + 6x + 9 = -5 + 9$$

or

$$x^2 + 6x + 9 = 4$$

By following this procedure, we produce a perfect square trinomial on the left side of the equation. The expression $x^2 + 6x + 9$ is a perfect square trinomial that can be expressed as $(x + 3)^2$. Therefore,

$$x^2 + 6x + 9 = 4$$

can be written $(x + 3)^2 = 4$

Now use the square root property,

$$x + 3 = \pm\sqrt{4}$$
$$x + 3 = \pm 2$$

Finally, solve for x by subtracting 3 from both sides of the equation.

$$x + 3 - 3 = -3 \pm 2$$
$$x = -3 \pm 2$$
$$x = -3 + 2 \quad \text{or} \quad x = -3 - 2$$
$$x = -1 \qquad\qquad x = -5$$

Thus, the solutions are -1 and -5. Check both solutions in the original equation.

Check:

$x = -1$	$x = -5$
$x^2 + 6x + 5 = 0$	$x^2 + 6x + 5 = 0$
$(-1)^2 + 6(-1) + 5 = 0$	$(-5)^2 + 6(-5) + 5 = 0$
$1 - 6 + 5 = 0$	$25 - 30 + 5 = 0$
$0 = 0$ true	$0 = 0$ true

To Solve a Quadratic Equation by Completing the Square

1. Use the multiplication (or division) property of equality if necessary to make the numerical coefficient of the squared term equal to 1.
2. Rewrite the equation with the constant by itself on the right side of the equation.
3. Take one-half the numerical coefficient of the first-degree term, square it, and add this quantity to both sides of the equation.
4. Replace the trinomial with its equivalent squared binomial.
5. Use the square root property.
6. Solve for the variable.
7. Check your answers in the original equation.

EXAMPLE 2 Solve the equation $x^2 - 8x + 15 = 0$ by completing the square.

Solution: $x^2 - 8x + 15 = 0$

$$x^2 - 8x = -15$$

Take half the numerical coefficient of the x term, square it, and add this product to both sides of the equation.

$$\frac{1}{2}(-8) = -4, \quad (-4)^2 = 16$$

Now add 16 to both sides of the equation

$$x^2 - 8x + 16 = -15 + 16$$
$$x^2 - 8x + 16 = 1$$
$$\text{or} \quad (x - 4)^2 = 1$$
$$x - 4 = \pm\sqrt{1}$$
$$x - 4 = \pm 1$$
$$x = 4 \pm 1$$
$$x = 4 + 1 \quad \text{or} \quad x = 4 - 1$$
$$x = 5 \qquad\qquad x = 3$$

A check will show that the solutions are 5 and 3.

EXAMPLE 3 Solve the equation $x^2 = 3x + 18$ by completing the square.

Solution: Place all terms except the constant on the left side of the equation.

$$x^2 = 3x + 18$$
$$x^2 - 3x = 18$$

Take half the numerical coefficient of the x term, square it, and add this product to both sides of the equation.

$$\frac{1}{2}(-3) = -\frac{3}{2}, \qquad \left(-\frac{3}{2}\right)^2 = \frac{9}{4}$$

$$x^2 - 3x + \frac{9}{4} = 18 + \frac{9}{4}$$
$$\left(x - \frac{3}{2}\right)^2 = 18 + \frac{9}{4}$$
$$\left(x - \frac{3}{2}\right)^2 = \frac{72}{4} + \frac{9}{4}$$
$$\left(x - \frac{3}{2}\right)^2 = \frac{81}{4}$$
$$x - \frac{3}{2} = \pm\sqrt{\frac{81}{4}}$$
$$x - \frac{3}{2} = \pm\frac{9}{2}$$
$$x = \frac{3}{2} \pm \frac{9}{2}$$
$$x = \frac{3}{2} + \frac{9}{2} \quad \text{or} \quad x = \frac{3}{2} - \frac{9}{2}$$
$$x = \frac{12}{2} = 6 \qquad\qquad x = -\frac{6}{2} = -3$$

The solutions are 6 and -3.

In the following examples we will not display some intermediate steps.

EXAMPLE 4 Solve the equation $x^2 - 6x + 1 = 0$.

Solution:

$$
\begin{aligned}
x^2 - 6x + 1 &= 0 \\
x^2 - 6x &= -1 \\
x^2 - 6x + 9 &= -1 + 9 \\
(x - 3)^2 &= 8 \\
x - 3 &= \pm\sqrt{8} \\
x - 3 &= \pm 2\sqrt{2} \\
x &= 3 \pm 2\sqrt{2}
\end{aligned}
$$

The solutions are $3 + 2\sqrt{2}$ and $3 - 2\sqrt{2}$.

EXAMPLE 5 Solve the equation $3m^2 - 9m + 6 = 0$ by completing the square.

Solution: To solve an equation by completing the square, the numerical coefficient of the squared term must be 1. Since the coefficient of the squared term is 3, we multiply both sides of the equation by $\frac{1}{3}$ to make the coefficient equal to 1.

$$
\begin{aligned}
3m^2 - 9m + 6 &= 0 \\
\frac{1}{3}(3m^2 - 9m + 6) &= \frac{1}{3}(0) \\
m^2 - 3m + 2 &= 0
\end{aligned}
$$

Now proceed as in earlier examples.

$$
\begin{aligned}
m^2 - 3m &= -2 \\
m^2 - 3m + \frac{9}{4} &= -2 + \frac{9}{4} \\
\left(m - \frac{3}{2}\right)^2 &= -\frac{8}{4} + \frac{9}{4} \\
\left(m - \frac{3}{2}\right)^2 &= \frac{1}{4} \\
m - \frac{3}{2} &= \pm\sqrt{\frac{1}{4}} \\
m - \frac{3}{2} &= \pm\frac{1}{2} \\
m &= \frac{3}{2} \pm \frac{1}{2}
\end{aligned}
$$

$$m = \frac{3}{2} + \frac{1}{2} \qquad \text{or} \qquad m = \frac{3}{2} - \frac{1}{2}$$

$$m = \frac{4}{2} = 2 \qquad\qquad m = \frac{2}{2} = 1$$

The solutions are 2 and 1.

Exercise Set 10.2

Solve by completing the square.

1. $x^2 + 2x - 3 = 0$
2. $x^2 - 6x + 8 = 0$
3. $x^2 - 4x - 5 = 0$
4. $x^2 + 8x + 12 = 0$
5. $x^2 + 3x + 2 = 0$
6. $x^2 + 4x - 32 = 0$
7. $x^2 - 2x - 8 = 0$
8. $x^2 - 9x + 14 = 0$
9. $x^2 = -6x - 9$
10. $x^2 + 5x + 4 = 0$
11. $x^2 = -5x - 6$
12. $x^2 = 2x + 15$
13. $x^2 + 9x + 18 = 0$
14. $x^2 - 9x + 18 = 0$
15. $x^2 = 15x - 56$
16. $x^2 = 3x + 28$
17. $-4x = -x^2 + 12$
18. $-x^2 - 3x + 40 = 0$
19. $x^2 + 2x - 6 = 0$
20. $x^2 - 4x + 2 = 0$
21. $6x + 6 = -x^2$
22. $x^2 - x - 3 = 0$
23. $-x^2 + 5x = -8$
24. $x^2 + 3x - 6 = 0$
25. $2x^2 + 4x - 6 = 0$
26. $2x^2 + 2x - 24 = 0$
27. $2x^2 + 18x + 4 = 0$
28. $2x^2 = 8x + 90$
29. $3x^2 + 33x + 72 = 0$
30. $4x^2 = -28x + 32$
31. $2x^2 + 10x - 3 = 0$
32. $3x^2 - 8x + 4 = 0$
33. $3x^2 + 6x = 6$
34. $2x^2 - x = 5$
35. $x^2 + 4x = 0$
36. $2x^2 - 6x = 0$
37. $2x^2 - 4x = 0$
38. $3x^2 = 9x$

39. When three times a number is added to the square of a number, the sum is 4. Find the number(s).

40. When five times a number is subtracted from two times the square of a number, the difference is 12. Find the number(s).

41. If the square of three more than a number is 9, find the number(s).

42. If the square of two less than an integer is 16, find the number(s).

43. The product of two positive numbers is 21. Find the two numbers if the larger is 4 greater than the smaller.

44. **(a)** What is a perfect square trinomial?
(b) Fill in the shaded area to make a perfect square trinomial and explain how you determined your answer.

$$x^2 + 8x \quad \square$$

45. In a perfect square trinomial, what is the relationship between the constant and the coefficient of the x term?

46. **(a)** Write a perfect square trinomial that has a term of $-12x$.
(b) Explain how you constructed your perfect square trinomial.

Cumulative Review Exercises

[6.5] **47.** Simplify $\dfrac{x^2}{x^2 - x - 6} - \dfrac{x - 2}{x - 3}$.

[7.4] **48.** Explain how you can determine if two equations represent parallel lines without graphing the equations.

[8.3] **49.** Solve the system of equations.

$$3x - 4y = 6$$
$$2x + y = 8$$

[9.5] **50.** Solve the equation $\sqrt{2x + 3} = 2x - 3$.

Group Activity/Challenge Problems

Solve by completing the square.

1. $x^2 + \frac{3}{5}x - \frac{1}{2} = 0.$

2. $x^2 - \frac{2}{3}x - \frac{1}{5} = 0.$

3. $3x^2 + \frac{1}{2}x = 4.$

4. $0.1x^2 + 0.2x - 0.54 = 0$

5. $-5.26x^2 + 7.89x + 15.78 = 0$

10.3 Solving Quadratic Equations by the Quadratic Formula

1. Solve quadratic equations by the quadratic formula.
2. Determine the number of solutions to a quadratic equation using the discriminant.

Tape 17

The Quadratic Formula

1 A method that can be used to solve any quadratic equation is the quadratic formula. It is the most useful and versatile method of solving quadratic equations.

The standard form of a quadratic equation is $ax^2 + bx + c = 0$, where a is the coefficient of the squared term, b is the coefficient of the first-degree term, and c is the constant.

Quadratic Equation in Standard Form	*Values of a, b, and c*		
$x^2 - 3x + 4 = 0$	$a = 1,$	$b = -3,$	$c = 4$
$-2x^2 + \frac{1}{2}x - 2 = 0$	$a = -2,$	$b = \frac{1}{2},$	$c = -2$
$3x^2 - 4 = 0$	$a = 3,$	$b = 0,$	$c = -4$
$5x^2 + 3x = 0$	$a = 5,$	$b = 3,$	$c = 0$
$-\frac{1}{2}x^2 + 5 = 0$	$a = -\frac{1}{2},$	$b = 0$	$c = 5$
$-12x^2 + 8x = 0$	$a = -12,$	$b = 8,$	$c = 0$

To Solve a Quadratic Equation by the Quadratic Formula

1. Write the equation in standard form, $ax^2 + bx + c = 0$, and determine the numerical values for a, b, and c.
2. Substitute the values for a, b, and c from step 1 in the quadratic formula below and then evaluate to obtain the solution.

The Quadratic Formula

$$x = \frac{-b \pm \sqrt{b^2 - 4ac}}{2a}$$

The Quadratic Formula can be derived by starting with the equation $ax^2 + bx + c = 0$ and solving the equation for x using completing the square. See Group Activity 6.

EXAMPLE 1 Solve the equation $x^2 + 2x - 8 = 0$ using the quadratic formula.

Solution: In this equation $a = 1$, $b = 2$, and $c = -8$. Substitute these values into the quadratic formula and then evaluate.

$$\begin{aligned} x &= \frac{-b \pm \sqrt{b^2 - 4ac}}{2a} \\ &= \frac{-(2) \pm \sqrt{(2)^2 - 4(1)(-8)}}{2(1)} \\ &= \frac{-2 \pm \sqrt{4 + 32}}{2} \\ &= \frac{-2 \pm \sqrt{36}}{2} \\ &= \frac{-2 \pm 6}{2} \end{aligned}$$

$$x = \frac{-2 + 6}{2} \quad \text{or} \quad x = \frac{-2 - 6}{2}$$

$$x = \frac{4}{2} = 2 \qquad x = \frac{-8}{2} = -4$$

Check:

$x = 2$		$x = -4$	
$x^2 + 2x - 8 = 0$		$x^2 + 2x - 8 = 0$	
$(2)^2 + 2(2) - 8 = 0$		$(-4)^2 + 2(-4) - 8 = 0$	
$4 + 4 - 8 = 0$		$16 - 8 - 8 = 0$	
$0 = 0$	true	$0 = 0$	true

COMMON STUDENT ERROR The **entire numerator** of the quadratic formula must be divided by $2a$.

Correct

$$x = \frac{-b \pm \sqrt{b^2 - 4ac}}{2a}$$

Incorrect

$$x = -b \pm \frac{\sqrt{b^2 - 4ac}}{2a}$$

$$x = \frac{-b}{2a} \pm \sqrt{b^2 - 4ac}$$

EXAMPLE 2 Solve the equation $6x^2 - x - 2 = 0$ using the quadratic formula.

Solution:

$$6x^2 - x - 2 = 0$$

$$a = 6, \quad b = -1, \quad c = -2$$

$$x = \frac{-b \pm \sqrt{b^2 - 4ac}}{2a}$$

$$= \frac{-(-1) \pm \sqrt{(-1)^2 - 4(6)(-2)}}{2(6)}$$

$$= \frac{1 \pm \sqrt{1 + 48}}{12}$$

$$= \frac{1 \pm \sqrt{49}}{12}$$

$$= \frac{1 \pm 7}{12}$$

$$x = \frac{1 + 7}{12} \quad \text{or} \quad x = \frac{1 - 7}{12}$$

$$x = \frac{8}{12} = \frac{2}{3} \qquad x = \frac{-6}{12} = -\frac{1}{2}$$

Check:

$$x = \frac{2}{3} \qquad\qquad x = -\frac{1}{2}$$

$$6x^2 - x - 2 = 0 \qquad\qquad 6x^2 - x - 2 = 0$$

$$6\left(\frac{2}{3}\right)^2 - \frac{2}{3} - 2 = 0 \qquad\qquad 6\left(-\frac{1}{2}\right)^2 - \left(-\frac{1}{2}\right) - 2 = 0$$

$$\overset{2}{\cancel{6}}\left(\frac{4}{\cancel{9}}\right) - \frac{2}{3} - 2 = 0 \qquad\qquad \overset{3}{\cancel{6}}\left(\frac{1}{\cancel{4}}\right) + \frac{1}{2} - 2 = 0$$

$$\frac{8}{3} - \frac{2}{3} - \frac{6}{3} = 0 \qquad\qquad \frac{3}{2} + \frac{1}{2} - \frac{4}{2} = 0$$

$$0 = 0 \quad \text{true} \qquad\qquad 0 = 0 \quad \text{true}$$

EXAMPLE 3 Solve the equation $2x^2 + 4x - 5 = 0$ using the quadratic formula.

Solution:

$$a = 2, \quad b = 4, \quad c = -5$$

$$x = \frac{-b \pm \sqrt{b^2 - 4ac}}{2a}$$

$$= \frac{-4 \pm \sqrt{(4)^2 - 4(2)(-5)}}{2(2)}$$

$$= \frac{-4 \pm \sqrt{16 + 40}}{4}$$

$$= \frac{-4 \pm \sqrt{56}}{4}$$

$$= \frac{-4 \pm 2\sqrt{14}}{4}$$

Now factor out 2 from both terms in the numerator; then divide out common factors as explained in section 9.4.

$$x = \frac{\overset{1}{\cancel{2}}(-2 \pm \sqrt{14})}{\underset{2}{\cancel{4}}}$$

$$x = \frac{-2 \pm \sqrt{14}}{2}$$

Thus, the solutions are

$$x = \frac{-2 + \sqrt{14}}{2} \quad \text{and} \quad x = \frac{-2 - \sqrt{14}}{2}$$

EXAMPLE 4 Solve the equation $x^2 = -4x + 6$ using the quadratic formula.

Solution: Write the equation in standard form.

$$x^2 + 4x - 6 = 0$$

$$a = 1, \qquad b = 4, \qquad c = -6$$

$$x = \frac{-b \pm \sqrt{b^2 - 4ac}}{2a}$$

$$= \frac{-4 \pm \sqrt{(4)^2 - 4(1)(-6)}}{2(1)}$$

$$= \frac{-4 \pm \sqrt{16 + 24}}{2}$$

$$= \frac{-4 \pm \sqrt{40}}{2}$$

$$= \frac{-4 \pm 2\sqrt{10}}{2}$$

$$= \frac{\overset{1}{\cancel{2}}(-2 \pm \sqrt{10})}{\underset{1}{\cancel{2}}}$$

$$= -2 \pm \sqrt{10}$$

$$x = -2 + \sqrt{10} \quad \text{or} \quad x = -2 - \sqrt{10}$$

COMMON STUDENT ERROR Many students solve quadratic equations correctly until the last step, where they make an error. Do not make the mistake of trying to simplify an answer that cannot be simplified any further. The following are answers that cannot be simplified, along with some common errors.

Answers That Cannot Be Simplified	*Incorrect*
$\dfrac{3 + 2\sqrt{5}}{2}$	$\dfrac{3 + 2\sqrt{5}}{2} = \dfrac{3 + \overset{1}{\cancel{2}}\sqrt{5}}{\underset{1}{\cancel{2}}} = 3 + \sqrt{5}$
$\dfrac{4 + 3\sqrt{5}}{2}$	$\dfrac{\overset{2}{\cancel{4}} + 3\sqrt{5}}{\underset{1}{\cancel{2}}} = 2 + 3\sqrt{5}$

Other examples of common errors of this type are illustrated in the Common Student Error box on page 509.

EXAMPLE 5 Solve the equation $x^2 = 4$ using the quadratic formula.

Solution: Write in standard form.

$$x^2 - 4 = 0$$

$$a = 1, \qquad b = 0, \qquad c = -4$$

$$x = \frac{-b \pm \sqrt{b^2 - 4ac}}{2a}$$

$$= \frac{-0 \pm \sqrt{0^2 - 4(1)(-4)}}{2(1)}$$

$$= \frac{\pm\sqrt{16}}{2} = \frac{\pm 4}{2} = \pm 2$$

Thus, the solutions are 2 and -2.

EXAMPLE 6 Solve the quadratic equation $2x^2 + 5x = -6$.

Solution:

$$2x^2 + 5x + 6 = 0$$

$$a = 2, \qquad b = 5, \qquad c = 6$$

$$x = \frac{-b \pm \sqrt{b^2 - 4ac}}{2a}$$

$$= \frac{-5 \pm \sqrt{(5)^2 - 4(2)(6)}}{2(2)}$$

$$= \frac{-5 \pm \sqrt{25 - 48}}{4}$$

$$= \frac{-5 \pm \sqrt{-23}}{4}$$

Since $\sqrt{-23}$ is not a real number, we can go no further. This equation has no real number solution. *When given a problem of this type, your answer should be "no real number solution." Do not leave the answer blank, and do not write 0 for the answer.*

The Discriminant

2 The expression under the square root sign in the quadratic formula is called the **discriminant.**

$$\underbrace{b^2 - 4ac}_{\text{discriminant}}$$

The discriminant can be used to determine the number of solutions to a quadratic equation.

When the discriminant is:

1. **Greater than zero,** $b^2 - 4ac > 0$, the quadratic equation has **two distinct real number solutions.**
2. **Equal to zero,** $b^2 - 4ac = 0$, the quadratic equation has **one real number solution.**
3. **Less than zero,** $b^2 - 4ac < 0$, the quadratic equation has **no real number solution.**

$b^2 - 4ac$	Number of Solutions
Positive	Two distinct real number solutions
0	One real number solution
Negative	No real number solution

EXAMPLE 7

(a) Find the discriminant of the equation $x^2 - 8x + 16 = 0$.
(b) Use the quadratic formula to find the solution.

Solution:

(a) $a = 1, \quad b = -8, \quad c = 16$

$$b^2 - 4ac = (-8)^2 - 4(1)(16) = 64 - 64 = 0$$

Since the discriminant equals zero, there is one real number solution.

(b)
$$x = \frac{-b \pm \sqrt{b^2 - 4ac}}{2a}$$
$$= \frac{-(-8) \pm \sqrt{0}}{2(1)}$$
$$= \frac{8 \pm 0}{2} = \frac{8}{2} = 4$$

The only solution is 4.

EXAMPLE 8 Without actually finding the solutions, determine if the following equations have two distinct real number solutions, one real number solution, or no real number solution.

(a) $2x^2 - 4x + 6 = 0$ **(b)** $x^2 - 5x - 8 = 0$ **(c)** $4x^2 - 12x = -9$

Solution: We use the discriminant of the quadratic formula to answer these equations.

(a) $b^2 - 4ac = (-4)^2 - 4(2)(6) = 16 - 48 = -32$

Since the discriminant is negative, this equation has no real number solution.

(b) $b^2 - 4ac = (-5)^2 - 4(1)(-8) = 25 + 32 = 57$

Since the discriminant is positive, this equation has two distinct real number solutions.

(c) First rewrite $4x^2 - 12x = -9$ as $4x^2 - 12x + 9 = 0$.

$$b^2 - 4ac = (-12)^2 - 4(4)(9) = 144 - 144 = 0$$

Since the discriminant is zero, this equation has one real number solution.

Now let us look at one of many application problems that may be solved using the quadratic formula.

EXAMPLE 9 Mr. Jackson is planning to plant a grass walkway of uniform width around his rectangular swimming pool, which measures 18 feet by 24 feet. How far will the walkway extend from the pool if Mr. Jackson has only enough seed to plant 2000 square feet of grass?

Solution: Let us make a diagram of the pool and grassy area (Fig. 10.2). Let $x =$ the uniform width of the grass area. Then the total length of the larger rectangular area is $2x + 24$. The total width of the larger rectangular area is $2x + 18$.

The grassy area can be found by subtracting the area of the pool from the larger rectangular area.

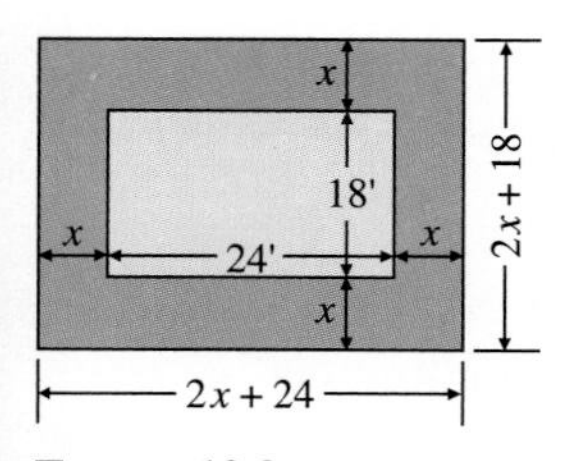

FIGURE 10.2

$$\text{Area of pool} = l \cdot w = (24)(18) = 432 \text{ square feet}$$

$$\begin{aligned}\text{Area of large rectangle} &= l \cdot w = (2x + 24)(2x + 18)\\ &= 4x^2 + 84x + 432 \text{ (pool plus grassy area)}\end{aligned}$$

$$\begin{aligned}\text{Grassy area} &= \text{area of large rectangle} - \text{area of pool}\\ &= 4x^2 + 84x + 432 - (432)\\ &= 4x^2 + 84x\end{aligned}$$

The total grassy area must be 2000 square feet.

$$\text{Grassy area} = 4x^2 + 84x$$
$$2000 = 4x^2 + 84x$$
$$\text{or} \quad 4x^2 + 84x - 2000 = 0$$
$$4(x^2 + 21x - 500) = 0$$
$$\frac{1}{\cancel{4}} \cdot \cancel{4}(x^2 + 21x - 500) = \frac{1}{4} \cdot 0$$
$$x^2 + 21x - 500 = 0$$

By the quadratic formula.

$$a = 1, \qquad b = 21, \qquad c = -500$$
$$x = \frac{-b \pm \sqrt{b^2 - 4ac}}{2a}$$
$$x = \frac{-21 \pm \sqrt{(21)^2 - 4(1)(-500)}}{2(1)}$$
$$= \frac{-21 \pm \sqrt{441 + 2000}}{2}$$
$$= \frac{-21 \pm \sqrt{2441}}{2}$$
$$\approx \frac{-21 \pm 49.41}{2}$$
$$x \approx \frac{-21 - 49.41}{2} \quad \text{or} \quad x \approx \frac{-21 + 49.41}{2}$$
$$x \approx \frac{-70.41}{2} \qquad x \approx \frac{28.41}{2}$$
$$x \approx -35.21 \qquad x \approx 14.21$$

Since lengths are positive, the only possible answer is $x \approx 14.21$. Thus, there will be a grass walkway about 14.2 feet wide all around the pool.

Helpful Hint

Notice in Example 9 that when we had the quadratic equation $4x^2 + 84x - 2000 = 0$ we factored out the common factor 4 to get

$$4x^2 + 84x - 2000 = 0$$
$$4(x^2 + 21x - 500) = 0$$

We then multiplied both sides of the equation by $\frac{1}{4}$ and used the quadratic equation $x^2 + 21x - 500 = 0$, where $a = 1$, $b = 21$, and $c = -500$, in the quadratic formula.

If all the terms in a quadratic equation have a common factor, it will be easier to factor it out first so that you will have smaller numbers when you use the quadratic formula. Consider the quadratic equation $4x^2 + 8x - 12 = 0$.

In this equation $a = 4$, $b = 8$, and $c = -12$. If you solve this equation with the quadratic formula, after simplification you will get the solutions -3 and 1. Try this and see. If you factor out 4 to get

$$4x^2 + 8x - 12 = 0$$
$$4(x^2 + 2x - 3) = 0$$

and then use the quadratic formula with the equation $x^2 + 2x - 3 = 0$, where $a = 1$, $b = 2$, and $c = -3$, you get the same solution. Try this and see.

Exercise Set 10.3

Determine whether each equation has two distinct real number solutions, one real number solution, or no real number solution.

1. $x^2 + 3x - 5 = 0$
2. $2x^2 + 6x + 3 = 0$
3. $3x^2 - 4x + 7 = 0$
4. $-4x^2 + x - 8 = 0$
5. $5x^2 + 3x - 7 = 0$
6. $2x^2 = 16x - 32$
7. $4x^2 - 24x = -36$
8. $5x^2 - 4x = 7$
9. $x^2 - 8x + 5 = 0$
10. $x^2 - 5x - 9 = 0$
11. $-3x^2 + 5x - 8 = 0$
12. $x^2 + 4x - 8 = 0$
13. $x^2 + 7x - 3 = 0$
14. $2x^2 - 6x + 9 = 0$
15. $4x^2 - 9 = 0$
16. $6x^2 - 5x = 0$

Use the quadratic formula to solve each equation. If the equation has no real number solution, so state.

17. $x^2 - 3x + 2 = 0$
18. $x^2 + 6x + 8 = 0$
19. $x^2 - 9x + 20 = 0$
20. $x^2 - 3x - 10 = 0$
21. $x^2 + 5x - 24 = 0$
22. $x^2 - 6x = -5$
23. $x^2 = 13x - 36$
24. $x^2 - 36 = 0$
25. $x^2 - 25 = 0$
26. $x^2 - 6x = 0$
27. $x^2 - 3x = 0$
28. $z^2 - 17z + 72 = 0$
29. $p^2 - 7p + 12 = 0$
30. $2x^2 - 3x + 2 = 0$
31. $2y^2 - 7y + 4 = 0$
32. $2x^2 - 7x = -5$
33. $6x^2 = -x + 1$
34. $4r^2 + r - 3 = 0$
35. $2x^2 - 4x - 1 = 0$
36. $3w^2 - 4w + 5 = 0$
37. $2s^2 - 4s + 3 = 0$
38. $x^2 - 7x + 3 = 0$
39. $4x^2 = x + 5$
40. $x^2 - 2x - 1 = 0$
41. $2x^2 - 7x = 9$
42. $-x^2 + 2x + 15 = 0$
43. $-2x^2 + 11x - 15 = 0$
44. $6x^2 + 5x + 9 = 0$
45. The product of two consecutive positive integers is 20. Find the two consecutive integers.
46. The length of a rectangle is 3 feet longer than its width. Find the dimensions of the rectangle if its area is 28 square feet.
47. The length of a rectangle is 3 feet smaller than twice its width. Find the length and width of the rectangle if its area is 20 square feet.
48. Lisa wishes to plant a uniform strip of grass around her pool. If her pool measures 20 feet by 30 feet and she has only enough seed to cover 336 square feet, what will be the width of the uniform strip? (See Example 9.)
49. The McDonald's garden is 30 feet by 40 feet. They wish to lay a uniform border of pine bark around their garden. How large a strip should they lay if they only have enough bark to cover 296 square feet?
50. **(a)** What is the discriminant? **(b)** Explain how the discriminant can be used to determine the number of real number solutions a quadratic equation has.
51. How many real number solutions does a quadratic equation have if the discriminant equals
    (a) -4 **(b)** 0 **(c)** $\frac{1}{2}$?
52. Without looking at your notes, write down the quadratic formula. You must learn this formula by memory.
53. Explain in your own words why a quadratic equation will have two real number solutions when the discriminant is greater than 0, one real number solution when the discriminant is equal to 0, and no real number solution when the discriminant is less than 0. Use the quadratic formula in explaining your answer.

CUMULATIVE REVIEW EXERCISES

[5.6, 10.2, 10.3]

Solve the following quadratic equations by (a) factoring, (b) completing the square, and (c) using the quadratic formula. If the equation cannot be solved by factoring, so state.

54. $x^2 - 13x + 42 = 0.$ **55.** $6x^2 + 11x - 35 = 0.$ **56.** $2x^2 + 3x - 4 = 0.$ **57.** $6x^2 = 54.$

Group Activity/ Challenge Problems

1. In Section 10.4 we will graph quadratic equations. We will learn that the graphs of quadratic equations are *parabolas*. The graph of the quadratic equation $y = x^2 - 2x - 8$ is illustrated below.

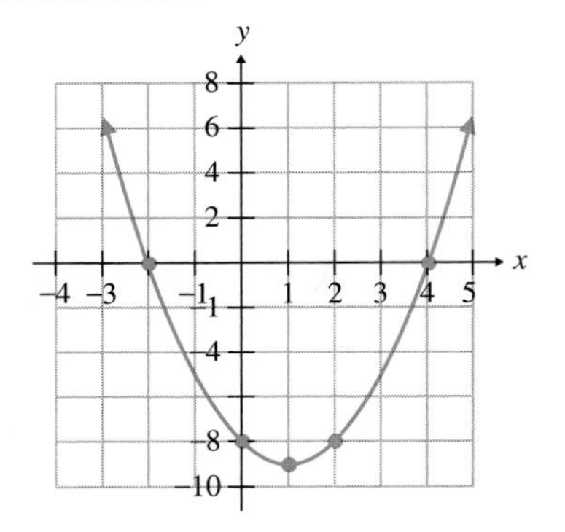

(a) Copy this graph in your notebook. Then graph the equation $y = 2x - 3$ on the same axes and estimate the points of intersection of the graphs.

(b) The graphs represent the system of equations

$$y = x^2 - 2x - 8$$
$$y = 2x - 3$$

Use substitution to obtain one quadratic equation in x. Then solve the quadratic equation.

(c) Does the answer to part (b) agree with the x coordinates of the points of intersection in part (a)?

(d) Use the values of x found in part (b) to find the solution to the system of equations.

Find all the values of c that will result in the equation having **(a)** two real number solutions **(b)** one real number solution **(c)** no real number solution.

2. $x^2 + 6x + c = 0$ **3.** $2x^2 + 3x + c = 0$ **4.** $-3x^2 + 6x + c = 0$

5. Farmer Justina Wells wishes to form a rectangular region along a river bank by constructing fencing on three sides, as illustrated in the diagram. If she has only 400 feet of fencing and wishes to enclose an area of 15,000 square feet, find the dimensions of the rectangular region.

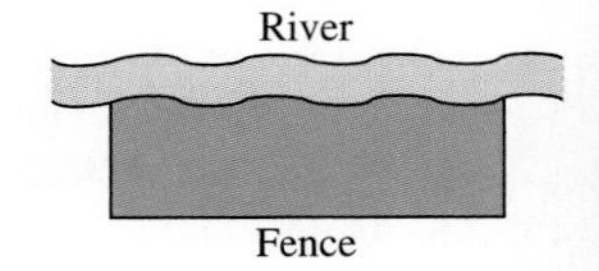

6. By completing the square, solve the following equation for x in terms of a, b, and c: $ax^2 + bx + c = 0.$

10.4 Graphing Quadratic Equations

Tape 17

1. Graph quadratic equations in two variables.
2. Find the coordinates of the vertex of a parabola.
3. Use symmetry to graph quadratic equations.
4. Find the x intercepts of the graph of a quadratic equation.

Graphing Quadratic Equations

1 In Section 7.3 we learned how to graph linear equations. In this section we graph quadratic equations of the form

$$y = ax^2 + bx + c, \qquad a \neq 0$$

The graph of every quadratic equation of this form will be a **parabola.** The graph of $y = ax^2 + bx + c$ will have one of the shapes indicated in Figure 10.3.

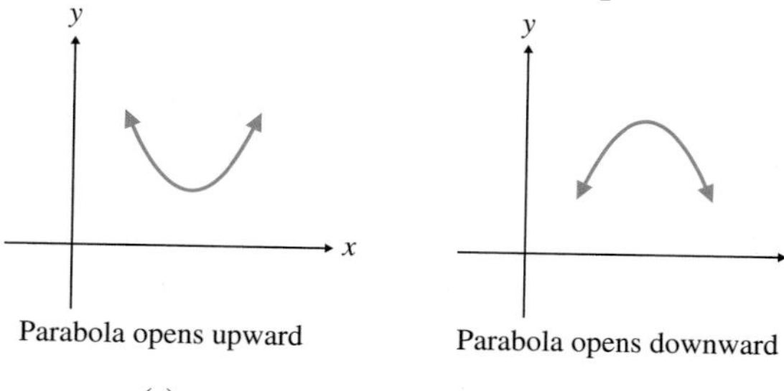

FIGURE 10.3

When a quadratic equation is in the form $y = ax^2 + bx + c$, the sign of a, the numerical coefficient of the squared term, will determine whether the parabola will open upward (Fig. 10.3a) or downward (Fig. 10.3b). When a is positive, the parabola will open upward, and when a is negative, the parabola will open downward. The **vertex** is the lowest point on a parabola that opens upward and the highest point on a parabola that opens downward (Fig. 10.4).

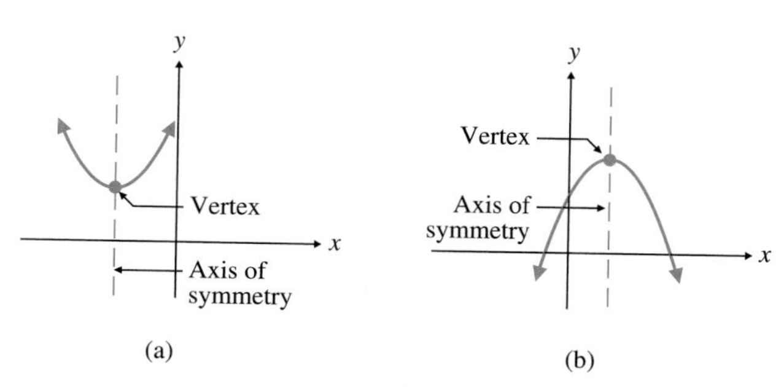

FIGURE 10.4

Graphs of quadratic equations of the form $y = ax^2 + bx + c$ have **symmetry** about a line through the vertex. This means that if we fold the paper along this imaginary line, called the **axis of symmetry,** the right and left sides of the graph will coincide.

One method that can be used to graph a quadratic equation is to plot it point by point. When determining points to plot, select values for x and determine the corresponding values for y.

EXAMPLE 1 Graph the equation $y = x^2$.

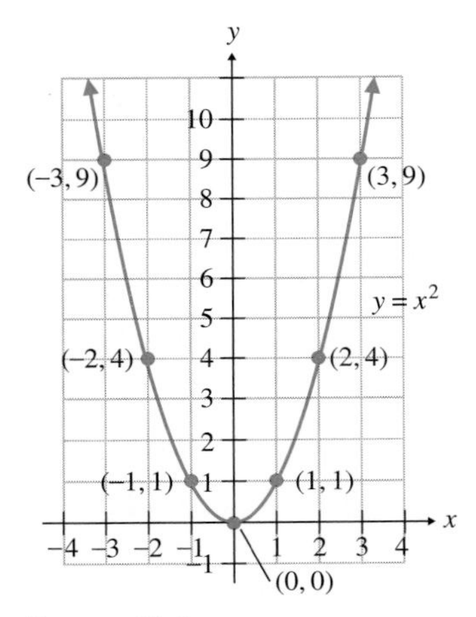

FIGURE 10.5

Solution: Since $a = 1$, which is positive, this parabola opens upward.

$$y = x^2$$

Let $x = 3$, $y = (3)^2 = 9$
Let $x = 2$, $y = (2)^2 = 4$
Let $x = 1$, $y = (1)^2 = 1$
Let $x = 0$, $y = (0)^2 = 0$
Let $x = -1$, $y = (-1)^2 = 1$
Let $x = -2$, $y = (-2)^2 = 4$
Let $x = -3$, $y = (-3)^2 = 9$

x	y
3	9
2	4
1	1
0	0
−1	1
−2	4
−3	9

Connect the points with a smooth curve (Fig. 10.5). Note how the graph is symmetric about the line $x = 0$ (or the y axis).

EXAMPLE 2 Graph the equation $y = -2x^2 + 16x - 24$.

Solution: Since $a = -2$, which is negative, this parabola opens downward.

$$y = -2x^2 + 16x - 24$$

Let $x = 0$, $y = -2(0)^2 + 16(0) - 24 = -24$
Let $x = 1$, $y = -2(1)^2 + 16(1) - 24 = -10$
Let $x = 2$, $y = -2(2)^2 + 16(2) - 24 = 0$
Let $x = 3$, $y = -2(3)^2 + 16(3) - 24 = 6$
Let $x = 4$, $y = -2(4)^2 + 16(4) - 24 = 8$
Let $x = 5$, $y = -2(5)^2 + 16(5) - 24 = 6$
Let $x = 6$, $y = -2(6)^2 + 16(6) - 24 = 0$
Let $x = 7$, $y = -2(7)^2 + 16(7) - 24 = -10$
Let $x = 8$, $y = -2(8)^2 + 16(8) - 24 = -24$

x	y
0	−24
1	−10
2	0
3	6
4	8
5	6
6	0
7	−10
8	−24

Note how the graph (Fig. 10.6) is symmetric about the line $x = 4$, which is dashed because it is not part of the graph. The vertex of this parabola is the point (4, 8). Since the y values are large, the y axis has been marked with 4-unit intervals to allow us to graph the points (0, −24) and (8, −24). The arrows on the ends of the graph indicate that the parabola continues indefinitely.

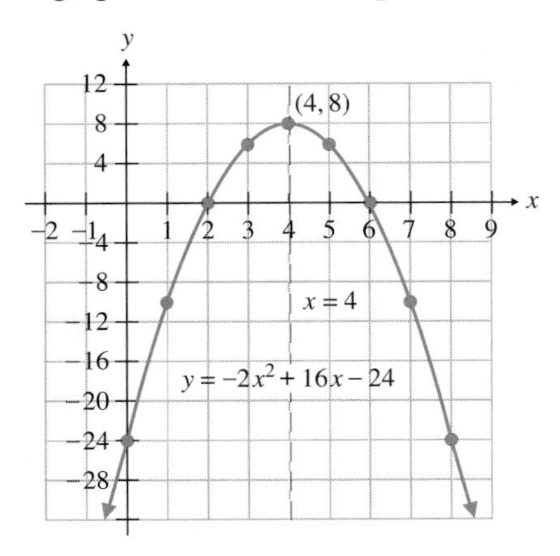

FIGURE 10.6

Finding the Vertex of a Parabola

 When graphing quadratic equations, how do we decide what values to use for x? When the location of the vertex is unknown, this is a difficult question to answer. When the location of the vertex is known, it becomes more obvious which values to use.

In Example 2, the axis of symmetry is $x = 4$, and the x coordinate of the vertex is also 4. For a quadratic equation in the form $y = ax^2 + bx + c$, both the axis of symmetry and the x coordinate of the vertex can be found by using the following formula:

Axis of Symmetry and x Coordinate of Vertex

$$x = \frac{-b}{2a}$$

In the quadratic equation in Example 2, $a = -2$, $b = 16$, and $c = -24$. Substituting these values in the formula for the axis of symmetry gives

$$x = \frac{-b}{2a} = \frac{-(16)}{2(-2)} = \frac{-16}{-4} = 4$$

Thus, the graph is symmetric about the line $x = 4$, and the x coordinate of the vertex is 4.

The y coordinate of the vertex can be found by substituting the value of the x coordinate of the vertex into the quadratic equation and solving for y.

$$\begin{aligned} y &= -2x^2 + 16x - 24 \\ &= -2(4)^2 + 16(4) - 24 \\ &= -2(16) + 64 - 24 \\ &= -32 + 64 - 24 \\ &= 8 \end{aligned}$$

Thus, the vertex is at the point (4, 8).

For a quadratic equation of the form $y = ax^2 + bx + c$, the y coordinate of the vertex can also be found by the following formula.

y Coordinate of Vertex

$$y = \frac{4ac - b^2}{4a}$$

For Example 2,

$$\begin{aligned} y &= \frac{4ac - b^2}{4a} \\ &= \frac{4(-2)(-24) - (16)^2}{4(-2)} \\ &= \frac{192 - 256}{-8} = \frac{-64}{-8} = 8 \end{aligned}$$

You may use the method of your choice to find the y coordinate of the vertex. Both methods result in the same value of y.

Use Symmetry to Graph Quadratic Equations

3 One method to use in selecting points to plot when graphing parabolas is to determine the axis of symmetry and the vertex of the graph. Then select nearby values of x on either side of the axis of symmetry. When graphing the equation, make use of the symmetry of the graph.

EXAMPLE 3

(a) Find the axis of symmetry of the graph of the equation $y = x^2 + 8x + 15$.

(b) Find the vertex of the graph.

(c) Graph the equation.

Solution: **(a)** $a = 1, \quad b = 8, \quad c = 15$

$$x = \frac{-b}{2a} = \frac{-8}{2(1)} = -4$$

The parabola is symmetric about the line $x = -4$. The x coordinate of the vertex is -4.

(b) Now find the y coordinate of the vertex. Substitute -4 for x in the quadratic equation.

$$y = x^2 + 8x + 15$$
$$y = (-4)^2 + 8(-4) + 15 = 16 - 32 + 15 = -1$$

The vertex is at the point $(-4, -1)$.

(c) Since the axis of symmetry is $x = -4$, we will select values for x that are greater than or equal to -4. It is often helpful to plot each point as it is determined. If a point does not appear to lie on the parabola, check it.

$$y = x^2 + 8x + 15$$

Let $x = -3$, $\quad y = (-3)^2 + 8(-3) + 15 = 0$

Let $x = -2$, $\quad y = (-2)^2 + 8(-2) + 15 = 3$

Let $x = -1$, $\quad y = (-1)^2 + 8(-1) + 15 = 8$

x	y
−3	0
−2	3
−1	8

These points are plotted in Figure 10.7a. The entire graph of the equation is illustrated in Figure 10.7b. Note how we use symmetry to complete the graph. The points $(-3, 0)$ and $(-5, 0)$ are each 1 horizontal unit from the axis of symmetry, $x = -4$. The points $(-2, 3)$ and $(-6, 3)$ are each 2 horizontal units from the axis of symmetry, and the points $(-1, 8)$ and $(-7, 8)$ are each 3 horizontal units from the axis of symmetry.

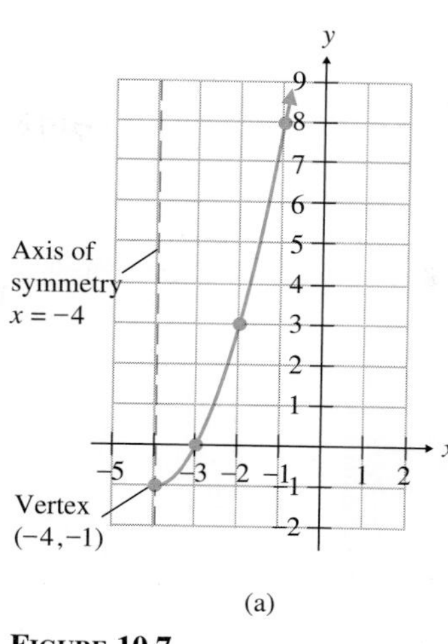

(a)

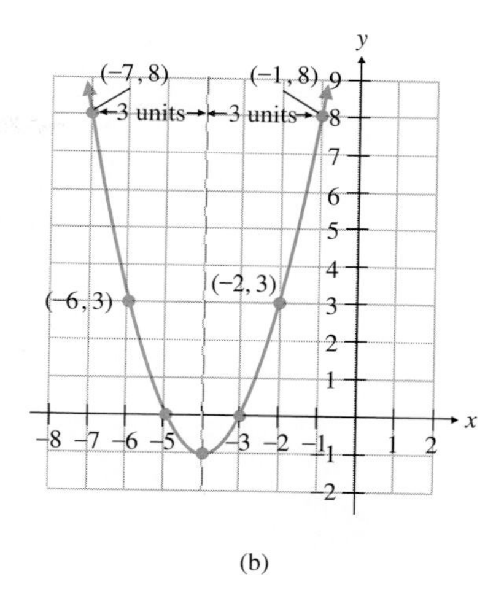

(b)

FIGURE 10.7

EXAMPLE 4 Graph the equation $y = -2x^2 + 3x - 4$.

Solution: $a = -2, \quad b = 3, \quad c = -4$

Since $a < 0$, this parabola will open downward.

$$\text{Axis of symmetry:} \quad x = \frac{-b}{2a}$$

$$x = \frac{-3}{2(-2)} = \frac{-3}{-4} = \frac{3}{4}$$

Since the x value of the vertex is a fraction, we will use the formula to find the y coordinate of the vertex.

$$y = \frac{4ac - b^2}{4a}$$

$$y = \frac{4(-2)(-4) - 3^2}{4(-2)} = \frac{32 - 9}{-8} = \frac{23}{-8} = -\frac{23}{8} \quad (\text{or } -2\frac{7}{8})$$

The vertex of this graph is at the point $\left(\frac{3}{4}, -\frac{23}{8}\right)$. Since the axis of symmetry is $x = \frac{3}{4}$, we will begin by selecting values of x that are greater than $\frac{3}{4}$.

$$y = -2x^2 + 3x - 4$$

Let $x = 1$, $\quad y = -2(1)^2 + 3(1) - 4 = -3$

Let $x = 2$, $\quad y = -2(2)^2 + 3(2) - 4 = -6$

Let $x = 3$, $\quad y = -2(3)^2 + 3(3) - 4 = -13$

x	y
1	-3
2	-6
3	-13

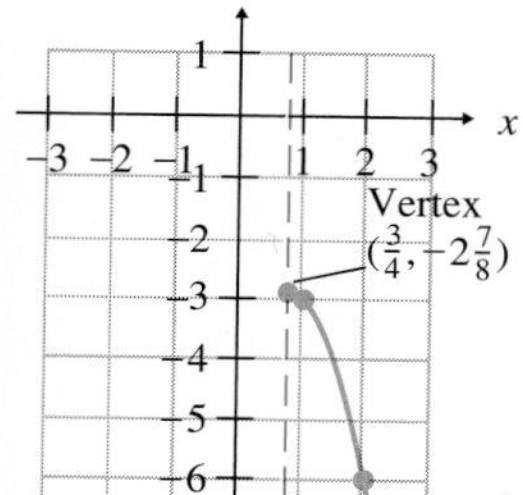
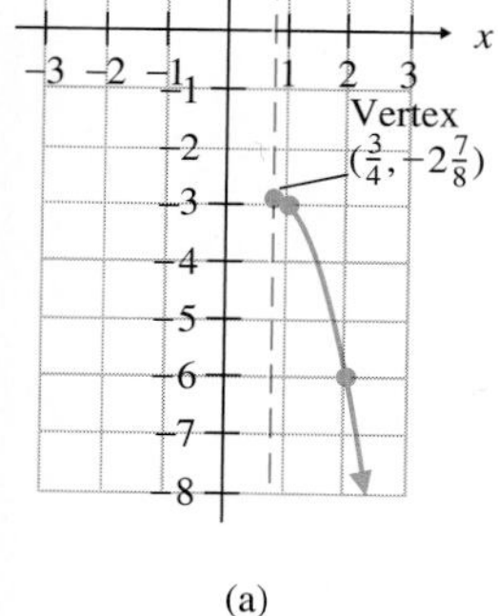

(a)

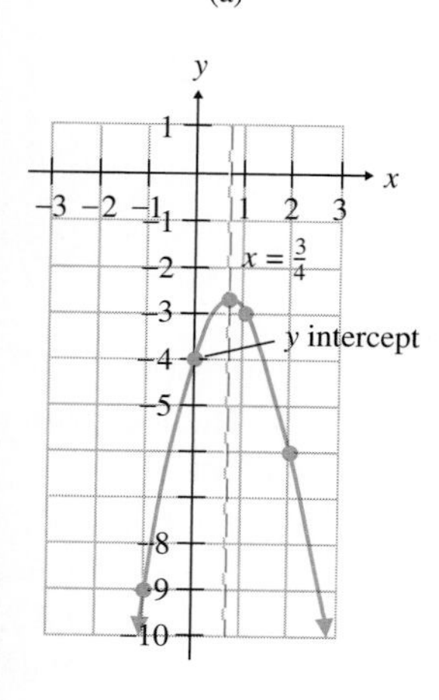

(b)

FIGURE 10.8

When the axis of symmetry is a fractional value, be very careful when constructing the graph. You should plot as many additional points as needed. In this example, when $x = 0$, $y = -4$, and when $x = -1$, $y = -9$. Figure 10.8a shows the points plotted on the right side of the axis of symmetry. Figure 10.8b shows the completed graph.

(a)

(b)

FIGURE 10.9 Not every shape that resembles a parabola is a parabola. For example, the St. Louis Arch (a) resembles a parabola, but it is not a parabola. However, the bridge over the Mississippi near Jefferson Barracks Missouri, which connects Missouri and Illinois (b), is á parabola.

Calculator Corner

EVALUATING QUADRATIC EXPRESSIONS

In this section when graphing we will often have to evaluate quadratic expressions to obtain values for y. Here we show how the polynomial $6x^2 - 3x + 4$ can be evaluated using a *scientific calculator* for the values $x = 5$ and $x = -5$.

Evaluate: $6x^2 - 3x + 4$ for $x = 5$

Evaluate: $6(5)^2 - 3(5) + 4$

6 [×] 5 [x^2] [−] 3 [×] 5 [+] 4 [=] 139

Evaluate: $6x^2 - 3x + 4$ for $x = -5$

Evaluate: $6(-5)^2 - 3(-5) + 4$

6 [×] 5 [+/−] [x^2] [−] 3 [×] 5 [+/−] [+] 4 [=] 169

Finding the x Intercepts of a Parabola

4 In Example 4 the graph crossed the y axis at $y = -4$. Recall from earlier sections that to find the y intercept we let $x = 0$ and solve for y. The location of the y intercept is often helpful when graphing quadratic equations.

To find the x intercept when graphing straight lines in Section 7.3, we set $y = 0$ and found the corresponding value of x. We did this because the value of y where a graph crosses the x axis is 0. We use the same procedure here when finding the x intercepts of a quadratic equation of the form $y = ax^2 + bx + c$. To find the x intercepts of the graph of $y = ax^2 + bx + c$ we set y equal to 0 and solve the resulting equation, $ax^2 + bx + c = 0$. The real number solutions of the equation $ax^2 + bx + c = 0$ will be the x intercepts of the graph $y = ax^2 + bx + c$. To solve equations of the form $ax^2 + bx + c = 0$ we can use factoring as explained in Section 5.6; completing the square, in Section 10.2; or the quadratic formula, in Section 10.3.

In Section 10.3 we mentioned that when solving quadratic equations of the form $ax^2 + bx + c = 0$, when the discriminant, $b^2 - 4ac$ is greater than zero, there are two distinct real number solutions; when it is equal to zero, there is one real number solution; and when it is less than zero, there is no real number solution.

A quadratic equation of the form $y = ax^2 + bx + c$ will have either two distinct x intercepts (Fig. 10.10a), one x intercept (Fig. 10.10b), or no x intercepts (Fig. 10.9c). The number of x intercepts may be determined by the discriminant.

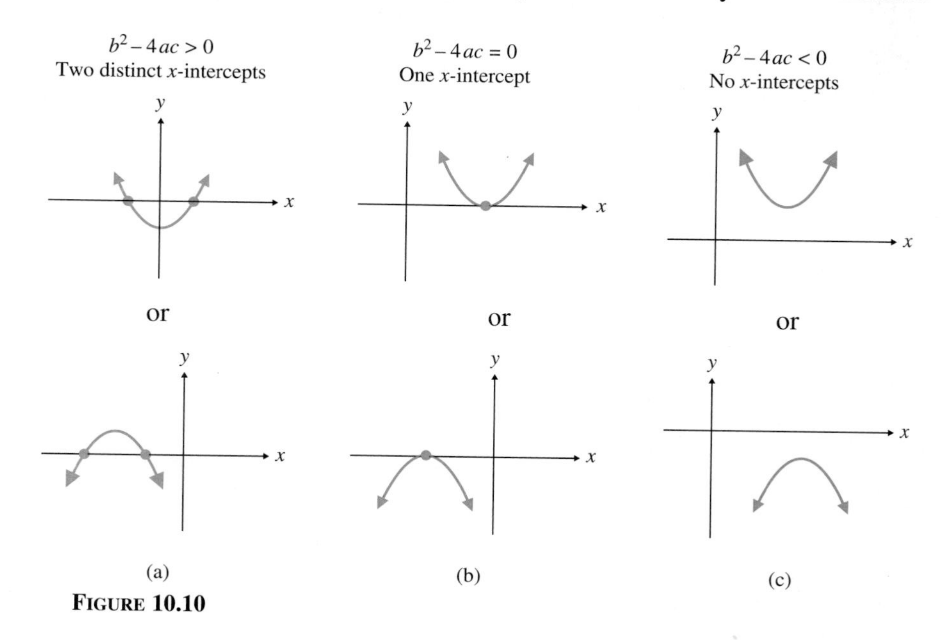

FIGURE 10.10

The x intercepts can be found graphically. They may also be found algebraically by setting y equal to 0 and solving the resulting equation, as in Example 5.

EXAMPLE 5

(a) Find the x intercepts of the graph of the equation $y = x^2 - 2x - 24$ by factoring, by completing the square, and by the quadratic formula.

(b) Graph the equation.

Solution:

(a) To find the x intercepts algebraically we set y equal to 0, and solve the resulting equation, $x^2 - 2x - 24 = 0$. We will solve this equation by all three algebraic methods.

Method 1: Factoring.

$$x^2 - 2x - 24 = 0$$
$$(x - 6)(x + 4) = 0$$
$$x - 6 = 0 \quad \text{or} \quad x + 4 = 0$$
$$x = 6 \qquad\qquad x = -4$$

Method 2: Completing the square.

$$x^2 - 2x - 24 = 0$$
$$x^2 - 2x = 24$$
$$x^2 - 2x + 1 = 24 + 1$$
$$(x - 1)^2 = 25$$
$$x - 1 = \pm 5$$
$$x = 1 \pm 5$$
$$x = 1 + 5 \quad \text{or} \quad x = 1 - 5$$
$$x = 6 \qquad\qquad x = -4$$

Method 3: Quadratic formula.

$$x^2 - 2x - 24 = 0$$
$$a = 1, \qquad b = -2, \qquad c = -24$$
$$x = \frac{-b \pm \sqrt{b^2 - 4ac}}{2a}$$
$$= \frac{-(-2) \pm \sqrt{(-2)^2 - 4(1)(-24)}}{2(1)}$$
$$= \frac{2 \pm \sqrt{4 + 96}}{2}$$
$$= \frac{2 \pm \sqrt{100}}{2}$$
$$= \frac{2 \pm 10}{2}$$
$$x = \frac{2 + 10}{2} \quad \text{or} \quad x = \frac{2 - 10}{2}$$
$$x = \frac{12}{2} = 6 \qquad\qquad x = \frac{-8}{2} = -4$$

Note that the same solutions, 6 and -4, were obtained by all three methods. The graph of the equation $y = x^2 - 2x - 24$ will have two distinct x intercepts, at 6 and -4.

(b) Since $a > 0$, this parabola opens upward.

$$\text{Axis of symmetry: } x = \frac{-b}{2a} = \frac{-(-2)}{2(1)} = \frac{2}{2} = 1$$

$$y = x^2 - 2x - 24$$

Let $x = 1$, $y = 1^2 - 2(1) - 24 = -25$

Let $x = 2$, $y = 2^2 - 2(2) - 24 = -24$

Let $x = 3$, $y = 3^2 - 2(3) - 24 = -21$

Let $x = 4$, $y = 4^2 - 2(4) - 24 = -16$

Let $x = 5$, $y = 5^2 - 2(5) - 24 = -9$

Let $x = 6$, $y = 6^2 - 2(6) - 24 = 0$

x	y
1	−25
2	−24
3	−21
4	−16
5	−9
6	0

Again we use symmetry to complete the graph (Fig. 10.11). The x-intercepts 6 and -4 agree with the answer obtained in part **(a)**.

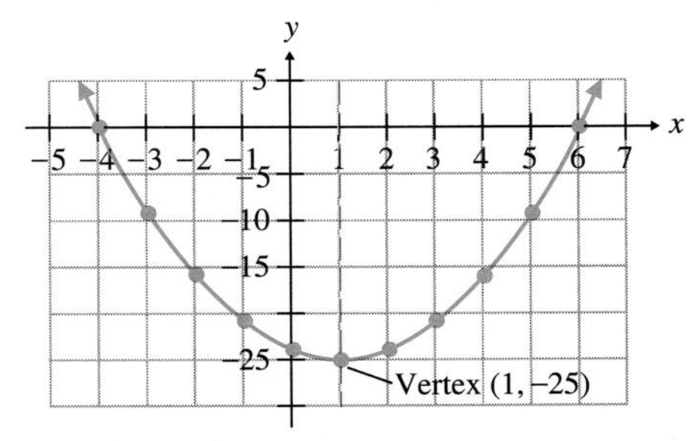

FIGURE 10.11

Exercise Set 10.4

Indicate the axis of symmetry, the coordinates of the vertex, and whether the parabola opens up or down.

1. $y = x^2 + 2x - 7$

2. $y = x^2 + 4x - 9$

3. $y = -x^2 + 5x - 6$

4. $y = 3x^2 + 6x - 9$

5. $y = -3x^2 + 5x + 8$

6. $y = x^2 + 3x - 6$

7. $y = -4x^2 - 8x - 12$

8. $y = 2x^2 + 3x + 8$

9. $y = 3x^2 - 2x + 2$

10. $y = -x^2 + x + 8$

11. $y = 4x^2 + 12x - 5$

12. $y = -2x^2 - 6x - 5$

Graph each quadratic equation and determine the x-intercepts, if they exist.

13. $y = x^2 - 1$

14. $y = x^2 + 4$

15. $y = -x^2 + 3$

16. $y = -x^2 - 2$

17. $y = x^2 + 2x + 3$

18. $y = x^2 + 4x + 3$

19. $y = x^2 + 2x - 15$

20. $y = -x^2 + 10x - 21$

21. $y = -x^2 + 4x - 5$

22. $y = x^2 + 8x + 15$

23. $y = x^2 - x - 12$

24. $y = x^2 - 5x + 4$

25. $y = x^2 - 6x + 9$

26. $y = x^2 - 6x$

27. $y = -x^2 + 5x$

28. $y = x^2 - 4x + 4$

29. $y = 2x^2 - 6x + 4$

30. $y = x^2 - 2x + 1$

31. $y = -x^2 + 11x - 28$

32. $y = 4x^2 + 12x + 9$

33. $y = x^2 - 2x - 15$

34. $y = x^2 - 5x + 4$

35. $y = -2x^2 + 7x - 3$

36. $y = 2x^2 + 3x - 2$

37. $y = -2x^2 + 3x - 2$

38. $y = -4x^2 - 6x + 4$

39. $y = 2x^2 - x - 15$

40. $y = 6x^2 + 10x - 4$

Using the discriminant, determine the number of x intercepts the graph of the equation will have. Do not graph the equation.

41. $y = 3x^2 - 6x + 4$

42. $y = -2x^2 - x + 5$

43. $y = 4x^2 - 6x - 7$

44. $y = x^2 - 6x + 9$

45. $y = 0.1x^2 + 2x - 3$

46. $y = -4.3x^2 + 5.7x$

47. $y = -\frac{1}{2}x^2 - \frac{3}{4}$

48. $y = 5x^2 - 13.2x + 9.3$

49. What are the coordinates of the vertex of a parabola?

50. What determines whether the graph of a quadratic equation of the form $y = ax^2 + bx + c$, $a \neq 0$, is a parabola that opens upward or downward? Explain your answer.

51. **(a)** What are the x-intercepts of a graph? **(b)** How can you find the x-intercepts of a graph algebraically?

52. How many x-intercepts will the graph of a quadratic equation have if the discriminant has a value of
(a) 5 (b) -2 (c) 0?

53. (a) When graphing a quadratic equation of the form $y = ax^2 + bx + c$, what is the equation of the vertical line about which the parabola will be symmetric?
(b) What is this vertical line called?

The graph of a quadratic equation of the form $y = ax^2 + bx + c$ is a parabola. The value of a (the coefficient of the squared term in the equation) and the vertex of the parabola, are given. Determine the number of x intercepts the parabola will have. Explain how you determined your answer.

54. $a = -2$, vertex at $(0, -3)$

55. $a = 5$, vertex at $(4, -3)$

56. $a = -3$, vertex at $(-4, 0)$

57. $a = -1$, vertex at $(2, -4)$

58. Will the equations below have the same x intercepts when graphed? Explain how you determined your answer.
$y = x^2 - 2x - 15$ and $y = -x^2 + 2x + 15$

CUMULATIVE REVIEW EXERCISES

[6.3] **59.** Add $\frac{3}{x+3} - \frac{x-2}{x-4}$

[2.5] **60.** Solve the equation $\frac{1}{3}(x + 6) = 3 - \frac{1}{4}(x - 5)$

[7.2] **61.** Graph $4x - 6y = 20$.

[7.6] **62.** Graph $y < 2$.

Group Activity/ Challenge Problems

1. In Section 7.6 we defined a function. Will all quadratic equations of the form $y = ax^2 + bx + c$, $a \neq 0$, be functions? Explain your answer.
2. Graph $f(x) = x^2 - 4x + 3$.
3. (a) How will the graphs of the equations below compare? Explain how you determined your answer. $y = x^2 - 2x - 8$ and $y = -x^2 + 2x + 8$
   (b) Graph $y = x^2 - 2x - 8$ and $y = -x^2 + 2x + 8$ on the same axes.
4. The equation $x^2 + y^2 = r^2$ is the equation of a circle whose center is at the origin and whose radius is r. Graph the equation $x^2 + y^2 = 4$.

Summary

GLOSSARY

Axis of symmetry (555): The imaginary line about which a graph is symmetric.

Parabola (555): The graph of a quadratic equation is a parabola.

Perfect square trinomial (539): A trinomial that can be expressed as the square of a binomial.

Standard form of a quadratic equation (534): $ax^2 + bx + c = 0$, $a \neq 0$.

Vertex of a parabola (555): The lowest point on a parabola that opens upward or the highest point on a parabola that opens downward.

IMPORTANT FACTS

Square root property: If $x^2 = a$, then $x = \sqrt{a}$ or $x = -\sqrt{a}$ (or $x = \pm\sqrt{a}$).

Quadratic formula: $x = \dfrac{-b \pm \sqrt{b^2 - 4ac}}{2a}$

Discriminant: $b^2 - 4ac$

If $b^2 - 4ac > 0$ the quadratic equation has two distinct real number solutions.

If $b^2 - 4ac = 0$ the quadratic equation has one real number solution.

If $b^2 - 4ac < 0$ the quadratic equation has no real number solution.

Coordinates of the vertex of a parabola:

$\left(\dfrac{-b}{2a}, \dfrac{4ac - b^2}{4a}\right).$

Review Exercises

[10.1] *Solve by using the square root property.*

1. $x^2 = 25$ **2.** $x^2 = 8$ **3.** $2x^2 = 12$

4. $x^2 + 3 = 9$ **5.** $x^2 - 4 = 16$ **6.** $2x^2 - 4 = 10$

7. $3x^2 + 8 = 32$ **8.** $(x - 3)^2 = 12$ **9.** $(2x + 4)^2 = 30$

10. $(3x - 5)^2 = 50$

[10.2] *Solve by completing the square.*

11. $x^2 - 10x + 16 = 0$ **12.** $x^2 - 8x + 15 = 0$ **13.** $x^2 - 14x + 13 = 0$

14. $x^2 + x - 6 = 0$ **15.** $x^2 - 3x - 54 = 0$ **16.** $x^2 = -5x + 6$

17. $x^2 + 2x - 5 = 0$ **18.** $x^2 - 3x - 8 = 0$ **19.** $2x^2 - 8x = 64$

20. $2x^2 - 4x = 30$ **21.** $4x^2 + 2x - 12 = 0$ **22.** $6x^2 - 19x + 15 = 0$

[10.3] *Determine whether each equation has two distinct real number solutions, one real number solution, or no real number solution.*

23. $3x^2 - 4x - 20 = 0$ **24.** $-3x^2 + 4x = 9$ **25.** $2x^2 + 6x + 7 = 0$

26. $x^2 - x + 8 = 0$ **27.** $x^2 - 12x = -36$ **28.** $3x^2 - 4x + 5 = 0$

29. $-3x^2 - 4x + 8 = 0$ **30.** $x^2 - 9x + 6 = 0$

Solve using the quadratic formula. If an equation has no real number solution, so state.

31. $x^2 - 9x + 14 = 0$ **32.** $x^2 + 7x - 30 = 0$ **33.** $x^2 = 7x - 10$

34. $5x^2 - 7x = 6$ **35.** $x^2 - 18 = 7x$ **36.** $x^2 - x - 30 = 0$

37. $6x^2 + x - 15 = 0$ **38.** $2x^2 + 4x - 3 = 0$ **39.** $-2x^2 + 3x + 6 = 0$

40. $x^2 - 6x + 7 = 0$ **41.** $3x^2 - 4x + 6 = 0$ **42.** $3x^2 - 6x - 8 = 0$

43. $2x^2 + 3x = 0$ **44.** $2x^2 - 5x = 0$

[10.1– 10.3] *Solve each quadratic equation using the method of your choice.*

45. $x^2 - 11x + 24 = 0$ **46.** $x^2 - 16x + 63 = 0$ **47.** $x^2 = -3x + 40$

48. $x^2 + 6x = 27$ **49.** $x^2 - 4x - 60 = 0$ **50.** $x^2 - x - 42 = 0$

51. $x^2 + 11x - 12 = 0$ **52.** $x^2 = 25$ **53.** $x^2 + 6x = 0$

54. $2x^2 + 5x = 3$ **55.** $2x^2 = 9x - 10$ **56.** $6x^2 + 5x = 6$

57. $x^2 + 3x - 6 = 0$ **58.** $3x^2 - 11x + 10 = 0$ **59.** $-3x^2 - 5x + 8 = 0$

60. $-2x^2 + 6x = -9$ **61.** $2x^2 - 5x = 0$ **62.** $3x^2 + 5x = 0$

0.4] *Indicate the axis of symmetry, the coordinates of the vertex, and whether the parabola opens upward or downward.*

63. $y = x^2 - 2x - 3$

64. $y = x^2 - 10x + 24$

65. $y = x^2 + 7x + 12$

66. $y = -x^2 - 2x + 15$

67. $y = x^2 - 3x$

68. $y = 2x^2 + 7x + 3$

69. $y = -x^2 - 8$

70. $y = -4x^2 + 8x + 5$

71. $y = -x^2 - x + 20$

72. $y = 3x^2 + 5x - 8$

Graph each quadratic equation, and determine the x-intercepts if they exist. If they do not exist, so state.

73. $y = x^2 + 6x$

74. $y = -2x^2 + 8$

75. $y = x^2 + 2x - 8$

76. $y = x^2 - x - 2$

77. $y = x^2 + 5x + 4$

78. $y = x^2 + 4x + 3$

79. $y = -2x^2 + 3x - 2$

80. $y = 3x^2 - 4x + 1$

81. $y = 4x^2 - 8x + 6$

82. $y = -3x^2 - 14x + 5$

83. $y = -x^2 - 6x - 4$

84. $y = 2x^2 + 5x - 12$

2–10.3] *Solve.*

85. The product of two positive integers is 88. Find the two integers if the larger one is 3 greater than the smaller.

86. The length of a rectangle is 5 feet less than twice its width. Find the length and width of the rectangle if its area is 63 square feet.

Practice Test

1. Solve the equation $x^2 + 1 = 21$.
2. Solve the equation $(2x - 3)^2 = 35$.
3. Solve by completing the square: $x^2 - 4x = 60$.
4. Solve by completing the square: $x^2 = -x + 12$.
5. Solve by the quadratic formula: $x^2 - 5x - 6 = 0$.
6. Solve by the quadratic formula: $2x^2 + 5 = -8x$.
7. Solve by the method of your choice: $3x^2 - 5x = 0$.
8. Solve by the method of your choice: $2x^2 + 9x = 5$.
9. Determine whether the following equation has two distinct real solutions, one real solution, or no real solution: $3x^2 - 4x + 2 = 0$.
10. Indicate the axis of symmetry, the coordinates of the vertex, and whether the graph opens upward or downward: $y = -x^2 + 3x + 8$.
11. Graph the following equation, and determine the x-intercepts, if they exist: $y = x^2 + 2x - 8$.
12. Graph the following equation, and determine the x-intercepts, if they exist: $y = -x^2 + 6x - 9$.
13. The length of a rectangle is 1 foot greater than three times its width. Find the length and width of the rectangle if its area is 30 square feet.

Cumulative Review Test

1. Evaluate $-x^2y + y^2 - 3xy$ when $x = -3$ and $y = 4$.
2. Solve the equation $\frac{1}{4}x + \frac{3}{5}x = \frac{1}{3}(x + 2)$.
3. Find the length of side x.

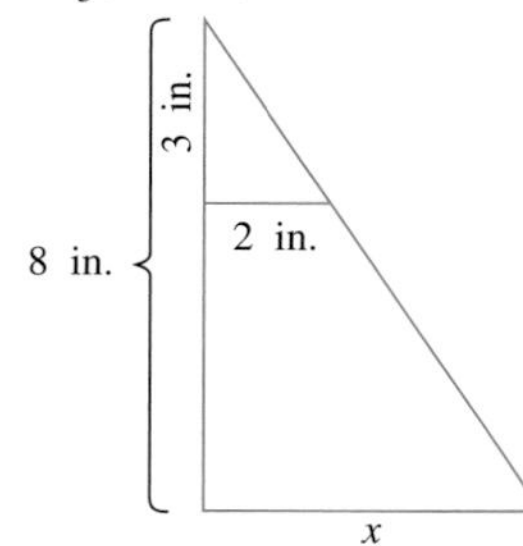

4. Solve the inequality $2(x - 3) \leq 6x - 5$ and graph the solution on the number line.
5. Solve the formula $A = \dfrac{m + n + P}{3}$ for P.
6. Simplify $(6x^2y^4)^3(2x^4y^5)^2$.
7. Divide $\dfrac{x^2 + 6x + 5}{x + 2}$.
8. Factor by grouping $2x^2 - 3xy - 4xy + 6y^2$.
9. Factor $4x^2 - 14x - 8$.
10. Add $\dfrac{4}{a^2 - 16} + \dfrac{2}{(a - 4)^2}$.
11. Solve the equation $x + \dfrac{48}{x} = 14$.
12. Graph the equation $3x + 5y = 10$.
13. Solve the system of equations by the addition method.

$$3x - 4y = 12$$
$$4x - 3y = 6$$

14. Simplify $\sqrt{\dfrac{3x^2y^3}{54x}}$.
15. Add $2\sqrt{28} - 3\sqrt{7} + \sqrt{63}$.
16. Solve the equation $\sqrt{x^2 + 5} = x + 1$.
17. Solve the equation $2x^2 - 4x - 5 = 0$ using the quadratic formula.
18. If 4 pounds of fertilizer can fertilize 500 square feet of lawn, how many pounds of fertilizer are needed to fertilize 3200 square feet of lawn?
19. The length of a rectangle is 3 feet less than three times its width. Find the width and length of the rectangle if its perimeter is 74 feet.
20. Willie jogs 3 miles per hour faster than she walks. She jogs for 2 miles and then walks for 2 miles. If the total time of her outing is 1 hour, find the rate at which she walks and jogs.

Appendices

A. Review of Decimals and Percent

B. Finding the Greatest Common Factor and Least Common Denominator

C. Geometry

Appendix A Review of Decimals and Percent

To Add or Subtract Numbers Containing Decimal Points

1. Align the numbers by the decimal points.
2. Add or subtract the numbers as if they were whole numbers.
3. Place the decimal point in the sum or difference directly below the decimal points in the numbers being added or subtracted.

Example 1 Add 4.6 + 13.813 + 9.02.

Solution:

$$\begin{array}{r} 4.600 \\ 13.813 \\ +\ 9.020 \\ \hline 27.433 \end{array}$$

Example 2 Subtract 3.062 from 25.9.

Solution:

$$\begin{array}{r} 25.900 \\ -\ 3.062 \\ \hline 22.838 \end{array}$$

To Multiply Numbers Containing Decimal Points

1. Multiply as if the factors were whole numbers.
2. Determine the total number of digits to the right of the decimal points in the factors.
3. Place the decimal point in the product so that the product contains the same number of digits to the right of the decimal as the total found in step 2. For example, if there are a total of three digits to the right of the decimal points in the factors, there must be three digits to the right of the decimal point in the product.

EXAMPLE 3 Multiply 2.34×1.9.

Solution:

$$\begin{array}{r l} 2.34 & \longleftarrow \text{two digits to the right of the decimal point} \\ \times\ 1.9 & \longleftarrow \text{one digit to the right of the decimal point} \\ \hline 2106 & \\ 234 & \\ \hline 4.446 & \longleftarrow \text{three digits to the right of the decimal point in the product} \end{array}$$

EXAMPLE 4 Multiply 2.13×0.02.

Solution:

$$\begin{array}{r l} 2.13 & \longleftarrow \text{two digits to the right of the decimal point} \\ \times\ 0.02 & \longleftarrow \text{two digits to the right of the decimal point} \\ \hline 0.0426 & \longleftarrow \text{four digits to the right of the decimal point in the product} \end{array}$$

Note that it was necessary to add a zero preceding the digit 4 in the answer in order to have four digits to the right of the decimal point.

To Divide Numbers Containing Decimal Points

1. Multiply both the dividend and divisor by a power of 10 that will make the divisor a whole number.
2. Divide as if working with whole numbers.
3. Place the decimal point in the quotient directly above the decimal point in the dividend.

To make the divisor a whole number, multiply *both* the dividend and divisor by 10 if the divisor is given in tenths, by 100 if the divisor is given in hundredths, by 1000 if the divisor is given in thousandths, and so on. Multiplying both the numerator and denominator by the same nonzero number is the same as multiplying the fraction by 1. Therefore, the value of the fraction is unchanged.

EXAMPLE 5 Divide $\dfrac{1.956}{0.12}$.

Solution: Since the divisor, 0.12, is twelve-hundredths, we multiply both the divisor and dividend by 100.

$$\frac{1.956}{0.12} \times \frac{100}{100} = \frac{195.6}{12.}$$

Now divide.

$$\begin{array}{r} 16.3 \\ 12.\overline{)195.6} \\ \underline{12} \\ 75 \\ \underline{72} \\ 3\,6 \\ \underline{3\,6} \\ 0 \end{array}$$

The decimal point in the answer is placed directly above the decimal point in the dividend. Thus, $1.956/0.12 = 16.3$.

EXAMPLE 6 Divide 0.26 by 10.4.

Solution: First, multiply both the dividend and divisor by 10.

$$\frac{0.26}{10.4} \times \frac{10}{10} = \frac{2.6}{104.}$$

Now divide.

$$\begin{array}{r} 0.025 \\ 104\overline{)2.600} \\ \underline{2\ 08} \\ 520 \\ \underline{520} \\ 0 \end{array}$$

Note that a zero had to be placed before the digit 2 in the quotient.

$$\frac{0.26}{10.4} = 0.025$$

Percent The word *percent* means "per hundred." The symbol % means percent. One percent means "one per hundred," or

$$1\% = \frac{1}{100} \quad \text{or} \quad 1\% = 0.01$$

EXAMPLE 7 Convert 16% to a decimal.

Solution: Since

$$\begin{aligned} 1\% &= 0.01 \\ 16\% &= 16(0.01) = 0.16 \end{aligned}$$

EXAMPLE 8 Convert 2.3% to a decimal.

Solution: $2.3\% = 2.3(0.01) = 0.023.$

EXAMPLE 9 Convert 1.14 to a percent.

Solution: To change a decimal number to a percent, we multiply the number by 100%.

$$1.14 = 1.14 \times 100\% = 114\%$$

Often you will need to find an amount that is a certain percent of a number. For example, when you purchase an item in a state or county that has a sales tax you must often pay a percent of the item's price as the sales tax. Examples 10 and 11 show how to find a certain percent of a number.

EXAMPLE 10 Find 12% of 200.

Solution: To find a percent of a number, use multiplication. Change 12% to a decimal number, then multiply by 200.

$$(0.12)(200) = 24$$

Thus, 12% of 200 is 24.

EXAMPLE 11 Monroe County in New York State charges an 8% sales tax.

(a) Find the sales tax on a stereo system that cost \$580.
(b) Find the total cost of the system, including tax.

Solution: **(a)** The sales tax is 8% of 580.

$$(0.08)(580) = 46.40$$

The sales tax is \$46.40
(b) The total cost is the purchase price plus the sales tax:

$$\text{Total cost} = \$580 + \$46.40 = \$626.40$$

Appendix B Finding the Greatest Common Factor and Least Common Denominator

Prime Factorization

In Section 1.2 we mentioned that to reduce fractions to their lowest terms you can divide both the numerator and denominator by the *greatest common factor* (GCF). One method to find the GCF is to use *prime factorization.* Prime factorization is the process of writing a given number as a product of prime numbers. *Prime numbers* are natural numbers, excluding 1, that can be divided by only themselves and 1. The first ten prime numbers are 2, 3, 5, 7, 11, 13, 17, 19, 23, and 29. Can you find the next prime number? If you answered 31, you answered correctly.

To write a number as a product of primes, we can use a *tree diagram.* Begin by selecting any two numbers whose product is the given number. Then continue factoring each of these numbers into prime numbers, as shown in Example 1.

EXAMPLE 1 Determine the prime factorization of the number 120.

Solution: We will use three different tree diagrams to illustrate the prime factorization of 120.

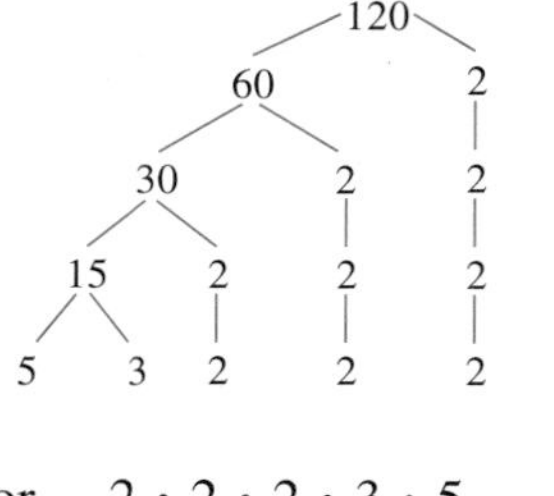

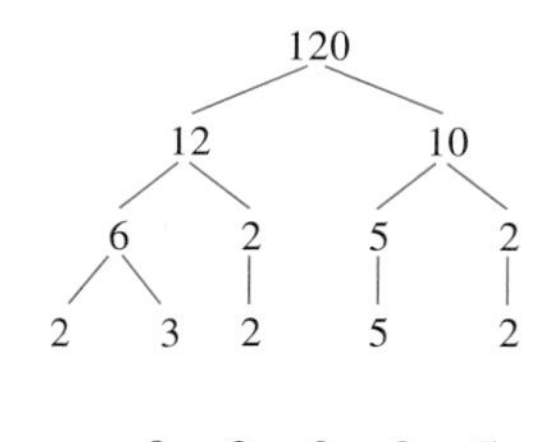

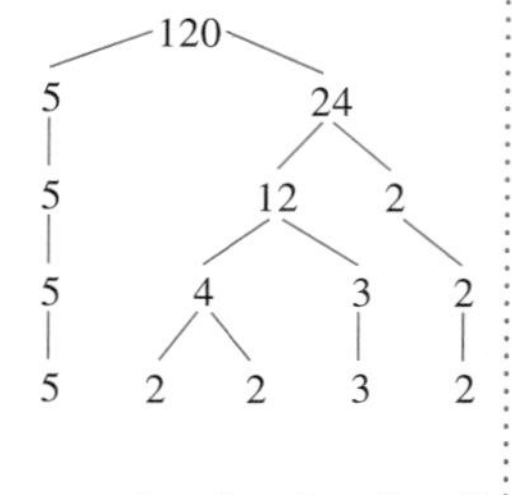

or $2 \cdot 2 \cdot 2 \cdot 3 \cdot 5$ or $2 \cdot 2 \cdot 2 \cdot 3 \cdot 5$ or $2 \cdot 2 \cdot 2 \cdot 3 \cdot 5$

Note that no matter how you start, if you do not make a mistake, you find that the prime factorization of 120 is $2 \cdot 2 \cdot 2 \cdot 3 \cdot 5$. There are other ways 120 can be factored but all will lead to the prime factorization $2 \cdot 2 \cdot 2 \cdot 3 \cdot 5$.

Greatest Common Factor

The greatest common factor (GCF) of two natural numbers is the greatest integer that is a factor of both numbers. We use the GCF when reducing fractions to lowest terms.

To Find the Greatest Common Factor of a Given Numerator and Denominator

1. Write both the numerator and the denominator as a product of primes.
2. Determine all the prime factors that are common to both prime factorizations.
3. Multiply the prime factors found in step 2 to obtain the GCF.

EXAMPLE 2 Consider the fraction $\frac{108}{156}$.

(a) Find the GCF of 108 and 156.

(b) Reduce $\frac{108}{156}$ to lowest terms.

Solution: **(a)** First determine the prime factorizations of both 108 and 156.

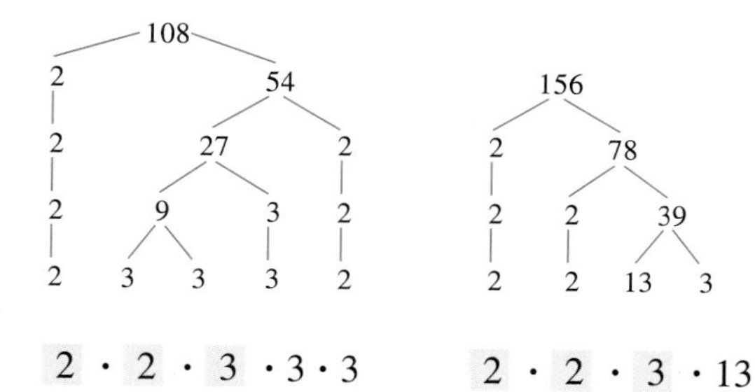

There are two 2's and one 3 common to both prime factorizations; thus

$$\text{GCF} = 2 \cdot 2 \cdot 3 = 12$$

The greatest common factor of 108 and 156 is 12. Twelve is the greatest integer that divides into both 108 and 156.

(b) To reduce $\frac{108}{156}$, we divide both the numerator and denominator by the GCF, 12.

$$\frac{108 \div 12}{156 \div 12} = \frac{9}{13}$$

Thus, $\frac{108}{156}$ reduces to $\frac{9}{13}$.

Least Common Denominator When adding two or more fractions, you must write each fraction with a common denominator. The best denominator to use is the *least common denominator* (LCD). The LCD is the smallest number that each denominator divides into. Sometimes the least common denominator is referred to as the *least common multiple* of the denominators.

To Find the Least Common Denominator of Two or More Fractions

1. Write each denominator as a product of prime numbers.
2. For each prime number, determine the maximum number of times that prime number appears in any of the prime factorizations.
3. Multiply all the prime numbers, found in Step 2. Include each prime number the maximum number of times it appears in any of the prime factorizations. The product of all these prime numbers will be the LCD.

Example 3 illustrates the procedure to determine the LCD.

EXAMPLE 3 Consider $\frac{7}{108} + \frac{5}{156}$.

(a) Determine the least common denominator.

(b) Add the fractions.

Solution: **(a)** We found in Example 2 that

$$108 = 2 \cdot 2 \cdot 3 \cdot 3 \cdot 3 \quad \text{and} \quad 156 = 2 \cdot 2 \cdot 3 \cdot 13$$

We can see that the maximum number of 2's that appear in either prime factorization is two (there are two 2's in both factorizations), the maximum number of 3's is three, and the maximum number of 13's is one. Multiply as follows:

$$2 \cdot 2 \cdot 3 \cdot 3 \cdot 3 \cdot 13 = 1404$$

Thus, the least common denominator is 1404. This is the smallest number that both 108 and 156 divide into.

(b) To add the fractions, we need to write both fractions with a common denominator. The best common denominator to use is the LCD. Since $1404 \div 108 = 13$, we will multiply $\frac{7}{108}$ by $\frac{13}{13}$. Since $1404 \div 156 = 9$, we will multiply $\frac{5}{156}$ by $\frac{9}{9}$.

$$\frac{7}{108} \cdot \frac{13}{13} + \frac{5}{156} \cdot \frac{9}{9} = \frac{91}{1404} + \frac{45}{1404} = \frac{136}{1404} = \frac{34}{351}$$

Thus, $\frac{7}{108} + \frac{5}{156} = \frac{34}{351}$.

Appendix C Geometry

This appendix introduces or reviews important geometric concepts. Table C.1 gives the names and descriptions of various types of angles.

Angles

TABLE C.1

Angle	Sketch of Angle
An **acute angle** is an angle whose measure is between 0° and 90°.	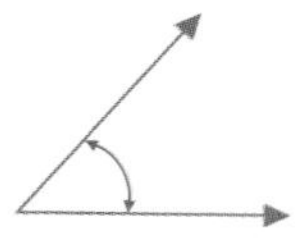
A **right angle** is an angle whose measure is 90°.	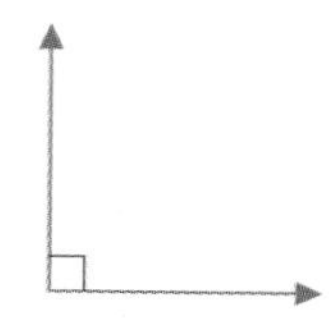
An **obtuse angle** is an angle whose measure is between 90° and 180°.	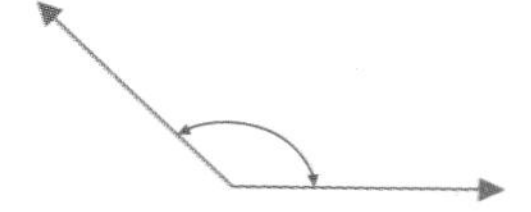
A **straight angle** is an angle whose measure is 180°.	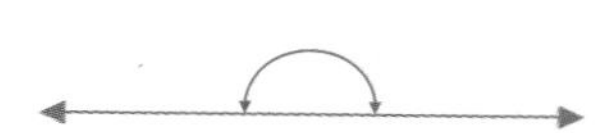
Two angles are **complementary angles** when the sum of their measures is 90°. Each angle is the complement of the other. Angles A and B are complementary angles.	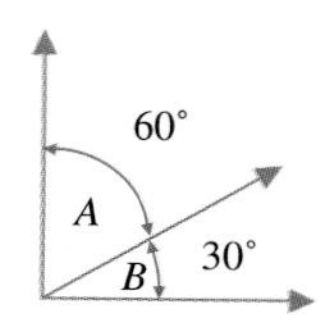
Two angles are **supplementary angles** when the sum of their measures is 180°. Each angle is the supplement of the other. Angles A and B are supplementary angles.	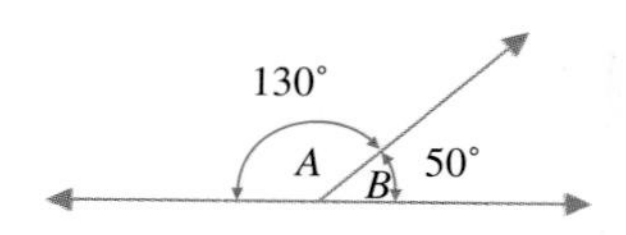

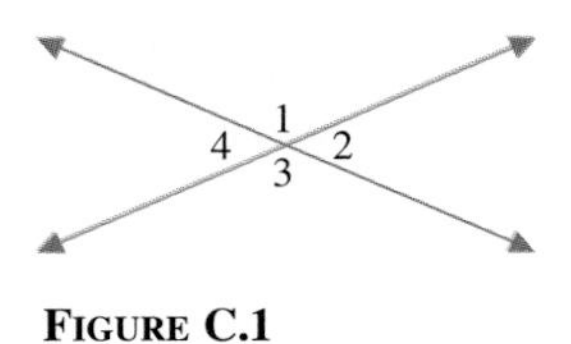

FIGURE C.1

When two lines intersect, four angles are formed as shown in Figure C.1. The pair of opposite angles formed by the intersecting lines are called **vertical angles.**

Angles 1 and 3 are vertical angles. Angles 2 and 4 are also vertical angles. *Vertical angles have equal measures.* Thus, angle 1, symbolized by $\angle 1$, is equal to angle 3, symbolized by $\angle 3$. We can write $\angle 1 = \angle 3$. Similarly, $\angle 2 = \angle 4$.

Parallel and Perpendicular Lines

Parallel lines are two lines in the same plane that do not intersect (Fig. C2). **Perpendicular lines** are lines that intersect at right angles (Fig. C3).

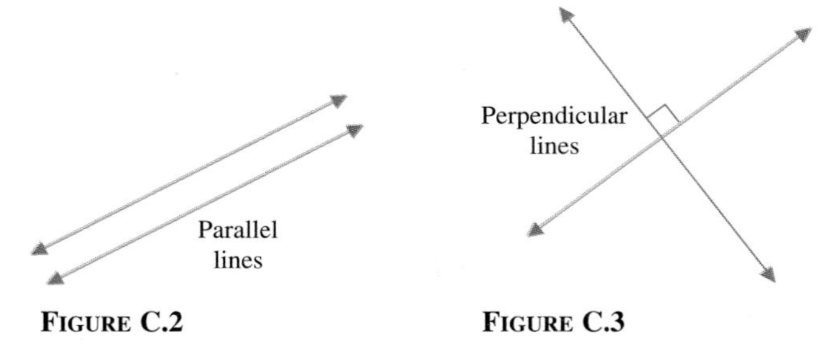

FIGURE C.2 **FIGURE C.3**

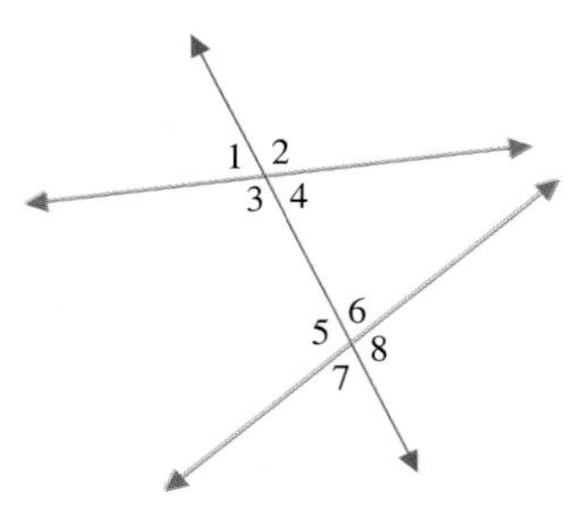

FIGURE C.4

A **transversal** is a line that intersects two or more lines at different points. When a transversal line intersects two other lines, eight angles are formed, as illustrated in Figure C.4. Some of these angles are given special names.

Interior angles: 3, 4, 5, 6

Exterior angles: 1, 2, 7, 8

Pairs of corresponding angles: 1 and 5; 2 and 6; 3 and 7; 4 and 8

Pairs of alternate interior angles: 3 and 6; 4 and 5

Pairs of alternate exterior angles: 1 and 8; 2 and 7

Parallel Lines Cut by a Transversal

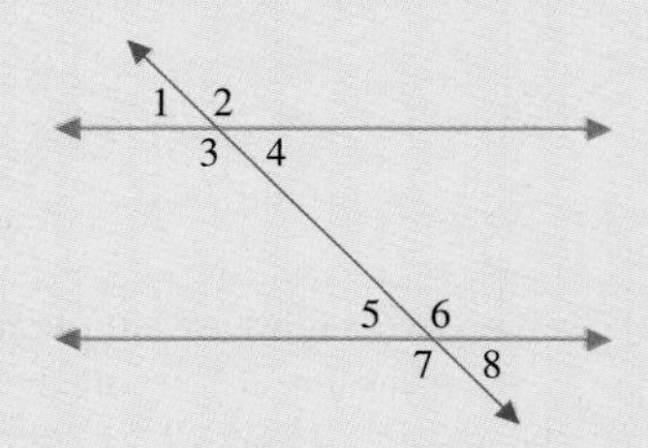

When two parallel lines are cut by a transversal:

1. Corresponding angles are equal ($\angle 1 = \angle 5$, $\angle 2 = \angle 6$, $\angle 3 = \angle 7$, $\angle 4 = \angle 8$).
2. Alternate interior angles are equal ($\angle 3 = \angle 6$, $\angle 4 = \angle 5$).
3. Alternate exterior angles are equal ($\angle 1 = \angle 8$, $\angle 2 = \angle 7$).

EXAMPLE 1 If line 1 and line 2 are parallel lines and $\angle 1 = 112°$, find the measure of angles 2 through 8.

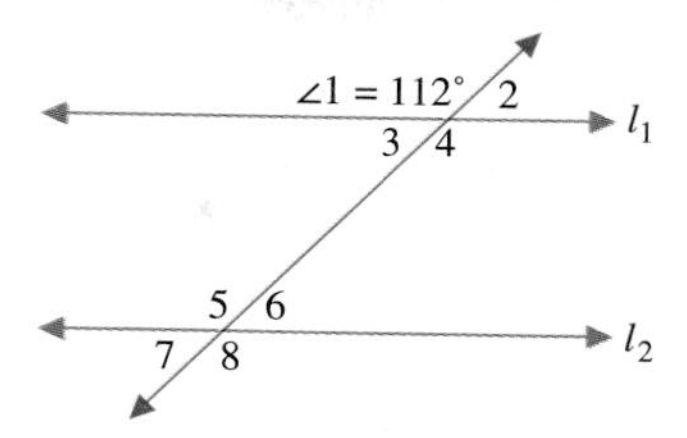

Solution: Angles 1 and 2 are supplementary. So $\angle 2$ is $180° - 112° = 68°$. Angles 1 and 4 are equal since they are vertical angles. Thus, $\angle 4 = 112°$. Angles 1 and 5 are corresponding angles. Thus, $\angle 5 = 112°$. It is equal to its verticle angle, $\angle 8$, so $\angle 8 = 112°$. Angles 2, 3, 6, and 7 are all equal and measure 68°.

Polygons A **polygon** is a closed figure in a plane determined by three or more line segments. Some polygons are illustrated in Figure C.5.

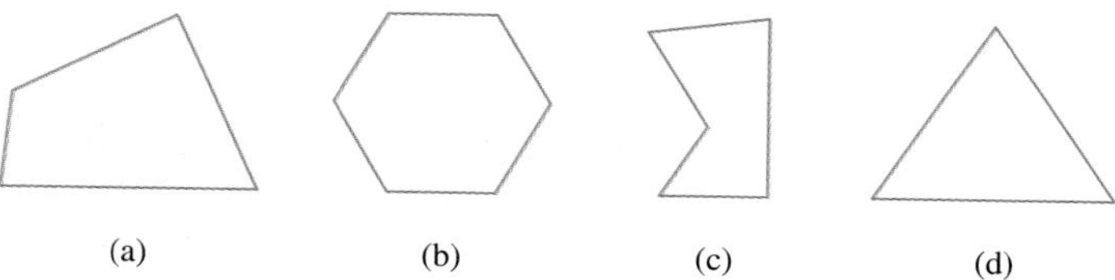

FIGURE C.5

A **regular polygon** has sides that are all the same length, and interior angles that all have the same measure. Figures C.5(b) and (d) are regular polygons.

Sum of the Interior Angles of a Polygon

The sum of the interior angles of a polygon can be found by the formula

$$\text{Sum} = (n - 2)180°$$

where n is the number of sides of the polygon.

EXAMPLE 2 Find the sum of the measures of the interior angles of **(a)** a triangle; **(b)** a quadrilateral (4 sides), **(c)** an octagon (8 sides).

Solution: **(a)** Since $n = 3$, we write

$$\begin{aligned}\text{Sum} &= (n - 2)180° \\ &= (3 - 2)180° = 1(180°) = 180°\end{aligned}$$

The sum of the measures of the interior angles in a triangle is 180°.

(b) $$\begin{aligned}\text{Sum} &= (n - 2)180° \\ &= (4 - 2)180° = 2(180°) = 360°\end{aligned}$$

The sum of the measures of the interior angles in a quadrilateral is 360°.

(c) $\text{Sum} = (n - 2)(180°) = (8 - 2)180° = 6(180°) = 1080°$

The sum of the measures of the interior angles in an octagon is 1080°.

Now we will briefly define several types of triangles in Table C.2.

Triangles

TABLE C.2

Triangle	Sketch of Triangle
An **acute triangle** is one that has three acute angles (angles of less than 90°).	
An **obtuse triangle** has one obtuse angle (an angle greater than 90°).	
A **right triangle** has one right angle (an angle equal to 90°). The longest side of a right triangle is opposite the right angle and is called the **hypotenuse.** The other two sides are called the **legs.**	
An **isosceles triangle** has two sides of equal length. The angles opposite the equal sides have the same measure.	
An **equilateral triangle** has three sides of equal length. It also has three equal angles that measure 60° each.	60° 60° 60°

When two sides of a *right triangle* are known, the third side can be found using the **Pythagorean theorem,** $a^2 + b^2 = c^2$, where a and b are the legs and c is the hypotenuse of the triangle. (See Chapter 9 for examples.)

Congruent and Similar Figures

If two triangles are **congruent,** it means that the two triangles are identical in size and shape. Two congruent triangles could be placed one on top of the other if we were able to move and rearrange them.

Two triangles are congruent if any one of the following statements is true.

1. Two angles of one triangle are equal to two corresponding angles of the other triangle, and the lengths of the sides between each pair of angles are equal. This method of showing that triangles are congruent is called the *angle, side, angle* method.

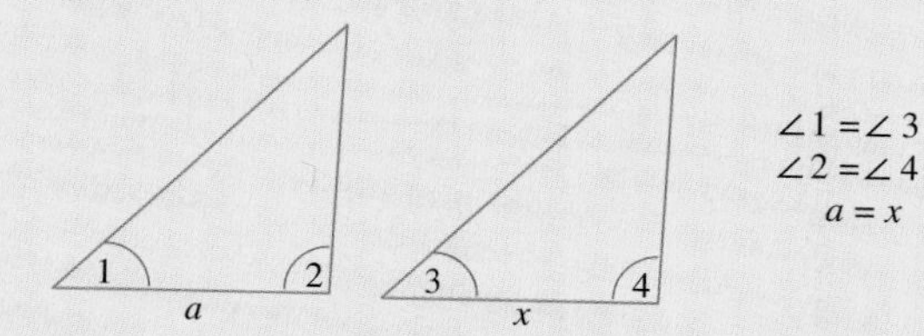

2. Corresponding sides of both triangles are equal. This is called the *side, side, side* method.

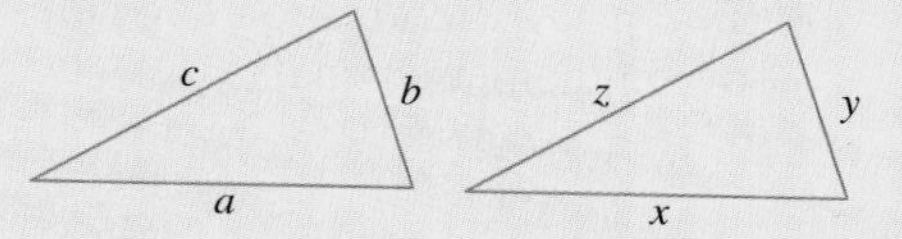

3. Two corresponding pairs of sides are equal, and the angle between them is equal. This is referred to as the *side, angle, side* method.

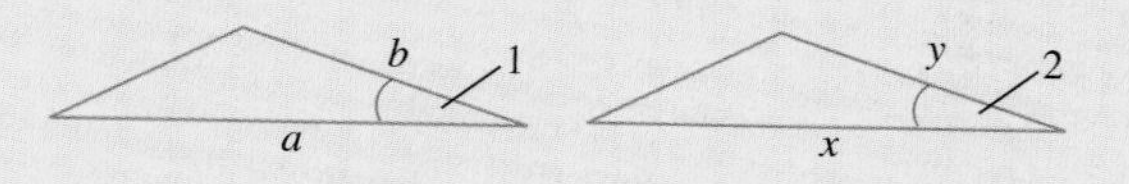

EXAMPLE 3 Determine if the two triangles are congruent.

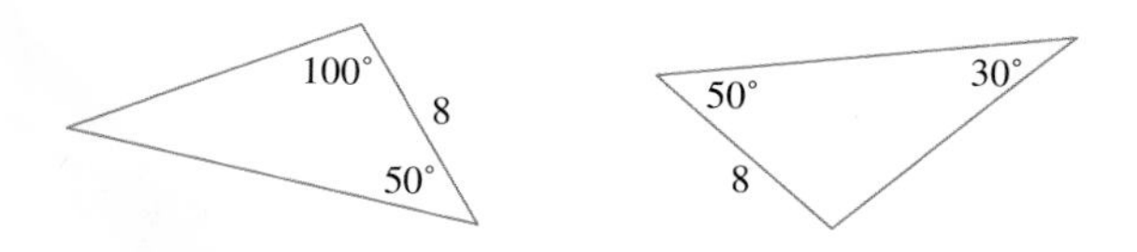

Solution: The unknown angle in the figure on the right must measure 100° since the sum of the angles of a triangle is 180°. Both triangles have the same two angles (100° and 50°), with the same length side between them, 8 units. Thus, these two triangles are congruent by the angle, side, angle method.

Two triangles are **similar** if all three pairs of corresponding angles are equal and corresponding sides are in proportion. Similar figures do not have to be the same size but must have the same general shape.

Two triangles are similar if any one of the following statements is true.

1. Two angles of one triangle equal two angles of the other triangle.

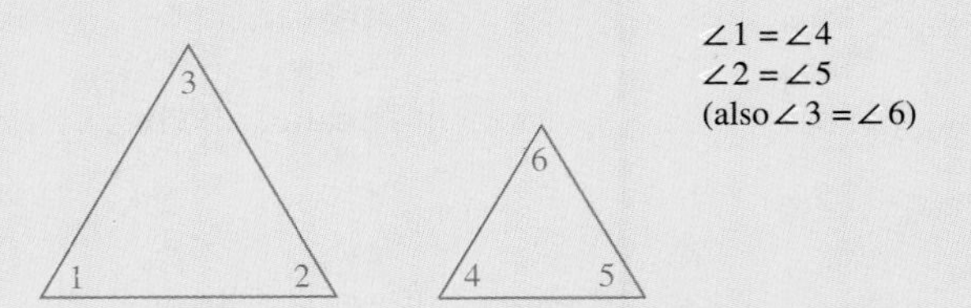

2. Corresponding sides of the two triangles are proportional.

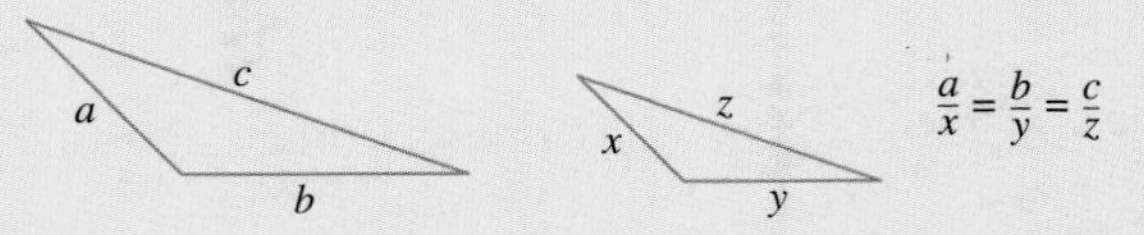

3. Two pairs of corresponding sides are proportional, and the angles between them are equal.

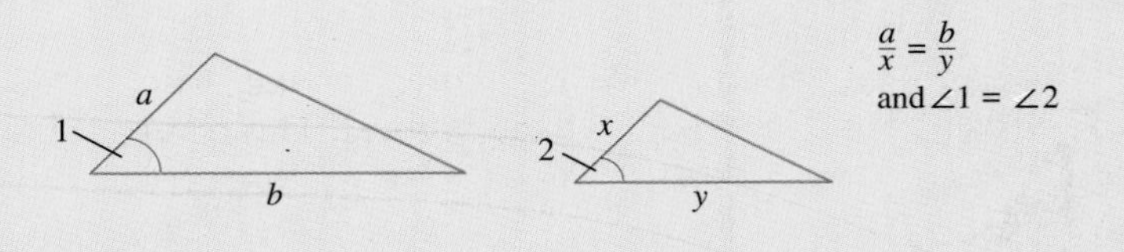

EXAMPLE 4 Are the triangles ABC and $AB'C'$ similar?

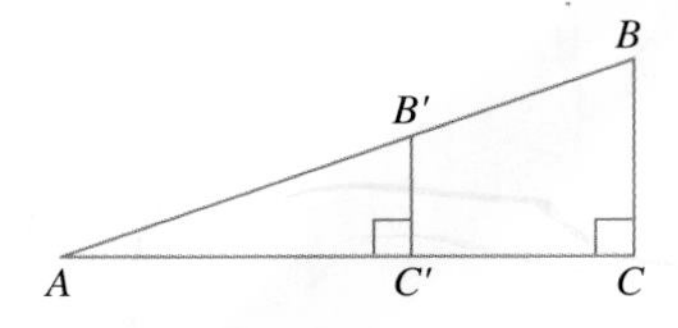

Solution: Angle A is common to both triangles. Since angle C and angle C' are equal (both 90°), then $\angle B$ and $\angle B'$ must be equal. Since the three angles of triangle ABC equal the three angles of triangle $AB'C'$, the two triangles are similar.

Answers

Chapter 1

Exercise Set 1.2

1. $\frac{1}{4}$ **3.** $\frac{2}{3}$ **5.** $\frac{1}{2}$ **7.** $\frac{3}{7}$ **9.** $\frac{5}{8}$ **11.** lowest terms **13.** $\frac{4}{3}$ or $1\frac{1}{3}$ **15.** $\frac{10}{7}$ or $1\frac{3}{7}$ **17.** (a) **19.** (b) **21.** $\frac{3}{8}$ **23.** $\frac{5}{14}$ **25.** $\frac{1}{12}$ **27.** $\frac{5}{4}$ or $1\frac{1}{4}$ **29.** $\frac{5}{16}$ **31.** 6 **33.** $\frac{4}{9}$ **35.** $\frac{16}{65}$ **37.** $\frac{19}{14}$ or $1\frac{5}{14}$ **39.** 12 **41.** 1 **43.** 2 **45.** $\frac{5}{7}$ **47.** $\frac{1}{6}$ **49.** 1 **51.** $\frac{7}{13}$ **53.** $\frac{37}{30}$ or $1\frac{7}{30}$ **55.** $\frac{1}{5}$ **57.** $\frac{4}{15}$ **59.** $\frac{3}{56}$ **61.** $\frac{29}{24}$ or $1\frac{5}{24}$ **63.** $\frac{9}{28}$ **65.** $\frac{93}{20}$ or $4\frac{13}{20}$ **67.** $\frac{23}{6}$ or $3\frac{5}{6}$ **69.** $\frac{52}{15}$ or $3\frac{7}{15}$ **71.** $\frac{81}{20}$ or $4\frac{1}{20}$ **73.** $\frac{5}{6}$ **75.** $\frac{11}{12}$ **77.** $20\frac{1}{4}$ yd **79.** $13\frac{11}{16}$ in **81.** $8\frac{7}{16}$ ft **83.** $11\frac{7}{8}$ ft **85.** 297 min or 4 hr 57 min **87.** $\frac{25}{16}$ or $1\frac{9}{16}$ in **89.** 5 mg **91. (a)** yes **(b)** $\frac{5}{8}$ in **93. (a)** $\frac{5}{16}$ **(b)** $\frac{20}{9}$ or $2\frac{2}{9}$ **(c)** $\frac{29}{24}$ or $1\frac{5}{24}$ **(d)** $\frac{11}{24}$ **95.** A general term for any collection of numbers, variables, grouping symbols, and operations. **97.** (b). In part (a) we have divided out common factors from *two* fractions.

Group Activity/Challenge Problems

1. (b) Rice and water 1 cup, salt $\frac{3}{8}$ tsp, butter $1\frac{1}{2}$ tsp **3.** 3 **5.** $\frac{427}{90}$ or $4\frac{67}{90}$

Exercise Set 1.3

1. $\{\ldots, -3, -2, -1, 0, 1, 2, 3, \ldots\}$ **3.** $\{1, 2, 3, 4, \ldots\}$ **5.** $\{1, 2, 3, 4, \ldots\}$ **7.** T **9.** T **11.** F **13.** F **15.** T **17.** T **19.** F **21.** F **23.** T **25.** T **27.** T **29.** F **31.** T **33.** T **35.** T **37.** F **39. (a)** 7, 9 **(b)** 7, 0, 9 **(c)** -6, 7, 0, 9 **(d)** -6, 7, 12.4, $-\frac{9}{5}$, $-2\frac{1}{4}$, 0, 9, 0.35 **(e)** $\sqrt{3}$, $\sqrt{7}$ **(f)** -6, 7, 12.4, $-\frac{9}{5}$, $-2\frac{1}{4}$, $\sqrt{3}$, 0, 9, $\sqrt{7}$, 0.35 **41. (a)** 5 **(b)** 5 **(c)** -300 **(d)** 5, -300 **(e)** $\frac{1}{2}$, $4\frac{1}{2}$, $\frac{5}{12}$, -1.67, 5, -300, $-9\frac{1}{2}$ **(f)** $\sqrt{2}$, $-\sqrt{2}$ **(g)** $\frac{1}{2}$, $\sqrt{2}$, $-\sqrt{2}$, $4\frac{1}{2}$, $\frac{5}{12}$, -1.67, 5, -300, $-9\frac{1}{2}$ **43.** $-\frac{2}{3}$, $\frac{1}{2}$, 6.3 **45.** $-\sqrt{7}$, $\sqrt{3}$, $\sqrt{6}$ **47.** -5, 0, 4 **49.** -13, -5, -1 **51.** 1.5, 3, $6\frac{1}{4}$ **53.** -7, 1, 5

Cumulative Review Exercises

56. $\frac{14}{3}$ **57.** $5\frac{1}{3}$ **58.** $\frac{49}{40}$ or $1\frac{9}{40}$ **59.** $\frac{70}{27}$ or $2\frac{16}{27}$

Group Activity/Challenge Problems

1. (a) no **(b)** Infinite set **3. (a)** and **(b)** an infinite number **5.** $A \cup B = \{a, b, c, d, g, h, i, j, m, p\}$, $A \cap B = \{b, c, d\}$

Exercise Set 1.4

1. 4 **3.** 15 **5.** 0 **7.** -8 **9.** -65 **11.** $2 < 3$ **13.** $-3 < 0$ **15.** $\frac{1}{2} > -\frac{2}{3}$ **17.** $0.2 < 0.4$ **19.** $-\frac{1}{2} > -1$ **21.** $4 > -4$ **23.** $-2.1 < -2$ **25.** $\frac{5}{9} > -\frac{5}{9}$ **27.** $-\frac{3}{2} < \frac{3}{2}$ **29.** $0.49 > 0.43$ **31.** $5 > -7$ **33.** $-0.006 > -0.007$

35. $-5 < -2$ **37.** $-\frac{2}{3} > -3$ **39.** $-\frac{1}{2} > -\frac{3}{2}$ **41.** $8 > |-7|$ **43.** $|0| < \frac{2}{3}$ **45.** $|-3| < |-4|$ **47.** $4 < |-\frac{9}{2}|$
49. $|-\frac{6}{2}| > |-\frac{2}{6}|$ **51.** $=$ **53.** $<$ **55.** $<$ **57.** 4, −4 **59.** 2, −2
61. The distance between 0 and the number on the number line.

Cumulative Review Exercises

62. $\frac{31}{24}$ or $1\frac{7}{24}$ **63.** {0, 1, 2, 3, . . .} **64.** {1, 2, 3, 4, . . .} **65. (a)** 5 **(b)** 5, 0 **(c)** 5, −2, 0 **(d)** 5, −2, 0, $\frac{1}{3}$, $-\frac{5}{9}$, 2.3
(e) $\sqrt{3}$ **(f)** 5, −2, 0, $\frac{1}{3}$, $\sqrt{3}$, $-\frac{5}{9}$, 2.3

Group Activity/Challenge Problems

1. less than **3.** 3, −3 **5. (a)** x **(b)** $-x$ **(c)** $|x| = \begin{cases} x, & x \geq 0 \\ -x, & x < 0 \end{cases}$

Exercise Set 1.5

1. −12 **3.** 40 **5.** 0 **7.** $-\frac{5}{3}$ **9.** $-\frac{3}{5}$ **11.** −0.63 **13.** $-3\frac{1}{5}$ **15.** 3.1 **17.** 7 **19.** 1 **21.** −6 **23.** 0 **25.** 0
27. −10 **29.** 0 **31.** −10 **33.** 0 **35.** −6 **37.** 3 **39.** −9 **41.** 9 **43.** −27 **45.** −44 **47.** −26 **49.** 5
51. −20 **53.** −31 **55.** 91 **57.** −140 **59.** −98 **61.** 266 **63.** −373 **65.** −452 **67.** −1300 **69.** −22
71. −3880 **73.** −1267 **75.** −2050 **77.** −14,559 **79.** −1215 **81.** −7494 **83.** 7458 **85.** True **87.** False
89. False **91.** \$174 **93.** \$1927 **95.** 54 ft **97.** increased by 92 million people

Cumulative Review Exercises

101. 1 **102.** $\frac{43}{16}$ or $2\frac{11}{16}$ **103.** $|-3| > 2$ **104.** $8 > |-7|$

Group Activity/Challenge Problems

1. −22 **2.** −10 **3.** 20 **4.** 55 **5.** 210 **6.** 5050

Exercise Set 1.6

1. 3 **3.** −1 **5.** 0 **7.** −3 **9.** −6 **11.** 6 **13.** −6 **15.** 6 **17.** −8 **19.** 2 **21.** 2 **23.** 9 **25.** 0
27. −18 **29.** −2 **31.** 0 **33.** 0 **35.** 0 **37.** −1 **39.** −5 **41.** −41 **43.** −3 **45.** −180 **47.** −110
49. −10 **51.** 220 **53.** 0 **55.** −46 **57.** −18 **59.** −16 **61.** 10 **63.** −2 **65.** 13 **67.** 0 **69.** −4
71. 11 **73.** 81 **75.** 99 **77.** −595 **79.** 847 **81.** 1712 **83.** 196 **85.** −448 **87.** 116.1 **89.** 0 **91.** −69
93. −1670 **95.** 97.32 **97.** 7 **99.** −2 **101.** −15 **103.** −2 **105.** 0 **107.** 43 **109.** −6 **111.** −9 **113.** 35
115. −21 **117.** −12 **119.** −3 **121.** 18 **123.** 12 **125.** −22 **127.** 326 boxes **129.** \$1246 **131.** 100°
133. (b) 8 **135. (a)** 148 mi **(b)** 12 mi

Cumulative Review Exercises

137. {. . . , −3, −2, −1, 0, 1, 2, 3, . . .} **138.** The set of rational numbers together with the set of irrational numbers forms the set of real numbers. **139.** $|-3| > -5$ **140.** $|-6| < |-7|$

Group Activity/Challenge Problems

1. −5 **3.** 50 **4. (a)** 7 units **(b)** 5 − (−2) **5.** 253 ft. **7.** 0

Exercise Set 1.7

1. 12 **3.** −9 **5.** −32 **7.** −9 **9.** 12 **11.** 36 **13.** 96 **15.** 81 **17.** −10 **19.** 36 **21.** 0 **23.** −1
25. −120 **27.** 84 **29.** −90 **31.** 360 **33.** $-\frac{3}{10}$ **35.** $\frac{14}{27}$ **37.** 4 **39.** $-\frac{15}{28}$ **41.** 3 **43.** 4 **45.** 4 **47.** −4
49. −18 **51.** 5 **53.** 6 **55.** 5 **57.** −1 **59.** −4 **61.** 0 **63.** 16 **65.** 0 **67.** −3 **69.** −6 **71.** 5 **73.** $-\frac{3}{4}$

75. $\frac{3}{80}$ **77.** 1 **79.** $-\frac{144}{5}$ **81.** -30 **83.** 9 **85.** 5 **87.** 60 **89.** 0 **91.** -45 **93.** -20 **95.** -1 **97.** 0
99. undefined **101.** 0 **103.** 0 **105.** 0 **107.** undefined **109.** -1440 **111.** -4 **113.** -16 **115.** -9
117. 2550 **119.** 17,052 **121.** 0 **123.** -199.5 **125.** undefined **127.** 0 **129.** -172.8 **131.** 7027.2
133. True **135.** True **137.** True **139.** False **141.** False **143.** True **145.** True
147. Each pair of negative numbers has a positive product.

Cumulative Review Exercises

149. $\frac{25}{7}$ or $3\frac{4}{7}$ **150.** -2 **151.** -3 **152.** 3

Group Activity/Challenge Problems

1. 81 **3.** $\frac{8}{27}$ **9.** 1

Exercise Set 1.8

1. 25 **3.** 8 **5.** 27 **7.** 216 **9.** -8 **11.** -1 **13.** 27 **15.** -36 **17.** 36 **19.** 16 **21.** 4 **23.** 16 **25.** -16
27. -64 **29.** 225 **31.** 80 **33.** 32 **35.** -75 **37.** x^2y^2 **39.** xy^3z **41.** y^2z^3 **43.** x^2y^2z **45.** a^2x^2y **47.** xy^2z^3
49. $3xy^2$ **51.** xxy **53.** $xyyy$ **55.** $xyyzzz$ **57.** $3 \cdot 3yz$ **59.** $2 \cdot 2 \cdot 2xxxy$ **61.** $(-2)(-2)yyyz$ **63.** 9, -9
65. 16, -16 **67.** 4, -4 **69.** 49, -49 **71.** 1, -1 **73.** $\frac{1}{4}$, $-\frac{1}{4}$ Calculator answers sometimes vary in the last digit displayed and some calculators do not display a 0 before a decimal number between -1 and 1. **75.** 243 **77.** -8 **79.** -32
81. $-15{,}625$ **83.** 1296 **85.** 592.704 **87.** -12.167 **89.** 0.0625 **91.** 0.0256 **93.** 0.197530864 **95.** False
97. False **99.** True **101.** True **103.** True **105.** Any nonzero number will be positive when squared.
107. Positive; an even number of negative numbers are being multiplied. **109.** Any real number except 0 raised to the zero power equals 1.

Cumulative Review Exercises

110. 18 **111.** -5 **112.** 10,364 **113.** $\frac{10}{3}$ or $3\frac{1}{3}$ **114.** 0

Group Activity/Challenge Problems

1. (a) 2^5 **(b)** 3^5 **(c)** 2^7 **(d)** x^{m+n} **3. (a)** 2^6 **(b)** 3^6 **(c)** 4^4 **(d)** x^{mn}

Exercise Set 1.9

1. 23 **3.** 8 **5.** 13 **7.** -10 **9.** 16 **11.** 29 **13.** -13 **15.** -2 **17.** 12 **19.** 10 **21.** 7 **23.** 36 **25.** 121
27. $\frac{1}{2}$ **29.** -5 **31.** 12 **33.** 9 **35.** 169 **37.** 156 **39.** 25 **41.** 129.81 **43.** 26.04 **45.** $\frac{71}{112}$ **47.** $\frac{1}{4}$ **49.** $\frac{170}{9}$
51. $[(6 \cdot 3) - 4] - 2$, 12 **53.** $9[[(20 \div 5) + 12] - 8]$, 72 **55.** $(\frac{4}{5} + \frac{3}{7}) \cdot \frac{2}{3}$, $\frac{86}{105}$ **57.** 2 **59.** 10 **61.** 3 **63.** -7
65. -25 **67.** 75 **69.** -20 **71.** 0 **73.** -5 **75.** 21 **77.** 33 **79.** -18 **81.** -3 **83.** 49 **85.** 5 **87.** 28
89. 4 **91.** -8 **93.** 20 **95.** -47 **97.** -50 **99.** 38 **103. (b)** -91

Cumulative Review Exercises

104. 144 **105. (a)** 25 **(b)** -25 **106.** 16 **107.** -16

Group Activity/Challenge Problems

1. 160 **2.** 177 **3.** -312 **5.** $12 - (4 - 6) + 10$ **7.** $30 + (15 \div 5) + 10 \div 2$

Exercise Set 1.10

1. Distributive property **3.** Commutative property of multiplication **5.** Distributive property **7.** Associative property of multiplication **9.** Distributive property **11.** $4 + 3$ **13.** $(-6 \cdot 4) \cdot 2$ **15.** $(y)(6)$ **17.** $1 \cdot x + 1 \cdot y$ or $x + y$

19. $3y + 4x$ **21.** $5(x + y)$ **23.** $3(x + 2)$ **25.** $3x + (4 + 6)$ **27.** $(x + y)3$ **29.** $4x + 4y + 12$ **31.** Commutative property of addition **33.** Distributive property **35.** Commutative property of addition **37.** Distributive property **39.** yes **41.** no

Cumulative Review Exercises

44. $\frac{49}{15}$ or $3\frac{4}{15}$ **45.** $\frac{23}{16}$ or $1\frac{7}{16}$ **46.** 45 **47.** -25

Group Activity/Challenge Problems

1. Commutative property of addition

Chapter 1 Review Exercises

1. $\frac{1}{2}$ **2.** $\frac{9}{25}$ **3.** $\frac{25}{36}$ **4.** $\frac{7}{6}$ or $1\frac{1}{6}$ **5.** $\frac{19}{72}$ **6.** $\frac{17}{15}$ or $1\frac{2}{15}$ **7.** $\{1, 2, 3, \ldots\}$ **8.** $\{0, 1, 2, 3, \ldots\}$ **9.** $\{\ldots, -3, -2, -1, 0, 1, 2, 3, \ldots\}$ **10.** {quotient of two integers, denominator not 0} **11.** {all numbers that can be represented on the real number line} **12. (a)** 3, 426 **(b)** 3, 0, 426 **(c)** $3, -5, -12, 0, 426$ **(d)** $3, -5, -12, 0, \frac{1}{2}, -0.62, 426, -3\frac{1}{4}$ **(e)** $\sqrt{7}$ **(f)** $3, -5, -12, 0, \frac{1}{2}, -0.62, \sqrt{7}, 426, -3\frac{1}{4}$ **13. (a)** 1 **(b)** 1 **(c)** $-8, -9,$ **(d)** $-8, -9, 1$ **(e)** $-2.3, -8, -9, 1\frac{1}{2}, 1, -\frac{3}{17}$ **(f)** $-2.3, -8, -9, 1\frac{1}{2}, \sqrt{2}, -\sqrt{2}, 1, -\frac{3}{17}$ **14.** $>$ **15.** $<$ **16.** $>$ **17.** $>$ **18.** $<$ **19.** $>$ **20.** $<$ **21.** $>$ **22.** $<$ **23.** $=$ **24.** 3 **25.** -9 **26.** 0 **27.** -5 **28.** -3 **29.** -6 **30.** -6 **31.** -5 **32.** 8 **33.** -2 **34.** -9 **35.** -10 **36.** 5 **37.** -5 **38.** 4 **39.** -12 **40.** 5 **41.** -4 **42.** -12 **43.** -7 **44.** 6 **45.** 6 **46.** -1 **47.** 9 **48.** -28 **49.** 27 **50.** -36 **51.** -6 **52.** $-\frac{6}{35}$ **53.** $-\frac{6}{11}$ **54.** $\frac{15}{56}$ **55.** 0 **56.** 48 **57.** 12 **58.** -70 **59.** -60 **60.** -24 **61.** 144 **62.** -5 **63.** -3 **64.** -4 **65.** 18 **66.** 0 **67.** 0 **68.** -8 **69.** 5 **70.** 9 **71.** $-\frac{3}{32}$ **72.** $-\frac{3}{4}$ **73.** $\frac{56}{27}$ **74.** $-\frac{35}{9}$ **75.** 1 **76.** 0 **77.** 0 **78.** Undefined **79.** Undefined **80.** Undefined **81.** 0 **82.** 24 **83.** -8 **84.** 1 **85.** 3 **86.** -8 **87.** 18 **88.** -2 **89.** -4 **90.** 10 **91.** 1 **92.** 15 **93.** -4 **94.** 16 **95.** 36 **96.** 729 **97.** 1 **98.** 81 **99.** 16 **100.** -27 **101.** -1 **102.** -32 **103.** $\frac{4}{49}$ **104.** $\frac{9}{25}$ **105.** $\frac{8}{125}$ **106.** x^2y **107.** xy^2 **108.** x^3y^2 **109.** y^2z^2 **110.** $2^2 \cdot 3^3xy^2$ **111.** $5 \cdot 7^2x^2y$ **112.** x^2y^2z **113.** xxy **114.** $xzzz$ **115.** $yyyz$ **116.** $2xxxyy$ **117.** -9 **118.** -16 **119.** -27 **120.** -16 **121.** 23 **122.** -2 **123.** 23 **124.** 22 **125.** 26 **126.** -19 **127.** -39 **128.** -3 **129.** -4 **130.** -60 **131.** 10 **132.** 20 **133.** 20 **134.** 114 **135.** 9 **136.** 14 **137.** 2 **138.** 26 **139.** 9 **140.** 0 **141.** -3 **142.** -11 **143.** -3 **144.** 21 **145.** 3 **146.** -335 **147.** 353.6 **148.** -2.88 **149.** 117.8 **150.** 78,125 **151.** 729 **152.** -74.088 **153.** 58 **154.** 1 **155.** Associative property of addition **156.** Commutative property of multiplication **157.** Distributive property **158.** Commutative property of multiplication **159.** Commutative property of addition **160.** Associative property of addition **161.** Commutative property of addition

Chapter 1 Practice Test

1. (a) 42 **(b)** 42, 0 **(c)** $-6, 42, 0, -7, -1$ **(d)** $-6, 42, -3\frac{1}{2}, 0, 6.52, \frac{5}{9}, -7, -1$ **(e)** $\sqrt{5}$ **(f)** $-6, 42, -3\frac{1}{2}, 0, 6.52, \sqrt{5}, \frac{5}{9}, -7, -1$ **2.** $<$ **3.** $>$ **4.** -12 **5.** -11 **6.** 16 **7.** -14 **8.** 8 **9.** -24 **10.** $\frac{16}{63}$ **11.** -2 **12.** -69 **13.** -2 **14.** 12 **15.** 81 **16.** $\frac{27}{125}$ **17.** $2^25^2y^2z^3$ **18.** $2 \cdot 2 \cdot 3 \cdot 3 \cdot 3xxxxyy$ **19.** 26 **20.** 10 **21.** 11 **22.** Commutative property of addition **23.** Distributive property **24.** Associative property of addition **25.** Commutative property of multiplication

Chapter 2

Exercise Set 2.1

1. $5x$ **3.** $-x$ **5.** $x + 9$ **7.** $3x$ **9.** $4x - 7$ **11.** $2x + 3$ **13.** $5x + 8$ **15.** $5x + 3y + 3$ **17.** $2x - 4$ **19.** $x + 8$ **21.** $-8x + 2$ **23.** $-2x + 11$ **25.** $3x - 6$ **27.** $-5x + 3$ **29.** $6y + 6$ **31.** $4x - 10$ **33.** $3x - 4$

35. $x + \frac{5}{12}$ **37.** $48.5x + 8.3$ **39.** $x + \frac{1}{8}y$ **41.** $-4x - 8.3$ **43.** $-2x + 7$ **45.** $7x - 16$ **47.** $7x - 1$ **49.** $x - 8$ **51.** $21.72x - 7.11$ **53.** $-\frac{23}{20}x - 5$ **55.** $2x + 12$ **57.** $5x + 20$ **59.** $-2x + 8$ **61.** $-x + 2$ **63.** $x - 4$ **65.** $\frac{1}{4}x - 3$ **67.** $-1.8x + 3$ **69.** $-x + 3$ **71.** $0.8x - 0.2$ **73.** $x - y$ **75.** $-2x + 6y - 8$ **77.** $14.26x - 10.58y + 8.28$ **79.** $x - 8y + \frac{1}{2}$ **81.** $x + 3y - 9$ **83.** $x - 4 - 2y$ **85.** $3x - 8$ **87.** $2x - 5$ **89.** $14x + 18$ **91.** $4x - 2y + 3$ **93.** $y + 3$ **95.** $7x + 3$ **97.** $x - 9$ **99.** $6x - 12$ **101.** $-x - 2$ **103.** $-x + 6$ **105.** $8x - 19$ **107.** $x + 6.8$ **109.** $3x$ **111.** $x + 15$ **113.** $0.2x + 4y + 0.4$ **115.** $-6x + 3y - 3$ **117.** $x - 5$ **119.** $\frac{1}{6}x - \frac{10}{3}$ **121.** The signs of all the terms inside the parentheses are changed when the parentheses are removed. **123. (a)** $2x^2, 3x, -5$; The terms are the part of the expression that are added or subtracted. **(b)** The factors of $2x^2$ are $1, 2, x, 2x, x^2$ and $2x^2$. Note that $1 \cdot 2x^2 = 2x^2$, $2 \cdot x^2 = 2x^2$ and $x \cdot 2x = 2x^2$. Expressions that are multiplied are factors of the product.

Cumulative Review Exercises

124. 7 **125.** -16 **127.** -12

Group Activity/Challenge Problems

1. $18x - 25y + 3$ **3.** $6x^2 + 5y^2 + 3x + 7y$ **5. (a)** $3x^2, -10x, 8$ **(b)** positive factors of $3x^2$: $1, 3, x, 3x, x^2, 3x^2$ **(c)** Factors of 8: $1, 2, 4, 8, -1, -2, -4, -8$

Exercise Set 2.2

1. solution **3.** not solution **5.** solution **7.** not solution **9.** solution **11.** solution **13.** $x + 5 = 8, x = 3$ **15.** $12 = x + 3, x = 9$ **17.** $10 = x + 7, x = 3$ **19.** $x + 6 = 4 + 11, x = 9$ **21.** 4 **23.** -10 **25.** -9 **27.** -61 **29.** 22 **31.** 43 **33.** -12 **35.** 72 **37.** -26 **39.** -58 **41.** 12 **43.** -12 **45.** 3 **47.** -9 **49.** 53 **51.** 3 **53.** 5 **55.** 1 **57.** 17 **59.** -26 **61.** -36 **63.** -47.5 **65.** 46.5 **67.** -21.58 **69.** 0 **71.** 720 **73. (a)** The number or numbers that make the equation a true statement. **(b)** To find the solutions to an equation. **77. (a)** Get the variable by itself on one side of the equation. **79.** Subtract 3

Cumulative Review Exercises

80. 18 **81.** -8 **82.** $2x - 13$ **83.** $10x - 32$

Group Activity/Challenge Problems

1. (a) yes **(b)** yes **(c)** yes **3. (a)** $2x = 8$ **(b)** $x = 4$ **5. (a)** $20 = 4x$ **(b)** $5 = x$ (or $x = 5$)

Exercise Set 2.3

1. $2x = 10, x = 5$ **3.** $6 = 3x, x = 2$ **5.** $2x = 5, x = \frac{5}{2}$ **7.** $4 = 3x, x = \frac{4}{3}$ **9.** 3 **11.** 8 **13.** -2 **15.** -12 **17.** 5 **19.** 3 **21.** $-\frac{3}{2}$ **23.** 6 **25.** 2 **27.** 49 **29.** $-\frac{1}{3}$ **31.** 6 **33.** $\frac{10}{13}$ **35.** 2 **37.** -1 **39.** $-\frac{3}{40}$ **41.** -75 **43.** 125 **45.** -35 **47.** 20 **49.** -5 **51.** 12 **53.** -16 **55.** -36 **57.** 6 **59.** -20.2 **61.** $-\frac{5}{4}$ **63.** 9 **65. (a)** $-a$ **(b)** -5 **(c)** 5 **67.** Divide by -2 **69.** Multiply by $\frac{1}{4}, \frac{3}{20}$

Cumulative Review Exercises

71. -4 **72.** 0 **73.** $-11x + 38$ **74.** -57

Group Activity/Challenge Problems

1. (a) 4 **(b)** $2x + 6 = 14$ **(c)** $x = 4$ **3. (a)** 1 **(b)** $6 = 2x + 4$ **(c)** $x = 1$

Exercise Set 2.4

1. $2x + 4 = 16, x = 6$ **3.** $30 = 2x + 12, x = 9$ **5.** $3x + 10 = 4, x = -2$ **7.** $5 + 3x = 12, x = \frac{7}{3}$ **9.** 3 **11.** -6 **13.** 5 **15.** $\frac{12}{5}$ **17.** -12 **19.** 3 **21.** $\frac{11}{3}$ **23.** $-\frac{19}{16}$ **25.** -2.9 **27.** 5 **29.** $-\frac{7}{6}$ **31.** 3 **33.** 20 **35.** 6.8 **37.** 32 **39.** 0 **41.** 0 **43.** -1 **45.** 0 **47.** $-\frac{7}{6}$ **49.** 1 **51.** -4 **53.** 4 **55.** 3 **57.** 0.6 **59.** -1 **61.** 6 **63.** 2.1 **65.** -3.6 **67.** Addition property **69. (b)** 5

Cumulative Review Exercises

70. $\frac{49}{40}$ or $1\frac{9}{40}$ **71.** 64 **72.** Isolate the variable on one side of the equation. **73.** Divide both sides of the equation by -4.

Group Activity/Challenge Problems

1. $\frac{35}{6}$ **2.** $\frac{4}{5}$ **3.** -4

4. (a) 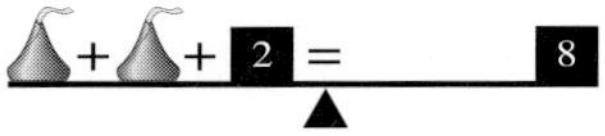 + + 2 = 8 **(b)** $2x + 2 = 8$ **(c)** \$3

5. (a) 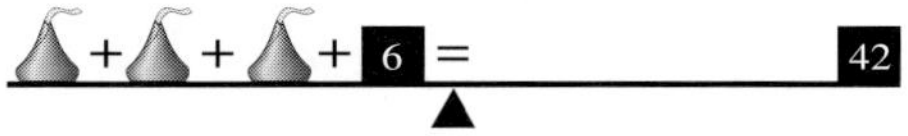 + + + 6 = 42 **(b)** $3x + 6 = 42$ **(c)** \$12 **7. (a)** $2x = x + 3$ **(b)** $x = 3$

9. (a) $2x + 3 = 4x + 2$ **(b)** $x = \frac{1}{2}$

Exercise Set 2.5

1. $2x = x + 6, x = 6$ **3.** $5 + 2x = x + 19, x = 14$ **5.** $5 + x = 2x + 5, x = 0$ **7.** $2x + 8 = x + 4, x = -4$ **9.** 5 **11.** 1 **13.** $\frac{3}{5}$ **15.** 3 **17.** 2 **19.** 1 **21.** 4.16 **23.** 4 **25.** $-\frac{17}{7}$ **27.** No solution **29.** $\frac{34}{5}$ **31.** -4 **33.** 25 **35.** All real numbers **37.** All real numbers **39.** 0 **41.** All real numbers **43.** $-\frac{112}{15}$ **45.** 14 **47.** $-\frac{5}{3}$ **49.** 12 **51.** 16 **53.** $-\frac{10}{3}$ **57.** You will obtain a false statement. **59. (b)** -8.

Cumulative Review Exercises

60. 0.131687243 **61.** Numbers or letters multiplied together are factors; numbers or letters added or subtracted are terms. **62.** $7x - 10$ **63.** $\frac{10}{7}$ **64.** -3

Group Activity/Challenge Problems

1. $\frac{1}{4}$ **3.** -4 **5. (a)** 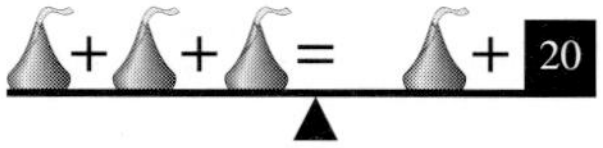 + + = + 20 **(b)** $3x = x + 20$ **(c)** 10 lbs

Exercise Set 2.6

1. 5:8 **3.** 2:1 **5.** 25:4 **7.** 5:3 **9.** 1:3 **11.** 6:1 **13.** 13:32 **15.** 8:1 **17. (a)** 5.3:3.6 **(b)** about 1.47:1 **19. (a)** 20,000:165 **(b)** about 121.21:1 **21.** 16 **23.** 45 **25.** -100 **27.** 5 **29.** -30 **31.** 6 **33.** 384 mi **35.** 7 gal **37.** 1.27 ft **39.** 260 in or 21.67 ft **41.** 24 tsp **43.** 340 trees **45.** 0.55 mL **47.** 96 sec or 1 min 36 sec **49.** about 261,200 thousand people or 261,200,000 people **51.** 4.75 ft **53.** 2.9 sq. yd **55.** 10.5 in **57.** 15.63 mi **59.** \$0.83 **61.** 5 points **63.** 2,033,898.3 lire **65.** 32 in **67.** 11.2 ft **69.** 5.6 in **71.** yes, her ratio is 2.12:1 **75.** You need a given ratio and one of the two parts of a second ratio.

Cumulative Review Exercises

76. Commutative property of addition **77.** Associative property of multiplication **78.** Distributive property **79.** $\frac{3}{4}$

Group Activity/Challenge Problems

1. (a) 750:10,000 **(b)** 0.075:1 **2.** $\frac{1}{3}$ cup flour, $\frac{2}{3}$ tsp nutmeg, $\frac{2}{3}$ tsp cinnamon, $\frac{1}{6}$ tsp salt, $1\frac{1}{3}$ tbsp butter, 1 cup sugar

3. 0.625 cc **5. (a)** about 140,920,000 people **(b)** about 48,780,000 people **(c)** about 92,140,000 people

EXERCISE SET 2.7

1. $x > 4$ 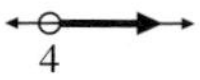 **3.** $x \geq -2$ 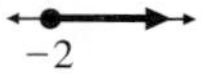 **5.** $x > -5$  **7.** $x < 10$

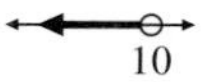

9. $x \leq -4$  **11.** $x > -\frac{3}{2}$ 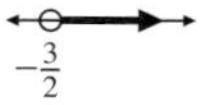 **13.** $x \leq 1$  **15.** $x < -3$

17. $x < \frac{3}{2}$ **19.** $x > \frac{35}{9}$ 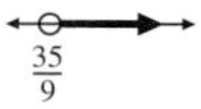 **21.** $x > -\frac{8}{3}$ 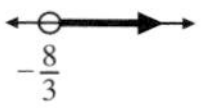 **23.** $x \leq -\frac{11}{3}$

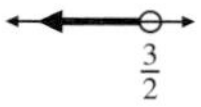

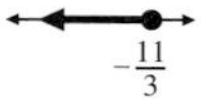

25. $x \geq -6$  **27.** $x < 1$ **29.** $x < 2$ **31.** All real numbers

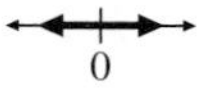

33. $x > \frac{3}{4}$  **35.** $x > \frac{23}{10}$ 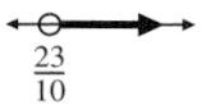 **37.** No solution 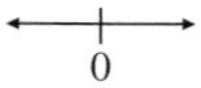 **39.** $x \geq -\frac{7}{11}$

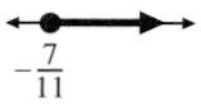

41. All real numbers **43.** The inequality has no solution. **45.** When multiplying or dividing by a negative number

CUMULATIVE REVIEW EXERCISES

47. -9 **48.** -25 **49.** $\frac{14}{5}$ or $2\frac{4}{5}$ **50.** 500 kwh

GROUP ACTIVITY/CHALLENGE PROBLEMS

1. $x \geq -2$ **4. (a)** \$10 **(c)** \$160

CHAPTER 2 REVIEW EXERCISES

1. $2x + 8$ **2.** $3x - 6$ **3.** $8x - 6$ **4.** $-2x - 8$ **5.** $-x - 2$ **6.** $-x + 2$ **7.** $-16 + 4x$ **8.** $18 - 6x$ **9.** $20x - 24$ **10.** $-6x + 15$ **11.** $36x - 36$ **12.** $-4x + 12$ **13.** $-3x - 3y$ **14.** $-6x + 4$ **15.** $-3 - 2y$ **16.** $-x - 2y + z$ **17.** $3x + 9y - 6z$ **18.** $-4x + 6y - 14$ **19.** $5x$ **20.** $7y + 2$ **21.** $-2y + 7$ **22.** $5x + 1$ **23.** $6x + 3y$ **24.** $-3x + 3y$ **25.** $6x + 8y$ **26.** $6x + 3y + 2$ **27.** $-x - 1$ **28.** $3x + 3y + 6$ **29.** 3 **30.** $-12x + 3$ **31.** $5x + 6$ **32.** $-2x$ **33.** $5x + 7$ **34.** $-10x + 12$ **35.** $5x + 3$ **36.** $4x - 4$ **37.** $22x - 42$ **38.** $3x - 3y + 6$ **39.** $-x + 5y$ **40.** $3x + 2y + 16$ **41.** 3 **42.** $-x - 2y + 4$ **43.** 2 **44.** -8 **45.** 11 **46.** -27 **47.** 2 **48.** $\frac{11}{2}$ **49.** -2 **50.** -3 **51.** 12 **52.** 1 **53.** 6 **54.** -3 **55.** $-\frac{21}{5}$ **56.** -5 **57.** -19 **58.** -1 **59.** $\frac{2}{3}$ **60.** $\frac{1}{5}$ **61.** $\frac{9}{2}$ **62.** $\frac{10}{7}$ **63.** -3 **64.** -1 **65.** -8 **66.** $-\frac{23}{5}$ **67.** -10 **68.** $-\frac{4}{3}$ **69.** No solution **70.** All real numbers **71.** $\frac{17}{3}$ **72.** $-\frac{20}{7}$ **73.** 3:4 **74.** 5:12 **75.** 1:1 **76.** 3 **77.** 3 **78.** 9 **79.** $\frac{135}{4}$ **80.** -4 **81.** -24 **82.** 36 **83.** 90 **84.** 40 in **85.** 1 ft **86.** $x \geq 2$  **87.** $x < 3$

88. $x \geq -\frac{12}{5}$ 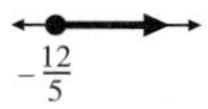 **89.** No solution 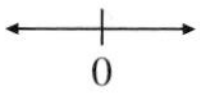 **90.** All real numbers 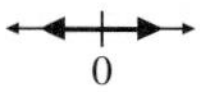 **91.** $x < -3$

92. $x \leq \frac{9}{5}$  **93.** $x > \frac{8}{5}$  **94.** $x < -\frac{5}{3}$ **95.** $x \leq \frac{5}{11}$

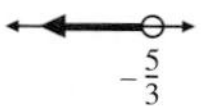

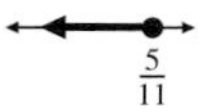

96. No solution 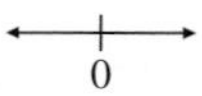 **97.** All real numbers  **98.** 240 calories **99.** 110 copies **100.** $6\frac{1}{3}$ in **101.** 9.45 ft. **102.** approximately \$0.3209 **103.** 57.3° **104.** 192 bottles

CHAPTER 2 PRACTICE TEST

1. $4x - 8$ **2.** $-x - 3y + 4$ **3.** $2x + 4$ **4.** $-x + 10$ **5.** $-6x + y - 6$ **6.** $7x - 5y + 3$ **7.** $8x - 1$ **8.** 4 **9.** -2 **10.** 2 **11.** -1 **12.** $-\frac{1}{7}$ **13.** No solution **14.** All real numbers **15.** -45

16. $x > -7$  **17.** $x \leq 12$ 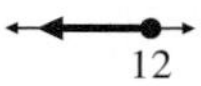 **18.** No solution 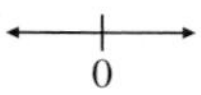 **19.** $\frac{32}{3}$ or $10\frac{2}{3}$ ft

20. 150 gal

CUMULATIVE REVIEW TEST

1. $\frac{16}{25}$ **2.** $\frac{1}{2}$ **3.** $>$ **4.** -6 **5.** -8 **6.** 7 **7.** 3 **8.** 1 **9.** Associative Property of Addition **10.** $10x + y$

11. $3x + 16$ **12.** 3 **13.** -40 **14.** 2 **15.** -1 **16.** 6 **17.** $x > 10$ 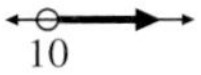

18. $x \geq -12$ 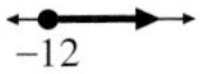

19. 158.4 lbs **20.** \$42

Chapter 3

EXERCISE SET 3.1

1. 25 **3.** 22 **5.** 126 **7.** 12.56 **9.** 10 **11.** 6 **13.** 1080 **15.** 56 **17.** 5 **19.** 60 **21.** 16 **23.** 6

25. $y = -2x + 8, 4$ **27.** $y = \dfrac{2x + 4}{6}, 4$ **29.** $y = \dfrac{-3x + 6}{2}, 0$ **31.** $y = \dfrac{4x - 20}{5}, -\dfrac{4}{5}$ **33.** $y = \dfrac{3x + 18}{6}, 3$

35. $y = \dfrac{-x + 8}{2}, 6$ **37.** $t = \dfrac{d}{r}$ **39.** $p = \dfrac{i}{rt}$ **41.** $d = \dfrac{C}{\pi}$ **43.** $b = \dfrac{2A}{h}$ **45.** $w = \dfrac{P - 2l}{2}$ **47.** $n = \dfrac{m - 3}{4}$

49. $b = y - mx$ **51.** $r = \dfrac{I - P}{Pt}$ **53.** $d = \dfrac{3A - m}{2}$ **55.** $b = d - a - c$ **57.** $y = \dfrac{-ax + c}{b}$ **59.** $h = \dfrac{V}{\pi r^2}$

61. 35 **63.** 10°C **65.** 95°F **67.** $P = 10$ **69.** $K = 4$ **71.** 30 **73.** \$1440 **75.** \$5000 **77.** 30 in **79.** 24 sq cm **81.** 50.24 sq in **83.** 25.12 in **85.** 8 ft **87. (a)** 62.1 ft **(b)** 124.2 ft **89.** 18,237.12 cu in

91. When you multiply a unit by the same unit you get a square unit. **93. (a)** $\pi = \dfrac{C}{2r}$ or $\pi = \dfrac{C}{d}$ **(b)** π or about 3.14

CUMULATIVE REVIEW EXERCISES

94. 0 **95.** 3:2 **96.** 1620 min or 27 hrs **97.** $x \leq -17$

GROUP ACTIVITY/CHALLENGE PROBLEMS

1. (a) $A = d^2 - \pi\left(\dfrac{d}{2}\right)^2$ **(b)** 3.44 sq ft **(c)** 7.74 sq ft **2. (a)** $V = 18x^3 - 3x^2$ **(b)** 6027 cu cm **(c)** $S = 54x^2 - 8x$ **(d)** 2590 sq cm **4. (a)** diameter: approximately 142 ft, radius: approximately 71 ft **(b)** approximately 134 ft **(c)** approximately 14,095 sq ft **(d)** approximately 1,691,455 cu ft

EXERCISE SET 3.2

1. $x + 5$ **3.** $4x$ **5.** $0.70x$ **7.** $0.10c$ **9.** $0.16p$ **11.** $6x - 3$ **13.** $\frac{3}{4}x + 7$ **15.** $2(x + 8)$ **17.** $4x$ **19.** $0.23x$

21. $8.20b$ **23.** $300n$ **25.** $25x$ **27.** $12x$ **29.** $16c$ **31.** $275x + 25y$ **33.** Six less than a number. **35.** One more than four times a number. **37.** Seven less than five times a number. **39.** Four times a number, decreased by two. **41.** Three times a number subtracted from two. **43.** Twice the difference of a number and one. **45.** Martin's salary is x; Eileen's salary is $x + 45$. **47.** One number is x; the second number is $x/3$. **49.** The first consecutive even integer is x; the second consecutive even integer is $x + 2$. **51.** One number is x; the second number is $x + 12$ (or $x - 12$). **53.** One number is x; the other number is $(x/2) + 3$. **55.** One number is x; the second number is $3x - 4$. **57.** The first consecutive odd integer is x; the second consecutive odd integer is $x + 2$. **59.** One number is x; the second number is $x - 0.15x$. **61.** $c, c - 0.10c$ **63.** $p, p - 0.50p$ **65.** $w, 2w$ **67.** $m, m + 0.15m$ **69.** $x + 5x = 18$ **71.** $x + (x + 1) = 47$

73. $2x - 8 = 12$ **75.** $\frac{1}{5}(x + 10) = 150$ **77.** $x + (2x - 8) = 1000$ **79.** $x + 0.08x = 92$ **81.** $x - 0.25x = 65$

83. $x - 0.20x = 215$ **85.** $x + (2x - 3) = 21$ **87.** $40t = 180$ **89.** $15y = 215$ **91.** $25q = 150$ **93.** Three more than a number is six. **95.** Three times a number, decreased by one, is four more than twice the number **97.** Four times the difference of a number and one is six. **99.** Six more than five times a number is the difference of six times the number and one **101.** The sum of a number and the number increased by four is eight **103.** The sum of twice a number and the number increased by three is five

Cumulative Review Exercises

107. 3.35 tsp **108.** $\frac{1}{6}$ cup **109.** 15 **110.** $y = \frac{3x - 6}{2}\left(\text{or } y = \frac{3}{2}x - 3\right)$, 6

Group Activity/Challenge Problems

1. (a) $86{,}400d + 3600h + 60m + s$ **(b)** 368,125 sec **3.** $30 = 6t$ **5.** $40{,}000y = 1{,}000{,}000$
7. $20 + 0.60m = 30 + 0.45m$

Exercise Set 3.3

1. $x + (x + 1) = 45$; 22, 23 **3.** $x + (x + 2) = 68$; 33, 35 **5.** $x + (3x - 5) = 43$; 12, 31
7. $x + (x + 2) + (x + 4) = 87$; 27, 29, 31 **9.** $(2x - 8) - x = 17$; 25, 42 **11.** $x + (5x + 3) = 69$; Vector, 11 mpg; Geo, 58 mpg **13.** $x + (15x + 6) = 22$; Mexico, 1; Canada, 21 **15.** $x - 0.06x = 65{,}800$; \$70,000
17. $20x = 15$; 75% **19.** $1500 + 2x = 3100$; 800 **21.** $2000 + 0.02x = 2400$; \$20,000 **23.** $0.08x = 60$; \$750
25. $x + 0.08x = 37{,}800$; \$35,000 **27.** $x + 0.07x = 1.50$; \$1.40 **29.** $x + 3x + 3x = 210{,}000$; each child, \$90,000; charity, \$30,000 **31.** $4000 - 300x = 2000$; about 6.67 yr. **33.** $x - 0.60x = 24$; 60 gal **35.** $15 + 5x = 65$; 10 mph
37. $x + 0.90x = 110{,}000$; about 57,895 people **39.** $x + 0.07x + 0.15x = 20$; \$16.39 **41.** $x - 0.10x - 20 = 250$; \$300 **43.** $x + 15x = 4$; $\frac{1}{4}$ or 0.25 gal oil and $3\frac{3}{4}$ or 3.75 gal gas **45.** $x - 0.01x = 2.362$; about 2.386 million
47. (a) about 42 months or 3.5 yr **(b)** Citibank **49. (a)** 18.75 mo or about 1.56 yr **(b)** Countrywide **51. (a)** 9.2 mo **(b)** \$325.50 per month

Cumulative Review Exercises

54. $\frac{17}{12}$ **55.** Associative property of addition **56.** Commutative property of multiplication **57.** Distributive property
58. 56 lb **59.** $b = 2M - a$

Group Activity/Challenge Problems

3. n
$4n$
$4n + 6$
$(4n + 6)/2 = 2n + 3$
$2n + 3 - 3 = 2n$

4. about 185.7% increase **6.** about 251,476.2 people per sq mi

Exercise Set 3.4

1. 9.5 in **3.** $A = 47°$, $B = 133°$ **5.** 50°, 60°, 70° **7.** 4m, 4m, 2m **9.** $w = 48$ ft, $l = 72$ ft
11. two smaller angles, 69°; two larger angles, 111° **13.** $w = 2\frac{2}{3}$ ft, $h = 4\frac{2}{3}$ ft **15.** $h = 3$ ft, $l = 9$ ft
17. The area remains the same. **19.** The volume becomes eight times as great.

Cumulative Review Exercises

22. $<$ **23.** $>$ **24.** -8 **25.** $-2x - 4y + 6$ **26.** $y = \frac{-2x + 9}{3}$ or $y = -\frac{2}{3}x + 3$, 1

Group Activity/Challenge Problems

2. $ac + ad + bc + bd$ **3. (a)** 342.56 ft **(b)** 192 ft

Chapter 3 Review Exercises

1. 12.56 **2.** 48 **3.** 20 **4.** 300 **5.** 240 **6.** 28.26 **7.** 113.04 **8.** 20 **9.** 21 **10.** -11 **11.** -8 **12.** 15
13. 4.5 **14.** $y = 2x - 12$, 8 **15.** $y = \frac{3x + 4}{2}$, 5 **16.** $y = \frac{3x - 5}{2}$, -7 **17.** $y = -3x - 10$, -10

18. $y = \frac{-3x - 6}{2}, 6$ **19.** $y = \frac{4x - 3}{3}, \frac{5}{3}$ **20.** $m = \frac{F}{a}$ **21.** $h = \frac{2A}{b}$ **22.** $t = \frac{i}{pr}$ **23.** $w = \frac{P - 2l}{2}$
24. $y = \frac{2x - 6}{3}$ **25.** $B = 2A - C$ **26.** $h = \frac{V}{\pi r^2}$ **27.** \$180 **28.** 6 in **29.** 29 and 33 **30.** 127 and 128
31. 38 and 7 **32.** \$8000 **33.** \$2000 **34.** \$650 **35. (a)** 166.7 mo or 13.9 yr **(b)** Mellon Bank
36. (a) 27.4 mo or 2.3 yr **(b)** yes **37.** 45°, 55°, 80° **38.** 30°, 40°, 150°, 140° **39.** $w = 15.5$ ft, $l = 19.5$ ft
40. 103 and 105 **41.** \$450 **42.** \$12,000 **43.** 42°, 50°, 88° **44.** 8 years **45.** 70°, 70°, 110°, 110°
46. (a) 500 copies **(b)** King Kopie by \$5

PRACTICE TEST

1. 18 ft **2.** 145 **3.** 100.48 **4.** $R = \frac{P}{I}$ **5.** $y = \frac{3x - 6}{2}$ **6.** $a = 3A - b$ **7.** $c = \frac{D - Ra}{R}$ or $c = \frac{D}{R} - a$
8. 56 and 102 **9.** 13, 14, and 15 **10.** \$16.39 **11.** 15, 30, 30 in **12.** 50°, 50°, 130°, 130°

Chapter 4

EXERCISE SET 4.1

1. x^6 **3.** y^3 **5.** 243 **7.** y^5 **9.** y^5 **11.** x^7 **13.** 25 **15.** y **17.** 1 **19.** y **21.** 1 **23.** 3 **25.** 1 **27.** 1
29. x^{10} **31.** x^{25} **33.** x^3 **35.** x^{12} **37.** x^{15} **39.** $1.69x^2$ **41.** x^2 **43.** $64x^6$ **45.** $-27x^9$ **47.** $8x^6y^3$
49. $73.96x^4y^{10}$ **51.** $-216x^9y^6$ **53.** $-x^{12}y^{15}z^{18}$ **55.** $\frac{x^2}{y^2}$ **57.** $\frac{x^3}{64}$ **59.** $\frac{y^5}{x^5}$ **61.** $\frac{216}{x^3}$ **63.** $\frac{27x^3}{y^3}$ **65.** $\frac{9x^2}{25}$
67. $\frac{64y^9}{x^3}$ **69.** $\frac{-27x^9}{64}$ **71.** $\frac{x^2}{y^4}$ **73.** $\frac{y^4}{x^7}$ **75.** $\frac{5x^2}{y^2}$ **77.** $\frac{1}{4x^2y}$ **79.** $\frac{7}{2x^5y^5}$ **81.** $\frac{-3y^4}{x^3z}$ **83.** $\frac{-3}{x^3y^2z}$ **85.** $\frac{8}{x^6}$
87. $27y^{12}$ **89.** 1 **91.** $\frac{x^4}{y^4}$ **93.** $\frac{z^{24}}{16y^{20}}$ **95.** $64x^6y^9$ **97.** x^4y^{12} **99.** $\frac{9}{16x^{12}y^4}$ **101.** $9x^2y^8$ **103.** $-18x^3y^9$
105. $8x^5y^8$ **107.** $10x^2y^7$ **109.** $4x^2y^2$ **111.** $3x^{14}y^{23}$ **113.** $54x^{17}y^{17}$ **115.** $x^{11}y^{13}$ **117.** $18x^{10}y^{28}$ **119.** cannot be simplified **121.** cannot be simplified **123.** cannot be simplified **125.** y^2 **127.** cannot be simplified **129.** $\frac{x^2}{y}$
131. 0 **133.** Negative, exponent is odd **135.** Positive, exponent is even

CUMULATIVE REVIEW EXERCISES

137. An equation of the form $ax + b = c$ **138.** A linear equation that has only one solution **139.** An equation that is true for all real numbers. **140.** $C = 18.84$ in, $A = 28.26$ sq in **141.** $y = \frac{2x - 6}{5}$ or $y = \frac{2}{5}x - \frac{6}{5}$

GROUP ACTIVITY/CHALLENGE PROBLEMS

1. $\frac{9x^2}{8y^3}$ **3.** 15 **5.** 3 **7.** 2, 5 **9.** 4, 11 **11.** 3, 2, 4

EXERCISE SET 4.2

1. $\frac{1}{x^2}$ **3.** $\frac{1}{4}$ **5.** x^3 **7.** x **9.** 16 **11.** $\frac{1}{x^6}$ **13.** $\frac{1}{y^{21}}$ **15.** $\frac{1}{x^8}$ **17.** 64 **19.** x^3 **21.** x^2 **23.** 9 **25.** $\frac{1}{x^2}$
27. y^3 **29.** $\frac{1}{x^4}$ **31.** 27 **33.** $\frac{1}{27}$ **35.** z^9 **37.** $\frac{1}{x^{25}}$ **39.** y^6 **41.** $\frac{1}{x^4}$ **43.** $\frac{1}{x^{15}}$ **45.** $\frac{1}{x^8}$ **47.** y^{10} **49.** 1 **51.** 1
53. 1 **55.** x^4 **57.** 1 **59.** $\frac{1}{4}$ **61.** $\frac{1}{36}$ **63.** x^3 **65.** $\frac{1}{9}$ **67.** 125 **69.** $\frac{1}{4}$ **71.** 1 **73.** $\frac{2y}{x}$ **75.** $\frac{1}{3x^3}$ **77.** $\frac{5x^4}{y}$
79. $\frac{1}{9x^4y^6}$ **81.** $\frac{y^9}{x^{15}}$ **83.** $\frac{15}{x^3}$ **85.** $\frac{6}{x}$ **87.** $\frac{-27}{x^2}$ **89.** $\frac{2x}{y^2}$ **91.** $\frac{15y}{x}$ **93.** $2x^5$ **95.** $\frac{4}{x^7}$ **97.** $\frac{x^3}{5}$ **99.** $\frac{6}{x^5y^2}$
101. $\frac{4}{x^{12}y^4}$ **103.** $\frac{x^5}{2y^6}$ **105. (a)** Yes, $a^{-1}b^{-1} = \frac{1}{a} \cdot \frac{1}{b} = \frac{1}{ab}$ **(b)** No, $a^{-1} + b^{-1} = \frac{1}{a} + \frac{1}{b} \neq \frac{1}{a + b}$

CUMULATIVE REVIEW EXERCISES

108. -18 **109.** 18 **110.** 6.67 oz **111.** 9, 28

GROUP ACTIVITY/CHALLENGE PROBLEMS

1. (a) and **(b)** $\frac{z^2}{9x^4y^6}$ **3.** -2 **5.** -2 **7.** -3 **9.** $\frac{2}{3}$ **11.** $\frac{11}{6}$

EXERCISE SET 4.3

1. 4.2×10^4 **3.** 9×10^2 **5.** 5.3×10^{-2} **7.** 1.9×10^4 **9.** 1.86×10^{-6} **11.** 9.14×10^{-6} **13.** 1.07×10^2 **15.** 1.53×10^{-1} **17.** 4200 **19.** 40,000,000 **21.** 0.0000213 **23.** 0.312 **25.** 9,000,000 **27.** 535 **29.** 35,000 **31.** 10,000 **33.** 120,000,000 **35.** 0.0153 **37.** 930 **39.** 320 **41.** 0.0021 **43.** 20 **45.** 4.2×10^{12} **47.** 4.5×10^{-7} **49.** 2×10^3 **51.** 2×10^{-7} **53.** 3×10^8 **55.** 9.2×10^{-5}, 1.3×10^{-1}, 8.4×10^3, 6.2×10^4 **57.** 3.2×10^7 seconds **59.** 8,640,000,000 cu ft **61. (a)** 2.0×10^8 lb **(b)** 200,000,000 lb **63. (a)** 1.8×10^{10} mi **(b)** 3.332×10^6 mi or 3,332,000 mi **65.** -8

CUMULATIVE REVIEW EXERCISES

67. 0 **68. (a)** $\frac{3}{2}$ **(b)** 0 **69.** 2 **70.** $\frac{-y^{12}}{64x^9}$

GROUP ACTIVITY/CHALLENGE PROBLEMS

1. (a) 1×10^6, 1×10^9, 1×10^{12} **(b)** 1000 days or about 2.74 yr **(c)** 1,000,000 days or about 2740 yr **(d)** 1,000,000,000 days or about 2,740,000 yr **(e)** 1000

EXERCISE SET 4.4

1. Monomial **3.** Monomial **5.** Binomial **7.** Trinominal **9.** Not polynomial **11.** Binomial **13.** Monomial **15.** Polynomial **17.** Trinomial **19.** Not polynomial **21.** First degree **23.** $2x^2 + x - 6$, second **25.** $-x^2 - 4x - 8$, second **27.** Third **29.** Second **31.** $-6x^3 + x^2 - 3x + 4$, third **33.** $5x^2 - 2x - 4$, second **35.** $-2x^3 + 3x^2 + 5x - 6$, third **37.** $6x + 1$ **39.** $-2x + 11$ **41.** $-x - 2$ **43.** $21x - 21$ **45.** $x^2 + 6.6x + 0.8$ **47.** $2x^2 + 8x + 5$ **49.** $5x^2 + x + 20$ **51.** $-3x^2 + x + \frac{17}{2}$ **53.** $5.4x^2 - 5x + 4$ **55.** $-2x^3 - 3x^2 + 4x - 3$ **57.** $3x^2 - 2xy$ **59.** $5x^2y - 3x + 2$ **61.** $7x - 1$ **63.** $x^2 + x + 16$ **65.** $-3x^2 + 2x - 12$ **67.** $7x^2 + 7x - 13$ **69.** $2x^3 - x^2 + 6x - 2$ **71.** $5x^3 - 7x^2 - 2$ **73.** $3xy + 3x + 3$ **75.** $x - 6$ **77.** $3x + 4$ **79.** -5 **81.** $-7x + 3$ **83.** $6x^2 + 7x - 8.5$ **85.** $8x^2 + x + 4$ **87.** $5x^2 - 2x + 7$ **89.** $2x^2 + 3x + 7.4$ **91.** $2x^3 - \frac{23}{5}x^2 + 2x - 2$ **93.** $9x^3 - x^2 - 5x - \frac{1}{5}$ **95.** $-x + 11$ **97.** $2x^2 - 9x + 14$ **99.** $-x^3 + 11x^2 + 9x - 7$ **101.** $3x + 17$ **103.** $4x + 7$ **105.** $5x^2 + 7x - 2$ **107.** $4x^2 - 5x - 6$ **109.** $4x^3 - 7x^2 + x - 2$ **111.** Sum of a finite number of terms of the form ax^n, where a is a real number and n is a whole number. **113. (a)** The exponent on the variable is the degree of the term. **(b)** It is the same as the degree of the highest degree term in the polynomial. **115.** Write the polynomial with exponents on the variable decreasing from left to right.

CUMULATIVE REVIEW EXERCISES

120. $|-4| < |-6|$ **121.** False **122.** True **123.** False **124.** False **125.** $\frac{y^3}{8x^9}$

GROUP ACTIVITY/CHALLENGE PROBLEMS

1. $-12x + 18$ **2.** $3x^2y - 13xy + 9xy^2 + 3x$ **3.** $8x^2 + 28x - 24$

EXERCISE SET 4.5

1. $3x^3y$ **3.** $30x^5y^7$ **5.** $-28x^6y^{15}$ **7.** $54x^6y^{14}$ **9.** $3x^6y$ **11.** $9x^5y^4$ **13.** $3x + 12$ **15.** $2x^2 - 6x$ **17.** $8x^2 - 24x$ **19.** $2x^3 + 6x^2 - 2x$ **21.** $-2x^3 + 4x^2 - 10x$ **23.** $-20x^3 + 30x^2 - 20x$ **25.** $24x^3 + 32x^2 - 40x$

27. $0.6x^2y + 1.5x^2 - 1.8xy$ **29.** $xy - y^2 - 3y$ **31.** $x^2 + 7x + 12$ **33.** $6x^2 + 3x - 30$ **35.** $4x^2 - 16$ **37.** $-6x^2 - 8x + 30$ **39.** $-2x^2 + x + 15$ **41.** $x^2 + 7x + 12$ **43.** $x^2 + 2x - 8$ **45.** $6x^2 + 23x + 20$ **47.** $6x^2 - x - 12$ **49.** $x^2 - 1$ **51.** $4x^2 - 12x + 9$ **53.** $-2x^2 + 5x + 12$ **55.** $-4x^2 + 2x + 12$ **57.** $x^2 - y^2$ **59.** $6x^2 - 5xy - 6y^2$ **61.** $8xy - 12x - 6y^2 + 9y$ **63.** $x^2 + 0.9x + 0.18$ **65.** $2x^2 + 5x + 2$ **67.** $x^2 - 16$ **69.** $4x^2 - 1$ **71.** $x^2 + 2xy + y^2$ **73.** $x^2 - 0.4x + 0.04$ **75.** $9x^2 - 25$ **77.** $0.16x^2 + 0.8xy + y^2$ **79.** $2x^3 + 10x^2 + 11x - 3$ **81.** $5x^3 - x^2 + 16x + 16$ **83.** $-14x^3 - 22x^2 + 19x - 3$ **85.** $18x^3 - 69x^2 + 54x - 27$ **87.** $6x^4 + 5x^3 + 5x^2 + 10x + 4$ **89.** $x^4 - 3x^3 + 5x^2 - 6x$ **91.** $3x^4 - 7x^3 - 7x^2 + 3x$ **93.** $a^3 + b^3$ **95.** Yes
97. No, it may contain either 2, 3, or 4 terms.
99. **(a)** $(x + 2)(2x + 1)$ or $2x^2 + 5x + 2$ **(b)** 54 sq ft. **(c)** 1 ft

Cumulative Review Exercises

100. 13 miles **101.** $\frac{1}{16y^4}$ **102.** **(a)** -216 **(b)** $\frac{1}{216}$ **103.** $-6x^2 - 2x + 8$

Group Activity/Challenge Problems

1. $2x^3\sqrt{5} + 5x^2 - \frac{x\sqrt{5}}{2}$ **3.** $6x^6 - 18x^5 + 3x^4 + 35x^3 - 54x^2 + 38x - 12$ **5.** $3x^2$ **6.** $-4xy^3$ **7.** $x + 2$
11. $x + 3, x - 2$

Exercise Set 4.6

1. $\frac{x^2 - 2x - 15}{x + 3} = x - 5$ or $\frac{x^2 - 2x - 15}{x - 5} = x + 3$ **3.** $\frac{2x^2 + 5x + 3}{2x + 3} = x + 1$ or $\frac{2x^2 + 5x + 3}{x + 1} = 2x + 3$
5. $\frac{4x^2 - 9}{2x + 3} = 2x - 3$ or $\frac{4x^2 - 9}{2x - 3} = 2x + 3$ **7.** $x + 2$ **9.** $x + 3$ **11.** $\frac{3}{2}x + 4$ **13.** $-3x + 2$ **15.** $3x + 1$
17. $3 + \frac{6}{x}$ **19.** $-\frac{3}{x} + 1$ **21.** $1 + \frac{2}{x} - \frac{3}{x^2}$ **23.** $-2x^3 + \frac{3}{x} + \frac{4}{x^2}$ **25.** $x^3 + 4x - \frac{3}{x^3}$ **27.** $3x^2 - 2x + 6 - \frac{5}{2x}$
29. $-x^2 - \frac{3}{2}x + \frac{2}{x}$ **31.** $3x^4 + x^2 - \frac{10}{3} - \frac{3}{x^2}$ **33.** $x + 3$ **35.** $2x + 3$ **37.** $2x + 4$ **39.** $x - 2$
41. $x + 5 - \frac{3}{2x - 3}$ **43.** $2x + 3$ **45.** $2x - 3 + \frac{2}{4x + 9}$ **47.** $4x - 3 - \frac{3}{2x + 3}$ **49.** $2x^2 + 3x - 1$
51. $3x^2 - 3x - 1 + \frac{6}{3x + 2}$ **53.** $x^2 + 3x + 9 + \frac{19}{x - 3}$ **55.** $x^2 + 3x + 9$ **57.** $2x^2 + x - 2 - \frac{2}{2x - 1}$
59. $-x^2 - 7x - 5 - \frac{8}{x - 1}$ **61.** $3x^2$ **63.** 7, 6, 4, 2

Cumulative Review Exercises

64. **(a)** 2 **(b)** 2, 0 **(c)** $2, -5, 0, \frac{2}{5}, -6.3, -\frac{23}{34}$ **(d)** $\sqrt{7}, \sqrt{3}$ **(e)** $2, -5, 0, \sqrt{7}, \frac{2}{5}, -6.3, \sqrt{3}, -\frac{23}{34}$ **65.** **(a)** 0 **(b)** undefined **66.** parentheses, exponents, multiplication or division (left to right), addition or subtraction (left to right) **67.** $-\frac{2}{3}$

Group Activity/Challenge Problems

1. $2x^2 - 3x + \frac{5}{2} - \frac{3}{2(2x + 3)}$ **3.** $-3x + 3 + \frac{1}{x + 3}$ **5.** $x^2 + 4x + 2$

Exercise Set 4.7

1. 5 gal/min **3.** 40 copies/min **5.** 2736 mph **7.** 250 cm^3/hr **9.** 1 mile **11.** 336 gal **13.** 3.5 hr **15.** Santa Fe Special, 52 mph; Amtrak, 82 mph **17.** **(a)** 35 mph, 40 mph **(b)** approximately 30.43 knots, approximately 34.78 knots

19. 2.24 ft/min. **21.** Chestnut, 4 mph; Midnight, 7 mph **23.** 5400 ft **25. (a)** 275 ft **(b)** 120 ft/min **27.** Road, 2.45 ft/day; Bridge, 1.25 ft/day **29.** \$3500 at 8%, \$5400 at 11% **31.** \$1875 at 10%, \$3125 at 6% **33.** Rate increased after 3 months, or in April. **35.** 9 ones, 3 tens **37.** 6 hours at \$6, 12 hours at \$6.50 **39.** 9 lb **41.** $1\frac{2}{3}L$ **43.** 11.1% **45.** 2.86 lb Family, 7.14 lb Spot Filler **47. (a)** 13 shares Mattel, 52 shares United **(b)** \$434

Cumulative Review Exercises

48. (a) $\frac{22}{13}$ or $1\frac{9}{13}$ **(b)** $\frac{35}{8}$ or $4\frac{3}{8}$ **49.** All real numbers **50.** $\frac{3}{4}$ or 0.75 **51.** $x \leq \frac{1}{4}$

Group Activity/Challenge Problems

1. (a) $52{,}800 = 4416t$ **(b)** approximately 11.96 hr **3.** 0.35 ft/sec

Chapter 4 Review Exercises

1. x^6 **2.** x^8 **3.** 243 **4.** 32 **5.** x^3 **6.** 1 **7.** 9 **8.** 16 **9.** $\frac{1}{x^2}$ **10.** x^5 **11.** 1 **12.** 3 **13.** 1 **14.** 1 **15.** $4x^2$ **16.** $27x^3$ **17.** $4x^2$ **18.** $-27x^3$ **19.** $16x^8$ **20.** $-x^{12}$ **21.** x^{12} **22.** $\frac{4x^6}{y^2}$ **23.** $\frac{27x^{12}}{8y^3}$ **24.** $24x^5$ **25.** $\frac{4x}{y}$ **26.** $12x^5y^2$ **27.** $9x^2$ **28.** $24x^7y^7$ **29.** $16x^8y^{11}$ **30.** $24x^{10}y^8$ **31.** $\frac{16x^6}{y^4}$ **32.** $\frac{x^9}{8y^9}$ **33.** $\frac{1}{x^3}$ **34.** $\frac{1}{y^7}$ **35.** $\frac{1}{25}$ **36.** x^3 **37.** x^7 **38.** 9 **39.** $\frac{1}{x^2}$ **40.** $\frac{1}{x^5}$ **41.** $\frac{1}{x^3}$ **42.** x^8 **43.** x^5 **44.** x^7 **45.** $\frac{1}{x^6}$ **46.** $\frac{1}{9x^8}$ **47.** $\frac{x^9}{64y^3}$ **48.** $\frac{-125}{x^6}$ **49.** $-8x^8$ **50.** $\frac{27y^3}{x^6}$ **51.** $\frac{x^4}{16y^6}$ **52.** $\frac{6}{x}$ **53.** $10x^2y^2$ **54.** $\frac{24y^2}{x^2}$ **55.** $\frac{6}{x^6y}$ **56.** $3y^5$ **57.** $\frac{3y}{x^3}$ **58.** $\frac{5x}{y^4}$ **59.** $\frac{4y^{10}}{x}$ **60.** $\frac{1}{2x^2y^5}$ **61.** 3.64×10^5 **62.** 1.64×10^6 **63.** 7.63×10^{-3} **64.** 1.76×10^{-1} **65.** 2.08×10^3 **66.** 3.14×10^{-4} **67.** 0.0042 **68.** 16,500 **69.** 970,000 **70.** 0.00000438 **71.** 0.914 **72.** 536 **73.** 4,600,000 **74.** 1260 **75.** 19.84 **76.** 340,000 **77.** 0.09 **78.** 0.00003 **79.** 1.2×10^9 **80.** 2.4×10^1 **81.** 9.2 **82.** 2×10^8 **83.** 3.4×10^{-3} **84.** 5×10^8 **85. (a)** about 28 times greater **(b)** 22,083,000 **86.** 3.0524×10^{15} mi **87.** Binomial, 1 **88.** Monomial, 0 **89.** $x^2 + 3x - 4$, Trinomial, second **90.** $4x^2 - x - 3$, Trinomial, second **91.** Binomial, second **92.** Not polynomial **93.** $-4x^2 + x$, Binomial, second **94.** Not polynomial **95.** $x^3 + 4x^2 - 2x - 6$, third **96.** $3x + 7$ **97.** $9x + 1$ **98.** $2x - 5$ **99.** $-3x^2 + 10x - 15$ **100.** $11x^2 - 2x - 3$ **101.** $x - 6.7$ **102.** $-2x + 2$ **103.** $9x^2 - \frac{5}{4}x + 4$ **104.** $6x^2 - 18x - 4$ **105.** $-5x^2 + 8x - 19$ **106.** $4x + 2$ **107.** $2x^2 - 4x$ **108.** $4.5x^3 - 13.5x^2$ **109.** $6x^3 - 12x^2 + 21x$ **110.** $-3x^3 + 6x^2 + x$ **111.** $24x^3 - 16x^2 + 8x$ **112.** $x^2 + 9x + 20$ **113.** $2x^2 - 2x - 12$ **114.** $16x^2 + 48x + 36$ **115.** $-6x^2 + 14x + 12$ **116.** $x^2 - 16$ **117.** $3x^3 + 7x^2 + 14x + 4$ **118.** $3x^3 + x^2 - 10x + 6$ **119.** $10x^3 - 19x^2 + 36x - 12$ **120.** $x + 2$ **121.** $x - 2$ **122.** $8x + 4$ **123.** $2x^2 + 3x - \frac{4}{3}$ **124.** $8x + 6 - \frac{4}{x}$ **125.** $4x^4 - 2x^3 + \frac{3}{2}x - \frac{1}{x}$ **126.** $-8x + 2$ **127.** $-2x - 1 - \frac{4}{x}$ **128.** $\frac{5}{2}x + \frac{5}{x} + \frac{1}{x^2}$ **129.** $x + 4$ **130.** $2x - 3$ **131.** $5x - 2 + \frac{2}{x + 6}$ **132.** $2x^2 + 3x - 4$ **133.** $2x^2 + x - 2 + \frac{2}{2x - 1}$ **134.** 4 hr **135.** 538.46 mph **136.** 2 hr **137.** \$4000 at 8%, \$8000 at $7\frac{1}{4}$% **138.** 1.2 *L* of 10%, 0.8 *L* of 5% **139.** 6.5 mph **140.** 4 hr **141.** 60 lb of \$3.50, 20 lb of \$4.10 **142.** 32¢ stamps, 36; 22¢ stamps, 4 **143.** older brother, 55 mph; younger brother, 60 mph **144.** 0.4*L*.

Chapter 4 Practice Test

1. $6x^6$ **2.** $27x^6$ **3.** $4x^3$ **4.** $\frac{x^3}{8y^6}$ **5.** $\frac{y^4}{5x^6}$ **6.** $\frac{1}{5x^3y^6}$ **7.** Trinomial **8.** Monomial **9.** Not polynomial **10.** $6x^3 - 2x^2 + 5x - 5$, third degree **11.** $3x^2 - 3x + 1$ **12.** $-2x^2 + 4x$ **13.** $3x^2 - x + 3$ **14.** $12x^3 - 6x^2 + 15x$ **15.** $8x^2 + 2x - 21$ **16.** $-12x^2 - 2x + 30$ **17.** $6x^3 - 4x^2 - 28x + 24$ **18.** $4x^2 + 2x - 1$

19. $-x + 2 - \frac{5}{3x}$ **20.** $4x + 5$ **21.** 57.14 hr **22.** 80 mph **23.** 20 L **24.** 1.26×10^9 **25.** 2×10^{-7}

Cumulative Review Test

1. -40 **2.** 20 **3.** All real numbers **4.** $x < -5$,  **5.** $w = \frac{v}{lh}$ **6.** $y = \frac{4x - 6}{3}$ or $y = \frac{4}{3}x - 2$
7. $6x^9$ **8.** $135x^8y^{13}$ **9.** $3x^2 - 2x - 5$, second degree **10.** $-4x^2 - 6x - 11$ **11.** $8x^2 - 3x + 9$ **12.** $x^2 - 3x - 7$
13. $6x^2 - 19x + 15$ **14.** $2x^3 - 6x^2 - 12x - 40$ **15.** $3x - 2 + \frac{8}{3x}$ **16.** $2x - 3$ **17.** \$3.33 **18.** 4
19. $w = 3$ ft, $l = 10$ ft **20.** 2 hr

Chapter 5

Exercise Set 5.1

1. $2^2 \cdot 3^2$ **3.** $2 \cdot 3^2 \cdot 5$ **5.** $2^3 \cdot 5^2$ **7.** 4 **9.** 12 **11.** 18 **13.** x **15.** $3x$ **17.** 1 **19.** xy **21.** x^3y^5 **23.** 5
25. x^2y^2 **27.** x **29.** $x + 3$ **31.** $2x - 3$ **33.** $3x - 4$ **35.** $x + 4$ **37.** $3(x + 2)$ **39.** $5(3x - 1)$
41. cannot be factored **43.** $3x(3x - 4)$ **45.** $2p(13p - 4)$ **47.** $2x(2x^2 - 3)$ **49.** $12x^8(3x^4 - 2)$ **51.** $3y^3(8y^{12} - 3)$
53. $x(1 + 3y^2)$ **55.** cannot be factored **57.** $4xy(4yz + x^2)$ **59.** $2xy^2(17x + 8y^2)$ **61.** $36xy^2z(z^2 + x^2)$
63. $y^3z^5(14 - 9x)$ **65.** $3(x^2 + 2x + 3)$ **67.** $3(3x^2 + 6x + 1)$ **69.** $4x(x^2 - 2x + 3)$ **71.** cannot be factored
73. $3(5p^2 - 2p + 3)$ **75.** $4x^3(6x^3 + 2x - 1)$ **77.** $xy(8x + 12y + 9)$ **79.** $(x + 3)(x + 4)$ **81.** $(7x - 4)(4x - 3)$
83. $(4x + 1)(2x + 1)$ **85.** $(4x + 1)(2x + 1)$ **87.** An expression written as a product of factors.

Cumulative Review Exercises

91. $-2x + 18$ **92.** 1 **93.** $\frac{1}{2}$ **94.** $h = \frac{2A}{b}$

Group Activity/Challenge Problems

1. $2(x - 3)[2x^2(x - 3)^2 - 3x(x - 3) + 2]$ **2.** $2x^2(2x + 7)(3x^3 + 2x - 1)$ **3.** $x^{1/3}(x + 2)(x + 3)$ **4.** $5x^{-1/2}(3x + 1)$
5. $(x + 3)(x + 2)$ **6. (a)** $3(0) + 3(1) + 3(2) + 3(3) + 3(4)$ **(b)** $3(0 + 1 + 2 + 3 + 4)$ **(c)** $3(10) = 30$
(d) $3(55) = 165$

Exercise Set 5.2

1. $(x + 3)(x + 4)$ **3.** $(x + 4)(x + 2)$ **5.** $(x + 5)(x + 2)$ **7.** $(x - 5)(x + 3)$ **9.** $(2x - 3)(2x + 3)$
11. $(3x + 1)(x + 3)$ **13.** $(2x - 1)(2x - 1) = (2x - 1)^2$ **15.** $(8x + 1)(x + 4)$ **17.** $(x + 1)(3x - 2)$
19. $(2x - 3)(x - 2)$ **21.** $(3x + 5)(5x - 3)$ **23.** $(x - 3y)(x + 2y)$ **25.** $(3x + y)(2x - 3y)$ **27.** $(2x - 5y)(5x - 6y)$
29. $(x + a)(x + b)$ **31.** $(x - 2)(y + 4)$ **33.** $(a + b)(a + 2)$ **35.** $(x + 5)(y - 1)$ **37.** $(4 - x)(3 + 2y)$
39. $(a^2 + 1)(a + 2)$ **41.** $(x^2 - 3)(x + 4)$ **43.** $2(x + 4)(x - 6)$ **45.** $4(x + 2)(x + 2) = 4(x + 2)^2$
47. $x(3x - 1)(2x + 3)$ **49.** $2(x + 4y)(x - 2y)$ **51.** $(x + 2)(y + 3)$ **53.** $(x + 5)(y + 6)$ **55.** $(a + b)(x + y)$
57. $(d + 3)(c - 4)$ **59.** $(a + b)(c - d)$ **61.** Determine if all terms have a common factor; if so, factor out the GCF.
63. $x^2 + 4x - 2x - 8$, multiply the factors using the FOIL method.

Cumulative Review Exercises

65. 7.14 years **66.** 21 days or more **67.** $5x^2 - 2x - 3 + \frac{5}{3x}$ **68.** $x + 3$

Group Activity/Challenge Problems

1. $x(3x^2 + 2)(x^2 - 5)$ **3.** $(3a - x)(6a + x^2)$ **5. (a)** $3x^2 + 6x + 4x + 8$ **(b)** $3x^2 + 10x + 8 = (3x + 4)(x + 2)$
7. (a) $2x^2 - 6x - 5x + 15$ **(b)** $2x^2 - 11x + 15 = (2x - 5)(x - 3)$ **9. (a)** $4x^2 - 20x + 3x - 15$
(b) $4x^2 - 17x - 15 = (4x + 3)(x - 5)$

EXERCISE SET 5.3

1. $(x + 2)(x + 1)$ **3.** $(x + 2)(x + 4)$ **5.** $(x + 4)(x + 3)$ **7.** cannot be factored **9.** $(y - 15)(y - 1)$
11. $(x + 3)(x - 2)$ **13.** $(r + 5)(r - 3)$ **15.** $(b - 9)(b - 2)$ **17.** cannot be factored **19.** $(a + 11)(a + 1)$
21. $(x + 15)(x - 2)$ **23.** $(x + 2)^2$ **25.** $(k + 3)^2$ **27.** $(x + 5)^2$ **29.** $(w - 15)(w - 3)$ **31.** $(x + 24)(x - 2)$
33. $(x - 5)(x + 4)$ **35.** $(y - 7)(y - 2)$ **37.** $(x + 16)(x - 4)$ **39.** $(x - 2)(x - 12)$ **41.** $(x + 8)(x - 10)$
43. $(x - 5)(x - 12)$ **45.** $(x + 2)(x + 28)$ **47.** $(x - 2y)^2$ **49.** $(x + 5y)(x + 3y)$ **51.** $2(x - 5)(x - 1)$
53. $5(x + 3)(x + 1)$ **55.** $2(x - 4)(x - 3)$ **57.** $x(x - 6)(x + 3)$ **59.** $2x(x + 7)(x - 4)$ **61.** $x(x + 2)^2$
63. Both +, the constant and x term are both positive, therefore the signs in the factors must both be positive.
65. One + the other −, the constant is negative, therefore one factor must contain a + sign and the other a − sign.
67. Both −, the constant is positive and the x term is negative, therefore the signs in the factors must both be negative.
69. $x^2 - 11x + 24$, multiply the factors and combine like terms. **71.** $2x^2 - 8xy - 10y^2$, multiply the factors and combine like terms. **73.** Multiply factors using the FOIL method.

CUMULATIVE REVIEW EXERCISES

75. 7 **76.** $2x^3 + x^2 - 16x + 12$ **77.** $3x + 2 - \dfrac{2}{x - 4}$ **78.** 19.6% **79.** $(x - 2)(3x + 5)$

GROUP ACTIVITY/CHALLENGE PROBLEMS

1. $(x + 0.4)(x + 0.2)$ **3.** $\left(x + \frac{1}{5}\right)\left(x + \frac{1}{5}\right)$ **5.** $-(x + 2)(x + 4)$ **7.** $(x + 20)(x - 15)$ **9.** $(x + 100)(x + 80)$
11. $(x + 100)(x - 80)$ **13.** $(x - 200)(x - 40)$

EXERCISE SET 5.4

1. $(2x + 3)(x + 2)$ **3.** $(2x + 3)(3x + 2)$ **5.** $(3x + 1)(x + 1)$ **7.** $(2x + 5)(x + 3)$ **9.** $(2x - 1)(2x + 3)$
11. $(5y - 3)(y - 1)$ **13.** cannot be factored **15.** $(4x + 1)(x + 3)$ **17.** cannot be factored **19.** $(5y - 1)(y - 3)$
21. $(2x + 5)(2x - 3)$ **23.** $(7x - 2)(x - 2)$ **25.** $(x - 1)(3x - 7)$ **27.** $(5z + 2)(z - 7)$ **29.** $(4x + 3)(2x - 1)$
31. $(5x - 1)(2x - 5)$ **33.** $(4x + 5)(2x - 3)$ **35.** $3(2x + 1)(x + 5)$ **37.** $2(3x + 5)(x - 1)$ **39.** $x(2x - 1)(3x + 4)$
41. $2x(2x + 3)(x - 1)$ **43.** $2x(3x + 5)(x - 1)$ **45.** $5(6x + 1)(2x + 1)$ **47.** $(2x + y)(x + 2y)$
49. $(2x - y)(x - 3y)$ **51.** $2(3x - y)(3x + 4y)$ **53.** $3(2x + 3y)(x - 4y)$ **55.** $6x^2 + x - 12$ **57.** $6x^2 + 21x + 15$
59. $2x^4 - x^3 - 3x^2$ **61. (a)** The quotient of the trinomial divided by the binomial is the second factor. **(b)** $6x + 11$
63. Multiply the factors together; the product should be the trinomial you started with.

CUMULATIVE REVIEW EXERCISES

65. $\dfrac{1}{2}$ **66.** $w = 3$ ft, $l = 8$ ft **67.** $12xy^2(3x^3y - 1 + 2x^4y^4)$ **68.** $(x - 9)(x - 6)$

GROUP ACTIVITY/CHALLENGE PROBLEMS

1. $(6x - 5)(3x + 4)$ **3.** $(5x - 8)(3x - 20)$ **5.** $4(6x + 5)(3x - 10)$

EXERCISE SET 5.5

1. $(x + 2)(x - 2)$ **3.** $(y + 5)(y - 5)$ **5.** $(x + 7)(x - 7)$ **7.** $(x + y)(x - y)$ **9.** $(3y + 4)(3y - 4)$
11. $4(4a + 3b)(4a - 3b)$ **13.** $(5x + 4)(5x - 4)$ **15.** $(z^2 + 9x)(z^2 - 9x)$ **17.** $9(x^2 + 3y)(x^2 - 3y)$
19. $(7m^2 + 4n)(7m^2 - 4n)$ **21.** $20(x + 3)(x - 3)$ **23.** $4(2x + 5y^2)(2x - 5y^2)$ **25.** $(x + y)(x^2 - xy + y^2)$
27. $(a - b)(a^2 + ab + b^2)$ **29.** $(x + 2)(x^2 - 2x + 4)$ **31.** $(x - 3)(x^2 + 3x + 9)$ **33.** $(a + 1)(a^2 - a + 1)$
35. $(2x + 3)(4x^2 - 6x + 9)$ **37.** $(3a - 4)(9a^2 + 12a + 16)$ **39.** $(3 - 2y)(9 + 6y + 4y^2)$
41. $(2x - 3y)(4x^2 + 6xy + 9y^2)$ **43.** $2(x - 3)(x + 2)$ **45.** $y(x + 4)(x - 4)$ **47.** $3(x + 1)^2$ **49.** $5(x + 3)(x - 1)$
51. $3(x + 3)(y - 2)$ **53.** $2(x + 6)(x - 6)$ **55.** $3y(x + 3)(x - 3)$ **57.** $3y^2(x + 1)(x^2 - x + 1)$
59. $2(x - 2)(x^2 + 2x + 4)$ **61.** $2(x + 4)(3x - 2)$ **63.** $x(3x + 2)(x - 4)$ **65.** $(x + 2)(4x - 3)$
67. $25(b + 2)(b - 2)$ **69.** $a^3b^2(a + 2b)(a - 2b)$ **71.** $3x^2(x - 3)^2$ **73.** $x(x^2 + 25)$ **75.** $(y^2 + 4)(y + 2)(y - 2)$

77. $5(2a - 3b)(a + 4b)$ **79.** $(2a - 3)(b + 2)$ **81.** $9(1 + y^2)(1 + y)(1 - y)$
83. (a) $a^3 + b^3 = (a + b)(a^2 - ab + b^2)$

Cumulative Review Exercises

85. All real numbers 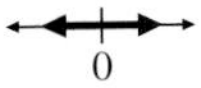 **86.** $y = \dfrac{2x - 6}{5}$ or $y = \dfrac{2}{5}x - \dfrac{6}{5}$ **87.** $\dfrac{8x^9}{27y^{12}}$ **88.** $\dfrac{1}{x^5}$

Group Activity/Challenge Problems

2. $(x^2 + 1)(x^4 - x^2 + 1)$ **3.** $(x^2 - 3y^3)(x^4 + 3x^2y^3 + 9y^6)$ **4.** $(x - 3 + 2y)(x - 3 - 2y)$

Exercise Set 5.6

1. 0, −2 **3.** 0, 9 **5.** $-\frac{5}{2}$, 3 **7.** 4, −4 **9.** 0, 12 **11.** 0, −2 **13.** −4, 3 **15.** 2, 10 **17.** 3, −6 **19.** −4, 6
21. 2, −21 **23.** −5, −6 **25.** 4, −2 **27.** 32, −2 **29.** 6, −3 **31.** $\frac{1}{3}$, 7 **33.** $\frac{2}{3}$, −1 **35.** −4, 3 **37.** $\frac{4}{3}$, $-\frac{1}{2}$
39. 2, 3 **41.** 0, 16 **43.** 6, −6 **45.** 3, −3 **47.** 9, −9 **49.** 10, 12 **51.** 4, 16 **53.** $w = 6$ in, $l = 9$ in
55. $w = 5$ ft, $l = 7$ ft **57.** 5 sec **59.** 30 water sprinklers **61. (a)** 650 **(b)** 10 **63. (a)** 66 **(b)** 11
65. You can eliminate the 3 by dividing both sides of the equation by 3 to get $(x - 4)(x + 5) = 0$.

Cumulative Review Exercises

66. −9 **67.** $-x^2 + 7x - 4$ **68.** $6x^3 + x^2 - 10x + 4$ **69.** $2x - 3$ **70.** $2x - 3$

Group Activity/Challenge Problems

1. $x^2 - 9x + 18 = 0$ **3.** $x^2 + 14x + 45 = 0$ **5.** $x^3 - 8x^2 + 15x = 0$ **7.** 20 and 50 **9.** 0, 2, −5

Chapter 5 Review Exercises

1. x^2 **2.** $3p$ **3.** 6 **4.** $4x^2y^2$ **5.** 1 **6.** $8x^2$ **7.** $2x - 7$ **8.** $x + 5$ **9.** $4(x - 4)$ **10.** $5(2x + 1)$
11. $4y(6y - 1)$ **12.** $5p^2(11p - 4)$ **13.** $6x^2y(4 + 3xy)$ **14.** $6xy(1 - 2x)$ **15.** $2(x^2 + 2x - 4)$
16. $6x^4y^2(10y^2 + x^5y - 3x)$ **17.** Cannot be factored **18.** $(x - 2)(5x + 3)$ **19.** $(3x - 2)(x - 1)$ **20.** $(2x + 1)(4x - 3)$
21. $(x + 2)(x + 4)$ **22.** $(x + 4)(x - 3)$ **23.** $(x + 7)(x - 7)$ **24.** $(2a - 1)(a - b)$ **25.** $(3x + 2)(y + 1)$
26. $(x - 2y)(x + 3)$ **27.** $(5x - 1)(x + 4)$ **28.** $(x + 4y)(5x - y)$ **29.** $(4x + 5y)(3x - 2y)$ **30.** $(3x - 2y)(4x + 5y)$
31. $(a + 1)(b - 1)$ **32.** $(3x + 2y)(x - 3y)$ **33.** $(4x + 3)(5x - 3)$ **34.** $(3x - 1)(2x + 3)$ **35.** $(x + 2)(x + 5)$
36. $(x - 5)(x - 3)$ **37.** $(x - 5)(x + 4)$ **38.** $(x + 5)(x - 4)$ **39.** $(x - 6)(x - 5)$ **40.** $(x - 8)(x - 7)$
41. Cannot be factored **42.** $(x + 8)(x + 3)$ **43.** $x(x + 4)(x + 1)$ **44.** $x(x - 8)(x + 5)$ **45.** $(x + 3y)(x - 5y)$
46. $4x(x + 5y)(x + 3y)$ **47.** $(2x - 1)(x + 3)$ **48.** $(3x + 1)(x + 4)$ **49.** $(4x - 5)(x - 1)$ **50.** $(5x + 2)(x - 3)$
51. $(2x - 3)(2x + 5)$ **52.** $(5x - 2)(x - 6)$ **53.** Cannot be factored **54.** $(6x + 1)(x + 5)$ **55.** $(2x - 3)(2x + 7)$
56. $(3x - 2)(2x + 5)$ **57.** $(4x + 5)(2x - 7)$ **58.** $(2x + 5)^2$ **59.** $x(3x - 2)^2$ **60.** $2x(3x + 1)(3x - 5)$
61. $(2x - 3y)(2x - 5y)$ **62.** $(8x + y)(2x - 3y)$ **63.** $(x + 4)(x - 4)$ **64.** $(x + 8)(x - 8)$ **65.** $4(x + 2)(x - 2)$
66. $9(3x + y)(3x - y)$ **67.** $(8x^2 + 9y^2)(8x^2 - 9y^2)$ **68.** $(4 + 5y)(4 - 5y)$ **69.** $(2x^2 + 3y^2)(2x^2 - 3y^2)$
70. $(10x^2 + 11y^2)(10x^2 - 11y^2)$ **71.** $(x - y)(x^2 + xy + y^2)$ **72.** $(x + y)(x^2 - xy + y^2)$ **73.** $(a + 2)(a^2 - 2a + 4)$
74. $(a - 1)(a^2 + a + 1)$ **75.** $(a + 3)(a^2 - 3a + 9)$ **76.** $(x - 2)(x^2 + 2x + 4)$ **77.** $(2x - y)(4x^2 + 2xy + y^2)$
78. $(3 - 2y)(9 + 6y + 4y^2)$ **79.** $3(3x^2 + 5y)(3x^2 - 5y)$ **80.** $2(x - 4y)(x^2 + 4xy + 16y^2)$ **81.** $(x - 5)(x - 10)$
82. $2(x - 4)^2$ **83.** $4(x + 3)(x - 3)$ **84.** $4(y + 4)(y - 4)$ **85.** $8(x + 3)(x - 1)$ **86.** $(x - 9)(x + 3)$
87. $(2x + 3)(2x - 5)$ **88.** $3(2x - 3)(x - 4)$ **89.** $8(x - 1)(x^2 + x + 1)$ **90.** $y(x - 3)(x^2 + 3x + 9)$
91. $y(x + 4)(x - 1)$ **92.** $3x(2x + 3)(x + 5)$ **93.** $(x + 3y)(x + 2y)$ **94.** $(2x - 5y)(x + 2y)$ **95.** $(2x - 5y)^2$
96. $(4y + 7z)(4y - 7z)$ **97.** $(a + 6)(b + 7)$ **98.** $y^5(4 + 5y)(4 - 5y)$ **99.** $2x(x + 4y)(x + 2y)$
100. $(2x - 3y)(3x + 7y)$ **101.** $2x(4x + 1)(4x + 3)$ **102.** $(y^2 + 1)(y + 1)(y - 1)$ **103.** 0, 5 **104.** −3, −4
105. 5, $-\frac{2}{3}$ **106.** 0, 3 **107.** 0, −4 **108.** 6, −4 **109.** −3, −5 **110.** −4, 2 **111.** −4, 3 **112.** −2, −5
113. 2, 4 **114.** 1, −2 **115.** $\frac{1}{4}$, $-\frac{3}{2}$ **116.** $\frac{1}{2}$, −8 **117.** 2, −2 **118.** $\frac{7}{6}$, $-\frac{7}{6}$ **119.** $\frac{3}{2}$, −8 **120.** $\frac{3}{2}$, $\frac{5}{2}$ **121.** 10, 11
122. 6, 8 **123.** 5, 8 **124.** $w = 7$ ft, $l = 9$ ft **125.** 5 in, 9 in **126.** 20

CHAPTER 5 PRACTICE TEST

1. $2x^2$ **2.** $3xy^2$ **3.** $4xy(x-2)$ **4.** $3x(8xy-2y+3)$ **5.** $(x+2)(x-3)$ **6.** $(3x+1)(x-4)$ **7.** $(5x-3y)(x-3y)$ **8.** $(x+4)(x+8)$ **9.** $(x+8)(x-3)$ **10.** $(x-5y)(x-4y)$ **11.** $2(x-5)(x-6)$ **12.** $x(2x-1)(x-1)$ **13.** $(3x+2y)(4x-3y)$ **14.** $(x+3y)(x-3y)$ **15.** $(x+3)(x^2-3x+9)$ **16.** $2, \frac{5}{2}$ **17.** $-2, -3$ **18.** $-5, 1$ **19.** $4, 9$ **20.** $l = 6$ m, $w = 4$ m

Chapter 6

EXERCISE SET 6.1

1. $x \neq 0$ **3.** $x \neq 6$ **5.** $x \neq 2, x \neq -2$ **7.** $x \neq 2, x \neq -8$ **9.** $\frac{1}{1+y}$ **11.** 4 **13.** $\frac{x^2+6x+3}{2}$ **15.** $x+1$ **17.** $\frac{x}{x-2}$ **19.** $\frac{x-3}{x-2}$ **21.** $2(x+1)$ **23.** -1 **25.** $-(x+2)$ **27.** $\frac{-(x+6)}{2x}$ **29.** $-(x+3)$ **31.** $3x+2$ **33.** $3x-4$ **35.** $2x-3$ **37.** $x+4$ **39.** $\frac{x-4}{x+4}$ **41.** x^2+2x+4 **43.** No real number will make the denominator 0. **45.** $x \neq 3$ since 3 would make the denominator 0. **47.** No **49.** The numerator and denominator have no common factors other than 1. **51.** There is no factor common to both terms in the numerator. **53.** $x+2$ **55.** $x^2+7x+12$

CUMULATIVE REVIEW EXERCISES

57. $y = x - 2z$ **58.** 28°, 58°, 94° **59.** $\frac{x^4}{9y^2}$ **60.** $9x^2-10x-17$

GROUP ACTIVITY/CHALLENGE PROBLEMS

1. (a) $x \neq -3, x \neq 2$ **(b)** $\frac{1}{x-2}$ **3. (a)** $x \neq 0, x \neq -5, x \neq \frac{3}{2}$ **(b)** $\frac{1}{x(2x-3)}$ **5.** 1

EXERCISE SET 6.2

1. $\frac{xy}{4}$ **3.** $\frac{80x^4}{y^6}$ **5.** $\frac{36x^9y^2}{25z^7}$ **7.** $\frac{-3x+2}{3x+2}$ **9.** 1 **11.** a^2-b^2 **13.** $\frac{x+3}{2(x+2)}$ **15.** $x+3$ **17.** $3x^2y$ **19.** $\frac{10z}{x}$ **21.** $6a^2b$ **23.** $\frac{3x^2}{2}$ **25.** $\frac{x}{x+6}$ **27.** $\frac{x-8}{x+2}$ **29.** $\frac{x+3}{x-1}$ **31.** 1 **33.** $6x^3y^3$ **35.** $\frac{5}{3ab^2c^2}$ **37.** $\frac{3y^2}{a^2}$ **39.** $\frac{32x^7}{35my^2}$ **41.** $\frac{1}{2}$ **43.** $3x$ **45.** 1 **47.** 1 **49.** $\frac{2}{3}$ **51.** $-\frac{1}{x}$ **53.** $\frac{4x}{3y}$ **55.** 1 **57.** $12x^2-x-6$ **59.** $\frac{x-1}{x+1}$ **61.** $\frac{3x+6}{3x+y}$ **63.** $\frac{-2x+3}{3x-2}$ **65.** $\frac{1}{x-2y}$ **67.** x^2-2x-8, numerator must be $(x-4)(x+2)$. **69.** x^2-x-12, denominator must be $(x-4)(x+3)$.

CUMULATIVE REVIEW EXERCISES

73. $20x^4y^5z^{11}$ **74.** $2x^2+x-2-\frac{2}{2x-1}$ **75.** $3(x-5)(x+2)$ **76.** $5, -2$

GROUP ACTIVITY/CHALLENGE PROBLEMS

1. $\frac{x-3}{x+3}$ **2.** $\frac{x-1}{x-3}$ **3.** Numerator must be $(x-2)(x-3)$ or x^2-5x+6. Denominator must be $(x-5)(x+4)$ or x^2-x-20.

EXERCISE SET 6.3

1. $\frac{2x-1}{6x}$ **3.** $\frac{x+3}{5}$ **5.** $\frac{x+3}{x}$ **7.** 2 **9.** $\frac{3x+3}{x-2}$ **11.** $\frac{2x+1}{2x^2}$ **13.** $\frac{1}{x-2}$ **15.** 0 **17.** $\frac{-4x-1}{x-7}$ **19.** $x-3$ **21.** $\frac{-1}{x-2}$ **23.** 1 **25.** $\frac{2x+9}{2x-9}$ **27.** $\frac{x-3}{x+5}$ **29.** $\frac{3}{4}$ **31.** $\frac{x-8}{x+8}$ **33.** $\frac{3x+2}{x-4}$ **35.** $\frac{6x+1}{x-8}$ **37. (b)** $\frac{4x-3-(2x-7)}{5x+4}$ **39. (b)** $\frac{4x+5-(-x^2+3x+6)}{x^2-6x}$ **41.** 3 **43.** $6x$ **45.** $10x$ **47.** x^2 **49.** $2x+3$ **51.** $x^2(x+1)$ **53.** $144x^3y$ **55.** $36x(x+5)$ **57.** $x(x+1)$ **59.** $180x^2y^3$ **61.** $6(x+4)(x-3)$ **63.** $(x+6)(x+5)$ **65.** $(x-8)(x+3)(x+8)$ **67.** $(x+3)(x-3)(x-1)$ **69.** $(x-2)(x+1)(x+3)$ **71.** $(x+2)^2$ **73.** $(x-5)(x-4)$ **75.** $(3x-2)(x+6)(3x-1)$ **77.** $(2x+1)^2(4x+3)$ **79.** signs change **81.** $(x+4)(x-3)$ **83.** $x^2+9x-10$, sum of numerators must be $5x-7$ **85.** $-4x^2-3x-9$, difference of numerators must be x^2+3x

CUMULATIVE REVIEW EXERCISES

86. $\frac{92}{45}$ or $2\frac{2}{45}$ **87.** $-\frac{1}{5}$ **88.** $1\frac{1}{8}$ or 1.125 oz **89. (a)** 75 hr **(b)** plan 2 **90.** $\frac{(x-2)^2}{(2x+1)(x+3)}$

GROUP ACTIVITY/CHALLENGE PROBLEMS

1. $\frac{-4x^2+8x+3}{x^2-9}$ **3.** $30x^{12}y^9$ **5.** $(x-4)(x+3)(x-2)$ **7.** $(2x+5)(x-2)(x+1)$

EXERCISE SET 6.4

1. $\frac{7}{2x}$ **3.** $\frac{3x+8}{2x^2}$ **5.** $\frac{2x^2-1}{x^2}$ **7.** $\frac{3x+10}{5x^2}$ **9.** $\frac{28x+15y}{20x^2y^2}$ **11.** $\frac{xy+x}{y}$ **13.** $\frac{9x+1}{6x}$ **15.** $\frac{5x^2+y^2}{xy}$ **17.** $\frac{4y-30x^2}{5x^2y}$ **19.** $\frac{8x-10}{x(x-2)}$ **21.** $\frac{11a+6}{a(a+3)}$ **23.** $\frac{-2x^2+4x+8}{3x(x+2)}$ **25.** $\frac{2}{x-3}$ **27.** $\frac{11}{x+5}$ **29.** $\frac{7x+1}{(x+1)(x-1)}$ **31.** $\frac{20x}{(x-5)(x+5)}$ **33.** $\frac{5x-12}{(x+3)(x-3)}$ **35.** $\frac{-2x+12}{(x+3)(x-3)}$ **37.** $\frac{11}{(x-3)(x-4)}$ **39.** $\frac{6x-8}{(x+4)(x-2)}$ **41.** $\frac{x^2+2x-3}{(x+2)(x+2)}$ **43.** $\frac{4x+18}{(x+5)(x-2)(x+3)}$ **45.** $\frac{4x-3}{(x+3)(x+2)(x-2)}$ **47.** $\frac{2x^2-9x+4}{(3x-1)(x+2)(2x+3)}$ **49.** $\frac{2x^2-3x-4}{(4x+3)(x+2)(2x-1)}$ **51. (b)** $\frac{x^2+x-9}{(x+2)(x-3)(x-2)}$

CUMULATIVE REVIEW EXERCISES

52. 810 **53.** $x > -8$ **54.** $4x-3-\frac{4}{2x+3}$ **55.** 1

GROUP ACTIVITY/CHALLENGE PROBLEMS

1. $\frac{x^2+5x-2}{(x+2)(x-2)}$ **3.** $\frac{2x-3}{2-x}$ **5.** $\frac{3x^3-6x^2+16x-8}{(x+2)(x-2)(x^2-2x+4)}$

EXERCISE SET 6.5

1. $\frac{8}{11}$ **3.** $\frac{57}{32}$ **5.** $\frac{25}{1224}$ **7.** $\frac{x^3y}{8}$ **9.** $6xz^2$ **11.** $\frac{xy+1}{x}$ **13.** $\frac{3}{x}$ **15.** $\frac{3y-1}{2y-1}$ **17.** $\frac{x-y}{y}$ **19.** $-\frac{a}{b}$ **21.** -1
23. $\frac{2(x+2)}{x^3}$ **25.** $a+b$ **27.** $\frac{a^2+b}{b(b+1)}$ **29.** $\frac{y-x}{x+y}$ **31.** $\frac{y^2(x+1)}{x^2(y+1)}$
33. A fraction that contains a fraction in its numerator or its denominator or in both.

CUMULATIVE REVIEW EXERCISES

35. All real numbers
36. An expression containing a finite number of terms of the form ax^n, for any real number a and any whole number n.
37. $48x^5y^{10}$ **38.** $\frac{x^2-9x+2}{(3x-1)(x+6)(x-3)}$

GROUP ACTIVITY/CHALLENGE PROBLEMS

1. (a) $\frac{2}{7}$ **(b)** $\frac{4}{13}$ **3.** $\frac{a^3b+a^2b^3-ab^2}{a^3-ab^3+b^2}$

EXERCISE SET 6.6

1. 6 **3.** 48 **5.** 30 **7.** -1 **9.** 1 **11.** 4 **13.** 3 **15.** $-\frac{1}{5}$ **17.** $\frac{1}{4}$ **19.** $\frac{14}{3}$ **21.** No solution **23.** 2 **25.** $-\frac{12}{7}$
27. No solution **29.** 8 **31.** -14 **33.** $-2, -3$ **35.** 4 **37.** $-\frac{5}{2}$ **39.** No solution **41.** 5 **43.** No solution
45. -3 **47.** 2 **49. (b)** $\frac{2}{3}$
51. (a) The problem on the right is an equation, while the one on the left is not. **(b)** Problem on left: write each fraction with the least common denominator $12(x-1)$; then combine numerators. Problem on right: multiply both sides of the equation by the LCD $12(x-1)$ to eliminate fractions; then solve the remaining equation.
(c) Problem on left: $\frac{x^2-x+12}{12(x-1)}$; Problem on right: $4, -3$

CUMULATIVE REVIEW EXERCISES

52. 30 rides **53.** 50°, 130° **54.** 75 min
55. Linear equation: $ax+b=c, a \neq 0$; quadratic equation $ax^2+bx+c=0, a \neq 0$

GROUP ACTIVITY/CHALLENGE PROBLEMS

1. 15 cm **4. (a)** 3

EXERCISE SET 6.7

1. $80 = \frac{1}{2}(h+6)h$; $b = 16$ cm, $h = 10$ cm **3.** $\frac{1}{x}+\frac{1}{3x}=\frac{4}{3}$; 1, 3 **5.** $\frac{1}{3}+\frac{1}{5}=\frac{1}{x}$; $\frac{15}{8}$ **7.** $\frac{10}{4+c}=\frac{6}{4-c}$; 1 mph
9. $\frac{d}{30}+\frac{d}{30}=\frac{1}{2}$; 7.5 mi **11.** $\frac{400}{r+30}=\frac{250}{r}$; 50 km/hr, 80 km/hr **13.** $\frac{2}{r}+\frac{2}{2r}=1$, walks 3 mph, jogs 6 mph
15. $\frac{d}{5}+\frac{3-d}{2}=0.9$; jogs 0.4 hr, walks 0.5 hr **17.** $\frac{d}{6}=\frac{d}{9}+0.5$; 9 mi **19.** $\frac{t}{3}+\frac{t}{6}=1$; 2 hr
21. $\frac{t}{8}+\frac{t}{5}=1$; $3\frac{1}{13}$ hr **23.** $\frac{t}{4}-\frac{t}{5}=1$; 20 min **25.** $\frac{3}{4}+\frac{3}{t}=1$; 12 hr **27.** $\frac{5}{12}+\frac{t}{15}=1$; $8\frac{3}{4}$ days
29. $\frac{t}{60}+\frac{t}{50}-\frac{t}{30}=1$; 300 hr **31.** $\frac{t}{300}+\frac{t}{100}+\frac{t}{200}=1$; $\frac{600}{11}$ or about 54.5 yr

Cumulative Review Exercises

33. 8 **34.** $-\frac{3}{2}x - \frac{9}{2}$ **35.** 1 **36.** $\frac{3(x^2 - 3x - 5)}{(2x + 3)(3x - 5)(3x + 1)}$

Group Activity/Challenge Problems

1. $\frac{1}{x - 3} = 2\left(\frac{1}{2x - 6}\right)$; all real numbers except 3 **3.** $\frac{b}{3} = \frac{b}{6} + 1.5$; 9 buckets

5. (a) 40 min **(b)** 10 min **(c)** $\frac{t}{40} + \frac{t}{10} = 1$; 8 min

Chapter 6 Review Exercises

1. $x \neq 4$ **2.** $x \neq 3, x \neq 4$ **3.** $x \neq \frac{3}{2}, x \neq 5$ **4.** $\frac{1}{1 - y}$ **5.** $x^2 + 4x + 12$ **6.** $3x + 2y$ **7.** $x + 4$ **8.** $x + 2$
9. $-(2x - 1)$ **10.** $\frac{x - 6}{x + 2}$ **11.** $\frac{3x + 4}{x - 4}$ **12.** $\frac{x - 8}{2x + 3}$ **13.** $\frac{8xy^2}{3}$ **14.** $6xz^2$ **15.** $\frac{16b^3c^2}{a^2}$ **16.** $-\frac{1}{2}$ **17.** $-2x$
18. 1 **19.** 36 **20.** $\frac{32z}{x^3}$ **21.** $\frac{3}{x - y}$ **22.** $\frac{1}{3(a + 3)}$ **23.** 1 **24.** $2x(x - 5y)$ **25.** $\frac{x - 2}{x + 2}$ **26.** 4 **27.** 9
28. $\frac{4}{x + 10}$ **29.** $3x + 2$ **30.** $3x + 4$ **31.** 24 **32.** $15x^2$ **33.** $30x^6y^4$ **34.** $x(x + 1)$ **35.** $(x + 2)(x - 3)$
36. $x(x + 1)$ **37.** $(x + y)(x - y)$ **38.** $x - 7$ **39.** $(x + 7)(x - 5)(x + 2)$ **40.** $\frac{3}{x}$ **41.** $\frac{24x + y}{4xy}$ **42.** $\frac{5x^2 - 12y}{3x^2y}$
43. $\frac{5x + 12}{x + 3}$ **44.** $\frac{a^2 + c^2}{ac}$ **45.** $\frac{7x + 12}{x(x + 3)}$ **46.** $\frac{-x - 4}{3x(x - 2)}$ **47.** $\frac{4x + 26}{(x + 5)^2}$ **48.** $\frac{2x - 8}{(x - 3)(x - 5)}$
49. $\frac{8x + 25}{(x + 3)(x + 4)}$ **50.** $\frac{7x + 12}{x + 2}$ **51.** $\frac{a - 2}{2x(a + 2)}$ **52.** $\frac{3x - 3}{(x + 3)(x - 3)}$ **53.** $\frac{(x + y)y^2}{2x^3}$
54. $\frac{4}{(x + 2)(x - 3)(x - 2)}$ **55.** $\frac{22x + 5}{(x - 5)(x - 10)(x + 5)}$ **56.** $\frac{x + y}{x}$ **57.** $\frac{16(x - 2y)}{3(x + 2y)}$ **58.** $a - 3$
59. $\frac{8(x - 2)}{x}$ **60.** $\frac{3x^2 - 7x + 2}{(x + 1)(x - 1)(3x - 5)}$ **61.** $\frac{34}{9}$ **62.** $\frac{55}{26}$ **63.** $\frac{5yz}{6}$ **64.** $\frac{16x^3z^2}{y^3}$ **65.** $\frac{xy + 1}{y^3}$ **66.** $\frac{xy - x}{x + 1}$
67. $\frac{4x + 2}{x(6x - 1)}$ **68.** $\frac{2}{x}$ **69.** a **70.** $\frac{2a + 1}{2}$ **71.** $\frac{x + 1}{-x + 1}$ **72.** $\frac{x^2(3 - y)}{y(y - x)}$ **73.** 9 **74.** 1 **75.** 6 **76.** 6
77. 52 **78.** -20 **79.** No solution **80.** 18 **81.** $\frac{1}{2}$ **82.** -6 **83.** -18 **84.** $2\frac{2}{9}$ hr **85.** $16\frac{4}{5}$ hr **86.** $\frac{5}{2}$, 10
87. Bus 80 km/hr, train 120 km/hr

Chapter 6 Practice Test

1. $\frac{2x^3z}{3y^3}$ **2.** $a + 3$ **3.** $\frac{x - 3y}{3}$ **4.** $\frac{4}{y(y + 5)}$ **5.** $\frac{7x - 2}{2y}$ **6.** $\frac{7x^2 - 6x - 11}{x + 3}$ **7.** $\frac{10x + 3}{2x^2}$ **8.** $\frac{-x + 10}{x + 2}$
9. $\frac{-1}{(x + 4)(x - 4)}$ **10.** $\frac{29}{10}$ **11.** $\frac{x^2(1 + y)}{y}$ **12.** 60 **13.** 12 **14.** $3\frac{1}{13}$ hr

Cumulative Review Test

1. 50 **2.** -8 **3.** $-\frac{5}{2}$ **4.** $\frac{27y^6}{x^9}$ **5.** $R = \frac{P - 2E}{3}$ **6.** $3x^2 - 11x + 4$ **7.** $12x^3 - 38x^2 + 39x - 15$
8. $(6a - 5)(a - 1)$ **9.** $5(2x^2 - x + 1)$ **10.** $(x - 6)(x - 4)$ **11.** $(3x + 2)(2x - 5)$ **12.** 4, $\frac{3}{2}$ **13.** $\frac{x + 3}{2x + 1}$
14. $\frac{x^2 - 8x - 12}{(x + 4)(x - 5)}$ **15.** $\frac{6x + 2}{(x - 5)(x + 2)(x + 3)}$ **16.** $-\frac{3}{2}$ **17.** 5 **18.** \$2000 **19.** 1.5 lb
20. 3 mi at 4 mph, 3 mi at 12 mph

Chapter 7

EXERCISE SET 7.1

1. (a) 53.5% **(b)** No, only percents are indicated **(c)** No **(d)** No **3. (a)** \$2.7 million **(b)** \$0.54 million **(c)** State Fair 1993 revenue: \$3.6 million

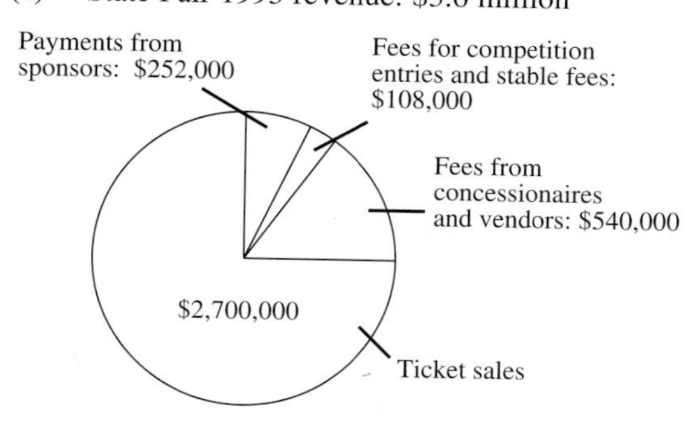

5. (a) 15 years **(b)** 23 years, five hundredths of a second **(d)** 9.82 sec **7. (a)** decreased by 14,464 **(b)** increased by 5554 **(c)** 1950: 67,148; 1991: 55,247 **9. (a)** 30,000 **(b)** 60,000 **(c)** 160,000 **(d)** about 960,000 **11. (a)** 31%, 20%, 18.5% **(b)** It has varied between 18.5% and 14% but has remained relatively stable. **(c)** It has varied between 31% and 35%. It increased up until 1990–1991, then began decreasing. **13. (a)** \$320,000 **(b)** after 1989 **(c)** \$450,000 **(d)** \$650,000 **(e)** 144% **(f)** 28.8% **15. (a)** 3 billion **(b)** 7 billion **(c)** 800 million **(d)** 1 billion **(e)** 2 billion **(f)** The world population is expected to increase from about 5 billion to 11 billion from 1990 to 2100, with the poorest accounting for most of the increase.

CUMULATIVE REVIEW EXERCISES

22. $\frac{18}{5}x - \frac{7}{6}$ **23.** $3x + 2 + \frac{10}{x - 2}$ **24.** $4(x + 2)(x - 2)$ **25.** $\frac{5}{12}$

GROUP ACTIVITY/CHALLENGE PROBLEMS

1. Disney's Income, 1992

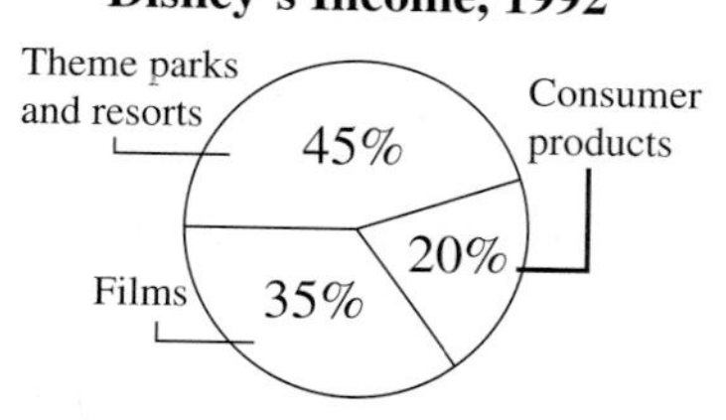

3. (a) 92%, 80%, 77%, 59%, 30%, 15%, 2%

b)

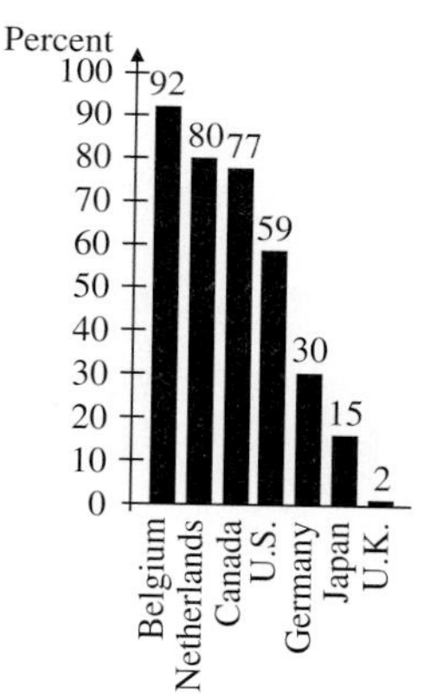

4.

Subscribers (millions) Cellular Telephone Subscribers

16, 14, 12, 10, 8, 6, 4, 2

84 85 86 87 88 89 90 91 92 93

Year

EXERCISE SET 7.2

1. I **3.** III **5.** II **7.** I **9.** III **11.** II **13.** $A(3, 1)$, $B(-3, 0)$, $C(1, -3)$, $D(-2, -3)$, $E(0, 3)$, $F(\frac{3}{2}, -1)$

15.

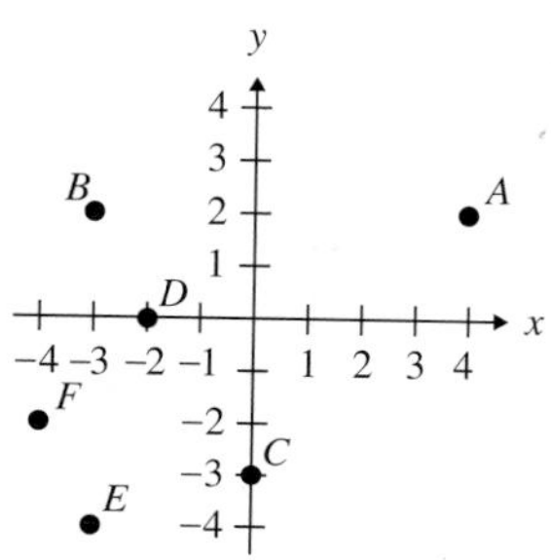

17.

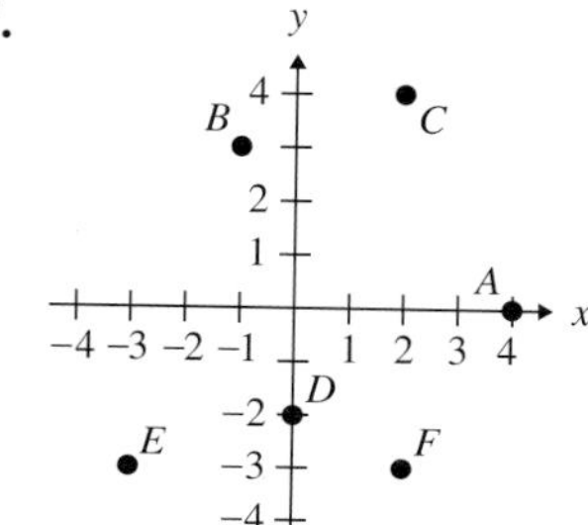

19. (3, 1) not on line

21. (a) (0, 1), (−1, 0), (2, 3)
(b)

23. (a) (2, 0), $(\frac{2}{3}, -2)$, $(\frac{4}{3}, -1)$
(b)

25. (a) $(1, \frac{1}{2})$, (0, 1), (−2, 2)
(b)

27. The x coordinate **29. (a)** x axis **(b)** y axis **31.** axis is singular, axes is plural
33. An illustration of the set of points whose coordinates satisfy the equation. **35.** a straight line **37.** $ax + by = c$

Cumulative Review Exercises

39. $-8x^2 - x + 4$ **40.** $6x^3 - 17x^2 + 22x - 15$ **41.** $(x + 3y)(x - 2)$ **42.** $\dfrac{-x - 5}{(x + 2)(x + 1)}$

Group Activity/Challenge Problems

1.

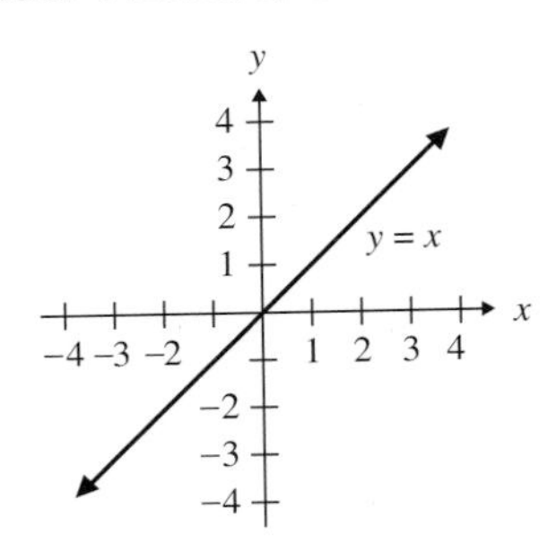

3.

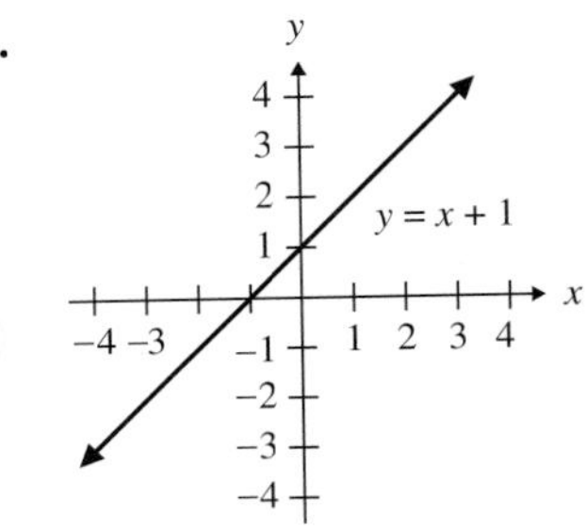

Exercise Set 7.3

1. 2 **3.** $\frac{11}{2}$ **5.** 3 **7.** −1 **9.** $\frac{8}{3}$ **11.** $-\frac{17}{2}$

13.

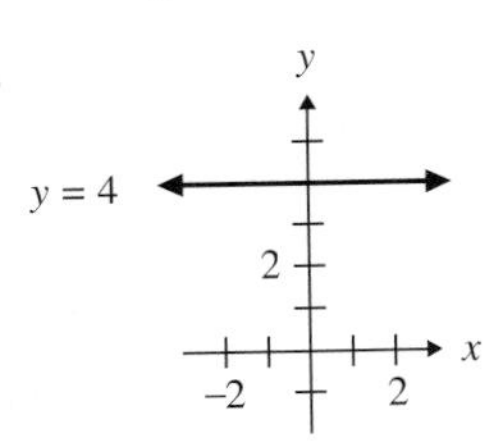

15.

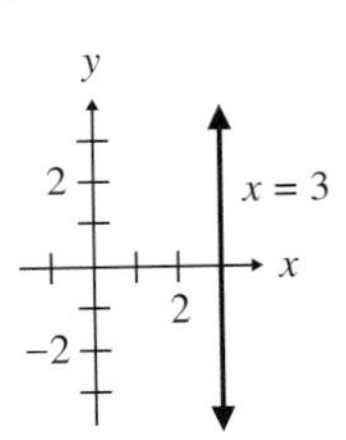

17.

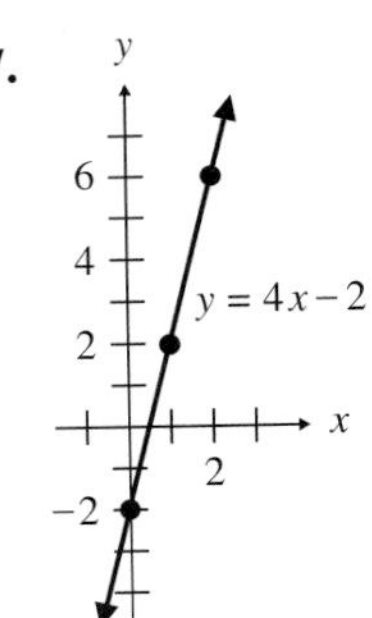

19.

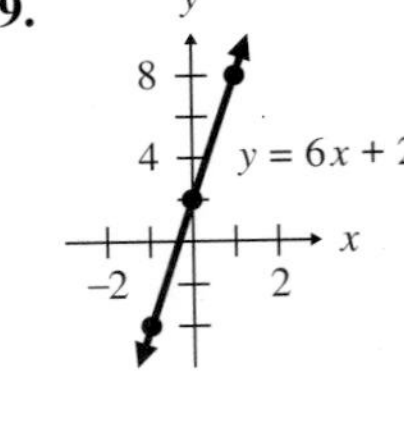

21.

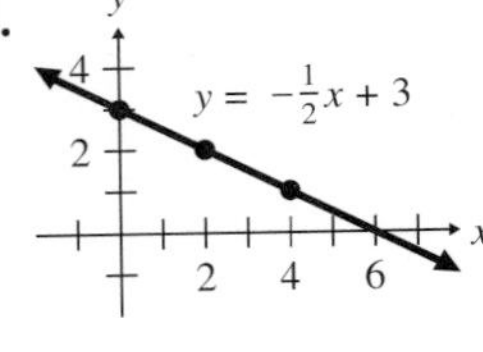

23.

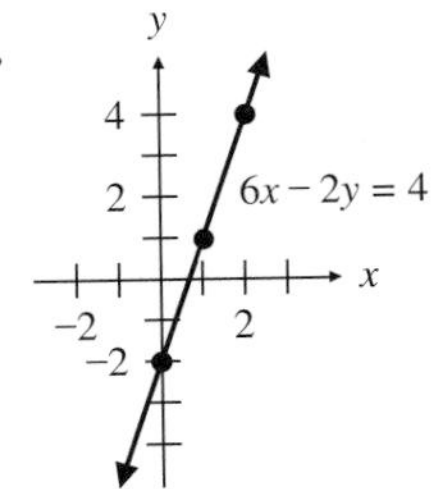

25.

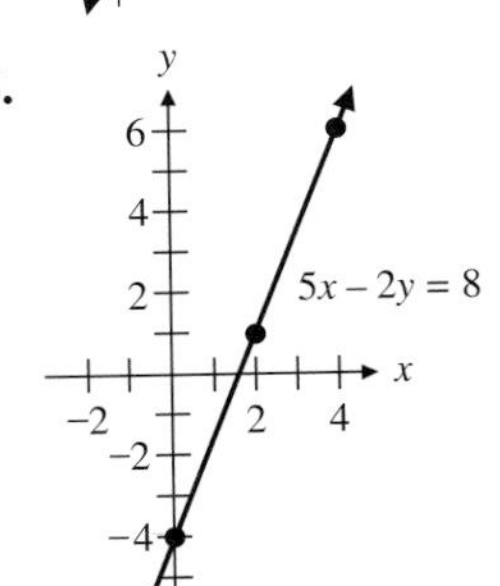

27.

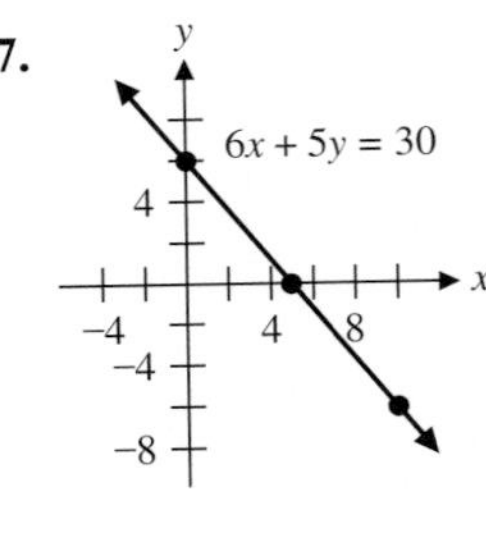

29.

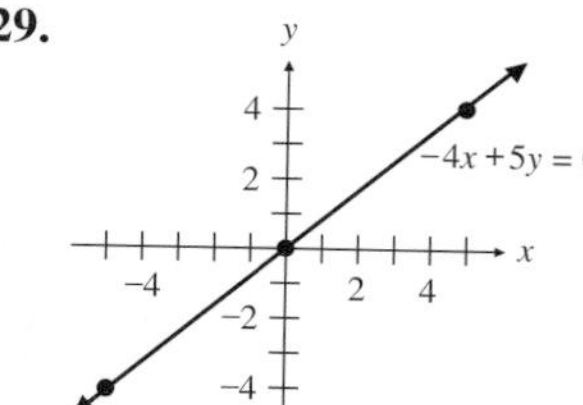

31.

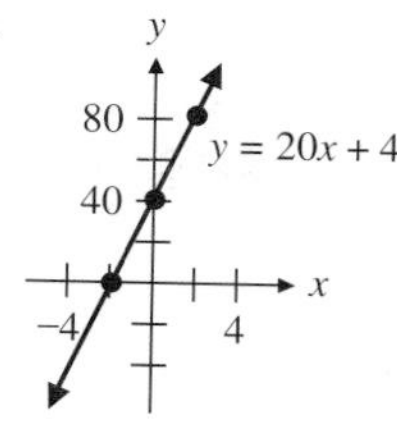

33.

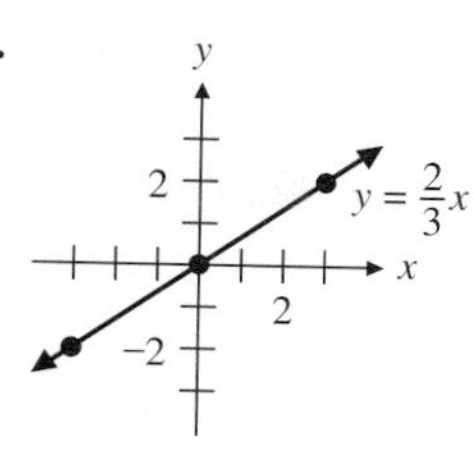

35.

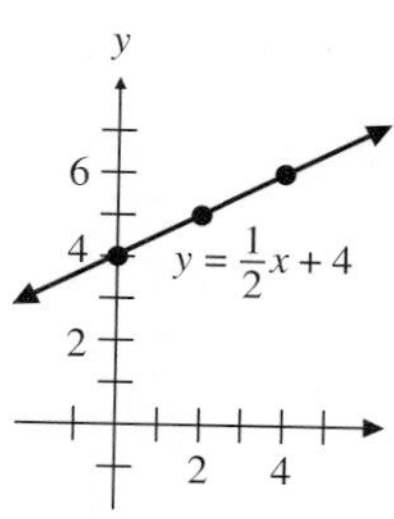

37.

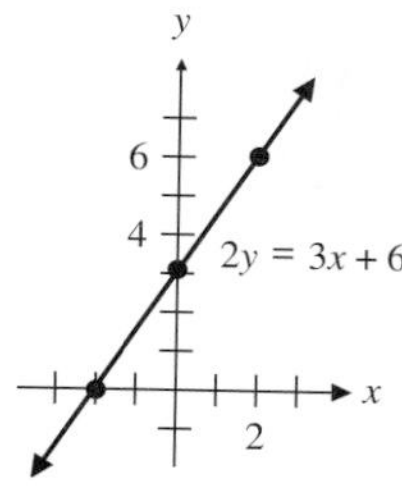

39.

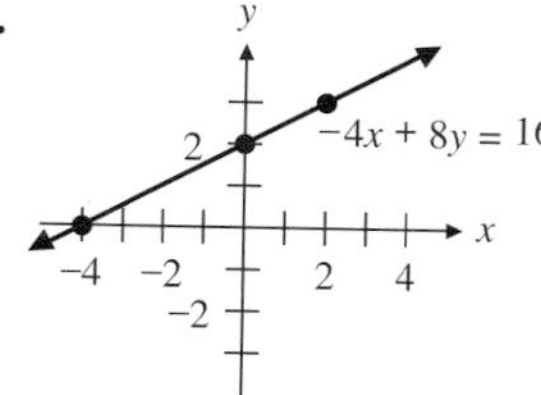

41.

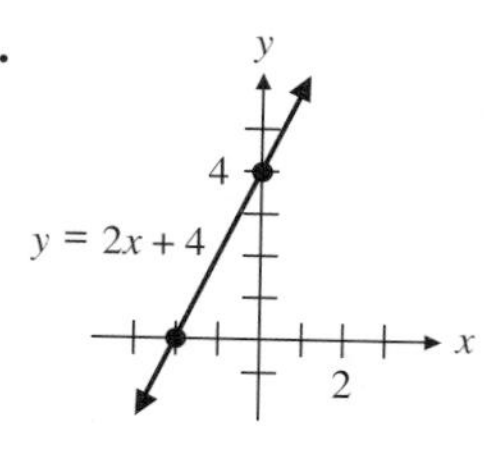

43.

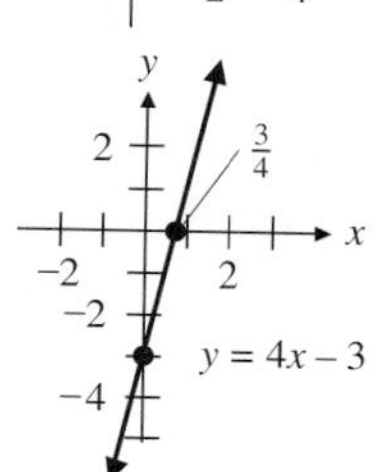

45.

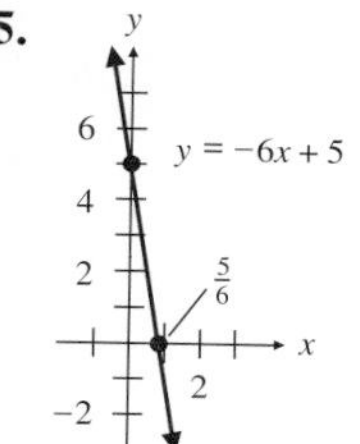

47.

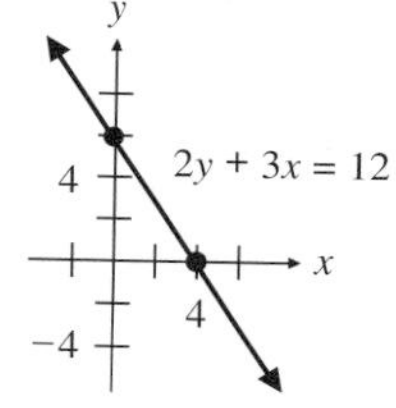

49.

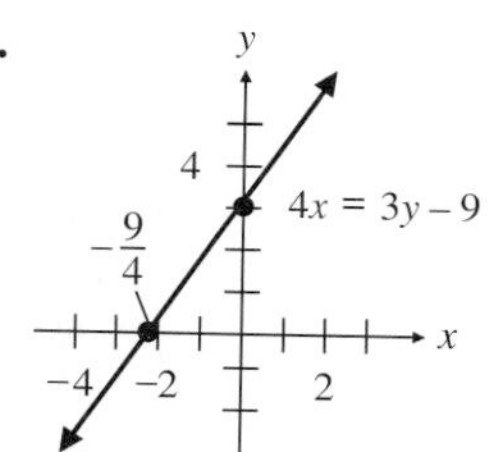

51.

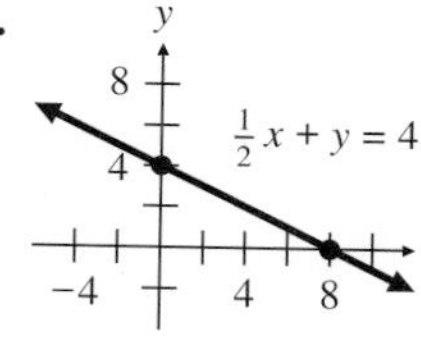

53.

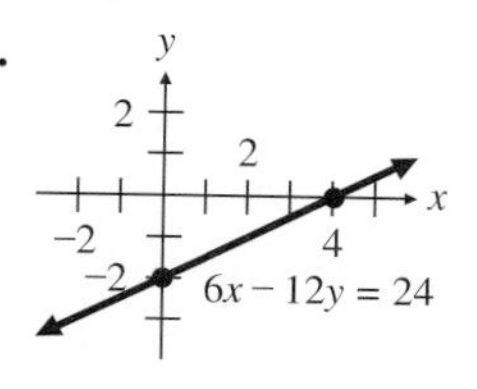

55.

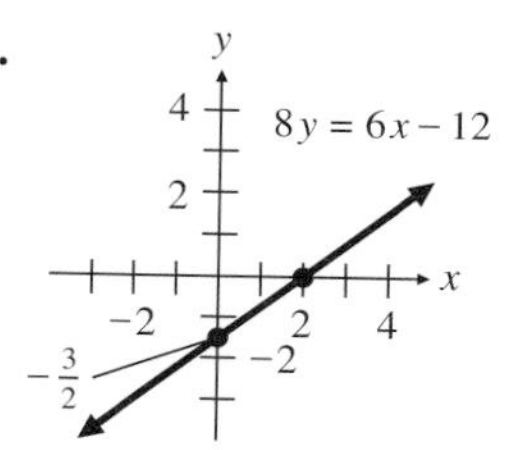

57.

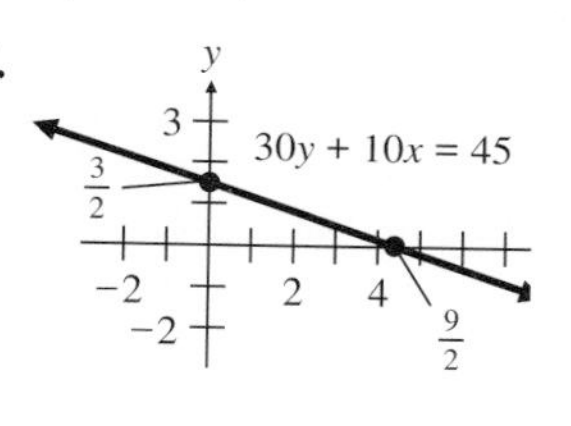

59.

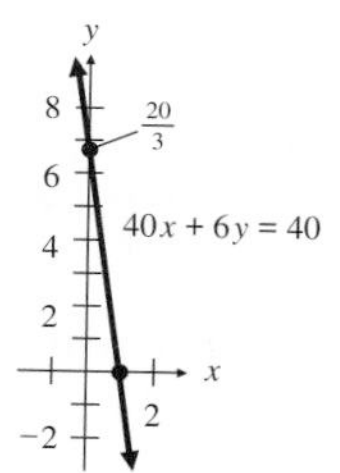

61.

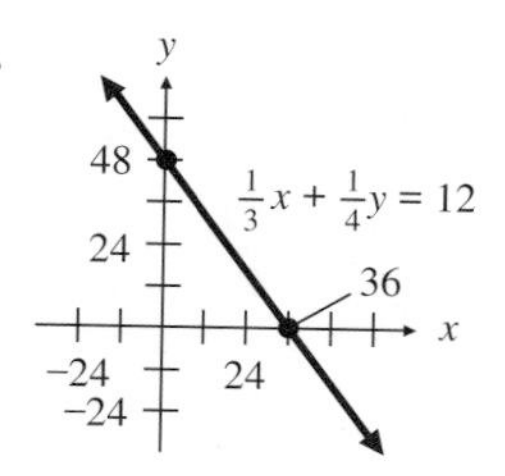

63.

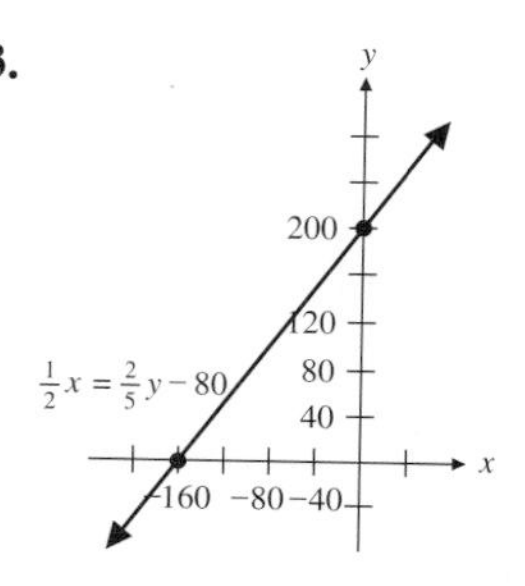

65. $x = -3$
67. $y = 3$

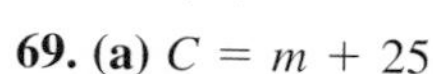

69. (a) $C = m + 25$ **(b)**

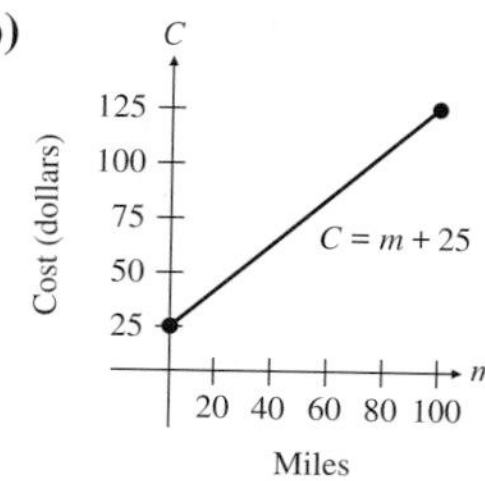

(c) \$75 **(d)** \$35

71. (a)

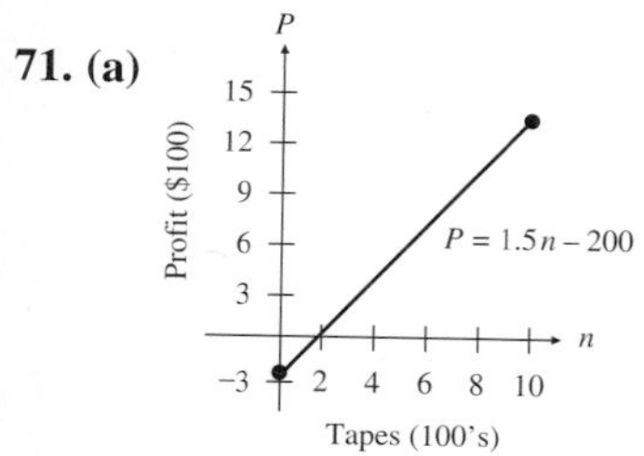

(b) \$550 **(c)** 800 tapes

73. x intercept: substitute 0 for y and find the corresponding value of x; y intercept: substitute 0 for x and find the corresponding value of y. **75.** a horizontal line **77.** You may not be able to read the exact answers from a graph. **79.** 5, 4 **81.** 6, 4

Cumulative Review Exercises

83. 5 **84.** 4.5 mph, 7.5 mph **85.** $\frac{1}{2}$, -12 **86.** 6, 8

Group Activity/Challenge Problems

1. (a)

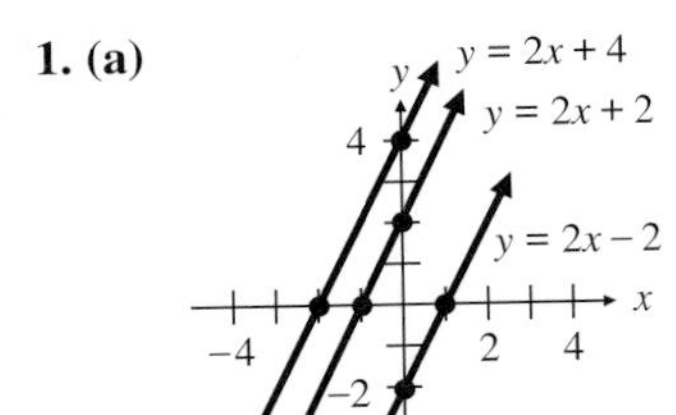

(b) They appear to be parallel lines

3. (a)

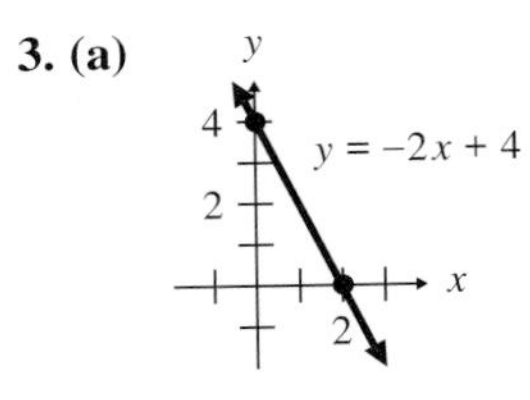

(b) $y = -2x + 4$

5.

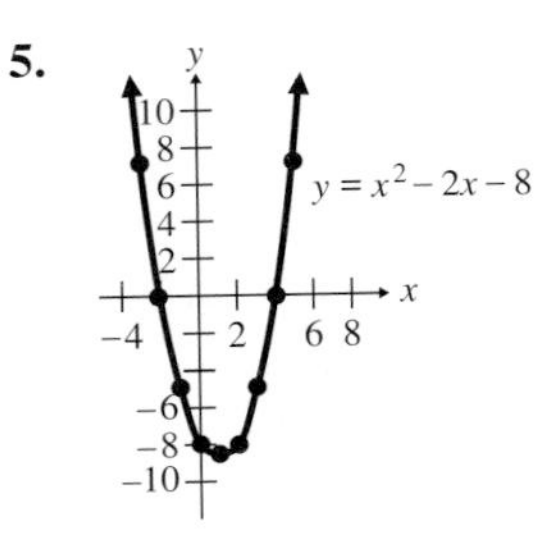

Exercise Set 7.4

1. 5 **3.** $\frac{1}{2}$ **5.** 0 **7.** 0 **9.** undefined **11.** $-\frac{5}{9}$ **13.** -2 **15.** $m = 2$ **17.** $m = -\frac{3}{2}$ **19.** $m = -\frac{3}{2}$ **21.** $m = \frac{7}{4}$
23. $m = 0$ **25.** slope is undefined
27. The ratio of the vertical change to the horizontal change between any two points on the line.
29. The values of y increase as the values of x increase.
31. Lines that rise from left to right have a positive slope. Lines that fall from left to right have a negative slope.
33. No, since we cannot divide by 0. We say that the slope of a vertical line is undefined.

Cumulative Review Exercises

34. (a) An equation that contains only one variable and that variable has an exponent of 1.
(b) $2x + 3 = 5x - 6$ (answers will vary)
35. (a) An equation that contains only one variable and the greatest exponent on that variable is 2.
(b) $x^2 + 2x - 3 = 0$ (answers will vary) **36. (a)** An equation that contains only one variable and one or more fractions.
(b) $\frac{x}{3} + \frac{x}{4} = 12$ (answers will vary) **37.** (a) An equation that contains two variables and the exponent on both variables is 1. **(b)** $y = 3x - 2$ (answers will vary)

Group Activity/Challenge Problems

1. $\frac{225}{68}$ **3.** 7

5. (a)

(b) AC, $m = \frac{3}{5}$

CB, $m = -2$

DB, $m = \frac{3}{5}$

AD, $m = -2$

(c) yes

EXERCISE SET 7.5

1.

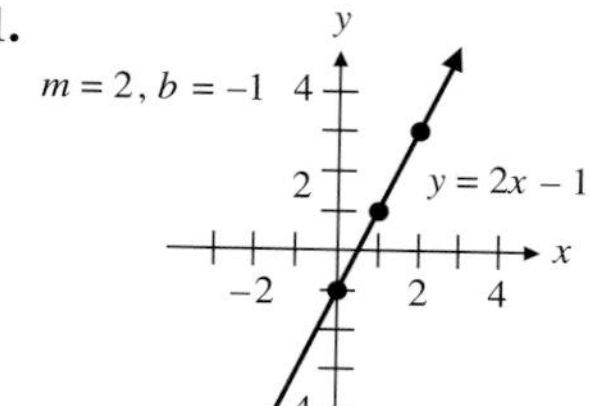

3.

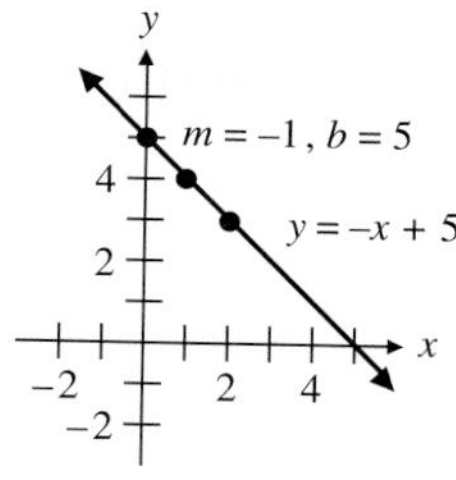

5.

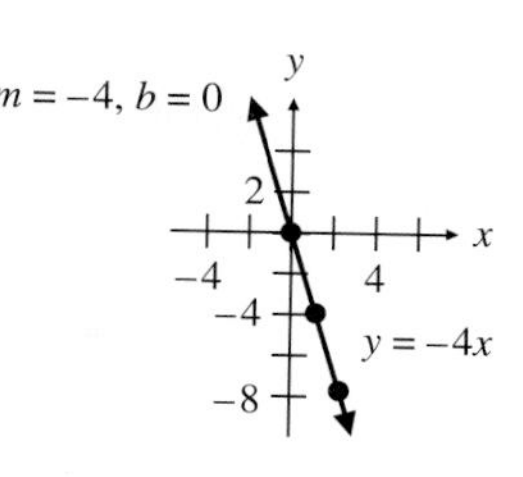

7.

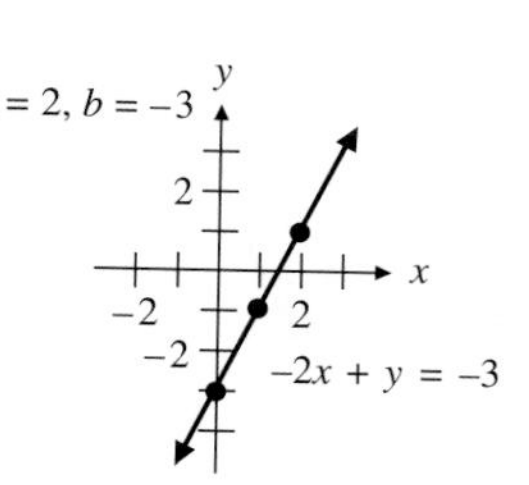

9.

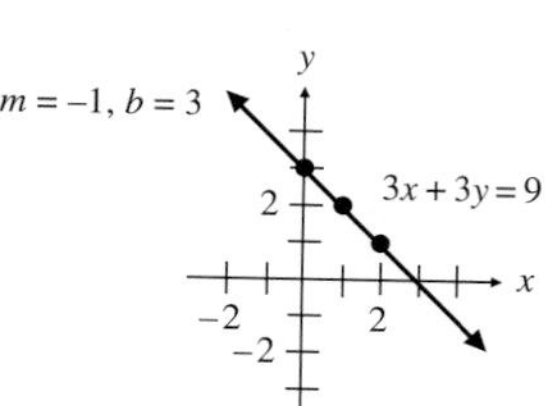

11.

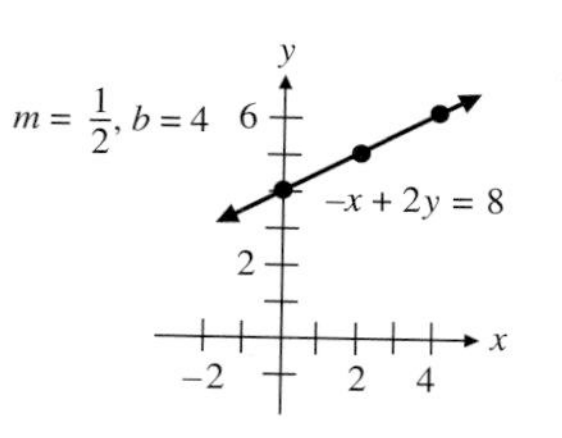

13.

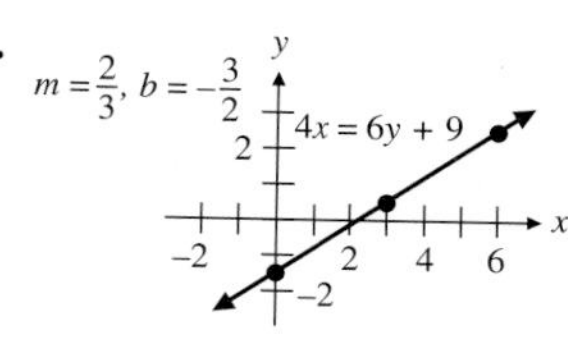

15.

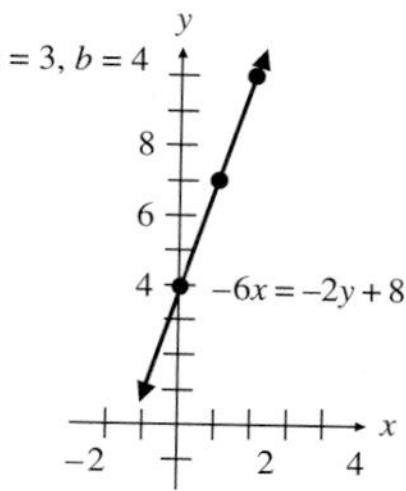

17.

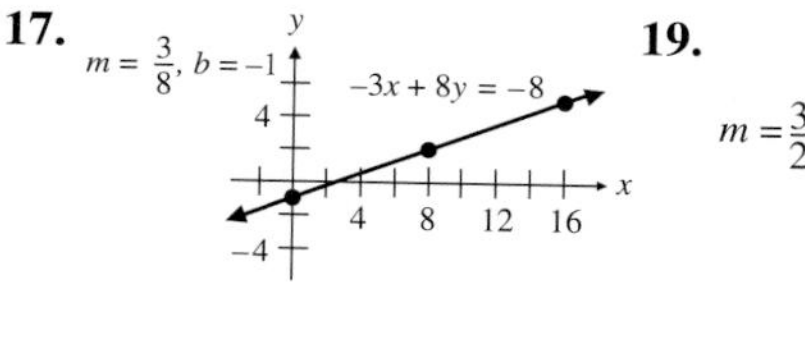

19.

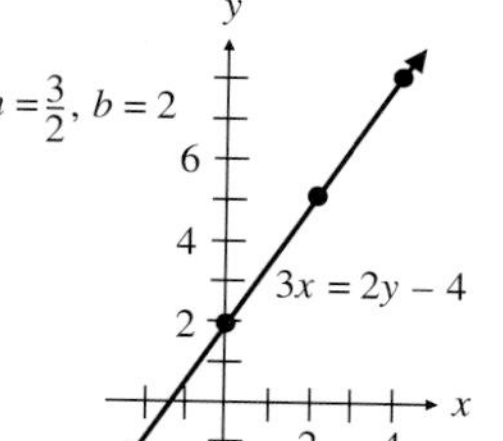

21. $y = 2x$
23. $y = -3x - 5$
25. $y = \frac{1}{3}x + 5$
27. $y = 2x - 1$
29. Yes
31. Yes
33. No
35. No
37. $y = 5x + 4$
39. $y = -2x - 3$
41. $y = \frac{1}{2}x - \frac{9}{2}$
43. $y = \frac{3}{5}x + 7$
45. $y = 3x + 10$
47. $y = -\frac{3}{2}x$
49. $y = \frac{1}{2}x - 2$
51. $y = 5.2x - 1.6$

53.

x	y
0	−2
2	1
4	4

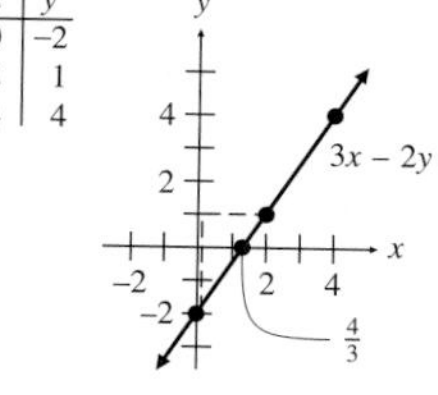

55.

x	y
0	2
3	4
−3	0

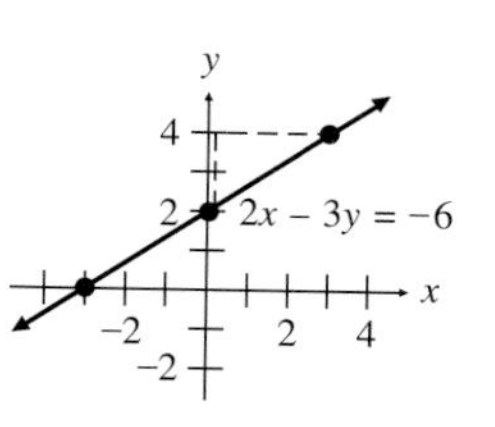

57. Compare their slopes. If the slopes are the same but their y intercepts are different then the lines are parallel.

59. **(a)** $ax + by = c$ **(b)** $y = mx + b$ **(c)** $y - y_1 = m(x - x_1)$

61. **(b)**

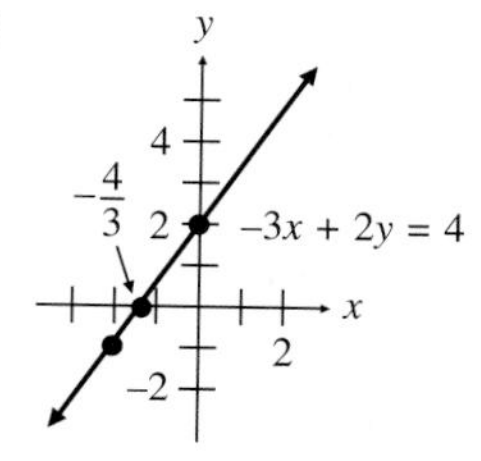

CUMULATIVE REVIEW EXERCISES

62. $3x^2 - 3x - 1 + \dfrac{6}{3x + 2}$ **63.** 1.5 L **64.** $\dfrac{x - 2}{2x + 1}$ **65.** $\dfrac{4(2x - 1)}{(x - 2)(x + 2)}$ **66.** 5, 6 **67.** $b = 8$ ft, $h = 9$ ft

GROUP ACTIVITY/CHALLENGE PROBLEMS

3. Yes **4.** $y = \dfrac{3}{4}x + 2$ **5.** $y = -\dfrac{2}{5}x + \dfrac{13}{10}$

EXERCISE SET 7.6

1. function, domain $\{1, 2, 3, 4, 5\}$, range $\{1, 2, 3, 4, 5\}$ **3.** relation, domain $\{1, 2, 3, 5, 7\}$, range $\{-2, 0, 2, 4, 5\}$
5. relation, domain $\{0, 1, 3, 5\}$, range $\{-4, -1, 0, 1, 2\}$ **7.** relation, domain $\{0, 1, 4\}$, range $\{-3, 0, 2, 5\}$

9. function, domain {0, 1, 2, 3, 4}, range {3} **11. (a)** {(1, 4), (2, 5), (3, 5), (4, 7)} **(b)** Function **13. (a)** {(−4, 5), (0, 7), (6, 9), (6, 3)} **(b)** Not function **15.** Function **17.** Function **19.** Not function **21.** Function **23.** Not function **25.** Function **27. (a)** 7 **(b)** −1 **29. (a)** 0 **(b)** 5 **31. (a)** 5 **(b)** 15 **33. (a)** 3 **(b)** 5

35.
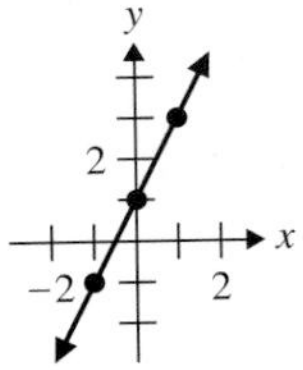

37.
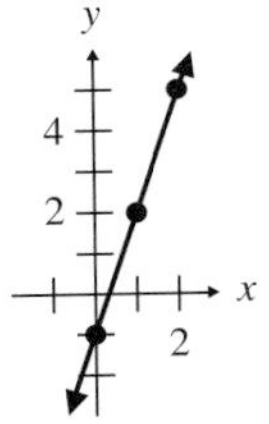

39.
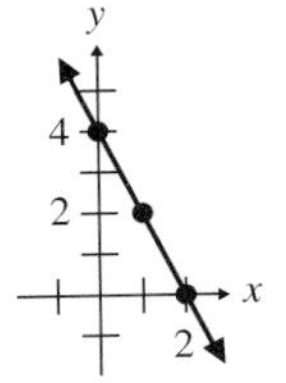

41.
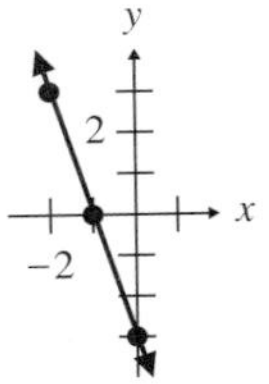

43. (a)
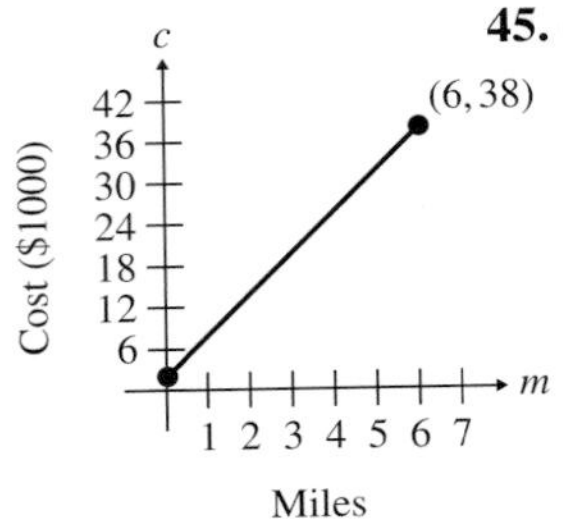

(b) $14,000

45. (a)
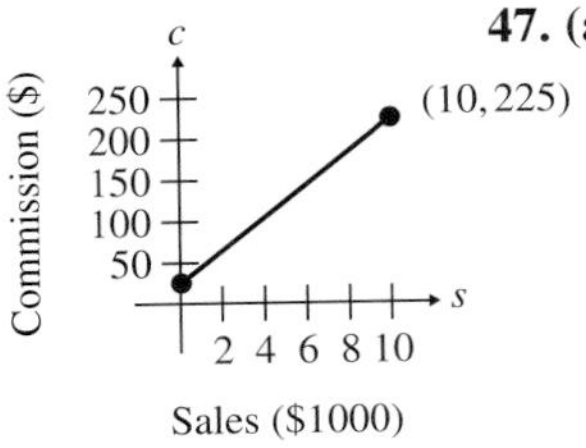

(b) $185

47. (a)
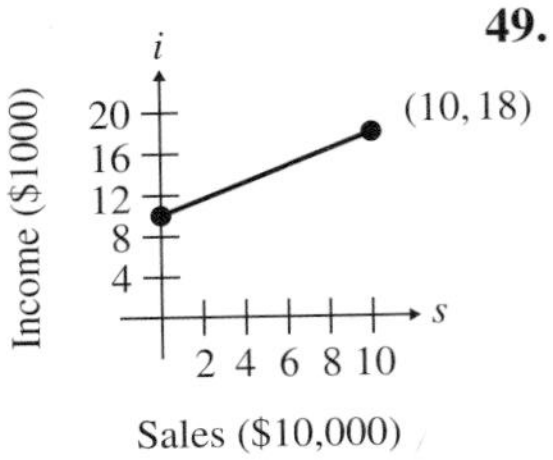

(b) $11,600

49. (a)
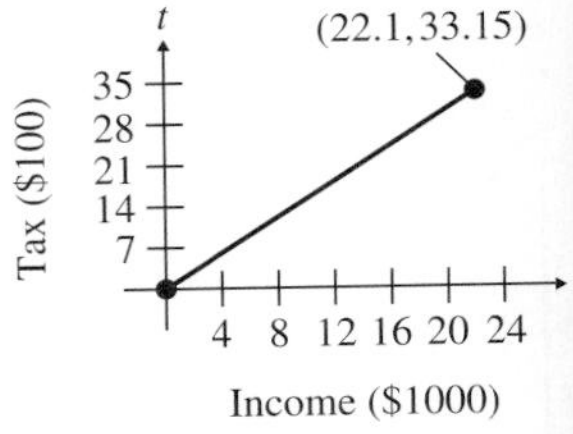

(b) $2250

51. A relation in which no two ordered pairs have the same first coordinate and a different second coordinate. **53.** The set of values that represent the dependent variable. **55.** No, each x must have a unique y for it to be a function. **57.** Yes, each x may have a unique value of y. **59.** yes **61.** no

Cumulative Review Exercises

63. $\frac{7}{2}$ **64.** $53.57 **65.** $x < -1$, 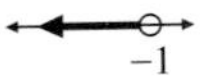 **66.** $w = \frac{A - 2l}{2}$

Group Activity/Challenge Problems

3. (a) 0 **(b)** 5

Exercise Set 7.7

1.
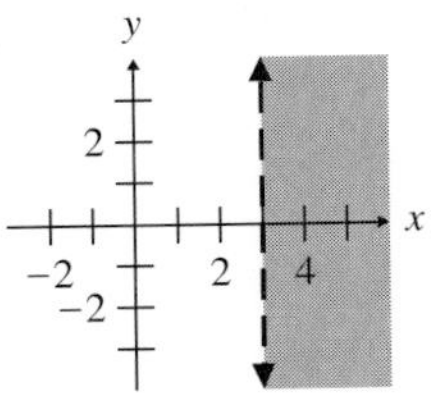

3.
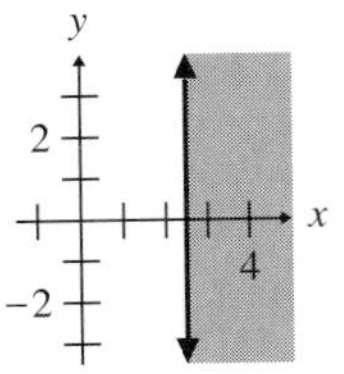

5.
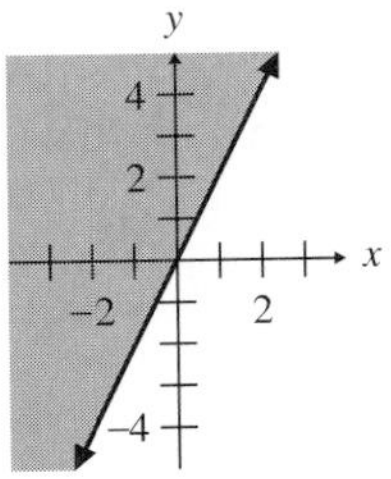

7.
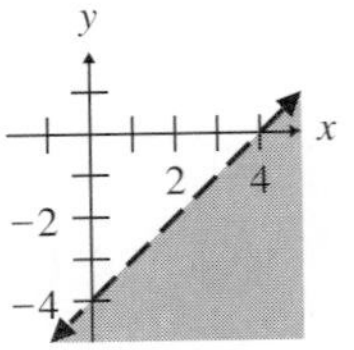

9.
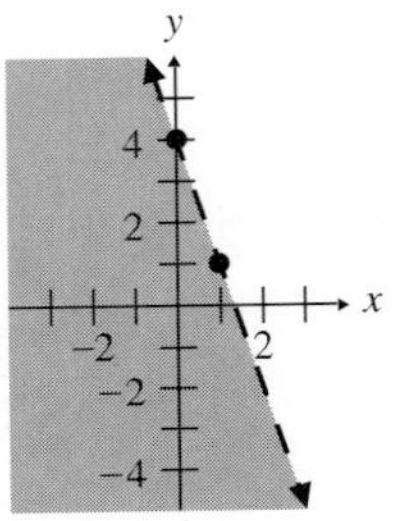

11.
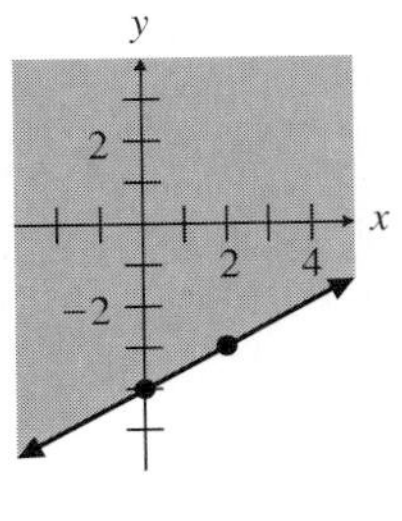

13.
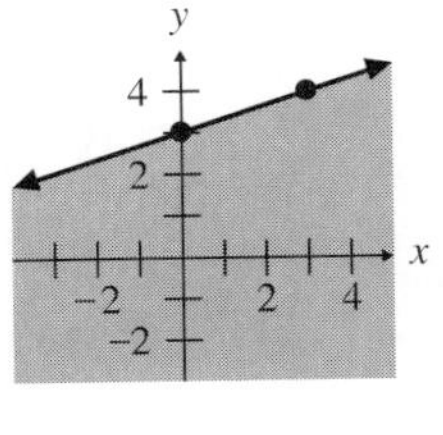

15.
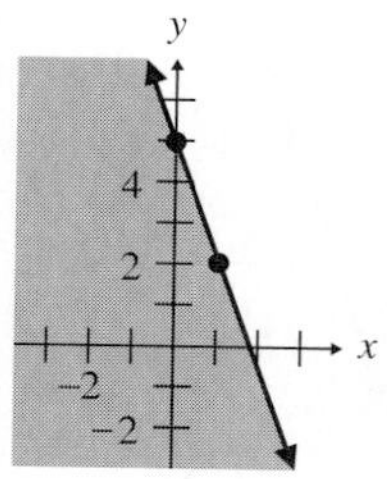

17.
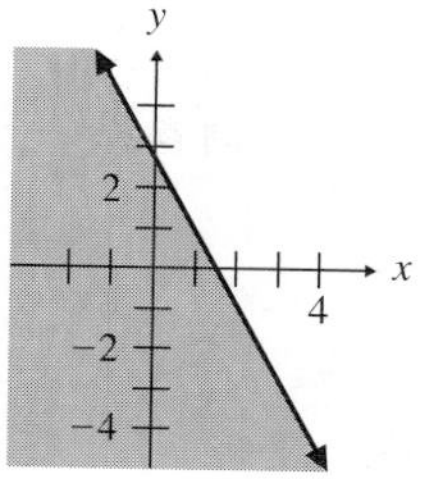

19.
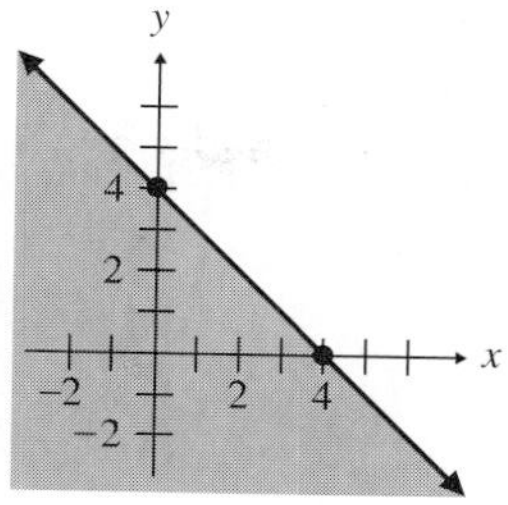

21.
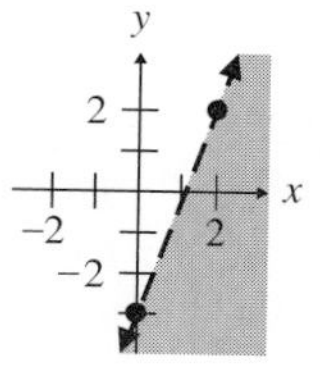

23.
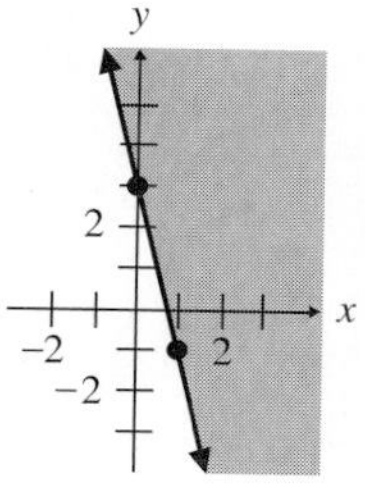

25. The points on the line satisfy the equal (=) part of the inequality.

CUMULATIVE REVIEW EXERCISES

27. $x > 3$, 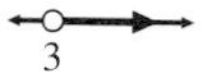

28. \$1250 **29.** $F = \frac{9}{5}C + 32$ **30.** $m = \frac{6}{5}$, $b = -\frac{9}{5}$

GROUP ACTIVITY/CHALLENGE PROBLEMS

4. (a) 1 **(b)** 2 **(c)** $x > 2$, 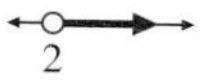

(d)
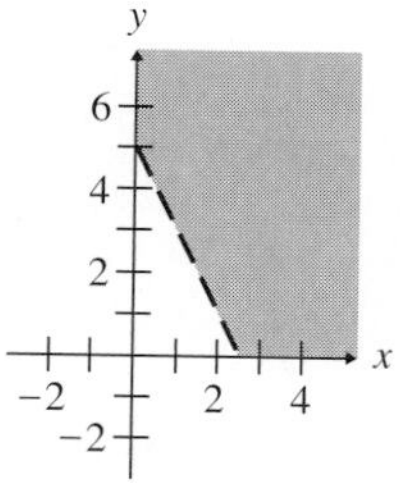

5.
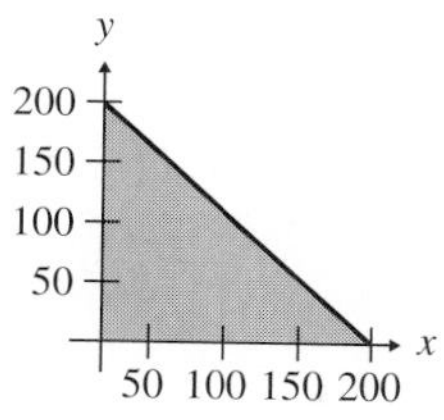

7. (a) $2x + y \le 10$

(b)
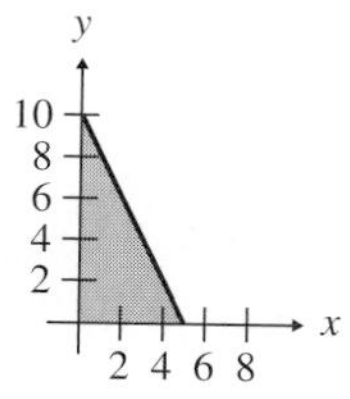

CHAPTER 7 REVIEW EXERCISES

1. (a) 38%
(b) 62%
(c) \$101.6 billion
(d) Major Cards' Amount Charged in 1993

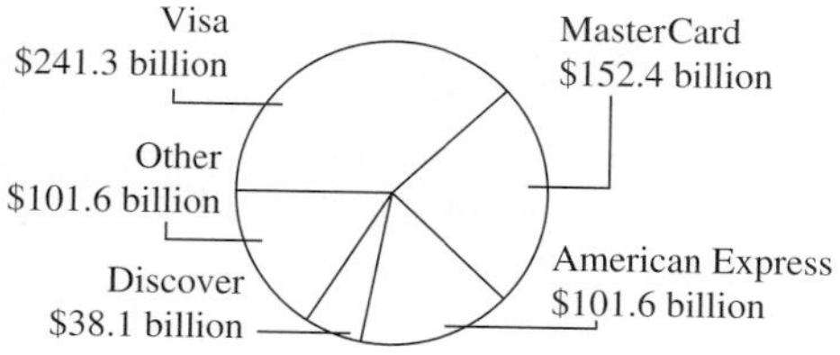

2. (a) 48% **(b)** 18% **(c)** greater, by about 30% **(d)** superstores
3. (a) 68%, 10% **(b)** about the third quarter of 1991 **(c)** about 94% **(d)** about 87% **4. (a)** about 135 per 100,000; about 350 per 100,000; increase of about 215 per 100,000 **(b)** 25%, 60% **(c)** 100% **(d)** 100%

5.
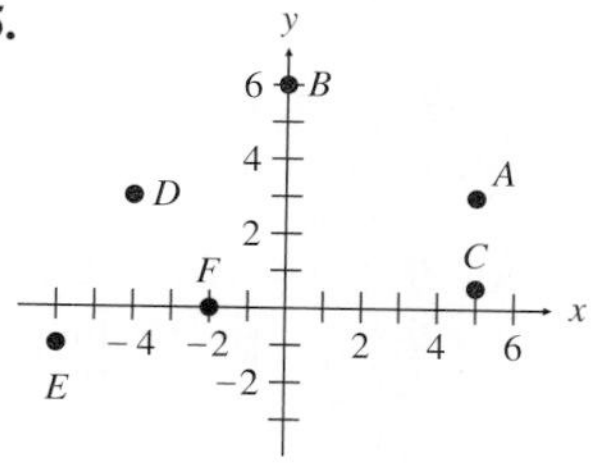

6. No
7. b, d
8. (a) -1 **(b)** -4 **(c)** $\frac{16}{3}$ **(d)** $\frac{8}{3}$

9.

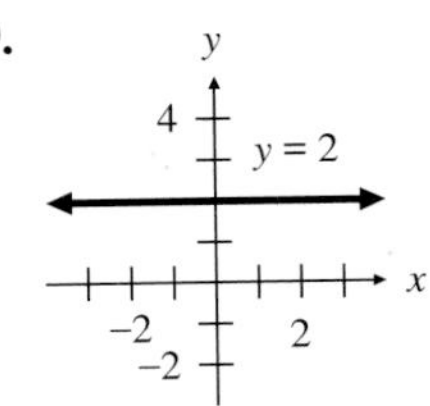

10.

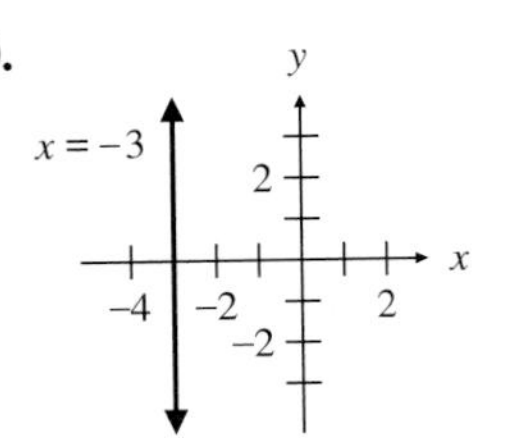

11.

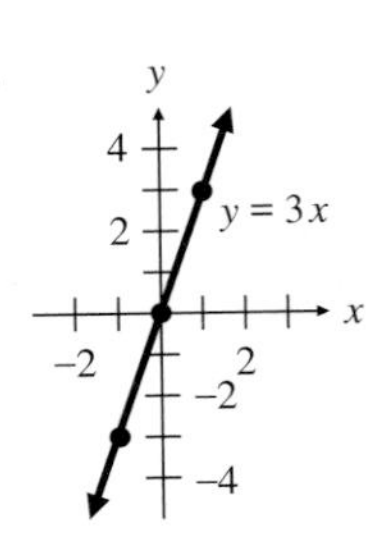

12.

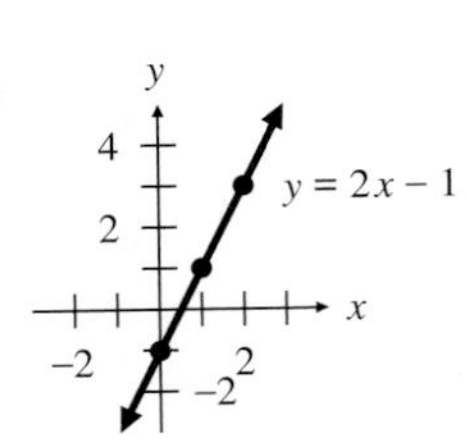

13.

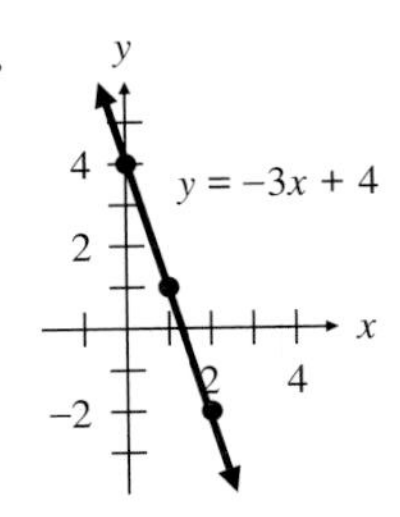

14.

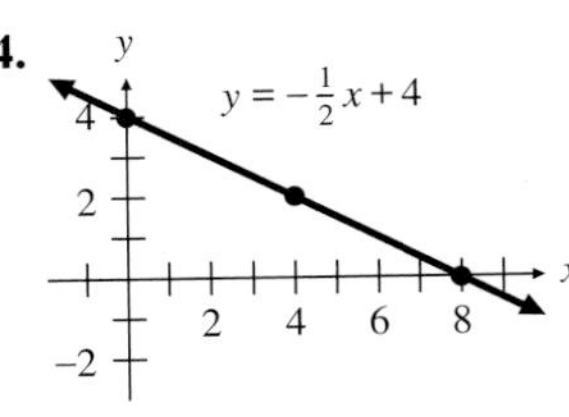

15.

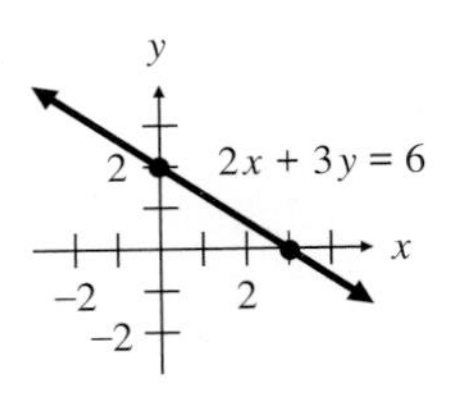

16.

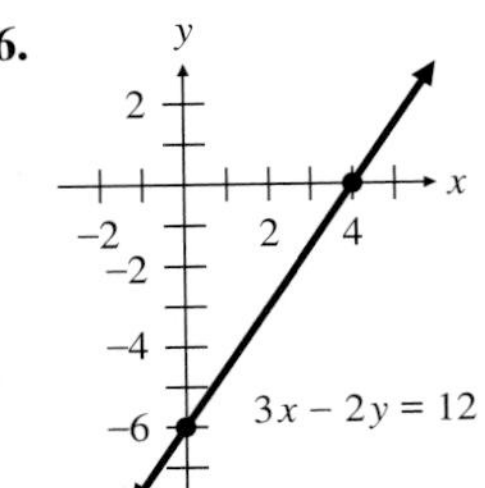

17.

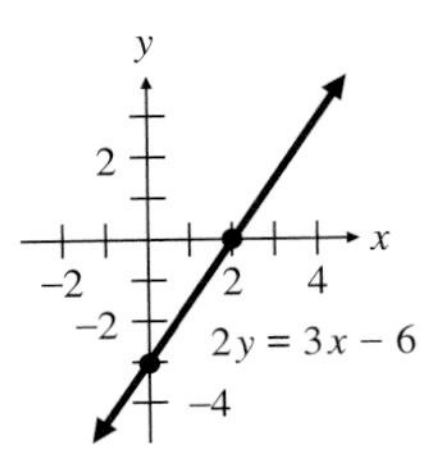

18.

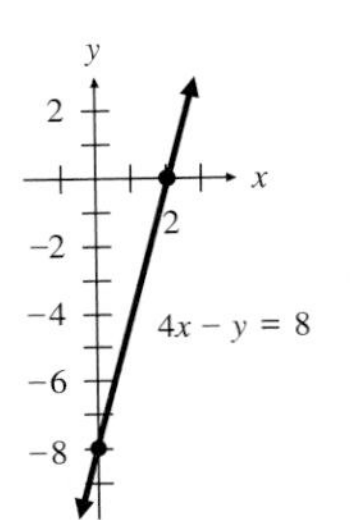

19.

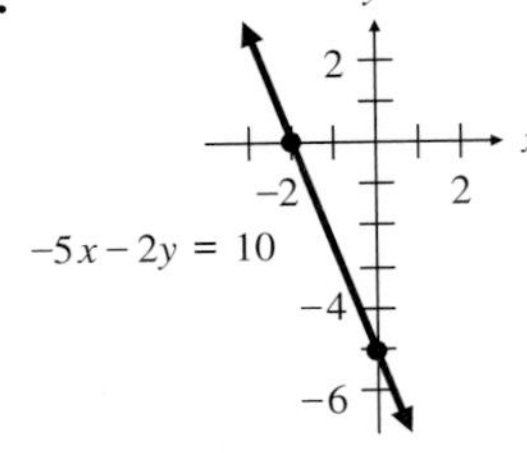

20.

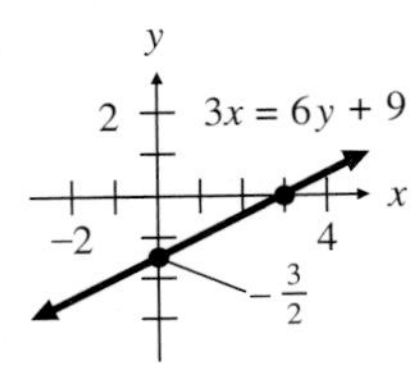

21.

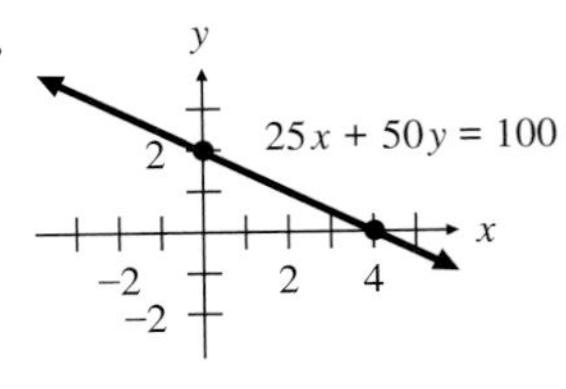

22.

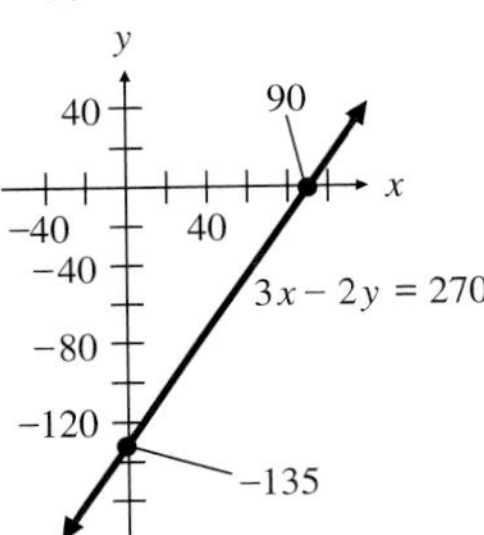

23.

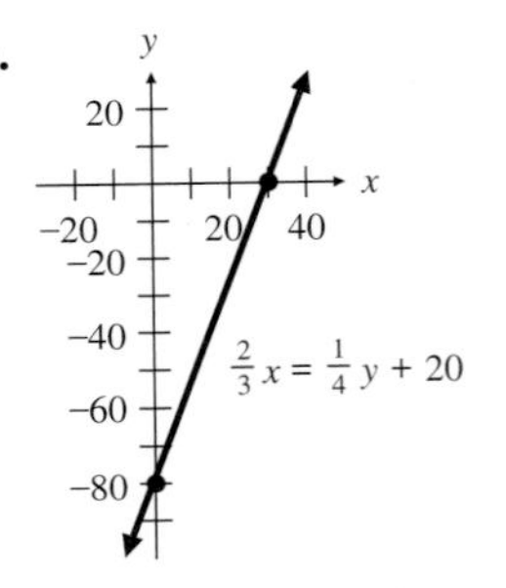

24. $-\frac{12}{5}$ **25.** $-\frac{1}{12}$ **26.** -2 **27.** 0 **28.** undefined **29.** The ratio of the vertical change to the horizontal change between any two points on the line. **30.** $-\frac{5}{7}$ **31.** $\frac{9}{4}$ **32.** $\frac{1}{4}$ **33.** $m = -1, b = 4$ **34.** $m = -4, b = \frac{1}{2}$ **35.** $m = -\frac{2}{3}, b = \frac{8}{3}$ **36.** $m = -\frac{1}{2}, b = \frac{3}{2}$ **37.** $m = \frac{3}{2}, b = 3$ **38.** $m = -\frac{3}{5}, b = \frac{12}{5}$ **39.** $m = -\frac{9}{7}, b = \frac{15}{7}$ **40.** Slope is undefined, no y intercept **41.** $m = 0, b = -3$ **42.** $y = 2x + 2$ **43.** $y = x - \frac{5}{2}$ **44.** $y = -\frac{1}{2}x + 2$ **45.** Yes **46.** No **47.** Yes **48.** Yes **49.** $y = 2x - 2$ **50.** $y = -3x + 2$ **51.** $y = -\frac{2}{3}x + 4$ **52.** $y = 2$ **53.** $x = 3$ **54.** $y = -2x - 4$ **55.** $y = -\frac{7}{2}x - 4$ **56.** $x = -4$
57. Function, domain $\{0, 1, 2, 4, 6\}$, range $\{-3, -1, 2, 4, 5\}$ **58.** Not function, domain $\{3, 4, 6, 7\}$, range $\{0, 1, 2, 5\}$
59. Not function, domain $\{3, 4, 5, 6\}$, range $\{-3, 1, 2\}$ **60.** Function, domain $\{-2, 3, 4, 5, 9\}$, range $\{-2\}$
61. (a) $\{(1, 3), (4, 5), (7, 2), (9, 2)\}$ **(b)** Function **62. (a)** $\{(4, 1), (6, 3), (6, 5), (8, 7)\}$ **(b)** Not function
63. Function **64.** Not function **65.** Function **66.** Function **67. (a)** 2 **(b)** -19 **68. (a)** 11 **(b)** -37
69. (a) -4 **(b)** -8 **70. (a)** 12 **(b)** 76

71.

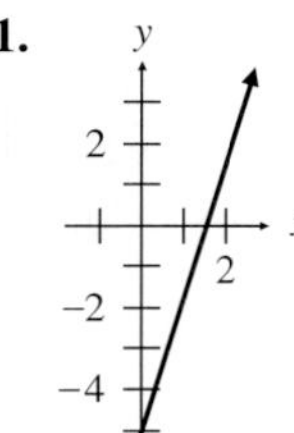

72.

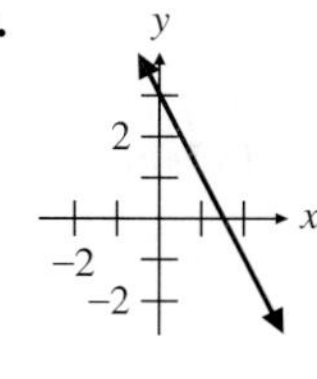

73.

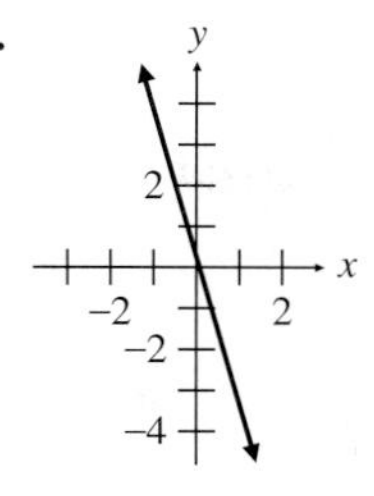

74.

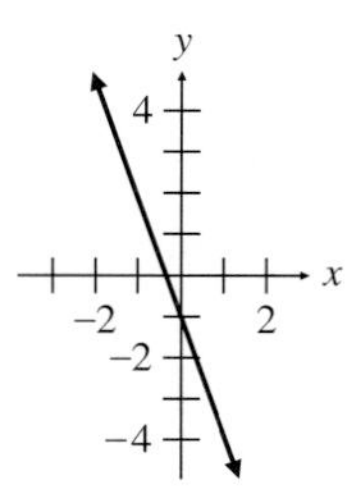

75. (a)

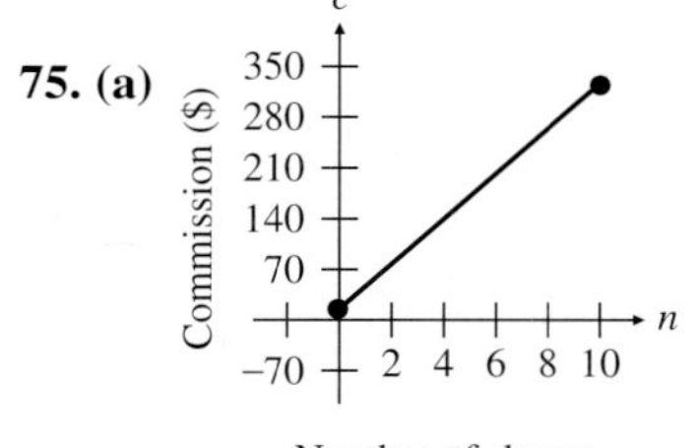

(b) \$55

76. (a)

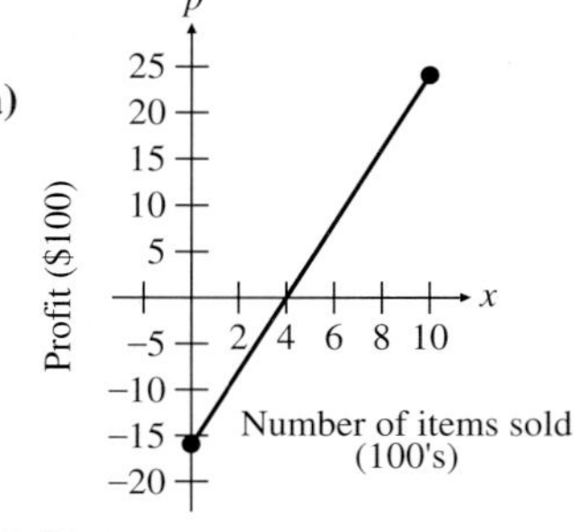

(b) \$0

77.

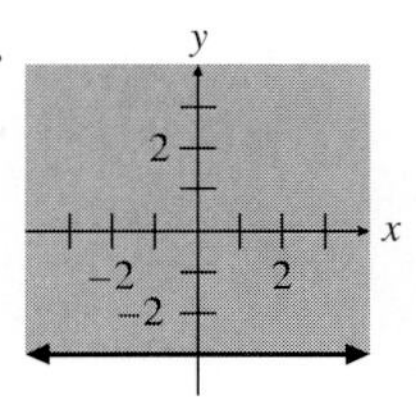

78.

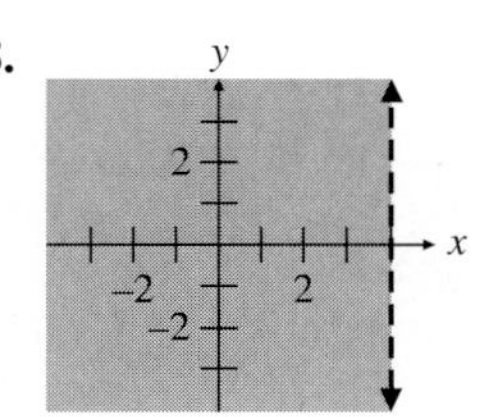

79.

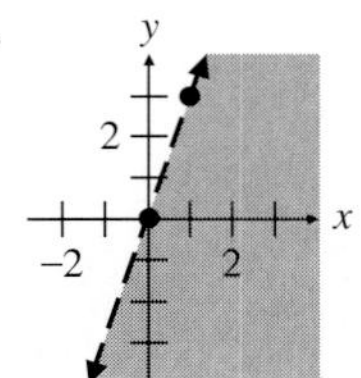

80.

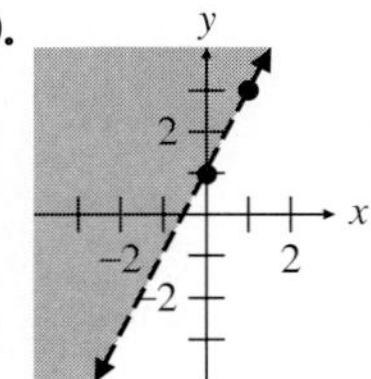

81.

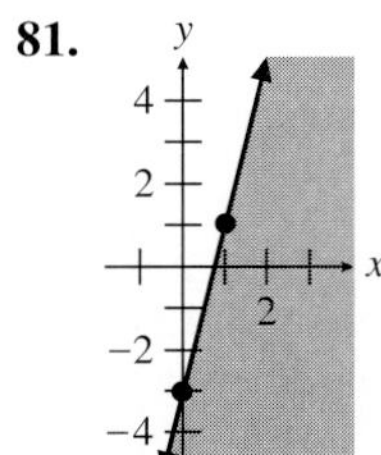

82.

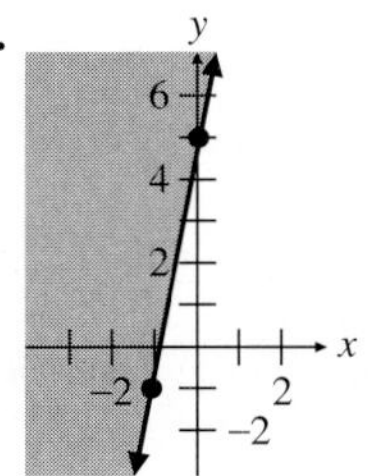

83.

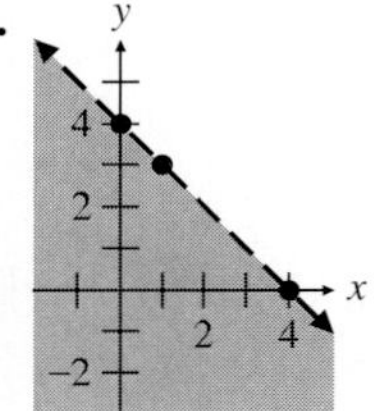

84.

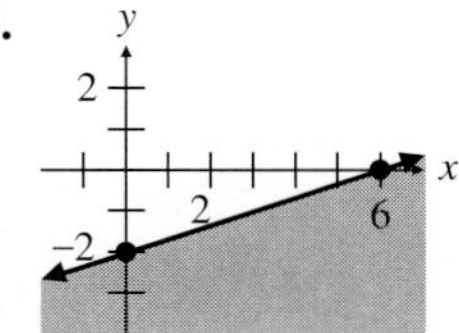

CHAPTER 7 PRACTICE TEST

1. (a) 36.7% **(b)** 5.7% **(c)** \$0.8112 billion or \$811.2 million **2.** a, b **3.** $-\frac{4}{3}$ **4.** $m = \frac{4}{9}$, $b = -\frac{5}{3}$ **5.** $y = -x - 1$

6. $y = 3x$ **7.** $y = -\frac{3}{7}x + \frac{2}{7}$ **8.** Yes, the slope of both lines is $\frac{3}{2}$ and y intercepts are different.

9.

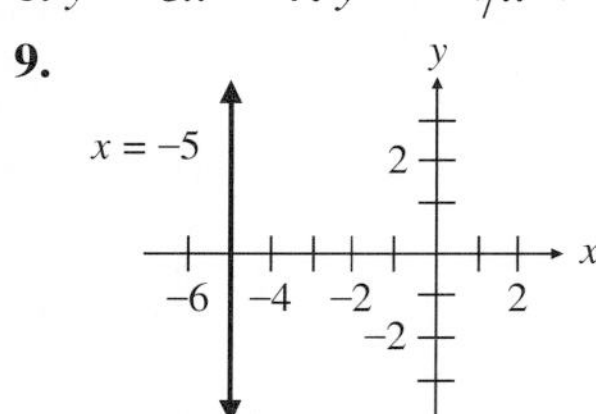

10.

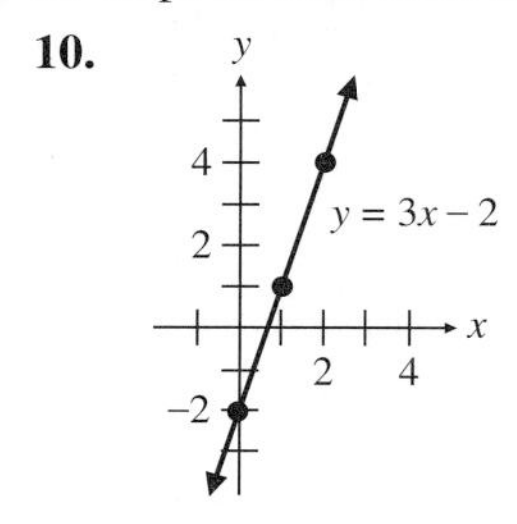

11.

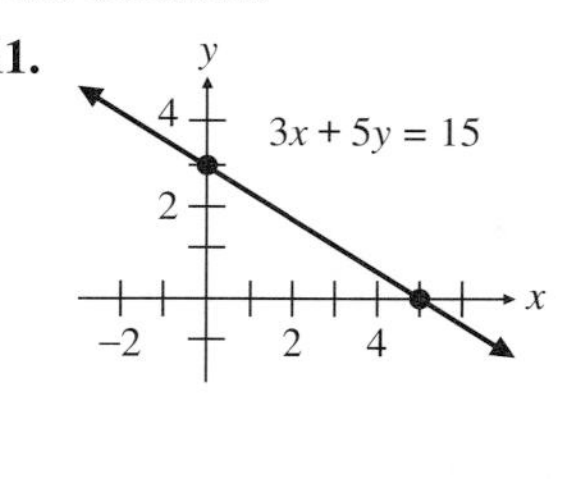

12.

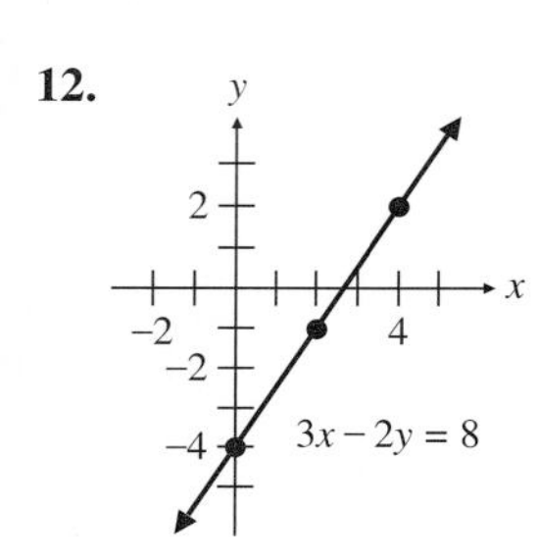

13. A relation in which no two ordered pairs have the same first element and a different second element.
14. (a) Not function **(b)** domain {1, 3, 5, 6}, range {−4, 0, 2, 3, 5}
15. (a) Function **(b)** Not function

16.

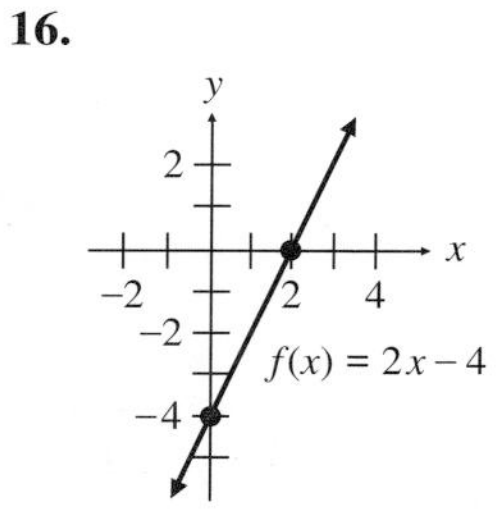

17.

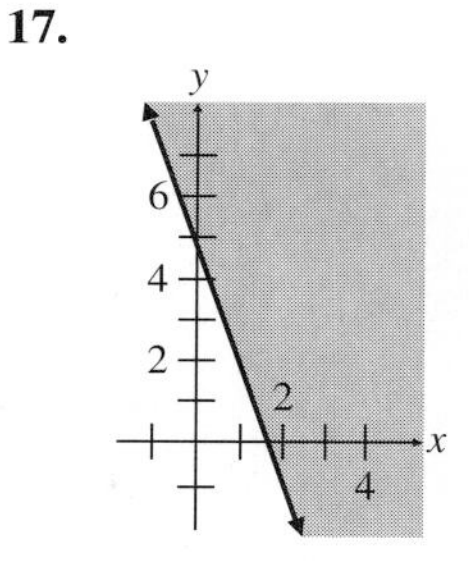

18.

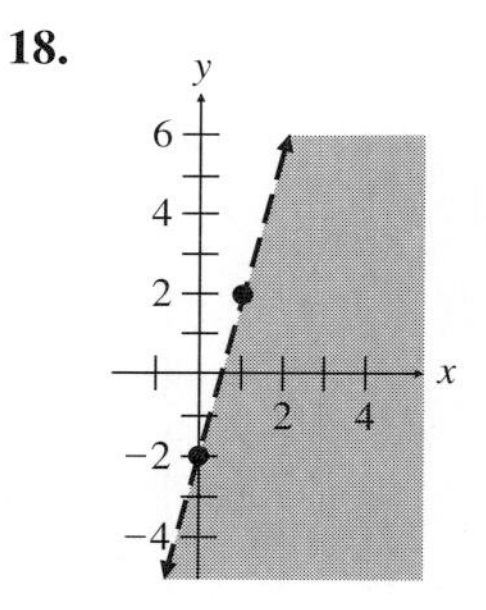

Chapter 8

EXERCISE SET 8.1

1. c **3.** a **5.** a **7.** a, c **9.** a, c **11.** Consistent—one solution **13.** Dependent—infinite number of solutions **15.** Consistent—one solution **17.** Dependent—infinite number of solutions **19.** One solution **21.** No solution **23.** One solution **25.** Infinite number of solutions **27.** No solution **29.** No solution

31. **33.**

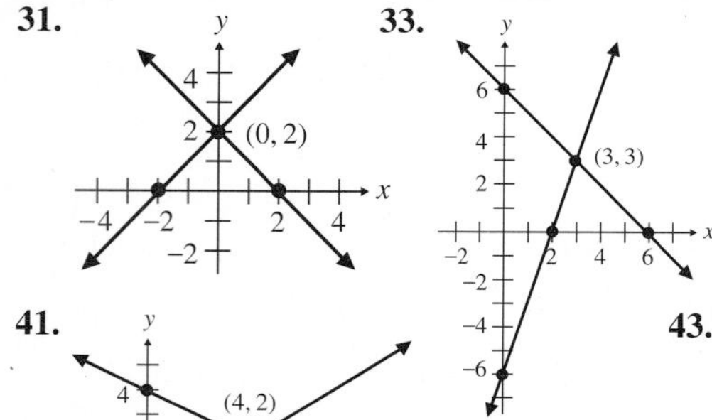

35.

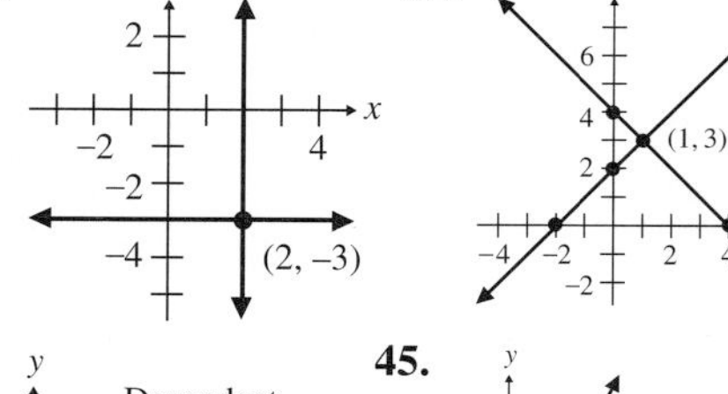

37. **39.**

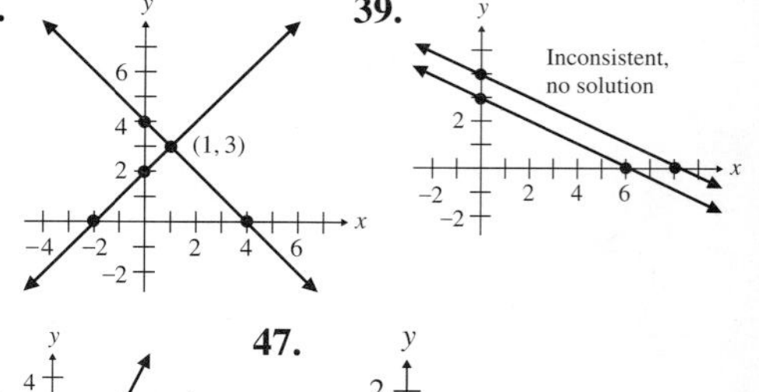

41.

43.

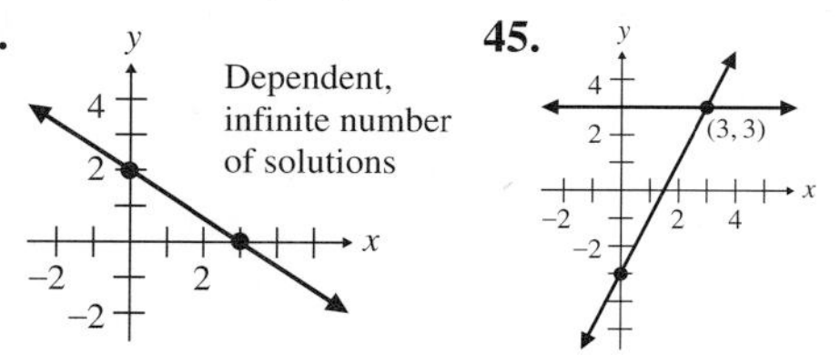

45. **47.**

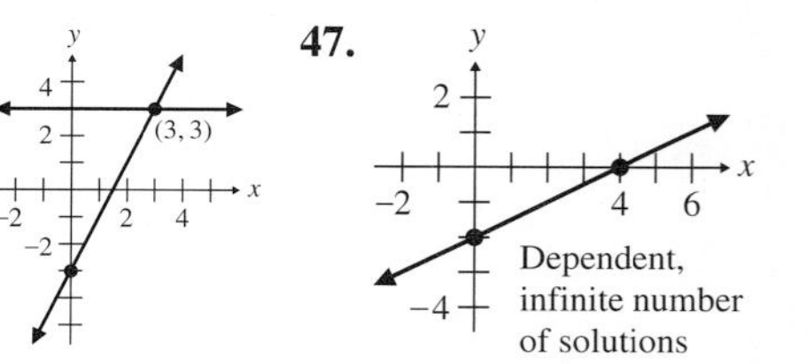

49.

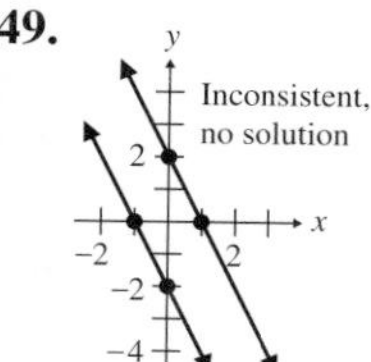

51.

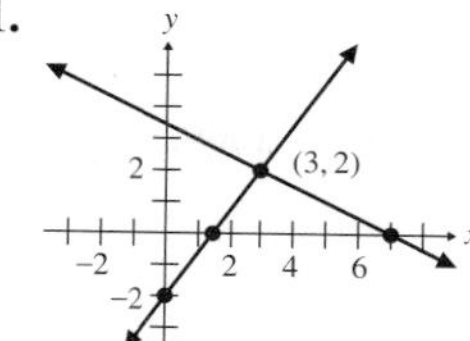

53.

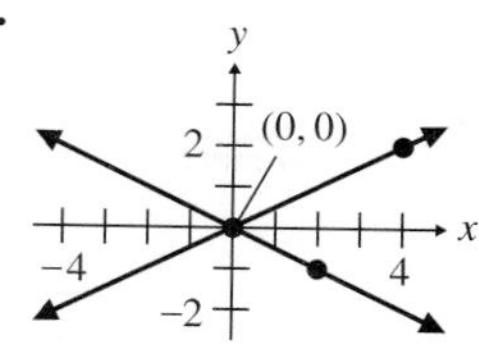

55.

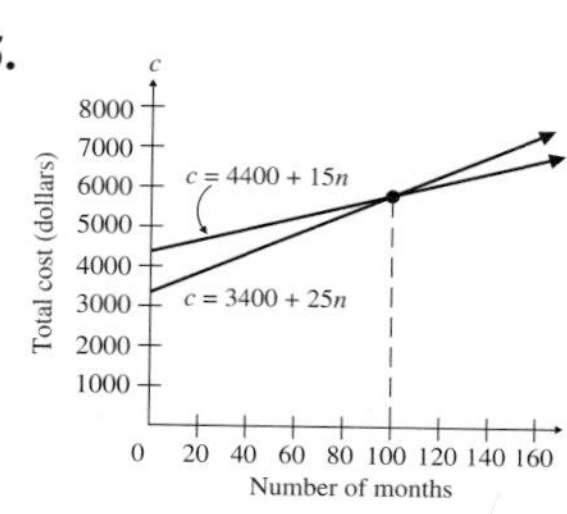

57.

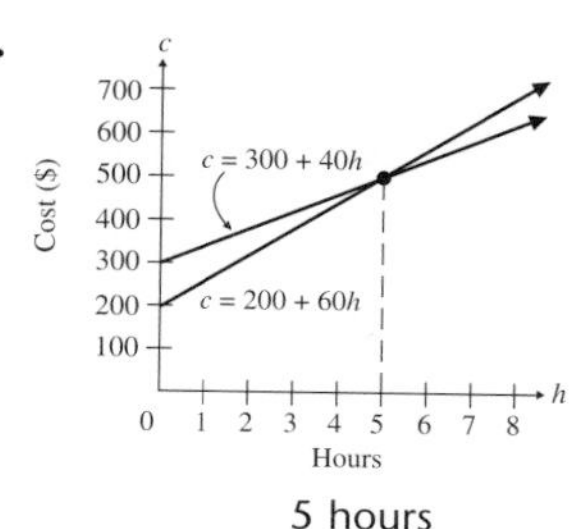

5 hours

59. The lines will be parallel since the slopes are the same but the y intercepts are different.

61. Write both equations in slope-intercept form and then compare their slopes and y intercepts. If the slopes are different there is one solution. If the slopes are the same but the y intercepts are different there is no solution. If both the slopes and y intercepts are the same there are an infinite number of solutions.

Cumulative Review Exercises

63. (a) 6 **(b)** 6, 0 **(c)** 6, -4, 0 **(d)** 6, -4, 0, $2\frac{1}{2}$, $-\frac{9}{5}$, 4.22 **(e)** $\sqrt{3}$, $-\sqrt{7}$ **(f)** 6, -4, 0, $\sqrt{3}$, $2\frac{1}{2}$, $-\frac{9}{5}$, 4.22, $-\sqrt{7}$

64. $(x + 3)(y - 4)$ **65.** $(2x + 3)(4x - 5)$ **66.** $\dfrac{-x - 6}{x + 3}$ **67.** -6

Group Activity/Challenge Problems

1. 0 **2.** an infinite number

Exercise Set 8.2

1. (2, 1) **3.** $(-1, -1)$ **5.** Inconsistent—no solution **7.** $(4, -9)$ **9.** Dependent—infinite number of solutions **11.** $(-1, 2)$ **13.** Dependent—infinite number of solutions **15.** (2, 1) **17.** Inconsistent—no solution **19.** $(-1, 0)$

21. $\left(-\dfrac{59}{37}, \dfrac{9}{37}\right)$ **23. (a)** 4 hr **(b)** 74°F **25. (a)** 3 hr **(b)** 295 mile marker

27. You will obtain a false statement, such as $3 = 0$.

Cumulative Review Exercises

29. -27 **30.** -27 **31.** -81 **32.** 81 **33.** 6 eggs **34.** $5\dfrac{1}{3}$ in. **35.** $\dfrac{9y^4}{x^2}$ **36.** $576x^{14}y^{19}$ **37.** 10

Group Activity/Challenge Problems

1. (a) $x = \dfrac{d - b}{a - c}$ **(b)** $x = 2$ **(c)** (2, 8) **3. (a)** (300, 1800) **(b)** x value remains the same, but the y value is halved. **(c)** (300, 900) **5.** $(4, 2, -1)$

Exercise Set 8.3

1. (6, 2) **3.** $(-2, 3)$ **5.** $(4, \frac{11}{2})$ **7.** Inconsistent—no solution **9.** $(-8, -26)$ **11.** $(4, -2)$ **13.** (5, 4)

15. $(\frac{3}{2}, -\frac{1}{2})$ **17.** Dependent—infinite number of solutions **19.** $(\frac{2}{5}, \frac{8}{5})$ **21.** Inconsistent—no solution **23.** (0, 0)

25. $(10, \frac{15}{2})$ **27.** $(\frac{31}{28}, -\frac{23}{28})$ **29.** $(\frac{20}{39}, -\frac{16}{39})$ **31.** $(\frac{14}{5}, -\frac{12}{5})$ **33.** You will obtain a false statement like $0 = 6$.

35. (b) (4, 1)

Cumulative Review Exercises

36. $\dfrac{64y^3}{x^6z^{12}}$ **37.** 22 days **38.** $\dfrac{2x+3}{2x+1}$ **39.** $\dfrac{x^2-18x-3}{(x+1)(x-1)(x-15)}$

Group Activity/Challenge Problems

3. $(8, -1)$ **4.** $(-\frac{105}{41}, \frac{447}{82})$ **6.** $(1, 2, 3)$

Exercise Set 8.4

1. $x + y = 26$, $x = 2y + 2$; 8, 18 **3.** $a + b = 90$, $b = a + 18$; $a = 36°$, $b = 54°$
5. $a + b = 180$, $a = b + 52$; $a = 116°$, $b = 64°$ **7.** $l = w + 100$, $2l + 2w = 800$; $l = 250$ ft, $w = 150$ ft
9. $x + y = 14$, $480x + 460y = 6560$; 6 Eagles, 8 Maple Leafs **11.** $B = 5M$, $37B + 75M = 7800$; 150 shares BancOne, 30 shares Microsoft **13.** $k + c = 4.5$, $k - c = 3.2$; 3.85 mph, kayak; 0.65 mph, current
15. $c = 50 + 0.85p$, $c = 100 + 0.80p$; \$1000 **17. (a)** $c = 0.40s$, $c = 1.00s - 1500$; \$2500
(b) $1500n = 6000$; 4 months **19.** $x + y = 12{,}500$, $0.1x + 0.0525y = 1200$; \$11,447.37 at 10%
21. $T = J + 4$, $3T = 3.2J$; Jill's boat, 60 mph; Teresa's boat, 64 mph
23. $U = D + 100$, $3U + 3D = 2700$; Delta, 400 mph; United, 500 mph **25.** $M = P + 0.3$, $5M = 8P$; 0.5 hour
27. $x + y = 10$, $0.25x + 0.50y = 0.40(10)$; $4L$ of 25%, $6L$ of 50% **29.** $x + y = 30$, $5x + 7y = 160$; 25 lb, \$5; 5 lb, \$7
31. $x + y = 8$, $12x + 6y = 10(8)$; $2\frac{2}{3}$ oz drink, $5\frac{1}{3}$ oz juice

Cumulative Review Exercises

33. 234 **34.** $\dfrac{1}{9x^8}$ **35. (a)** yes, third **(b)** yes, fourth **(c)** No, a polynomial cannot have a negative exponent on the variable. **36.** $2x - 3 + \dfrac{2}{x+4}$

Group Activity/Challenge Problems

1. 45 min **3.** $9t = d$, $5t = d - \frac{1}{2}$; 1.125 mi
4. $x + y = 300$, $0.7x + 0.4y = 0.6(300)$, or $[0.3x + 0.6y = 0.4(300)]$; 200 g of first alloy, 100 g of second alloy

Exercise Set 8.5

1.
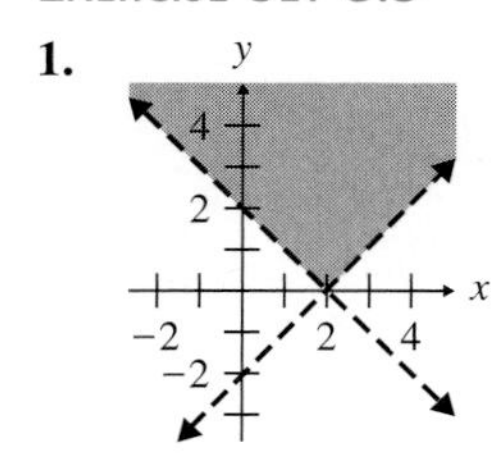

3.
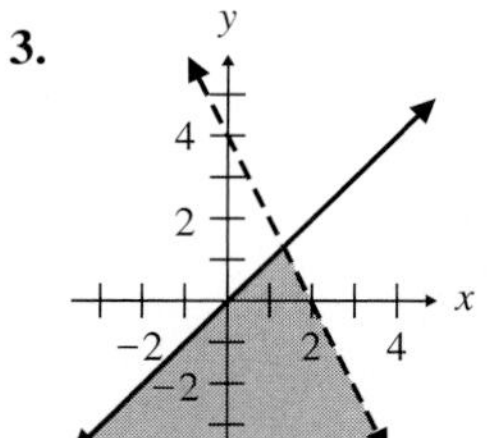

5.
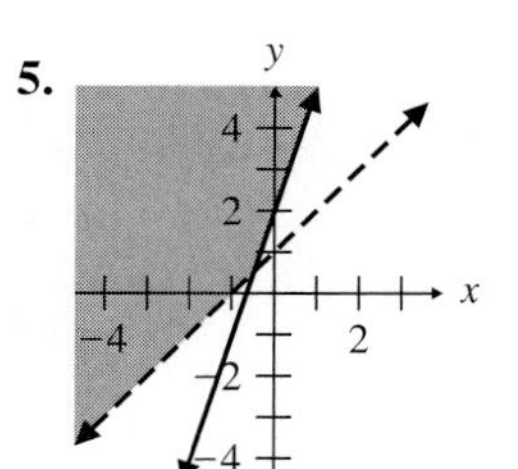

7.
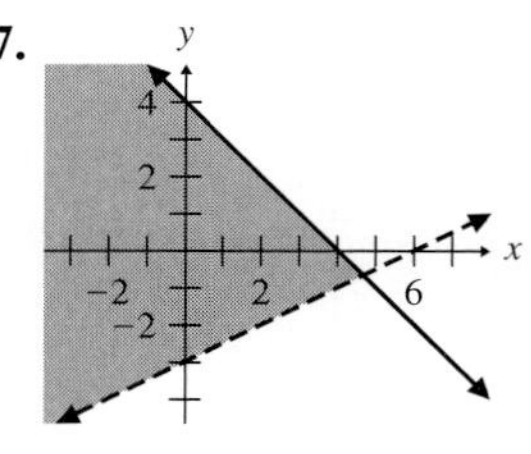

9.
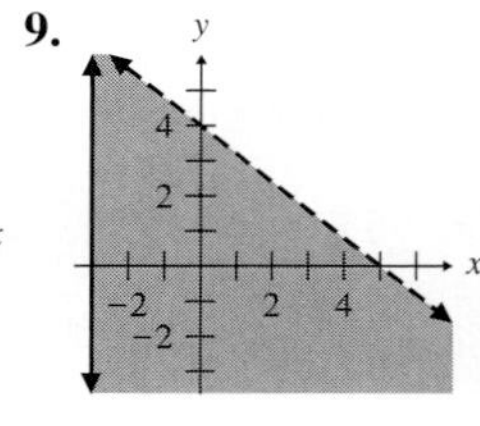

11.
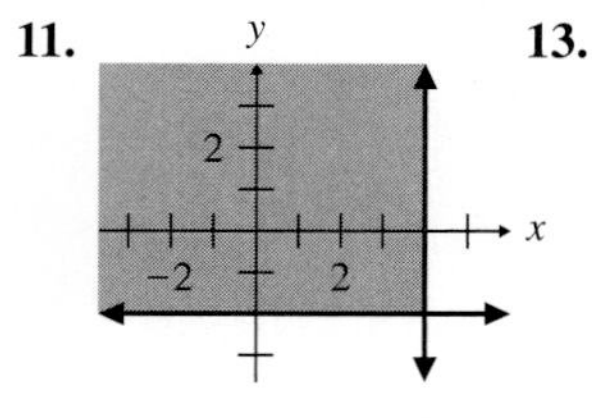

13.
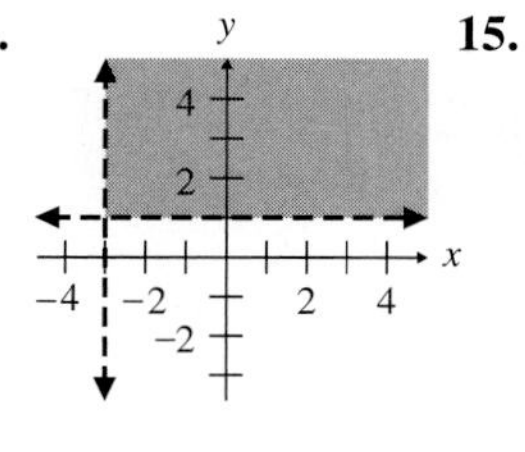

15.
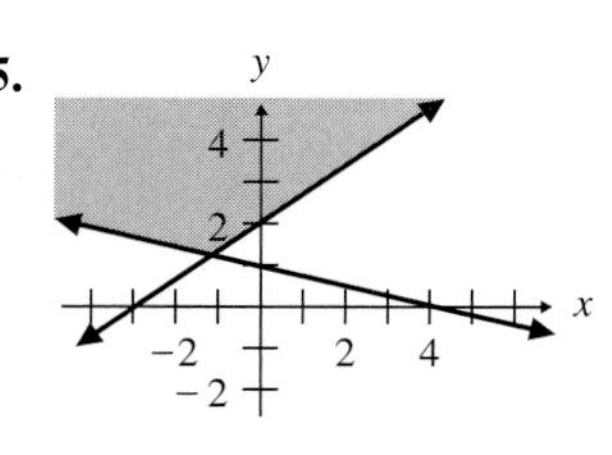

17. No

Cumulative Review Exercises

18. $3x^3 - 10x^2 + 29x - 28$ **19.** $(x - 3)(y + 1)$ **20.** $(x - 6)(x - 7)$ **21.** $(3x - 2)(2x + 1)$

GROUP ACTIVITY/CHALLENGE PROBLEMS

1.

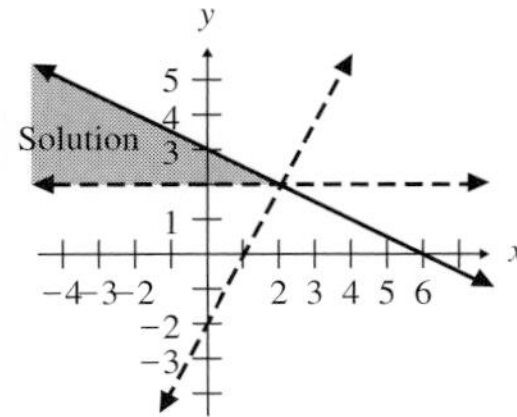

3.

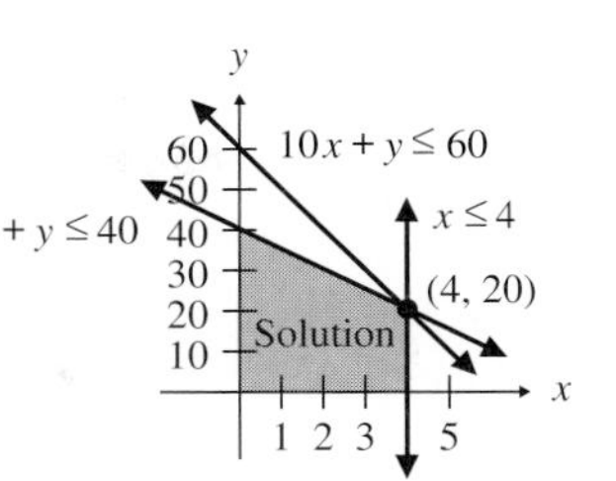

CHAPTER 8 REVIEW EXERCISES

1. c **2.** a **3.** Consistent, one **4.** Inconsistent, none **5.** Dependent, infinite number of solutions **6.** Consistent, one **7.** No solution **8.** One solution **9.** Infinite number of solutions **10.** One solution

11.

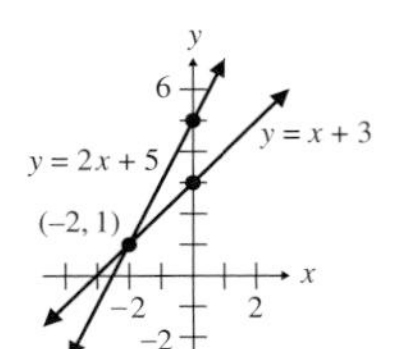

12.

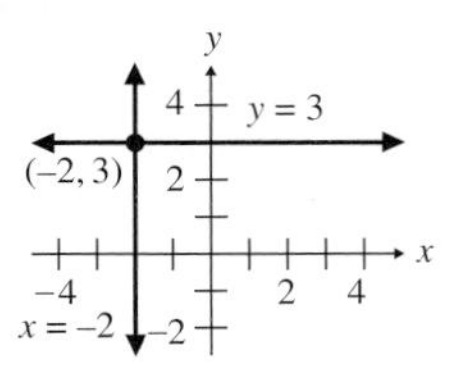

13.

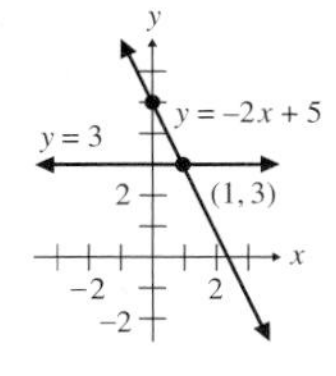

14.

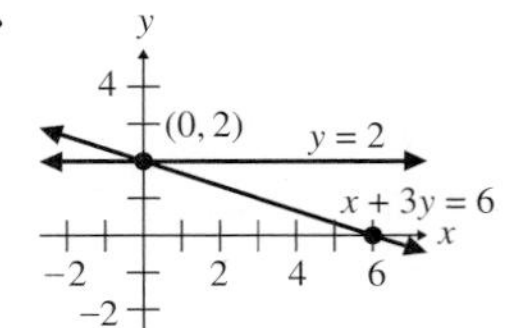

15.

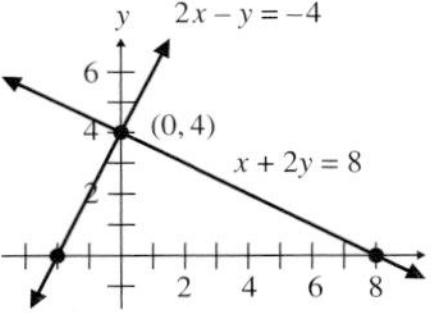

16.

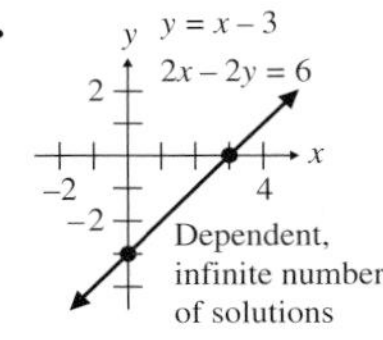

17.

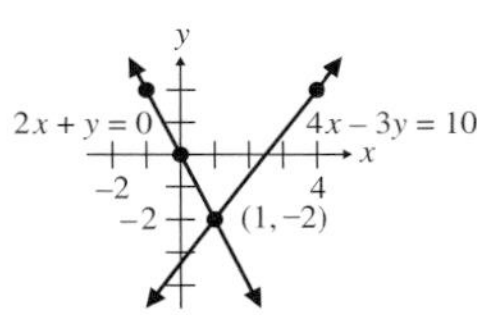

18.

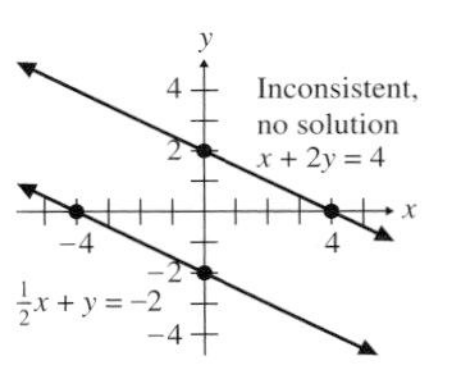

19. (5, 2) **20.** (−3, 2) **21.** (2, 1) **22.** (5, 4) **23.** (5, 2) **24.** (−18, 6) **25.** No solution

26. Infinite number of solutions **27.** $(\frac{5}{2}, -1)$ **28.** $(\frac{20}{9}, \frac{26}{9})$ **29.** (8, −2) **30.** (1, −2) **31.** (26, −16) **32.** (−7, 19)

33. $(\frac{32}{13}, \frac{8}{13})$ **34.** $(-1, \frac{13}{3})$ **35.** No solution **36.** $(\frac{7}{5}, \frac{13}{5})$ **37.** Infinite number of solutions **38.** $(-\frac{78}{7}, -\frac{48}{7})$

39. $x + y = 48$, $y = 2x - 3$; 17, 31 **40.** $p + w = 600$, $p - w = 530$; 565 mph, plane; 35 mph, wind **41.** $c = 20 + 0.50$ m, $c = 35 + 0.40$ m; 150 mi **42.** $x + y = 16{,}000$, $0.04x + 0.06y = 760$; \$10,000 at 4%, \$6000 at 6% **43.** $A = R + 6$, $5A + 5R = 600$; Ron 57 mph, Audra 63 mph **44.** $G + A = 40$; $0.6G + 0.45A = 20.25$; 15 lb of Green Turf, 25 lb of Agway **45.** $x + y = 6$; $0.3x + 0.5y = 0.4(6)$; 3 liters of each

46.

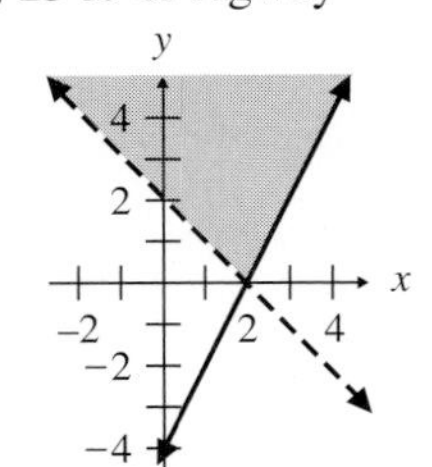

47.

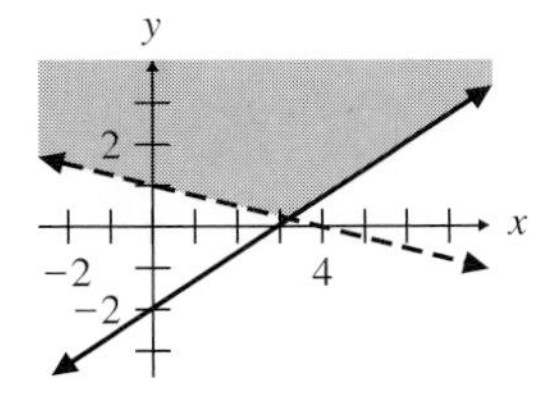

48.

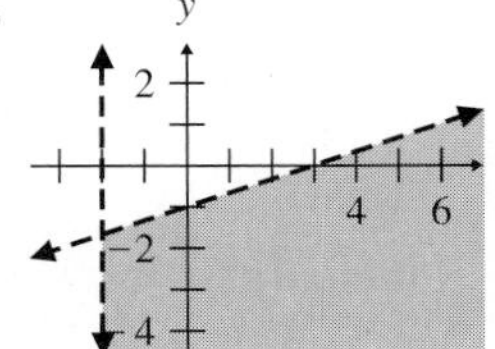

49.

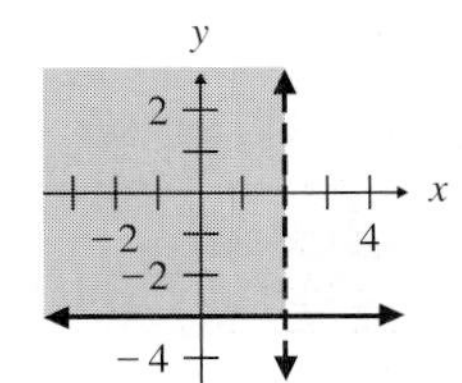

CHAPTER 8 PRACTICE TEST

1. b **2.** Inconsistent—no solution **3.** Consistent—one solution **4.** Dependent—infinite number of solutions **5.** No solution **6.** One solution

7.

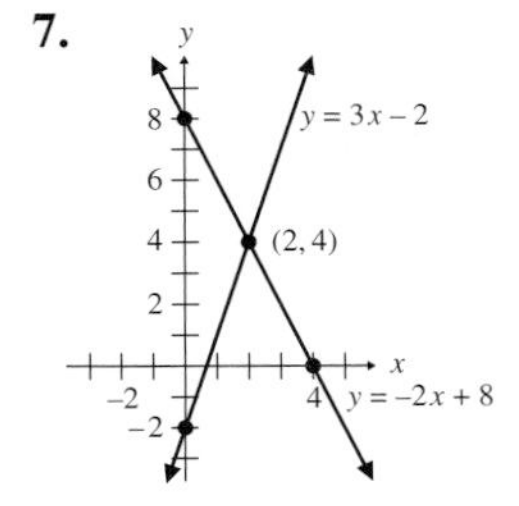

8.

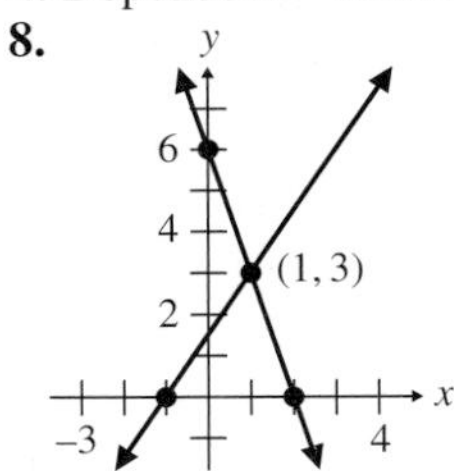

9. $\left(\frac{7}{2}, -\frac{5}{2}\right)$ **10.** (3, 1) **11.** (5, −5) **12.** $\left(\frac{44}{19}, \frac{48}{19}\right)$ **13.** $c = 40 + 0.08x$, $c = 45 + 0.03x$; 100 mi
14. $x + y = 20, 6x + 4.5y = 5(20)$; $13\frac{1}{3}$ lb of peanuts, $6\frac{2}{3}$ lb of cashews.
15.

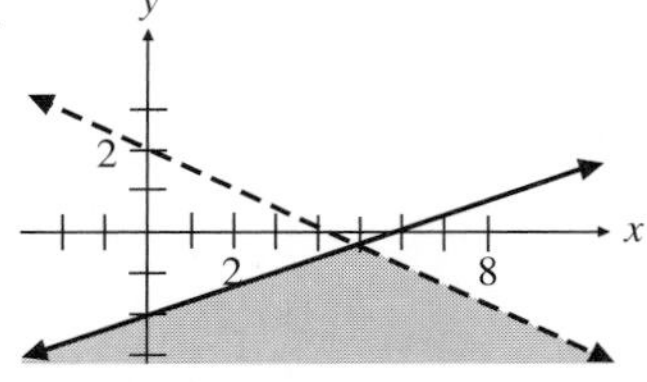

Cumulative Review Test

1. $\frac{8}{3}$ **2.** 2 **3.** $3, \frac{4}{3}$ **4.** $\frac{85}{4}$ **5.** $\frac{1}{2}$ **6.** $\frac{20}{3} \approx 6.67$ in **7.** $2x^{23}y^{17}$ **8.** $(3x - 4)(2x - 1)$ **9.** $\dfrac{x - 33}{(x + 3)(x - 3)(x - 6)}$
10. $\dfrac{(x - 4)(x + 4)}{(2x - 3)(x + 3)}$ **11.**

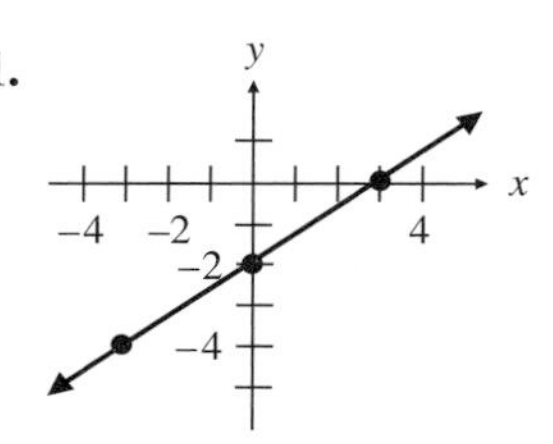

12.

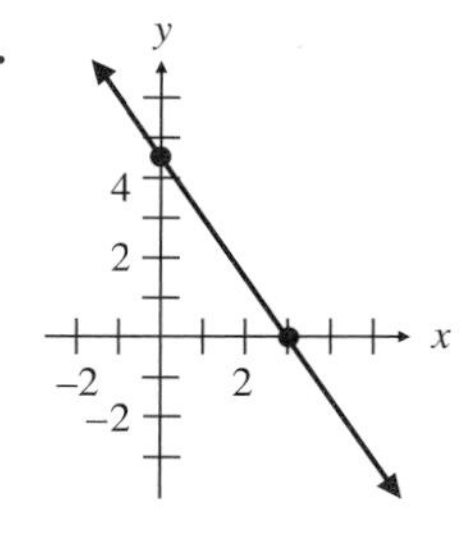

13.

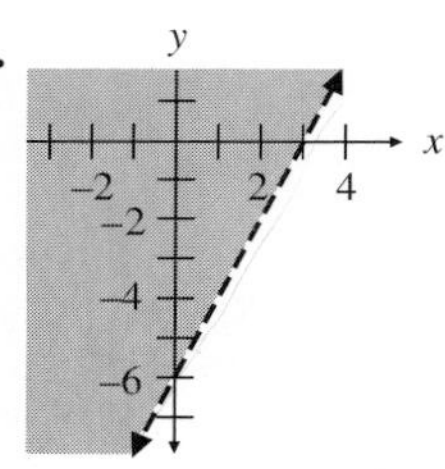

14. No solution, the lines are parallel (same slope) **15.**

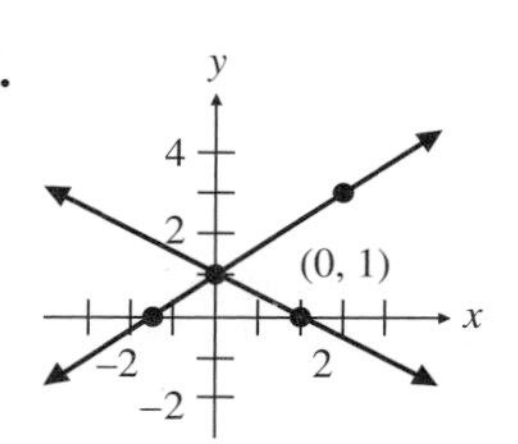

16. $\left(\frac{34}{11}, \frac{-8}{11}\right)$ **17.** 60 min **18. (a)** \$500,000 **(b)** PCR **19.** $6\frac{2}{3}$ *L* of 20%, $3\frac{1}{3}$ *L* of 35% **20.** $\frac{300}{11} \approx 27.3$ min

Chapter 9

Exercise Set 9.1

1. 1 **3.** 0 **5.** −9 **7.** 11 **9.** −4 **11.** 12 **13.** 13 **15.** −1 **17.** 9 **19.** −11 **21.** $\frac{1}{2}$ **23.** $\frac{3}{4}$ **25.** $-\frac{2}{5}$
27. $\frac{6}{7}$
The answers to exercises 29–39 may differ slightly due to variations in calculators.
29. 2.8284271 **31.** 3.8729833 **33.** 8.9442719 **35.** 9 **37.** 9.8488578 **39.** 1.7320508 **41.** True **43.** True
45. False **47.** True **49.** False **51.** True **53.** $7^{1/2}$ **55.** $(17)^{1/2}$ **57.** $(5x)^{1/2}$ **59.** $(12x^2)^{1/2}$ **61.** $(19xy^2)^{1/2}$
63. $(40x^3)^{1/2}$ **65.** It is nonnegative. The square root of a negative number is not a real number.
67. No real number when squared will be a negative number.

Cumulative Review Exercises

68. −6 **69.** −1 **70.** $\frac{30}{11}$ or $2\frac{8}{11}$ hr **71.** $\frac{4}{11}$ **72.** 16

GROUP ACTIVITY/CHALLENGE PROBLEMS

1. **(a)** Yes **(b)** No **(c)** No **(d)** Yes **(e)** No **3.** **(a)** Yes **(b)** Yes **(c)** a **5.** $x^{3/2}$ **7.** $(2x)^{8/4}$ which equals $(2x)^2$ or $4x^2$

EXERCISE SET 9.2

1. 4 **3.** $2\sqrt{2}$ **5.** $4\sqrt{6}$ **7.** $4\sqrt{2}$ **9.** $4\sqrt{10}$ **11.** $4\sqrt{3}$ **13.** $6\sqrt{3}$ **15.** $2\sqrt{39}$ **17.** 16 **19.** 30 **21.** y^3 **23.** xy^2 **25.** $x^4y^6\sqrt{x}$ **27.** $ab^2\sqrt{c}$ **29.** $x\sqrt{3x}$ **31.** $5xy\sqrt{2y}$ **33.** $10y^2z^6\sqrt{2y}$ **35.** $9qb\sqrt{3bc}$ **37.** $8xz^2\sqrt{2xyz}$ **39.** $5x^2\sqrt{10yz}$ **41.** $2\sqrt{6}$ **43.** $3\sqrt{6}$ **45.** $15\sqrt{2}$ **47.** $x\sqrt{15}$ **49.** $2x^2\sqrt{10x}$ **51.** $6xy^2\sqrt{2x}$ **53.** $3xy^2\sqrt{6xy}$ **55.** $3xy^3\sqrt{10yz}$ **57.** $6x^3y^5$ **59.** $4x$ **61.** $13x^4y^6$ **63.** $20x^2$ **65.** 4 **67.** exponent on x, 6; on y, 5 **69.** coefficient, 8; exponent on x, 12; on y, 7 **73.** Radicands cannot be negative. **75.** **(b)** $x^4\sqrt{x}$ **77.** **(a)** There can be no perfect square factors or any exponents greater than 1 in the radicand. **(b)** $5x^2\sqrt{3x}$

CUMULATIVE REVIEW EXERCISES

78. $\dfrac{3x+1}{x-1}$ **79.** $y = -\frac{1}{2}x + \frac{3}{2}, m = -\frac{1}{2}, b = \frac{3}{2}$ **80.**

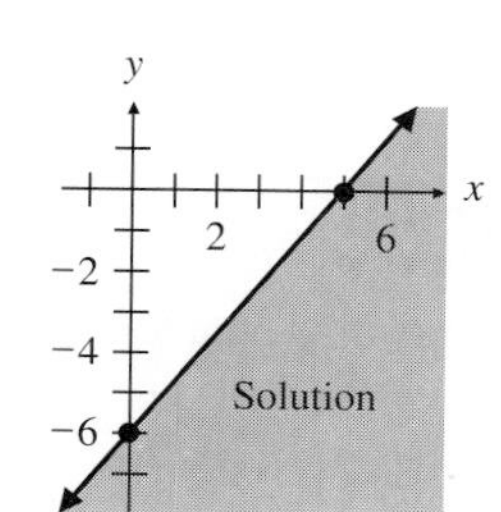

81. $(\frac{2}{11}, -\frac{15}{11})$

GROUP ACTIVITY/CHALLENGE PROBLEMS

4. **(a)** $a - b$ **(b)** 3

EXERCISE SET 9.3

1. 2 **3.** 3 **5.** 3 **7.** $\frac{1}{5}$ **9.** $\frac{3}{7}$ **11.** $\frac{1}{7}$ **13.** $2x\sqrt{5}$ **15.** $\sqrt{3xy}$ **17.** $\dfrac{\sqrt{5}}{3y}$ **19.** $\dfrac{1}{xy^2}$ **21.** $2x^2$ **23.** $\dfrac{4}{5xy}$ **25.** $\dfrac{y\sqrt{5}}{z}$ **27.** $6x^5y^8$ **29.** $\dfrac{3\sqrt{2}}{2}$ **31.** $\sqrt{2}$ **33.** $\dfrac{\sqrt{10}}{2}$ **35.** $\dfrac{\sqrt{10}}{5}$ **37.** $\dfrac{\sqrt{5}}{5}$ **39.** $\dfrac{x\sqrt{2}}{2}$ **41.** $\dfrac{x\sqrt{2}}{4}$ **43.** $\dfrac{x^2\sqrt{5}}{5}$ **45.** $\dfrac{x^3\sqrt{15y}}{15y}$ **47.** $\dfrac{x\sqrt{y}}{2y}$ **49.** $\dfrac{\sqrt{6}}{5x^2y^2z}$ **51.** $\dfrac{3\sqrt{5x}}{xy^2}$ **53.** 1. No perfect square factors in any radicand. 2. No radicand contains a fraction. 3. No square roots in any denominator. **55.** The radicand contains a fraction, $\dfrac{\sqrt{3}}{3}$. **57.** Cannot be simplified. **59.** Simplified to $x\sqrt{2}$. **61.** Cannot be simplified. **65.** **(b)** $\dfrac{a\sqrt{b}}{b}$

CUMULATIVE REVIEW EXERCISES

66. $3(x+4)(x-8)$ **67.** $\dfrac{1}{x+1}$ **68.** 4, 6 **69.**

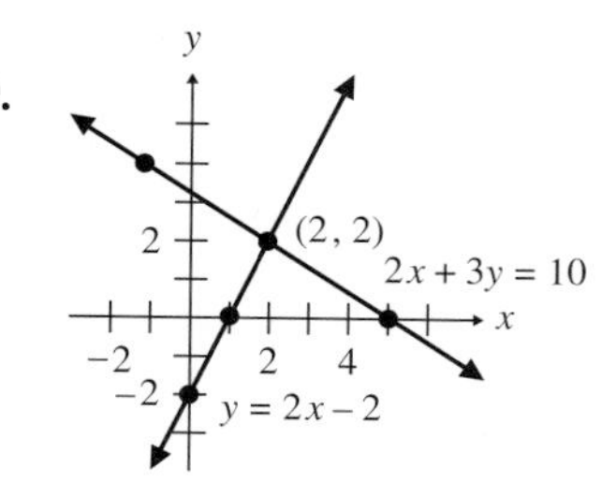

Group Activity/Challenge Problems

2. $\dfrac{2(\sqrt{6}-\sqrt{3})}{3}$ **4.** $64x^{10}$ **5.** $8x$ **7.** $2x$

Exercise Set 9.4

1. $3\sqrt{6}$ **3.** $-2\sqrt{7}$ **5.** $5-4\sqrt{3}$ **7.** $5\sqrt{x}$ **9.** $3\sqrt{x}$ **11.** $3+2\sqrt{y}$ **13.** $x+\sqrt{x}+4\sqrt{y}$ **15.** $3-2\sqrt{x}$ **17.** $5-8\sqrt{z}$ **19.** $2\sqrt{2}-2\sqrt{3}$ **21.** $4\sqrt{2}$ **23.** $7\sqrt{5}$ **25.** $16\sqrt{2}$ **27.** $-30\sqrt{3}+20\sqrt{5}$ **29.** $25\sqrt{10}-36\sqrt{5}$ **31.** $64-4\sqrt{6}$ **33.** 7 **35.** 31 **37.** $x-9$ **39.** $6-x^2$ **41.** $x-y^2$ **43.** $4x-9y$ **45.** $8-4\sqrt{3}$ **47.** $3\sqrt{5}-6$ **49.** $-2\sqrt{2}+2\sqrt{3}$ **51.** $\dfrac{-8\sqrt{5}-16\sqrt{2}}{3}$ **53.** $\dfrac{12-2\sqrt{x}}{36-x}$ **55.** $\dfrac{24+6\sqrt{y}}{16-y}$ **57.** $\dfrac{4\sqrt{x}+4y}{x-y^2}$ **59.** $\dfrac{x\sqrt{x}-x\sqrt{y}}{x-y}$ **61.** $\dfrac{\sqrt{5x}-x}{5-x}$ **63.** $2(2-\sqrt{3})$ **65.** $2(1+4\sqrt{5})$ **67.** Cannot be simplified **69.** $\dfrac{2(3+5\sqrt{3})}{3}$ **71.** $\dfrac{-1+8\sqrt{5}}{5}$ **73.** $\dfrac{3(1-2\sqrt{2})}{5}$ **75.** a^2-b **77.** $a-b$ **79.** When they have the same radicands.

Cumulative Review Exercises

81. 18 **82.** $\frac{9}{2}, -4$ **83.** $\dfrac{-2x-3}{(x+2)(x-2)}$ **84.** 30 minutes

Group Activity/Challenge Problems

2. Perimeter: $4(\sqrt{5}+\sqrt{3})$ or about 15.87; Area: $8+2\sqrt{15}$ or about 15.75 **3.** Perimeter: $4\sqrt{6}$ or about 9.80; Area: 2 **5.** 48

Exercise Set 9.5

1. 64 **3.** No solution **5.** 4 **7.** No solution **9.** 4 **11.** 4 **13.** No solution **15.** 7 **17.** 7 **19.** 2 **21.** 9 **23.** $\frac{14}{3}$ **25.** 1 **27.** 1, 9 **29.** No solution **31.** -2 **33.** 5, 6 **35.** No solution **37.** An equation that contains a variable in a radicand. **39.** because there may be extraneous roots

Cumulative Review Exercises

41.

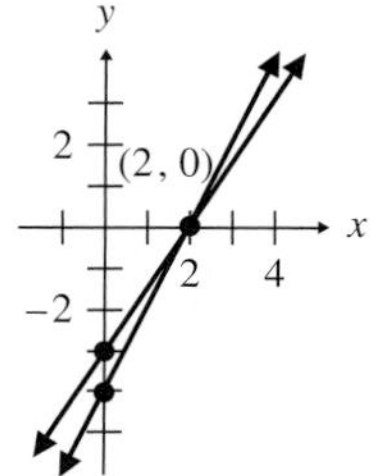

42. (2, 0) **43.** (2, 0) **44.** Boat 8 mph, current 4 mph

Group Activity/Challenge Problems

1. 9 **3.** 9 **5. (a)**

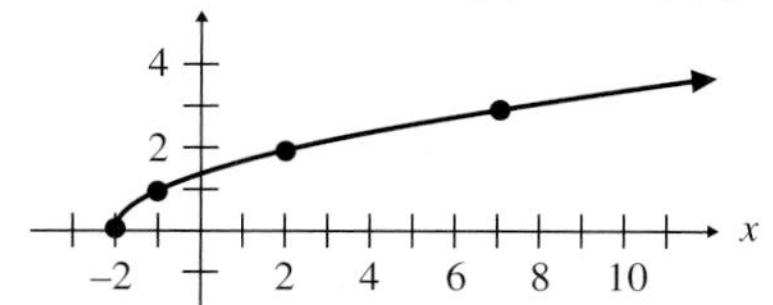

(b) No **(c)** yes

EXERCISE SET 9.6

1. $\sqrt{119} \approx 10.91$ **3.** $\sqrt{164} \approx 12.81$ **5.** $\sqrt{175} \approx 13.23$ **7.** $\sqrt{41} \approx 6.40$ **9.** $\sqrt{149} \approx 12.21$ **11.** $\sqrt{128} \approx 11.31$ **13.** $\sqrt{17{,}240.89} \approx 131.30$ yd **15.** $\sqrt{18.25} \approx 4.27$ m **17.** $\sqrt{10} \approx 3.16$ **19.** $\sqrt{288} \approx 16.97$ **21.** $\sqrt{144} = 12$ in **23.** $\sqrt{25.48} \approx 5.05$ ft **25.** $\sqrt{225} = 15$ in **27.** $\sqrt{5120} \approx 71.55$ ft/sec **29.** $\sqrt{80{,}000} \approx 282.84$ ft/sec **31.** $6.28\sqrt{1.25} \approx 7.02$ sec **33.** $6.28\sqrt{0.3125} \approx 3.51$ sec **35.** $0.2(\sqrt{1418})^3 \approx 10{,}679.34$ days **37.** $\sqrt{19.5(6{,}370{,}000)} \approx 11{,}145.18$ m/sec

CUMULATIVE REVIEW EXERCISES

38. $x > 6$ **39.** $\frac{x^4}{4y^3}$ **40.** $-2x^3 + 4x^2 - 6x - 4$ **41.** $\frac{5}{3}x^2 - 3x + 2 - \frac{4}{3x} - \frac{1}{x^2}$

GROUP ACTIVITY/CHALLENGE PROBLEMS

1. 9 in by 12 in **3.** 3.25 ft

EXERCISE SET 9.7

1. 2 **3.** -2 **5.** 2 **7.** 3 **9.** -1 **11.** 4 **13.** $3\sqrt[3]{2}$ **15.** $2\sqrt[3]{2}$ **17.** $3\sqrt[3]{3}$ **19.** $2\sqrt[4]{2}$ **21.** $2\sqrt[3]{5}$ **23.** x **25.** y^3 **27.** x^4 **29.** y **31.** x^5 **33.** 16 **35.** 8 **37.** 1 **39.** 27 **41.** 27 **43.** 625 **45.** $\frac{1}{9}$ **47.** $\frac{1}{32}$ **49.** $x^{7/3}$ **51.** $x^{4/3}$ **53.** $y^{15/4}$ **55.** $y^{21/4}$ **57.** $x^{2/3}$ **59.** x **61.** x^4 **63.** x^2 **65.** Both equal 4 **67. (b)** $\sqrt[8]{x^5}$ or $(\sqrt[8]{x})^5$

CUMULATIVE REVIEW EXERCISES

69. -42 **70.** $(3x - 4)(x - 8)$ **71.**

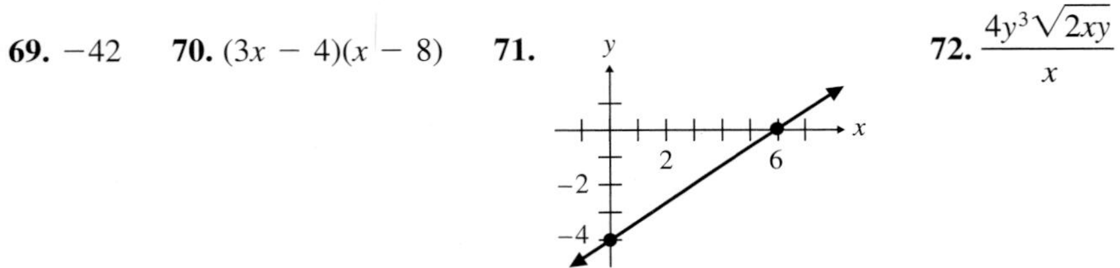

72. $\frac{4y^3\sqrt{2xy}}{x}$

GROUP ACTIVITY/CHALLENGE PROBLEMS

1. xy **3.** $\sqrt[4]{2}$ **5. (b)** $\frac{\sqrt[3]{4}}{2}$ **7.** $\frac{\sqrt[3]{x}}{x}$ **9.** $\frac{4\sqrt[3]{25y^2}}{5y^2}$

CHAPTER 9 REVIEW EXERCISES

1. 5 **2.** 6 **3.** -9 **4.** $8^{1/2}$ **5.** $(26x)^{1/2}$ **6.** $(20xy^2)^{1/2}$ **7.** $4\sqrt{2}$ **8.** $2\sqrt{11}$ **9.** $3x^2y^2\sqrt{5x}$ **10.** $5x^2y^3\sqrt{5}$ **11.** $x^2z\sqrt{15xyz}$ **12.** $4b^2c^2\sqrt{3ac}$ **13.** $4\sqrt{6}$ **14.** $5x$ **15.** $6x\sqrt{y}$ **16.** $5xy\sqrt{3}$ **17.** $10xy^3\sqrt{y}$ **18.** $2xy^2\sqrt{6xy}$ **19.** 4 **20.** $\frac{1}{5}$ **21.** $\frac{1}{2}$ **22.** $\frac{3\sqrt{5}}{5}$ **23.** $\frac{\sqrt{15x}}{6}$ **24.** $\frac{\sqrt{6x}}{6}$ **25.** $\frac{x\sqrt{2}}{2}$ **26.** $\frac{x^2\sqrt{2x}}{4}$ **27.** $\frac{y\sqrt{15}}{x^2}$ **28.** $\frac{x\sqrt{2y}}{y^2}$ **29.** $\frac{3\sqrt{5x}}{2x^2y}$ **30.** $\frac{z\sqrt{14x}}{7x}$ **31.** $-3(1 - \sqrt{2})$ **32.** $\frac{15 + 5\sqrt{6}}{3}$ **33.** $\frac{2\sqrt{3} - \sqrt{3x}}{4 - x}$ **34.** $\frac{2\sqrt{x} + 10}{x - 25}$ **35.** $\frac{\sqrt{5x} - \sqrt{15}}{x - 3}$ **36.** $4\sqrt{3}$ **37.** $-\sqrt{2}$ **38.** $-2\sqrt{x}$ **39.** 0 **40.** $\sqrt{2}$ **41.** $12\sqrt{10}$ **42.** $-10\sqrt{2}$ **43.** $26\sqrt{2}$ **44.** $16\sqrt{3} + 20\sqrt{5}$ **45.** 81 **46.** No solution **47.** 39 **48.** 8 **49.** 9 **50.** 4 **51.** 0 **52.** 7 **53.** 3 **54.** 10 **55.** $\sqrt{88} \approx 9.38$ **56.** $\sqrt{12} \approx 3.46$ **57.** $\sqrt{61} \approx 7.81$ **58.** $\sqrt{135} \approx 11.62$ ft **59.** $\sqrt{261} \approx 16.16$ in **60.** 10 **61.** $\sqrt{153} \approx 12.37$ **62.** 11 ft **63.** 2 **64.** -3 **65.** 2 **66.** $2\sqrt[3]{2}$ **67.** $2\sqrt[3]{3}$ **68.** $2\sqrt[4]{2}$ **69.** $2\sqrt[3]{6}$ **70.** $3\sqrt[3]{2}$

71. $2\sqrt[4]{6}$ **72.** x^5 **73.** x^4 **74.** y^4 **75.** 4 **76.** 4 **77.** $\frac{1}{9}$ **78.** 16 **79.** $\frac{1}{8}$ **80.** 125 **81.** $x^{5/3}$ **82.** $x^{10/3}$
83. $y^{9/4}$ **84.** $x^{5/2}$ **85.** $y^{11/2}$ **86.** $x^{7/4}$ **87.** x **88.** $\sqrt[3]{x^2}$ **89.** x^3 **90.** x^2 **91.** x^2 **92.** x^2 **93.** x^3 **94.** x^6

Chapter 9 Practice Test

1. $(3xy)^{1/2}$ **2.** $x + 3$ **3.** $4\sqrt{6}$ **4.** $2x\sqrt{3}$ **5.** $4x^2y^2\sqrt{2y}$ **6.** $4xy\sqrt{3x}$ **7.** $5x^2y^2\sqrt{3y}$ **8.** $\frac{1}{5}$ **9.** $\dfrac{y}{4x}$ **10.** $\dfrac{\sqrt{2}}{2}$
11. $\dfrac{2\sqrt{5x}}{5}$ **12.** $\dfrac{2\sqrt{15x}}{3xy}$ **13.** $-3(2 - \sqrt{5})$ **14.** $\dfrac{6\sqrt{x} + 18}{x - 9}$ **15.** $11\sqrt{3}$ **16.** $-3\sqrt{x}$ **17.** 76 **18.** 4, 8
19. $\sqrt{106} \approx 10.30$ **20.** $\sqrt{74} \approx 8.60$ **21.** $\dfrac{1}{81}$ **22.** x^5

Chapter 10

Exercise Set 10.1

1. 4, − 4 **3.** 10, −10 **5.** 6, −6 **7.** $\sqrt{10}, -\sqrt{10}$ **9.** $2\sqrt{2}, -2\sqrt{2}$ **11.** 2, −2 **13.** $\sqrt{17}, -\sqrt{17}$ **15.** 3, − 3
17. $\sqrt{15}/2,\ -\sqrt{15}/2$ **19.** $\sqrt{195}/5, -\sqrt{195}/5$ **21.** 1, −3 **23.** 7, −1 **25.** 2, −10 **27.** $1 + 2\sqrt{3}, 1 - 2\sqrt{3}$
29. $-6 + 2\sqrt{5}, -6 - 2\sqrt{5}$ **31.** 3, −7 **33.** 19, −1 **35.** $(-3 + 3\sqrt{2})/2, (-3 - 3\sqrt{2})/2$
37. $(-1 + 2\sqrt{5})/4, (-1 - 2\sqrt{5})/4$ **39.** $(6 + 3\sqrt{2})/2, (6 - 3\sqrt{2})/2$ **41.** $w = 2\sqrt{10} \approx 6.32$ ft, $l = 4\sqrt{10} \approx 12.65$ ft
43. $x^2 = 36$, other answers are possible **45.** 9, need an equation equivalent to $x^2 = 36$ **47.** $ax^2 + bx + c = 0, a \neq 0$

Cumulative Review Exercises

49. $2(2x + 3)(x - 4)$ **50.** No solution **51.** \$2000 at 6%, \$8000 at 8% **52.** $y = 4x - 1$

Group Activity/Challenge Problems

1. $s = \sqrt{A}$ **3.** $r = \sqrt{\dfrac{A}{\pi}}$ **5.** $d = \sqrt{\dfrac{k}{I}}$ **6.** $r = \sqrt{\dfrac{A}{p}} - 1$

Exercise Set 10.2

1. 1, −3 **3.** 5, −1 **5.** −2, −1 **7.** 4, −2 **9.** −3 **11.** −2, −3 **13.** −3, −6 **15.** 7, 8 **17.** 6, −2
19. $-1 + \sqrt{7}, -1 - \sqrt{7}$ **21.** $-3 + \sqrt{3}, -3 - \sqrt{3}$ **23.** $(5 + \sqrt{57})/2, (5 - \sqrt{57})/2$ **25.** 1, −3
27. $(-9 + \sqrt{73})/2, (-9 - \sqrt{73})/2$ **29.** −8, −3 **31.** $(-5 + \sqrt{31})/2, (-5 - \sqrt{31})/2$ **33.** $-1 + \sqrt{3}, -1 - \sqrt{3}$
35. 0, −4 **37.** 0, 2 **39.** 1, −4 **41.** 0, −6 **43.** 3, 7
45. The constant is the square of half the coefficient of the x term.

Cumulative Review Exercises

47. $\dfrac{4}{(x + 2)(x - 3)}$
48. Write the equations in slope-intercept form and compare the slopes. If the slopes are the same and the y-intercepts are different, the equations represent parallel lines. **49.** $(\frac{38}{11}, \frac{12}{11})$ **50.** 3

Group Activity/Challenge Problems

1. $(-3 + \sqrt{59})/10, (-3 - \sqrt{59})/10$ **3.** $(-1 + \sqrt{193})/12, (-1 - \sqrt{193})/12$ **5.** $0.75 + \sqrt{3.5625},\ 0.75 - \sqrt{3.5625}$

Exercise Set 10.3

1. Two real number solutions **3.** No real number solution **5.** Two real number solutions **7.** One real number solution
9. Two real number solutions **11.** No real number solution **13.** Two real number solutions

15. Two real number solutions **17.** 1, 2 **19.** 4, 5 **21.** −8, 3 **23.** 4, 9 **25.** 5, −5 **27.** 0, 3 **29.** 3, 4 **31.** $(7 + \sqrt{17})/4, (7 - \sqrt{17})/4$ **33.** $\frac{1}{3}, -\frac{1}{2}$ **35.** $(2 + \sqrt{6})/2, (2 - \sqrt{6})/2$ **37.** No real number solution **39.** $\frac{5}{4}, -1$ **41.** $\frac{9}{2}, -1$ **43.** $3, \frac{5}{2}$ **45.** 4, 5 **47.** $w = 4$ ft, $l = 5$ ft **49.** 2 ft **51. (a)** none **(b)** one **(c)** two

Cumulative Review Exercises

54. 6, 7 **55.** $\frac{5}{3}, -\frac{7}{2}$ **56.** Cannot be solved by factoring; $\frac{-3 + \sqrt{41}}{4}, \frac{-3 - \sqrt{41}}{4}$ **57.** 3, −3

Group Activity/Challenge Problems

1. (a)

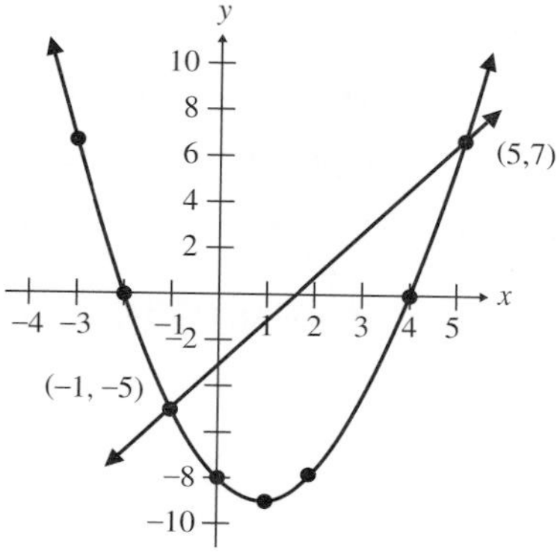

(5,7)

(−1, −5)

(b) −1, 5 **(c)** yes **(d)** (−1, −5), (5, 7) **2. (a)** $c < 9$ **(b)** $c = 9$ **(c)** $c > 9$ **3. (a)** $c < \frac{9}{8}$ **(b)** $c = \frac{9}{8}$ **(c)** $c > \frac{9}{8}$ **5.** 300 ft. long by 50 ft. wide, or 100 ft. long by 150 ft. wide

Exercise Set 10.4

1. $x = -1, (-1, -8)$, up **3.** $x = \frac{5}{2}, (\frac{5}{2}, \frac{1}{4})$, down **5.** $x = \frac{5}{6}, (\frac{5}{6}, \frac{121}{12})$, down **7.** $x = -1, (-1, -8)$, down **9.** $x = \frac{1}{3}, (\frac{1}{3}, \frac{5}{3})$, up **11.** $x = -\frac{3}{2}, (-\frac{3}{2}, -14)$, up

13. x-intercepts 1, −1

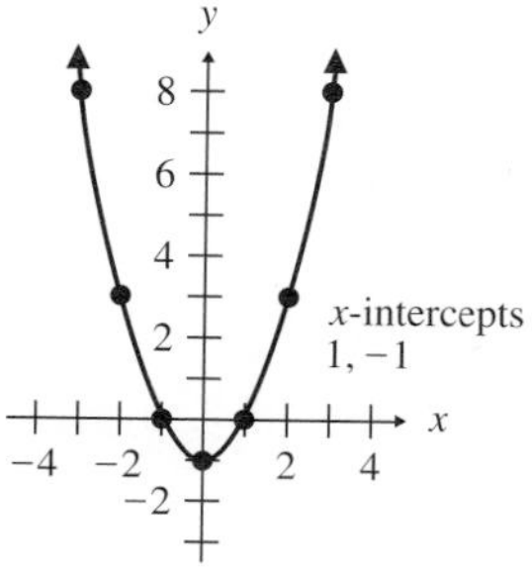

15. x-intercepts $\sqrt{3}, -\sqrt{3}$

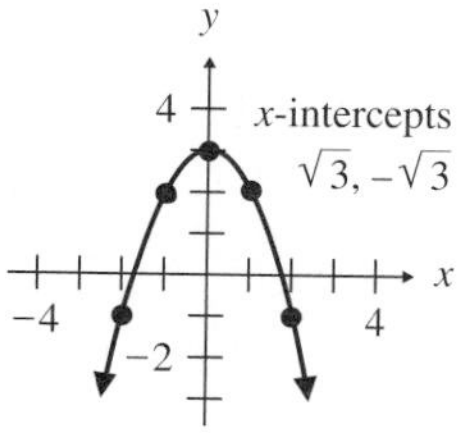

17. No x-intercepts

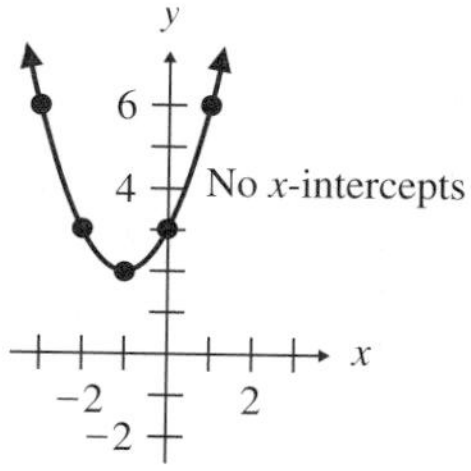

19. x-intercepts 3, −5

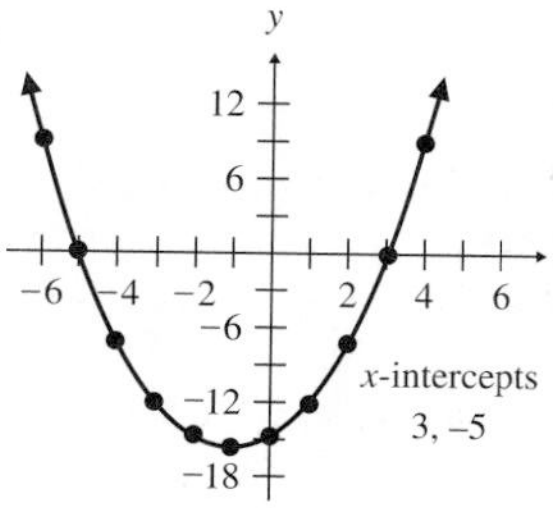

21. No x-intercepts

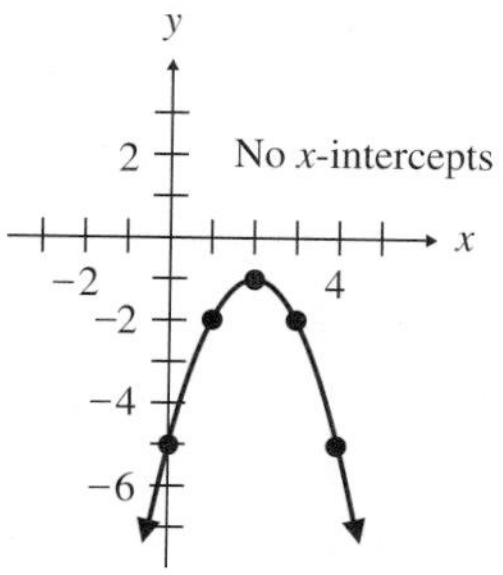

23. x-intercepts −3, 4

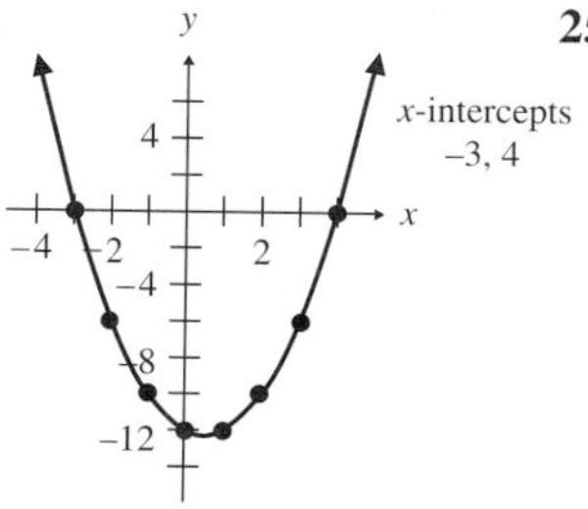

25. x-intercept 3

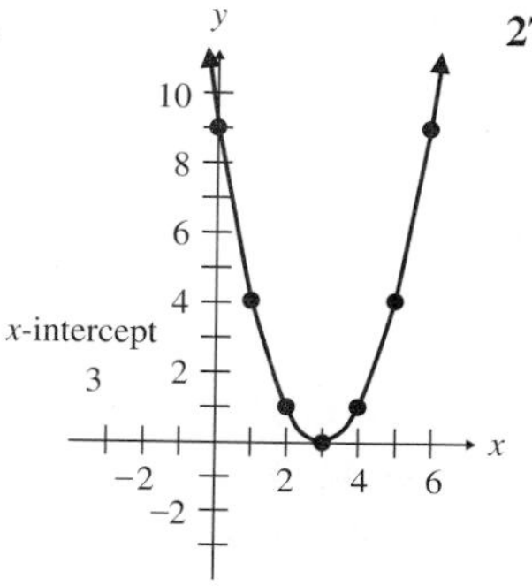

27. x-intercepts 0, 5

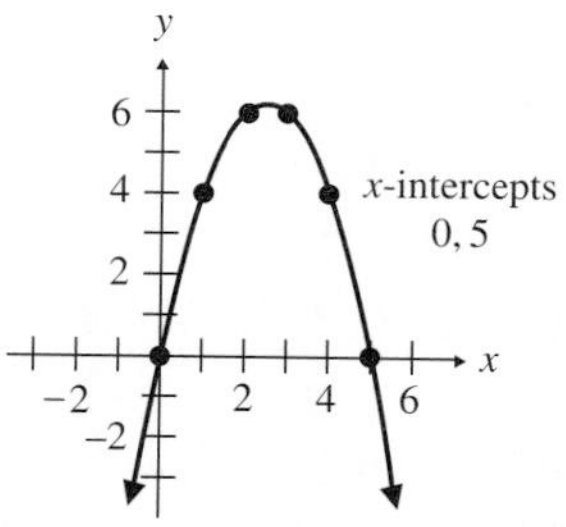

29.

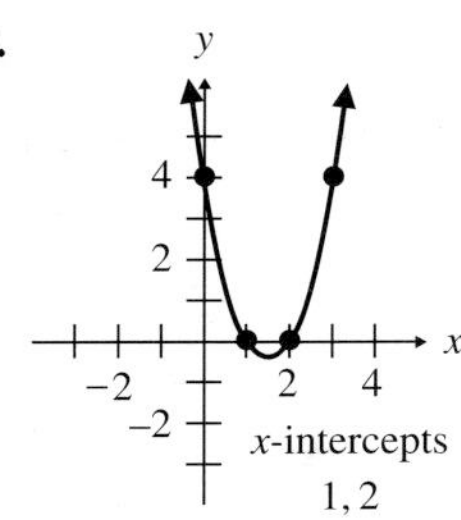

31.

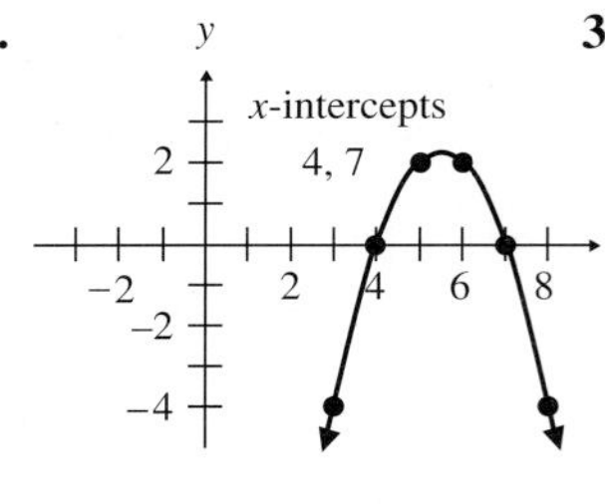

33.

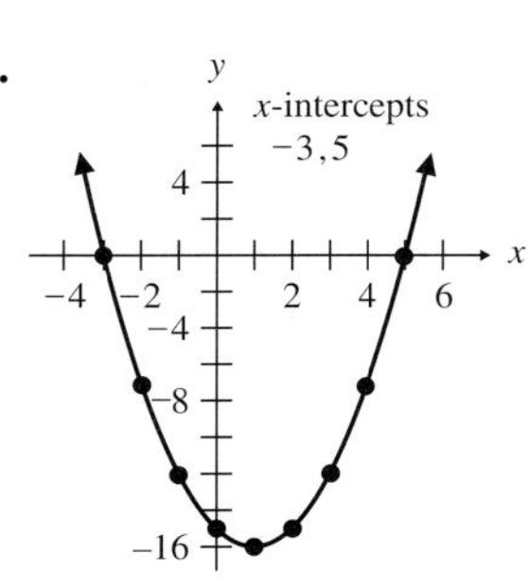

35.

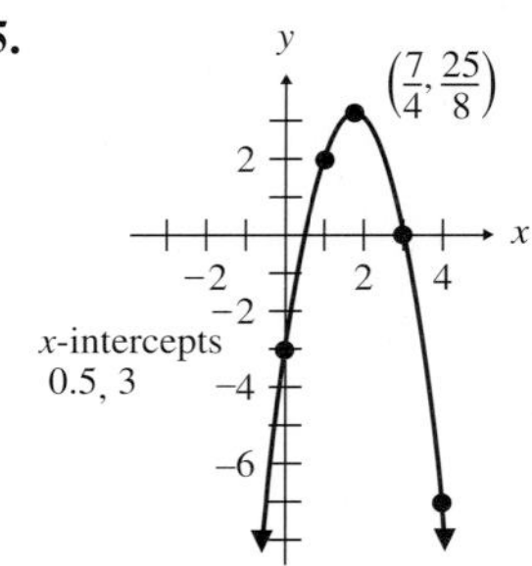

37.

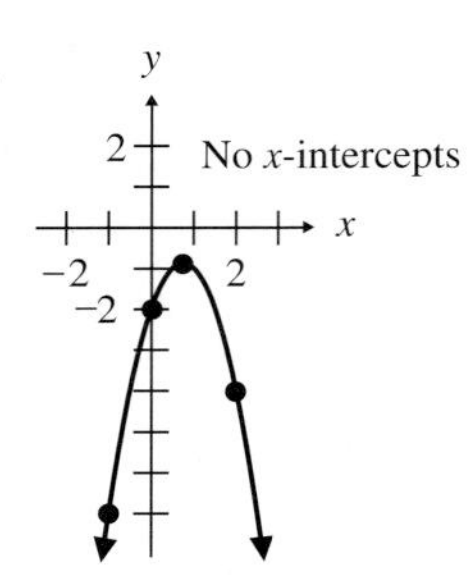

39.

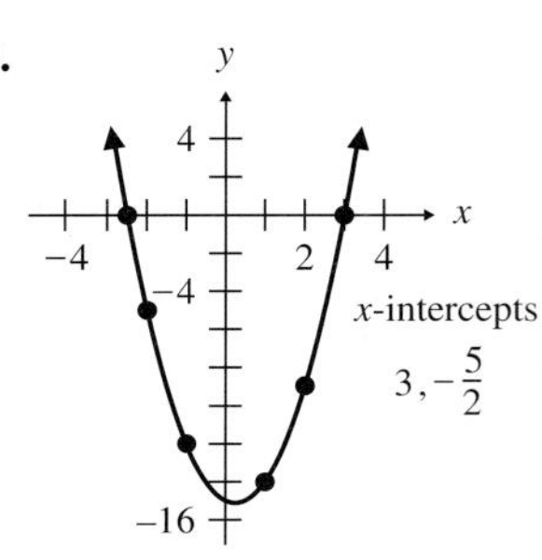

41. None **43.** Two **45.** Two **47.** None

49. $\left(\dfrac{-b}{2a}, \dfrac{4ac - b^2}{4a}\right)$

51. (a) The values of x where the graph crossses the x axis. **(b)** Let $y = 0$ and solve the resulting equation for x.

53. (a) $x = \dfrac{-b}{2a}$, **(b)** The axis of symmetry.

55. Two, the vertex is below the x axis and the parabola opens upward.

57. None, the vertex is below the x axis and the parabola opens downward

Cumulative Review Exercises

59. $\dfrac{-x^2 + 2x - 6}{(x + 3)(x - 4)}$ **60.** $\dfrac{27}{7}$ **61.**

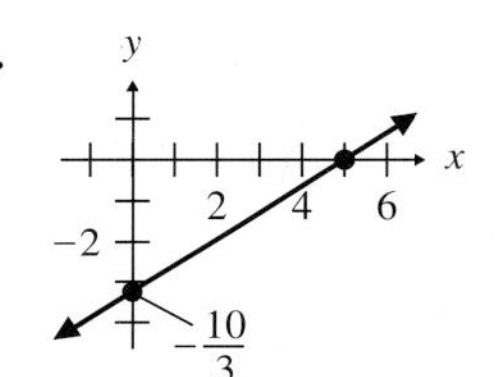

62.

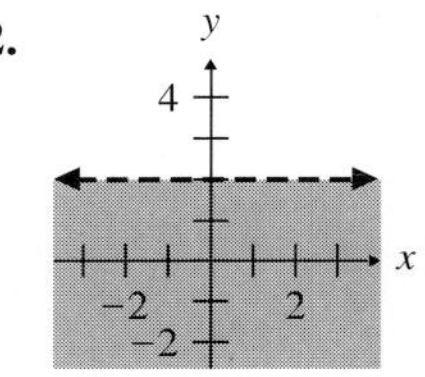

Group Activity/Challenge Problems

2.

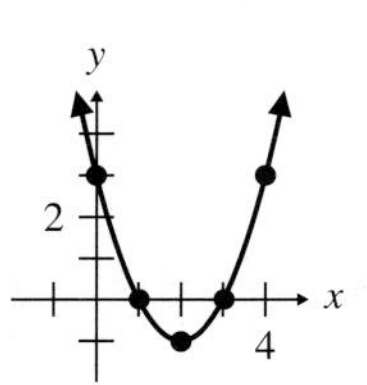

Chapter 10 Review Exercises

1. $5, -5$ **2.** $2\sqrt{2}, -2\sqrt{2}$ **3.** $\sqrt{6}, -\sqrt{6}$ **4.** $\sqrt{6}, -\sqrt{6}$ **5.** $2\sqrt{5}, -2\sqrt{5}$ **6.** $\sqrt{7}, -\sqrt{7}$ **7.** $2\sqrt{2}, -2\sqrt{2}$ **8.** $3 + 2\sqrt{3}, 3 - 2\sqrt{3}$ **9.** $(-4 + \sqrt{30})/2, (-4 - \sqrt{30})/2$ **10.** $(5 + 5\sqrt{2})/3, (5 - 5\sqrt{2})/3$ **11.** $2, 8$ **12.** $3, 5$ **13.** $1, 13$ **14.** $2, -3$ **15.** $9, -6$ **16.** $1, -6$ **17.** $-1 + \sqrt{6}, -1 - \sqrt{6}$ **18.** $(3 + \sqrt{41})/2, (3 - \sqrt{41})/2$

19. $-4, 8$ **20.** $5, -3$ **21.** $\frac{3}{2}, -2$ **22.** $\frac{5}{3}, \frac{3}{2}$ **23.** Two real number solutions **24.** No real number solution **25.** No real number solution **26.** No real number solution **27.** One real number solution **28.** No real number solution **29.** Two real number solutions **30.** Two real number solutions **31.** $2, 7$ **32.** $-10, 3$ **33.** $2, 5$ **34.** $2, -\frac{3}{5}$ **35.** $-2, 9$ **36.** $6, -5$ **37.** $\frac{3}{2}, -\frac{5}{3}$ **38.** $(-2 + \sqrt{10})/2, (-2 - \sqrt{10})/2$ **39.** $(3 + \sqrt{57})/4, (3 - \sqrt{57})/4$ **40.** $3 + \sqrt{2}, 3 - \sqrt{2}$ **41.** No real number solution **42.** $(3 + \sqrt{33})/3, (3 - \sqrt{33})/3$ **43.** $0, -\frac{3}{2}$ **44.** $0, \frac{5}{2}$ **45.** $3, 8$ **46.** $7, 9$ **47.** $5, -8$ **48.** $-9, 3$ **49.** $-6, 10$ **50.** $7, -6$ **51.** $1, -12$ **52.** $5, -5$ **53.** $0, -6$ **54.** $\frac{1}{2}, -3$ **55.** $\frac{5}{2}, 2$ **56.** $\frac{2}{3}, -\frac{3}{2}$ **57.** $(-3 + \sqrt{33})/2, (-3 - \sqrt{33})/2$ **58.** $2, \frac{5}{3}$ **59.** $1, -\frac{8}{3}$ **60.** $(3 + 3\sqrt{3})/2, (3 - 3\sqrt{3})/2$ **61.** $0, \frac{5}{2}$ **62.** $0, -\frac{5}{3}$ **63.** $x = 1, (1, -4)$, up **64.** $x = 5, (5, -1)$, up **65.** $x = -\frac{7}{2}, (-\frac{7}{2}, -\frac{1}{4})$, up **66.** $x = -1, (-1, 16)$, down **67.** $x = \frac{3}{2}, (\frac{3}{2}, -\frac{9}{4})$, up **68.** $x = -\frac{7}{4}, (-\frac{7}{4}, -\frac{25}{8})$, up **69.** $x = 0, (0, -8)$, down **70.** $x = 1, (1, 9)$, down **71.** $x = -\frac{1}{2}, (-\frac{1}{2}, \frac{81}{4})$, down **72.** $x = -\frac{5}{6}, (-\frac{5}{6}, -\frac{121}{12})$, up

73. x-intercepts 0, −6

74. x-intercepts 2, −2

75. x-intercepts −4, 2

76. x-intercepts −1, 2

77. x-intercepts −1, −4

78.

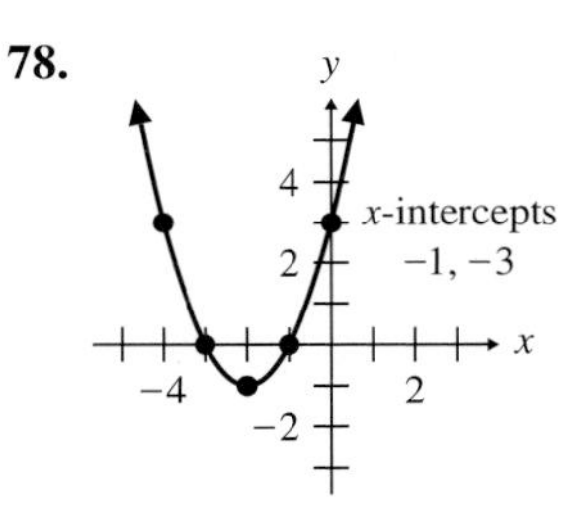

79.

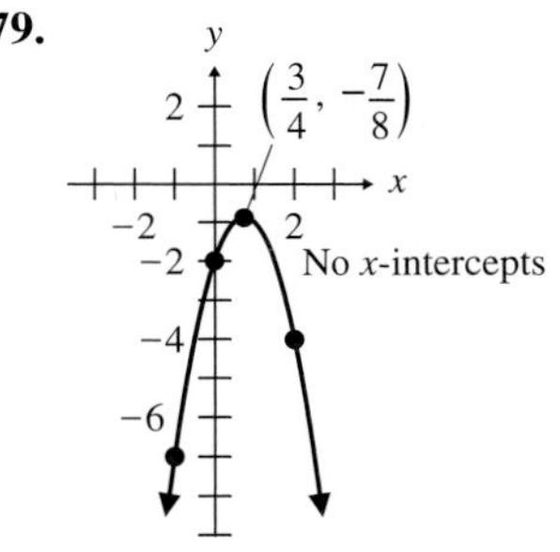

80.

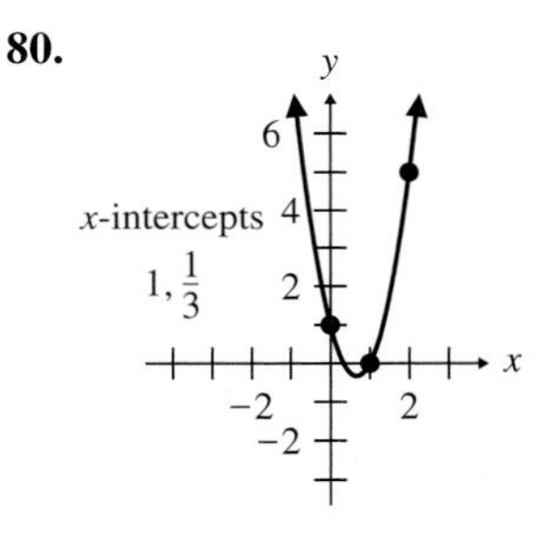

81.

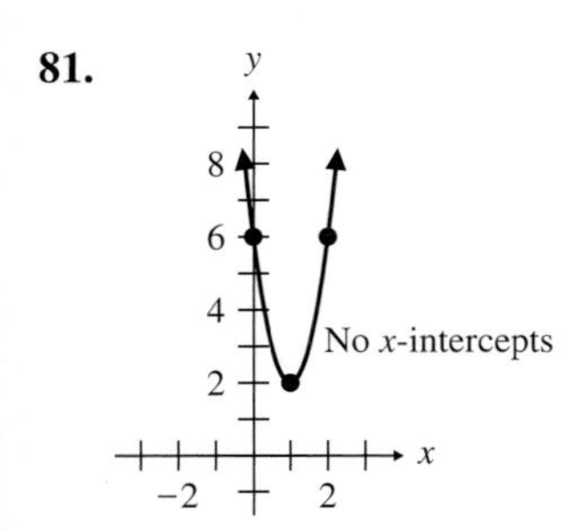

82.

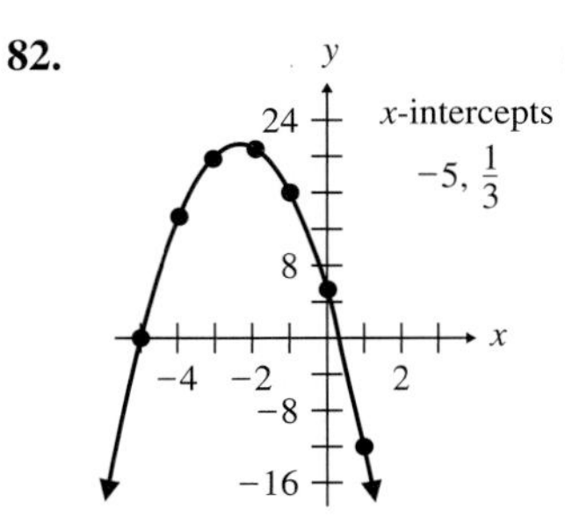

83.

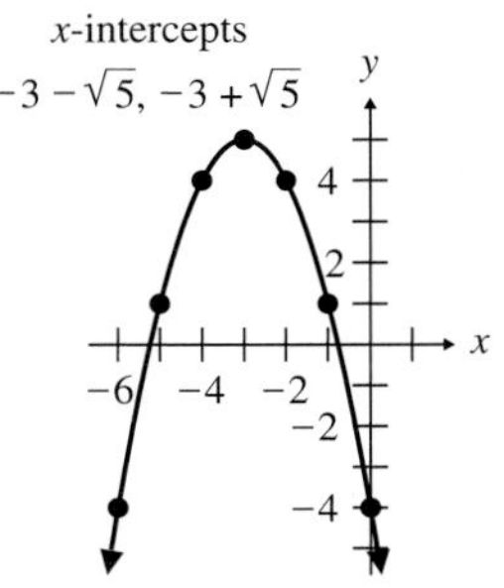

84.

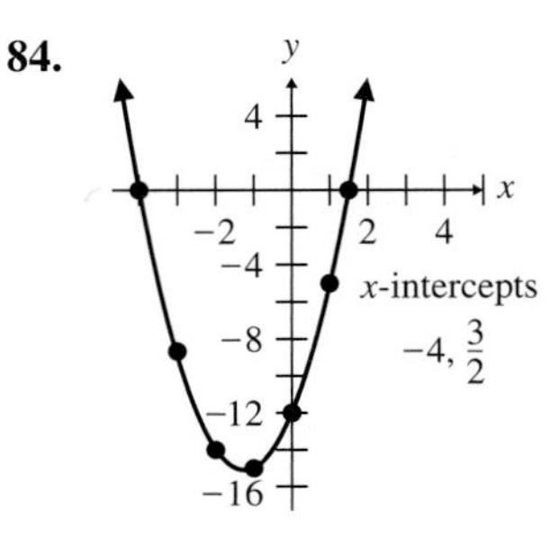

85. 8, 11 **86.** $w = 7$ ft, $l = 9$ ft

CHAPTER 10 PRACTICE TEST

1. $2\sqrt{5}, -2\sqrt{5}$ **2.** $(3 + \sqrt{35})/2, (3 - \sqrt{35})/2$ **3.** $10, -6$ **4.** $-4, 3$ **5.** $6, -1$

6. $(-4+\sqrt{6})/2, (-4-\sqrt{6})/2$ **7.** $0, \frac{5}{3}$ **8.** $-5, \frac{1}{2}$ **9.** No real number solution **10.** $x=\frac{3}{2}, (\frac{3}{2}, \frac{41}{4})$, down

11.

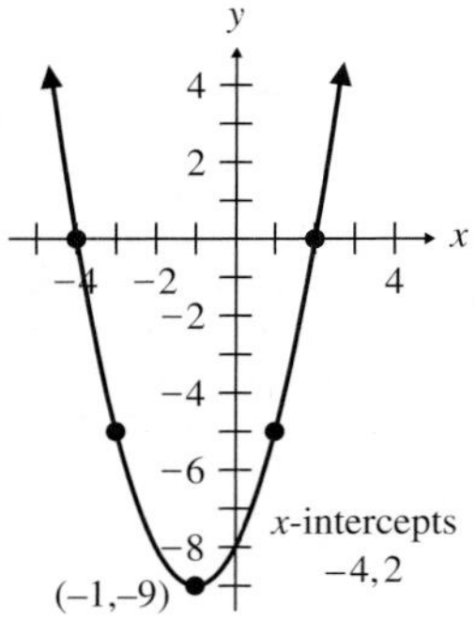

12.

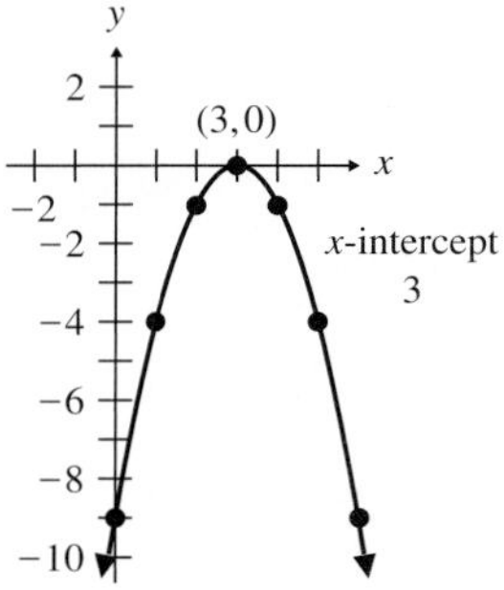

13. $w=3$ ft, $l=10$ ft

Cumulative Review Test

1. 16 **2.** $\frac{40}{31}$ **3.** $5\frac{1}{3}$ in **4.** $x \geq -\frac{1}{4}$

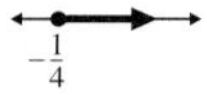

5. $P=3A-m-n$ **6.** $864x^{14}y^{22}$ **7.** $x+4-\dfrac{3}{x+2}$

8. $(x-2y)(2x-3y)$ **9.** $2(2x+1)(x-4)$ **10.** $\dfrac{6a-8}{(a+4)(a-4)^2}$ **11.** 6, 8 **12.**

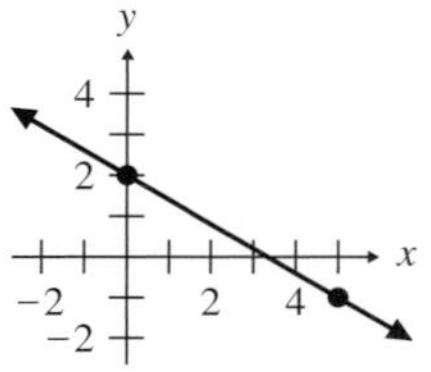

13. $(-\frac{12}{7}, -\frac{30}{7})$ **14.** $\dfrac{y\sqrt{2xy}}{6}$ **15.** $4\sqrt{7}$ **16.** 2 **17.** $(2+\sqrt{14})/2,\ (2-\sqrt{14})/2$ **18.** 25.6 lb

19. Width = 10 ft; length = 27 ft **20.** walks, 3 mph; jogs, 6 mph

Index

Photo Credits

	Title Page	Gerald French/FPG International
Chapter 1	Chapter Opener	Frans Lanting/Minden Pictures
	Calculator Corner	Allen R. Angel
	Exercise Set 1.5, Problem 94	Allen R. Angel
	Exercise Set 1.6, Problem 130	DPA/SP/The Image Works
Chapter 2	Chapter Opener	Frans Lanting/Minden Pictures
	Section 2.6, Example 12	George Haling/Photo Researchers
	Exercise Set 2.6, Problem 39	Allen R. Angel
	Exercise Set 2.6, GA Problem 4	Leonard Lee Rue/Monkmeyer Press
Chapter 3	Chapter Opener	Deborah Davis/PhotoEdit
	Exercise Set 3.1, Problem 87	Allen R. Angel
	Exercise Set 3.1, GA Problem 4	Allen R. Angel
	Section 3.3, Example 13	Dr. Morley, Read/Science Photo Library/Photo Researchers
	Exercise Set 3.3, Problem 35	Mark Burnett/Photo Researchers
	Exercise Set 3.3, GA Problem 6	H. Donnezant/Photo Researchers
Chapter 4	Chapter Opener	NASA Headquarters
	Exercise Set 4.3, GA Problem 3	NASA Headquarters
	Exercise Set 4.7, Problem 4	Larry Keenan Assoc./The Image Bank
	Exercise Set 4.7, Problem 23	Steve Dunwell/The Image Bank
Chapter 5	Chapter Opener	Jeff Greenberg/dMRp Photo Researchers
	Exercise Set 5.2, Problem 65	Pascal Quittemelle/Stock Boston
	Exercise Set 5.3, Problem 78	Blair Seitz/Photo Researchers
	Section 5.6, Example 10	Tracy Knauer/Photo Researchers
Chapter 6	Chapter Opener	G. C. Kelley/Photo Researchers
	Exercise Set 6.6, GA Problem 1	Tony Freeman/PhotoEdit
	Section 6.7, Example 4	Stephen Marks/The Image Bank
	Exercise Set 6.7, Problem 9	Janeart Ltd./The Image Bank
	Exercise Set 6.7, Problem 17	Steve Hansen/Stock Boston
Chapter 7	Chapter Opener	John Griffin/The Image Works
	Exercise Set 7.1, Problem 5	Auscape-Agence Vandystadt/Photo Researchers
	Exercise Set 7.1, Problem 7	Bullaty Lomeo/The Image Bank

	Exercise Set 7.1, Problem 15	Jeffrey Dunn/Stock Boston
	Calculator Corner	Courtesy, Texas Instruments
	Exercise Set 7.4, GA Problem 4	Ben Blankenburg/Stock Boston
	Section 7.6, Example 6	Jeff Greenberg/dMRp Photo Researchers
	Exercise Set 7.6, Problem 43	Van Etten/Monkmeyer Press
Chapter 8	Chapter Opener	Jim Corwin/Stock Boston
	Exercise Set 8.1, Problem 57	Das/Monkmeyer Press
	Section 8.4, Example 2	Mike Mazzaschi/Stock Boston
	Exercise Set 8.4, Problem 10	Peter Skinner/Photo Researchers
	Exercise Set 8.4, GA Problem 3	Tony Freeman/PhotoEdit
Chapter 9	Chapter Opener	NASA Headquarters
	Exercise Set 9.4, Problem 84	Michael P. Gadomsk/Photo Researchers
	Section 9.6, Example 3	Mark Burnett/Photo Researchers
	Section 9.6, Example 5	Jim Charles/Photo Researchers
	Section 9.6, Example 6	Richard Megna/Fundamental Photographs
Chapter 10	Chapter Opener	George Malave/Stock Boston
	Section 10.4 Arch	Joe Sohm/Chromosohn/Photo Researchers
	Section 10.4 Bridge	Illinois Department of Transportation
	Practice Test, Problem 20	Allen R. Angel

Chapter 7 Graphing Linear Equations

Linear equation in two variables: $ax + by = c$

A **graph** is an illustration of the set of points whose coordinates satisfy the equation.

Every **linear equation** of the form $ax + by = c$ will be a straight line when graphed.

To find the y intercept (where the graph crosses the y axis) set $x = 0$ and solve for y.

To find the x intercept (where the graph crosses the x axis) set $y = 0$ and solve for x.

$$\text{slope } (m) = \frac{\text{change in } y}{\text{change in } x} = \frac{y_2 - y_1}{x_2 - x_1}$$

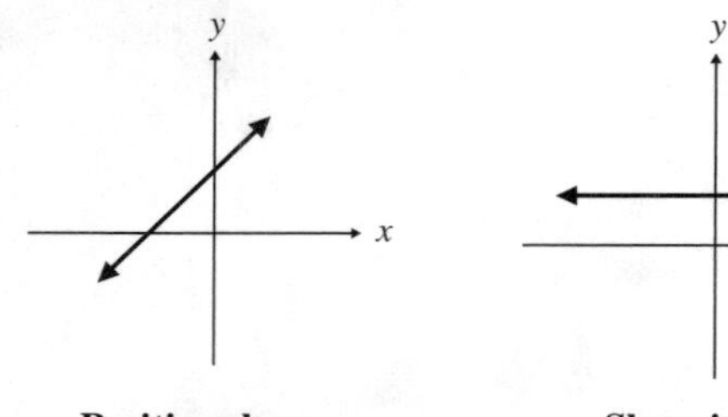

Positive slope **Slope is 0**

Negative slope

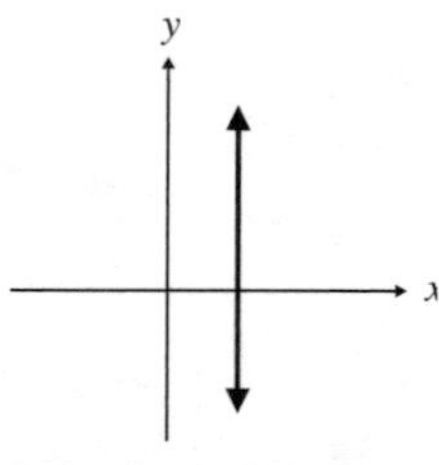

Slope is undefined

LINEAR EQUATIONS

Standard form of a linear equation: $ax + by = c$

Slope-intercept form of a linear equation: $y = mx + b$, where m is the slope and b is the y intercept.

Point-slope form of a linear equation: $y - y_1 = m(x - x_1)$ where m is slope and (x_1, y_1) is a point on the line.

A **relation** is any set of ordered pairs.

A **function** is a relation in which no two ordered pairs have the same first coordinate and a different second coordinate.

Chapter 8 Systems of Linear Equations

The **solution** to a system of linear equations is the ordered pair or pairs that satisfy all equations in the system. A system of linear equations may have no solution, exactly one solution, or an infinite number of solutions.

Consistent system
Exactly 1 solution
(Nonparallel lines)

y

Solution

x

Line 1 Line 2

(a)

One solution

Inconsistent system
No solution
(Parallel lines)

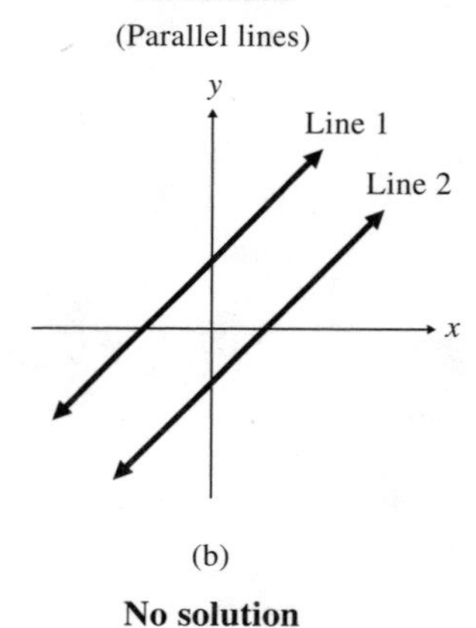

(b)

No solution

Dependent system
Infinite number of solutions
(Same line)

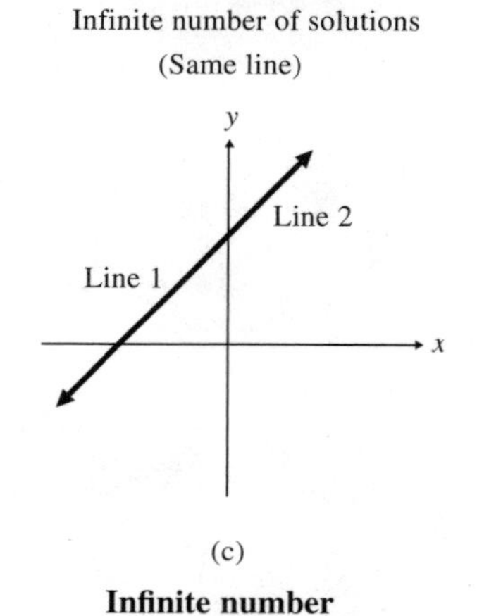

(c)

Infinite number of solutions

A system of linear equations may be solved graphically, or algebraically by the substitution method or by the addition (or elimination) method.